Disciplinary Approaches to Aging

General Editor

Donna Lind Infeld, Ph. D.
George Washington Univesity

A ROUTLEDGE SERIES

Contents of the Collection

Disciplinary Approaches to Aging

Volume 1

Biology of Aging

Edited with introductions by

Donna Lind Infeld, Ph. D.

George Washington University

Routledge
Taylor & Francis Group
NEW YORK AND LONDON

First published by Garland Publishing, Inc.

This edition published 2013 by Routledge

Routledge
711 Third Avenue
New York
NY 10017

Routledge
2 Park Square, Milton Park
Abingdon
Oxon, OX14 4RN

Routledge is an imprint of the Taylor & Francis Group, an informa business

10 9 8 7 6 5 4 3 2 1

Library of Congress Cataloging-in-Publication Data

Disciplinary approaches toward aging / edited with introductions by Donna Lind Infeld.
p. cm.
Includes bibliographical references.
Contents: v. 1. Biology of aging—v. 2. Psychology of aging—v. 3. Sociology of aging—v. 4. Anthropology of aging—v. 5. Economics of aging—v. 6. Politics, policy, aging.
ISBN 0-415-93895-3 (set : alk paper)—ISBN 0-415-93896-1 (vol. 1 : alk. paper) — ISBN 0-415-93897-X (vol. 2 : alk. paper)—ISBN 0-415-93898-8 (vol. 3 : alk. paper)—ISBN 0-415-93899-6 (vol. 4 : alk. paper)—ISBN 0-415-93900-3 (vol. 5 : alk. paper)—ISBN 0-415-93901-1 (vol. 6 : alk. paper)
1. Gerontology. 2. Aged. 3. Aging. I. Infeld, Donna Lind.
HQ1061.D557 2002
305.26—dc21 2002020627

SET ISBN 9780415938952
POD ISBN 9780415870665
Vol1 9780415938969
Vol2 9780415938976
Vol3 9780415938983
Vol4 9780415938990
Vol5 9780415939003
Vol6 9780415939010

Contents

Cellular and Molecular Biology of Aging

Age-Related Changes in Human Physiology

Pathophysiology and Aging

The Aging Brain

Cancer and Aging

Population Biology

Future Issues in the Biology of Aging

Series Introduction

Life expectancy around the world has been increasing dramatically over the last several decades. Global average life expectancy has now reached 66 years (U.N. 2000). This trend has resulted in a large, and still growing, elderly population. There are currently well over 550 million people in the world who are over age 60, and this number is expected to approach 1.2 billion by 2025 (AoA 1998).

Increasing longevity is a result of numerous factors. Chief among them are improved nutrition and sanitation, higher levels of healthy behaviors (e.g., reduced smoking), and improved medical care. Underlying these changes at the population level are higher levels of education, income, and knowledge about health, trends that are continuing to spread throughout the world.

However, this lengthening life expectancy and increasing size of the older population is not occurring evenly. In more developed regions of the world, including Europe, Northern America, Australia/New Zealand and Japan, the 14 percent of the population that was 65 or older in 2000 is expected to increase to 19 percent by 2020 (U.N. 2001). Life expectancy in the United States, now 76.9 years, is almost 30 years longer than it was in 1900 (AoA 2001). The age group 65 and over makes up over 12 percent U.S. of the population and is expected to increase to 20 percent by 2030 (Ibid). One way I like to represent this proportion to students is to compare it to the current 17.6 percent of the population of Florida that is age 65 and older. In other words, there will be a greater proportion of older people throughout the United States than currently live in Florida. If you have ever visited Florida, you will understand why this is hard to comprehend.

In less developed regions of the world, including Africa, Asia (excluding Japan), Latin America, the Caribbean, Melanesia, Micronesia and Polynesia, the growth rate of the elderly population is even greater than it is in more developed regions. The projected increase of people over age 60 is from 5 percent in 2000 to over 7 percent by 2020 (ibid.). This is increase of two percentage points represents a faster rate of growth than that occurring in the more developed parts of the world.

In addition to the growth of the older population, the older population itself is getting older. The oldest old population (age 80 and older) is growing at an even faster rate than the young-old (age 60–79), with an increase of the 60 and older age group worldwide from 11 percent to 19 percent projected for 2050 (U.N. 2000).

These numbers are so large and changing so rapidly that the consequences to societies and to individuals' lives are almost impossible to comprehend. Gerontology, the study of aging, is the field that attempts to undertake this daunting task. Further extension of the life span, and more importantly adding quality to the added years, will be based on our knowledge of the multi-faceted components of aging. These components include new levels of understanding and intervention in the normal processes of aging (biology), mood, morale, and life-satisfaction (psychology), productive and rewarding roles in retirement and widowhood (sociology), cultural norms and supports for aging members of communities (anthropology), distribution of resources, particularly income opportunities for retired persons (economics) and the ability to shape public policy to assure equity of access to public resources (political science). These challenges are at the core of the disciplines that contribute to the field of gerontology.

GERONTOLOGY

The term gerontology is based on the Greek word *geron*, which means old men. It is the field that involves the study of the many dimensions of aging. It is a relatively young field. The Gerontological Society of America, the primary American organization devoted to promoting the scientific study of aging, was founded in 1945, and the behavioral and social sciences section was established in 1956 (www.geron.org/history.htm). Since the 1960s, the study of aging has grown dramatically and there has been extensive theoretical development and growing research applying a wide range of disciplinary approaches to examining the phenomenon of aging.

Gerontology is a multidisciplinary field. Numerous disciplines are interested in and make unique contributions to understanding aging. What is less clear is the extent to which gerontology can be considered interdisciplinary, that is reflecting integration among disciplines in the study of aging. While some studies have interdisciplinary research questions and methods, most research on age-related topics continues to be either disciplinary or multidisciplinary. In other words, research is generally grounded in a single discipline or else involves a team of researchers exploring a range of discipline-based issues. However, in some cases it achieves significant integration of theoretical concepts and approaches within an individual study.

I come from a multidisciplinary gerontological background. My aging career started with support from the Administration on Aging as an undergraduate trainee at Portland State University. My degree was in Psychology with a minor in Sociology. My doctoral work at the Florence Heller

Graduate School for Advanced Studies of Social Welfare at Brandeis University continued this multidisciplinary approach involving coursework in economics and political science as well as continued work in gerontology.

In the twenty-five years since completing my formal education, my professional positions have involved teaching aging courses to students in the applied fields of health services management and policy, public health, and public administration. As a result of this education and these academic settings, I view my background as strong on breadth and short on depth in the basic disciplines that contribute to understanding aging. This lack of depth was a motivating factor for creating this anthology. In preparing for classes in my gerontology overview course I found myself trying to stay up to date with the latest disciplinary work in numerous fields. It was my goal to produce a set of publications to help faculty members and other professionals in the field of aging to keep abreast of these developments.

DEVELOPMENT OF THE ANTHOLOGY

The series is designed to provide an integrated set of foundational articles from the primary disciplines interested in the study of aging. Each volume includes articles that provide an historical overview of theoretical perspectives in that discipline. Each volume also explores major concepts and issues within the discipline, highlights recent research findings, and provides directions for future research

I decided to organize this anthology along traditional disciplinary lines for two reasons. First, most academic training in gerontology continues to be organized along disciplinary lines. Thus, disciplinary-based readings can be more easily integrated into courses and training. Second, despite the inter/multidisciplinary nature of much research in aging, most theoretical developments continue to emerge from disciplinary homes.

University courses and textbooks about aging can be found in all major academic disciplines. According to the Association for Gerontology in Higher Education, there are more than 500 campuses in the United States that offer aging courses in more than 30 disciplines and fields of study (www.aghe.org/natdbase.htm). For example, The University of Southern California, arguably among the most extensive gerontology programs in the United States, offers 37 undergraduate and 35 graduate courses thorough its Leonard Davis School of Gerontology. Disciplines represented include biology, physiology, psychology, sociology, demography, policy, and ethics (www.usc.edu/dept/gero/ldsg/index.htm).

In addition to these traditional disciplinary approaches, numerous professional fields offer courses related to aging, many of which are grounded on disciplinary theories and applied research. Training in fields such as social work, nursing, counseling, recreation therapy, and health care administration depend on the multi-disciplinary literature on aging. Further, international attention to aging is expanding rapidly. As a result there is broad and growing demand for accessible and coherent gerontological literature.

This six-volume anthology on Disciplinary Approaches to Aging is designed to help respond to the need for more gerontological literature. The volumes include 1) Biology, 2) Psychology, 3) Anthropology, 4) Sociology, 5) Economics, and 6) Political Science. The series draws from a wide array of published sources from the United States and around the world. While most articles were selected from academic journals, a broad range of sources was reviewed including professional publications, book chapters, web sites and government reports and documents.

The process of searching for articles was complex and varied across the disciplines. The criteria I used in making selections were:

Overview or review articles rather than reports of single research studies

My goal was to provide articles that would bring readers up to date with current developments in each discipline. Review articles were easier to identify in biology because several medical literature databases provide a search option for them. In each of the other disciplines I had to look through hundreds of abstracts and articles to identify those that provide a broad overview of the topic or field.

Recent publication

Almost all of the articles were published within the last five years, with the large majority less than two years old. Recent publication supports the anthology's objective of focusing on *current* developments in each discipline.

Broadest possible array of authors and publications

In several disciplines I could have produced a whole volume from articles published in a single journal or written by a single author. However, my objective was to represent the widest possible range of publications that address issues of aging. Therefore, no authors or publications have more than one article in any volume of the anthology.

Inclusion of publications from around the world

Priority was given to articles that contribute to the understanding of current *international* theoretical and research developments contributing to disciplinary approaches to aging as part of the effort to represent the breadth of the field.

With these criteria in mind, I applied multiple search engines and databases, focusing first on the academic and professional journals they catalogue. I started with broad terms (e.g., anthropology and aging/old age/aged/elderly) but also searched a range of subtopics (culture, ethnography, etc.). For some of the volumes there was ample material identified using these steps. Next I had to make difficult choices, again using the criteria identified above. For several volumes the initial search processes did not provide adequate coverage of work in the discipline. I then turned to other sources such as the web and book chapters to round out the coverage.

Despite these efforts, I know that there are some gaps in the selections included here. When I was overwhelmed with the choices and decisions, I tried to keep in mind the overall goal of creating a resource for faculty and other professionals to use to provide the basis for developing and updating lectures in a wide range of aging-related courses. To supplement the readings, a list of additional references is provided in each volume to direct readers to additional resources and recent examples of research conducted in that field.

The multi-disciplinary nature of aging research and theory presented a particular challenge when sorting and organizing the articles into discipline-specific volumes. Not only do many major research studies include, for example, psychological as well as sociological and economic data, but concepts often cut across disciplinary lines as well. Concepts such as norms and roles, whose homes are in sociology, are also prevalent in psychological and anthropological research. Further, it was not always easy to decide where to place an article when a sociologist was publishing in an anthropological journal or visa versa. And, of course, research in social psychology belongs in both sociology and psychology. The most significant area of overlap was between economics and political science, more specifically, public policy. In fact much of the economic research on aging could have been just as appropriately considered to be public policy research. Despite these challenges, I feel the resulting compilation meets the goal of being a collection of articles that captures the major issues, concepts, and research directions of each disciplinary field.

As I struggled with making the many decisions involved in producing this anthology, it became painfully clear that knowledge simply does not fit into neat and distinct boxes. Thanks to Mark Georgiev, Associate Editor at Routledge, who reminded me that there is no perfect anthology. Without his support, encouragement, and prompting this project would not have been completed. My thanks also go to Lauren Block who tried to make the product as perfect as possible. Her help was invaluable throughout every step. Thanks also to Marcel and Amanda for their never-ending love and support. In the end, producing *Disciplinary Approaches to Aging* was an education for me and I hope it will be a useful resource for my colleagues around the world.

Donna Lind Infeld, Ph.D.

REFERENCES

Administration on Aging (November 9, 1998). International Activities of the Administration on Aging. Retrieved March 4, 2002 from *http://www.aoa.gov/Factsheets/international.html.*

Administration on Aging (December 21, 2001). A Profile of Older Americans: 2001. Retrieved March 4, 2002, from *http://www.aoa.gov/aoa/STATS/profile/profile.pdf.*

Association for Gerontology in Higher Education. Retrieved April 1, 2002, from *http://www.aghe.org/natdbase.thm.*

Gerontological Society of America. Retrieved April 1, 2002, from *http://www.geron.org/history.htm.*

United Nations/Division for Social Policy Development, (May 24, 2000). The Aging of the World Population. Retrieved March 4, 2002, from *http://www.un.org/esa/socdev/ageing/agewpop.htm.*

United Nations, Population Division (2001). Population Division of the Department of Economic and Social Affairs of the United Nations Secretariat, World Population Prospects: The 2000 Revision. Retrieved March 4, 2002, from *http://esa.un.org/unpp/index.asp?Panel=3*

University of Southern California. Leonard Davis School of Gerontology. Retrieved April 2, 2002, from *http://usc.edu/dept/gero/ldsg/index.htm.*

Volume Introduction

Exciting things are happening in the understanding of the basic mechanisms of the biology of aging. Periodically these developments make their way to the newspapers and television, but these brief spotlights do not allow a thorough understanding of the mechanisms explored and the implications of the findings. It is the objective of this volume to select current materials that provide an overview of research in the various areas of biology and physiology of aging.

The Gerontological Society of America has, on occasion, sponsored a pre-conference workshop entitled "Biology for Non Biologists." This workshop presents basic concepts in biogerontology and discusses how the new findings refine our views of the aging process. Its purpose is to communicate biological research findings in a way that non-biologists can understand. Articles for this volume of *Disciplinary Approaches to Aging* were selected with same idea in mind. While the articles are written by leading scientists in the field, they are not reports of findings from individual studies. Rather they are overview articles that synthesize the current understanding about a range of aspects of the underlying mechanisms of biophysical aging.

I am not a biologist. So why did I include biology in this series? I could have argued for a series on "Social Science Approaches to Aging." However, biology is the fundamental mechanism that drives the changes that result in problems and issues for the other disciplines — so it could not be ignored. For example, biological changes in the brain are clearly associated with changes in behavior. Thus we need to understand the biology of aging to have a baseline for understanding the psychology of aging.

When I first started teaching an overview course on aging I invited a biologist to give the lecture on biology of aging. Unfortunately, too often the students were unable to understand what the guest expert was trying to explain because, as a scientist, he did not understand how little the students understood of basic biology. I decided that I should be knowledgeable enough to give my own overview lecture on the biology of aging. With the help of works like Leonard Hayflick's *How and Why We Age* (1994; 1996) and with print materials and videos available from the National Institute on Aging (www.nia.nih.gov), I was able to become conversant with the major theories about why we age.

There are two underlying questions that drive the study of biology of aging:

1. Why do human bodies change over time?
2. What is normal aging?

Numerous theories have been developed to try to answer the question of why people age. Over time it has become increasingly clear that there will likely not be one single answer but rather a combination of causes resulting in the multifaceted process we call aging. It is also difficult to answer the question of what constitutes normal aging. Conditions that are viewed as diagnosable illnesses may result from gradual age-related processes. Therefore, one must question what changes constitute normal aging as opposed to disease. Our understanding of the answer to this question is also changing. In general, the biological process of human aging can be viewed as a combination of decreasing physiological functioning and increasing vulnerability to disease.

There are some anatomical manifestations of aging common to all species, such as thinning and graying of hair and changes in pigmentation such as age spots. However, more significant than the similarities in aging processes are dramatically different rates at which these age changes occur across, and even within, a given species. There is also significant variation in the rate of age-related changes that occur within a given individual.

An important contribution to understanding variability in aging was made by the Successful Aging Study (Rowe and Kahn, 1998), funded by the MacArthur Foundation. This multidisciplinary study involved longitudinal examinations of relatively high-functioning men and women in their 70s. Sixteen scientists focused a range of studies on the components of successful aging. The studies addressed questions such as why some people in their 80–90s function well while others do not? The work identified three major components of successful aging: the ability to engage with life, avoid disease, and maintain cognitive and physical functioning. The study provides evidence that all of these components can be modified, thereby increasing the likelihood of successful aging. The article by Seeman et al. (2001), included in this volume, is a product of the Successful Aging Study. It examines "*Allostatic load as a marker of cumulative biological risk.*" The findings of the Successful Aging Study have contributed to increasing interest in a range of efforts to promote healthy aging.

The most significant ongoing study of aging in the United States is the Baltimore Longitudinal Studies of Aging (BLSA). The BLSA is a division of the National Institute on Aging of the National Institutes of Health in the U.S. Department of Health and Human Services. The study started in 1958 with 50 non-hospitalized men. Women were added to the study in 1978.

Now there are over 2,400 subjects; however, as a result of the voluntary nature of the sample, subjects do not reflect the general population of the United States. For example, recruitment of an adequate number of minorities is an ongoing challenge and volunteers tend to be more highly educated than the general population. Participants undergo a wide array of physical and psychological tests, spending two-and-a-half days under study every year or two, depending on their age. Because of its ability to examine a large group of subjects over a long period of time, the BLSA provides a unique opportunity to explore normal age changes. A theme across the findings of the BLSA is that there is often more variability of performance between subjects of the same age than between age groups. In this volume, Sorkin et al. (1999) and Haman et al. (2001) present findings from the BLSA on physiological changes that occur with aging.

The National Institute on Aging, one of the twenty-five institutes and centers of the U.S. National Institutes of Health (NIH), in addition to operating the BLSA, supports a wide range of biological and biomedical research on aging. Through its extramural and intramural programs, including international cooperative research initiatives, NIA is a major driving force in biology of aging (www.nia.nih.gov/research/). Five articles included in this volume were produced with NIA funding. Other Institutes of the NIA also support aging-related research and two such projects are reported on in this volume.

In addition to the significant contribution of biologists in the United States, many other countries are also producing important research in this area. Countries represented in this volume include: Canada (Murray 1996), England (Mann 1997), France (Flanigan et al. 1998), Germany (Graeber et al. 1998), Israel and Russia (Kobliansky et al. 1995), Scotland (Charlesworth 2000), and Switzerland (Saurat 1995).

A wide range of biological research endeavors, supported by government agencies and foundations around the world, have moved theories of aging beyond the early approaches, such as the "Wear and Tear Theory," to new levels of understanding including attention to the role of genes and telemeres in determining the rate of aging. Researchers are investigating areas from caloric restriction to genetic engineering, with some biologists predicting extension of the human life span beyond 120 years (Walford, 1996). Clearly, these are exciting times in the biology of aging and it is important for non-biologists to understand the direction of these developments in order to examine the implications of changes and plan for the future.

REFERENCES

Hayflick, L. 1994. *Why We Age*. Ballantine Books.

National Institute on Aging. Research Programs. Retrieved April 2, 2002, from *http://www.nia.nih.gov/research*.

Rowe, J. W. and R. L. Kahn. 1998. *Successful Aging: The MacArthur Foundation Study*. Pantheon.

Walford, R. 2000. Beyond the 120 Year Diet. NY: Four Walls, Eight Windows.

Additional References

Barzilai, N., & Gupta, G. (1999). Revisiting the role of fat mass in the life extension induced by caloric restriction. *Journals of Gerontology Series A-Biological Sciences & Medical Sciences, 54*(3), B89–96; discussion B97–88.

Blazekovic-Milakovic, S., Kern, J., & Kulenovic, M. (2000). Health status as geneologic burden in aging process. *Collegium Antropologicum, 24*(1), 79–89.

Bonneux, L., Barendregt, J. J., & Van der Maas, P. J. (1998). The expiry date of man: a synthesis of evolutionary biology and public health. *Journal of Epidemiology & Community Health, 52*(10), 619–623.

Butler, R. N. (2000). The realities of aging. *Geriatrics, 55*(5), 3–4.

Campisi, J. (2001). From cells to organisms: can we learn about aging from cells in culture? *Experimental Gerontology, 36*(4–6), 607–618.

Carlson, J. C., & Riley, J. C. (1998). A consideration of some notable aging theories. *Experimental Gerontology, 33*(1–2), 127–134.

Chen, C. P. (1997). Genetics of Alzheimer's disease: just how is molecular biology going to help grandma? *Singapore Medical Journal, 38*(3), 100–101.

Cohen, H. J. (2000). In search of the underlying mechanisms of frailty. *Journals of Gerontology Series A-Biological Sciences & Medical Sciences, 55*(12), M706–708.

Comai, L. (1999). The nucleolus: a paradigm for cell proliferation and aging. *Brazilian Journal of Medical & Biological Research, 32*(12), 1473–1478.

Cristofalo, V. J. (1996). Ten years later: what have we learned about human aging from studies of cell cultures? *Gerontologist, 36*(6), 737–741.

Ershler, W. B., & Longo, D. L. (1997). The biology of aging: the current research agenda. *Cancer, 80*(7), 1284–1293.

Finch, C. E., & Austad, S. N. (2001). History and prospects: symposium on organisms with slow aging. *Experimental Gerontology, 36*(4–6), 593–597.

Ford, K. G., Darling, D., Souberbielle, B., & Farzaneh, F. (2000). Protein transduction: a new tool for the study of cellular ageing and senescence. *Mechanisms of Ageing & Development, 121*(1–3), 113–121.

Fossel, M. (2000). Human aging and progeria. *Journal of Pediatric Endocrinology, 13*(Suppl 6), 1477–1481.

Fukagawa, N. K. (1999). Aging: is oxidative stress a marker or is it causal? *Experimental Biology & Medicine, 222*(3), 293–298.

Gershon, H., & Gershon, D. (2000). Paradigms in aging research: a critical review and assessment. *Mechanisms of Ageing & Development, 117*(1–3), 21–28.

Globerson, A. (2001). Biogerontology research in Israel. *Experimental Gerontology, 36*(1), 3–19.

Gutteridge, J. M., & Halliwell, B. (2000). Free radicals and antioxidants in the year 2000. A historical look to the future. *Annals of the New York Academy of Sciences, 899*, 136–147.

Hayflick, L. (1996). *How and Why We Age*. New York: Ballantine Books.

Holloszy, J. O. (2000). The biology of aging. *Mayo Clinic Proceedings, 5*(Suppl), S3–8; discussion S8–9.

Holmes, D. J., Fluckiger, R., & Austad, S. N. (2001). Comparative biology of aging in birds: an update. *Experimental Gerontology, 36*(4–6), 869–883.

Kanungo, M. S., Gupta, S., & Upadhyay, R. (1997). Molecular biology of ageing. *Indian Journal of Medical Research, 106*, 413–422.

Kawas, C., Gray, S., Brookmeyer, R., Fozard, J., & Zonderman, A. (2000). Age-specific incidence rates of Alzheimer's disease: the Baltimore Longitudinal Study of Aging. *Neurology, 54*(11), 2072–2077.

Kirkwood, T. B., & Austad, S. N. (2000). Why do we age? *Nature, 408*(6809), 233–238.

Klapper, W., Parwaresch, R., & Krupp, G. (2001). Telomere biology in human aging and aging syndromes. *Mechanisms of Ageing & Development, 122*(7), 695–712.

Manton, K. G. (1999). Dynamic paradigms for human mortality and aging. *Journals of Gerontology Series A-Biological Sciences & Medical Sciences, 54*(6), B247–254.

Martin, G. M. (2000). Some new directions for research on the biology of aging. *Annals of the New York Academy of Sciences, 908*, 1–13.

Miller, R. A. (1997). When will the biology of aging become useful? Future landmarks in biomedical gerontology. *Journal of the American Geriatrics Society, 45*(10), 1258–1267.

Oehmichen, M., & Meissner, C. (2000). Natural death. *Gerontology, 46*(2), 105–110.

Oulton, R., & Harrington, L. (2000). Telomeres, telomerase, and cancer: life on the edge of genomic stability. *Current Opinion in Oncology, 12*(1), 74–81.

Paolisso, G., Barbieri, M., Bonafe, M., & Franceschi, C. (2000). Metabolic age modelling: the lesson from centenarians. *European Journal of Clinical Investigation, 30*(10), 888–894.

Perls, T. (2001). Genetic and phenotypic markers among centenarians. *Journals of Gerontology Series A-Biological Sciences & Medical Sciences, 56*(2), M67–70.

Raha, S., & Robinson, B. H. (2000). Mitochondria, oxygen free radicals, disease and ageing. *Trends in Biochemical Sciences, 25*(10), 502–508.

Robert, L., & Peterszegi, G. (1998). Aging and matrix biology. *Pathologie et Biologie, 46*(7), 491–495.

Roth, G. S., Ingram, D. K., Black, A., & Lane, M. A. (2000). Effects of reduced energy intake on the biology of aging: the primate model. *European Journal of Clinical Nutrition, 54*(Suppl 3), S15–20.

Rothschild, H., & Jazwinski, S. M. (1998). Human longevity determinant genes. *Journal of the Louisiana State Medical Society, 150*(6), 272–274.

Rowe, J. W., & Kahn, R. L. (2000). Successful aging and disease prevention. *Advances in Renal Replacement Therapy, 7*(1), 70–77.

Shay, J. W., & Wright, W. E. (2000). Hayflick, his limit, and cellular ageing. *Nature Reviews Molecular Cell Biology, 1*(1), 72–76.

Shock, N. W., & Baker, G. T., 3rd. (1990). Gerontology: the scientific study of aging, past and future. *Journal of Gerontology, 45*(2), B31.

Skulachev, V. P. (2001). The programmed death phenomena, aging, and the Samurai law of biology. *Experimental Gerontology, 36*(7), 995–1024.

Slavkin, H. C. (1998). And we all lived happily ever after: understanding the biological controls of aging. *Journal of the American Dental Association, 129*(5), 629–633.

Strauss, E. (2001). Longevity. Growing old together. *Science, 292*(5514), 41–43.

Wallace, D. C., Brown, M. D., Melov, S., Graham, B., & Lott, M. (1998). Mitochondrial biology, degenerative diseases and aging. *Biofactors, 7*(3), 187–190.

Warner, H. R. (2001). Support for basic gerontological research in the USA. *Experimental Gerontology, 36*(3), 403–412.

Williams, G. C. (1999). The 1999 Crafoord Prize lectures. The Tithonus error in modern gerontology. *Quarterly Review of Biology, 74*(4), 405–415.

CHAPTER 1

Defining Aging

How old would you be if you didn't know how old you was?

—Satchel Paige

ONE of our earliest realizations as children is that most of the things that we see or know, including humans, animals, and inanimate objects, change or deteriorate with the passage of time. Some take aging for granted, accepting it as the fate of most things with which they are familiar and a normal consequence of the passage of time. Others spend a substantial part of their time in activities designed to arrest or thwart the unwelcome effects of aging on themselves and their surroundings. Many of the things in our homes or workplaces need to be protected, repaired, and renewed with time. Other things, as they age, come to be considered more beautiful and desirable.

With material objects, most of the ravages of time are the result of oxidation—the combination of the molecules that compose the objects with atmospheric oxygen. In iron-containing objects we call the phenomenon rust, but the avidity of oxygen for other material things also produces changes that we associate with their aging. Other age changes can be attributed to temperature changes, water, radiation, trauma, microorganisms, or insects. If we own pets, we see changes in them. We are literally surrounded with living and inanimate things that are continuously changing with time. Aging is an omnipresent part of our material and living world, and eventually, we realize that the phenomenon is occurring even within ourselves.

Aging is the only fatal affliction that all of us share. Scientists long scorned conducting research in the field of biological aging because it was deemed uninteresting or fundamentally intractable. Happily, this attitude has changed in recent years, and a major effort has been mounted in many countries to understand the aging process. But in order to understand the process, we first need to define the changes that we call aging. After all, as the French aphorist put it, "Most troubles of mankind are a matter of grammar."

CHRONOLOGICAL VERSUS BIOLOGICAL AGE

Aging defies easy definition, at least in biological terms. Aging is not merely the passage of time. It is the manifestation of biological events that occur over a span of time. There is no perfect definition of aging but, as with love and beauty, most of us know it when we experience it or see it. We all recognize an old person when we see one, and some of us are quite good at estimating chronological ages. However, subjective determinations based on appearances are frequently wrong, and, more importantly, age in years does not directly correlate with biological age. What is needed is a measurement of something biological that changes as a function of increasing age—a measurement that distinguishes biological age from chronological age. The distinction is crucial.

Scientific knowledge advances only when we can measure or count things in some way. We can determine the age of woody trees by counting the annual rings and the age of some fish by counting the layers on their scales, but we have no reliable measurement to determine the biological age of humans and most other animals. Of course, with humans we can use birth certificates, but, even if they are reliable, they specify only a single point in time. Chronological age measures how much time, how many years, have elapsed since that point. It tells us when to celebrate birthdays and what number to put on forms when asked our age, but for gerontologists aging is chronological only in the legal or social sense. Time itself produces no biological effects. Events occur *in* time but not *because of* its passage. The biological events that follow birth happen at different times and occur at different rates in each of us.

When we remark with surprise that someone "looks young" (or old) for his or her chronological age, we are observing that we all age biologically at *different rates*. Gerontologists have much good scientific evidence that this apparent difference is real. It is likely that age changes begin in different parts of the body at different times and that the rate of annual change varies among various cells, tissues, and organs, as well as from person to person. Unlike the passage of time, biological aging, sometimes called functional aging, defies easy measurement. What we would like to have is one or a few measurable biological changes that mirror all other biological age changes without reference to the passage of time, so that we could say, for example, that someone who is chronologically eighty years old is biologically, or functionally, sixty years old. This kind of measurement would help explain why one eighty-year-old has so many more youthful qualities than does another eighty-year-old, who may be biologically eighty or even ninety years old.

In terms of aging, we resemble a clock shop. Each of our many tissues and organs behaves like an independent clock ticking at a rate different from that of the others. Because of this, a person of a given chronological age could be considerably younger or older *biologically*, depending on the average of how quickly or slowly all his or her clocks are ticking. Knowing our biological age would be much more informative than knowing our chronological age, but, alas, we have no way of measuring it.

That is not for lack of trying. Gerontologists have tracked various biological measurements for humans over time with the hope that the changes found could predict the rate of aging. They have measured hair color, ear length, grip strength, heart function, exercise capacity, and dozens of other variables. None of these studies has established an infallible measurement of biological age, for at least one important reason: There is too much individual variability in the potential markers of aging. Even at birth humans have an enormous range of difference in most variables that can be measured. The differences are frequently unrelated to age. It is as if each of us begins the aging race with our feet on a different starting line. For example, women are generally smaller than men and therefore have a reduced vital capacity. (Vital capacity is measured as the volume of air one can forcibly exhale after taking a deep breath.) Yet, women generally live longer than men, so measurement of vital capacity is obviously useless as a marker for the rate of biological or functional aging.

A better approach to determining biological or functional age might lie in measuring the nearness of death as a function of some earlier biological change. Of course, the population studied must consist entirely of people who died from a known cause, and the biological change must be a function of aging, not illness or lifestyle. Retrospective analyses on human populations have successfully determined some variables that have value in predicting impending death. This kind of study is called ***risk factor analysis***. Smokers, for example, have been shown by this method to have a shorter life expectation than nonsmokers and those who quit smoking. This is valuable information, but it does not tell us anything meaningful about the rate of aging. Smokers do not age faster than nonsmokers; they simply die sooner on average.

Even if there were an accurate way to determine biological age, knowing a person's biological age might be a mixed blessing. It could give rise to some bizarre birthday celebrations. Think what a deliriously happy occasion it would be if you celebrated your fiftieth chronological birthday knowing that you were only forty years old biologically. However, it would be an unhappy party if you were chronologically fifty and biologically sixty! Husbands and wives might discover that, although one is chronologically older than the other, their biological ages are reversed. We might even find adult children who are biologically older than their parents! Should retirement age and life insurance premiums be based on biological, not chronological, age? If so, some people might be forced to retire or to increase their life insurance premiums at age fifty, others at age eighty, or later. These seemingly odd situations will become more likely if a reliable biological marker is ever discovered that can demonstrate someone to be much older or younger biologically than he or she is chronologically.

We do have fairly reliable biological markers for age in some other animals and in plants. Still, these markers tell us not about the rate of aging, but only the passage of time. The two are related, of course, but the passage of time does not measure the rate of aging. The most familiar example of a biological marker in the plant kingdom is the growth rings that are laid down annually by woody trees. Time-dependent rings or markings have also been found in some animals. In mammals, markings on teeth of seals and the horns of sheep, goats, and caribou have been used to determine chronological age. The earwax of whales also has time-dependent growth zones. No time-dependent markers have been found in birds, which is one reason bird-banding techniques are used. Growth zones are found in the bones of snakes and other reptiles and in the annual layers that form the scales of many fish species. The number of fin rays has been used to measure the age of catfish and sturgeon. Other signs that occur in some animals regularly over time include markings on shells and on otoliths, which are tiny stones found in the balancing organ of the inner ear.

Humans have none of these age markers, which is why we must rely on birth certificates, other written documents, hearsay, or memory to establish chronological age.

LONGEVITY, AGING, AND DEATH

Any serious consideration of biological aging must define all three of the phenomena that characterize the finitude of life. These are longevity, aging, and death.

Longevity is the period of time that an animal can be expected to live, given the best of circumstances. For newborn humans, average longevity (life expectation) in developed countries is about seventy-five years and maximum longevity (life span) is about one hundred fifteen years. The essential question about longevity is, Why do we live as long as we do?

Aging represents losses in normal function that occur after sexual maturation and continue up to the time of maximum longevity for members of a species. The essential question about aging is, Why do we grow old?

Death is the final event that ends life. It is not, as one wag said, nature's way of telling you to slow down. The essential question about death is, of course, Why do we die?

Death need not be related to aging at all. For example, the death of a child by accident or infectious disease is unrelated to aging. Death is linked to aging in the sense that, with age, the probability of death for all members of a species increases. The qualification "all members of a species" is important because certain subgroups may die before they have an opportunity to age. For example, the steep rise in the death rate for young men has nothing to do with the aging process; it has everything to do with automobile accidents and, in some communities, homicide.

HOW OLD ARE YOU—REALLY?

This question might seem silly. Most of us know our date of birth as well as we know our own name. However, in many less-developed countries dates of birth are often unknown. In China, on the other hand, even illiterate people can provide their precise date of birth. The Chinese, however, consider themselves to be one year old on the day of their birth and determine age according to a lunar calendar, which makes it very easy to miscalculate the ages of the Chinese according to the Western solar calendar or of Westerners according to the Chinese calendar. (There are 29.5 days in a lunar month. Thus, a twelve-month lunar year rapidly falls out of synchronization with the 365.25-day solar year. To compensate, the Chinese insert an extra lunar month at irregular intervals, which makes the year in which a month is added a leap year.)

Similar uncertainties about age occur when we ponder what it is in each of us that is one year older at each birthday. Ordinarily we think not about the aging of our cells, but rather about how we relate to our family and friends or to what we perceive that society expects of someone our age. But if you say figuratively that you are not the same person you were five or ten years ago, you are quite literally right. All of us are composed of billions of individual cells and the products that cells make. Most of the cells present in our body today were not present five or ten years ago. In fact, some were not present yesterday.

There are at least three kinds of tissue in which cells are in a constant state of division: the skin, the cells lining the digestive tract, and the cells that make red and white blood cells. Many of the cells in these tissues divide every day, and the older cells are simply discarded. Old skin cells at the outermost edge of our skin are ultimately sloughed into the environment. Old gut cells are sloughed into the digestive tract, (yes, we cannibalize ourselves!) and old blood cells are literally eaten by other cells called phagocytes. Cells in other tissues are replaced by new cells (a process called turnover) on a longer time scale.

Two important kinds of tissue are composed of cells that do not divide or turn over. These are nerve cells (found throughout the body, and including the brain) and skeletal and heart muscle cells. Most of the brain and muscle cells that you will possess for life are present at birth or soon thereafter and do not divide further. Muscles, hearts and brains become larger after birth partly because the cells that form them increase in size (not in number) and partly because *other* cells surrounding the neurons and muscle cells divide and contribute to size increase.

If many of your cells turn over in less than ten years, then how old are you, *really*? After all, the cells that turned over during the last ten years are gone and new ones may have replaced them several times, so your present cells may be younger than the cells that were present a decade ago. Most of your cells do not accompany you from the cradle to the grave, so you really are not the same person that you were even a few years ago, either figuratively or literally. The best answer to the question "How old am I?" is that you are as old as your oldest cells—those neurons and skeletal muscle cells you were born with and still have. When you celebrate your next birthday, accuracy demands that you celebrate the birthday only of your nerve and muscle cells and the cell products that haven't been replaced!

But before you plan the bizarre celebration of the birthday of your neurons and muscle cells, I will add another complication. Although we are all composed of individual cells and their products, the cells themselves consist of smaller units called molecules, which in turn are composed of atoms. The complication is that molecules may turn over, or be replaced, as a result of normal metabolic processes, without the individual cells in which they are packaged being replaced. What happens in an old cell is analogous to what might happen in an old car. If the car has had *every* part replaced it is not the same old car. If the cell has had *every* part replaced it is not the same old cell. The neurons you were born with might appear today to be the same cells, but in reality many of the molecules that composed them when you were born (except DNA) may have been replaced with new molecules. So nondividing cells may not be the same cells you were born with after all!

It is probable that turnover in nondividing cells does occur to some extent, although except for DNA we have no proof that it is complete. If most of the molecules in your nerve and muscle cells turn over, then your birthday celebration is going to be tricky for two reasons. First, if all but a few of the molecules in all of your cells have turned over, then you are literally a different person today. Second, all of your molecules, whether they have turned over or not, are composed of more fundamental units called atoms, most of which have been the same since our planet formed. You and I simply represent unique rearrangements of ancient atoms that are themselves billions of years old. We are really composed of billion-year-old atoms; we might actually claim to be immortal! In that sense we are all billion-year-olds no matter when we were born, and celebrating birthdays is absurd.

In partial compensation for the loss of birthday celebrations consider this benefit: The atoms in our bodies may have been part of the body of someone else long since dead. This is the only scientific basis for believing that we, the living, represent a form of reincarnation. When we die our atoms will dissipate into the environment, and some, perhaps, will become part of another human in a continuing pattern of recycling atoms. You could argue that this is scientific evidence for life after death; our atoms are immortal but we, as individuals, are not. Upon death our concept of self is gone forever, regardless of where our atoms came from or where they will go.

Attempting to precisely define the age of a living organism can obviously boggle the mind. About the best you can do, if you insist on painstaking accuracy, is to celebrate the birth of all of your dead cells because, although they no longer exist, those cells generated the cell lineages whose progeny you are today.

PHYSIOLOGICAL REVIEWS
Vol. 78, No. 2, April 1998
Printed in U.S.A.

The Free Radical Theory of Aging Matures

KENNETH B. BECKMAN AND BRUCE N. AMES

Department of Molecular and Cell Biology, University of California, Berkeley, California

Beckman, Kenneth B., and Bruce N. Ames. The Free Radical Theory of Aging Matures. *Physiol. Rev.* 78: 547–581, 1998.—The free radical theory of aging, conceived in 1956, has turned 40 and is rapidly attracting the interest of the mainstream of biological research. From its origins in radiation biology, through a decade or so of dormancy and two decades of steady phenomenological research, it has attracted an increasing number of scientists from an expanding circle of fields. During the past decade, several lines of evidence have convinced a number of scientists that oxidants play an important role in aging. (For the sake of simplicity, we use the term *oxidant* to refer to all "reactive oxygen species," including $O_2^{-}\cdot$, H_2O_2, and •OH, even though the former often acts as a reductant and produces oxidants indirectly.) The pace and scope of research in the last few years have been particularly impressive and diverse. The only disadvantage of the current intellectual ferment is the difficulty in digesting the literature. Therefore, we have systematically reviewed the status of the free radical theory, by categorizing the literature in terms of the various types of experiments that have been performed. These include phenomenological measurements of age-associated oxidative stress, interspecies comparisons, dietary restriction, the manipulation of metabolic activity and oxygen tension, treatment with dietary and pharmacological antioxidants, in vitro senescence, classical and population genetics, molecular genetics, transgenic organisms, the study of human diseases of aging, epidemiological studies, and the ongoing elucidation of the role of active oxygen in biology.

I. INTRODUCTION

The study of aging, by nature multidisciplinary, has been characterized by a dizzying variety of theories, a huge phenomenological literature, and the absence of firmly established primary causes. The diverse life histories of animal species, which manifest aging in very different ways, has been an obstacle to testing unified theories. For experimental gerontology to provide more than a catalog of age-related changes, it has been necessary for biologists to define the alterations that are common to most old cells, tissues, and animals, while simultaneously respecting that there is not a single phenomenon of aging or a single cause. This has taken some time, and from the outside it may have appeared that the field has been mired in phenomenology. Perhaps for this reason, the study of aging was until recently avoided by molecular biologists, who naturally favored clear-cut phenomena. As the molecular details of development, cancer, and immunology yielded to modern tools during the 1970s and 1980s, the field of aging lagged, and mechanisms responsible for aging failed to emerge.

Throughout this time, though, there has never been a shortage of unified theories attempting to reduce aging to something more tractable. In fact, gerontologists have been prolific in this regard (83, 205). Whereas some researchers have believed that a small number of random, deleterious mechanisms could explain degenerative senescence, others have opted for theories of "programmed" aging, in which senescence is the final destination in a developmental pathway. In the course of these debates, a number of scientists have rallied around a set of ideas called the free radical theory of aging: loosely, the belief that damage by reactive oxygen species is critical in determining life span. This theory inspired many experiments in which evidence of oxidative damage in aged animals was sought.

In the last 10 years or so, the nature of aging research has changed dramatically; one might say that the field has entered early adulthood. The tools of molecular biology are now sophisticated and accessible enough that researchers within gerontology have adopted them. At the same time, molecular biologists situated on the edge of aging research have made inroads and have discovered that fruit flies and nematodes are amenable to the study of aging. Also, medical researchers investigating human diseases of aging, such as Alzheimer's disease (AD) and inherited progerias, have overcome long-standing roadblocks.

It has been gratifying, therefore, that many of the preliminary studies in what might be called "molecular gerontology" lend credibility to the free radical theory. Results from disparate experimental systems have recently shown that oxygen radicals play a role in degenerative senescence, and the pace of discoveries is quickening. The likely result of this collision of scientific approaches will be the unraveling of the physiological tangle of aging, and it seems safe to say that one of the important knots will turn out to be oxidative stress.

However, there is a danger that in the excitement of theoretical confirmation, certain nuances are lost. For instance, the revelation that oxygen radicals may be involved in neurodegeneration does not mean that oxidative stress determines life span. The free radical theory has sought not only to explain the mechanisms of degenerative senescence, but it has also attempted to explain differences between species' life spans in terms of oxidants. So, although many recent studies indicate that oxygen radicals play some kind of role in aging, only a small number of these support the more ambitious version of the free radical theory. On the other hand, there is no reason to cling to such a stringent version of the free radical theory, and it is becoming apparent that whether or not they determine life span, oxygen radicals are certainly important players in aging's pathophysiology. In other words, the scope of the free radical theory of aging should include aging-associated oxidative stress in general, rather than limiting itself to those oxidative events that may determine life span. In fact, many current articles indicate that such a blurring of distinctions has already occurred and that as it is commonly used, "free radical theory" encompasses a broad set of ideas. Therefore, our first purpose is to delineate these different conceptions of the free radical theory, as a prerequisite to its critical evaluation.

Because of the recent popularity of free radical research, a large number of reviews have addressed various aspects of the interplay between oxidants and aging (6,

15, 18, 33, 51, 57, 58, 60, 65, 82, 85, 87, 103, 106, 123–126, 130, 161, 166, 196, 203, 227, 257, 273, 275, 288, 293, 297, 307, 312, 315, 335, 338, 348, 353, 357). Rather than merely updating this literature, our aim is to provide a systematic categorization of the types of experiments that have been performed. The phenomenon and study of aging are incredibly diverse, encompassing organisms from rotifers to mammals and techniques from physiology to genetics. Although it is precisely the broad sweep of evidence that lends the free radical theory its appeal, the menagerie of animals and techniques sometimes obscures the logic. By breaking the literature down into smaller pieces (a practice we find necessary ourselves), we hope to make it easier for readers to judge the theory. Moreover, by imposing a structure, we aim to highlight novel and definitive approaches, because it is these that will replace the phenomenology of past decades.

In this review, then, we briefly outline the evolution of the free radical theory and then delineate the different areas of evidence. We focus on recent experiments and point to the areas that we feel are most likely to provide future insights. The way in which we have categorized the literature is outlined in the table of contents (sects. IV-XVII) and in Table 1. Although any such system is somewhat arbitrary, we hope that ours will make it easier both to assimilate the existing literature and to envision future experiments. In writing a review on as broad a topic as the free radical theory, we have been forced to limit both the content and the number of references cited. Although we have done our best to include recent work, omissions were inevitable. We apologize to all authors whose work we have not managed to include, and direct readers to other recent reviews for material we have left out.

II. AN OVERVIEW OF THE FREE RADICAL THEORY OF AGING

A. Origins of the Free Radical Theory

In 1956, Denham Harman suggested that free radicals produced during aerobic respiration cause cumulative oxidative damage, resulting in aging and death. He noted parallels between the effects of aging and of ionizing radiation, including mutagenesis, cancer, and gross cellular damage (120, 128). At the time, it had recently been discovered that radiolysis of water generates hydroxyl radical (•OH) (319), and early experiments using paramagnetic resonance spectroscopy had identified the presence of •OH in living matter (45). Harman (120) therefore hypothesized that endogenous oxygen radical generation occurs in vivo, as a by-product of enzymatic redox chemistry. He ventured that the enzymes involved would be those "involved in the direct utilization of molecular oxygen, particularly those containing iron." Finally, he hypothesized that traces of iron and other metals would catalyze oxidative reactions in vivo and that peroxidative chain reactions were possible, by analogy to the principles of in vitro polymer chemistry. All of these predictions have been confirmed during the past 40 years.

The theory gained credibility with the identification in 1969 of the enzyme superoxide dismutase (SOD) (204), which provided the first compelling evidence of in vivo generation of superoxide anion (O_2^-•), and from the subsequent elucidation of elaborate antioxidant defenses (351). The use of SOD as a tool to locate subcellular sites of O_2^-• generation led to a realization that buttressed the free radical theory, namely, that mitochondria are a principal source of endogenous oxidants (37). Gerontologists had long observed that species with higher metabolic rates have shorter maximum life span potential (MLSP); they age faster (268). In fact, it had been proposed at the turn of the century that energy consumption per se was responsible for senescence, a concept referred to as the "rate of living" hypothesis (246, 268, 295). The realization that energy consumption by mitochondria may result in O_2^-• production linked the free radical theory and the rate of living theory irrevocably: a faster rate of respiration, associated with a greater generation of oxygen radicals, hastens aging. By now, the two concepts have essentially merged.

B. Sources of Oxidants

Ground-state diatomic oxygen ($^3\Sigma g^- O_2$ or more commonly, O_2), despite being a radical species and the most important oxidant in aerobic organisms, is only sparingly reactive itself due to the fact that its two unpaired electrons are located in different molecular orbitals and possess "parallel spins." As a consequence, if O_2 is simultaneously to accept two electrons, these must both possess antiparallel spins relative to the unpaired electrons in O_2, a criterion which is not satisfied by a typical pair of electrons in atomic or molecular orbitals (which have opposite spins according to the Pauli exclusion principle). As a result, O_2 preferentially accepts electrons one at a time from other radicals (such as transition metals in certain valences). Thus, in vivo, typical two- or four-electron reduction of O_2 relies on coordinated, serial, enzyme-catalyzed one-electron reductions, and the enzymes that carry these out typically possess active-site radical species such as iron. One- and two-electron reduction of O_2 generates O_2^-• and hydrogen peroxide (H_2O_2), respectively, both of which are generated by numerous routes in vivo, as discussed below. In the presence of free transition metals (in particular iron and copper), O_2^-• and H_2O_2 together generate the extremely reactive hydroxyl radical (•OH). Ultimately, •OH is assumed to be the species responsible for initiating the oxidative destruction of biomolecules. In addition to O_2^-•, H_2O_2, and •OH, two energetically ex-

TABLE 1. *The strengths and weaknesses of approaches to the testing of the free radical theory of aging*

Experimental Approach (Review Section)	Strengths of the Approach	Weaknesses of the Approach
Oxidative phenomenology (Sect. IV)	Simplicity; large existing data set; elucidation of the basic biochemistry of oxidative stress; technical foundation for other approaches; negative result may be instructive	Results are merely correlative (oxidative damage may be a consequence of aging); negative results, considered uninteresting, may fail to appear in the literature
Interspecies comparisons (Sect. V)	Specific testable predictions; deviations from predictions may refine theory; use of different species avoids conclusions that are species or strain specific; negative results may be instructive	Logistical problems in animal handling; quantitative comparisons complicated by qualitative interspecies differences in oxidative defenses and/or repair
Dietary restriction (Sect. VI)	Very well established model of life span extension; straightforward methodology; probably relevant to human aging and cancer	Results are somewhat correlative (alterations in oxidative stress may be consistent with but incidental to a more fundamental cellular switch)
Manipulation of rate of living (Sect. VII)	A direct test of rate of living theory, with specific testable predictions; straightforward methodology	To date, limited to relatively simple organisms, such as invertebrates
Manipulation of oxygen concentration (Sect. VIII)	A direct test of rate of living theory, with specific testable predictions; straightforward methodology	Results hard to interpret. Are positive results (life span attenuation by hyperoxia, extension by hypoxia) relevant to normoxic aging? Not applicable to most mammals
Supplementation with dietary antioxidants (Sect. IX)	(Potentially) a test of free radical theory	Results hard to interpret, due to unknown fates of supplements, potential adverse effects, complexity of overall antioxidant defenses
Administration of pharmacological antioxidants (Sect. X)	(Potentially) a test of free radical theory	Results hard to interpret, due to unknown fates of supplements, potential adverse effects, complexity of overall antioxidant defenses
In vitro senescence (Sect. XI)	A very well-established model system; relevance to human cancer (and perhaps to degenerative senescence in vivo)	Relevance to aging in vivo not yet clear
Classical and population genetics (Sect. XII)	"Awesome" power of genetics; very well established methods linked to a vast body of species-specific information; concurrent genome projects may forge interspecies links between gerontogenes	Currently limited to invertebrate model organisms (yeast, *C. elegans*, *Drosophila*)
Molecular genetics (Sect. XIII)	Powerful techniques for measuring mutational events (quantitatively and qualitatively); mutational fingerprinting to identify oxidative stress; simple methods applicable to wide array of species	In addition to causing cancer, physiological relevance of somatic mutations (and other genetic abnormalities) is not obvious
Transgenic organisms (Sect. XIV)	Immensely powerful tools; a direct test of free radical theory; interbreeding of transgenic animals for sophisticated in vivo research; established methods for widely divergent animals (*Caenorhabditis elegans*, *Drosophila*, mice)	Somewhat expensive and time consuming; dynamic nature of oxidative stress may complicate interpretations of phenotypes
Sporadic degenerative diseases (Sect. XV)	Results (if positive) provide mechanistic insights into the biology of oxidative pathophysiology and evidence of etiological relevance	Lack of animal models; disease may be of strictly human relevance (not a general phenomenon of aging)
Inherited degenerative diseases (Sect. XVI)	Results (if positive) provide mechanistic insights into the biology of oxidative pathophysiology and evidence of etiological relevance	Usually lack of an animal model; phenomenon (often severe) may not be of relevance either to most humans, nor to aging in general
Epidemiology (Sect. XVII)	Large data sets permit subtle species-wide trends to be detected; public health and policy relevance (preventative medicine); may be facilitated by sequencing of the human genome	Currently expensive, long term, and difficult

cited species of O_2 termed "singlet oxygens" can result from the absorption of energy (for instance, from ultraviolet light). Designated by the formulas $^1\Delta gO_2$ and $^1\Sigma g^+O_2$, both of these species differ from the triplet ground state ($^3\Sigma g^-O_2$) in having their two unpaired electrons in opposite spins, thereby eliminating the spin restriction of ground-state O_2 and enabling greater reactivity. The chemistry of oxygen and its derivatives has been extensively discussed elsewhere (115, 342). Because all of these species ($O_2^-\cdot$, H_2O_2, $\cdot OH$, $^1\Delta gO_2$, and $^1\Sigma g^+O_2$), by different routes, are involved in oxygen's toxicity, we will collectively refer to them as "oxidants."

It is now beyond doubt that oxidants are generated in vivo and can cause significant harm (20, 37, 60, 91, 112, 351). There are numerous sites of oxidant generation, four of which have attracted much attention: mitochondrial electron transport, peroxisomal fatty acid metabolism, cytochrome *P*-450 reactions, and phagocytic cells (the "respiratory burst"). Before a discussion of the potential contributions of different sources of oxidants, it is worthwhile briefly to outline them.

In the textbook scheme of mitochondrial respiration, electron transport involves a coordinated four-electron reduction of O_2 to H_2O, the electrons being donated by NADH or succinate to complexes I and II, respectively, of the mitochondrial electron transport chain (ETC). Ubiquinone (coenzyme Q, or UQ), which accepts electrons from complexes I (NADH dehydrogenase) and II (succinate dehydrogenase), undergoes two sequential one-electron reductions to ubisemiquinone and ubiquinol (the Q cycle), ultimately transferring reducing equivalents to the remainder of the electron transport chain: complex III (UQ-cytochrome *c* reductase), cytochrome *c*, complex IV (cytochrome-*c* oxidase), and finally, O_2 (115). However, it appears mitochondrial electron transport is imperfect, and one-electron reduction of O_2 to form $O_2^-\cdot$ occurs. The spontaneous and enzymatic dismutation of $O_2^-\cdot$ yields H_2O_2, so a significant by-product of the actual sequence of oxidation-reduction reactions may be the generation of $O_2^-\cdot$ and H_2O_2.

How much $O_2^-\cdot$ and H_2O_2 do mitochondria generate? In classic experiments during the 1970s, measurements of H_2O_2 generation by isolated mitochondria indicated that it is maximal when ADP is limiting and the electron carriers are consequently reduced ("state 4" respiration) (26). Estimates of state 4 H_2O_2 generation by pigeon and rat mitochondrial preparations amounted to 1–2% of total electron flow (26, 229). One problem with this estimate of mitochondrial H_2O_2 generation is its reliance on the use of buffer saturated with air (20% O_2). In vivo, the partial pressure of O_2 is ~5%, so these calculations may overestimate the flux of oxidants in vivo. Even disregarding the use of air-saturated buffer, the initial estimate of percentage ETC flux leading to H_2O_2 can be challenged on the grounds that in these experiments the concentrations of substrates fed to mitochondria were higher than occurs physiologically (118, 146). When H_2O_2 is measured with more physiological concentrations, the flux is ~10-fold lower (118), and experiments using subcellular fractions of SOD-deficient *Escherichia coli* suggest in vivo leakage of 0.1% from the respiratory chain (146).

What proportion of mitochondrial H_2O_2 ultimately derives from ETC $O_2^-\cdot$ generation? Unfortunately, the measurement of $O_2^-\cdot$ generation by intact mitochondria is prevented by the presence of mitochondrial SOD (mSOD). Therefore, the isolation of submitochondrial particles from which mSOD has been removed (by sonication of the intact organelles followed by extensive washing) was used for the detection of ETC $O_2^-\cdot$. In these experiments, stoichiometric estimates of the ratio of $O_2^-\cdot$ generation (by submitochondrial particles) to H_2O_2 generation (by the intact organelles) fell between 1.5 and 2.1 (24, 25, 69, 89, 191); because two $O_2^-\cdot$ molecules dismutate (either spontaneously or with the help of mSOD) to form one molecule of H_2O_2, such results suggest that virtually all mitochondrial H_2O_2 may originate as $O_2^-\cdot$ (27). Moreover, because most cellular H_2O_2 originates from mitochondria, $O_2^-\cdot$ from the ETC may be a cell's most significant source of oxidants (37).

In a recent discussion of the classic in vitro work (88), some of the original experimenters take issue with the idea that free $O_2^-\cdot$ exists in mitochondria as a result of normal flux through the ETC. They point out that in addition to having removed mSOD from mitochondria, the sonication they employed also resulted in the loss of cytochrome *c*, which rapidly scavenges $O_2^-\cdot$ in vitro and is present in mitochondria at local concentrations from 0.5 to 5 mM. In mitochondria, they argue, mSOD and cytochrome *c* rapidly scavenge $O_2^-\cdot$ (in the matrix and intermembrane spaces, respectively). More to the point, the authors stress that unless the ETC was poisoned with inhibitors such as antimycin A, $O_2^-\cdot$ generation was not detected in their experiments (89, 191). Arguing that mSOD should act to increase the rate of $O_2^-\cdot$ generation in vivo (by accelerating product removal by dismutation to H_2O_2), they suggest that the actual role of mSOD in vivo may be to increase H_2O_2 generation (with $O_2^-\cdot$ as a rapidly consumed intermediate) (88). Ultimately, there remains a good deal of uncertainty surrounding the mechanisms, quantity, and meaning of mitochondrial $O_2^-\cdot$ generation in vivo (228), a mystery which is deepened by recent reports of enzymatic nitric oxide (NO·) generation in mitochondria (C. Giulivi and C. Richter, personal communication). Because $O_2^-\cdot$ and NO· react to form the oxidant peroxynitrite ($ONOO^-$), mitochondrial $O_2^-\cdot$ generation may soon need to be considered in the light of its ability to destroy NO· and form $ONOO^-$, as discussed in section XVIII*B*.

A second source of oxygen radicals is peroxisomal β-oxidation of fatty acids, which generates H_2O_2 as a by-product. Peroxisomes possess high concentrations of catalase, so it is unclear whether or not leakage of H_2O_2 from peroxisomes contributes significantly to cytosolic oxidative stress under normal circumstances. However, a class of nonmutagenic carcinogens, the peroxisome proliferators, which increase the number of hepatocellular peroxisomes and result in liver cancer in rodents, also cause oxidative stress and damage (7, 157, 177, 230). Interestingly, during the regeneration of the liver after partial hepatectomy, there exist peroxisomes that do not stain for catalase activity (232), hinting that during rapid cell proliferation, oxidant leakage from peroxisomes may be enhanced.

Microsomal cytochrome *P*-450 enzymes metabolize xenobiotic compounds, usually of plant origin, by catalyzing their univalent oxidation or reduction. Although these reactions typically involve NADPH and an organic substrate, some of the numerous cytochrome *P*-450 isozymes directly reduce O_2 to $O_2^{-}\cdot$ (105, 168) and may cause oxidative stress. An alternative route for cytochrome *P*-450-mediated oxidation involves redox cycling, in which substrates accept single electrons from cytochrome *P*-450 and transfer them to oxygen. This generates $O_2^{-}\cdot$ and simultaneously regenerates the substrate, allowing subsequent rounds of $O_2^{-}\cdot$ generation (115). Although it is unclear to what extent cytochrome *P*-450 side reactions proceed under normal conditions, it is possible that such chronic $O_2^{-}\cdot$ generation by cytochrome *P*-450 is the price animals pay for their ability to detoxify acute doses of toxins (6).

Finally, phagocytic cells attack pathogens with a mixture of oxidants and free radicals, including $O_2^{-}\cdot$, H_2O_2, NO•, and hypochlorite (38, 220, 262). Although the massive generation of oxidants by immune cells differs from the above three sources of free radicals to the extent that it is the result of pathogenesis, it is nevertheless a normal and unavoidable consequence of innate immunity. Chronic inflammation is therefore unique among the endogenous sources of oxidants, because it is mostly preventable (49, 231, 245).

In addition to these four sources of oxidants, there exist numerous other enzymes capable of generating oxidants under normal or pathological conditions, often in a tissue-specific manner (115). To give a single relevant example, the deamination of dopamine by monoamine oxidase generates H_2O_2, in some neurons, and has been implicated in the etiology of Parkinson's disease (80). Finally, the widespread catalytic generation of NO•, achieved by various isozymes of nitric oxide synthase and central to processes as diverse as vascular regulation, immune responses, and long-term potentiation, increases the potential routes for destructive oxidative reactions (187). The interaction between $O_2^{-}\cdot$ and NO• results in $ONOO^-$, which is a powerful oxidant.

As originally articulated by Harman (120), the free radical theory of aging did not distinguish between these different sources of oxidants. However, the rate of living hypothesis clearly singled out mitochondrial $O_2^{-}\cdot$ and H_2O_2 generation, since it is the mitochondrial respiration rate that negatively correlates with MLSP. Also, as many other established sources of oxidants are tissue specific (associated with hepatic, neuronal, and other specialized functions), they are less likely to explain aging across a broad range of species. For this reason, mitochondrial $O_2^{-}\cdot$ and H_2O_2 have captured the lion's share of attention. However, it may turn out that for some age-associated disorders, nonmitochondrial oxidants are critical. In the expanded sense of the free radical theory, any oxidants, mitochondrial or not, may play a role. Therefore, despite the great number of intracellular sources of oxidant that have been identified in a qualitative way, in terms of ranking their relative importance, the field is in its infancy.

C. Targets of Oxidants

What are the targets of endogenous oxidants? The three main classes of biological macromolecules (lipids, nucleic acids, and proteins) are susceptible to free radical attack, and there is plentiful evidence that all suffer oxidative damage in vivo. Although it is well beyond the scope of this review to treat the biochemistry of oxidative damage in any great depth, the area has been expertly reviewed (115). A synopsis of the better known pathways of oxidative damage, however, is warranted; the most familiar end products are described here.

The earliest research on the destruction of biological molecules by oxidants involved lipids (109). Food chemists have long understood that the rancidity of fats results from peroxidative chain reactions in lipids ("autoxidation"); a lipid hydroperoxyl radical abstracts a hydrogen atom from the double bond of a neighboring unsaturated lipid, forming a hydroperoxide and an alkyl radical, the latter which combines with O_2 to regenerate a lipid hydroperoxyl radical capable of initiating another round of oxidation. Ultimately, intramolecular reactions and decomposition yield cyclic endoperoxides and unsaturated aldehydes, the latter of which are reactive and may act as mutagens (194) or inactivate enzymes (39, 322), or operate as endogenous fixatives, reacting with proteins and nucleic acids to form heterogeneous cross-links (42). Moreover, a primary effect of lipid peroxidation is decreased membrane fluidity, which alters membrane properties and can significantly disrupt membrane-bound proteins (324).

Oxidative damage to nucleic acids includes adducts of base and sugar groups, single- and double-strand breaks in the backbone, and cross-links to other molecules. The spectrum of adducts in mammalian chromatin oxidized in vitro and in vivo includes more than 20 known products, including damage to all four bases and thymine-tyrosine cross-links (70, 71, 113). The electrochemical properties of the adduct 8-oxo-guanine (oxo8gua) and the deoxynucleoside 8-oxo-2,7-dihydro-2′-deoxyguanosine (oxo8dG), which have permitted the coupling of extremely sensitive electrochemical detection to high-performance liquid chromatography (HPLC), have resulted in hundreds of studies of its formation, accumulation, and excretion (17). The identification of specific enzymatic repair of oxidative lesions has recently provided both proof of the significance of oxidative DNA damage as well as tools to manipulate the load of damage in vivo by genetic knockout (17, 23, 78, 192, 266, 291).

The oxidation of proteins is less well characterized, but several classes of damage have been documented, including oxidation of sulfhydryl groups, reduction of disulfides, oxidative adduction of amino acid residues close to metal-binding sites via metal-catalyzed oxidation, reactions with aldehydes, protein-protein cross-linking, and peptide fragmentation (317, 318). A particularly intriguing recent development has been the realization that a number of enzymes possessing active-site iron-sulfur clusters are acutely sensitive to inactivation by O_2^-• (86, 176). For example, *E. coli* aconitase is inactivated by O_2^-• with a rate constant of 10^9 $M^{-1} \cdot s^{-1}$ (95, 96). Mammalian mitochondrial aconitase is inactivated in vitro and in vivo by treatments that increase mitochondrial O_2^-• generation, such as growth under hyperbaric conditions (97, 98). Because aconitase participates in the citric acid cycle, its inhibition would be expected to have pleiotropic effects. Moreover, the mechanism of aconitase inhibition by O_2^-• has been demonstrated to involve the release of free iron from the enzyme (86). Free iron atoms catalytically exacerbate oxygen stress (see below), and it has been proposed that superoxide's genotoxicity is a function of its ability to liberate protein-bound iron (159, 184).

Unlike lipids and nucleic acids, proteins represent a very diverse target for oxidative damage. Although protein oxidation has been demonstrated at the level of the peptide backbone and amino acids, there has been relatively little scrutiny of differences between proteins in their sensitivities. A detailed quantitative comparison of bovine serum albumin and glutamine synthase has shown susceptible residues of the former (methionine and the aromatic amino acid residues) to be oxidized about twice as fast as those on the latter, implicating all four levels of protein structure in relative susceptibility (19). A study of the oxidation sensitivities of a various cloned K^+ channels from T lymphocytes, cardiac cells, and neurons revealed that whereas five of the cloned channels were highly sensitive to oxidation, an equal number were resistant (74). Differential sensitivities raise the possibility that the loss of homeostasis that is a hallmark of aging could result from the selective oxidation of proteins.

In the context of aging, a particularly relevant aspect of oxygen's toxicity is its promotion by some metals and by elevated O_2 partial pressure. Iron and copper catalyze the homolytic cleavage of ROOH (the Fenton reaction), leading to the generation of •OH (115). It is •OH that is the most reactive oxidant, reacting at diffusion-limited rates. The catalytic properties of iron and copper explain why cells possess metal-chelating proteins such as ferritin and transferrin, which reduce the concentration of redox-active metals (114, 211). In humans, the body's content of iron increases with age (in men throughout their lives, and in women after menopause), and it has been suggested that this accumulation may increase the risk of oxidative damage with age (169, 331). Finally, oxidative stress in vivo is aggravated by increasing O_2 partial pressure, due to a more pronounced flux of mitochondrial O_2^-• (37). Consequently, the manipulation of O_2 partial pressure is a relatively simple tool that has been used to test the free radical theory.

D. Antioxidant Defenses

Cells are equipped with an impressive repertoire of antioxidant enzymes, as well as small antioxidant molecules mostly derived from dietary fruits and vegetables (5, 351). These include *1)* enzymatic scavengers such as SOD, which hastens the dismutation of O_2^-• to H_2O_2, and catalase and glutathione peroxidase (GPX), which convert H_2O_2 to water; *2)* hydrophilic radical scavengers such as ascorbate, urate, and glutathione (GSH); *3)* lipophilic radical scavengers such as tocopherols, flavonoids, carotenoids, and ubiquinol; *4)* enzymes involved in the reduction of oxidized forms of small molecular antioxidants (GSH reductase, dehydroascorbate reductase) or responsible for the maintenance of protein thiols (thioredoxin reductase); and *5)* the cellular machinery that maintains a reducing environment (e.g., glucose-6-phosphate dehydrogenase, which regenerates NADPH). The complement of defenses deployed differs not only between organisms or tissues, but even between cellular compartments. For instance, GPX plays an important role in mammals but is absent from flies and nematodes (298, 330), and there exist in humans three forms of SOD (cytosolic Cu,Zn-SOD, mitochondrial Mn-SOD, and extracellular SOD), encoded and regulated independently (91).

As far as supporting the free radical theory of aging is concerned, the universality of antioxidant defenses is good news. Although the nature of these defenses varies between species, the presence of some type of antioxidant defense is universal. In fact, some antioxidants, such as SOD, are very highly conserved. Clearly, an indifference to oxygen free radicals is inconsistent with life, underlining the centrality of oxidative damage. Moreover, the fact that antioxidant defenses are not uniform has been incorporated into the free radical theory; differences in antioxidant defenses between species have been put forth to explain differences in life span. Although there is something uncomfortably ad hoc in these two different interpretations of the data, they are not inconsistent. Whereas aerobic life requires organisms to cope with oxidation to some extent, different evolutionary pressures appear to have selected for more or less investment in these defenses, as is discussed in section III.

Finally, a persistent problem in testing the free radical theory is that antioxidants are both parallel (different antioxidants can play similar roles, e.g., catalase and GPX) and serial (enzymes operate in tandem to decompose radicals to harmless products, e.g., SOD and catalase). Consequently, measurements of individual antioxidant activities do not have great relevance. In fact, as is discussed below, measurements of age-related changes in individual antioxidants have led to conflicting results (297). For this reason, aggregate assays have been devised, such as the susceptibility of crude cellular homogenates to in vitro oxidation by ionizing radiation (2, 300). Although these assays do not provide any information about the specific mechanisms of defense, they conveniently measure overall effectiveness.

E. Repair of Oxidative Damage

Unlike defenses against oxidants, which have been extensively characterized, the machinery for repairing oxidative damage is relatively unexplored. Nevertheless, it is clear that cells repair oxidized lipids (e.g., phospholipase A_2 cleaves lipid peroxides from phospholipids; Ref. 239), oxidized nucleic acids (e.g., glycosylases specifically recognize and excise oxidized bases from double-stranded DNA; Refs. 23, 56a, 56b, 325), and oxidized proteins (104, 240, 261, 316). The comparative biochemistry of cellular repair is very fertile ground for the free radical theory.

F. Synthesis: Interaction of Oxidant Generation, Oxidative Damage, and Repair

The existence of multiple intracellular sources of oxidants and complex defenses has led to refinements of the free radical theory. For example, it is clear that the metabolism of oxygen radicals is dynamic, with damage resulting from an increase in oxidant generation or a decrease in antioxidant defenses. Consequently, a difference in life span between species or individuals could be due to different rates of living, or to different "rates of scavenging" (57, 82). The picture has been further complicated by the discovery of specific enzymatic repair of oxidative damage, leading to "repair" or "fidelity" versions of the theory, in which life span is determined by the failure to correct oxidative damage (137, 239).

The relationship among these three components of oxidative stress: oxidant generation, antioxidant protection, and repair of oxidative damage, and the way in which they have been investigated in testing the free radical theory, is illustrated schematically in Figure 1. Increases in oxidant generation, and decreases in antioxidant protection and repair systems, are among the theory's testable predictions and have been examined both as a function of age in individuals of the same species, as well as between species of differing MLSP.

Finally, an extremely important (if experimentally recalcitrant) aspect of the interactions between oxidants, antioxidants, and repair are feedback loops, positive and negative, between them. Antioxidant defenses and cellular repair systems have been shown to be induced in response to oxidative challenges (67, 119, 320) and are of course potential targets of oxidative destruction (145). Also, the generation of oxidants may be enhanced by the malfunctioning of oxidatively damaged molecules (28, 303). Therefore, with the examination of Figure 1, it is not difficult to envision ways in which primary oxidative destruction of any target (e.g., the components of the mitochondrial ETC, scavenging enzymes such as SOD, or DNA repair enzymes) might promote further oxidative damage in what is frequently called a "catastrophic vicious cycle."

Although such cycles are intuitively appealing, their documentation awaits future work and will be extremely difficult from a technical standpoint. An alternative to lab-based approaches, namely, the computational modeling of these complex interactions in what has been termed a "Network Theory of Ageing," is being pursued by theoretical gerontologists (170). Ultimately, a question of obvious importance is whether or not such cycles, if they exist, could be broken, and modeling may help pinpoint weak links for therapeutic intervention.

III. REFINEMENTS AND COROLLARIES OF THE FREE RADICAL THEORY

As the free radical theory has gained ground, it has incorporated other ideas. For example, as mentioned above, the rate of living hypothesis dovetailed with the free radical theory once mitochondrial free radical generation was confirmed. Three other ideas that have been influential are the evolutionary concept of antagonistic pleiotropy, the somatic mutation theory of aging, and the mitochondrial theory of aging.

A. Oxidants and Evolutionary Theories of Aging

The intracellular generation of oxidants capable of limiting life span may appear paradoxical. It seems reasonable to expect that natural selection might have devised aerobic cells that do not leak toxic by-products. Evolutionary biologists have contributed to the free radical theory by suggesting why physiologically harmful generation of oxygen radicals occurs. They have argued that natural selection favors genes that act early in life and increase reproduction, rather than genes that act to preserve nongerm cells (the "disposable soma"), a principle called "antagonistic pleiotropy" (162–164, 265, 341).

The concept of antagonistic pleiotropy stresses that in the wild, reproductive success is principally a function of external factors. With the exception of modern-day humans, individuals do not usually die of old age, but are eaten, parasitized, or out-competed by others. Therefore, preserving the cells of the disposable soma, otherwise known as the body, may be disadvantageous if it detracts

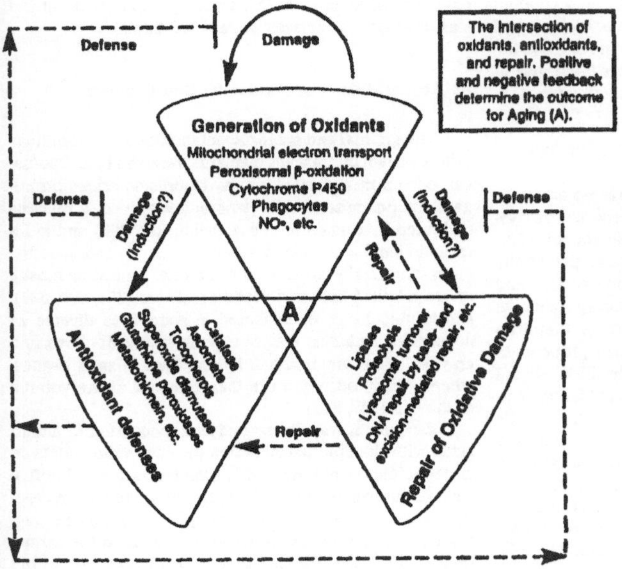

FIG. 1. The ultimate outcome of oxidative stress is a function of *1*) oxidant generation, *2*) antioxidant defenses, and *3*) repair of oxidative damage. Bolds arrows denote oxidative damage, and dashed arrows denote routes for its prevention or repair. Because of the ways in which these processes may interact, multiple positive- and negative-feedback loops are possible. Aging (A) is situated at intersection of these processes. In testing the free radical theory, changes in *processes 1–3* have been measured both as a function of age and as a function of species' maximum life span potential. The similarity of the figure to the international emblem of radiation is not a coincidence; the free radical theory has its roots in radiation biology.

from more pressing problems. The toxicity of respiratory burst oxidants, for instance, cannot easily be eliminated by evolution, since this would result in death from childhood infection. Similarly, an investment in improved antioxidant defenses maximizes fitness only if the resources are not better invested in strength, beauty, speed, or cunning.

In terms of natural selection, the tremendous cost of death before reproductive age, the constantly compounding probability of death from external threats, and the cost of failing to reproduce all ensure that selective pressure is strongest at young ages. Any novel mutations that decrease oxidative damage first have to satisfy the criteria of youthful reproduction. In short, selective pressure to compete effectively at an early age may guarantee a certain degree of O_2 toxicity and work against the conservation of the soma in the long run.

B. Oxidants and the Somatic Mutation Theory of Aging

The somatic mutation theory holds that the accumulation of DNA mutations is responsible for degenerative senescence (23, 79, 212, 218, 332). In the case of cancer, which results from both point mutations in oncogenes and the loss of tumor suppressor gene function (often by deletion), the role of mutations is unquestionable (5). It remains to be seen whether or not the argument is valid for nonproliferative senescence. For instance, whereas significant age-related increases in somatic mutations in a reporter transgene (*lacZ*) have been measured in a mitotic tissue of transgenic mice (the liver), no increase was detected in the largely postmitotic brain of the same animals (71a), suggesting that neurodegeneration, at least, is unlikely to be the result of accumulated somatic mutations in nuclear DNA. Moreover, the accumulation of mutations in the liver tissues was not dramatic, suggesting that mutagenesis may be of little functional consequence to mitotic tissues as well (338b). In light of these data, what evidence is there that somatic mutations are related to aging? A compelling argument for the somatic mutation theory of aging was provided years ago in the discovery that DNA repair ability correlates with species-specific life span (127a), a phenomenon that has recently been reconfirmed (52a). However, it has been noted that DNA repair, which is necessary for the prevention of tumorigenesis, is necessary but not sufficient for longevity (52a). Ultimately, arguments about the physiological significance of somatic mutations hinge on how disruptive a given mutational burden is to a cell or animal; with current methods, this is an unanswerable question.

In any case, it has been demonstrated in numerous studies with prokaryotes, yeast, and mammalian cells that oxidants are mutagens, against which cells actively protect their genetic material (81, 108). Although it is not yet clear what fraction of mutations can be attributed to oxidative damage, the characterization and cloning of defense genes against oxidative mutagenesis (17), and the development of in vivo mutagenesis assays (198), has finally opened up avenues for definitive experiments.

C. Oxidants and Mitochondrial Theories of Aging

The mitochondrion has also long attracted attention as one of the cell's weak links, an organelle whose dysfunction has profound negative pleiotropic effects (193). Mitochondria supply ATP and also sequester potentially toxic Ca^{2+}, yet because of their generation of $O_2^-\bullet$ and H_2O_2, they are on the front lines of respiratory oxidative stress. The idea that the mitochondrion is therefore uniquely vulnerable was embraced early on by proponents of the free radical theory (121). In the early 1980s, Miquel and colleagues (84, 212, 213) proposed that oxidative damage to mitochondrial DNA (mtDNA) in postmitotic cells would lead to mutations and blocks to replication, and consequently to mitochondrial dysfunction and physiological decline. This "mtDNA mutation hypothesis of aging," which incorporates free radicals, somatic mutations, and the central role of mitochondria in homeostasis, is presently under intense scrutiny (8, 11, 21, 50, 110, 222, 223, 238, 258–260, 276, 288, 336, 339).

IV. OXIDATIVE PHENOMENOLOGY: AGE-ASSOCIATED TRENDS

The phenomenological approach to the free radical theory has involved looking for traces of oxidative damage in vivo. Phenomenology is not well suited to critically testing the free radical theory, since the data (which are voluminous and generally supportive) mainly represent correlations. Documented increases in oxidative damage, no matter how impressive, may be a consequence of a primary nonoxidative event. Nevertheless, phenomenology is the foundation upon which more powerful experiments depend, since the analytical methods developed for it have been used to compare species, genetic mutants, and populations with differing life spans. In fact, almost all of the biomarkers of oxidative stress described in this section have been found to accumulate at a faster rate in short-lived species, and in many cases, this rate correlates with O_2 consumption. Familiarity with the most frequently measured end points is a prerequisite to assessing the free radical theory.

If oxidative damage is a significant cause of cellular degeneration, then one expects to see more of it in older individuals. Oxidative damage has been described in terms of "accumulation, modification, and depletion": accumulation of end products of oxidative damage (such as lipofuscin), modification of existing structures (such as oxidative adducts in DNA), and depletion (such as the loss of enzymatic activity or reduced thiols).

A. Accumulation of Oxidative End Products

The gradual and steady accumulation of intracellular yellow-brown fluorescent pigments, referred to as lipofuscin, occurs in numerous phyla. Lipofuscin arises prominently in postmitotic cells (where, it is argued, it remains undiluted by rounds of cell division; Ref. 296) and is located in small granules in secondary lysosomes. Lipofuscin is structurally complex and variable, consisting mostly of cross-linked lipid and protein residues (251, 296, 327), and is ubiquitous, documented in species as diverse as nematodes, fruit flies, rats, bees, crab-eating monkeys, and crayfish. Most important, it is abundant in aged tissues, where it may occupy more than one-half of the volume of the cell (347, 348).

Early on, it was discovered that incubation of amino acids with the lipid peroxidation product malonaldehyde under acidic conditions leads to the formation of lipofuscin-like fluorophores (42). The plausibility of such a reaction, given the contents and low pH of lysosomes, suggested that lipid peroxidation in vivo leads to the formation of lipofuscin (324). Other in vitro studies of lipid peroxidation have since uncovered a great number of routes to fluorescent, cross-linked products via promiscuous oxidative chemistry, which suggests that lipofuscin is a biomarker of lipid peroxidation (160, 296, 348).

Despite extensive in vitro experiments, it is not known with certainty how lipofuscinogenesis occurs in vivo, nor how lipofuscin comes to accumulate with age. Lipid peroxidation could occur throughout the cell and be followed by lysosomal phagocytosis and cross-linking of peroxidative by-products, in which case an age-related increase in lipofuscin content could be seen as the result of oxidative damage. Alternatively, an age-associated decline in lysosomal activity (due to something besides oxidation) might increase the residence time of phagocytosed material enough to enhance lipofuscinogenesis in situ from constant amounts of peroxides (135). In support of the latter possibility, infusion into rat brains of lysosomal proteinase inhibitors leads to the rapid accumulation of lipofuscin-like granules (149). This scenario suggests that lipofuscinogenesis may be a consequence, not a cause, of aging.

Experiments with cultured cardiac myocytes have established roles for both oxidative damage and lysosomal turnover (28). Lipofuscin accumulates in these cells in culture, and growth under increasing O_2 partial pressure from 5 to 40% markedly enhances its accumulation (306). Inclusion of iron in the growth medium further increases lipofuscinogenesis, and the iron chelator desferal depresses it, suggesting that Fenton reaction-generated •OH is an initiator (200). Finally, antioxidants inhibit lipofuscin formation in cultured cardiomyocytes, whereas lysosomal protease inhibitors increase it (199, 201, 202).

Even if oxidative damage is primarily responsible for depositing lipofuscin in the lysosomes of senescing animals, is it more than a biomarker of aging? It has been theorized that lipofuscin accumulation is likely to impair autophagy, as more lysosomal volume is occupied by the indigestible material (28). Because lysosomes are responsible for the recycling of materials and organelles, their failure may include the following: *1*) a delay in mitochondrial turnover (with a concomitant decrease in mitochondrial efficiency or an increase in mitochondrial oxidant generation), *2*) an accumulation of oxidatively modified proteins and lipids in the cytosol awaiting degradation (potentially aggravating cytosolic lipid peroxidation), *3*) an accumulation of lipofuscin-bound iron in a redox-active form (which might promote further intralysosomal lipid peroxidation), and *4*) the disruption of lysosomal membranes (and the spillage of hydrolytic enzymes into the cytosol). Although these speculations (28) remain to be substantiated, it has been shown that when treated with sublethal doses of H_2O_2, cultured cells display lysosomal disruption and leakage of the lysosomal compartment into the cytosol (29). Also, it has been demonstrated that the sensitivity of cultured primary hepatocytes to oxidation, which was associated with a loss of GSH and an influx of Ca^{2+}, was prevented by the iron chelator desferal (235). What is intriguing about these results is the fact that whereas desferal stabilized the lysosomes, it did not prevent the loss of GSH or the increase in intracellular calcium, so it may be that it is lysosomal leakage per se, rather than peroxidative damage, which is the actual lethal step in this model of oxidative killing.

B. Steady-State Levels of Oxidative Modification

Unlike cytosolic proteins whose half-lives are measured in minutes or hours, some extracellular proteins are rarely recycled, and oxidative modification of these old macromolecules occurs. A class of fluorescent cross-linked molecules that is distinct from lipofuscin forms on long-lived proteins such as collagen and lens crystallin (216). These modifications are initiated by the reaction of reducing sugars with free amino groups (glycation), a chemical sequence that is unrelated to oxidation and results in a molecule known as an Amadori product. Further nonoxidative rearrangements result in stable, cross-linked advanced glycation end products (AGEs) (35), whose absolute abundance appears to be an excellent biomarker of age (217). Recently, it was discovered that oxidation is one fate of the Amadori product. Pentosidine, the name given to a cross-link involving arginine, lysine, and pentose moieties, is one such "glycoxidation product," the formation of which requires the presence of O_2 (326). It appears as if Amadori products themselves are a source of H_2O_2 in vitro, which then accelerates glucose-mediated fluorogenic collagen cross-linking in a catalase-sensitive fashion, although it is unclear to what extent this occurs in vivo (77, 217). As is the case with other AGEs, the tissue burden of pentosidine is elevated in diabetics, as a consequence of hyperglycemia.

Pentosidine has been found to accumulate as a function of age in shrews, rats, dogs, cows, pigs, monkeys, and humans, yielding equivalently shaped curves in all cases (284). It is not clear, however, how glycooxidative modifications might contribute to degeneration. It has been proposed that cross-linking in cartilage is related to its decreased elasticity and relative resistance to proteolysis in old animals (9). However, the absolute amount of collagen pentosidine cross-links attained at death is much higher in long-lived than in short-lived species: 6–7 pmol/mg in 3.5-yr-old shrews, 15–18 pmol/mg in 25-yr-old monkeys, and 50–100 pmol/mg in 90-yr-old humans (284). In other words, it appears that the rate of pentosidine accumulation may merely be a measure of more rapid oxidative damage in short-lived species rather than an actual cause of dysfunction.

An intriguing twist to this story has been the cloning of a specific cellular receptor for AGE, called RAGE (receptor for AGE), that belongs to the immunoglobulin superfamily and is expressed by mononuclear cells and the vascular endothelium (277, 278). One of the effects of AGE binding by RAGE is the generation (in mononuclear cells) of intracellular oxidants, the activation of the oxidant-sensitive transcription factor NFκB, and the induction of downstream events linked to atherogenesis (279, 280). As discussed in section xv*B*, it has recently been shown that RAGE, which is highly expressed by microglial cells in the brain (142), is a receptor for amyloid β-peptide (Aβ). The RAGE binding of Aβ results in oxidant generation, implicated in the etiology of AD (293, 344).

Several amino acid residues in proteins are susceptible to oxidative modification, forming side chain carbonyl derivatives (317). The development of sensitive methods for the analysis of protein carbonyls by Stadtman and coworkers (181) enabled them to study oxidative modification in human brain tissue and cultured fibroblasts, and in rat liver. They found a two- to threefold rise in protein carbonyl content between young and old age, an increase from 10 to ~30% of the total protein pool (315). The increase was exponential and correlated well with decreased activity of the oxidation-sensitive enzyme glucose-6-phosphate dehydrogenase (G-6-PD). In comparison, the rise in protein oxidation in the mongolian gerbil was less dramatic, increasingly significantly in brain, heart, and testis, but not in kidney. As in human tissues, trends for the activity of G-6-PD correspond to the increased damage, falling in brain and heart but not in kidney (300). Similar results have been reported in an insect model. An age-associated 2.5-fold increase in the protein carbonyl content of old versus young houseflies has also been documented (299), and as in humans, the increase occurs exponentially during the life span. The similarity of the degree and pattern of increase in insects and mammals is striking, considering the enormous difference in their MLSP (40 days vs. 100 yr). Moreover, protein carbonyl levels increase similarly in mitochondrial extracts from the thoracic flight muscles of these animals (303). Mitochondrial aconitase is particularly prone to oxidative modification during aging in vivo and was identified by the immunoblotting of housefly mitochondrial protein extracts with a monoclonal antibody designed to detect protein carbonyls (343b). Carbonylation of this key citric acid cycle enzyme increased in parallel with a decline in its activity.

Somewhat stronger evidence that protein oxidation may play a causative role in senescence comes from comparisons of "crawlers" versus "fliers" of the same aging cohort. Although the two groups share the same chronological age, crawlers are phenotypically senescent individuals that have lost the ability to fly and have a shorter remaining average life span than do fliers (e.g., 9.0 days vs. 13.3 days for 10-day-old crawlers and fliers, respectively). The protein carbonyl content of crawlers was 29% higher than that of fliers (299), reflecting their greater phenotypic age, as was the degree of carbonyl modification of mitochondrial (but not cytosolic) aconitase (343b). Humans suffering from Werner's syndrome, a disease characterized by premature senescence, are individuals whose phenotypic aging is also accelerated, and they too appear to have more extensive protein oxidation. Fibroblasts from Werner's patients of all ages have a level of protein carbonyls equivalent to that in 80-yr-old controls (233). In a creative study attempting to correlate protein oxidation to a physiologically relevant end point, it was shown that in old mice, interanimal variation in protein carbonyl content of two different areas of the brain (cerebral cortex vs. cerebellum) was associated with parallel interanimal variation in memory and motor function deficits (90).

Are protein carbonyls physiologically relevant, or are they merely markers? What are the actual consequences of protein modification? Unfortunately, there are few quantitative data with which to answer this question, although qualitative data exist. The fate of oxidized proteins may depend on the form of damage. For example, metal-catalyzed oxidation of G-6-PD by iron/citrate results in a thermolabile enzyme that is a better substrate for proteolysis than is the native enzyme (93). Rapid turnover of metal-oxidized G-6-PD may therefore proceed efficiently. On the other hand, G-6-PD modification by 4-hydroxy-2-nonenal, a lipid peroxidation product, also inactivates the enzyme but does not render the enzyme thermolabile or increase its degradation by proteases (322). This difference exists despite the fact that in both cases, the same lysine residue is affected. To make matters more complex, the cross-linking of G-6-PD multimers by 4-hydroxy-2-nonenal (which predictably results in a product with lipofuscin-like fluorescence) produces a molecular species that actually inhibits the multicatalytic protease (92). The physiological cost of protein oxidation is presently an unknown quantity.

The appearance of protein-bound 3,4-dihydroxyphenylalanine (DOPA) on •OH-damaged proteins has been characterized; when converted to a quinone, protein-bound DOPA can undergo redox cycling, generating O_2^-•. It has therefore been proposed that protein oxidation may contribute to the progression of aging not merely by the loss of protein function, but also by an acceleration of the flux of oxidants (61, 63, 64, 69, 101, 102).

The oxidative modification of DNA has also been studied in animals of different ages, with conflicting results. Although some studies have reported a modest increase in specific oxidative adducts, single-strand breaks, and abasic sites, others have been negative (23, 132, 156, 221, 337). The failure to detect an age-related increase in oxidative adducts by the analytical chromatographic techniques typically employed may have been due to the difficulty of working close to the limit of sensitivity (17). In fact, it has become apparent that the measurement of the adduct oxo8dG is frequently plagued by artifacts (29a, 44b, 127c, 156a, 248a) and that these may have compromised some published experiments. Of particular concern are measurements of oxo8dG in mtDNA (16), which have generally been higher than in nuclear DNA, but which may be particularly prone to artifacts associated with the analysis of small samples (16, 127c). Moreover, it is noteworthy that even among the highly variable published estimates of oxo8dG in mtDNA are values that are equivalent to the lowest measured values of oxo8dG in nuclear DNA (131a). Because of the small number of studies of mtDNA and the high variability between the measured values, it is not yet possible to conclude that mtDNA is, in fact, more heavily oxidized than nDNA. Encouragingly, alternative PCR-based methods for measuring oxidative damage have recently been used to compare oxidation of mtDNA and nDNA by exogenous oxidants, with the result that the former appears more sensitive than the latter (270a, 343a), although these studies could not quantify baseline values of damage. With methodological improvements, future experiments may be more conclusive. For instance, the use of single-cell gel electrophoresis (the comet assay) to measure single-strand breaks and abasic sites in whole rat hepatocytes in situ revealed a statistically significant 1.5-fold increase in old rats compared with young rats (131) (although this experiment did not distinguish between oxidative and nonoxidative damage).

In any case, even if the burden of oxidative adducts

does increase with age, there is virtually no information about the likely effect of oxidative DNA damage in vivo, apart from the knowledge that it leads to mutations and cancer. The fact that there is active DNA repair in postmitotic tissues (in which the danger of mutation due to replication is nonexistent), and that such repair is often targeted to transcribed regions of the genome, suggests that DNA damage itself interferes with gene expression and is not tolerated (116, 117). This important question deserves more attention.

C. Oxidative Depletion of Biochemical Pools

The oxidative depletion of molecules with increasing age has not been well documented in senescent animals, since the destruction of molecules does not often leave traces; luckily, some pathways of oxidative damage do leave biochemical fingerprints. The loss of integrity of lipid bilayers due to peroxidation is one of the most salient effects of oxidative damage (324) and results in the generation of aldehydes and alkanes. Unfortunately, these are not easily measured, the widespread use of the simple and nonspecific thiobarbituric acid test notwithstanding (109). Nevertheless, countless studies have reported an increase in thiobarbituric acid-reactive substances (TBARS) with age. Combined with other more reliable assays, these studies have demonstrated that there is a greater degree of lipid peroxidation in older animals (209). The measurements of exhaled ethane and pentane is a technique that has the advantage of being applicable to humans (165). Unlike lipofuscin and TBARS, which measure the size of a pool of destroyed molecules and require a tissue biopsy, the assay of exhaled hydrocarbons measures the rate of damage and is noninvasive. Breath pentane has been found to increase significantly with age in humans, suggesting that increased lipid turnover occurs with age because of peroxidation (161a, 208, 356). Refinement of the technique and elimination of the associated artifacts (314) should facilitate further testing of the free radical theory in humans.

The loss of activity of several oxygen-sensitive enzymes (G-6-PD, glutamate synthetase) has been reported in mammalian models of aging (315). In houseflies, a decline in G-6-PD, glutamate synthetase, and alcohol dehydrogenase activities has also been documented and coincides with a dramatic loss of protein sulfhydryls (3). Another commonly reported age-related loss is an increase in the ratio of oxidized to reduced glutathione, which may reflect a disruption of the cell's redox state (215, 273).

D. Age-Associated Trends in Antioxidant Defenses and Repair

What is the cause of age-related oxidative damage? It could result from less active antioxidant defenses and repair, but studies that have measured age-related changes in antioxidant defenses have generated conflicting results. Recent measurements of antioxidants in mongolian gerbils (300) and mice (215) are representative of the types of patterns that have been uncovered in many other studies (65, 248, 254, 263, 274, 301, 302, 305, 329). In various tissues of gerbils, there was not a consistent pattern of change; increases in SOD and decreases in GSH were observed, whereas GPX was equivalent at different ages and catalase increased or decreased, depending on the tissues and the age at analysis. In mouse brain, on the other hand, significant decreases in SOD, catalase, and GSH reductase were observed, although GPX levels were unchanged.

Another complication is that defenses are induced in response to stress. Therefore, a higher level may indicate better protection, or alternatively, greater need for antioxidant defenses due to an increase in oxidant generation. Studies of antioxidants in rat heart and skeletal muscles illustrate this point. In heart, decreases in cytosolic SOD and GPX and increases in mitochondrial SOD and GPX were noted in older animals, and several indexes of oxidative damage were also elevated (151). From these results, it was concluded that although overall myocardial antioxidant defenses were weakened in the older animals, they were induced in mitochondria as a compensatory response. In skeletal muscles, in contrast, increases were observed in both cytosolic and mitochondrial forms of all of the enzymes studied (150), despite the fact that indexes of lipid peroxidation were again elevated; in this case, it was concluded that both cytosolic and mitochondrial antioxidants were induced. The credibility of these hypotheses is not in question, but it is hard to see how they could be disproved. When these and similar studies of age-related antioxidant levels are combined, what remains is a confusing assemblage of ambiguous trends.

Of course, interactions between antioxidants are complex, which aggravates the problem. To avoid the problems posed by assays of individual antioxidants, aggregate measures of antioxidant defenses have been devised. A crude but integrative measure of antioxidant defenses, for instance, is the susceptibility of a homogenate to induced oxidation. X-irradiation of a whole body homogenate of houseflies results in a linear, dose-dependent increase in protein carbonyls. When homogenates of old and young flies are compared, the rate of induction of protein carbonyls by X-irradiation is 45% higher in 14- than 5-day-old flies. This suggests that the antioxidant defenses in older flies are less able to cope with oxidative stress. Moreover, the activity of G-6-PD, an enzyme known to be sensitive to oxidation, decreases upon X-irradiation of living flies, and does so to a greater extent in old than young animals (2). When this assay was applied to the gerbil samples described above, in which no overall change in antioxidants was seen, a clear difference between young and old tissues emerged. Whereas 6 krad of X-irradiation induced a 20–38% increase in protein carbonyls in 5-mo-old animals, it induced a 152–211% increase in 26-mo-old animals (300). Similarly, although synaptosomes from young and old mice contain equivalent amounts of ATP and GSH, those of old mice were far more sensitive to GSH depletion by the diethyl maleate than those from young mice (197). Lastly, reperfusion injury is a well-established model of oxidative stress associated with the reestablishment of blood flow following ischemia, and it causes greater oxidative damage to heart tissues of old rats than young ones (192a). The use of a polyclonal antiserum specific for adducts between lipid peroxidation end products and proteins detected such covalent modifications of mitochondrial proteins from old but not young animals, which was associated with a more dramatic loss of respiratory capacity in the former. Whereas the baseline mitochondrial respiratory parameters (before ischemia-reperfusion) did not differ between young and old animals, the administration of a physiologically relevant stress revealed a probable age-related decline in antioxidant defenses.

Another alternative to measuring absolute levels of antioxidants in old versus young animals is to investigate the ability of animals of different ages to induce antioxidants, an approach that has been applied to the analysis of SOD in the nematode *Caenorhabditis elegans* (59). Whereas in young animals challenge with hyperoxia or the redox cycling compound plumbagin resulted in an increase in SOD activity, in middle-aged or old animals it actually resulted in a net loss of activity.

What about repair of oxidative damage? Does its activity decrease with age? The bulk of evidence suggests that there is probably not an overall age-associated change in the intrinsic ability of cells to degrade damaged proteins (94, 270). Although a dramatic decrease in the activity of the oxidized protein-specific alkaline protease has been reported in old rat hepatocytes (318), no change in this activity was measured in the heart or brain of 25- versus 5-mo-old gerbils (300). In a separate work, a 50% age-related decline in a single activity (peptidylglutamyl-peptide hydrolase activity) of the hepatic multicatalytic protease was associated with its selective sensitivity (relative to the multicatalytic protease's other activities) to metal-catalyzed oxidation, suggesting that resistance to oxidants may (logically) characterize the proteases responsible for degrading damaged proteins (46).

Although the intrinsic protease activity may not decrease with age, there is evidence that repair of oxidized proteins may be less easily induced in response to an oxidative insult in old animals. For instance, exposure of young and old rats to 100% O_2 increased the content of protein carbonyls in both groups over a 48-h period. Between 48 and 54 h of exposure, however, alkaline protease activity was induced in young animals, with a corresponding decrease in protein carbonyls to initial levels. In old animals, on the other hand, no increase in activity was observed, and protein carbonyl levels continued to rise throughout the time course (318).

There is circumstantial evidence from mutagenesis studies that either antioxidant defenses or repair of oxidative DNA damage (or both) is less efficient in old mice. The induction of somatic mutations in mice by γ-irradiation is from 2.3- to 3.6-fold higher in old than in young animals, depending on the dose (99). The induction of mutations in young and old animals was reduced by feeding the animals a cocktail of dietary antioxidants, confirming that oxidants played a mutagenic role in these experiments. Therefore, the more pronounced induction of mutations in older mice is indirect evidence of decreased antioxidant defenses and repair (99). Later experiments employing peripheral lymphocytes from young and old human subjects resulted in similar results (100). The ability of human peripheral lymphocytes to repair oxidative DNA damage induced by H_2O_2 has also been found to be less efficient in cells from older donors (14).

Altogether, the results above suggest that older cells may be less able to prevent oxidative damage from occurring, and less effective at removing the damage once it has occurred. There is a clear need for more and better data about age-related trends in defenses and repair.

E. Age-Associated Trends in Oxidant Generation

The accumulation of oxidative damage could also result from an age-associated increase in the primary generation of oxidants, and some research suggests that this is the case. Generation of H_2O_2 and $O_2^{\bullet -}$ by isolated mitochondria and submitochondrial particles from 25-mo-old gerbils, for instance, is ~150–200% that of 5-mo-old animals (300), and that of aged rat heart (305) and brain (93a) has also been reported to be elevated. On the other hand, a recent study that paid specific attention to maintaining physiological substrate concentrations during in vitro mitochondrial incubations failed to detect an age-related increase in mitochondria H_2O_2 output (118). Recent measurements of oxidant generation in carefully isolated rat hepatocytes from young and old animals, employing an intracellular dye that fluoresces upon oxidation, have confirmed under conditions that preserve cellular integrity that cellular oxidant generation appears to increase (110). A similar increase in oxidant generation documented in flight muscle mitochondria of houseflies was associated with increases in the activities of every measured component of the electron transport chain except for the content of UQ (308), suggesting that an imbalance in electron transport may be the cause of aberrant reduction of oxygen. In another experiment, a population of chronologically identical 12-day-old flies was separated into phenotypically older crawlers and younger fliers as described above; mitochondrial H_2O_2 generation was twice as high in the crawlers than in the fliers, reflecting their greater phenotypic age.

Interestingly, oxidative damage to mitochondrial membranes and proteins has itself been implicated in enhanced oxidant generation; exposure of isolated mitochondria to the free radical generator 2,2-azobis(2-aminopropane)dihydrochloride or the cross-linking agent glutaraldehyde resulted in mitochondria that were more able to generate H_2O_2 when fed exogenous substrate (308). When these in vitro studies are combined with the above evidence of increased oxidative damage, the picture that emerges is a potential "vicious cycle" of oxidative damage and oxidant generation.

V. INTERSPECIES COMPARISONS

All of the data discussed in the previous section identify age-related differences within a given species. A complementary, and in some ways more powerful approach, is to compare species that have different MLSP. Comparative biochemistry and physiology, inspired by the rate of living hypothesis, has played a key role in establishing the free radical theory. Several representative studies are described below.

A. Oxidative Damage and Maximum Life Span Potential

The accumulation of cardiac lipofuscin in the monkey correlates with its cumulative O_2 consumption, starting at sexual maturation (226). Comparisons between rodent, dog, and primate species also revealed a close correlation between specific metabolic rate and lipofuscin accumulation (224, 225). The accumulation of the glycooxidation product pentosidine is similar. Short-lived species such as the shrew (MLSP = 3.5 yr) or rat (MLSP = 3 yr), with high specific metabolic rates, accumulate pentosidine in collagen at a rate of ~0.3–0.5 pmol/mg annually. Longer-lived primates such as monkeys and humans accumulate pentosidine at about one-tenth as fast (284).

Interspecies comparisons of protein oxidation mirror these results. The protein carbonyl content of 15-day-old individuals of five species of fly correlated negatively with mean life span; the longest-lived species (*Drosophila melanogaster*, mean life span = 65.5 days) had roughly one-third the level of protein oxidation as shortest-lived species (*Phaenecia sericata*, mean life span = 29.5 days) (310). The content of protein carbonyls was 20% higher in the heart, and ~50% higher in the brain, of 3.5-mo-old individuals of *Mus musculus* (MLSP = 3.5 yr) than in 3.5-mo-old individuals of *Peromyscus leucopus* (MLSP = 8 yr) (304).

The excretion of the oxo8gua and oxo8dG in urine is a reflection of whole body oxidative hits to DNA (287). The validity of its use to measure in vivo oxidative hits is strengthened by recent study of 33 women, in which both O_2 consumption and excretion of oxo8dG were measured. There was a highly significant positive correlation (P = 0.00007, r = 0.64) between O_2 consumption and excretion of adducts (188). When the urinary output of the oxidative DNA repair products oxo8gua thymine glycol and thymidine glycol were compared in mice, rats, and humans, they were found to correlate with species-specific metabolic rate (1, 287).

B. Antioxidant Defenses and Maximum Life Span Potential

One version of the free radical theory proposes that the differences in life span between species are due to species-specific antioxidant capacity. Early work by Cutler (57) tested this association: MLSP correlated positively with SOD but negatively with catalase and GPX (57). Recently, other groups have revisited this hypothesis, with similar results. Comparisons between SOD, catalase, GPX, and GSH in brain, liver, and heart from mice, rats, guinea pigs, rabbits, pigs, and cows were carried out and revealed that *1*) SOD and catalase activities correlated positively with MLSP, *2*) GSH activity correlated negatively with MLSP, and *3*) GPX correlated positively with MLSP in the brain and negatively in the liver and heart (309).

Interspecies comparisons between more distantly related vertebrate classes found either no correlation or a negative correlation between antioxidants and MLSP. For instance, in the liver of fish, frogs, birds, and mammals, strongly significant negative correlations were found for catalase, GPX, and GSH, whereas all other measured antioxidants (SOD, GSH reductase, ascorbate, urate, GSH) failed to correlate with MLSP (190). When the same antioxidants were measured in brain and lung tissues of the same species, virtually identical results were obtained (12, 247).

Comparisons between two very closely related species of mice, the house mouse (*M. musculus*) and the white-footed mouse (*P. leucopus*), which lives twice as long, have revealed higher levels of SOD, catalase, and GPX in brain and heart extracts of the latter, with the differences in GPX being the most dramatic. Also, when brain homogenates of the two species were challenged with X-irradiation, protein carbonyls accumulated at a threefold higher rate in shorter-lived *M. musculus* than longer-lived *P. leucopus* (304).

Together, these results suggest that the evolution of longevity has not been associated with a clear-cut increase in antioxidant capacity, at least across the broad sweep of evolution. (The comparisons between *M. mus-*

culus and *P. leucopus* invite the speculation that the more recent radiation of long-lived species may be associated with the coordinate upregulation of antioxidant defenses.) In any case, what emerged from the comparative biochemistry of free radical defenses is a fascinating paradox: birds, which typically have much longer long life spans than rodents, have lower activities of antioxidants and higher metabolic rates. This paradox has inspired the interspecies comparisons of mitochondrial oxidant generation discussed in section V.

C. Generation of Reactive Oxygen Species and Maximum Life Span Potential

The rate of living hypothesis identified an inverse correlation between metabolic rate and life span, and the free radical theory supplied a convenient mechanistic link: the mitochondrial generation of oxidants. Appealing as this theory is, one of the problems with the rate of living hypothesis is the conspicuous lack of fit of a few groups, notably birds and primates. Both of these groups live longer than their specific metabolic rate would predict. As a result, the total lifetime oxygen consumption of these groups is quite a bit higher than other groups (e.g., during their MLSP, pigeons and canaries consume 465 and 1,222 liters O_2/g, respectively, whereas mice, rats, guinea pigs, and trout consume 77, 28, 48, and 26 liters O_2, respectively) (247, 248). However, it cannot be assumed that the generation of oxidants is a direct function of metabolic rate, since mitochondria of different species (or even tissues) may exhibit different rates of intrinsic oxidant generation. These lines of reasoning by two laboratories have led to a similar line of experimentation, the results of which suggest that longevity is associated with a lesser capacity for mitochondrial oxidant generation (13, 172, 173, 304, 310).

For example, isolated pigeon mitochondria from brain, liver, or heart consume from two- to threefold as much oxygen as isolated rat mitochondria from the same tissues. However, pigeon mitochondria generate only one-third to one-half as much H_2O_2 as rat mitochondria under the same conditions. As a result, the calculated percentage of O_2 converted to $O_2^-\cdot/H_2O_2$ by mitochondria from lung, liver, and brain is ~10-fold lower in pigeons than in rats. This lower rate of oxidant generation corresponds with the roughly 10-fold longer life span of pigeons (13). A second, independent comparison of mitochondrial oxidant generation by pigeons and rats found similar results (172). Detailed respiratory comparisons have shown that generation of oxidants from both complex I and complex III of the mitochondrial electron transport chain is lower in pigeon than rat (130a).

In fact, the earliest interspecies study of mitochondrial oxidant production compared liver mitochondria of five mammalian species and thoracic muscle mitochondria of two species of fly (311), and also found that longer MLSPs were associated with lower mitochondrial oxidant generation. The mitochondria from flies produced from 6-fold more to 300-fold more oxidants than those from mammals. In a more recent comparison of heart tissues of eight diverse mammalian species of widely varying life span, submitochondrial particles of the short-lived species were found to generate more $O_2^-\cdot$ than those of long-lived species, a property correlated directly with the concentration of CoQ_9 (coenzyme Q possessing 9 isoprene units in its isoprenoid tail) and inversely with CoQ_{10} (although experiments intended to demonstrate a direct relationship between CoQ_9:CoQ_{10} ratio and $O_2^-\cdot$ generation failed to do so) (178a).

Interestingly, it appears as if the concept of intrinsic mitochondrial radical generation may also partially explain longevity differences between more closely related species. Measurements of $O_2^-\cdot$ and H_2O_2 generation by isolated mitochondria from the mouse species *M. musculus* and *P. leucopus* revealed that the former, which live half as long as the latter, also generate 48–74% more $O_2^-\cdot$ and 300–500% more H_2O_2. [As is the case with rats and pigeons, long-lived *P. leucopus* has the higher specific metabolic rate of the two species (304).] Finally, a comparison of five species of dipteran fly whose mean life spans vary by twofold also showed that the rate of generation of mitochondrial $O_2^-\cdot$ and H_2O_2 correlated negatively with life span (310).

Together, interspecies comparisons of oxidative damage, antioxidant defenses, and oxidant generation provide some of the most compelling evidence that oxidants are one of the most significant determinants of life span.

VI. DIETARY RESTRICTION

Limiting the dietary intake of mammals is a well-established way to extend life span (340). Interestingly, early expectations that dietary restriction (DR) would lower metabolic rate per se have not been confirmed, and so if DR attenuates oxidative damage, it is not via a simple reduction in oxygen consumption (312, 339a). Rather, there is now convincing evidence that dietary restriction may act in part by decreasing oxidative stress and increasing antioxidant defenses and repair (352). For example, mice whose caloric intake was restricted by 40% (DR) compared with ad libitum (AL) fed controls, and who lived 43% longer on average, had a less rapid accumulation of protein carbonyls in brain, heart, and kidney. Moreover, the generation of H_2O_2 and $O_2^-\cdot$ by mitochondria and submitochondrial particles from DR animals was lower than from AL animals, and also increased less with advancing age. In this study, however, DR did not appear to enhance antioxidant defenses (305). In another study, 40% DR of rats increased the activities of SOD, catalase, and GPX at older ages; free radical damage, as measured by TBARS and lipofuscin accumulation, was correspondingly lower (253). In a third study, focusing on mitochondrial oxidant generation and membrane properties of synaptosomal preparations, 40% DR reduced mitochondrial oxidant generation in both young and old samples and prevented the age-dependent decline in membrane fluidity, despite the fact that increases in the cholesterol-to-phospholipid ratio were common to both groups of animals (43). Other experiments have been recently reviewed (335, 352).

Dietary restriction also delays cancer incidence, which may be a reflection of fewer mutations induced by oxidants in DR animals. Dietary restriction has been shown to strengthen DNA repair (111), although few studies have focused specifically on repair of oxidative lesions. Not only caloric restriction, but also restriction of protein intake, can forestall the occurrence of cancer in rodent models. Although the mechanism by which protein restriction operates is not clear, enhanced protection against cellular degeneration, including oxidative damage, is a likely candidate. It is therefore interesting that the level of protein carbonyls in animals fed a low-protein diet was significantly lower than in animals fed standard lab food and that treatment of the animals with γ-irradiation induced a greater increase in protein carbonyls in high- than in low-protein animals (350).

VII. MANIPULATION OF RATE OF LIVING

The metabolic activity of caged houseflies can be reduced by limiting the space available for flight (low activity = LA vs. high activity = HA). It is expected that the LA flies should live longer due to lower O_2 consumption and generation of oxidants, and this is in fact observed. LA flies exhibit more than a twofold increase in both mean life span and MLSP. Measurements of protein carbonyls in 14-day-old LA and HA flies, which have not yet begun to die, revealed an elevation of more than 50% in HA flies relative to LA flies in both whole body extracts (299) and in mitochondria (303).

Under unfavorable conditions, the nematode *C. elegans* may undergo a morphological transformation to form a "dauer larva," a resting stage that does not feed, exhibits altered metabolic activity, and can survive for several times the MLSP of adult nematodes before returning to the adult stage. Interestingly, dauer larvae have also been found to have increased SOD activity (178).

VIII. MANIPULATION OF OXYGEN TENSION

Another experimental manipulation that modulates life span is growth under different O_2 tensions. The growth of houseflies in an atmosphere of 100% O_2, for example, markedly reduces their mean and MLSPs, and also increases the rate of accumulation of protein carbonyls in whole body extracts (299) and in isolated mitochondria (303). Similarly, elevated atmospheric O_2 decreased, and subnormal oxygen increased, the mean and maximum life spans of nematodes (138). Of course, the principal drawbacks of this type of experiment are its inapplicability to mammals and the fact that elevations or decreases in ambient O_2 in such species will likely be confounded by the overt pathology of hypo- or hyperoxia.

IX. SUPPLEMENTATION WITH DIETARY ANTIOXIDANTS

As one of the most intuitive approaches to testing the free radical theory, nutritional supplementation has been attempted with numerous species and compounds, with the result that mean life span has been extended in some instances (122). In mammals, results have been mixed. To cite recent studies, life-long oral supplementation of rats with UQ was without effect on mortality (189), whereas supplementation of the senescence accelerated mouse strain (see sect. XII) with "βCATECHIN," a commercial supplement with antioxidant activities (174, 349) resulted in life extension (175). Negative results from the feeding of antioxidants have been rationalized by arguing that many of the antioxidants fed to mammals interfere with mitochondrial respiration, and so the failure of antioxidant trials to extend MLSP may have been due to their toxicity (127). In a recent short-term feeding experiment, in contrast, the thiol donor *N*-acetyl-cysteine was found to markedly improve mitochondrial ETC complex activity in rats (214) and short-term feeding with a complex antioxidant-containing plant extract attenuated oxidative damage in rat mitochondria (272a). Experiments aimed at extending life span should probably be preceded by such short-term trials, to document the uptake, distribution, and biochemical action of the compounds.

Although no large-scale human nutritional intervention trials have been aimed specifically at the study of aging, a few have been undertaken in the hopes that dietary antioxidants might help prevent cancer. Three similar and heavily scrutinized studies are the α-tocopherol β-carotene (ATBC) prevention study (4), the physicians' health study (129), and the β-carotene and retinol efficacy trial (CARET) (236), all of which were designed to assess the effect of antioxidants on lung cancer risks. Although the details of these enormous experiments are not important to this discussion, the overall result was instructive: there was either no decrease or a modest increase in lung incidence and mortality with β-carotene administration.

These negative results, as well as negative results described above using laboratory rodents, should not be taken as evidence that the free radical theory of aging is flawed. In fact, they prove merely that a complex organism like a human or rodent is unlikely to respond predictably to crude manipulations such as supplementation with one or a small number of compounds (22), as well as that a single end point (such as lung cancer) is not equivalent to aging. Since the time when the results of these human trials were announced, a number of explanations have been forwarded to explain the paradoxical promotion of cancer by β-carotene (252), which indicates that the ultimate value of these trials may be a more precise understanding of this antioxidant. Ultimately, what the entire experience should teach the field of molecular gerontology is that at least until the biochemistry of dietary and cellular antioxidants is better understood, dietary trials in laboratory animals or humans will remain unreliable tests of the free radical theory; molecular gerontology should focus on more instructive experiments.

X. ADMINISTRATION OF PHARMACOLOGICAL ANTIOXIDANTS

Dramatic antiaging effects have been observed by Floyd and co-workers (234) with the free radical spin trap *N*-tert-butyl-α-phenylnitrone (PBN). Initial studies used PBN as a spin trap to measure radical generation during ischemia-reperfusion injury of gerbil brain (234). During reperfusion, the appearance of protein carbonyls correlated well with the loss of glutamine synthetase activity and the appearance of PBN-dependent nitroxide radical. Interestingly, it was also discovered that pretreatment with PBN attenuated protein oxidation and loss of glutamine synthetase activity and reduced lethality.

In a subsequent study that investigated the effect of PBN on spontaneous protein oxidation, it was discovered that twice-daily administration of PBN to old gerbils (32 mg/kg) reversed age-related oxidative damage (34). For example, during normal aging of gerbils, the level of brain protein carbonyls increased 185%, and glutamine synthetase activity declined by ~35%. A marked drop in neutral protease activity was also observed. Old animals administered PBN for 2 wk, however, experienced a reversal of these trends: protein carbonyls dropped, and glutamine synthetase and neutral protease activities were virtually restored. This effect was reversible, since over a 2-wk period after cessation of PBN treatment, the oxidation of proteins increased, and the activities of glutamine synthetase and neutral protease fell. Most impressive, however, was the performance of the PBN-treated old gerbils in a test of short-term memory; whereas the old gerbils were twice as likely to make errors as young gerbils, the old animals treated with PBN performed as well as the young (34).

Although these preliminary results held great promise, not only for the study of aging, but also for its therapeutic treatment, subsequent experiments reported that the results were irreproducible (31, 32). Further follow-up work from a third laboratory confirmed the initial findings in gerbil brain, but at the same time found no effect of PBN in lowering the level of protein carbonyls in gerbil heart or mouse brain (75). The chronic treatment of aged rats with PBN was found to reverse age-related cognitive impairment (294). Finally, the administration of PBN from age 20 wk throughout the life of a short-lived strain of mouse [the senescence accelerated mouse (SAM)] increased its life span from 42 to 56 wk (an increase of 33%) (73). As is argued (in sect. IX) to be the case for experiments with dietary intervention, maximum worth will be derived from pharmacological experiments when they include not only end-point mortality measurements but more focused studies of cellular effects.

XI. IN VITRO SENESCENCE AND OXIDANTS

The growth of normal diploid fibroblast in vitro is limited; after a specific number of population doublings (PDs) characteristic of the species, tissue, and age of the donor (the "Hayflick limit," named for the phenomenon's discoverer), the cells cease replication, which culminates in a viable, nondividing population with characteristic "senescent" morphology (56). Research on such "replicative senescence," and the vigorous debate about its relevance to organismal aging, have been recently reviewed (30, 55, 56). For the purposes of this discussion, a few points are worth reiterating. *1*) There are numerous correlations between organismal MLSP and replicative senescence. Fibroblasts from species with shorter MLSP exhibit a shorter in vitro life span than cells from more long-lived species. Similarly, fibroblasts from older individuals possess a shorter in vitro replicative potential than those from younger individuals (30), although this generalization has been questioned (55). Moreover, cells from patients with Werner's syndrome, a disease of accelerated aging (see sect. XVII), also complete fewer PDs before "crisis," when replication ceases (30). *2*) In vitro senescence may be relevant to organismal life span in at least two ways. First, as one of the likely components of tumor suppression, cellular senescence is necessary (but not sufficient) for extended life span. In other words, a long life span (as in humans, for instance) requires the absence of cancer at younger ages, and this is achieved in part by growth arrest of preneoplastic cells by mechanisms that operate during in vitro senescence (30, 269). Support for this assertion are the observations that fibroblasts from rodents (which have a significant cancer incidence after 2–3 yr of age) spontaneously immortalize in vitro, whereas fibroblasts from humans (who are relatively free of cancer until much later ages) do not (30), and that immortalization is associated with inactivation of p53 and Rb, which are frequently mutated in neoplasms (107). Second, senescent cells appear to occur in vivo, as judged by enzymatic markers characterized in vitro (68). It has been suggested that even at a low frequency, such cells may exert a disproportionate deleterious effect on tissue physiology by dominant effects of aberrant cellular regulation (e.g., expression of inappropriate cytokines) (30). *3*) In vitro senescence has been closely associated with the shortening of telomeric DNA, at least in humans (56). In human somatic cells, the average length of telomeric DNA shortens by ~50 bp/PD in vitro, and by ~15 bp/yr in vivo (30).

Leaving aside the question of the relevance of replicative senescence, is there any evidence that oxidative damage contributes to it? As early as the mid 1970s, this hypothesis was tested, by culturing human fibroblasts under conditions of low O_2 partial pressures and supplementing culture medium with the antioxidant α-tocopherol, with somewhat equivocal results (10, 241, 242). During the 1980s, a handful of studies documented the increased sensitivity of late-passage cells to oxidant challenge (139, 140) and noted that hyperoxic exposure could shorten in vitro life span (140). Later studies, looking to PD-dependent changes in antioxidant defenses as a potential explanation for the increased sensitivity of late-passage cells to oxidants, failed to account for these observations (250), although an impaired ability of late-passage cells to reinitiate DNA synthesis after hyperbaric O_2 exposure was established (141). More recently, a loss of oxo8gua glycosylase activity with increasing passage number has been reported (133), suggesting that decreased repair may be involved.

Recently, the issue has been revisited, and it was shown that a short (2 h) exposure to H_2O_2 (200 μM) induces an abrupt senescent-like growth arrest in human diploid fibroblasts (40). Intriguingly, cells that resumed cell division were found to have a markedly attenuated life span; a short-term oxidative stress had resulted in a long-term loss of PDs. The inhibition of H_2O_2 growth arrest by a metal chelator suggested that a Fenton-derived •OH radical was involved, and it was proposed that oxidative DNA damage and its sensing by cell cycle checkpoint controls may be involved in the phenotype (40). Indeed, measurements of DNA oxidation indicated that senescent cells suffer greater oxidative DNA damage than young cells and that the addition of the antioxidant PBN to the culture medium prolonged the life span in a dose-dependent fashion (41). In these and other studies, it was shown that culture under low oxygen tension extended fibroblast life span (41, 355) and that the inhibition of catalase decreased it (355), implicating endogenous H_2O_2 in replicative senescence.

Interestingly, lipofuscin accumulates in fibroblasts in vitro as well as in vivo, and culture under mildly hyperoxic conditions (40% O_2 partial pressure) decreases life span and stimulates lipofuscin accumulation (333). Direct feeding of synthetic lipofuscin particles to fibroblasts, which phagocytose and accumulate the material, simulates senescence (333). Most intriguing, perhaps, is the revelation that 40% O_2 culture resulted in a rapid loss of telomeric DNA as well as other cytogenetic alterations characteristic of senescent cells, and a permanent loss of replicative capacity reminiscent of the growth arrest of cells with a pulse of H_2O_2 (330a, 334). The measurement of DNA single-strand breaks in telomeric sequences of such cells revealed an increase under hyperoxia (334), confirming the idea (40) that DNA damage by oxidants may be involved in replicative senescence in vitro and extending the model to embrace the theory of telomere shortening (334). Interestingly, it has also been recently shown that the repair of ultraviolet light-induced pyrimidine dimers in telomeric DNA is lower in fibroblasts from an old donor than a young donor (171).

Together, these experiments are consistent with the hypothesis that oxidative DNA damage, acting to shorten telomeric sequences and thereby induce DNA damage-responsive growth arrest controls (107), plays a role in in vitro senescence. Two recent studies permit a further speculation. It has been shown that overexpression of the oncogenic ras gene in primary cells, which initially serves as a strong mitogenic stimulus, rapidly induces premature senescence in a p53- and p16-dependent fashion (286). Interestingly, it has also been recently reported that ras induces cellular oxidant generation (147). It is tempting to suggest that the "aggressive mitogenic stimulus" (286) of oncogenic ras, credited with inducing senescence, may be oxidative in nature. There may be a role for oxidants in replicative senescence.

XII. CLASSICAL AND POPULATION GENETICS

In comparison to the measurement of age-related changes, genetic approaches have the potential to provide more definitive information about aging, by supplying model organisms (mutant strains with altered life span) that differ at a small number of loci. Many of the geneticist's favorite organisms have been found to be amenable to studies of aging and have furnished evidence of the role of oxidants (196, 289).

A. *Caenorhabditis elegans* Genetics

The emergence of *C. elegans* as a convenient metazoan genetic model has attracted scientists seeking to discover genes controlling life span, or "genontogenes" (152, 158). Several long-lived mutant strains of *C. elegans* have been identified, including *age* mutants (the originally identified phenotype being extended life span), *daf* mutants (identified by abnormalities in the formation of dauer larvae), and *clk* (clock) mutants [which exhibit a pleiotropic behavioral deceleration (slower feeding, defecation, and movement)] (158). These two classes of mutant have collapsed into a single category with the cloning of *age-1* and its identification as an allele of *daf-23* (219). Strains of *age-1* live 65% longer on average and have a MLSP 110% longer than wild type (153). The trait is recessive, and biochemical studies have revealed that strains carrying *age-1* alleles have enhanced oxidative defenses. For example, when the sensitivities of wild-type and *age-1* strains to H_2O_2 were compared, it was found that the 50% effective lethal dose (LD_{50}) of the wild-type remained constant, whereas the LD_{50} of the *age-1* mutant increased with age, an indication of increasing resistance to oxidants (178). Similar comparisons showed that *age-1* mutants are more resistant to $O_2^{-}\cdot$ as well (330). In both cases, greater resistance to oxidants is associated with elevated activities of SOD and catalase. Whereas in wild-type animals these activities remained constant throughout the life span, in *age-1* strains they increased with advancing age, such that the difference between the mutants and the wild type grew increasingly pronounced at old ages (178, 330). In addition to being resistant to oxidants, *age-1* strains have increased heat tolerance, so it is likely that enhanced ability to cope with all environmental stresses, including oxidative ones, is the result of the loss-of-function mutation (185).

A comparative study of *daf-2* and *age-1* mutants revealed that they lie in a common pathway. Life extension by either requires the action of wild-type *daf-16*, and their effects on life extension are not additive (72), and so it was reasoned that they work in tandem. Their cloning revealed that *daf-2* is a member of the insulin receptor family (160a) and that *age-1* is a catalytic subunit of phosphatidylinositol 3-kinase (219), suggesting that together they transduce an insulin-like metabolic signal with pleiotropic effects on metabolism, stress resistance, and dauer formation (127b, 291a). The more recent cloning of *daf-16*, revealing that it is a member of the hepatocyte nuclear factor 3 (HNF-3)/forkhead family of transcriptional regulators (183a), has provided insight into the entire pathway, since in humans, insulin signaling antagonizes HNF-3. The implications, then, are *1*) that signaling through *daf-2* and *age-1* ultimately may serve to antagonize the control of a genetic program controlled by *daf-16*, and *2*) that this pathway may be relevant to human aging, since it appears to have been conserved (183a). Of course, the connection of these proteins to the aerobic respiration and the rate of living, via insulin-mediated metabolic control, has escaped no one's notice (183a).

The cloning of the *clk-1* gene also implicates respiration rate in *C. elegans* life span, albeit indirectly. The gene is homologous to a yeast gene called *CAT5/COO7* (79a), which is necessary for the genetic control of the shift from anaerobic to aerobic metabolism (127b), as well as the biosynthesis of the mitochondrial electron transport component UQ (252a). It has been proposed that the decelerated metabolic rate in *Clk-1* mutants of *C. elegans* may result in a correspondingly slower rate of oxidant generation (127b), although the authors of this work note that so far data are merely consistent with such a proposal.

A mutant that is in some ways the opposite of *age-1* resulted from a screen for sensitivity to $O_2^{-}\cdot$ in a chemically mutagenized population of *C. elegans*. The resulting mutant, *mev-1*, is hypersensitive to both paraquat and hyperoxia. The activity of SOD in *mev-1* was found to be roughly one-half of wild type, and the average life span was reduced by ~35% (148). Lipofuscin-like fluorescence accumulated more rapidly in *mev-1* mutants that in wild-type animals (144). The life span of both *mev-1* and wild-type nematodes was lengthened under hypoxia and shortened by hyperoxia, but the effect was more dramatic for the mutants (138).

B. *Drosophila* Genetics

According to the evolutionary concept of antagonistic pleiotropy, natural selection favors traits that benefit organisms during their period of reproduction. Therefore, by imposing a strong artificial selection that consistently rewards females capable of late reproduction, one ought to be able to select against those pleiotropic traits that are most antagonistic to long-term survival. After a suitable number of populations, a long-lived phenotype should emerge from a large, heterogeneous population.

In the early 1980s, Rose (264) followed precisely this strategy using *Drosophila* thereby validating the concept of antagonistic pleiotropy and also providing five independently selected lines (referred to as lines O1-O5) for further experimentation. The strength of this model is that, unlike *age-1*, lines O1-O5 are not single gene mutants, but rather populations in which any number of allelic changes will have occurred. Hence, biochemical alterations that are found to have arisen independently in many of these lines are likely to reflect the type of complex polymorphisms that confer increased life span to long-lived species such as *Homo sapiens*.

Therefore, it is significant that in a comparison of the lines O1-O5 with lines B1-B5 (control lines not subjected to selection), all of the long-lived populations had acquired a higher frequency of a relatively rare SOD allele that is associated with higher in vitro activity (328). In *Drosophila*, there exist two electrophoretic alleles of SOD, designated F (with lower in vitro activity) and S (with higher in vitro activity). Whereas the genotypes of all B flies tested were FF (no S alleles were detected), both FS and SS individuals existed in all five O lines, such that the average allele frequency of S ranged from 0.10 to 0.28 [0.25 ± 0.05 (SE)]. In all five independent instances of laboratory evolution, SOD activity had increased.

C. Rodent Genetics

The senescence-accelerated mouse (SAM) is an increasingly important model in gerontology (244); the senescence-prone strain (SAM-P) has a mean life span of 9 mo, compared with 13 mo for a senescence-resistant (SAM-R) strain. In liver tissues, the activity of mitochondrial SOD in the SAM-P strain was about one-half that in the SAM-R strain, a result that was consistent at all ages examined (244), which could explain the more rapid age-related increase in lipid peroxidation seen in this tissue (244) as well as independent reports of enhanced oxidative stress in SAM-P mice (186).

A potentially very useful animal model for testing the free radical theory of aging is the "S strain" of Wistar rat, selected for sensitivity to cataractogenesis by galactose. A heritable increase in cellular hexose uptake by cells from the S strain is associated with an increased intracellular generation of oxidants, increased endogenous lipid peroxidation, mitochondrial dysfunction, an increased age-specific incidence of degenerative diseases, and earlier aging than the galactose-resistant R strain (271, 272). Moreover, SOD and catalase activities in the blood of S rats are one-half that in R rats, which may have explained the increased in protein oxidative damage that was also seen in the S strain (345).

XIII. MOLECULAR GENETICS

A. Oxidants and Nuclear Somatic Mutations

Oxidants are twice removed from the somatic mutation theory of aging, since two requirements need to be satisfied for the relevance of oxidants to be ensured: *1*) oxidants must account for a significant proportion of mutations, and *2*) mutations must account for a significant proportion of the phenomena of aging. The second criterion is complex and has been recently discussed (71a, 195, 218, 338b); suffice it to say that to the extent that cancer is a degenerative disease of age (6), there is little doubt that mutations are relevant. What remains to be seen is whether or not the age-related degenerative changes that do not involve neoplasia are also the result of mutations (79) (or epigenetic alterations; Ref. 136).

With the assumption that they are, the question then remains, To what extent are spontaneous mutations due to oxidants? Some of the best evidence so far of the relevance of oxidative mutagenesis in vivo is an argument by analogy. In *E. coli*, mutation rates can be elevated from 10- to 1,000-fold by the loss of mutator genes *mutM*, *mutT*, and *mutY*, whose gene products are involved in repair and prevention of oxidation of guanine in DNA (210). Significantly, homologous genes and novel genes with analagous activities have been found in a number of species including humans (17, 192, 266), which forcefully makes the case that oxidants are significant mutagens in vivo. There is, moreover, plentiful direct evidence of oxidative mutagenesis in eukaryotic cells, such as the demonstration that disruption of the metallothionein gene in tissue culture, which may abrogate the cell's ability to chelate redox active metals such as iron, results in a 5- to 10-fold increase in spontaneous mutagenesis (267). Also, mutagenesis in cells from humans with Fanconi's anemia, a condition associated with oxidative stress, displayed a higher mutation frequency than controls and a marked increase in mutation frequency in response to the elevation of O_2 partial pressure (183).

Despite these results, the proportion of mutations due to oxidants in the presence of wild-type antimutators is still an unknown quantity. Recent experiments have illustrated that dietary supplementation with antioxidants, which reduces irradiation-induced mutagenesis, appears to reduce the spontaneous mutation rate as well (99, 100), suggesting that antioxidant scavenging in vivo is incomplete. The quantitative contribution of oxidants in somatic mutagenesis is an open (and important) question.

B. Oxidants and Mitochondrial Somatic Mutations

A burst of activity was stimulated by the report that a large deletion of the mitochondrial genome (ΔmtDNA) accumulates exponentially with age in humans and by the speculation that this may be the work of mitochondrial oxidants (53). Since this first detection of ΔmtDNA in normal elderly humans, dozens of studies have confirmed the finding, detecting the same "common" deletion in humans as well as different deletions in other species, including *C. elegans* (206), mice (44, 323), rats (76), and rhesus monkeys (179). Increased frequencies of ΔmtDNA in brain samples have been associated with Huntington's disease and AD, degenerative diseases of aging that affect millions of people (48, 143).

In most of these cases, a significant accumulation of the deletions occurs only at advanced ages and is associated primarily with postmitotic tissues. This mosaicism of mtDNA deletion (313), which is reminiscent of the pattern the maternally transmitted mitochondrial diseases, is hypothesized to be partly due to the lack of proliferative competition in postmitotic tissues (50, 336). Whereas in dividing tissues cells that inherit a relatively greater percentage of ΔmtDNA molecules by unequal mitotic assortment stand to compete less well in subsequent rounds of cell division, this "phenotypic sieving" will not operate in a nondividing tissue.

There are reasons to believe that the deletion of mtDNA may stem from oxidative damage (51). For one, the steady-state burden of oxidative adducts in mtDNA has been measured to be 16-fold higher in rat mtDNA than in nuclear DNA (16, 260). Moreover, quantitative polymerase chain reaction analyses of ΔmtDNA accumulation in different areas of the brain have consistently revealed highest frequencies of areas of high metabolic activity (47, 54). Despite the promise of these results, it is not self-evident that the observed accumulation of ΔmtDNA represents more than a biomarker (182). For one, the highest frequencies of ΔmtDNA detected, even at the oldest ages and in the most deletion-prone tissues, are generally no higher than 0.1%. Because in maternally transmitted mitochondrial genetic disorders symptoms are often absent even when more than one-half of the mtDNA molecules carry deletions (336), it is hard to see how such low frequencies could exert an effect. On the other hand, because there may be numerous deletions existing independently, measurements of a single one may merely represent the "tip of the iceberg" of mtDNA damage (54). The use of "long polymerase chain reaction" to amplify the entire mitochondrial genome has revealed the presence of numerous mtDNA rearrangements in human skeletal muscle (207). On the other hand, a recent attempt to measure an increase in somatic point mutations in human skeletal muscle was unsuccessful (243).

The analysis of progressively smaller bundles of cells has shown that the ΔmtDNA frequency is unevenly distributed (282). For example, when a muscle homogenate of rhesus monkey fibers is analyzed, the detected frequency of ΔmtDNAs is 0.02–0.1%. However, restriction of the sample size to 10-fiber bundles decreases the number of positive reactions but increases the measured frequency of ΔmtDNAs in positive reactions to 4.6–13.2%. In other words, a frequency of 0.1% could represent 100% deleted molecules in 1 cell of 1,000 (180). At any given point in time, only a small number of cells may carry deletions, but if these cells continually die as a result of a failure of oxidative phosphorylation, and other deletions arise de novo, then the low steady-state frequency of deletions could obscure a high rate of ΔmtDNA-associated cell death.

Finally, it has been proposed that a vicious cycle of increasing oxidative damage and stress of the type discussed in section *IIF* is particularly likely in the case of mtDNA, since it encodes components of the mitochondrial electron transport chain (52). This hypothesis is supported by the fact that in statistical terms, a random mutation of mtDNA is most likely to affect complex I, combined with the observation that a complex I poison (MPTP) triggers neuronal cell death in mice (52). It remains to be shown that known somatic mtDNA mutations do, in fact, result in increased oxidant generation.

XIV. TRANSGENIC ORGANISMS

A. Transgenic *Drosophila*

The modern acid test of a biological theory is the generation of a transgenic organism, and recent efforts to genetically engineer long-lived *Drosophila* by bolstering antioxidant defenses have been successful. The first such efforts involved the introduction of a bovine cDNA for Cu,Zn-SOD into *Drosophila*, which resulted in a significant increase in SOD activity as well as resistance to oxidative stresses and a small, statistically significant increase in mean life span (85). Whereas a repeat of this procedure with the *Drosophila* Cu,Zn-SOD gene resulted in only marginal effects on resistance to oxidants, and failed to increase life span, the introduction of single copies of *Drosophila* SOD and catalase resulted in a number of strains with significantly extended mean and maximum life spans (237).

As expected, the augmentation of antioxidant activities in the transgenic flies led to significant improvements over controls in numerous age-related indexes of oxidative metabolism: *1*) the accumulation of protein carbonyls was reduced; *2*) the rate of oxygen consumption at middle age and old age was higher; *3*) the ~50% loss of G-6-PD activity with age was reduced to ~10%; *4*) the age-related doubling of the load of oxo8dG was virtually abolished, as was the roughly twofold increase in sensitivity to the induction of oxidative DNA damage; and *5*) the age-related threefold increase in mitochondrial oxidant generation was reduced by about one-half (237, 298). In short, the predictions of the free radical theory were confirmed.

B. Transgenic Mice

Ultimately, transgenic mice are the molecular gerontologist's Holy Grail, and it is not until similar experiments are performed in mice that the free radical theory will have been satisfactorily tested. Although such experiments will clearly be exciting, work that has already been performed with mice transgenic for SOD suggests that their phenotypes may not be straightforward. As has been stressed, antioxidant defenses are interactive and coordinately regulated, and it has become apparent that oxidants play important roles as signaling molecules (286). As a consequence, up- or downregulating a single antioxidant may disrupt an internal balance and either exacerbate oxidative stress or otherwise compromise cellular physiology. Indeed, the phenotype of Cu,Zn-SOD-overexpressing mice appears to mimic the pathology of Down's syndrome (for which they may be a model, due to the localization of Cu,Zn-SOD to the trisomic chromosome 21; Ref. 36). Moreover, it has been recently discovered that a complete knockout of Cu,Zn-SOD has a phenotypic effect only on a small subset of motor neurons thought to be involved in amyotrophic lateral sclerosis (255). Even more surprising, the recent elimination of the enzyme GPX in a transgenic knockout mouse model resulted in generally normal, fertile mice without overt signs either of oxidative stress or increased sensitivity to oxidants

(134). Together, these results indicate that the promising results with *Drosophila* may be difficult to replicate, although no matter what the results, antioxidant overexpressing strains will clearly be extremely useful. By crossing animals with different transgenes, it will be possible to dissect antioxidant interactions and their effects on aging.

XV. SPORADIC DEGENERATIVE DISEASES

In the form of the specific age-related diseases in elderly humans, there exists a natural laboratory of oxidative stress. Indeed, investigation of the etiologies of human degenerative disease implicate oxidative stress in cancer (5, 6), atherosclerosis (112, 343), diabetes (326), and neurodegeneration (15, 18, 293), to name a few.

For example, the role of oxidants in AD has been a particularly active area of research in recent years. It has been known for some time that the formation of amyloid deposits, a hallmark feature of the clinical diagnosis, is associated with aberrant processing and folding of a protein called amyloid β-precursor protein (AβPP) to form the amyloid plaques (Aβ) characteristic of the disease (283). In recent years, it was discovered that not only the in vitro formation of synthetic Aβ from AβPP can be accelerated by the presence of oxygen but also that Aβ, which is cytotoxic when fed to neurons in vitro, appears to stimulate the generation of oxidant (293). In fact, amyloid peptides themselves generate oxidants in vitro (130). As discussed above, it has also been recently discovered that a cellular receptor (on microglial cells and endothelial cells) responsible for transducing the cytotoxic signal from Aβ and inducing oxidant generation is none other than RAGE (344). Amyloid plaques have also been found to bind redox-active iron, providing another potential positive feedback loop for oxidative damage (292a). Although we do not have space to go into the details, what is relevant here is not so much the specifics but the realization that in numerous ways, oxidants act as effectors in the progression of the disease. In other words, research into the etiology of AD (193b, 203a, 292b) [and also Parkinson's disease (15, 44a, 145a), amyotrophic lateral sclerosis (346), and dozens of other age-related disorders] is revealing that cytotoxicity very frequently involves oxidative stress.

Even if oxidants turn out to be essential effectors of AD pathology, can they initiate the process? Recently, it was claimed that a higher degree of heteroplasmy of mtDNA is associated with individuals with AD versus controls (62), consistent with a genetic predisposition based on mitochondrial inheritance. What is intriguing in this work is the reported association of these AD polymorphisms with higher intracellular oxidant generation by mitochondria isolated from the patients (62). However, recent work by two groups has independently revealed an artifact in the unusual protocol used for mtDNA extraction in these experiments: a nuclear pseudogene, rather than mtDNA itself, was amplified, and so the conclusions from these studies are invalid (131b, 336a).

XVI. INHERITED DEGENERATIVE DISEASES

Werner's syndrome (WS) is a segmental progeroid syndrome of late-onset accelerated aging, associated with an increased incidence of cancer, attenuated replicative senescence of cells in vitro, and, interestingly, an elevation of oxidized protein in isolated fibroblasts (315, 354). The cloning (354) of the gene responsible for the syndrome (*WRN*) does not suggest an oxidative etiology, however, as the protein product WRN appears to be a helicase (106a). *WRN* is homologous to the RecQ helicase of *E. coli*, which is involved in unwinding DNA during repair; hence, it has been proposed that the loss of helicase activity in WS results in a deficiency in recombination-mediated repair in humans cells (106a, 291a). Inefficient or dysregulated DNA repair, in turn, may cause the chromosomal instability, which is one of the disease's hallmarks and which may be the root of some or all of the diseases associated with the syndrome (106a, 291a).

Although there is consequently no obvious causal link between mutations in *WRN* and oxidative damage, the increased protein oxidation in WS fibroblasts (315) allows the speculation that measurable oxidative damage may result from inefficient recombination-dependent repair, chromosomal instability, and an ensuing disruption of genetic programs. If this turns out to be the case, the oxidative phenomenology of WS will serve as an important reminder that the measurement of oxidative damage does not imply oxidative causation! [On the other hand, if normal aging, like WS, results in part from chromosomal instability, it may be that oxidative genotoxicity plays a central role (sect. XIII*A*). In other words, it may be that oxidative damage may both result from and result in chromosomal instability.]

XVII. EPIDEMIOLOGY OF OXIDANTS AND ANTIOXIDANTS

Nutritional epidemiology has suggested that dietary antioxidants are crucially involved in the prevention of degenerative disease (22), and global epidemiological studies have pointed to geographical and economic antioxidant deficiencies in their localized occurrence (6). For instance, the loss of immune function with aging is amenable to therapeutic treatment with antioxidants (208a); the fact that both antioxidant status and immunological decline vary between individuals (209) suggests that there may exist heritable difference in antioxidant capacities, linked to differences in immune function. To date, however, the epidemiological study of senescence is merely a theoretical possibility and awaits powerful tools before it can be affordably attempted.

XVIII. SUMMARY

A. Are Oxidants Responsible for Aging?

We have outlined what we see to be the major experimental approaches to testing the free radical theory of aging. In Table 1, the strengths and weaknesses of different modes of experimentation are assessed. How powerful are different types of experiment? Has the free radical theory been supported?

Before a discussion of the merits of individual experiments, it should be stressed that if one is to judge them by their abilities to falsify the free radical theory, they are all bound to fail, for the simple reason that the free radical theory is very hard to falsify. The absence of a predicted outcome (for instance, life span extension by antioxidant supplementation) may often be explained by (justifiable) ad hoc reasoning. The physiology of oxidative stress, being a complex interaction of endogenous and exogenous factors, generally permits the "explanation" of negative results. How then shall one judge the different types of experiments and the theory itself? In fact, as the results discussed above illustrate, the momentum gathering behind the free radical theory is not due to any single experiment or approach, but rather derives from the extraordinarily multidisciplinary nature of current research. Although no single line of reasoning alone permits definitive conclusions, together they present a compelling case. [This philosophy, that the study of aging should employ a combination of different approaches, has been convincingly espoused before (83).]

Age-related oxidative phenomenology continues to provide evidence that senescence is associated with increased oxidant generation, a decline in the robustness of defenses and repair, and an accumulation of end products of oxidative damage (sect. IV). Although these trends are only correlations, the study of age-related changes has been rewarded with surprises such as the receptor for AGE (RAGE), which turns out to be involved in the etiology of AD and evidence that lysosomal lipofuscin may directly disrupt cells in vivo and in vitro. In fact, oxidative phenomenology has focused efforts on understanding the interplay between the components of oxidative stress outlined in Figure 1. The search for evidence of age-related DNA and protein oxidation has stimulated research into DNA repair, proteolytic salvage pathways, mitochondrial defenses, and so on.

Interspecies comparisons have better potential directly to test the free radical theory (sect. V). As predicted, differences between long- and short-lived species have been documented in all components of oxidative stress. Perhaps most instructive, however, have been results that initially appeared contradictory, such as the fact that birds are an outlier group in rate of living calculations. Attempts to account for the discrepancy have led to promising ad hoc hypotheses about the role of mitochondrial oxidant generation in aging. Unfortunately, relatively few researchers are familiar with or prepared to handle a variety of experimental animals, despite the research potential of a diverse menagerie.

In contrast, the model of dietary restriction has appeal precisely because it is so well established, homogeneous, and relevant to specific human diseases (sect. VI). Although early predictions that DR would result in a lower metabolic rate have proved unfounded (312), the discovery that oxidative defenses are enhanced by DR have strengthened both the free radical theory and the concept of antagonistic pleiotropy (158). Although, currently, results from DR studies are merely consistent with the free radical theory, future experiments may reveal specific pathways whereby DR upregulates defenses and repair.

Manipulation of metabolic activity and oxygen concentration are interesting systems, if of somewhat questionable relevance to normal aging (sects. VII and VIII). One drawback of such experiments is that they have generally been limited to invertebrates; the manipulation of metabolism in mammals is less straightforward. In the future, transgenic mouse models may allow a specific manipulation of metabolic rate in a consistent genetic background.

Attempts to manipulate life span with nutritional and pharmacological antioxidants obviously hold great appeal, as direct and intuitive tests of the theory (sects. IX and X). Although until recently they have not quite lived up to their promise (at least in mammals), this may have been due to a lack of the appropriate compounds. The serendipitous discovery of PBN's pharmacological properties appears to have opened up new avenues.

The study of fibroblast aging in vitro might appear the least likely area of research to support the free radical theory, being a phenomenon of replication per se, rather than of chronological aging (sect. XI). Nevertheless, the results discussed above not only suggest that oxidants may play a role, but they also have led to novel hypotheses about the dynamics of telomeric DNA in vivo. Future work in this area should be exciting.

The utility of genetics and transgenic animals in the study of aging, on the other hand, does not need elaboration (sects. XII and XIV). At present, the chief limitation to genetic studies is that fact that the adult forms of *C. elegans* and *Drosophila* are postmitotic. Molecular genetics (sect. XIII), which in the study of aging has primarily centered around the concept of somatic mutagenesis, is a field with particularly powerful tools that are generally applicable to any species. It is therefore very well suited to comparative studies, as illustrated by the hunt for mtDNA deletions in a wide range of species. Moreover, because mutational spectra provide information about the primary mutagenic event (256), molecular genetic studies may find the "smoking gun" of oxidative mutagenesis.

Finally, the push to understand human sporadic and inherited diseases through biochemical, genetic, and epidemiological methods may ultimately provide the most detailed information about age-related oxidative stress, as is becoming apparent in the cases of cancer, atherosclerosis, and neurodegenerative disorders (sects. xv–xvii). The complete sequencing of the human genome, and modern high-throughput genetic methods, are poised to provide an unprecedented amount of genetic information about a biological sample of many millions of individuals whose life histories (medical records) are accessible. For instance, an approach to the genetic study of aging that is only now attracting mainstream interest is the study of "human longevity outliers," or centenarians (292). Technical advances in the identification and analysis of polymorphisms, fueled by the Human Genome project, may permit the study of oxidative stress-associated alleles in these extraordinary individuals. An example of such an approach is the identification of a coding mtDNA polymorphism (a C-to-A transversion at mtDNA nucleotide position 5178 in the NADH dehydrogenase subunit 2 gene) in a larger percentage of centenarians (62% with 5178A) than controls (45% with 5178A) (322a). In this study, it was also found that as the age of randomly selected hospital inpatients and outpatients increased, the percentage carrying the 5178C polymorphism increased, suggesting that such individuals are at higher risk of degenerative disease. The authors hypothesize that the 5178A polymorphism, which is rare in all human populations except the Japanese, may be responsible both for the longevity of Japanese centenarians and, therefore, for the general longevity of the Japanese population as a whole. They suggest that such a polymorphism, affecting the mitochondrion, may have an effect on oxidative stress.

The answer to the question of whether or not the free radical theory has yet been confirmed depends on one's conception of what the theory predicts. In its broader sense ("oxidants contribute significantly to the process of degenerative senescence"), the theory has clearly been validated. In the more strict sense of the theory ("oxidants determine MLSP"), the data are not yet conclusive, although a large body of consistent data has been generated.

B. Nitric Oxide

A topic of great importance, which we have here largely omitted, is the role of nitric oxide (NO•) both as a damaging oxidant in its own right and as a signaling molecule that is susceptible to oxidative scavenging. NO•, the chemical previously known as endothelium-derived relaxation factor, is generated enzymatically by isozymes of nitric oxide synthase and is involved in vascular tone, innate immunity, and neuronal signal transduction. Moreover, NO reacts with O_2^-• to form peroxynitrite ($ONOO^-$), itself a powerful oxidant. Therefore, not only must nitric oxide synthase be seen as a potential source of damaging oxidants (249, 281), but O_2^-• may now be considered a physiologically relevant scavenger of the free radical signaling molecule NO• (58).

C. Oxidants and Apoptosis

Another topic we have decided to omit, because of space limitations and because this is an extremely fast-moving and frequently reviewed area, is the role of oxidants and mitochondria in apoptosis. Suffice it to mention that in the previous year there have appeared dozens of experiments documenting either the loss of mitochondrial membrane potential (355a), the release of mitochondrial cytochrome *c* (255a) [or an "apoptosis inducing factor" (355a)] into the cytosol, or the generation of oxidants by mitochondria during apoptosis (355a). Combined with the mitochondrial localization of the Bcl-2 family of anti-apoptosis proteins, and copious evidence that oxidant challenges can induce apoptosis (74a, 154, 290a), these experiments have shown that mitochondria are central to the process of cell suicide (255a, 355a). Despite this general consensus, however, a number of disagreements exist about the details of the mitochondrion's role in the process (149a) and even greater uncertainty about the role of oxidants. An equally complex problem will be defining the role of apoptosis itself in aging (338a). For instance, by eliminating neuronal cells, apoptosis may contribute to neurodegeneration, yet effective apoptosis is also clearly critical in preventing (age-related) tumorigenesis. In short, both the machinery and impact of apoptosis are still incompletely understood; conclusions about the interplay among oxidants, apoptosis, and aging will have to await further experimentation (338a).

D. Regulatory Roles for Oxidants and Antioxidants: Signal Transduction

The discovery that NO• could serve as a signaling molecule has foreshadowed a more general recognition that radical species are more than by-products. As mentioned in section xviii*B*, the interaction between O_2^-• and NO• indicates that both may be regulatory molecules (58). In other ways, the relationship between oxidants and signaling is surfacing. For instance, the nitration of tyrosine residues on a synthetic peptide substrate of cell cycle kinases destroyed its ability to be phosphorylated, which illustrates a mechanism for an individual amino acid adduct to exert an amplified effect (167). At a more physiological level, it has been shown that the sensitivity of hippocampal brain slices to muscarinic acetylcholine receptor agonists, which falls in an age-related fashion, can be restored by various antioxidant treatments and is potentiated (in old animals) by NO•-generating systems (155).

Perhaps most intriguing have been discoveries, such as the potential regulation of aconitases by O_2^{-}• (95, 98) and the realization that cell signaling via the Ras family of tyrosine kinases involves oxidants (147), that suggest that oxidants play a role in signal transduction by design. These discoveries strengthen the free radical theory for the following reason: if the regulated, enzymatic generation of oxidants plays a role in signaling pathways ("legitimate oxidants"), then even a small increase in oxidant generation by nonregulated pathways ("illegitimate oxidants") may have an exaggerated impact, by mimicking the legitimate process. The arrival of "redox regulation" as a viable field of inquiry, as evidenced by numerous topical recent reviews (66, 285, 321), may usher in a period when oxidants are seen as molecules of central physiological importance, in much the same way as are (for instance) lipid metabolites. As far as acceptance of the free radical theory is concerned, such a paradigm shift will probably lead to de facto acceptance of its broader conception (the idea that oxidants play a role in the process of aging). After all, if it turns out that oxidants and oxidative reactions are central to physiology, then how could they not be central to aging as well? If the free radical theory once lacked appeal because oxidants appeared to be stoichiometrically minor by-products, then those days may soon be over.

On the other hand, in its more narrow conception ("oxidants define MLSP"), the free radical theory still lacks experimental confirmation. A hypothetical example will illustrate this point. Even if mitogenic stimulation by Ras, for instance, were to involve oxidant generation in mice and humans, and the disregulation of Ras signaling by illegitimate oxidants in aged individuals were shown to be important in degenerative senescence, the different life spans of these organisms still beg an explanation; it still would remain to be explained why the process of oxidative disregulation takes so much longer to occur in humans than in mice. In conclusion, despite the growing concensus that the oxidants and oxygen free radicals are involved in degenerative senescence, countless mechanistic questions remain to be uncovered, as well as a central outstanding unknown: Do oxidants determine maximum life span in humans and other organisms?

We thank Caleb Finch for critically reading the manuscript and Irwin Fridovich for offering numerous detailed and useful comments. K. B. Beckman thanks Dr. I. Philipson for numerous useful discussions.

B. N. Ames was supported by National Cancer Institute Outstanding Investigator Grant CA39910 and National Institute of Environmental Health Sciences Grant ES-01896.

In writing a review on as broad a topic as the free radical theory, we have been forced to limit both the content and the number of references cited. Although we have done our best to include recent work, omissions were inevitable. We apologize to all authors whose work we have not managed to include and direct readers to other recent reviews for material we have left out.

Address for reprint requests: K. B. Beckman, Dept. of Molecular and Cell Biology, 401 Barker Hall, Univ. of California, Berkeley, CA 94720-3202.

REFERENCES

1. ADELMAN, R., R. L. SAUL, AND B. N. AMES. Oxidative damage to DNA: relation to species metabolic rate and life span. *Proc. Natl. Acad. Sci. USA* 85: 2706–2708, 1988.
2. AGARWAL, S., AND R. S. SOHAL. Relationship between aging and susceptibility to protein oxidative damage. *Biochem. Biophys. Res. Commun.* 194: 1203–1206, 1993.
3. AGARWAL, S., AND R. S. SOHAL. Aging and protein oxidative damage. *Mech. Ageing Dev.* 75: 11–19, 1994.
4. ALBANES, D., O. P. HEINONEN, J. K. HUTTUNEN, P. R. TAYLOR, J. VIRTAMO, B. K. EDWARDS, J. HAAPAKOSKI, M. RAUTALAHTI, A. M. HARTMAN, AND J. PALMGREN. Effects of alpha-tocopherol and beta-carotene supplements on cancer incidence in the Alpha-Tocopherol Beta-Carotene Cancer Prevention Study. *Am. J. Clin. Nutr.* 62: 1427S-1430S, 1995.
5. AMES, B. N., L. S. GOLD, AND W. C. WILLETT. The causes and prevention of cancer. *Proc. Natl. Acad. Sci. USA* 92: 5258–5265, 1995.
6. AMES, B. N., M. K. SHIGENAGA, AND T. M. HAGEN. Oxidants, antioxidants, and the degenerative diseases of aging. *Proc. Natl. Acad. Sci. USA* 90: 7915–7922, 1993.
7. ARNAIZ, S. L., M. TRAVACIO, S. LLESUY, AND A. BOVERIS. Hydrogen peroxide metabolism during peroxisome proliferation by fenofibrate. *Biochim. Biophys. Acta* 1272: 175–180, 1995.
8. ARNHEIM, N., AND G. CORTOPASSI. Deleterious mitochondrial DNA mutations accumulate in aging human tissues. *Mutat. Res.* 275: 157–167, 1992.
9. BAKER, G. T. D., AND R. L. SPROTT. Biomarkers of aging. *Exp. Gerontol.* 23: 223–239, 1988.
10. BALIN, A. K., D. B. GOODMAN, H. RASMUSSEN, AND V. J. CRISTOFALO. The effect of oxygen and vitamin E on the life span of human diploid cells in vitro. *J. Cell Biol.* 74: 58–67, 1977.
11. BANDY, B., AND A. J. DAVISON. Mitochondrial mutations may increase oxidative stress: implications for carcinogenesis and aging? *Free Radical Biol. Med.* 8: 523–539, 1990.
12. BARJA, G., S. CADENAS, C. ROJAS, M. LOPEZ-TORRES, AND R. PEREZ-CAMPO. A decrease of free radical production near critical targets as a cause of maximum longevity in animals. *Comp. Biochem. Physiol. B Biochem. Mol. Biol.* 108: 501–512, 1994.
13. BARJA, G., S. CADENAS, C. ROJAS, R. PEREZ-CAMPO, AND M. LOPEZ-TORRES. Low mitochondrial free radical production per unit O_2 consumption can explain the simultaneous presence of high longevity and high aerobic metabolic rate in birds. *Free Radical Res.* 21: 317–327, 1994.
14. BARNETT, Y. A., AND C. M. KING. An investigation of antioxidant status, DNA repair capacity and mutation as a function of age in humans. *Mutat. Res.* 338: 115–128, 1995.
15. BEAL, M. F. Aging, energy, and oxidative stress in neurodegenerative diseases. *Ann. Neurol.* 38: 357–366, 1995.
16. BECKMAN, K. B., AND B. N. AMES. Detection and quantification of oxidative adducts of mitochondrial DNA. *Methods Enzymol.* 264: 442–453, 1996.
17. BECKMAN, K. B., AND B. N. AMES. Oxidative decay of DNA. *J. Biol. Chem.* 272: 19633–19636, 1997.

18. BENZI, G., AND A. MORETTI. Are reactive oxygen species involved in Alzheimer's disease? *Neurobiol. Aging.* 16: 661–674, 1995.
19. BERLETT, B. S., R. L. LEVINE, AND E. R. STADTMAN. Comparison of the effects of ozone on the modification of amino acid residues in glutamine synthetase and bovine serum albumin. *J. Biol. Chem.* 271: 4177–4182, 1996.
20. BEYER, W., J. IMLAY, AND I. FRIDOVICH. Superoxide dismutases. *Prog. Nucleic Acids Res.* 40: 221–253, 1991.
21. BITTLES, A. H. Evidence for and against the causal involvement of mitochondrial DNA mutation in mammalian ageing. *Mutat. Res.* 275: 217–225, 1992.
22. BLOCK, G. Are clinical trials really the answer? *Am. J. Clin. Nutr. 62, Suppl.*: 1517S–1520S, 1995.
23. BOHR, V. A., AND R. M. ANSON. DNA damage, mutation and fine structure DNA repair in aging. *Mutat. Res.* 338: 25–34, 1995.
24. BOVERIS, A., AND E. CADENAS. Mitochondrial production of superoxide anions and its relationship to the antimycin insensitive respiration. *FEBS Lett.* 54: 311–314, 1975.
25. BOVERIS, A., E. CADENAS, AND A. O. STOPPANI. Role of ubiquinone in the mitochondrial generation of hydrogen peroxide. *Biochem. J.* 156: 435–444, 1976.
26. BOVERIS, A., AND B. CHANCE. The mitochondrial generation of hydrogen peroxide. General properties and effect of hyperbaric oxygen. *Biochem. J.* 134: 707–716, 1973.
27. BOVERIS, A., N. OSHINO, AND B. CHANCE. The cellular production of hydrogen peroxide. *Biochem. J.* 128: 617–630, 1972.
28. BRUNK, U. T., C. B. JONES, AND R. S. SOHAL. A novel hypothesis of lipofuscinogenesis and cellular aging based on interactions between oxidative stress and autophagocytosis. *Mutat. Res.* 275: 395–403, 1992.
29. BRUNK, U. T., H. ZHANG, H. DALEN, AND K. OLLINGER. Exposure of cells to nonlethal concentrations of hydrogen peroxide induces degeneration-repair mechanisms involving lysosomal destabilization. *Free Radical Biol. Med.* 19: 813–822, 1995.
29a. CADET, J., T. DOUKI, AND J.-L. RAVANAT. Artifacts associated with the measurement of oxidized DNA bases. *Environ. Health Perspect.* 105: 1034–1039, 1997.
30. CAMPISI, J. Replicative senescence: an old lives' tale? *Cell* 84: 497–500, 1996.
31. CAO, G., AND R. G. CUTLER. Protein oxidation and aging. I. Difficulties in measuring reactive protein carbonyls in tissues using 2,4-dinitrophenylhydrazine. *Arch. Biochem. Biophys.* 320: 106–114, 1995.
32. CAO, G., AND R. G. CUTLER. Protein oxidation and aging. II. Difficulties in measuring alkaline protease activity in tissues using the fluorescamine procedure. *Arch. Biochem. Biophys.* 320: 195–201, 1995.
33. CARNEY, J. M., C. D. SMITH, A. M. CARNEY, AND D. A. BUTTERFIELD. Aging- and oxygen-induced modifications in brain biochemistry and behavior. *Ann. NY Acad. Sci.* 738: 44–53, 1994.
34. CARNEY, J. M., P. E. STARKE-REED, C. N. OLIVER, R. W. LANDUM, M. S. CHENG, J. F. WU, AND R. A. FLOYD. Reversal of age-related increase in brain protein oxidation, decrease in enzyme activity, and loss in temporal and spatial memory by chronic administration of the spin-trapping compound *N-tert*-butyl-alpha-phenylnitrone. *Proc. Natl. Acad. Sci. USA* 88: 3633–3636, 1991.
35. CERAMI, A. Hypothesis. Glucose as a mediator of aging. *J. Am. Geriatr. Soc.* 33: 626–634, 1985.
36. CHAN, P. H., C. J. EPSTEIN, Y. LI, T. T. HUANG, E. CARLSON, H. KINOUCHI, G. YANG, H. KAMII, S. MIKAWA, AND T. KONDO. Transgenic mice and knockout mutants in the study of oxidative stress in brain injury. *J. Neurotrauma* 12: 815–824, 1995.
37. CHANCE, B., H. SIES, AND A. BOVERIS. Hydroperoxide metabolism in mammalian organs. *Physiol. Rev.* 59: 527–605, 1979.
38. CHANOCK, S. J., B. J. EL, R. M. SMITH, AND B. M. BABIOR. The respiratory burst oxidase. *J. Biol. Chem.* 269: 24519–24522, 1994.
39. CHEN, J. J., AND B. P. YU. Alterations in mitochondrial membrane fluidity by lipid peroxidation products. *Free Radical Biol. Med.* 17: 411–418, 1994.
40. CHEN, Q., AND B. N. AMES. Senescence-like growth arrest induced by hydrogen peroxide in human diploid fibroblast F65 cells. *Proc. Natl. Acad. Sci. USA* 91: 4130–4134, 1994.
41. CHEN, Q., A. FISCHER, J. D. REAGAN, L. J. YAN, AND B. N. AMES. Oxidative DNA damage and senescence of human diploid fibroblast cells. *Proc. Natl. Acad. Sci. USA* 92: 4337–4341, 1995.
42. CHIO, K. S., AND A. L. TAPPEL. Synthesis and characterization of the fluorescent products derived from malonaldehyde and amino acids. *Biochemistry* 8: 2821–2826, 1969.
43. CHOI, J. H., AND B. P. YU. Brain synaptosomal aging: free radicals and membrane fluidity. *Free Radical Biol. Med.* 18: 133–139, 1995.
44. CHUNG, S. S., R. WEINDRUCH, S. R. SCHWARZE, D. I. MCKENZIE, AND J. M. AIKEN. Multiple age-associated mitochondrial DNA deletions in skeletal muscle of mice. *Aging* 6: 193–200, 1994.
44a. COHEN, G., R. FAROOQUI, AND N. KESLER. Parkinson disease: a new link between monoamine oxidase and mitochondrial electron flow. *Proc. Natl. Acad. Sci. USA* 94: 4890–4894, 1997.
44b. COLLINS, A., J. CADET, B. EPE, AND C. GEDIK. Problems in the measurement of 8-oxoguanine in human DNA. Report of a workshop, DNA oxidation, held in Aberdeen, UK, 19–21 January, 1997. *Carcinogenesis* 18: 1833–1836, 1997.
45. COMMONER, B., J. TOWNSEND, AND G. E. PAKE. Free radicals in biological materials. *Nature* 174: 689–691, 1954.
46. CONCONI, M., L. I. SZWEDA, R. L. LEVINE, E. R. STADTMAN, AND B. FRIGUET. Age-related decline of rat liver multicatalytic proteinase activity and protection from oxidative inactivation by heat-shock protein 90. *Arch. Biochem. Biophys.* 331: 232–240, 1996.
47. CORRAL-DEBRINSKI, M., T. HORTON, M. T. LOTT, J. M. SHOFFNER, M. F. BEAL, AND D. C. WALLACE. Mitochondrial DNA deletions in human brain: regional variability and increase with advanced age. *Nature Genet.* 2: 324–329, 1992.
48. CORRAL-DEBRINSKI, M., T. HORTON, M. T. LOTT, J. M. SHOFFNER, A. C. MCKEE, M. F. BEAL, B. H. GRAHAM, AND D. C. WALLACE. Marked changes in mitochondrial DNA deletion levels in Alzheimer brains. *Genomics* 23: 471–476, 1994.
49. CORREA, P., AND V. W. CHEN. Gastric cancer. *Cancer Surv.* 20: 55–76, 1994.
50. CORTOPASSI, G., AND Y. LIU. Genotypic selection of mitochondrial and oncogenic mutations in human tissue suggests mechanisms of age-related pathophysiology. *Mutat. Res.* 338: 151–159, 1995.
51. CORTOPASSI, G., Y. LIU, AND T. HUTCHIN. Degeneration of human oncogenes and mitochondrial genes occurs in cells that exhibit age-related pathology. *Exp. Gerontol.* 31: 253–265, 1996.
52. CORTOPASSI, G., AND E. WANG. Modelling the effects of age-related mtDNA mutation accumulation: complex I deficiency, superoxide and cell death. *Biochim. Biophys. Acta* 1271: 171–176, 1995.
52a. CORTOPASSI, G., AND E. WANG. There is substantial agreement among interspecies estimates of DNA repair activity. *Mech. Ageing Dev.* 91: 211–218, 1996.
53. CORTOPASSI, G. A., AND N. ARNHEIM. Detection of a specific mitochondrial DNA deletion in tissues of older humans. *Nucleic Acids Res.* 18: 6927–6933, 1990.
54. CORTOPASSI, G. A., D. SHIBATA, N. W. SOONG, AND N. ARNHEIM. A pattern of accumulation of a somatic deletion of mitochondrial DNA in aging human tissues. *Proc. Natl. Acad. Sci. USA* 89: 7370–7374, 1992.
55. CRISTOFALO, V. J. Ten years later: what have we learned about human aging from studies of cell cultures? *Gerontologist* 36: 737–741, 1996.
56. CRISTOFALO, V. J., AND R. J. PIGNOLO. Replicative senescence of human fibroblast-like cells in culture. *Physiol. Rev.* 73: 617–638, 1993.
56a. CROTEAU, D. L., AND V. A. BOHR. Repair of oxidative damage to nuclear and mitochondrial DNA in mammalian cells. *J. Biol. Chem.* 272: 25409–25412, 1997.
56b. CUNNINGHAM, R. P. DNA repair: caretakers of the genome? *Curr. Biol.* 7: R576–R579, 1997.
57. CUTLER, R. G. Antioxidants and aging. *Am. J. Clin. Nutr.* 53, *Suppl.*: 373S–379S, 1991.
58. DARLEY-USMAR, V., H. WISEMAN, AND B. HALLIWELL. Nitric oxide and oxygen radicals: a question of balance. *FEBS Lett.* 369: 131–135, 1995.
59. DARR, D., AND I. FRIDOVICH. Adaptation to oxidative stress in young, but not in mature or old, *Caenorhabditis elegans*. *Free Radical Biol. Med.* 18: 195–201, 1995.

60. DAVIES, K. J. Oxidative stress: the paradox of aerobic life. *Biochem. Soc. Symp.* 61: 1–31, 1995.
61. DAVIES, M. J., S. FU, AND R. T. DEAN. Protein hydroperoxides can give rise to reactive free radicals. *Biochem. J.* 305: 643–649, 1995.
62. DAVIS, R. E. M. S., C. HERRNSTADT, S. S. GHOSH, E. FAHY, L. A. SHINOBU, D. GALASKO, L. J. THAL, M. F. BEAL, N. HOWELL, AND W. D. PARKER. Mutations in mitochondrial cytochrome *c* oxidase genes segregate with late-onset Alzheimer's disease. *Proc. Natl. Acad. Sci. USA* 94: 4526–4531, 1997.
63. DEAN, R. T., J. GEBICKI, S. GIESEG, A. J. GRANT, AND J. A. SIMPSON. Hypothesis: a damaging role in aging for reactive protein oxidation products? *Mutat. Res.* 275: 387–393, 1992.
64. DEAN, R. T., S. GIESEG, AND M. J. DAVIES. Reactive species and their accumulation on radical-damaged proteins. *Trends Biochem. Sci.* 18: 437–441, 1993.
65. DE LA PAZ, M. A., J. ZHANG, AND I. FRIDOVICH. Antioxidant enzymes of the human retina: effect of age on enzyme activity of macula and periphery. *Curr. Eye Res.* 15: 273–278, 1996.
66. DEMPLE, B. Redox signaling and gene control in the *Escherichia coli* soxRS oxidative stress regulon—a review. *Gene* 179: 53–57, 1996.
67. DENNOG, C., A. HARTMANN, G. FREY, AND G. SPEIT. Detection of DNA damage after hyperbaric oxygen (HBO) therapy. *Mutagenesis* 11: 605–609, 1996.
68. DIMRI, G. P., X. LEE, G. BASILE, M. ACOSTA, G. SCOTT, C. ROSKELLEY, E. E. MEDRANO, M. LINSKENS, I. RUBELJ, O. PEREIRA-SMITH, M. PEACOCKE, AND J. CAMPISI. A biomarker that identifies senescent human cells in culture and in aging skin in vivo. *Proc. Natl. Acad. Sci. USA* 92: 9363–9367, 1995.
69. DIONISI, O., T. GALEOTTI, T. TERRANOVA, AND A. AZZI. Superoxide radicals and hydrogen peroxide formation in mitochondria from normal and neoplastic tissues. *Biochim. Biophys. Acta* 403: 292–300, 1975.
70. DIZDAROGLU, M. Measurement of radiation-induced damage to DNA at the molecular level. *Int. J. Radiat. Biol.* 61: 175–183, 1992.
71. DIZDAROGLU, M. Oxidative damage to DNA in mammalian chromatin. *Mutat. Res.* 275: 331–342, 1992.
71a. DOLLE, M. E., H. GIESE, C. L. HOPKINS, H. J. MARTUS, J. M. HAUSDORFF, AND J. VIJG. Rapid accumulation of genome rearrangements in liver but not in brain of old mice. *Nature Genet.* 17: 431–434, 1997.
72. DORMAN, J. B., B. ALBINDER, T. SHROYER, AND C. KENYON. The age-1 and daf-2 genes function in a common pathway to control the life span of *Caenorhabditis elegans*. *Genetics* 141: 1399–1406, 1995.
73. DUBEY, A., M. J. FORSTER, AND R. S. SOHAL. Effect of the spin-trapping compound *N-tert*-butyl-alpha-phenylnitrone on protein oxidation and life span. *Arch. Biochem. Biophys.* 324: 249–254, 1995.
74. DUPRAT, F., E. GUILLEMARE, G. ROMEY, M. FINK, F. LESAGE, M. LAZDUNSKI, AND E. HONORE. Susceptibility of cloned K^+ channels to reactive oxygen species. *Proc. Natl. Acad. Sci. USA* 92: 11796–11800, 1995.
74a. DYPBUKT, J. M., M. ANKARCRONA, M. BURKITT, A. SJOHOLM, K. STROM, S. ORRENIUS, AND P. NICOTERA. Different prooxidant levels stimulate growth, trigger apoptosis, or produce necrosis of insulin-secreting RINm5F cells. The role of intracellular polyamines. *J. Biol. Chem.* 269: 30553–30560, 1994.
75. EDAMATSU, R., A. MORI, AND L. PACKER. The spin-trap *N-tert*-alpha-phenyl-butylnitrone prolongs the life span of the senescence accelerated mouse. *Biochem. Biophys. Res. Commun.* 211: 847–849, 1995.
76. EDRIS, W., B. BURGETT, O. C. STINE, AND C. R. FILBURN. Detection and quantitation by competitive PCR of an age-associated increase in a 4.8-kb deletion in rat mitochondrial DNA. *Mutat. Res.* 316: 69–78, 1994.
77. ELGAWISH, A., M. GLOMB, M. FRIEDLANDER, AND V. M. MONNIER. Involvement of hydrogen peroxide in collagen cross-linking by high glucose in vitro and in vivo. *J. Biol. Chem.* 271: 12964–12971, 1996.
78. EPE, B. DNA damage profiles induced by oxidizing agents. *Rev. Physiol. Biochem. Pharmacol.* 127: 223–249, 1996.
79. EVANS, D. A., J. P. BURBACH, AND F. W. VAN LEEUWEN. Somatic mutations in the brain: relationship to aging? *Mutat. Res.* 338: 173–182, 1995.
79a. EWBANK, J. J., T. M. BARNES, B. LAKOWSKI, M. LUSSIER, H. BUSSEY, AND S. HEKIMI. Structural and functional conservation of the *Caenorhabditis elegans* timing gene clk-1 (see comments). *Science* 275: 980–983, 1997.
80. FAHN, S., AND G. COHEN. The oxidant stress hypothesis in Parkinson's disease: evidence supporting it. *Ann. Neurol.* 32: 804–812, 1992.
81. FEIG, D. I., T. M. REID, AND L. A. LOEB. Reactive oxygen species in tumorigenesis. *Cancer Res.* 54, *Suppl.*: 1890s–1894s, 1994.
82. FEUERS, R. J., R. WEINDRUCH, AND R. W. HART. Caloric restriction, aging, and antioxidant enzymes. *Mutat. Res.* 295: 191–200, 1993.
83. FINCH, C. E. *Longevity, Senescence, and The Genome.* Chicago, IL: Univ. of Chicago Press, 1990.
84. FLEMING, J. E., J. MIQUEL, S. F. COTTRELL, L. S. YENGOYAN, AND A. C. ECONOMOS. Is cell aging caused by respiration-dependent injury to the mitochondrial genome? *Gerontology* 28: 44–53, 1982.
85. FLEMING, J. E., I. REVEILLAUD, AND A. NIEDZWIECKI. Role of oxidative stress in *Drosophila* aging. *Mutat. Res.* 275: 267–279, 1992.
86. FLINT, D. H., J. F. TUMINELLO, AND M. H. EMPTAGE. The inactivation of Fe-S cluster containing hydrolyases by superoxide. *J. Biol. Chem.* 268: 22369–22376, 1993.
87. FLOYD, R. A., AND J. M. CARNEY. Free radical damage to protein and DNA: mechanisms involved and relevant observations on brain undergoing oxidative stress. *Ann. Neurol.* 32, *Suppl.*: S22–S27, 1992.
88. FORMAN, H. J., AND A. AZZI. On the virtual existence of superoxide anion in mitochondria: thoughts regarding its role in pathophysiology. *FASEB J.* 11: 374–375, 1997.
89. FORMAN, H. J., AND J. A. KENNEDY. Role of superoxide radical in mitochondrial dehydrogenase reactions. *Biochem. Biophys. Res. Commun.* 60: 1044–1050, 1974.
90. FORSTER, M. J., A. DUBEY, K. M. DAWSON, W. A. STUTTS, H. LAL, AND R. S. SOHAL. Age-related losses of cognitive function and motor skills in mice are associated with oxidative protein damage in the brain. *Proc. Natl. Acad. Sci. USA* 93: 4765–4769, 1996.
91. FRIDOVICH, I. Superoxide radical and superoxide dismutases. *Annu. Rev. Biochem.* 64: 97–112, 1995.
92. FRIGUET, B., E. R. STADTMAN, AND L. I. SZWEDA. Modification of glucose-6-phosphate dehydrogenase by 4-hydroxy-2-nonenal. Formation of cross-linked protein that inhibits the multicatalytic protease. *J. Biol. Chem.* 269: 21639–21643, 1994.
93. FRIGUET, B., L. I. SZWEDA, AND E. R. STADTMAN. Susceptibility of glucose-6-phosphate dehydrogenase modified by 4-hydroxy-2-nonenal and metal-catalyzed oxidation to proteolysis by the multicatalytic protease. *Arch. Biochem. Biophys.* 311: 168–173, 1994.
93a. GABBITA, S. P., D. A. BUTTERFIELD, K. HENSLEY, W. SHAW, AND J. M. CARNEY. Aging and caloric restriction affect mitochondrial respiration and lipid membrane status: an electron paramagnetic resonance investigation. *Free Radical Biol. Med.* 23: 191–201, 1997.
94. GAFNI, A. Altered protein metabolism in aging. *Annu. Rev. Gerontol. Geriatr.* 10: 117–131, 1990.
95. GARDNER, P. R., AND I. FRIDOVICH. Superoxide sensitivity of the *Escherichia coli* aconitase. *J. Biol. Chem.* 266: 19328–19333, 1991.
96. GARDNER, P. R., AND I. FRIDOVICH. Inactivation-reactivation of aconitase in *Escherichia coli*. A sensitive measure of superoxide radical. *J. Biol. Chem.* 267: 8757–8763, 1992.
97. GARDNER, P. R., D. D. NGUYEN, AND C. W. WHITE. Aconitase is a sensitive and critical target of oxygen poisoning in cultured mammalian cells and in rat lungs. *Proc. Natl. Acad. Sci. USA* 91: 12248–12252, 1994.
98. GARDNER, P. R., I. RAINERI, L. B. EPSTEIN, AND C. W. WHITE. Superoxide radical and iron modulate aconitase activity in mammalian cells. *J. Biol. Chem.* 270: 13399–13405, 1995.
99. GAZIEV, A. I., A. PODLUTSKY, B. M. PANFILOV, AND R. BRADBURY. Dietary supplements of antioxidants reduce hprt mutant frequency in splenocytes of aging mice. *Mutat. Res.* 338: 77–86, 1995.
100. GAZIEV, A. I., G. R. SOLOGUB, L. A. FOMENKO, S. I. ZAICHKINA,

N. I. KOSYAKOVA, AND R. J. BRADBURY. Effect of vitamin-antioxidant micronutrients on the frequency of spontaneous and in vitro gamma-ray-induced micronuclei in lymphocytes of donors: the age factor. *Carcinogenesis* 17: 493–499, 1996.
101. GEBICKI, J. M. Protein hydroperoxides as new reactive oxygen species. *Redox Report* 3: 99–110, 1997.
102. GIESEG, S. P., J. A. SIMPSON, T. S. CHARLTON, M. W. DUNCAN, AND R. T. DEAN. Protein-bound 3,4-dihydroxyphenylalanine is a major reductant formed during hydroxyl radical damage to proteins. *Biochemistry* 32: 4780–4786, 1993.
103. GILCHREST, B. A., AND V. A. BOHR. Aging processes, DNA damage, and repair. *FASEB J.* 11: 322–330, 1997.
104. GIULIVI, C., R. E. PACIFICI, AND K. J. DAVIES. Exposure of hydrophobic moieties promotes the selective degradation of hydrogen peroxide-modified hemoglobin by the multicatalytic proteinase complex, proteasome. *Arch. Biochem. Biophys.* 311: 329–341, 1994.
105. GOEPTAR, A. R., H. SCHEERENS, AND N. P. VERMEULEN. Oxygen and xenobiotic reductase activities of cytochrome *P450*. *Crit. Rev. Toxicol.* 25: 25–65, 1995.
106. GOTO, S., AND A. NAKAMURA. Age-associated, oxidatively modified proteins: a critical evaluation. *Age* 20: 81–89, 1997.
106a. GRAY, M. D., J. C. SHEN, A. S. KAMATH-LOEB, A. BLANK, B. L. SOPHER, G. M. MARTIN, J. OSHIMA, AND L. A. LOEB. The Werner syndrome protein is a DNA helicase. *Nature Genet.* 17: 100–103, 1997.
107. GREIDER, C. W. Mammalian telomere dynamics: healing, fragmentation shortening and stabilization. *Curr. Opin. Genet. Dev.* 4: 203–211, 1994.
108. GROLLMAN, A. P., AND M. MORIYA. Mutagenesis by 8-oxoguanine: an enemy within. *Trends Genet.* 9: 246–249, 1993.
109. GUTTERIDGE, J. M., AND B. HALLIWELL. The measurement and mechanism of lipid peroxidation in biological systems. *Trends Biochem. Sci.* 15: 129–135, 1990.
110. HAGEN, T. M., D. L. YOWE, J. C. BARTHOLOMEW, C. M. WEHR, K. L. DO, J. Y. PARK, AND B. N. AMES. Mitochondrial decay in hepatocytes from old rats: membrane potential declines, heterogeneity and oxidants increase. *Proc. Natl. Acad. Sci. USA* 94: 3064–3069, 1997.
111. HALEY, Z. V., AND A. RICHARDSON. Effect of dietary restriction on DNA repair and DNA damage. *Mutat. Res.* 295: 237–245, 1993.
112. HALLIWELL, B. Free radicals, reactive oxygen species, and human disease: a critical evaluation with special reference to atherosclerosis. *Br. J. Exp. Pathol.* 70: 737–757, 1989.
113. HALLIWELL, B., AND M. DIZDAROGLU. The measurement of oxidative damage to DNA by HPLC and GC/MS techniques. *Free Radical Res. Commun.* 16: 75–87, 1992.
114. HALLIWELL, B., AND J. M. GUTTERIDGE. Oxygen free radicals and iron in relation to biology and medicine: some problems and concepts. *Arch. Biochem. Biophys.* 246: 501–514, 1986.
115. HALLIWELL, B. H., AND J. M. C. GUTTERIDGE. *Free Radicals in Biology and Medicine*. Oxford, UK: Oxford Univ. Press, 1989.
116. HANAWALT, P. C. DNA repair comes of age. *Mutat. Res.* 336: 101–113, 1995.
117. HANAWALT, P. C., P. GEE, L. HO, R. K. HSU, AND C. J. KANE. Genomic heterogeneity of DNA repair. Role in aging? *Ann. NY Acad. Sci.* 663: 17–25, 1992.
118. HANSFORD, R. G., B. A. HOGUE, AND V. MILDAZIENE. Dependence of H_2O_2 formation by rat heart mitochondria on substrate availablity and donor age. *J. Bioenerg. Biomembr.* 29: 89–95, 1997.
119. HARDMEIER, R., H. HOEGER, S. FANG-KIRCHER, A. KHOSCHSORUR, AND G. LUBEC. Transcription and activity of antioxidant enzymes after ionizing irradiation in radiation-resistant and radiation-sensitive mice. *Proc. Natl. Acad. Sci. USA* 94: 7572–7576, 1997.
120. HARMAN, D. Aging: a theory based on free radical and radiation chemistry. *J. Gerontol.* 2: 298–300, 1956.
121. HARMAN, D. The biologic clock: the mitochondria? *J. Am. Geriatr. Soc.* 20: 145–147, 1972.
122. HARMAN, D. The aging process. *Proc. Natl. Acad. Sci. USA* 78: 7124–7128, 1981.
123. HARMAN, D. Free radical theory of aging. *Mutat. Res.* 275: 257–266, 1992.
124. HARMAN, D. Free radical involvement in aging. Pathophysiology and therapeutic implications. *Drugs Aging* 3: 60–80, 1993.
125. HARMAN, D. Free-radical theory of aging. Increasing the functional life span. *Ann. NY Acad. Sci.* 717: 1–15, 1994.
126. HARMAN, D. Aging and disease: extending functional life span. *Ann. NY Acad. Sci.* 786: 321–336, 1996.
127. HARMAN, D. H. Free radical theory of aging: effects of antioxidants on mitochondrial function. *Age* 10: 58–61, 1987.
127a. HART, R. W., AND R. B. SETLOW. Correlation between deoxyribonucleic acid excision-repair and life-span in a number of mammalian species. *Proc. Natl. Acad. Sci. USA* 71: 2169–2173, 1974.
127b. HEKIMI, S., B. LAKOWSKI, T. M. BARNES, AND J. J. EWBANK. Molecular genetics of life span in *C. elegans*: how much does it teach us? *Trends Genet.* 14: 14–20, 1998.
127c. HELBOCK, H. J., K. B. BECKMAN, M. K. SHIGENAGA, P. W. WALTER, A. A. WOODALL, H. C. YEO, AND B. N. AMES. DNA oxidation matters: the HPLC-electrochemical detection assay of 8-oxo-deoxyguanosine and 8-oxo-guanine. *Proc. Natl. Acad. Sci. USA* 95: 288–293, 1998.
128. HEMPELMANN, L. H., AND J. G. HOFFMAN. Practical aspects of radiation injury. *Annu. Rev. Nucl. Sci.* 3: 369–389, 1953.
129. HENNEKENS, C. H., J. E. BURING, J. E. MANSON, M. STAMPFER, B. ROSNER, N. R. COOK, C. BELANGER, F. LAMOTTE, J. M. GAZIANO, P. M. RIDKER, W. WILLETT, AND R. PETO. Lack of effect of long-term supplementation with beta carotene on the incidence of malignant neoplasms and cardiovascular disease. *N. Engl. J. Med.* 334: 1145–1149, 1996.
130. HENSLEY, K., D. A. BUTTERFIELD, N. HALL, P. COLE, R. SUBRAMANIAM, R. MARK, M. P. MATTSON, W. R. MARKESBERY, M. E. HARRIS, M. AKSENOV, M. AKSENOVA, J. F. WU, AND J. M. CARNEY. Reactive oxygen species as causal agents in the neurotoxicity of the Alzheimer's disease-associated amyloid beta peptide. *Ann. NY Acad. Sci.* 786: 120–134, 1996.
130a. HERRERO, A., AND G. BARJA. Sites and mechanisms responsible for the low rate of free radical production of heart mitochondria in the long-lived pigeon. *Mech. Ageing Dev.* 98: 95–111, 1997.
131. HIGAMI, Y., I. SHIMOKAWA, T. OKIMOTO, AND T. IKEDA. An age-related increase in the basal level of DNA damage and DNA vulnerability to oxygen radicals in the individual hepatocytes of male F344 rats. *Mutat. Res.* 316: 59–67, 1994.
131a. HIGUCHI, Y., AND S. LINN. Purification of all forms of HeLa cell mitochondrial DNA and assessment of damage to it caused by hydrogen peroxide treatment of mitochondria or cells. *J. Biol. Chem.* 270: 7950–7956, 1995.
131b. HIRANO, M., A. SHTILBANS, R. MAYEUX, M. M. DAVIDSON, S. DIMAURO, J. A. KNOWLES, AND E. A. SCHON. Apparent mtDNA heteroplasmy in Alzheimer's disease patients and in normals due to PCR amplification of nucleus-embedded mtDNA pseudogenes. *Proc. Natl. Acad. Sci. USA* 94: 14894–14899. 1997.
132. HIRANO, T., R. YAMAGUCHI, S. ASAMI, N. IWAMOTO, AND H. KASAI. 8-Hydroxyguanine levels in nuclear DNA and its repair activity in rat organs associated with age. *J. Gerontol.* 51: 303–307, 1996.
133. HIRANO, T., Y. YAMAGUCHI, H. HIRANO, AND H. KASAI. Age-associated change of 8-hydroxyguanine repair activity in cultured human fibroblasts. *Biochem. Biophys. Res. Commun.* 214: 1157–1162, 1995.
134. HO, Y.-S., J.-L. MAGNENAT, R. T. BRONSON, J. CAO, M. GARGANO, M. SUGAWARA, AND C. D. FUNK. Mice deficient in cellular glutathione peroxidase develop normally and show no increased sensitivity to hyperoxia. *J. Biol. Chem.* 272: 16644–16651, 1997.
135. HOCHSCHILD, R. Lysosomes, membranes and aging. *Exp. Gerontol.* 6: 153–166, 1971.
136. HOLLIDAY, R. Mutations and epimutations in mammalian cells. *Mutat. Res.* 250: 351–363, 1991.
137. HOLMES, G. E., C. BERNSTEIN, AND H. BERNSTEIN. Oxidative and other DNA damages as the basis of aging: a review. *Mutat. Res.* 275: 305–315, 1992.
138. HONDA, S., N. ISHII, K. SUZUKI, AND M. MATSUO. Oxygen-dependent perturbation of life span and aging rate in the nematode. *J. Gerontol.* 48: 57–61, 1993.
139. HONDA, S., AND M. MATSUO. The sensitivity to hyperbaric oxygen

of human diploid fibroblasts during ageing in vitro. *Mech. Ageing Dev.* 12: 31–37, 1980.

140. HONDA, S., AND M. MATSUO. Shortening of the in vitro life span of human diploid fibroblasts exposed to hyperbaric oxygen. *Exp. Gerontol.* 18: 339–345, 1983.

141. HONDA, S., AND M. MATSUO. Lack of recovery from oxygen-induced damage to colony formation and DNA synthesis in senescent human diploid fibroblasts. *Mech. Ageing Dev.* 40: 81–87, 1987.

142. HORI, O., J. BRETT, T. SLATTERY, R. CAO, J. ZHANG, J. X. CHEN, M. NAGASHIMA, E. R. LUNDH, S. VIJAY, D. NITECKI, AND A. L. ET. The receptor for advanced glycation end products (RAGE) is a cellular binding site for amphoterin. Mediation of neurite outgrowth and co-expression of rage and amphoterin in the developing nervous system. *J. Biol. Chem.* 270: 25752–25761, 1995.

143. HORTON, T. M., B. H. GRAHAM, D. M. CORRAL, J. M. SHOFFNER, A. E. KAUFMAN, M. F. BEAL, AND D. C. WALLACE. Marked increase in mitochondrial DNA deletion levels in the cerebral cortex of Huntington's disease patients. *Neurology* 45: 1879–1883, 1995.

144. HOSOKAWA, H., N. ISHII, H. ISHIDA, K. ICHIMORI, H. NAKAZAWA, AND K. SUZUKI. Rapid accumulation of fluorescent material with aging in an oxygen-sensitive mutant mev-1 of *Caenorhabditis elegans*. *Mech. Ageing Dev.* 74: 161–170, 1994.

145. HU, J. J., N. DUBIN, D. KURLAND, B.-L. MA, AND G. C. ROUSH. The effects of hydrogen peroxide on DNA repair activities. *Mutat. Res.* 336: 193–201, 1995.

145a. HUNOT, S., B. BRUGG, D. RICARD, P. P. MICHEL, M. P. MURIEL, M. RUBERG, B. A. FAUCHEUX, Y. AGID, AND E. C. HIRSCH. Nuclear translocation of NF-kappaB is increased in dopaminergic neurons of patients with Parkinson disease. *Proc. Natl. Acad. Sci. USA* 94: 7531–7536, 1997.

146. IMLAY, J. A., AND I. FRIDOVICH. Assay of metabolic superoxide production in *Escherichia coli*. *J. Biol. Chem.* 266: 6957–6965, 1991.

147. IRANI, K., Y. XIA, J. L. ZWEIER, S. J. SOLLOTT, C. J. DER, E. R. FEARON, M. SUNDARESAN, T. FINKEL, AND P. J. GOLDSCHMIDT-CLERMONT. Mitogenic signaling mediated by oxidants in ras-transformed fibroblasts. *Science* 275: 1649–1652, 1997.

148. ISHII, N., K. TAKAHASHI, S. TOMITA, T. KEINO, S. HONDA, K. YOSHINO, AND K. SUZUKI. A methyl viologen-sensitive mutant of the nematode *Caenorhabditis elegans*. *Mutat. Res.* 237: 165–171, 1990.

149. IVY, G. O., F. SCHOTTLER, J. WENZEL, M. BAUDRY, AND G. LYNCH. Inhibitors of lysosomal enzymes: accumulation of lipofuscin-like dense bodies in the brain. *Science* 226: 985–987, 1984.

149a. JACOBSON, M. D. Reactive oxygen species and programmed cell death. *Trends Biochem. Sci.* 21: 83–86, 1996.

150. JI, L. L., D. DILLON, AND E. WU. Alteration of antioxidant enzymes with aging in rat skeletal muscle and liver. *Am. J. Physiol.* 258 (*Regulatory Integrative Comp. Physiol.* 27): R918–R923, 1990.

151. JI, L. L., D. DILLON, AND E. WU. Myocardial aging: antioxidant enzyme systems and related biochemical properties. *Am. J. Physiol.* 260 (*Regulatory Integrative Comp. Physiol.* 29): R386–R392, 1991.

152. JOHNSON, T. E., AND G. J. LITHGOW. The search for the genetic basis of aging: the identification of gerontogenes in the nematode *Caenorhabditis elegans*. *J. Am. Geriatr. Soc.* 40: 936–945, 1992.

153. JOHNSON, T. E., P. M. TEDESCO, AND G. J. LITHGOW. Comparing mutants, selective breeding, and transgenics in the dissection of aging processes of *Caenorhabditis elegans*. *Genetica* 91: 65–77, 1993.

154. JOHNSON, T. M., Z. X. YU, V. J. FERRANS, R. A. LOWENSTEIN, AND T. FINKEL. Reactive oxygen species are downstream mediators of p53-dependent apoptosis. *Proc. Natl. Acad. Sci. USA* 93: 11848–11852, 1996.

155. JOSEPH, J. A., R. VILLALOBOS-MOLINA, N. DENISOVA, S. ERAT, N. JIMENEZ, AND J. STRAIN. Increased sensitivity to oxidative stress and the loss of muscarinic receptor responsiveness in senescence. *Ann. NY Acad. Sci.* 786: 112–119, 1996.

156. KANEKO, T., S. TAHARA, AND M. MATSUO. Nonlinear accumulation of 8-hydroxy-2'-deoxyguanosine, a marker of oxidized DNA damage, during aging. *Mutat. Res.* 316: 277–285, 1996.

156a. KASAI, H. Analysis of a form of oxidative DNA damage, 8-hydroxy-2'-deoxyguanosine, as a marker of cellular oxidative stress during carcinogenesis. *Mutat. Res.* 387: 147–163, 1997.

157. KASAI, H., Y. OKADA, S. NISHIMURA, M. S. RAO, AND J. K. REDDY. Formation of 8-hydroxydeoxyguanosine in liver DNA of rats following long-term exposure to a peroxisome proliferator. *Cancer Res.* 49: 2603–2605, 1989.

158. KENYON, C. Ponce d'elegans: genetic quest for the fountain of youth. *Cell* 84: 501–504, 1996.

159. KEYER, K., A. S. GORT, AND J. A. IMLAY. Superoxide and the production of oxidative DNA damage. *J. Bacteriol.* 177: 6782–6790, 1995.

160. KIKUGAWA, K., T. KATO, AND A. HAYASAKA. Formation of dityrosine and other fluorescent amino acids by reaction of amino acids with lipid hydroperoxides. *Lipids* 26: 922–929, 1991.

160a. KIMURA, K. D., H. A. TISSENBAUM, Y. LIU, AND G. RUVKUN. Daf-2, an insulin receptor-like gene that regulates longevity and diapause in *Caenorhabditis elegans*. *Science* 277: 942–946, 1997.

161. KING, C. M., AND Y. A. BARNETT. Oxidative stress and human ageing (Abstract). *Biochem. Soc. Trans.* 23: 375S, 1995.

161a. KING, C. M. H. E. BRISTOW-CRAIG, E. S. GILLESPIE, AND Y. A. BARNETT. In vivo antioxidant status, DNA damage, mutation and DNA repair capacity in cultured lymphocytes from healthy 75- to 80-year-old humans. *Mutat. Res.* 377: 137–147, 1997.

162. KIRKWOOD, T. B. Comparative life spans of species: why do species have the life spans they do? *Am. J. Clin. Nutr.* 55, *Suppl.*: 1191S–1195S, 1992.

163. KIRKWOOD, T. B., AND M. R. ROSE. Evolution of senescence: late survival sacrificed for reproduction. *Philos. Trans. R. Soc. Lond B Biol. Sci.* 332: 15–24, 1991.

164. KIRKWOOD, T. B. L. Evolution and aging. *Nature* 270: 301–304, 1977.

165. KNEEPKENS, C. M., G. LEPAGE, AND C. C. ROY. The potential of the hydrocarbon breath test as a measure of lipid peroxidation. *Free Radical Biol. Med.* 17: 127–160, 1994.

166. KNIGHT, J. A. The process and theories of aging. *Ann. Clin. Lab. Sci.* 25: 1–12, 1995.

167. KONG, S. K., M. B. YIM, E. R. STADTMAN, AND P. B. CHOCK. Peroxynitrite disables the tyrosine phosphorylation regulatory mechanism: lymphocyte-specific tyrosine kinase fails to phosphorylate nitrated cdc2(6–20)NH_2 peptide. *Proc. Natl. Acad. Sci. USA* 93: 3377–3382, 1996.

168. KOOP, D. R. Oxidative and reductive metabolism by cytochrome *P450 2E1*. *FASEB J.* 6: 724–730, 1992.

169. KOSTER, J. F., AND W. SLUITER. Is increased tissue ferritin a risk factor for atherosclerosis and ischaemic heart disease? *Br. Heart J.* 73: 208–209, 1995.

170. KOWALD, A., AND T. B. KIRKWOOD. A network theory of ageing: the interactions of defective mitochondria, aberrant proteins, free radicals and scavengers in the ageing process. *Mutat. Res.* 316: 209–236, 1996.

171. KRUK, P. A., N. J. RAMPINO, AND V. A. BOHR. DNA damage and repair in telomeres: relation to aging. *Proc. Natl. Acad. Sci. USA* 92: 258–262, 1995.

172. KU, H. H., U. T. BRUNK, AND R. S. SOHAL. Relationship between mitochondrial superoxide and hydrogen peroxide production and longevity of mammalian species. *Free Radical Biol. Med.* 15: 621–627, 1993.

173. KU, H. H., AND R. S. SOHAL. Comparison of mitochondrial prooxidant generation and anti-oxidant defenses between rat and pigeon: possible basis of variation in longevity and metabolic potential. *Mech. Ageing Dev.* 72: 67–76, 1993.

174. KUMARI, M. V., T. YONEDA, AND M. HIRAMATSU. Scavenging activity of "beta catechin" on reactive oxygen species generated by photosensitization of riboflavin. *Biochem. Mol. Biol. Int.* 38: 1163–1170, 1996.

175. KUMARI, M. V., T. YONEDA, AND M. HIRAMATSU. Effect of "beta CATECHIN" on the life span of senescence accelerated mice (SAM-P8 strain). *Biochem. Mol. Biol. Int.* 41: 1005–1011, 1997.

176. KUO, C. F., T. MASHINO, AND I. FRIDOVICH. Alpha,beta-dihydroxyisovalerate dehydratase. A superoxide-sensitive enzyme. *J. Biol. Chem.* 262: 4724–4727, 1987.

177. LAKE, B. G. Mechanisms of hepatocarcinogenicity of peroxisome-

proliferating drugs and chemicals. *Annu. Rev. Pharmacol. Toxicol.* 35: 483–507, 1995.
178. LARSEN, P. L. Aging and resistance to oxidative damage in *Caenorhabditis elegans. Proc. Natl. Acad. Sci. USA* 90: 8905–8909, 1993.
178a. LASS, A., S. AGARWAL, AND R. S. SOHAL. Mitochondrial ubiquinone homologues, superoxide radical generation, and longevity in different mammalian species. *J. Biol. Chem.* 272: 19199–19204, 1997.
179. LEE, C. M., S. S. CHUNG, J. M. KACZKOWSKI, R. WEINDRUCH, AND J. M. AIKEN. Multiple mitochondrial DNA deletions associated with age in skeletal muscle of rhesus monkeys. *J. Gerontol.* 48: 201–205, 1993.
180. LEE, C. M., R. WEINDRUCH, AND J. M. AIKEN. Age-associated alterations of the mitochondrial genome. *Free Radical Biol. Med.* 22: 1259–1269, 1997.
181. LEVINE, R. L., D. GARLAND, C. N. OLIVER, A. AMICI, I. CLIMENT, A. G. LENZ, B. W. AHN, S. SHALTIEL, AND E. R. STADTMAN. Determination of carbonyl content in oxidatively modified proteins. *Methods Enzymol.* 186: 464–478, 1990.
182. LEZZA, A. M., D. BOFFOLI, S. SCACCO, P. CANTATORE, AND M. N. GADALETA. Correlation between mitochondrial DNA 4977-bp deletion and respiratory chain enzyme activities in aging human skeletal muscles. *Biochem. Biophys. Res. Commun.* 205: 772–779, 1994.
183. LIEBETRAU, W., T. M. RUNGER, B. E. MEHLING, M. POOT, AND H. HOEHN. Mutagenic activity of ambient oxygen and mitomycin C in Fanconi's anaemia cells. *Mutagenesis* 12: 69–77, 1997.
183a. LIN, K., J. B. DORMAN, A. RODAN, AND C. KENYON. daf-16: an HNF-3/forkhead family member that can function to double the life-span of *Caenorhabditis elegans. Science* 278: 1319–1322, 1997.
184. LIOCHEV, S. I., AND I. FRIDOVICH. The role of $O_2^{\bar{}}\cdot$ in the production of ·OH: in vitro and in vivo. *Free Radical Biol. Med.* 16: 29–33, 1994.
185. LITHGOW, G. J., T. M. WHITE, S. MELOV, AND T. E. JOHNSON. Thermotolerance and extended life-span conferred by single-gene mutations and induced by thermal stress. *Proc. Natl. Acad. Sci. USA* 92: 7540–7544, 1995.
186. LIU, J., AND A. MORI. Age-associated changes in superoxide dismutase activity, thiobarbituric acid reactivity and reduced glutathione level in the brain and liver in senescence accelerated mice (SAM): a comparison with ddY mice. *Mech. Ageing Dev.* 71: 23–30, 1993.
187. LIU, R. H., AND J. H. HOTCHKISS. Potential genotoxicity of chronically elevated nitric oxide: a review. *Mutat. Res.* 339: 73–89, 1995.
188. LOFT, S., A. ASTRUP, B. BUEMANN, AND H. E. POULSEN. Oxidative DNA damage correlates with oxygen consumption in humans. *FASEB J.* 8: 534–537, 1994.
189. LONNROT, K., T. METSA-KETELA, AND H. ALHO. The role of coenzyme Q-10 in aging: a follow-up study on life-long oral supplementation Q-10 in rats. *Gerontology* 41, *Suppl.* 2: 109–120, 1995.
190. LOPEZ-TORRES, M., R. PEREZ-CAMPO, C. ROJAS, S. CADENAS, AND G. BARJA. Maximum life span in vertebrates: relationship with liver antioxidant enzymes, glutathione system, ascorbate, urate, sensitivity to peroxidation, true malondialdehyde, in vivo H_2O_2, and basal and maximum aerobic capacity. *Mech. Ageing Dev.* 70: 177–199, 1993.
191. LOSCHEN, G., A. AZZI, C. RICHTER, AND L. FLOHE. Superoxide radicals as precursors of mitochondrial hydrogen peroxide. *FEBS Lett.* 42: 68–72, 1974.
192. LU, R., H. M. NASH, AND G. L. VERDINE. A mammalian DNA repair enzyme that excises oxidatively damaged guanines maps to a locus frequently lost in lung cancer. *Curr. Biol.* 7: 397–407, 1997.
192a. LUCAS, D. T., AND L. I. SZWEDA. Cardiac reperfusion injury: aging, lipid peroxidation, and mitochondrial dysfunction. *Proc. Natl. Acad. Sci. USA* 95: 510–514, 1998.
193. LUFT, R. The development of mitochondrial medicine. *Proc. Natl. Acad. Sci. USA* 91: 8731–8738, 1994.
193a. MARBOIS, B. N., AND C. F. CLARKE. The COQ7 gene encodes a protein in *Saccharomyces cerevisiae* necessary for ubiquinone biosynthesis. *J. Biol. Chem.* 271: 2995–3004, 1996.
193b. MARKESBERY, W. R. Oxidative stress hypothesis in Alzheimer's disease. *Free Radical Biol. Med.* 23: 134–147, 1997.
194. MARNETT, L. J., H. K. HURD, M. C. HOLLSTEIN, D. E. LEVIN, H. ESTERBAUER, AND B. N. AMES. Naturally occurring carbonyl compounds are mutagens in *Salmonella tester* strain TA104. *Mutat. Res.* 148: 25–34, 1985.
195. MARTIN, G. M. Somatic mutagenesis and antimutagenesis in aging research. *Mutat. Res.* 350: 35–41, 1996.
196. MARTIN, G. M., S. N. AUSTAD, AND T. E. JOHNSON. Genetic analysis of ageing: role of oxidative damage and environmental stresses. *Nature Genet.* 13: 25–34, 1996.
197. MARTINEZ, M., M. L. FERRANDIZ, A. DIEZ, AND J. MIQUEL. Depletion of cytosolic GSH decreases the ATP levels and viability of synaptosomes from aged mice but not from young mice. *Mech. Ageing Dev.* 84: 77–81, 1995.
198. MARTUS, H. J., M. E. DOLLE, J. A. GOSSEN, M. E. BOERRIGTER, AND J. VIJG. Use of transgenic mouse models for studying somatic mutations in aging. *Mutat. Res.* 338: 203–213, 1995.
199. MARZABADI, M. R., C. JONES, AND J. RYDSTROM. Indenoindole depresses lipofuscin formation in cultured neonatal rat myocardial cells. *Mech. Ageing Dev.* 80: 189–197, 1995.
200. MARZABADI, M. R., R. S. SOHAL, AND U. T. BRUNK. Effect of ferric iron and desferrioxamine on lipofuscin accumulation in cultured rat heart myocytes. *Mech. Ageing Dev.* 46: 145–157, 1988.
201. MARZABADI, M. R., R. S. SOHAL, AND U. T. BRUNK. Effect of alpha-tocopherol and some metal chelators on lipofuscin accumulation in cultured neonatal rat cardiac myocytes. *Ann. Cell Pathol.* 2: 333–346, 1990.
202. MARZABADI, M. R., R. S. SOHAL, AND U. T. BRUNK. Mechanisms of lipofuscinogenesis: effect of the inhibition of lysosomal proteinases and lipases under varying concentrations of ambient oxygen in cultured rat neonatal myocardial cells. *APMIS* 99: 416–426, 1991.
203. MATSUO, M. Oxygen dependency of life-span in the nematode. *Comp. Biochem. Physiol. A Physiol.* 105: 653–658, 1993.
203a. MATTSON, M. P. Advances fuel Alzheimer's conundrum. *Nature Genet.* 17: 254–256, 1997.
204. McCORD, J. M., AND I. FRIDOVICH. Superoxide dismutase. An enzymic function for erythrocuprein (hemocuprein). *J. Biol. Chem.* 244: 6049–6055, 1969.
205. MEDVEDEV, Z. A. An attempt at a rational classification of theories of ageing. *Biol. Rev. Camb. Philos. Soc.* 65: 375–398, 1990.
206. MELOV, S., G. J. LITHGOW, D. R. FISCHER, P. M. TEDESCO, AND T. E. JOHNSON. Increased frequency of deletions in the mitochondrial genome with age of *Caenorhabditis elegans. Nucleic Acids Res.* 23: 1419–1425, 1995.
207. MELOV, S., J. M. SHOFFNER, A. KAUFMAN, AND D. C. WALLACE. Marked increase in the number and variety of mitochondrial DNA rearrangements in aging human skeletal muscle. *Nucleic Acids Res.* 23: 4122–4126, 1995.
208. MENDIS, S., P. A. SOBOTKA, AND D. E. EULER. Pentane and isoprene in expired air from humans: gas-chromatographic analysis of single breath. *Clin. Chem.* 40: 1485–1488, 1994.
208a. MEYDANI, S. N., M. MEYDANI, J. B. BLUMBERG, L. S. LEKA, G. SIBER, R. LOSZEWSKI, C. THOMPSON, M. C. PEDROSA, R. D. DIAMOND, AND B. D. STOLLAR. Vitamin E supplementation and in vivo immune response in healthy elderly subjects. A randomized controlled trial. *JAMA* 277: 1380–1386, 1997.
209. MEYDANI, S. N., D. WU, M. S. SANTOS, AND M. G. HAYEK. Antioxidants and immune response in aged persons: overview of present evidence. *Am. J. Clin. Nutr.* 62, *Suppl.*: 1462S–1476S, 1995.
210. MILLER, J. H. The relevance of bacterial mutators to understanding human cancer. *Cancer Surv.* 28: 141–153, 1996.
211. MINOTTI, G., AND S. D. AUST. The role of iron in the initiation of lipid peroxidation. *Chem. Phys. Lipids.* 44: 191–208, 1987.
212. MIQUEL, J. An update on the mitochondrial-DNA mutation hypothesis of cell aging. *Mutat. Res.* 275: 209–216, 1992.
213. MIQUEL, J., A. C. ECONOMOS, J. FLEMING, AND J. J. JOHNSON. Mitochondrial role in cell aging. *Exp. Gerontol.* 15: 575–591, 1980.
214. MIQUEL, J., M. L. FERRANDIZ, E. DE JUAN, I. SEVILA, AND M. MARTINEZ. *N*-acetylcysteine protects against age-related decline of oxidative phosphorylation in liver mitochondria. *Eur. J. Pharmacol.* 292: 333–335, 1995.
215. MO, J. Q., D. G. HOM, AND J. K. ANDERSEN. Decreases in protective enzymes correlates with increased oxidative damage in the aging mouse brain. *Mech. Ageing Dev.* 81: 73–82, 1995.
216. MONNIER, V. M., AND A. CERAMI. Nonenzymatic browning in vivo:

possible process for aging of long-lived proteins. *Science* 211: 491–493, 1981.
217. MONNIER, V. M., M. GLOMB, A. ELGAWISH, AND D. R. SELL. The mechanism of collagen cross-linking in diabetes: a puzzle nearing resolution. *Diabetes* 45, *Suppl.* 3: S67–S72, 1996.
218. MORLEY, A. A. The somatic mutation theory of ageing. *Mutat. Res.* 338: 19–23, 1995.
219. MORRIS, J. Z., H. A. TISSENBAUM, AND G. RUVKUN. A phosphatidylinositol-3-OH kinase family member regulating longevity and diapause in *Caenorhabditis elegans*. *Nature* 382: 536–539, 1996.
220. MOSLEN, M. T. Reactive oxygen species in normal physiology, cell injury and phagocytosis. *Adv. Exp. Med. Biol.* 366: 17–27, 1994.
221. MULLAART, E., P. H. LOHMAN, F. BERENDS, AND J. VIJG. DNA damage metabolism and aging. *Mutat. Res.* 237: 189–210, 1990.
222. MULLER, H. J. Mitochondria and ageing. *Brain Pathol.* 2: 149–158, 1992.
223. NAGLEY, P., I. R. MACKAY, A. BAUMER, R. J. MAXWELL, F. VAILLANT, Z. X. WANG, C. ZHANG, AND A. W. LINNANE. Mitochondrial DNA mutation associated with aging and degenerative disease. *Ann. NY Acad. Sci.* 673: 92–102, 1992.
224. NAKANO, M., AND S. GOTOH. Accumulation of cardiac lipofuscin depends on metabolic rate of mammals. *J. Gerontol.* 47: 126–129, 1992.
225. NAKANO, M., T. MIZUNO, AND S. GOTOH. Accumulation of cardiac lipofuscin in crab-eating monkeys (*Macaca fasicularis*): the same rate of lipofuscin accumulation in several species of primates. *Mech. Ageing Dev.* 66: 243–248, 1993.
226. NAKANO, M., T. MIZUNO, H. KATOH, AND S. GOTOH. Age-related accumulation of lipofuscin in myocardium of Japanese monkey (*Macaca fuscata*). *Mech. Ageing Dev.* 49: 41–48, 1989.
227. NOHL, H. Involvement of free radicals in ageing: a consequence or cause of senescence. *Br. Med. Bull.* 49: 653–667, 1993.
228. NOHL, H., L. GILLE, K. SCHONHEIT, AND Y. LIU. Conditions allowing redox-cycling ubisemiquinone in mitochondria to establish a direct redox couple with molecular oxygen. *Free Radical Biol. Med.* 20: 207–213, 1996.
229. NOHL, H., AND D. HEGNER. Do mitochondria produce oxygen radicals in vivo? *Eur. J. Biochem.* 82: 563–567, 1978.
230. OCKNER, R. K., R. M. KAIKAUS, AND N. M. BASS. Fatty-acid metabolism and the pathogenesis of hepatocellular carcinoma: review and hypothesis. *Hepatology* 18: 669–676, 1993.
231. OHSHIMA, H., AND H. BARTSCH. Chronic infections and inflammatory processes as cancer risk factors: possible role of nitric oxide in carcinogenesis. *Mutat. Res.* 305: 253–264, 1994.
232. OIKAWA, I., AND P. M. NOVIKOFF. Catalase-negative peroxisomes: transient appearance in rat hepatocytes during liver regeneration after partial hepatectomy. *Am. J. Pathol.* 146: 673–687, 1995.
233. OLIVER, C. N., B. W. AHN, E. J. MOERMAN, S. GOLDSTEIN, AND E. R. STADTMAN. Age-related changes in oxidized proteins. *J. Biol. Chem.* 262: 5488–5491, 1987.
234. OLIVER, C. N., P. E. STARKE-REED, E. R. STADTMAN, G. J. LIU, J. M. CARNEY, AND R. A. FLOYD. Oxidative damage to brain proteins, loss of glutamine synthetase activity, and production of free radicals during ischemia/reperfusion-induced injury to gerbil brain. *Proc. Natl. Acad. Sci. USA* 87: 5144–5147, 1990.
235. OLLINGER, K., AND U. T. BRUNK. Cellular injury induced by oxidative stress is mediated through lysosomal damage. *Free Radical Biol. Med.* 19: 565–574, 1995.
236. OMENN, G. S., G. E. GOODMAN, M. D. THORNQUIST, J. BALMES, M. R. CULLEN, A. GLASS, J. P. KEOGH, F. L. MEYSKENS, JR., B. VALANIS, J. H. WILLIAMS, JR., S. BARNHART, M. G. CHERNIACK, C. A. BRODKIN, AND S. HAMMAR. Risk factors for lung cancer and for intervention effects in CARET, the beta-carotene and retinol efficacy trial. *J. Natl. Cancer Inst.* 88: 1550–1559, 1996.
237. ORR, W. C., AND R. S. SOHAL. Extension of life-span by overexpression of superoxide dismutase and catalase in *Drosophila melanogaster*. *Science* 263: 1128–1130, 1994.
238. OZAWA, T. Mechanism of somatic mitochondrial DNA mutations associated with age and diseases. *Biochim. Biophys. Acta* 1271: 177–189, 1995.
238a. OZAWA, T. Genetic and functional changes in mitochondria associated with aging. *Physiol. Rev.* 77: 425–464, 1997.
239. PACIFICI, R. E., AND K. J. DAVIES. Protein, lipid and DNA repair systems in oxidative stress: the free-radical theory of aging revisited. *Gerontology* 37: 166–180, 1991.
240. PACIFICI, R. E., Y. KONO, AND K. J. DAVIES. Hydrophobicity as the signal for selective degradation of hydroxyl radical-modified hemoglobin by the multicatalytic proteinase complex, proteasome. *J. Biol. Chem.* 268: 15405–15411, 1993.
241. PACKER, L., AND K. FUEHR. Low oxygen concentration extends the life span of cultured human diploid cells. *Nature* 267: 423–425, 1977.
242. PACKER, L., AND J. R. SMITH. Extension of the life span of cultured normal human diploid cells by vitamin E: a reevaluation. *Proc. Natl. Acad. Sci. USA* 74: 1640–1641, 1977.
243. PALLOTTI, F., X. CHEN, E. BONILLA, AND E. A. SCHON. Evidence that specific mtDNA point mutations may not accumulate in skeletal muscle during normal human aging. *Am. J. Hum. Genet.* 59: 591–602, 1996.
244. PARK, J. W., C. H. CHOI, M. S. KIM, AND M. H. CHUNG. Oxidative status in senescence-accelerated mice. *J. Gerontol.* 51: 337–345, 1996.
245. PARSONNET, J., G. D. FRIEDMAN, D. P. VANDERSTEEN, Y. CHANG, J. H. VOGELMAN, N. ORENTREICH, AND R. K. SIBLEY. *Helicobacter pylori* infection and the risk of gastric carcinoma. *N. Engl. J. Med.* 325: 1127–1131, 1991.
246. PEARL, R. *The Rate of Living.* London: Univ. of London Press, 1928.
247. PEREZ-CAMPO, R., M. LOPEZ-TORRES, C. ROJAS, S. CADENAS, AND G. BARJA. Longevity and antioxidant enzymes, non-enzymatic antioxidants and oxidative stress in the vertebrate lung: a comparative study. *J. Comp. Physiol. B Biochem. Syst. Environ. Physiol.* 163: 682–689, 1994.
248. PEREZ-CAMPO, R., M. LOPEZ-TORRES, C. ROJAS, S. CADENAS, G. BARJA, AND D. QUIROGA. Lung glutathione reductase induction in aging catalase-depleted frogs correlates with early survival throughout the life span. *Mech. Ageing Dev.* 67: 115–127, 1993.
248a. PFLAUM, M., O. WILL, AND B. EPE. Determination of steady-state levels of oxidative DNA base modifications in mammalian cells by means of repair endonucleases. *Carcinogenesis* 18: 2225–2231, 1997.
249. PODEROSO, J. J., M. C. CARRERAS, C. LISDERO, N. RIOBO, F. SCHOPFER, AND A. BOVERIS. Nitric oxide inhibits electron transfer and increases superoxide radical production in rat heart mitochondria and submitochondrial particles. *Arch. Biochem. Biophys.* 328: 85–92, 1996.
250. POOT, M. Oxidants and antioxidants in proliferative senescence. *Mutat. Res.* 256: 177–189, 1991.
251. PORTA, E. A. Advances in age pigment research. *Arch. Gerontol. Geriatr.* 12: 303–320, 1991.
252. POTTER, J. D. Beta-carotene and the role of intervention studies. *Cancer Lett.* 114: 329–331, 1997.
252a. PROFT, M., P. KOTTER, D. HEDGES, N. BOJUNGA, AND K. D. ENTIAN. CAT5, a new gene necessary for derepression of gluconeogenic enzymes in *Saccharomyces cerevisiae*. *EMBO J.* 14: 6116–6126, 1995.
253. RAO, G., E. XIA, M. J. NADAKAVUKAREN, AND A. RICHARDSON. Effect of dietary restriction on the age-dependent changes in the expression of antioxidant enzymes in rat liver. *J. Nutr.* 120: 602–609, 1990.
254. RAO, G., E. XIA, AND A. RICHARDSON. Effect of age on the expression of antioxidant enzymes in male Fischer F344 rats. *Mech. Ageing Dev.* 53: 49–60, 1990.
255. REAUME, A. G., J. L. ELLIOTT, E. K. HOFFMAN, N. W. KOWALL, R. J. FERRANTE, D. F. SIWEK, H. M. WILCOX, D. G. FLOOD, M. F. BEAL, R. H. BROWN, JR., R. W. SCOTT, AND W. D. SNIDER. Motor neurons in Cu/Zn superoxide dismutase-deficient mice develop normally but exhibit enhanced cell death after axonal injury. *Nature Genet.* 13: 43–47, 1996.
255a. REED, J. C. Cytochomre *c*: can't live with it—can't live without it. *Cell* 91: 559–562, 1997.
256. REID, T. M., D. I. FEIG, AND L. A. LOEB. Mutagenesis by metal-induced oxygen radicals. *Environ Health Perspect.* 102, *Suppl.* 3: 57–61, 1994.
257. REITER, R. J., M. I. PABLOS, T. T. AGAPITO, AND J. M. GUER-

RERO. Melatonin in the context of the free radical theory of aging. *Ann. NY Acad. Sci.* 786: 362–378, 1996.
258. RICHTER, C. Reactive oxygen and DNA damage in mitochondria. *Mutat. Res.* 275: 249–255, 1992.
259. RICHTER, C. Oxidative damage to mitochondrial DNA and its relationship to ageing. *Int. J. Biochem. Cell Biol.* 27: 647–653, 1995.
260. RICHTER, C., J. W. PARK, AND B. N. AMES. Normal oxidative damage to mitochondrial and nuclear DNA is extensive. *Proc. Natl. Acad. Sci. USA* 85: 6465–6467, 1988.
261. RIVETT, A. J. Intracellular protein degradation. *Essays Biochem.* 25: 39–81, 1990.
262. ROBINSON, J. M., AND J. A. BADWEY. Production of active oxygen species by phagocytic leukocytes. *Immunol. Ser.* 60: 159–178, 1994.
263. RODRIGUEZ-MARTINEZ, M. A., AND A. RUIZ-TORRES. Homeostasis between lipid peroxidation and antioxidant enzyme activities in healthy human aging. *Mech. Ageing Dev.* 66: 213–222, 1992.
264. ROSE, M. R. Laboratory evolution of postponed senescence in *Drosophila melanogaster. Evolution* 38: 1004–1010, 1984.
265. ROSE, M. R., AND C. E. FINCH. Genetics and evolution of aging. In: *Contemporary Issues in Genetics and Evolution*. Dordrecht, Germany: Kluwer, 1994, vol 3.
266. ROSENQUIST, T. A., D. O. ZHARKOV, AND A. P. GROLLMAN. Cloning and characterization of a mammalian 8-oxoguanine DNA glycosylase. *Proc. Natl. Acad. Sci. USA* 94: 7429–7434, 1997.
267. ROSSMAN, T. G., E. I. GONCHAROVA, A. NÁDAS, AND N. DOLZHANSKAYA. Chinese hamster cells expressing antisense to metallothionein become spontaneous mutators. *Mutat. Res.* 373: 75–85, 1997.
268. RUBNER, M. *Das problem der Lebensdauer und seine Beziehungen zu Wachstum und Ernèhrung*. Munich: Oldenburg, 1908.
269. SAGER, R. Senescence as a mode of tumor suppression. *Environ. Health Perspect.* 93: 59–62, 1991.
270. SAHAKIAN, J. A., L. I. SZWEDA, B. FRIGUET, K. KITANI, AND R. L. LEVINE. Aging of the liver: proteolysis of oxidatively modified glutamine synthetase. *Arch. Biochem. Biophys.* 318: 411–417, 1995.
270a. SALAZAR, J. M., AND B. VAN HOUTEN. Preferential mitochondrial DNA injury caused by glucose oxidase as a steady generator of hydrogen peroxide in human fibroblasts. *Mutat. Res.* 385: 139–149, 1997.
271. SALGANIK, R. I., I. G. SHABALINA, N. A. SOLOVYOVA, N. G. KOLOSOVA, V. N. SOLOVYOV, AND A. R. KOLPAKOV. Impairment of respiratory functions in mitochondria of rats with an inherited hyperproduction of free radicals. *Biochem. Biophys. Res. Commun.* 205: 180–185, 1994.
272. SALGANIK, R. I., N. A. SOLOVYOVA, S. I. DIKALOV, O. N. GRISHAEVA, L. A. SEMENOVA, AND A. V. POPOVSKY. Inherited enhancement of hydroxyl radical generation and lipid peroxidation in the S strain rats results in DNA rearrangements, degenerative diseases, and premature aging. *Biochem. Biophys. Res. Commun.* 199: 726–733, 1994.
272a. SASTRE, J., A. MILLAN, J. GARCIA DE LA ASUNCION, R. PLA, G. JUAN, F. V. PALLARDO, E. O'CONNOR, J. A. MARTIN, M.-T. DROY-LEFAIX, AND J. VINA. A Ginko Biloba extract (EGb761) prevents mitochondrial aging by protecting against oxidative stress. *Free Radical Biol. Med.* 24: 298–304, 1998.
273. SASTRE, J., F. V. PALLARDO, AND J. VINA. Glutathione, oxidative stress, and aging. *Age* 19: 129–139, 1996.
274. SAWADA, M., AND J. C. CARLSON. Biochemical changes associated with the mechanism controlling superoxide radical formation in the aging rotifer. *J. Cell. Biochem.* 44: 153–165, 1990.
275. SCHAPIRA, A. H. Oxidative stress in Parkinson's disease. *Neuropathol. Appl. Neurobiol.* 21: 3–9, 1995.
276. SCHAPIRA, A. H., AND J. M. COOPER. Mitochondrial function in neurodegeneration and ageing. *Mutat. Res.* 275: 133–143, 1992.
277. SCHMIDT, A. M., O. HORI, J. BRETT, S. D. YAN, J. L. WAUTIER, AND D. STERN. Cellular receptors for advanced glycation end products. Implications for induction of oxidant stress and cellular dysfunction in the pathogenesis of vascular lesions. *Arterioscler. Thromb.* 14: 1521–1528, 1994.
278. SCHMIDT, A. M., O. HORI, R. CAO, S. D. YAN, J. BRETT, J. L. WAUTIER, S. OGAWA, K. KUWABARA, M. MATSUMOTO, AND D. STERN. RAGE: a novel cellular receptor for advanced glycation end products. *Diabetes* 45, *Suppl.* 3: S77–S80, 1996.
279. SCHMIDT, A. M., O. HORI, J. X. CHEN, J. F. LI, J. CRANDALL, J. ZHANG, R. CAO, S. D. YAN, J. BRETT, AND D. STERN. Advanced glycation end products interacting with their endothelial receptor induce expression of vascular cell adhesion molecule-1 (VCAM-1) in cultured human endothelial cells and in mice. A potential mechanism for the accelerated vasculopathy of diabetes. *J. Clin. Invest.* 96: 1395–1403, 1995.
280. SCHMIDT, A. M., S. D. YAN, J. BRETT, R. MORA, R. NOWYGROD, AND D. STERN. Regulation of human mononuclear phagocyte migration by cell surface-binding proteins for advanced glycation end products. *J. Clin. Invest.* 91: 2155–2168, 1993.
281. SCHULZ, J. B., R. T. MATTHEWS, D. R. HENSHAW, AND M. F. BEAL. Neuroprotective strategies for treatment of lesions produced by mitochondrial toxins: implications for neurodegenerative diseases. *Neuroscience* 71: 1043–1048, 1996.
282. SCHWARZE, S. R., C. M. LEE, S. S. CHUNG, E. B. ROECKER, R. WEINDRUCH, AND J. M. AIKEN. High levels of mitochondrial DNA deletions in skeletal muscle of old rhesus monkeys. *Mech. Ageing Dev.* 83: 91–101, 1995.
283. SELKOE, D. J. Cell biology of the amyloid beta-protein precursor and the mechanism of Alzheimer's disease. *Annu. Rev. Cell Biol.* 10: 373–403, 1994.
284. SELL, D. R., M. A. LANE, W. A. JOHNSON, E. J. MASORO, O. B. MOCK, K. M. REISER, J. F. FOGARTY, R. G. CUTLER, D. K. INGRAM, G. S. ROTH, AND V. M. MONNIER. Longevity and the genetic determination of collagen glycoxidation kinetics in mammalian senescence. *Proc. Natl. Acad. Sci. USA* 93: 485–490, 1996.
285. SEN, C. K., AND L. PACKER. Antioxidant and redox regulation of gene transcription. *FASEB J.* 10: 709–720, 1996.
286. SERRANO, M., A. W. LIN, M. E. McCURRACH, D. BEACH, AND S. W. LOWE. Oncogenic ras provokes premature cell senescence associated with accumulation of p53 and p16INK4a. *Cell* 88: 593–602, 1997.
287. SHIGENAGA, M. K., C. J. GIMENO, AND B. N. AMES. Urinary 8-hydroxy-2'-deoxyguanosine as a biological marker of in vivo oxidative DNA damage. *Proc. Natl. Acad. Sci. USA* 86: 9697–9701, 1989.
288. SHIGENAGA, M. K., T. M. HAGEN, AND B. N. AMES. Oxidative damage and mitochondrial decay in aging. *Proc. Natl. Acad. Sci. USA* 91: 10771–10778, 1994.
289. SHMOOKLER REIS, R. J., AND R. H. N. EBERT. Genetics of aging: current animal models. *Exp. Gerontol.* 31: 69–81, 1996.
290. SHOFFNER, J. M., M. D. BROWN, A. TORRONI, M. T. LOTT, M. F. CABELL, S. S. MIRRA, M. F. BEAL, C. C. YANG, M. GEARING, AND R. SALVO. Mitochondrial DNA variants observed in Alzheimer disease and Parkinson disease patients. *Genomics* 17: 171–184, 1993.
290a. SLATER, A. F., C. S. NOBEL, AND S. ORRENIUS. The role of intracellular oxidants in apoptosis. *Biochim. Biophys. Acta* 1271: 59–62, 1995.
291. SLUPSKA, M. M., C. BAIKALOV, W. M. LUTHER, J. H. CHIANG, Y. F. WEI, AND J. H. MILLER. Cloning and sequencing a human homolog (hMYH) of the *Escherichia coli* mutY gene whose function is required for the repair of oxidative DNA damage. *J. Bacteriol.* 178: 3885–3892, 1996.
291a. SMEAL, T., AND L. GUARENTE. Mechanisms of cellular senescence. *Curr. Opin. Genet. Dev.* 7: 281–287, 1997.
292. SMITH, D. W. E. Centenarians: human longevity outliers. *Gerontologist* 37: 200–207, 1997.
292a. SMITH, M. A., P. L. HARRIS, L. M. SAYRE, AND G. PERRY. Iron accumulation in Alzheimer disease is a source of redox-generated free radicals. *Proc. Natl. Acad. Sci. USA* 94: 9866–9868, 1997.
292b. SMITH, M. A., G. PERRY, P. L. RICHEY, L. M. SAYRE, V. E. ANDERSON, M. F. BEAL, AND N. KOWALL. Oxidative damage in Alzheimer's. *Nature* 382: 120–121, 1996.
293. SMITH, M. A., L. M. SAYRE, V. M. MONNIER, AND G. PERRY. Radical ageing in Alzheimer's disease. *Trends Neurosci.* 18: 172–176, 1995.
294. SOCCI, D. J., B. M. CRANDALL, AND G. W. ARENDASH. Chronic antioxidant treatment improves the cognitive performance of aged rats. *Brain Res.* 693: 88–94, 1995.
295. SOHAL, R. S. Metabolic rate and life span. In: *Cellular Aging: Concepts and Mechanisms*, edited by R. Witler. Basel: Karger, 1976, p. 25–40.
296. SOHAL, R. S. *Age Pigments*. Amsterdam: Elsevier, 1981.

297. SOHAL, R. S. The free radical hypothesis of aging: an appraisal of the current status. *Aging Clin. Exp. Res.* 5: 3–17, 1993.
298. SOHAL, R. S., A. AGARWAL, S. AGARWAL, AND W. C. ORR. Simultaneous overexpression of copper- and zinc-containing superoxide dismutase and catalase retards age-related oxidative damage and increases metabolic potential in *Drosophila melanogaster*. *J. Biol. Chem.* 270: 15671–15674, 1995.
299. SOHAL, R. S., S. AGARWAL, A. DUBEY, AND W. C. ORR. Protein oxidative damage is associated with life expectancy of houseflies. *Proc. Natl. Acad. Sci. USA* 90: 7255–7259, 1993.
300. SOHAL, R. S., S. AGARWAL, AND B. H. SOHAL. Oxidative stress and aging in the Mongolian gerbil (*Meriones unguiculatus*). *Mech. Ageing Dev.* 81: 15–25, 1995.
301. SOHAL, R. S., L. ARNOLD, AND W. C. ORR. Effect of age on superoxide dismutase, catalase, glutathione reductase, inorganic peroxides, TBA-reactive material, GSH/GSSG, NADPH/NADP$^+$ and NADH/NAD$^+$ in *Drosophila melanogaster*. *Mech. Ageing Dev.* 56: 223–235, 1990.
302. SOHAL, R. S., L. A. ARNOLD, AND B. H. SOHAL. Age-related changes in antioxidant enzymes and prooxidant generation in tissues of the rat with special reference to parameters in two insect species. *Free Radical Biol. Med.* 9: 495–500, 1990.
303. SOHAL, R. S., AND A. DUBEY. Mitochondrial oxidative damage, hydrogen peroxide release, and aging. *Free Radical Biol. Med.* 16: 621–626, 1994.
304. SOHAL, R. S., H. H. KU, AND S. AGARWAL. Biochemical correlates of longevity in two closely related rodent species. *Biochem. Biophys. Res. Commun.* 196: 7–11, 1993.
305. SOHAL, R. S., H. H. KU, S. AGARWAL, M. J. FORSTER, AND H. LAL. Oxidative damage, mitochondrial oxidant generation and antioxidant defenses during aging and in response to food restriction in the mouse. *Mech. Ageing Dev.* 74: 121–133, 1994.
306. SOHAL, R. S., M. R. MARZABADI, D. GALARIS, AND U. T. BRUNK. Effect of ambient oxygen concentration on lipofuscin accumulation in cultured rat heart myocytes—a novel in vitro model of lipofuscinogenesis. *Free Radical Biol. Med.* 6: 23–30, 1989.
307. SOHAL, R. S., AND W. C. ORR. Relationship between antioxidants, prooxidants, and the aging process. *Ann. NY Acad. Sci.* 663: 74–84, 1992.
308. SOHAL, R. S., AND B. H. SOHAL. Hydrogen peroxide release by mitochondria increases during aging. *Mech. Ageing Dev.* 57: 187–202, 1991.
309. SOHAL, R. S., B. H. SOHAL, AND U. T. BRUNK. Relationship between antioxidant defenses and longevity in different mammalian species. *Mech. Ageing Dev.* 53: 217–227, 1990.
310. SOHAL, R. S., B. H. SOHAL, AND W. C. ORR. Mitochondrial superoxide and hydrogen peroxide generation, protein oxidative damage, and longevity in different species of flies. *Free Radical Biol. Med.* 19: 499–504, 1995.
311. SOHAL, R. S., I. SVENSSON, AND U. T. BRUNK. Hydrogen peroxide production by liver mitochondria in different species. *Mech. Ageing Dev.* 53: 209–215, 1990.
312. SOHAL, R. S., AND R. WEINDRUCH. Oxidative stress, caloric restriction, and aging. *Science* 273: 59–63, 1996.
313. SOONG, N. W., D. R. HINTON, G. CORTOPASSI, AND N. ARNHEIM. Mosaicism for a specific somatic mitochondrial DNA mutation in adult human brain. *Nature Genet.* 2: 318–323, 1992.
314. SPRINGFIELD, J. R., AND M. D. LEVITT. Pitfalls in the use of breath pentane measurements to assess lipid peroxidation. *J. Lipid Res.* 35: 1497–1504, 1994.
315. STADTMAN, E. R. Protein oxidation and aging. *Science* 257: 1220–1224, 1992.
316. STADTMAN, E. R. Role of oxidized amino acids in protein breakdown and stability. *Methods Enzymol.* 258: 379–393, 1995.
317. STADTMAN, E. R., AND C. N. OLIVER. Metal-catalyzed oxidation of proteins. Physiological consequences. *J. Biol. Chem.* 266: 2005–2008, 1991.
318. STARKE-REED, P. E., AND C. N. OLIVER. Protein oxidation and proteolysis during aging and oxidative stress. *Arch. Biochem. Biophys.* 275: 559–567, 1989.
319. STEIN, G., AND J. WEISS. Chemical effects of ionizing radiations. *Nature* 161: 650, 1948.
320. STORZ, G., L. A. TARTAGLIA, S. B. FARR, AND B. N. AMES. Bacterial defenses against oxidative stress. *Trends Genet.* 6: 363–368, 1990.
321. SUZUKI, Y. J., H. J. FORMAN, AND A. SEVANIAN. Oxidants as stimulators of signal transduction. *Free Radical Biol. Med.* 22: 269–285, 1997.
322. SZWEDA, L. I., K. UCHIDA, L. TSAI, AND E. R. STADTMAN. Inactivation of glucose-6-phosphate dehydrogenase by 4-hydroxy-2-nonenal. Selective modification of an active-site lysine. *J. Biol. Chem.* 268: 3342–3347, 1993.
322a. TANAKA, M., J.-S. GONG, J. ZHANG, M. YONEDA, AND K. YAGI. Mitochondrial genotype associated with longevity. *Lancet* 351: 185–186, 1998.
323. TANHAUSER, S. M., AND P. J. LAIPIS. Multiple deletions are detectable in mitochondrial DNA of aging mice. *J. Biol. Chem.* 270: 24769–24775, 1995.
324. TAPPEL, A. L. Lipid peroxidation and fluorescent molecular damage to membranes. In: *Pathobiology of Cell Membranes*, edited by B. F. Trump and A. U. Arstila. New York: Academic, 1975, p. 145–170.
325. TCHOU, J., H. KASAI, S. SHIBUTANI, M. H. CHUNG, J. LAVAL, A. P. GROLLMAN, AND S. NISHIMURA. 8-Oxoguanine (8-hydroxyguanine) DNA glycosylase and its substrate specificity. *Proc. Natl. Acad. Sci. USA* 88: 4690–4694, 1991.
326. THORPE, S. R., AND J. W. BAYNES. Role of the Maillard reaction in diabetes mellitus and diseases of aging. *Drugs Aging* 9: 69–77, 1996.
327. TSUCHIDA, M., T. MIURA, AND K. AIBARA. Lipofuscin and lipofuscin-like substances. *Chem. Phys. Lipids* 44: 297–325, 1987.
328. TYLER, R. H., H. BRAR, M. SINGH, A. LATORRE, J. L. GRAVES, L. D. MUELLER, M. R. ROSE, AND F. J. AYALA. The effect of superoxide dismutase alleles on aging in *Drosophila*. *Genetica* 91: 143–149, 1993.
329. UEJIMA, Y., Y. FUKUCHI, S. TERAMOTO, R. TABATA, AND H. ORIMO. Age changes in visceral content of glutathione in the senescence accelerated mouse (SAM). *Mech. Ageing Dev.* 67: 129–139, 1993.
330. VANFLETEREN, J. R. Oxidative stress and ageing in *Caenorhabditis elegans*. *Biochem. J.* 292: 605–608, 1993.
330a. VAZIRI, H., M. D. WEST, R. C. ALLSOPP, T. S. DAVISON, Y. S. WU, C. H. ARROWSMITH, G. G. POIRIER, AND S. BENCHIMOL. ATM-dependent telomere loss in aging human diploid fibroblasts and DNA damage lead to the post-translational activation of p53 protein involving poly(ADP-ribose) polymerase. *EMBO J.* 16: 6018–6033, 1997.
331. VERCELLOTTI, G. M. A balanced budget—evaluating the iron economy. *Clin. Chem.* 42: 657, 1996.
332. VIJG, J., AND J. A. GOSSEN. Somatic mutations and cellular aging. *Comp. Biochem. Physiol B Biochem. Mol. Physiol.* 104: 429–437, 1993.
333. VON ZGLINICKI, T., E. NILSSON, W. D. DOCKE, AND U. T. BRUNK. Lipofuscin accumulation and ageing of fibroblasts. *Gerontology* 41, *Suppl.* 2: 95–108, 1995.
334. VON ZGLINICKI, T., G. SARETZKI, W. DOCKE, AND C. LOTZE. Mild hyperoxia shortens telomeres and inhibits proliferation of fibroblasts: a model for senescence? *Exp. Cell Res.* 220: 186–193, 1995.
335. WACHSMAN, J. T. The beneficial effects of dietary restriction: reduced oxidative damage and enhanced apoptosis. *Mutat. Res.* 350: 25–34, 1996.
336. WALLACE, D. C., J. M. SHOFFNER, I. TROUNCE, M. D. BROWN, S. W. BALLINGER, D. M. CORRAL, T. HORTON, A. S. JUN, AND M. T. LOTT. Mitochondrial DNA mutations in human degenerative diseases and aging. *Biochim. Biophys. Acta* 1271: 141–151, 1995.
336a. WALLACE, D. C., C. STUGARD, D. MURDOCH, T. SCHURR, AND M. D. BROWN. Ancient mtDNA sequences in the human nuclear genome: a potential source of errors in identifying pathogenic mutations. *Proc. Natl. Acad. Sci. USA* 94: 14900–14905, 1997.
337. WANG, Y. J., Y. S. HO, M. J. LO, AND J. K. LIN. Oxidative modification of DNA bases in rat liver and lung during chemical carcinogenesis and aging. *Chem. Biol. Interact.* 94: 135–145, 1995.
338. WARNER, H. R. Superoxide dismutase, aging, and degenerative disease. *Free Radical Biol. Med.* 17: 249–258, 1994.
338a. WARNER, H. R., R. J. HODES, AND K. POCINKI. What does cell

death have to do with aging? *J. Am. Geriatr. Soc.* 45: 1140–1146, 1997.

338b. WARNER, H. R., AND T. E. JOHNSON. Parsing age, mutations and time. *Nature Genet.* 17: 368–370, 1997.

339. WEI, Y. H. Mitochondrial DNA alterations as aging-associated molecular events. *Mutat. Res.* 275: 145–155, 1992.

339a. WEINDRUCH, R., AND R. S. SOHAL. Seminars in medicine of the Beth Israel Deaconess Medical Center. Caloric intake and aging. *N. Engl. J. Med.* 337: 986–994, 1997.

340. WEINDRUCH, R., R. L. WALFORD, S. FLIGIEL, AND D. GUTHRIE. The retardation of aging in mice by dietary restriction: longevity, cancer, immunity and lifetime energy intake. *J. Nutr.* 116: 641–654, 1986.

341. WILLIAMS, G. C., AND R. M. NESSE. The dawn of Darwinian medicine. *Q. Rev. Biol.* 66: 1–22, 1991.

342. WOOD, P. M. The potential diagram for oxygen at pH 7. *Biochem. J.* 253: 287–289, 1988.

343. YAGI, K., S. KOMURA, Y. SASAGURI, K. YOSHINO, AND N. OHISHI. Atherogenic change in the thoracic aorta of the senescence-accelerated mouse. *Atherosclerosis* 118: 233–236, 1995.

343a. YAKES, F. M., AND B. VAN HOUTEN. Mitochondrial DNA damage is more extensive and persists longer than nuclear DNA damage in human cells following oxidative stress. *Proc. Natl. Acad. Sci. USA* 94: 514–519, 1997.

343b. YAN, L. J., R. L. LEVINE, AND R. S. SOHAL. Oxidative damage during aging targets mitochondrial aconitase. *Proc. Natl. Acad. Sci. USA* 94: 11168–11172, 1997.

344. YAN, S. D., X. CHEN, J. FU, M. CHEN, H. ZHU, A. ROHER, T. SLATTERY, L. ZHAO, M. NAGASHIMA, J. MORSER, A. MIGHELI, P. NAWROTH, D. STERN, AND A. M. SCHMIDT. RAGE and amyloid-beta peptide neurotoxicity in Alzheimer's disease. *Nature* 382: 685–691, 1996.

345. YELINOVA, V., Y. GLAZACHEV, V. KHRAMTSOV, L. KUDRYASHOVA, V. RYKOVA, AND R. SALGANIK. Studies of human and rat blood under oxidative stress: changes in plasma thiol level, antioxidant enzyme activity, protein carbonyl content, and fluidity of erythrocyte membrane. *Biochem. Biophys. Res. Commun.* 221: 300–303, 1996.

346. YIM, M. B., J. H. KANG, H. S. YIM, H. S. KWAK, P. B. CHOCK, AND E. R. STADTMAN. A gain-of-function of an amyotrophic lateral sclerosis-associated Cu,Zn-superoxide dismutase mutant: an enhancement of free radical formation due to a decrease in K_m for hydrogen peroxide. *Proc. Natl. Acad. Sci. USA* 93: 5709–5714, 1996.

347. YIN, D. *Age Pigments and the Biochemical Basis of Their Formation.* Linkuping, Sweden: Linkuping University, 1995.

348. YIN, D. Biochemical basis of lipofuscin, ceroid, and age pigment-like fluorophores. *Free Radical Biol. Med.* 21: 871–888, 1996.

349. YONEDA, T., M. HIRAMATSU, M. SAKAMOTO, K. TOGASAKI, M. KOMATSU, AND K. YAMAGUCHI. Antioxidant effects of "beta catechin." *Biochem. Mol. Biol. Int.* 35: 995–1008, 1995.

350. YOUNGMAN, L. D., J. Y. PARK, AND B. N. AMES. Protein oxidation associated with aging is reduced by dietary restriction of protein or calories (published erratum appears in *Proc. Natl. Acad. Sci. USA* 89: 11107, 1992). *Proc. Natl. Acad. Sci. USA* 89: 9112–9116, 1992.

351. YU, B. P. Cellular defenses against damage from reactive oxygen species. *Physiol. Rev.* 74: 139–162, 1994.

352. YU, B. P. Aging and oxidative stress: modulation by dietary restriction. *Free Radical Biol. Med.* 5: 651–668, 1996.

353. YU, B. P., AND R. YANG. Critical evaluation of the free radical theory of aging. A proposal for the oxidative stress hypothesis. *Ann. NY Acad. Sci.* 786: 1–11, 1996.

354. YU, C. E., J. OSHIMA, Y. H. FU, E. M. WIJSMAN, F. HISAMA, R. ALISCH, S. MATTHEWS, J. NAKURA, T. MIKI, S. OUAIS, G. M. MARTIN, J. MULLIGAN, AND G. D. SCHELLENBERG. Positional cloning of the Werner's syndrome gene. *Science* 272: 258–262, 1996.

355. YUAN, H., T. KANEKO, AND M. MATSUO. Relevance of oxidative stress to the limited replicative capacity of cultured human diploid cells: the limit of cumulative population doublings increases under low concentrations of oxygen and decreases in response to aminotriazole. *Mech. Ageing Dev.* 81: 159–168, 1995.

355a. ZAMZAMI, N., T. HIRSCH, B. DALLAPORTA, P. X. PETIT, AND G. KROEMER. Mitochondrial implication in accidental and programmed cell death: apoptosis and necrosis. *J. Bioenerg. Biomembr.* 29: 185–193, 1997.

356. ZARLING, E. J., S. MOBARHAN, P. BOWEN, AND S. KAMATH. Pulmonary pentane excretion increases with age in healthy subjects. *Mech. Ageing Dev.* 67: 141–147, 1993.

357. ZOROV, D. B. Mitochondrial damage as a source of diseases and aging: a strategy of how to fight these. *Biochim. Biophys. Acta* 1275: 10–15, 1996.

in vivo *14:* 29-34 (2000)

Review

Cell Senescence in Human Aging: A Review of the Theory

MICHAEL FOSSEL

Clinical Professor of Medicine, Michigan State University, Box 630, Ada, Michigan 49301, U.S.A.

Abstract. *Cell senescence limits cell divisions in normal somatic cells and may play a central role in age-related diseases. In this review, we examine the theory that cell senescence underlies and is a pivotal event in human aging and age-related diseases. The model remains completely consistent with what we currently know of cell biology, cell senescence, and the pathology of all human age-related diseases including atherosclerosis, osteoarthritis, immune senescence, skin aging, Alzheimer's dementia, and cancer. Although there is in vitro and ex vivo data to support the model, there is currently no in vivo data supporting the theory. Such data is probably obtainable using available methods and, if confirmatory, would not only have stunning therapeutic consequences, but significant commercial consequences as well.*

Introduction

First described only four decades ago(1) and considered radical at the time(2), cell senescence is now accepted as a keystone of cell biology. The question of what role cell senescence plays in human aging and age-related diseases, however, remains unanswered(3) and not without theoretical complications(4). Until now, the question has remained moot, since we lacked the technology to either answer the question or institute any of the putative clinical interventions based on such a possibility(5). Recently, however, we have become capable of not only preventing and - to an extent- resetting the "clock" of cellular senescence(6), but we have made tentative in-roads in doing so in a clinically meaningful manner(7). Clearly cell senescence has implications for clinical medicine, both in terms of cancer(8) and increased cell senescence during aging or the accelerated cell aging which occurs during bone marrow transplantation(9). Over the next several years, our knowledge of cell senescence will doubtless increase, as will our understanding of the role of cell senescence in aging and age-related diseases.

Clinically, age-related diseases are seldom considered to share an underlying or unifying pathology. Age related diseases are thought of as being related not in a causal, but only in an epidemiological sense: their incidence increases with cohort age. More than a century ago, a similar lack of association was erroneously assumed for the group of diseases that we now regard as infectious(10). The lack of a basic understanding of mechanism prevented any comprehension of their commonality. In the case of aging, an analogous process may pertain as we fail to understand the fundamental mechanisms that underlie a common pathology.

Recent articles suggest that age-related diseases may have their shared etiology in cell senescence(11). As was the case of infectious disease in the nineteenth century, there is currently a lack of understanding of exactly *how* cell senescence might underlie age-related disease. The lack of any coherent, published discussion of the basic model has resulted in both naive criticism and unwarranted disbelief without reference to the data.

For example, in discussing age-related disease, some(12) have argued that *cell senescence* cannot play a role in myocardial disease because myocardial cells do not senesce. This argument appears sound, although substituting cholesterol for cell senescence exposes the lack of logical framework in the argument, viz., "*Cholesterol* cannot play a role in myocardial disease because myocardial cells do not *accumulate cholesterol.*" The lack of accumulation of cholesterol (or the lack of cell senescence) in myocardial cells is moot: the pathology (in both cases) occurs within the arterial walls which supply such myocardial cells. We not only have good reason to believe that cholesterol "causes" (i.e., plays a prominent role in the pathological cascade of) myocardial disease, but find that its accumulation within subendothelial cells of the coronary arterial walls is quite sufficient to cause death without any direct reference to primary pathology in the myocardial cells themselves. "Straw man" arguments regarding non-dividing cells have no intellectual validity as critiques of cell senescence in aging.

In this same vein, the cell senescence model of aging does not claim that *all* (or even most) cells divide or senesce in adult organisms (and clearly many cell phenotypes do not), only that those cells *which do divide* (and therefore senesce) have significant detrimental effects on other cells whether they divide or not. In the case of atherosclerosis, the

Key Words: Aging, cell senescence, telomeres, telomerase, degenerative disease, aging disease, cancer.

0258-851X/2000 $2.00+.40

senescence of arterial endothelial cells may be quite sufficient to cause death due to myocardial infarction or aortic aneurysm without any senescence in myocardial cells themselves(13). In the case of Alzheimer's dementia, microglial cell senescence may occur and may result in dysfunction of neurons which do not themselves divide or senesce. Criticisms based on the non-division of myocardial cells or neurons are specious and irrelevant to the pathology.

Such naive criticisms do a disservice both to the complexity of the pathology which is known to occur and (more pointedly here) to the plausible hypothesis that cell senescence underlies most (perhaps even all) age-related diseases. The minatory remarks made above regarding spurious and trivial arguments are warranted if we wish to understand the actual pathology underlying age-related disease, putatively that of cell senescence. Theoretical falsification of the model has proven naive and erroneous in cases such as atherosclerosis and dementia; this does not mean that theoretical falsification cannot occur, but rather that - to date - the model remains internally consistent, consistent with known data, and even elegant.

A more pertinent issue is the extent to which we might validate or falsify the model experimentally. The strength of any theory lies in making testable predictions. What is the actual data in support of cell senescence as a fundamental and primary player in the complex cascade of pathology that underlies age-related diseases? In the case of cell senescence, current supportive data is largely, although not exclusively, circumstantial. Nonetheless, the model makes specific and testable predications that we are now technically capable of validating. Specifically, cell senescence theory of aging predicts that if we can prevent or reverse cell senescence, then we can equally prevent or reverse age-related diseases. Within practical but currently unknown limits, to the extent that we can alter cell senescence, we will ameliorate clinical disease.

The model

The model is simple to state: cell senescence underlies organismal aging. Note that this is *not* the same as saying that cell senescence "causes" aging and is certainly not the same as saying that telomere shortening causes aging. The complexity of assigning causation in a biological system results in avoidable confusion and baseless disagreement. For example, does hypercholesterolemia "cause" atherosclerosis or does hypertension "cause" atherosclerosis? What about diabetes, genetic errors in lipid metabolism, arterial wall viral infection, tobacco use, etc? And which is the "real cause" of myocardial disease? Discussions which assign causation in a simplistic, monolithic sense founder and risk incoherence in biologic systems. Classic notions of cause and effect remain valid and useful but only if we give due respect to the complexity of interactions involved in such systems[14]. If we discuss causation in simplistic, unrealistic terms, we are prone to remain ignorant. Hypercholesterolemia may "cause" atherosclerosis without implying that hyperglycemia does not or that atherosclerosis cannot occur in its absence. Though generally present in patients with atherosclerosis, hypercholesterolemia is actually neither necessary nor sufficient to the disease process, yet this does not invalidate its causative role in many patients. Clinical advice regarding dietary and pharmaceutical intervention to lower serum cholesterol remains useful and accurately predictive of clinically benefits in the bulk of patients.

In biologic systems, rather than a simple chain of "domino" events, causation is more accurately a "web" of interacting factors. Discussions of biological and clinical cause would do well to reflect a sophisticated and realistic view of such complexity. Although the system may be dauntingly complex, the complexity need not prevent us from achieving a utilitarian understanding. Clinically - and practically - the key question should be "what is the most efficient intervention to ameliorate the course of the disease?" This addresses the heart of our common sense notion of causation, *i.e.*, can we significantly alter the outcome of the system? In this regard, the cell senescence model predicts that altering cell senescence would efficiently and effectively alter the disease process and this proposition is testable.

Consider atherosclerosis as a classic example of an age-related disease in which cell senescence is a fundamental "cause" of the pathology. The causes of atherosclerosis have been clarified progressively over the past century and there is a widespread consensus that we "understand" the pathology in at least general terms. In outline, the process of human atherogenesis begins with several "causes" in the arterial lumen. These include hypertension and shear stress, toxins (including nicotine), metabolic and physiological stressors (including hyperglycemia and hypercholesterolemia), and various implicated biological agents (including several viruses and bacteria such as Chlamydia). All of these collectively and independently can cause damage to the arterial vessel wall resulting in the standard, classic pathology of atheroma formation. At the microscopic level we see inflammation, cholesterol deposition, foam cells, smooth muscle proliferation, thrombus formation, and atheromatous growth. These changes in the vessel wall - and especially in the subendothelial layers - result in a narrowing lumen and a gradually elevating risk of both thrombotic occlusion and myocardial ischemia, as well as catastrophic infarct to the myocardium supplied by such pathologic coronary vessels.

But how do intralumenal causes result in subendothelial effects, ultimately the "cause" of ischemic effects in myocardium? Invoking cell senescence allows us to address this question, explain what might otherwise be seen as inexplicable sources of atherogenesis (e.g., Hutchinson-Gilford progeria), and does not invalidate any of our current understanding of the disease processes.

The cell senescence model of aging suggests that endothelial cells are damaged by intralumenal factors, resulting in endothelial cell loss with consequent cell replacement through cell division, itself resulting in telomere shortening, altered gene expression, and thereby, cell senescence. Theoretically, it is the altered gene expression - the concrete expression of cell senescence - of endothelial cells which "causes" the pathology in subendothelial cells. This model provides a linkage between our knowledge of epidemiology as the "input" and our knowledge of the pathology as an "output"; it does so through our growing understanding of the cell biology occurring in the vessel wall. There is no reason to expect that cell senescence would occur in myocardial cells, but every reason to expect it in the endothelial cells lining the aorta and coronary arteries.

Defining cell senescence[15] is problematic and ambiguous. Is cell senescence simply population doublings[16], telomere length[17], or expression of a single marker gene, such as Beta galactosidase[18]? While the most practical definition may prove to be the pattern of gene expression[19], telomere length as a marker has at least the sole value of being easily quantified. Using such a measure as a marker for endothelial cell senescence, we might predict that telomere lengths should reflect and correlate with arterial wall pathology. This is precisely what we find experimentally: telomeres shorten in areas of high shear stress and do so early in both atherogenesis and aneurysm formation[20]. Likewise, gene expression alters predictably[21]. Although merely correlational, this data is certainly consistent and predictable within the model: cell senescence does occur where and when it should to "cause" atherosclerosis. Endothelial cells senesce *pari passu* with subendothelial atherogenesis.

The result of increasing endothelial cell senescence is an alteration of cell-cell interactions, specifically upon subendothelial cells. And again, this is precisely what we find experimentally[22]. It would appear that cell senescence could certainly play a fundamental role in the pathology that we have long thought we understood. Tobacco, viral infections, shear stress, etc may still be "causes" of atherosclerosis, but their effects may be mediated through and incumbent upon the changes of gene expression and cell function that define cell senescence within endothelial cells lining the arterial vessel wall.

Note again the lack of relevance of the fact that myocardial cells do not divide: cell senescence does not "age" heart cells, but kills them indirectly through "aging" the endothelial cells of the coronary vessels which supply myocardial cells. One need not use a gun; starvation works quite as effectively. Cell senescence "causes" myocardial infarction not through myocardial cell senescence but through endothelial cell senescence and does it so reliably that atherosclerosis has become the leading cause of death in most developed nations.

The cell senescence model of aging is equally applicable throughout other aging organ systems and it has two primary and practical virtues. First, like any good theory, it is falsifiable. Second, if accurate, it will provide a highly effective point of clinical intervention. Bluntly, the clinical implications are withiout prior medical parallel. Since atherosclerosis is a leading cause of death, consider the supportive data and clinical implications for atherosclerosis alone. The model predicts that if we accelerate cell senescence we will accelerate atherosclerosis and if we slow or reverse cell senescence, we will slow or reverse atherosclerosis.

In Hutchinson-Gilford progeria the overwhelming cause of death is atherosclerosis[23] and the genetic effect appears to be expressed as early cell senescence at least in some cell lines[24]. These children lack the usual litany of risk factors - high cholesterol high blood glucose, high blood pressure, and tobacco use - yet suffer from impressive and classic atherogenesis as well as its clinical outcomes: strokes and myocardial infarctions[25]. Their average age at death is 12.7 years, almost without exception due to vascular disease. In this example, accelerated cell senescence and emphatically accelerated atherosclerosis go hand-in-hand.

The obvious and more interesting prediction is that the opposite may also be true. If we can slow, prevent, or reverse cell senescence could we, *pari passu*, alter the clinical outcome? If we can reverse cell senescence, as we know we can in the laboratory[26], can we prevent or cure atherosclerosis? In short, if we "telomerize" endothelial cells, can we prevent death due to heart disease, strokes, and aneurysms? The cell senescence model predicts that this is precisely what we might do[27]. We can effectively treat the clinical diseases of aging by treating the aging cells that underlie these diseases.

The issue regarding the efficiency of cell senescence as a therapeutic target is worth more detailed discussion. *In vitro*, we can reset and indefinitely postpone cell senescence, but, as yet, have not tried to do so *in vivo*. Potentially, therapies based directly upon cell senescence should be more effective and generically applicable than would current "piecemeal" therapies based upon individual risk factors such as hypercholesterolemia or based on secondary clinical "outcomes", such as coronary artery bypass grafting or angioplasty. The greater efficiency and efficacy of cell senescence treatments for atherosclerosis results because cell senescence functions as a single "common denominator" in atherogenesis, though which all other risk factors and "causes" must play out This "funneling" effect of cell senescence within the complex pathology underlying age-related diseases makes therapies aimed at reversing or preventing cell senescence unique and efficient points of therapeutic intervention. All "causes" of atherosclerosis share the common pathway of cell senescence and this pathway would be the more effective single point of clinical intervention.

In the case of atherosclerosis, not only would cell senescence therapy be applicable to children with Hutchinson-Gilford progeria (in which nothing else is currently effective), but equally effective in patients with any of the myriad of risk factors which trigger cell senescence in coronary and aortic vessels and which result in clinical disease and atherosclerotic death. The unique and fundamental place of cell . senescence in aging pathology makes it crucial that we test the theory. Whether cell senescence actually underlies atherogenesis remains to be shown, but the model is theoretically and experimentally consistent with all we know of atherogenesis and human pathology. It would be a therapeutic target without clinical parallel with potential for unprecedented efficacy and universal applicability. Many theories of aging are difficult to falsify; many have little potential value for clinical medicine or pharmacologic intervention. In contrast and in fact, the therapeutic, commercial and economic implications of the cell senescence theory of aging are so stunning and unexpected that it demands validation or falsification as rapidly as possible.

Nor do the implications rest solely within the purview of atherosclerosis. The cell senescence model applies equally well in other aging organ systems and in other age-related diseases. Cell senescence occurs in the skin[28], the immune system[29], joints[30], etc. The question is not whether or not it could be a fundamental "cause" within the pathologic cascades in each aging organ, but whether or not in fact it is such a cause. Let us consider some of these other organ systems and specific age-related diseases.

In the case of the central nervous system - Alzheimer's dementia - even circumstantially supportive data is not yet obtainable. There is no current evidence of cell senescence to match the evidence in atherosclerotic vessels. And just as some critics inappropriately cite the lack of myocardial cell division to criticize the model - forgetting where and how the pathology actually occurs - so too do they cite the lack of neuronal division in the adult mammalian brain as invalidating cell senescence as a cause of Alzheimer's dementia. Reality is not quite so simple.

The question is not "do neurons divide and senesce?", but rather "are there dividing cells whose senescence causes pathology in the involved neurons?" The obvious candidate for cell senescence is the astroglial cell[31]. These not only divide, but. have been implicated in the pathology and inflammation of Alzheimer's dementia[32]. Although such an hypothesis is consistent, fascinating, and falsifiable, there is no data that currently supports it. That cell senescence could cause Alzheimer's disease remains unassailable; whether it does so or not is completely unknown, with out even the benefit of circumstantial evidence. The sole exception to this lack of suggestive evidence is that trisomy 21[33] patients show early cell senescence in leukocyte lines (probably explaining their early immune senescence). If astrocytes, a related cell line, also show early cell senescence this might explain the early dementias which trisomy 21 patients are also prone to.

In neither of thesc two aging systems - the arterial walls or the central nervous system - are we capable of directly testing the cell senescence model. This is unfortunate because these two systems represent the most clinically pressing causes of age-related disease. Nor are they alone in this combination of importance and experimental frustration in our attempts to alter cell senescence experimentally *in vivo*. Aging joints likewise have been implicated in cell senescence as we would expect them to if cell senescence underlies osteoarthritis[34], but are similarly difficult to approach experimentally.

There are, however, other organ systems which age and which show evidence of cell senescence and, more importantly, some of these systems are technically approachable, allowing us the possibility of testing the theory. While we cannot yet prevent or reverse cell senescence *in vivo* within arterial walls, the brain, or joints, we are capable of doing so *ex vivo* in the immune system and *in vivo* in the skin.

The immune system is a remarkably easy system for transfection not only because we could manipulate it experimentally, but because we already do so on a regular basis without intending to. Bone marrow transplants, performed as a part of cancer treatments in an effort to "rescue" the hematopoetic system after whole body radiation or chemotherapy, result in substantial telomere shortening and rapid cell senescence in the transplanted leukocytes, as a resultof their necessary clonal expansion. The average bone marrow transplant causes the equivalent of four decades of cell senescence, perhaps the single most important limitation to serial marrow transplants[35]. Cell senescence has clinical consequences. Currently, we are technically capable of transfecting leukocytic stem cells with human telomerase (actually hTERT, the catalytic sub-unit of telomerase), resetting cell senescence, and replacing them *ex vivo*. That this would result in significant effects on immunosenescence is probable, unknown, and technically feasible. We could test the immune portion of the cell senescence theory of aging effectively and - technically speaking - with relatively ease. Likely clinical benefits to such an intervention are vouchsafed by the higher rates of death due to infectious disease in the elderly, in which immunosenescence probably plays a prominent role[36]. But the ethical and practical questions remain almost insurmountable: 1) is there a sufficient likelihood of clinical benefit to healthy, elderly volunteers to justify the risk, and 2) is there sufficient time to perform such an ex vivo experiment in the elderly patient dying of sepsis who might otherwise benefit from enhanced immune function? The risk to the former is not only the usual (and uncertain) risk of gene transfection, but the additional risk of constitutive telomerase expression with its potential (and likewise uncertain) risk of enhanced growth for what might otherwise be self-limited malignant cells (see below). The second problem remains as well: those patients who stand the most to gain from telomerase transfection to leukocyte stem cells and the consequent reconstitution of effective immune function are likely to either die within a matter of hours (long before immune reconstitution can occur) or be recovered and discharged by the time the cells have even been transfected in the laboratory, let alone returned to the patient. The upshot is that immune reconstitution *via* telomerase transfection *ex vivo* is technically feasible, but has not yet been attempted for want of appropriate experimental subjects.

The dermis is an approachable organ system and is likely to see the first *in vivo* test of cell senescence. Currently, there is little data to directly support the hypothesis. We know that a significantpercentage of cells show senescence in the elderly; we do not know that these senescent cells have clinical importance in explaining dermatologic aging. Nonetheless, there are a number of new findings which add credence to the hypothesis, including gene array findings which not only show that senescent cells have reliably altered pattems of gene expression, but that these patterns may be reliably and entirely reset by transfection with telomerase[37]. Even more intriguing and relevant are the findings that we cannot only reset cell senescence *in vitro*, but we can reconstitute dermis itself.

Transplantation of human fibroblasts and keratinocytes onto immunodeficient mice results in reconstitution of anatomically normal skin, complete with dermal/epidermal convolutions, etc. Using young human cells, the reconstituted tissue is grossly, microscopically, and genetically (gene array analysis) typical of young human skin. Using senescent cells, the tissue is typical of old human skin. However, if we transfect these cells with human telomerase, the resultant tissue is again typical of young human skin. The gross, microscopic, and gene expression characteristics are all those of young skin (38). In a surprising "end run", this experiment suggests that not only does cell senescence underlie the process of dermal aging, but *that it is reversible*. Whatever the complexity of dermal aging, whatever the percentage of senescent cells in older skin, whatever the cell-cell interactions which result in clinical skin aging, the process not only appears to depend upon cell senescence, but it is therapeutically malleable.

To what extent can this be extended to *in vivo* human experimentation? Gene transfection is advancing, but is far from a mature technology[39]. Nonetheless, the potential of cell senescence is sufficient, that dermal telomerase transfection is worth consideration in several cases, especially in Hutchinson-Gilford progeria and perhaps normal senescent skin. In the former case, the impact of the skin on clinical well-being is minimal: these children die of atherosclerosis, not senescent skin. Nonetheless. we are not technicallv capable of transfecting endothelial cells *in vivo* and transfection of dermal cells *in vivo* might prove that it was worth pursuing effective vectors for endothelial cells in these patients. In the latter case, normal elderly patients do suffer from dermal senescence and occasionally die as a result. Decubitus ulcers, poor wound healing, and the thinner dermal barrier (to infection and heat loss), all contribute to the morbidity and ultimate mortality of the elderly. While more attention (and appropriately so) is placed on atherosclerosis and cancer as a cause of death, and Alzheimer's dementia as a cause of morbidity, dermal senescence is not merely a cosmetic or trivial issue. The complications of dermal senescence may well justify direct transfection of human telomerase in such patients, both to provide a clinical benefitand to answer the question of what role cell senescence plays in clinical aging in skin.

A final disease, associated with cell senescence but in a notably different fashion, is cancer. Over the past few decades, our knowledge of cancer has increased immeasurably, albeit with as yet moderate clinical impact. It has become clear that both intercellular phenomenon (e.g., angiogenesis and immune surveillance) and intracellular phenomenon (the detection and repair of damaged DNA, apoptosis, the cell-cycle braking system[40], and telomerase expression) play a role in carcinogenesis. The relatively recent understanding that telomere shortening probably acts as a "dead man switch" to prevent the limitless replication of malignant cells[41], has prompted the suggestion that telomerase inhibitors might provide a superbly effective therapy[42]. Indeed this is being actively and appropriately pursued at both the research and clinical levels. Suffice it to say that telomerase inhibitors have great promise in that they offer to distinguish between normal and malignant cells, but that there is, as yet, no deliverable molecule which acts to inhibit telomerase safely *in vivo*. Nor is their potential without caveat Clearly a number of normal cell lines do express telomerase at least transiently[43] and a small percentage of (largely, but not exclusively benign) human tumors do not[44]. Nonetheless, telomere expression itself does not confer malignancy upon human cells[45] and the research progresses apace.

Conclusions

The cell senescence theory of aging is remarkably different from most current theories in two eminently practical respects: it is testable and it promises a remarkable and obtainable therapeutic return. Bluntly, we are at the point where we could test the theory, use it clinically, and obtain substantial commercial benefit from therapies based upon the model. Although we are currently unable to either test or derive clinical or commercial gains from the theory in the most relevant organ systems (endothelial aging in atherosclerotic vessels and glial aging in the CNS), there are at least two systems that deserve close and immediate consideration: the immune system and the skin. In both of these cases, we are (at least marginally) technically capable of transfecting the relevant cells with potential experimental confirmation and clinical utility. Cell senescence theory is not only supported by and consistent with all available experimental and clinical data in aging human organ systems, but it is testable and has almost immediate clinical relevance to human aging and age-related disease.

References

1 Hayflick L, Moorhead PS: The limited *in vitro* lifetime of human diploid cell strains. Experimental Aging Research *25:* 585-621, 1961.
2 Hayflick L: How and Why We Age. New York, Ballantine Books, 1994.

3 Banks DA, Fossel M: Telomeres, cancer, and aging, altering the human lifespan. JAMA 278: 1345-1348, 1997. Fossel M: Implications of Recent Work in Telomeres and Cell Senescence. Journal of Anti-Aging Medicine 1: 39-43, 1998.
4 Blasco MA, DePinho RA, Greider CW, Hande MP, Lansdorp PM, Lee HW, Samper E: Telomere shortening and tumor formation by mouse cells lacking telomerase RNA. Cell 91: 25-34, 1997. Rudolph KL, Chang S, Lee HW, Blasco M, Gotlieb GJ, Greider C, DePinho RA: Longevity, Stress Response, and Cancer in Aging Telomerase-Deficient Mice. Cell 96: 701-712, 1999. Cristofalo VJ, Allen RG, Pignolo R J, Martin BG, Beck JC: Relationship between donor age and the replicative lifespan of human cells in culture: A reevaluation. Proc Natl Acad Sci USA 95: 10614-10619, 1998.
5 Fossel M: Reversing Human Aging New York, William Morrow and Company, 1996.
6 Lange, T: Telomeres and senescence: ending the debate. Science 279: 334-335, 1998. Bodnar AG, Chiu C, Frolkis M, Harley CB, Holt SE, Lichtsteiner S, Morin GB, Ouellette M, Shay JW, Wright WE: Extension of Life-Span by Introduction of Telomerase into Normal Human Cells. Science 279: 349-352, 1998. Benchimol S, Vaziri H: Reconstruction of telomerase activity in normal cells leads to elongation of telomeres and extended replicative life span. Current Biology 8: 279-282, 1998. Vaziri H: Extension of Life Span in Normal Human Cells by Telomerase Activation: A Revolution in Culhlrul Senescence. Journal of Anti-Aging Medicine 1: 125-130, 1998.
7 Funk W, et al: Experimental Cell Research; submitted for publication, 1999.
8 Harley CB, Kim NW, Prowse KR, et al: Telomerase, cell immortality, and cancer. Cold Spring Harb Symp Ouant Biol 59: 307-315, 1994. Harley CB: Telomerase and cell immortality in humans. Program and abstracts of the 89th Annual Meeting of the American Association for Cancer Research. March 28-April 1, 1998; New Orleans, La, 1998. Hiyama E, Hiyama K, Yokoyama T, Matsuura Y, Piatyszek MA, Shay JW: Correlating telomerase activity levels with human neuroblastnma outcomes. Nat Med 1: 249-55, 1995. Hiyama E, Hiyama K, Ohtsu K, Yamaoka H, Ichikawa T, Shay JW, Yokoyama T: Telomerase activity in neuroblastoma: is it a prognostic indicator of clinical behaviour? Eur J Cancer 33: 1932-1936, 1997. Shay JW, Gazdar AF: Telomeres in the early detection of cancer. J Clin Pathol 50: 106-109, 1997. Shay JW, Bacchetti S: A survey of telomerase in human cancer. Eur J Cancer 33: 787-791, 1997. Maitra A, Yashima K, Rathi A, Timmons CF, Rogers BB, Shay JW, Gazdar AF: The RNA component of telomerase as a marker of biologic potential and clinical outcome in childhood neuroblastic tumors. Cancer 85: 741-749, 1999.
9 Wynn RF, Cross MA, Hatton C, et al: Accelerated telomere shortening in young recipients of allogenic bone-marrow transplants. Lancet 351: 178-181, 1998.
10 de Kruif P: Microbe Hunters. New York, Harcourt Brace & Company, 1954. Porter R: The Greatest Benefit to Mankind. New York, WW Norton Company, 1997.
11 Banks & Fossel op cit Fossel M: Telomerase and the Aging Cell; Implicatioqls for Human Health. JAMA 279: 1732-1735, 1998.
12 Clark W: Means fo an End. New York, Oxford University Press, 1999.
13 Chang E, Harley CB: Telomere length as a measure of replicative histories in human vascular tissues. Proc Natl Acad Sci USA 92: 1190-1194, 1995. Cooper LT, Cooke JP, Dzau VJ: The vasculopathy of aging. J Gerontol 49: 191-196, 1994.
14 Bierman EL: Aging and Atherosclerosis. In Principles of Geriatric Medicine and Gerontology (Hazzard WR, Bierman EL, Blass JP, Ettinger WH Jr, Halter JB, eds) New York, McGraw Hill Third Edition, 1994, pp 509-516.
15 Fossel M. Cell Senescence in Human Aging New York, Oxford Press, in preparation.
16 Hayflick, 1994, op cit
17 Harley CB, Futcher AB, Greider CW: Telomeres shorten during aging of human fibroblasts. Nature 345: 458-460, 1990.
18 Dimri GP, Campisi J: Molecular and cell biology of replicative senescence. Cold Spring Harbor Symp Ouant Biol 54: 67-73, 1994. Dimri GP, Lee X, Basile G, Acosta M, Scott G, Roskelley C, Medrano E, Linskens M, Rubelj I, Pereira-Smith O, Peacock M, Campisi J: A biomarker that identifies senescent human cells in culture and in aging skin in vivo. Proc Natl Acad Sci USA 92: 9363-9367, 1995. Campisi J: The role of cellular senescence in skin aging. J Investig Dermatol Symp Proc 3: 1-5, 1998.
19 Fossel in preparation, op cit.
20 Chang & Harley, 1995, op cit.
21 Kumazaki T: Modulation of gene expression during aging of human vascular endothelial cells. Hiroshima J Med Sci 42: 97, 1993.
22 Cooper et al 1994, op cit.
23 Fossel 1996, op cit. Brown WT, Zebrower M, Kieras FJ: Progeria: a genetic disease model of premature aging. In Genetic Effects on Aging II(Harrison DE, ed). Caldwell, NJ, Telford Press, 1990, Chapter 29.
24 Allsopp RC et al: Telomere length predicts replicative capacity of human fibroblasts. Proc Nat Acad Sci USA; 89:10114-10118, 1992.
25 Brown, 1990, op cit.
26 Bodnar et al, 1998, op cit Vaziri 1998, op cit Benchimol & Vaziri 1998, op cit.
27 Fossel 1996, op cit Fossel 1998, op cit.
28 Campisi 1998, op cit.
29 Effros RB, Valenzuela HF: Immunosenescence: analysis and genetic modulation of replicative senescence in T cells. Journal of Anti-Aging Medicine 1: 305-313, 1998.
30. Fossel 1998, op cit.
31. Fossel 1996, op cit.
32. Mayeux R, Schofield PW: Alzheimer's Disease. In Principles of Geriatric Medicine and Gerontology (Hazzard WR, Bierman EL, Blass JP, Ettinger WH Jr, Halter JB, eds) New York, McGraw Hill Third Edition, 1994, chapter 92.
33 Vaziri H, et al: Loss of telomeric DNA during aging of normal and trisomy 21 human leukocytes". American Journal of Human Genetics, 52: 661, 1993.
34 Fossel 1998, op cit.
35 Chiu C-P, Dragowska W, Kim NW, Vaziri H, Yui J, Thomas TE, et al: Differential expression of telomerase actiyity in hematopoietic progenitors from adult human bone marrow. Stem Cells 14: 239-248, 1996. Shay J: Accelerated telomere shortening in bone marrow recipients. Lancet 351: 153-154, 1998. Moore MA: Stem cell proliferation: ex vivo and in vivo observations. Stem Cells 15: 239-248, 1997. Wynn, 1998, op cit.
36 Effros, 1998, op cit.
37 Shelton DS, Chang E, Whittier PS, Choi D and Funk WD: Microarray analysis of replicative senescence. Current Biology, submitted for publication, 1999.
38 Funk, 1999, op cit.
39 Friedmann T: The Development of Human Gene Therapy. Cold Spring Harbor N.Y., Cold Spring Harbor Laboratory Press, 1999.
40 Jacks T, Weinberg RA: Cell-cycle control and its watchman. Nature 381: 643-644, 1996. Sherr CJ: Cancer Cell Cycles. Science 274: 1672-1677, 1996.
41 Shay & Gazdar, 1997, op cit. DePinho RA, de Lange T: Unlimited Mileage from Telomerase? Science 283: 947-950, 1999.
42 Counter CM, Hirte HW, Bacchetti S, Harley CB: Telomerase activity in human ovarian carcinoma. Proc Natl Acad Sci USA 91: 2900-2904, 1994. Fossel 1996, op cit Banks & Fossel 1997, op cit.
43 Avilion AA: Human telomerase RNA and telomerase activity in immortal cell lines and tumor tissues. Cancer Research 56: 645-650, 1996. Fossel 1998, op cit Ramirez RD, Wright WE, Shay JW, Taylor RS: Telomerase activity concentrates in the mitotically active segments of human hair follicles. J Invest Dermatol 108: 113-117, 1997. Wright WE, Piatyszek MA, Rainey WE, Byrd W, Shay JJ: Telomerase activity in human germline and embryonic tissues and cells. Dev Genet 18: 173-179, 1998. Englehardt M, et al: Telomerase regulation, cell cycle, and telomere stability in primitive hematopoetic cells. Blood 90: 182, 1997. Chiu, 1996, op cit.
44 Shay & Bacchetti, 1997, op cit.
45 Jiang X, Jimenez G, Chang E, Frolkis M, Kusler B, Sage M, Beeche M, Bodnar AG, Wahl GM, Tlsty TD, Chiu C Telomerase expression in human somatic cells does not induce changes associated with a transformed phenotype. Nature Genetics 21: 111-114, 1999.

Received September 16, 1999
Accepted November 1, 1999

Symposium: Calorie Restriction: Effects of Body Composition, Insulin Signaling and Aging

The Role of Fat Depletion in the Biological Benefits of Caloric Restriction[1,2]

Nir Barzilai[3] and Ilan Gabriely

Department of Medicine, Divisions of Geriatrics, Endocrinology and the Diabetes Research and Training Center, Albert Einstein College of Medicine, Bronx, NY 10461

ABSTRACT One of the most robust observations in the biology of aging is that caloric restriction (CR) extends life in a variety of species. Although CR results in substantial decrease in fat mass, the role of fat in life extension was considered minimal. Indeed, in the fields of obesity and diabetes, the amount of fat has been directly implicated in the metabolic consequences. Since it became apparent that fat is a massive endocrine tissue, some of its roles have been recently revised. Many of the systemic effects of CR can now be explained by the chronic effects related to decreased plasma levels of peptides, cytokines, complement factors and substrates that are produced in fat. Most of the benefits of CR on the neuroendocrine system and those related to the improvement in glucose homeostasis can be attributed to a decrease in adipose cells and their products. If all or most of the life-extending benefits of CR can be attributed to decreased fat stores, the expression of specific candidate substrates and proteins may be explored and manipulated in searching for the most powerful adipose-dependent signals that modulate life expectancy. J. Nutr. 131: 903S–906S, 2001.

KEY WORDS: • caloric restriction • fat mass • visceral fat • obesity • insulin sensitivity • fat-derived peptides

Although change in body weight, specifically marked reduction in fat mass (FM)[4], is the main phenotypic expression of caloric restriction (CR) in animals, only a few studies in the last decades have examined the causal relationship of FM to different metabolic disorders. The most compelling reason for investigating the role of FM in life extension is the fact that CR has overwhelming effects on a variety of organs and systems. In both animals and humans, elevated total body FM causes an increase in morbidity and mortality that may be attributed to a variety of pathological factors, among which insulin resistance is most important. Peripheral and hepatic insulin resistance is characterized by marked increases in fasting and postprandial insulin levels and, consequently, impaired glucose tolerance. Moreover, although insulin resistance is common in aging, it has been shown to be determined by FM in both rats and humans. Because CR reverses the increase in plasma insulin, glucose and glycosylated hemoglobin to youthful levels, it has been hypothesized that glucose and insulin mediate aging processes (the glucose and insulin hypothesis). A recent study showed a dramatic increase in life span in a nematode with a mutation of an insulin-like receptor complex gene, further supporting the importance of the insulin-signaling pathway in longevity (1). As will be discussed, increased body weight and FM in humans results in insulin resistance and is closely associated with the development of dyslipidemia, hypertension, diabetes and coronary heart disease, leading to early death.

FM, CR *and peripheral insulin sensitivity*

Most studies express glucose uptake in terms of kg of body weight, which does not correct for large amounts of relatively inactive adipose tissue and results in an underestimation of glucose uptake by lean body mass (LBM). Thus, while most studies demonstrated a decline in glucose utilization, body mass index (BMI) and FM were inversely correlated with glucose infusion rates during hyperinsulinemic clamp studies, independent of age (2,3). Indeed, insulin sensitivity in a large number of lean older individuals was similar to that of weight-matched young individuals (4). A recent multi-center study demonstrated that FM, and not age, caused decreased peripheral insulin sensitivity. Thus, FM has an overwhelming effect on peripheral insulin sensitivity with aging. Increased visceral fat (VF) was shown to be associated with a decrease in skeletal muscle insulin sensitivity measured by leg balance studies and

[1] Presented at the symposium "Calorie Restriction: Effects on Body Composition, Insulin Signaling and Aging" as part of the Experimental Biology 2000 meeting held April 15–18, 2000 in San Diego, California. This symposium was sponsored by the American Society for Nutritional Sciences and the American Society for Clinical. The proceedings of this symposium are published as a supplement to *The Journal of Nutrition*. Guest Editor for this supplement publication was Barbara Hansen, Obesity and Diabetes Research Center, School of Medicine, University of Maryland, Baltimore, Maryland.

[2] Supported by grants from the National Institutes of Health (KO8-AG00639, R29-AG15003, RO1-AG18381 to N.B.), the Core laboratories of the Albert Einstein Diabetes Research and Training Center (DK 20541). N.B. is a recipient of the Paul Beeson Physician Faculty Scholar in Aging Award.

[3] To whom correspondence should be addressed.
E-mail: barzilai@aecom.yu.edu

[4] Abbreviations used: FM, fat mass; CR, caloric restriction; LBM, lean body mass; BMI, body mass index; VF, visceral fat; HGP, hepatic glucose production; AT, angiotensinogen.

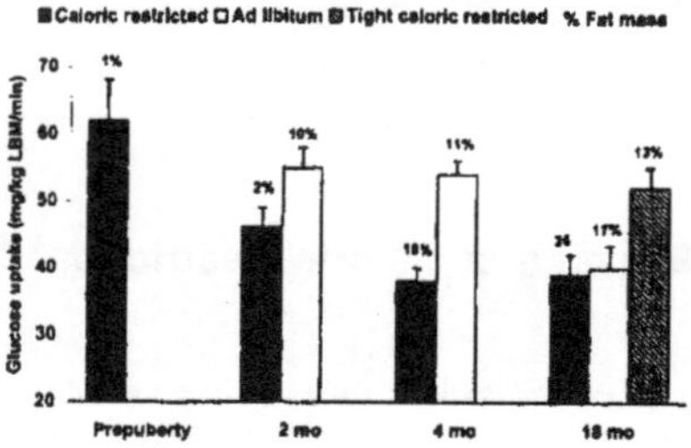

FIGURE 1 The effects of aging on insulin-mediated glucose uptake in ad libitum–fed and chronically calorie-restricted rats. Glucose uptake was markedly decreased after puberty and did not decrease with aging after late adulthood, when expressed in terms of lean body mass. CR restored insulin action when FM was decreased to below 13–14% of body weight. When tight CR was applied and FM decreased to 13%, glucose uptake increased dramatically. Values represent the mean ± SE.

a decrease in insulin responsiveness in BMI-matched women with upper versus lower body obesity (5,6). While improved insulin action is typical for CR and may be involved in mechanisms affecting longevity, impaired insulin action in humans is associated with morbidity and mortality. As in humans, rodents display an increase in basal plasma glucose and insulin levels with aging (7). The advantage in studying animal models is the ability to intervene by diet, drugs or surgery in a relatively homogeneous obese group and to relate the physiologic and molecular aspects of these changes.

We examined the relationship between FM, peripheral insulin and age in rats (8). We studied body composition and rates of (LBM) glucose uptake in a group of 2-, 4-, 6-, 12- and 18-mo-old rats. The results showed a significant gradual increase in FM from 2 to 18 mo of age (316% increase), with a high correlation between body weight, FM and epididimal fat. The rate of insulin-stimulated glucose uptake decreased significantly (~14%) from 2 to 4 mo of age. Surprisingly, we found that there was no further change in insulin responsiveness in rats that were much older. Analyzing the data using a best fitted model of two regression lines joined in a knot (spline curve model) demonstrated that when body weight was ~350 g or ~14% fat, no change in glucose uptake was noted with further increased body weight. This analysis suggests that the effect of body weight was saturable for the decrease in glucose uptake. Using the same statistical methodology to analyze the relationship between glucose uptake and FM and epididimal fat showed similar results (8). Because this saturable effect occurred at a relatively young age, it indicates that aging per se is not associated with further decrease in peripheral insulin sensitivity. We next set to prove this concept by examining whether a cause-effect relationship exists between FM and age-related decrease in peripheral insulin responsiveness (plasma insulin levels of ~100 pmol/L). We used the same age-group animals as the above ad libitum fed and compared them with chronically CR rats (Fig. 1). When CR young and older adult rats (2–4 mo) were studied, their insulin responsiveness significantly increased. At this time, their FM decreased to below 13% of their body weight. Interestingly, insulin responsiveness in older calorie-restricted animals did not improve; however, their FM percentage was higher (17%). We recently studied an additional group of more CR [50% food intake (18 mo; $n = 5$)] to decrease their FM to ~13%. This resulted in a dramatic improvement in insulin responsiveness that has not been noted when CR resulted in a decrease from 27 to 17% in FM (Fig. 1). Saturating effects of FM on peripheral insulin action had been previously demonstrated in Pima Indians (9), suggesting that once a certain critical level of fat accretion is achieved, further alterations in glucose metabolism are difficult to detect. In addition, this suggests that chronic benefits of weight reduction may be obtained not by losing any weight but by losing weight below a certain amount of fat. Furthermore, this suggests that FM accumulation, rather than aging, is responsible for the age-dependent insulin resistance.

CR *affects hepatic insulin sensitivity*

It was previously demonstrated that fasting plasma insulin levels are increased in both humans and rodents with aging. This is due to impaired suppression of hepatic glucose production (HGP) requiring significant portal hyperinsulinemia (10). However, suppression of HGP by insulin was significantly impaired in younger subjects with increased abdominal obesity (11). To better examine this question, we used chronic CR to maintain FM and VF at youthful levels and to abolish the overwhelming negative effect of fat on hepatic insulin action in 8- and 20-mo-old Brown Norway Fisher rats ($n = 344$). Using the pancreatic clamp technique, we measured their basal HGP and by inducing mild physiological hyperinsulinemia, we measured the ability of insulin to suppress HGP (Fig. 2). The results demonstrated that basal HGP was similar in all groups. However, insulin-induced suppression of HGP was significantly increased in the calorie-restricted (8- and 20-mo-old) rats compared with their ad libitum–fed controls (46 and 39%, respectively). In parallel, the percentage of FM and VF was significantly and proportionally lower in the CR groups. Importantly, rats were chronically calorie-restricted from young adulthood to maintain body composition at youthful levels. This study demonstrates that insulin action on modulating HGP is not impaired by aging per se but by age-related increases in FM and VF. This experimental manipulation provided evidence to support the hypothesis that the capacity of insulin to regulate hepatic glucose flux is not impaired with aging (12).

Fat distribution with aging and its metabolic effects

Increased VF is a common and typical change in body composition with aging (13,14). The increase in VF can be

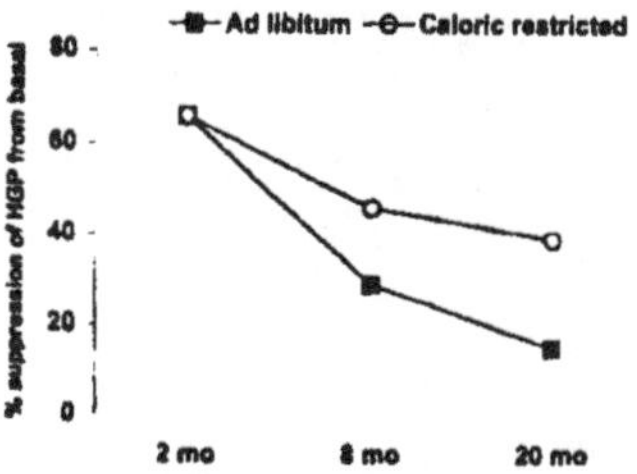

FIGURE 2 Suppression of HGP from basal in ad libitum–fed and calorie-restricted rats. Physiological hyperinsulinemia induced a more prominent suppression of HGP (with age) in calorie-restricted rats compared with ad libitum–fed rats; however, maximal suppression could not be achieved.

detected surgically (increased omental fat) or by CT scan or magnetic resonance imaging. However, the most common epidemiological tools are the observation of a pattern of upper body obesity, increased waist circumference and increased waist to hip ratio. Most important, waist to hip ratio in older men and women is increased independently of BMI (13). CT scan studies in men and women of all ages revealed a significant inverse correlation between the ratio of subcutaneous to VF tissue with age (14) and an increased accumulation of fat between the muscles of the abdominal wall in older compared with younger subjects (15). Thus, older men and women show an increase in VF, although the BMI may be normal and LBM may be decreased (16). VF has multi-factorial associations with aging and may determine longevity in humans by influencing atherosclerotic mechanisms (17). Numerous studies in subjects of all ages have demonstrated that increased VF is associated with increases in fasting and postprandial plasma insulin levels and impaired glucose tolerance (18,19).

This knowledge provided us with the opportunity to study specifically one small fraction of fat to prove the effects that fat has on the paradigm of ad libitum–fed and CR rats. Thus, removal of VF may provide some cause-effect relationship in the understanding of their metabolic impact. To directly examine whether VF modulates hepatic insulin action, we randomized moderately obese Sprague-Dawley rats either to surgical removal of the epididymal and perinephric fat pads or to sham operation and then studied them 3 wk later (20). Total VF was ~fourfold increased in the sham-operated group. However, whole body FM was not significantly different. While plasma glucose, FFA, glycerol and glucagon were similar, plasma insulin levels were decreased by one half in rats with surgically removed VF. The rate of insulin infusion needed to maintain plasma glucose levels at baseline during the hepatic-pancreatic clamp was dramatically decreased (~twofold) after VF removal. These results demonstrate that the hepatic responses to either endogenous (basal) or exogenous (pancreatic clamp) insulin were markedly enhanced by removing VF. The heightened hepatic insulin sensitivity was supported by the decrease in the gene expression of both glucose-6-phosphatase and phosphoenolpyruvate carboxykinase by physiological hyperinsulinemia only in the rats without VF. This improvement in hepatic insulin sensitivity was also supported by ~70% decrease in the plasma levels of insulin-like growth factor binding protein-1, a marker of insulin's transcription regulation in the liver. The marked increase in hepatic insulin action after surgical removal of selective intra-abdominal fat depots indicates a cause-effect relationship between them and explains the basal hyperinsulinemia in aging. Subsequently, we generated similar decreases in VF using leptin or β-adrenoreceptor agonists and demonstrated that hepatic insulin action was equally improved in all rats studied. While we defined specific effects to specific fat tissue (VF), these studies have also supported the general concept that fat has an overall metabolic effect (20).

Fat-derived peptides, their metabolic actions and risk factors for diseases

In the last several years, various peptides have been identified and shown to have the following characteristics: they are produced either primarily or also by fat tissue, secreted and measured in the plasma and have a distinct action on other tissues or organs. Additionally, manipulation at the level of FM (pharmacologically, CR or surgical removal of fat) induces important changes in these peptides action. The fat-derived protein TNF-α has been suggested to be directly involved in the development of insulin resistance in obesity by its effects on insulin signaling (21,22). We studied TNF-α gene expression in a variety of fat depots after surgical extraction of VF and compared the results with gene expression in sham-operated rats with intact VF (20). Most impressively, the gene expression of TNF-α in the subcutaneous fat (the largest fat depot) was decreased by ~fourfold to fivefold in rats without VF. Marked (45%) decreases in TNF-α gene expression have been demonstrated in fat obtained from obese humans who lost weight (23) and after dietary manipulations in obese mice. Leptin, the most investigated fat-derived peptide, also has tremendous implications in glucose metabolism (24); during physiological hyperinsulinemia (insulin clamp), leptin markedly enhanced the action of insulin on inhibiting HGP (~fourfold lower in leptin than in the control groups). This may have been through its effect on decreasing VF. However, leptin had an additional effect in improving peripheral insulin action; the rate of glucose uptake in the leptin group was 52 and 33% higher than in saline-administered and pair-fed groups, respectively. Interestingly, leptin may be responsible for many of the neuroendocrine adaptations to starvation and CR; during weight loss or fasting, both humans and animals (25,26) exhibit low levels of leptin gene expression and plasma levels. These conditions are associated with a marked neuroendocrine adaptation, including increased steroid stress hormones, decreased reproductive function, decreased thyroid function and delayed puberty (27). Angiotensinogen (AT), another peptide [expressed in liver and adipose tissue (28)], may also play an important role in the relationship between obesity, hypertension and hyperinsulinemia (the main components of syndrome X). The hormonal control of AT synthesis and secretion from adipocytes has been shown to be upregulated with glucocorticoids and TNF-α. However, in vitro studies have shown that insulin exerts an inhibitory effect on AT production (29). To examine the role of increased FM on AT, we followed AT expression after 3 h of hyperglycemia while maintaining plasma insulin concentration at basal (using somatostatin infusion) in lean and obese rats. The relative increase in AT expression was ~threefold in both groups (Fig. 3); however, the obese rats had ~fourfold increased FM compared with the lean group. Additionally, the obese rats also demonstrated resistance to the inhibitory effect of insulin on AT expression when examined in an hyperinsulinemic clamp.

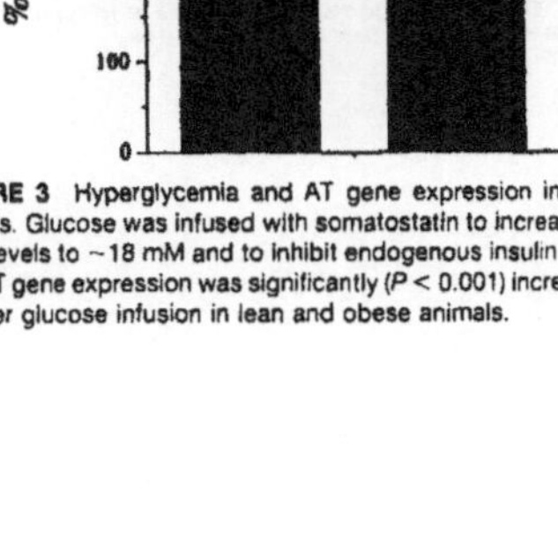

FIGURE 3 Hyperglycemia and AT gene expression in lean and obese rats. Glucose was infused with somatostatin to increase plasma glucose levels to ~18 mM and to inhibit endogenous insulin secretion, for 3 h. AT gene expression was significantly ($P < 0.001$) increased from basal after glucose infusion in lean and obese animals.

These findings demonstrate a potential pathological link between fat and hypertension. Plasminogen activator inhibitor-1 is also secreted from fat tissue (30), is increased with obesity, is associated with decreased fibrinolytic activity in the blood and is a risk factor for acute coronary events in humans. In fact, moderate CR in elderly obese subjects decreased plasminogen activator inhibitor-1 levels dramatically (31). In addition to cytokines, complement factors (D, C3, B) are also secreted by fat tissue and have a role in controlling immune responses (32,33). Newly developed differentiating techniques have suggested the existence of numerous other proteins with mRNA encoded in adipocytes and specific proteins formed and secreted from the adipocytes that achieve physiologically adequate plasma levels. For example, the protein designated Acrp30 is a novel serum protein similar to c1q, is produced exclusively in adipocytes, has plasma levels in ng/mL range, binds preferentially to muscle, has a cytokine tertiary structure and is regulated both acutely and chronically by insulin (34). Modulation of risk factors for diseases, which are nutritionally regulated in fat tissue, may play a role in the longevity of the fat-depleted CR animals.

We hypothesize that because evolution was typically associated with hunger and not with ad libitum feeding, the saturation of systems is easily achieved in human obesity and in ad libitum–fed animals. Fat depletion (by means of CR, pharmacologically or direct removal) has metabolic consequences on peripheral and hepatic insulin action, i.e., induces a reduction in peripheral and hepatic insulin resistance, an effect that seems to be independent of aging. In addition, although the accumulation of VF represents a risk for a variety of metabolic alterations (and manipulations at the level of VF may provide beneficial effects), it further proves the important role of fat in whole body homeostasis. In obesity, excess accumulation of FM may account not only for the perturbations in glucose metabolism, but also for a variety of additional risk factors for atherosclerotic cardiovascular and other unrelated diseases.

LITERATURE CITED

1. Kimura, K., Tissenbaum, H., Liu, Y. & Ruvkun, G. (1997) daf-2, an insulin receptor-like gene that regulates longevity and diapause in *Caenorhabditis elegans*. Science 277: 942–946.

2. Boden, G., Chen, X., DeSantis, R. A. & Kendrick, Z. (1993) Effects of age and body fat on insulin resistance in healthy men. Diabetes Care 16: 728–733.

3. O'Shaughnessy, I. M., Kasdort, G. M., Hottman, R. G. & Kalkhott, R. R. (1992) Does aging intensity the insulin resistance of human obesity. J. Clin. Endocrinol. Metab. 74: 1075–1081.

4. Ferrannini, E., Vichi, S., Beck-Nielsen, H., Laakso, M., Paolisso, G. & Smith, U. (1996) European Group for the Study of Insulin Resistance (EGIR). Insulin action and age. Diabetes 45: 947–953.

5. Colberg, S. R., Simoneau, J. A., Thaete, F. L. & Kelley, D. E. (1995) Skeletal muscle utilization of free fatty acids in women with with visceral obesity. J. Clin. Invest. 95: 1427–1428.

6. Peiris, A. N., Struve, M. F., Mueller, R. A., Lee, M. B. & Kissebah, A. H. (1988) Glucose metabolism in obesity: influence of body fat distribution. J. Clin. Endocrinol. Metab. 67: 760–767.

7. Barzilai, N. & Rossetti, L. (1995) The relationship between changes in body composition and insulin responsiveness in models of the aging rat. Am. J. Physiol. 269: E591–E597.

8. Barzilai, N., Banerjee, S., Hawkins, M., Chang, C. J., Chen, W. & Rossetti, L. (1998) The Effect of age-dependent increase in fat mass on peripheral insulin action is saturable. J. Gerontol. 53A: B141–B146.

9. Bogardus, C., Lillia, S., Mott, D. M., Hollenbeck, C. & Reaven, G. (1985) Relationship between degree of obesity and in vivo insulin action in man. Am. J. Physiol. 248: E286–E291.

10. Fink, R. I., Kolterman, O. G., Griffin, J. & Olefsky, J. M. (1983) Mechanisms of insulin resistance in aging. J. Clin. Invest. 71: 1523–1535.

11. Jackson, R. A., Hawa, M. I., Roshania, R. D., Sim, B. M., DiSilvio, L. & Jaspan, J. B. (1988) Influence of aging on hepatic and peripheral glucose metabolism in humans. Diabetes 37: 119–129.

12. Gupta, G., She, L., Ma, X. M., Cases, A. C., Yang, X. M., Hu, M., Wu, J., Rossetti, L., Barzilai, N. (2000) Ability of insulin to modulate hepatic glucose production in aging rats is impaired by fat accumulation. Am. J. Physiol. Endocrinol. Metab. 278: E985–E991.

13. Shimokata, H., Tobin, J. D., Muller, D. C., Elahi, D., Coon, P. J. & Andres, R. (1989) Studies in the distribution of body fat: effects of age sex, and obesity. J. Gerontol. 44: M66–M73.

14. Enzi, G., Gasparo, M., Binodetti, P.R., Fiore, D., Semisa, M. & Zurlo, F. (1986) Subcutaneous and visceral fat distribution according to sex, age, and overweight, evaluated by computed tomography. Am. J. Clin. Nutr. 44: 739–746.

15. Borkan, G. A., Hults, D. E., Gerzof, S. G., Robbins, A. H. & Silbert, C. K. (1983) Age changes in body composition revealed by computed tomography. J. Gerontol. 38: 673–677.

16. Norris, A. H., Lundy, T. & Shock, N. W. (1963) Trends in indices of body composition in men between ages 30–70 years. Ann. N. Y. Acad. Sci. 110: 623–639.

17. Larson, B. (1992) Regional obesity as a health hazard in men-prospective studies. Acta Med. Scand. Suppl. 723: 45–51.

18. Kissebah, A. H. (1991) Insulin resistance in visceral obesity [Review]. Int. J. Obes. 15(suppl. 2): 109–115.

19. Bjorntorp, P. (1991) Metabolic implications of body fat distribution [Review]. Diabetes Care 14: 1132–1143.

20. Barzilai, N., She, L., Liu, B. Q., Vuguin, P., Cohen, P., Wang, J. & Rossetti, L. (1999) Surgical removal of visceral fat reverses hepatic insulin resistance. Diabetes 48: 94–98.

21. Hotamisligil, G. S., Shargill, N. S. & Spiegelman, B. M. (1993) Adipose expression of tumor necrosis factor-alpha: direct role in obesity-linked insulin resistance. Science 259: 87–91.

22. Hotamisligil, G. S., Arner, P., Caro, J. F., Atkinson, R. L. & Spiegelman, B. M. (1995) Increased adipose tissue expression of tumor necrosis factor-alpha in human obesity and insulin resistance. J. Clin. Invest. 95: 2409–2415.

23. Hotamisligil, G. S., Peraldi, P., Budavari, A., Ellis, R., White, M. F. & Spiegelman, B. M. (1996) IRS-1-mediated inhibition of insulin receptor tyrosine kinase activity in TNF-alpha- and obesity-induced insulin resistance. Science 271: 665–668.

24. Flier, J. S. Leptin expression and action: new experimental paradigms. (1997) Proc. Natl. Acad. Sci. USA 94: 4242–4245.

25. Boden, G., Chen, X., Mozzoli, M. & Ryan, I. (1996) Effect of fasting on serum leptin in normal human subjects. J. Clin. Endocrinol. Metab. 81: 3419–3423.

26. Ahren, B., Mansson, S., Gingerich, R. & Havel, P. (1997) Regulation of plasma leptin in mice: influence of age, high-fat diet, and fasting. Am. J. Physiol. 273: R113–R120.

27. Masoro, E. (1995) Glucocorticoids and aging. Aging 7: 407–413.

28. Engeli, S., Gorzelniak, K., Kreutz, R., Runkel, N., Distler, A. & Sharma, A. M. (1999) Co-expression of renin-angiotensin system genes in human adipose tissue. J. Hypertens. 17: 555–560.

29. Aubert, J., Safnova, I., Negrel, R. & Ailhaud, G. (1998) Insulin down-regulates angiotensinogen gene expression and angiotensinogen secretion in cultured adipose cells. Biochem. Biophys. Res. Commun. 250: 77–82.

30. Alessi, M. C., Peiretti, F., Morange, P., Henry, M., Nalbone, G., & Juhan-Vague, I. (1997) Production of plasminogen activator inhibitor 1 by human adipose tissue: possible link between visceral fat accumulation and vascular disease. Diabetes 46: 860–867.

31. Calles-Escandon, J., Ballor, D., Harvey-Berino, J., Ades, P., Tracy, R. & Sobel, B. (1996) Amelioration of the inhibition of fibrinolysis in elderly, obese subjects by moderate energy intake restriction. Am. J. Clin. Nutr. 64: 7–11.

32. White, R. T., Damm, D., Hancock, N., Rosen, B. S., Lowell, B. B., Usher, P., Flier, J. S. & Spiegelman, B. M. (1992) Human adipsin is identical to complement factor D and is expressed at high levels in adipose tissue. J. Biol. Chem. 267: 9210–9213.

33. Roubenoff, R. (1997) Inflammatory and hormonal mediators of cachexia. J. Nutr. 127: 1014S–1016S.

34. Scherer, P. E., Williams, S., Fogliano, M., Baldini, G. & Lodish, H. F. (1995) A novel serum protein similar to C1q, produced exclusively in adipocytes. J. Biol. Chem. 270: 26746–26749.

Perspectives

Anecdotal, Historical and Critical Commentaries on Genetics

Edited by James F. Crow and William F. Dove

Fisher, Medawar, Hamilton and the Evolution of Aging

Brian Charlesworth

Institute for Cell, Animal and Population Biology, University of Edinburgh, Edinburgh EH9 3JT, United Kingdom

THE idea that senescent decline in the performance of biological systems must have an evolutionary basis traces back almost to the beginnings of evolutionary biology (ROSE 1991, Chap. 1), although Darwin does not seem to have discussed the problem. At first sight, the nearly universal existence of senescence in species of multicellular organisms is paradoxical, given that natural selection supposedly causes the evolution of increased, not decreased, fitness. As discussed by COMFORT (1979), many biologists have, therefore, taken the view that senescence reflects an inevitable process of damage accumulation with age, and indeed an analog of senescence can be seen in complex machines such as cars (GAVRILOV and GAVRILOVA 1991). But unicellular organisms, such as bacteria, which propagate simply by binary fission, and the germ lines of multicellular organisms, have been able to propagate themselves without senescence over billions of years, showing that biological systems are capable of ongoing repair and maintenance and so can avoid senescence at the cellular level. Senescence cannot, therefore, just be an unavoidable cumulative result of damage. The large amount of variation among different species in their rates of senescence also clearly indicates that aging is subject to variation and selection (COMFORT 1979; FINCH 1990; ROSE 1991; WACHTER and FINCH 1997). This conclusion is backed up by the existence of both quantitative genetic variation and major gene mutations affecting the rate of aging (FINCH 1990; ROSE 1991; WACHTER and FINCH 1997).

Modern evolutionary theory has demonstrated that, in species with a clearcut distinction between parent and offspring, senescence is a virtually inevitable result of the fact that genes that affect survival or fecundity only early in life have a greater selective impact than genes whose effects are manifest only late in life. The purpose of this *Perspectives* article is to trace the history of this idea, with especial emphasis on William HAMILTON's (1966) classic paper and its influence on subsequent work. This was motivated by Hamilton's untimely death earlier this year and the fact that his work on the evolution of senescence has probably received less attention than his other seminal contributions to evolutionary theory.

Within modern evolutionary genetics, the first discussion of the evolution of senescence was that of FISHER (1930), in the context of his famous concepts of the "Malthusian parameter" and "reproductive value." Fisher considered a sexually reproducing, age-structured population reproducing in continuous time, with a probability of survival to age x of $l(x)$, and a rate of production $m(x)$ of offspring of the same sex as their parent at age x. He pointed out that such a population reaches an asymptotic exponential rate of population increase, r, which is given by the (single) real root of the equation

$$\int_0^\infty e^{-rx} l(x)\, m(x)\, dx = 1. \qquad (1)$$

Fisher called r the Malthusian parameter of the population, in reference to Malthus's preoccupation with the supposedly inevitable exponential increase in numbers of the human species. Although not mentioned by Fisher, this result traces back to the work of EULER (1760) and SHARPE and LOTKA (1911). r is now commonly referred to as the "intrinsic rate of increase" of the populations, and Equation 1 is often called the Euler-Lotka equation.

Fisher drew attention to the quantity defined by the relation

$$v(x) = \frac{e^{rx}}{l(x)} \int_x^\infty e^{-ry} l(y)\, m(y)\, dy. \qquad (2)$$

This is the reproductive value of individuals of age x and measures their contribution to the future ancestry of a population growing at rate r, normalized to a value of unity at the time of conception. FISHER (1930) stated (p. 27), "The direct action of Natural Selection must be proportional to this contribution." He showed a curve of

Author e-mail: brian.charlesworth@ed.ac.uk

Genetics 156: 927–931 (November 2000)

reproductive value for Australian women, which rises with advancing age during childhood, reaches a maximum around 19, and then declines steadily, reaching zero at close to 50, when most women have reached menopause. He commented (p. 29),

> It is probably not without significance in this connexion that the death rate in Man takes a course generally inverse to the curve of the reproductive value. The minimum of the death rate curve is at twelve, certainly not far from the primitive maximum of the reproductive value; it rises more steeply for infants, and less steeply for the elderly than the curve of reproductive value falls, points which qualitatively we should anticipate, if the incidence of natural death had been to a large extent moulded by the effects of differential survival.

These ideas greatly influenced MEDAWAR (1946, 1952) when he was formulating the first explicit model of the evolution of aging. Medawar took it for granted that Fisher's reproductive value measures the relative effectiveness of selection at age x. Given this premise, a hypothetical mutant gene that increases survival over a small time interval at an age when reproductive value is high would thus have a higher net effect on fitness than a gene acting at an age when reproductive value is low. Since reproductive value declines over much of adult life, this leads to the expectation that selection will be more effective in improving performance early in adult life than late in life. He pointed out that this means that deleterious alleles with effects restricted to late stages of life would equilibrate at higher frequencies at mutation-selection balance than alleles that act earlier, the process now referred to as the "mutation-accumulation" theory of aging (ROSE 1991, Chap. 4). Late-acting deleterious mutations are, of course, familiar to medical geneticists. HALDANE (1941, pp. 192–194) had previously suggested that there would be selection for modifiers that postpone the age of onset of the effects of such mutations, and Medawar laid considerable stress on this possibility. However, the relevant selection pressure is of the order of the mutation rate to the trait in question, and so postponement is unlikely to be a major factor in the evolution of senescence (CHARLESWORTH 1994, p. 200).

Medawar also pointed out that alleles with positive effects on performance early in life, but with negative effects because of physiological trade-offs later on, are more likely to be established by selection than alleles with the opposite pattern. This idea was more fully developed by WILLIAMS (1957) and is now known as the "antagonistic pleiotropy" theory of aging (ROSE 1991, Chap. 4). Both of these mechanisms could cause an initially nonsenescent life history, in which mortality rates are independent of age, to evolve gradually to a state in which death rates increase with age, the commonly used demographic criterion for senescence (COMFORT 1979; FINCH 1990). The relative importance of mutation accumulation *vs.* antagonistic pleiotropy in the evolution of aging is still an unsettled issue (ROSE 1991; WACHTER and FINCH 1997).

HAMILTON (1966) noted that it is fallacious to use reproductive value as a measure of the effectiveness of selection as a function of age, and that a different measure should be used. He started with the Fisherian assumption that r is an appropriate measure of net fitness, so that a mutant gene whose effect on survival or reproductive succcess increases the value of r over that for the current population will be favored by selection. He noted that it is possible to make explicit calculations of the effects on r of small changes in survival or fecundity at a given age, by the method of implicit partial differentiation of Equation 1. In the context of Fisher's continuous-time model, if the integral of the fecundity rate between ages x and $x + \delta x$ is changed by a small amount $\delta m(x)$, the associated change in r as δx approaches zero is given by

$$\delta r = \frac{\delta m(x) e^{-rx} l(x)}{T}, \qquad (3)$$

where T is a measure of the generation time of the population, given by

$$T = \int_0^\infty x e^{-rx} l(x)\, m(x)\, dx. \qquad (4)$$

A similar treatment can be applied to the age-specific mortality rate, defined as

$$\mu(x) = -\frac{d \ln l(x)}{dx}. \qquad (5)$$

The change in r associated with a small change, $\delta\mu(x)$, in the integral of the mortality rate between ages x and $x + \delta x$ is given by

$$\delta r = -\frac{\delta\mu(x) \int_x^\infty e^{-ry} l(y)\, m(y)\, dy}{T}. \qquad (6)$$

Neither of these formulae corresponds to reproductive value, as given by Equation 2, and they have rather different implications for the relation between the age of effect of a gene and its impact on fitness. If the population is stationary in size or growing, as must be the case in the long term if it is not doomed to extinction, Equation 3 implies that, all else being equal, there is always a greater selective premium on early rather than late reproduction, since $l(x)$ declines with age. This is not predicted from the reproductive value curve, which increases during infancy, and it reflects the fact that a gene whose effect on fecundity occurs late in life may be removed by death of its carrier before this effect is expressed.

Similarly, Equation 6 implies that selection is indifferent to the timing of gene effects on age-specific mortality during infancy and that its intensity always decreases with age during adulthood. Again, this is quite different from the pattern predicted by reproductive value; the difference arises from the fact that reproductive value is conditioned on an individual having survived to age x and discounts the amount of population growth that occurs over a time period x, whereas Equation 6 measures the expected fitness effect of a change in mortality at age x for individuals censused at conception. HAMILTON (1966) pointed out that these differences are nontrivial. For example, if fecundity increases exponentially with age during adulthood, reproductive value also increases exponentially, so that its use would lead to the conclusion that selection opposes senescence. In contrast, Equation 6 implies that there is always a selective premium on early survival, given the monotonic decrease in the magnitude of its right-hand side with age, although the rate of decline of the intensity of selection with age is greatly slowed if fecundity increases with age.

On the basis of these results, Hamilton proposed that the more rapid incorporation of favorable mutations with early effects on survival or fecundity than mutations with effects later in life would cause an initially nonsenescent life history to evolve in the direction of relatively high mortality rates and low fecundity late in life, without having to postulate any harmful mutations or tradeoff effects. This explanation for the evolution of senescence has not been widely accepted, since (as Hamilton himself noted) it does not seem capable of explaining the evidently pathological aspects of many aspects of aging (ROSE 1991, pp. 70–71). Instead, most applications of Hamilton's formulae to the evolution of senescence have applied the same basic results to the mutation-accumulation and antagonistic pleiotropy theories (CHARLESWORTH 1994, Chap. 5).

Hamilton also pointed out that the oversimplified model of changes to mortality or fecundity at just one age can easily be extended, by calculating the net change in r owing to small changes in vital statistics at a whole range of ages. Functional relations among fecundity and mortality rates, reflecting resource allocation or physiological constraints, can also be included in such calculations, although he himself did not do this. The inclusion of such constraints has led to the development of elaborate models of life-history evolution, which attempt to predict optimal patterns of age-specific reproduction, growth, and survival and to relate comparative data on life histories to the predictions of these models (*e.g.*, STEARNS 1992; CHARLESWORTH 1994, Chap. 5; MCNAMARA and HOUSTON 1996). HAMILTON'S (1966) method of calculating the sensitivity of r to age-specific changes in vital statistics, later applied to the equivalent matrix model of discrete-time populations (DEMETRIUS 1969; GOODMAN 1971; CASWELL 1989), is at the core of this enterprise. Somewhat ironically, reproductive value reappears in optimization models as a weighting function for the effect of a change in fecundity at a given age on mortality at that age (SCHAFFER 1974), and optimal life histories can be viewed as maximizing reproductive value at each age (SCHAFFER 1974; CHARLESWORTH 1994, pp. 237–238).

Hamilton's analysis left one important gap, however. This concerns the validity of assuming that the Malthusian parameter r is indeed the correct measure of fitness for an age-structured population, in the sense that it accurately predicts the effect of selection on gene frequencies. No justification of this was provided by Fisher, who seems simply to have taken it for granted, as did most people who pioneered the theory of life-history evolution (*e.g.*, LEWONTIN 1965) and many distinguished theoretical population geneticists such as Kimura (*e.g.*, KIMURA 1958), for whom continuous-time models offered technical convenience.

It is easy to define r for a particular genotype, as the solution to Equation 1 for a (hypothetical) population consisting entirely of individuals with the set of $l(x)$ and $m(x)$ values characteristic of the genotype in question; this is presumably what Fisher had in mind as the Malthusian parameter of a given genotype. It is also easy to see that, with competition among clonally reproducing genotypes, the genotype with the highest r will outcompete the rest, since this situation is simply equivalent to a set of populations growing at different rates. It is less easy to see how to model a sexually reproducing diploid population in which each parent produces a mixture of genotypes, especially as changes in genotype frequencies induced by selection must cause continual changes in age structure (MORAN 1962, p. 90; CHARLESWORTH 1970).

There is, thus, no obvious guarantee that the use of r as a fitness measure gives correct results. In fact, although Fisher characteristically made no reference to their work, HALDANE (1927) and NORTON (1928) had made great progress toward solving this problem. HALDANE (1927) derived integral equations to represent the effects of selection on genotype frequencies in an age-structured population. He analyzed them by making the simplifying assumption that selection was weak and population growth was slow. Using some rather cumbersome algebra, he was able to obtain an approximate expression for the rate of change of gene frequency, in terms of the differences in lifetime expectations of offspring among genotypes. He returned to this problem at the end of his life (HALDANE 1962).

NORTON'S (1928) paper is one of the most profound papers in both demography and population genetics. Harry Norton was a mathematician at Trinity College, Cambridge (UK), and a member of the Bloomsbury group of British intellectuals. *Eminent Victorians* was dedicated to him by Lytton Strachey. He is mentioned in Strachey's biography as the only person in the group who could hold his own with Bertrand Russell and John Maynard Keynes (HOLROYD 1971). He had earlier anticipated HALDANE'S (1924) paper on the rate of change of gene frequency under selection, in a set of calculations published as an appendix to PUNNETT'S (1915) book on mimicry. Using integral equations similar to Haldane's, which Norton had derived independently in 1910 (HALDANE 1927), he examined the asymptotic properties of a diploid, randomly mating population segregating for a single locus with two alleles. Under some simplifying assumptions, notably random mating with respect to age and genotype and no sex differences in vital statistics, the genotype with the highest r value will supplant the others if there is directional selection. With heterozygote advantage in r, a polymorphism is maintained; with heterozygote disadvantage, polymorphism is eliminated. He also showed that, with heterozygote advantage, the population ultimately approaches the neighborhood of a fixed gene frequency.

Later work, reviewed by CHARLESWORTH (1994, Chaps. 3 and 4), has extended these pioneering analyses of age-structured populations. Sex differences in vital statistics, nonrandom mating, density-dependent modification of mortality and fecundity rates, and the effects of spatially and temporally fluctuating environments have all been included in the models. In the simplest case described above, when these complications can be neglected, we now know that the initial rate of increase of a nonrecessive rare mutant allele introduced into a large population is governed by its effect on the intrinsic rate of increase, even if its effect is arbitrarily large. In general, however, differences among genotypes in intrinsic rates of increase can be used only as approximate predictors of the rate of change of allele frequency under selection, although the approximation is very good if selection is weak.

Equilibrium frequencies of genotypes under the standard scenarios of population genetic models can, however, be calculated from equations that are of exactly the same form as those of the familiar discrete-generation models of deterministic population genetics (CROW and KIMURA 1970), such that the Wrightian fitness weight for a genotype, w_i, is replaced by the expression

$$w_i = \int_0^\infty e^{-rx} l_i(x)\, m_i(x)\, dx, \qquad (7)$$

where r is the growth rate of the population as a whole, and $l_i(x)$ and $m_i(x)$ are the vital statistics for individuals of genotype i. If selection is weak, differences in the w_i among genotypes are approximately proportional to differences in the corresponding genotypic intrinsic rates, where the constant of proportionality is equal to the value of T for some standard genotype (CHARLESWORTH 1994, p. 178).

It is interesting to note that the equilibrium gene frequency predicted from Equation 7 in the case of heterozygote advantage at a single locus with two alleles corresponds to the value that NORTON (1928) showed is approached asymptotically. An exact analysis of KIMURA's (1958) use of Malthusian parameters also enables Equation 7 to be recovered (CHARLESWORTH 1970). Bill Hamilton would probably have regarded these results as minor technicalities, but it is satisfying that the assumptions that underly the use of r in explaining the evolution of senescence and life-history patterns can be made explicit and analyzed in population genetic terms.

Our understanding of the evolution of senescence is, at one level, very complete; we know that senescence is an evolutionary response to the diminishing effectiveness of selection with age and that this explains many aspects of the comparative biology of senescence (WILLIAMS 1957; ROSE 1991; CHARLESWORTH 1994; RICKLEFS 1998). On the other hand, it is at present hard to be sure which of the two most likely important mechanisms by which this property of selection influences senescence (accumulation of late-acting deleterious mutations or fixation of mutations with favorable early effects and deleterious late effects) plays the more important role, especially as these are not mutually exclusive possibilities. If senescence does not take its toll, perhaps a future *Perspectives* by this author will provide an update on this problem.

I thank Deborah Charlesworth for her comments on the manuscript.

LITERATURE CITED

CASWELL, H., 1989 *Matrix Population Models*. Sinauer, Sunderland, MA.

CHARLESWORTH, B., 1970 Selection in populations with overlapping generations. I. The use of Malthusian parameters in population genetics. Theor. Popul. Biol. **1**: 352–370.

CHARLESWORTH, B., 1994 *Evolution in Age-structured Populations*. Cambridge University Press, Cambridge, UK.

COMFORT, A., 1979 *The Biology of Senescence*. Churchill Livingstone, Edinburgh, UK.

CROW, J. F., and M. KIMURA, 1970 *An Introduction to Population Genetics Theory*. Harper & Row, New York.

DEMETRIUS, L., 1969 The sensitivity of population growth rate to perturbations in the life-cycle components. Math. Biosci. **4**: 129–136.

EULER, L., 1760 Recherches générales sur la mortalité et la multiplication. Mém. Acad. R. Sci. Bell. Lett. **16**: 144–164.

FINCH, C. E., 1990 *Longevity, Senescence, and the Genome*. University of Chicago Press, Chicago, IL.

FISHER, R. A., 1930 *The Genetical Theory of Natural Selection*. Oxford University Press, Oxford, UK.

GAVRILOV, L. A., and N. S. GAVRILOVA, 1991 *The Biology of Life-Span: A Quantitative Approach*. Harwood, Chur, Switzerland.

GOODMAN, L. A., 1971 On the sensitivity of the population growth rate to changes in the age-specific birth and death rate. Theor. Popul. Biol. **2**: 339–354.

HALDANE, J. B. S., 1924 A mathematical theory of natural and artificial selection, I. Trans. Camb. Philos. Soc. **23**: 19–41.

HALDANE, J. B. S., 1927 A mathematical theory of natural and artificial selection, IV. Proc. Camb. Philos. Soc. **23**: 607–615.

HALDANE, J. B. S., 1941 *New Paths in Genetics*. Allen & Unwin, London.

HALDANE, J. B. S., 1962 Natural selection in a population with annual breeding but overlapping generations. J. Genet. **58**: 122–124.

HAMILTON, W. D., 1966 The moulding of senescence by natural selection. J. Theor. Biol. **12**: 12–45.

HOLROYD, M. L., 1971 *Lytton Strachey: A Biography*. Holt, Rinehart, Winston, New York.

KIMURA, M., 1958 On the change of population mean fitness by natural selection. Heredity **12**: 145–167.

LEWONTIN, R. C., 1965 Selection for colonizing ability, pp. 77–94 in *The Genetics of Colonizing Species*, edited by H. G. BAKER and G. L. STEBBINS. Academic Press, New York.

McNamara, J. M., and A. D. Houston, 1996 State-dependent life histories. Nature **380:** 215–221.

Medawar, P. B., 1946 Old age and natural death. Mod. **1:** 30–56.

Medawar, P. B., 1952 *An Unsolved Problem of Biology*. H. K. Lewis, London.

Moran, P. A. P., 1962 *The Statistical Processes of Evolutionary Theory*. Oxford University Press, Oxford.

Norton, H. T. J., 1928 Natural selection and Mendelian variation. Proc. Lond. Math. Soc. **28:** 1–45.

Punnett, R. C., 1915 *Mimicry in Butterflies*. Cambridge University Press, Cambridge, UK.

Ricklefs, R. E., 1998 Evolutionary theories of aging: confirmation of a fundamental prediction with implications for the genetic basis and evolution of lifespan. Am. Nat. **152:** 24–44.

Rose, M. R., 1991 *The Evolutionary Biology of Aging*. Oxford University Press, Oxford.

Schaffer, W. M., 1974 Selection for optimal life histories: the effect of age structure. Ecology **55:** 291–303.

Sharpe, F. R., and A. J. Lotka, 1911 A problem in age distribution. Philos. Mag. **21:** 435–438.

Stearns, S. C., 1992 *The Evolution of Life Histories*. Oxford University Press, Oxford.

Wachter, K. W., and C. E. Finch, 1997 *Between Zeus and the Salmon: The Biodemography of Longevity*. National Academy Press, Washington, DC.

Williams, G. C., 1957 Pleiotropy, natural selection and the evolution of senescence. Evolution **11:** 398–411.

Journal of Gerontology: BIOLOGICAL SCIENCES
2000, Vol. 55A, No. 3, B117–B123

Principles of Animal Use for Gerontological Research

Richard A. Miller[1] and Nancy L. Nadon[2]

[1]Department of Pathology and Geriatrics Center, University of Michigan School of Medicine, Institute of Gerontology, and Ann Arbor VA Medical Center, Michigan.
[2]National Institute on Aging, Bethesda, Maryland.

This essay presents some practical advice and suggestions for those who wish to use mice and rats in experiments on the biology of aging. Ten principles set forth guidance on choice of ages, choice of stocks, the importance of specific pathogen–free status, the uses of necropsy data, the dangers of pooling samples from different individuals, planning ahead for loss of aged mice to death and disease, the use of cost-adjusted power calculations, and the dangers of inferring causal associations from correlated age effects.

THIS article is intended to convey some rules-of-thumb for investigators just starting to think about the design of experiments on aging using mice and rats. The principles stated below reflect the opinions of the authors, based on years of experience in rodent-based work in gerontology and molecular biology. However, to avoid excessive circumlocution of the "in my opinion it may be helpful to" variety, the style is deliberately imperative, modeled on examples set by writers of style manuals (1) and columns of advice for the lovelorn. In addition the word "mice" will be used throughout to mean "mice and rats," except in those cases where mice are different from rats.

First Principle: Don't use mice that are too old.

Many beginners ask for the oldest available mice for their initial age-effect study under the presumption that the old ones will show bigger differences from young controls, and thus produce significant results quicker. This is rarely a good idea for several reasons:

(a) Old mice are usually sick (even if not quite dead yet). If a trait differs from young controls only in the last 10% of the cohort to drop off, then it's hard to be confident that the change is a result of aging rather than of the advanced disease or diseases most typical in the stock under study. After all, the aging process, which creates decrepit old mice from healthy, fit, young ones, takes many months to do this, and age-dependent changes in many cells, tissue, and organ systems can usually be demonstrated well before the median survival time for the species or stock. If your assay shows no change at 18, 22, 26, or 30 months of age (in a stock with a median survival of 24 months), then demonstrating a change in 34-month-old animals may well be due to sickness per se. Judicious selection of ages for initial exploratory work may depend on the specific characteristics of the stock to be used, and stocks with median survivals of 22 months or of 30 months may call for adjustments of ages selected for initial examination.

(b) Old mice are very expensive, particularly if you want them to be disease-free. The problem is that the real production cost of mice rises not linearly with chronologic age, but instead in proportion to the mortality rate, i.e., as an exponential function of age. If half your mice live to age 24 months, then producing a single 24-month-old mouse requires you to pay someone to house two mice for 24 months, one of which has just died. If only 10% of the mice survive to age 32 months, then the real cost of each 32-month-old mouse is the cost of raising 10 mice for anywhere from 18 to 32 months to get the one alive at 32 months. And then that one mouse, when you do the necropsy, may well turn out to have advanced neoplasia.

To illustrate the projected costs, at one well-known Midwestern Medical Center, animal users are charged $0.58/cage/day for cages of four mice. At this price it costs $106 to grow a mouse for 2 years; but because half the mice die, the cost of a live 2-year-old mouse is twice as high, or about $212. Because half of the 2-year-old mice are found to have advanced neoplasia even at a cursory necropsy, the cost of a more-or-less tumor-free 2-year-old mouse is another twofold higher, or $414. The nominal cost of a 32-month-old mouse (at $0.58/cage/day) is $140, but adjusting for attrition and disease gives a real cost closer to $1,400 each.

The National Institute on Aging (NIA) Office of Biological Resources provides highly subsidized animals, even with the recent price increase; for a 2-year-old C57BL/6 mouse, for example, the cost charged to investigators is a mere $72, well under the local production cost. The real subsidy (production cost minus cost to you) rises exponentially with age, however, and therefore it costs much more to raise a very old mouse in one's own facility than it does to buy them from NIA. Although NIA continues to raise these very old mice, despite their exceptionally high cost, in order to provide investigators maximal flexibility in designing their experimental protocols, investigators who do use this scarce resource are still confronted with the problems imposed by their very high rates of concurrent illness.

Second Principle: Don't use mice that are too young.

The beginner also begs for the youngest possible controls, often just weaned, again on the grounds that these are more likely to show big differences from old mice. The catch is that these mice are no more typical of "young adults" than, say, your typical 9-year-old human person.

Even 2–3-month-old mice, the biological equivalent of teenagers and college freshmen, i.e., technically postpubertal but hardly adult, are still in the throes of complex maturational changes we may wish to distinguish from the aging process. Although differences between 2-month-old and 5-month-old mice may be well worth studying, and in some cases may be highly relevant to aging (thymic involution, for example, is well advanced by 5 months of age), a conservative strategy might be to use mice aged 4–6 months as the "young" control group in experimental comparisons.

Third Principle: Don't use too few age groups.

It is certainly cheaper to do your initial survey experiment with only two age groups, e.g., 6 and 24 months, but there are major risks of missing interesting effects, and equal risks of reporting positive results whose significance would be routinely misinterpreted without data on intermediate ages. Figure 1 shows some hypothetical situations of this kind, in which the inclusion of one or more intermediate age groups radically alters the picture that would have been obtained by study of extreme ages only.

The take-home message is: if you start out knowing nothing about the effects of age on your measurement of interest, try an initial survey from young adult to the median survival age, say 6, 12, 18, and 24 months of age for most strains. If there really is no change between 6 and 24 months of age, then you're welcome to look at older animals, but remember that the older they get, the more likely it is that any effect you're seeing reflects the diseases and debilities that are likely to kill the mice in a few more months.

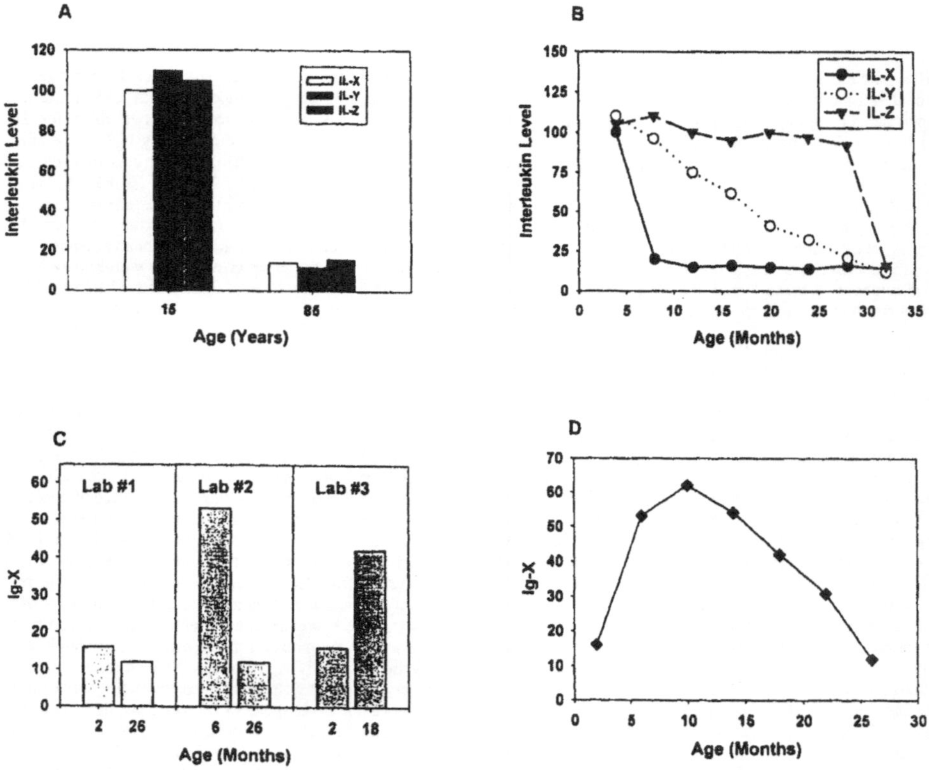

Figure 1. Some reasons to include middle-aged animals in a survey experiment. Panel A shows the hypothetical results of a study looking at three interleukins in the blood of young and old rats. The authors conclude that all three interleukins decline in parallel with aging—perhaps they share some common control mechanism on which one could base a Program Project application? Panel B shows a more comprehensive survey of the same topic conducted at another laboratory: IL-X drops early in adult life (a maturational effect), IL-Y shows a progressive decline through the life span, and IL-Z drops only when the animals get ill in very old age. Panel C shows three hypothetical reports of age effects on Ig-X levels in serum. Lab 1 says there's no change; Lab 2 says the levels go down; Lab 3 says the levels go up. Panel D shows the real hypothetical data set from which each lab drew its misleading conclusion: use of a wider and higher resolution set of age groups would have avoided much embarrassment and needless confusion.

Fourth Principle: The mice must be specific pathogen–free, and you have to be able to prove it.

Specific pathogen–free (SPF) colonies are *not* free of all disease, and are *not* free of all infectious agents, but they are free of a well-defined group of known murine (or ratoid) pathogens that routinely plague the old-fashioned "conventional" mouse colony. The most common mouse pathogens endemic in conventional colonies are Sendai virus, Coronavirus (mouse hepatitis virus), and pinworm, whereas rat colonies are most often infected with sialodacryoadenitis virus, Sendai, and *Mycoplasma pulmonis* (R. Dysko, DVM, personal communication). The methods required for maintenance of an SPF rodent vivarium are beyond the scope of this article, and are well reviewed elsewhere (2). Key steps include: never allowing in animals, from any source, that have not been proven to be SPF; never taking animals out of the colony and allowing them to return; never allowing visitors, even site visitors and deans, to enter the colony; never letting anyone into the colony who does not need to be there; and the routine but obsessive use of precautions like gloves, gowns, and shoe covers. Some facilities use an expensive barrier system, in which each room has a clean door (through which supplies and clean personnel enter) and a dirty door (through which used cages leave). Others have almost equal luck with commercially available single-cage filter bonnets that greatly reduce the risk of cage-to-cage transmission of airborne microbes.

These operating procedures, although necessary, are not sufficient to earn the right to refer to your colony as SPF; the colony must be proven to be free of key pathogens by routine testing of surveillance mice, for example on a quarterly basis. A minimal surveillance program involves introducing several cages of new mice, without filter bonnets, into each room every three months, using a stock (CD-1, for example) known to be susceptible to many common pathogens. These mice are then tested after 90 days, a period of time sufficient to allow them to become infected by anything in circulation, *and* to have developed antibodies to the infectious agent. The mice are then euthanized, examined carefully for evidence of intestinal and external parasites, and their sera tested (often by a commercial laboratory) for evidence of antiviral antibodies specific for the agents of interest. Periodic histopathological analysis is also recommended. If all tests come back negative, then you can call the colony SPF. Positive responses should induce a panicky feeling and a vigorous retesting effort, and repeat positives usually require that the entire affected colony be discarded and rederived.

If this surveillance system is not practicable in a given institutional vivarium, a useful alternative is to keep a small number of weanlings from breeding cage until they are about 3 months old and then to send these to a commercial facility for testing. Because the breeding pairs are usually long-term residents of the facility, and newborn pups are particularly sensitive to infection, this procedure is more sensitive than buying mice and allowing them simply to reside in the colony for a few months without contact with the local residents.

The optimal situation for maintaining an SPF colony combines the use of filter bonnets with the use of sentinel mice. Because filters work very well in preventing spread of airborne pathogens, exposure of the sentinel animals to potentially infectious agents requires a procedure in which the used bedding from a pool of cages be thoroughly mixed, and then added to the cages containing the sentinels. (A policemouse's life is not a happy one.) Sentinels who put up with this treatment for several months are then volunteered for necropsy and serological analysis. In these circumstances a stray positive result may not require sacrifice of the entire colony, because it is more likely that the infection has been confined to one or two cages; detailed follow-up studies may document good health for the majority of mice in the room. To be effective, this system requires good record-keeping in order to trace all cages that a sentinel has had contact with.

Why go through this hassle? "After all," the scientist stuck with a conventional colony might rationalize, "people are not free of all infectious agents; I'm just trying to more closely mimic the real world situation." The basic problem is that the intensity, variety, and prevalence of infection in any given conventional colony may well change from month to month and year to year, and is likely to differ greatly from one colony to another. Because many infections can alter a mouse's immune, hepatic, endocrine, digestive, pulmonary, and neurological responses, studies carried out on conventional colonies can prove very difficult to reproduce in another, or even in the same, laboratory. In some cases allegations of age effects on variables of interest have proven to occur only in conventional colonies (3), and are thus likely to reflect unsuspected influences of one or more uncharacterized infectious agents than of aging itself. Most effects of this kind doubtless go undetected, because few workers routinely use mice from two distinct colonies, one conventional and the other SPF. It seems likely, however, that many of the unnerving conflicts among reports in the gerontological literature may reflect variations in colony pathogen status.

Successful maintenance of an SPF colony also requires sufficient discipline to prevent the importation of new mouse stocks from uncertified suppliers. A well-run colony will usually permit unfettered importation of mice from only a very small number of commercial vendors, vendors that routinely submit clean bills of health with all shipments. A request for permission to bring in animals from an uncertified vendor or another research institution should trigger a process in which the sender is required to document the health status of the animals, and in which even allegedly clean animals are kept in a separate quarantine facility (or building) until tested locally for pathogens before they or their offspring are introduced into the general population. Importation of a stock that cannot be proven SPF ordinarily requires long-term quarantine or rederivation of the stock by caesarian delivery and foster nursing.

Two common mistakes: Scenario 1—the vivarium manager tells you it's an SPF colony, because the facility only buys from SPF suppliers. So why spend the money to test this? Four years ago, however, your technician visited a pet store on the way to work and every cage has had Sendai for four years—your laboratory mice are about as SPF as the ones in your basement. Test quarterly, and you can proudly report the clean bill of health in every paper and every grant proposal.

Scenario 2—you have a conventional colony, and you're not proud of it, but you buy the SPF mice from NIA and let them sit in the colony for just a week or two before use. This approach, which is remarkably common, just about guarantees that the mice used in your tests are infected with something; blessed with an SPF upbringing, they have no protective antibody titers, and are thus sitting ducks for whatever virus happens to be in your colony during their initiation into the tough realities of real world infection. Even if the mice will be housed in the animal facility short-term, it is well worthwhile to use an SPF facility or at least a quarantine room and filter bonnets.

It is not a good idea to try to sneak past the problem in Scenario 2 by using the mice the day they arrive off the delivery van. Shipping is very tough on mice, and the stress has an impact on adrenal size, steroid hormone levels, immunity, and other organ systems. No one would write a protocol that began, "Prior to their use in experiments, the mice were placed in a shipping container without access to their usual sources of food and water, and then flown 1000 miles in a dark, cold, noisy plane, followed by interstate truck shipment. . . ." It's a good idea to let the mice sit after arrival, for at least a week or preferably two, before use. But do it in an SPF colony.

Fifth Principle: Don't bet the farm on C57BL/6 mice. Don't bet it on F344 rats, either.

Inbred strains, developed over the last 80 years for their usefulness in transplantation and cancer research, have become the de facto standard strains for research in most other areas, including aging. Sixty percent of the rats ordered from the NIA colony are of the F344 stock, despite the well-known problems with this strain (4), and 40% of the mice used are C57BL/6. The F1 hybrid mice, which in some respects have much to recommend them when compared to inbred animals, account for only 15% of the mouse orders. This is a shame, whose implications for aging research have been reviewed in detail elsewhere (4–6). In brief, the case against inbreds includes the following counts:

- All mice in an inbred stock are genetically identical. It's therefore impossible to be certain that conclusions based on an inbred stock will apply equally to any other inbred stock without doing the study all over again.
- Inbred stocks are not only homogeneous, they are also weird, debilitated, and short lived. Creation of an inbred stock involves forced homozygosity at all loci. This is a highly selective process, because the inbreeding process frequently creates genotypes that impair viability and fertility; in fact, most brother-sister-mated families eventually die out, with the few surviving families becoming the "standard" inbred lines we all know and love. F1 hybrids created by a cross between two different inbred lines are almost invariably longer-lived than either of the two parents (7), consistent with the notion that the homozygous condition produces an animal of lower quality. Many inbred lines are famed for properties that clearly count as strain-specific oddities: the chronic renal disease and high lymphoma incidence of the F344 rat, the 100% incidence of thymic lymphoma in the AKR/J mouse, the near 100% incidence of reticulum cell sarcoma in SJL/J mice, and many other similar peculiarities. Table 1 shows a series of anecdotes; the take-home message is that individual inbred lines may have a very high incidence of lesions that are rarely seen in other inbred lines. These obvious illnesses, and other idiosyncrasies less obvious to the naked eye, could in principle wreak havoc on the process by which general conclusions are inferred from a limited data set. In some cases, of course, genetic identity is critical to the experimental plan; these situations include protocols that involve transfer of tissues from one mouse to another, and those where the goal of the study involves analyses of interstrain variation. But in many other instances the use of an inbred strain reflects mere custom rather than a careful decision among alternatives. To read more along these lines, check out R. Weindruch's article (5).

So what is a gerontologist to do? There are a number of possible pathways through the current difficulties:

(a) If you have to use genetically homogeneous animals, prefer F1 hybrids to inbred mice. Although each individual F1 stock is genetically uniform, at least you've ducked the homozygosity problem, and F1 mice are in general longer-lived, hardier beasts.

(b) Replicate key findings in multiple stocks. Once you've invested three years in proving something in CB6F1 mice, it may seem a waste of time to spend another few months checking the main points in two other F1 lines, but in the long run this may be more productive than spending the rest of your career chasing a finding that turns out to apply only to CB6F1.

(c) Consider the use of an animal stock with controlled heterogeneity, such as mice bred by a four-way cross (e.g., CB6F1 mothers crossed to C3D2F1 fathers). Such a cross yields an arbitrarily large group of full sibs; no two mice are genetically identical, but each mouse shares half of its genetic code with any other randomly chosen animal in the pool. There is a slowly growing literature demonstrating the usefulness of such heterogeneous lines in aging research, and the NIA Office of Biological Resources plans to add such mice to their contract colonies soon.

Table 1. Disease Incidence in Inbred Rodents: Some Sobering Anecdotes

Strain	Disease	Age (months)	Incidence	Reference
C3H/He	Hepatoma	14	85%	8
A/He	Pulmonary adenoma	18	90%	8
BALB/c	Lymphoma	13	44%	8
SJL/J	Reticulum cell sarcoma	13	91%	8
C57/BR	Pituitary tumor	"old"	33%	8
BALB/c female	Ovarian granulosa tumor	??	76%	8
F344 male	Leydig cell tumor	24	99%	8
F344	Glomerulonephropathy	27	> 90%	9
(A × B6)F1	At least one neoplasm	18–24	25%	10
		30–35	70%	10
		36–41	94%	10

Moral 1: If you need 25 "disease-free" old inbred or hybrid mice, you had better obtain more than 25 mice.

Moral 2: Aged animals of strain A may differ in known and unknown ways from those in strain B, due in part to disease effects.

Sixth Principle: Do at least a quickie autopsy on each old mouse; if you can afford it, pay a pro to do a gross necropsy.

Even a pathological novice can take a quick look: Is the spleen three times normal size? Are there little white bumps in the lung or liver? Battle-scars, probably infected, on the backs and legs? Skipping this precaution may make it easier to call the mice "apparently healthy" in the materials and methods section, but at the cost of making the results much harder to replicate and interpret. Tumors are common in aged rodents, and small tumors may not pose a problem for many experiments. However, as a rule, the simplest, and in some ways the best, plan is to toss out the data from mice that have large lesions—though of course this can be a very expensive rule to follow, particularly if you've ignored the First Principle.

Having a look yourself is better than not looking, but better still is to give the animal, after you've taken the tissue you need, to a veterinary pathologist or technician and ask them to have a look. This is fairly inexpensive, you get the written report back in a few weeks, and then you can go back through your notes and discard the data that came from the ones later found to have a serious illness.

Best of all is a histopathological autopsy. A thorough job costs \$50–\$100 per case; a quick microscopic look at the obvious lesions can cut the cost to \$25 or so. This is very important if you are characterizing a new model (does the drug you administer to prevent neurodegeneration increase the incidence of liver abnormalities?). But if your main goal is to eliminate data from sick mice, the gross inspection is often adequate and much cheaper.

An alternate, more informative, approach would be to consider whether the presence of a specific form of illness modifies the age effect or treatment effect under study. Limited statistical power—are the differences between tumor-ridden and tumor-free mice big enough to be worth separate analysis—and the difficulties of deciding the extent to which similar disease states can safely be lumped together can greatly complicate this variety of analysis. In any case it is important for the investigator to state explicitly the criteria used for elimination (or stratification) of individual animals, and the proportion of animals that met the inclusion criteria.

At what ages is the yield worth the cost? This will vary from strain to strain, and if you find that 95% of your 16-month-old mice are free of lesions that might compromise your interpretation, then you may want to skip paying for the gross inspections and do it yourself. As a rule of thumb, we try to get a gross necropsy on mice over 18 months and rats 24 months or older. Your mileage may vary.

Seventh Principle: Don't pool unless you absolutely have to.

There are two potential problems with pooling cells and tissues. The first is that pooling rapidly increases the chance that the cells or tissues under study contain abnormal cells from a diseased subject. In a study of immunity, for example, the inclusion of even a fairly small proportion of lymphoma cells in a pool of otherwise normal cells may strongly influence the result of the analysis. If sick mice make up one third of the population, each pool of five mice is very likely (83%) to include at least one diseased animal.

The second issue is a statistical one: the assessment of a statistical hypothesis ("mice aged 18 months express, on average, more of this gene than mice aged 6 months") depends on the number of individual mice, *or pools of mice*, tested independently. Thus an experiment in which a pool of 20 young mice is compared to a pool of 20 old mice has no greater statistical power than an experiment comparing one young to one old animal, i.e., none at all. The fewer mice used in each pool (ideally one mouse per pool), the more statistical power is achieved for the available mouse budget. Putting in the extra effort to miniaturize your test system to the point where it can be performed with material from a single mouse pays off handsomely in the long run.

Eighth Principle: Buy extra old mice to compensate for death and disease.

If you need 20 mice aged 24 months, don't buy 20 mice aged 24 months, because when you get around to using them 2 months later you'll have 18 live ones, of which only 10 will be free of visible lesions at necropsy. If you want 20 mice at age 24 months, buy 30 mice at age 22 months; use them 2–6 weeks later and discard the ones with lesions.

Ninth Principle: Do a cost-adjusted power analysis and save a bundle.

OK, you know how to do a power analysis. You call the local statistician, and indicate that you're trying to figure out if male mice have more muscles than female mice, and that the three males you've tested so far have 1200 μg worth of the muscle in question, with a standard deviation of 200 μg, and that you'd consider it worth knowing if the sex difference were as great as 200 μg, so how many mice do you need to use? Would 10 of each group do the trick? The statistician plugs the values into a secret program, and tells you that if you use a $p < .05$ criterion for significance (and who wouldn't?), then with $N = 10$ per group, you've got only a 58% chance of getting a significant result if the real difference is 200 μg. To get an 80% chance of detecting a difference of 200 μg, you're going to need to include 17 mice in each group. You then go out and buy some extra mice.

OK, next you want to know if the size of this muscle varies with age rather than with gender. So you do the same calculation, and get the same result: to get 80% power for detecting a difference between young and old of 200 μg, given the same assumptions above, you're going to need 17 young mice and 17 old mice. You go out and buy these 34 mice; actually you buy 17 young and more than 17 old, because some of the old animals will have to be discarded when you find out they have tumors.

Mistake: you've just wasted some money. The power analysis was done to calculate the minimum number of animals, but what you really want to do is get the maximal amount of statistical power per dollar spent. Because the old mice cost a lot more than the young ones, the cheapest way to get this statistical power is to buy slightly fewer old mice, and a good deal more young ones. If, for example, the cost of studying each young mouse (purchase cost plus cost of doing the assay) is \$18, and the cost of studying each old mouse is \$106, then the optimal solution is to buy 29 young mice and 12 old ones. If you buy 17 of each, you'll spend \$2108; if you buy

29 plus 12, you spend \$1797. You can allocate the \$311 to the next experiment, or give your tech a well-earned raise.

Actually, the real savings can be very high indeed, particularly if you have to grow the mice yourself rather than obtain them from the NIA's highly subsidized colonies, and particularly if you count in the cost of the mice you couldn't use because they had serious disease. The real cost of a 2-year-old mouse is not \$106, but the \$212 you spent to grow two of them, the one that died last week and one you've still got, or often the \$424 you need to spend to get a tumor-free mouse. At \$424 per old mouse, the minimal cost is to study 50 young and 10 old mice, at a cost of \$5242 and a savings of \$2272. At the NIA subsidized cost, you'll pay a mere \$72 a head for the old mice, or \$144 per tumor-free old mouse, and \$22 per young one; so you'll save a mere \$454—less cost, and less savings than the home-grown variety, but still worth the cost of writing down the following formulas, which were derived by Andrzej Galecki of the University of Michigan's Geriatrics Center and Institute of Gerontology:

$$N_Y = 0.5 \cdot N_e \cdot [1 + \mathrm{SQRT}(C_O/C_Y)]$$

$$N_O = 0.5 \cdot N_e \cdot [1 + \mathrm{SQRT}(C_Y/C_O)]$$

Here's what to do: first do the regular old power analysis, that tells you how many mice you'll need if you use equal *numbers* of mice in each group. This was $N_e = 17$ mice per group in the example shown above. Then figure out how much it costs to do the assay for each young mouse, including purchase costs, supply costs, tech time, overhead; this number is C_Y, the cost per young mouse. Calculate C_O, the cost per old mouse, in the same way, and be sure to throw in the adjustment for the number of mice you'll need to discard for disease. Then plug in the values and calculate N_Y, the number of young mice to buy, and N_O, the number of old mice to buy.

Tenth Principle: Don't misinterpret artifactual correlations due to age effects.

This isn't really a principle of animal use, because it applies with equal force to studies of human aging, but the error is so common that it's worth noting in this context. Consider, for example, a hypothetical situation in which an investigator is interested in testing the hypothesis that the relatively low ability of T cells from old mice to secrete interleukin-2 (IL-2) contributes to their low ability to make antibodies when injected with tumor cells. The investigator tests a group of mice, containing equal numbers of old and young animals, for both IL-2 and antibody production, and obtains the results shown in the left panel of Figure 2. He interprets this as good support for the idea that low IL-2 levels are indeed associated with poor antibody production, and because he knows from prior work that T cells from old mice do indeed make less IL-2 than cells from young mice, he concludes that the low IL-2 levels may underlie the poor antibody production.

The error in this inference is that the correlation between IL-2 and antibody production could well reflect the common influence on aging on both outcomes, rather than any direct connection between IL-2 production and antibody responses. The right panel of Figure 2 shows the same data set, but with triangles used to indicate the data from the old mice and circles to show data from young donors. *Within* each group, there is no correlation between IL-2 and antibody, and the impression of a correlation conveyed by the left panel of Figure 2 reflects the age influence on both traits. The older mice may also, compared to young animals, have more cataracts, weaker muscles, and a preference for classical music over grunge rock; plots of these measures against IL-2 production in a mixed-age group will also show excellent correlations that do not tell us much about causal relationships.

Summary

- Use youngish mice for your "old" groups, and use older young adults instead of adolescents as young controls. Use some mice in the middle range, too.
- Make sure they're SPF, and free of the most obvious tumors and other significant diseases.
- Try not to pool unless absolutely necessary.

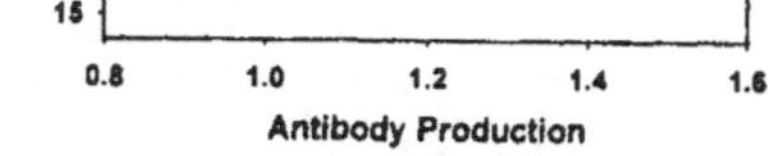

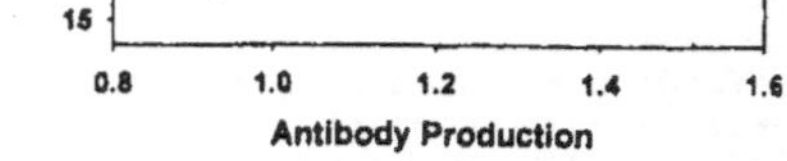

Figure 2. Misleading inferences from age-confounded correlations. This hypothetical data set represents an attempt to test the idea that the low levels of IL-2 production seen in old mice contribute to poor antibody production in these mice. Unsophisticated investigators might conclude, from the left panel, that there is indeed a strong correlation between IL-2 production (already known to decline with age) and antibody production, and thus claim that their hypothesis is supported. The right panel, showing old and young individuals separately, reveals the fallacy: the correlation results from the high age-sensitivity of both measured traits. Any two traits strongly influenced by aging would generate a similar correlation, even if neither trait had any direct mechanistic relation to the other.

• Don't put clean SPF rodents into a conventional colony; the power of prayer, though redoubtable in some earthly domains, does not always fully prevent infections in rodents.

• Shake the inbred habit—this is not your father's rodent. Pick F1s if you have to use genetically homogeneous stocks, and make sure to confirm your key findings in multiple strains.

ACKNOWLEDGMENTS

The preparation of this article was supported, in part, by NIA grants AG08808 and AG13283. The ideas presented represent the opinions of the authors, but were formulated in part during conversations with many colleagues, particularly David Harrison, Andrzej Galecki, Robert Dysko, and Bennett Cohen.

Address correspondence to Richard A. Miller, MD, PhD, The Geriatrics Center, University of Michigan, CCGCB, Room 5316, Ann Arbor, MI 48109-0940. E-mail:millerr@umich.edu

REFERENCES

1. Fowler HW. *A Dictionary of Modern English Usage*. 2nd ed. New York, NY: Oxford University Press; 1965.
2. Clough G. Suggested guidelines for the housing and husbandry of rodents for aging studies. *Neurobiol Aging*. 1991;12:653–658.
3. Florini JR. Limitations of interpretation of age-related changes in hormone levels: illustration by effects of thyroid hormones on cardiac and skeletal muscle. *J Gerontol Biol Sci*. 1989;44:B107–B109.
4. Weindruch R, Masoro EJ. Concerns about rodent models for aging research. *J Gerontol Biol Sci*. 1991;46:B87–B88.
5. Weindruch R. Animal models. In Masoro EJ, ed. *Handbook of Physiology Section 11: Aging*. New York: Oxford University Press; 1995: 37–52.
6. Miller RA, Austad S, Burke D, et al. Exotic mice as models for aging research: polemic and prospectus. *Neurobiol Aging*. 1999;20:217–231.
7. Smith GS, Walford RL, Mickey MR. Lifespan and incidence of cancer and other diseases in selected long-lived inbred mice and their F1 hybrids. *J Natl Cancer Inst*. 1967;50:1195–1213.
8. Altman PL, Katz DD. *Inbred and Genetically Defined Strains of Laboratory Animals. Part 1: Mouse and Rat*. Bethesda, MD: Federation of American Societies for Experimental Biology; 1979.
9. Maeda H, Gleiser CA, Masoro EJ, et al. Nutritional influences on aging of Fischer 344 rats: II. Pathology. *J Gerontol*. 1985;40:671–688.
10. Wolf NS, Giddens WE, Martin GM. Life table analysis and pathologic observations in male mice of a long-lived hybrid strain (A_f × C57BL/6)F_1. *J Gerontol Biol Sci*. 1988;43:B71–B78.

Received April 9, 1999
Accepted August 25, 1999
Decision Editor: Jay Roberts, PhD

ELSEVIER

Mutation Research 447 (2000) 117–135

Fundamental and Molecular Mechanisms of Mutagenesis

www.elsevier.com/locate/molmut
Community address: www.elsevier.com/locate/mutres

Mutation Research frontier

Somatic mutations and aging: a re-evaluation

J. Vijg *

Cancer Therapy and Research Center and University of Texas Health Science Center, San Antonio, TX 78229, USA

Received 1 September 1999; received in revised form 9 September 1999; accepted 30 September 1999

Abstract

Aging has been explained in terms of an accumulation of mutations in the genome of somatic cells, leading to tissue atrophy and neoplasms, as well as increased loss of function. Recent advances in transgenic mouse modeling and genomics technology have created, for the first time, the opportunity to begin testing this theory. In this paper the existing evidence for a possible role of somatic mutation accumulation in aging will be re-evaluated on the basis of the evolutionary logic of aging and recent insights in genome structure and function. New strategies for investigating the relationship between genome instability, mutation accumulation and aging will be discussed. © 2000 Elsevier Science B.V. All rights reserved.

Keywords: Somatic mutation; Aging; DNA repair; Transgenic and knockout mice; Genome instability

1. Historical background

Aging can be defined in practical terms as a series of time-related processes occurring in the adult individual that ultimately bring life to a close. To some extent, these processes can be understood as the coordinated action of the products of multiple genes. However, aging also has a major stochastic component, which includes effects of endogenous and environmental factors inflicting damage and subsequent mutations to DNA. Somatic mutation theories of aging were originally based on the observation that late irradiation damage in animals, like generalized atrophy and neoplasia, appeared to resemble premature aging [1]. Since radiation was known to induce mutations, it was proposed that aging could be the result of life-long exposure to low, natural levels of background radiation and other environmental agents capable of inflicting damage to DNA [2,3]. A major argument in support of a causal relation between somatic mutations and aging was the role of DNA as the primary informational biomolecule and the irreversibility of mutations, i.e., changes in sequence information. In view of the difficulties in quantification and characterization of mutations in vivo, i.e., in different organs and tissues of mammals, the concept did not evolve much beyond the theoretical state. Instead, attention switched to comparative studies of DNA repair capacity in species with different life spans [4].

Specific arguments against somatic mutations as a major cause of aging were raised in an early stage. Studies specifically addressing the question whether

* Department of Basic Research, Cancer Therapy and Research Center, 8122 Datapoint Drive, Suite 700, San Antonio, TX 78229, USA. Tel.: +1-210-949-7529; fax: +1-210-692-7502.
E-mail address: jvijg@saci.org (J. Vijg).

0027-5107/00/$ - see front matter
PII: S0027-5107(99)00202-X

radiation hastens typical changes seen in un-irradiated aging animals, then, indicated no complete similarity [5]. However, to obtain a definite answer extreme dose fractionation might be necessary and still, X-rays alone may not mimic the natural background of DNA damaging agents. Results from experiments in which survival of haploid and diploid wasps of the genus Habrobracon were compared after normal aging and X-irradiation seemed to indicate that mutations could not be a major cause of aging [6]. Indeed, although X-irradiation reduced life span of the haploid variant more than that of its diploid counterpart, their normal aging rate was the same. Hence, deleterious recessive mutations are clearly not the cause of aging in this species. However, these results do not account for dominant mutations [7], effects of X-rays other than mutations or the possibility of physiological differences between diploids and haploids. A third argument, used against somatic mutations as a cause of aging, was that impossibly high spontaneous mutation rates were considered necessary in order to have some functional impact [8]. Although this also mainly applies to recessive mutations, the question if DNA alterations ever reach the critical level to seriously impair the physiology of an organism is of genuine importance and has thus far not been adequately addressed.

Since the early days when somatic mutations were first considered as a possible cause of aging, significant progress has been made in understanding the structure and function of the higher animal genome and how a large variety of physical and chemical agents act upon it. There is important new information about the multiple enzymatic systems that process DNA, including DNA repair. Finally, the emergence of a whole new repertoire of methods for genome analysis and manipulation has provided the opportunity to directly test various aspects of somatic mutagenesis as a proximate cause of aging.

2. Evolutionary perspectives

In a world without aging the chance of dying would be the same at any age. Yet, this is not the case as exemplified by the rectangular survival curves of human or animal cohorts in a protected environment. A short period of relatively high mortality at early age is followed by a period of virtually no change, until at about 30%–40% of the average life span mortality starts to rise exponentially, resulting in the extinction of the cohort. The point of extinction, i.e., the age at death of the last survivor, can be considered as the maximum life span of the strain or species. A popular explanation of the phenomenon of aging is to see it as a genetically determined program that has some benefit for the group or species. Group selection, however, is rare and it is difficult to see a selective advantage in a process that creates worn-out individuals. Indeed, it has been argued that in view of the ample capacity of natural systems for regeneration and maintenance, aging, which results in decreased fitness, should not occur at all. To ultimately understand the role of somatic mutations as one of the possible proximate causes of aging it is important to define the evolutionary factors that determine why the process of aging is so intricately associated with most life forms.

2.1. Why do we age?

From an evolutionary perspective, aging is caused by the absence of selection against genes whose harmful effects occurred only after the oldest age of reproduction [9]. For most species this is easy to understand, since during aging many individuals are killed by forces other than senescence, resulting in increased evolutionary ignorance of genes with adverse affects later in life. By the same token, there is diminished selection for genes with a positive effect at late age. Hence, as a post-reproductive process aging is a by-product of evolution [10]. It is genetically determined, but in a non-adaptive way without any hidden benefit for the individual, group or species. (It still is occasionally proposed that aging has potential evolutionary advantages, and that the aging process may have been the subject of positive selection [11].)

In terms of the mutational variants driving the aging process, Martin [12] distinguished private and public mutations. Private mutations simply accumulate in the germ line because there is no selection against them (by genetic drift) and could be responsible for a number of adverse effects at late age, including age-related diseases, such as Huntington's disease and familial forms of Alzheimer's disease. The proximate causes of such manifestations of the aging process could involve a variety of mechanisms that may be unique to the individual, group or species.

Public mutations with negative effects at old age may have arisen in the form of pleiotropic mutations with a beneficial effect early in life [13]. Such alleles, selected for enhanced reproductive fitness, may modulate aging processes and life span in a way common to many species. The proximate mechanisms through which public mutations act could, for example, involve metabolic by-products with an antagonistic, negative effect at late ages. Indeed, such general processes as oxidative metabolism of glucose have been implicated in aging and a variety of common diseases through the production of reactive oxygen species and glycation end products [12]. DNA mutations could be an important molecular end point of such metabolism-related damaging events. Indeed, endogenous oxidative DNA damage has been considered as a significant factor in the initiation and progression of cancer, a major age-related disease [14,15]. Although unambiguous data that establish such a link is currently lacking [16], it is nevertheless conceivable that intrinsic aging processes specified by public aging genes predispose individuals to most of the common age-related diseases. Although mutation accumulation and antagonistic pleiotropy are now widely accepted to explain the evolutionary origin of aging, several questions still need to be addressed. For example, recent demographic findings indicating leveling off of late-age mortality need to be reconciled with the theoretical expectations [17].

In theories about the genetic control of aging ample attention is given to the concept of longevity assurance genes. The proposal that such genes exist is based on the recognition of stochastic processes causing somatic damage during aging. Accumulation of somatic damage as a cause of aging is predicted by the disposable soma theory of aging [18]. The disposable soma theory emphasizes a trade-off in the allocation of resources to either reproductive efforts or somatic maintenance and defense. Evolutionary theory would not predict a maximization of somatic maintenance and defense beyond levels that are strictly necessary to reach reproductive age, but individual variation is likely to be present. Hence, the expectation of allelic variation at the many loci harboring genes controlling cellular maintenance and repair, which could contribute to the observed variation in life span among individuals, strains and species. Examples of genes modulating aging rate in this way are those specifying antioxidant defense systems, genome stability systems, immune systems, apoptosis, but also degeneracy and redundancy, as well as cell and tissue regenerative capability.

2.2. Deleterious mutations in evolution and aging

Apart from adaptive mutagenesis, mutations can be expected to be almost always deleterious except from the rare few which offer some selective advantage. It is because of these rare few that mutations are essential as the ultimate source of genetic variation upon which evolution depends. Too many mutations, however, will reduce fitness and can lead to population extinction. According to the hypothesis known as Muller's ratchet [19], asexual lineages will tend to lose mutation-free genomes (due to genetic drift) and inevitably suffer from loss of viability. This is especially relevant for small populations and relatively easy to demonstrate in lower organisms such as *Escherichia coli*. For example, Kibota and Lynch [20] demonstrated that deleterious mutations of small effect escaped selection in *E. coli* lines with repeated population bottlenecks, resulting in decreased fitness. Elena and Lenski [21], also in *E. coli*, demonstrated that randomly introduced mutations interact to negatively affect fitness, i.e., the relationship between mutation number and decreased fitness was nearly log-linear. In small populations recombination processes associated with sexual reproduction appear to reduce mutational loads and to promote the accumulation of advantageous mutations [19,22]. Indeed, senescence-like phenomena have been observed in asexually propagated protozoa lineages, and were attributed to the accumulation of deleterious mutations [23].

In the somatic cells of multicellular organisms deleterious mutations are expected to accumulate, not unlike the situation in asexual unicellular lineages. In diploid organisms recessive mutations are masked, which limits adverse effects. Nevertheless, dominant and co-dominant mutations will also accumulate and could eventually result in a senescent phenotype.

3. Mutation accumulation in somatic cells of higher organisms

The ultimate source of somatic mutations, apart from replication errors per se, is the continuous introduction of chemical lesions in the DNA by a variety of environmental and endogenous physical, chemical and biological agents. Breaks of the sugar-phosphate backbone are spontaneously induced, e.g., under the influence of body heat, apurinic and apyrimidinic sites arise as a consequence of DNA hydrolysis and naturally present methyl donors cause non-enzymatic methylation [24]. As mentioned above, a major source of age-associated endogenous DNA damage could be oxygen free radicals, which are by-products of oxidative phosphorylation and several other biological and physiological processes [12,14,25]. In addition, a variety of exogenous agents, such as ultraviolet (UV) light, ionizing radiation and a range of chemicals in food, etc., can induce DNA damages in their target tissues, e.g., inter and intra-chromosomal cross links, DNA single and double strand breaks, bulky adducts and smaller adducts.

In spite of a wealth of recent methods to detect specific DNA lesions, it has been difficult to accurately assess spontaneous DNA damage loads in vivo in various organs and tissues during aging and rule out artifacts due to sample manipulation as the source of damage [24]. It seems likely that under normal circumstances the steady state level of spontaneous DNA lesions is very low. For example, Le et al. [26] reported less than 4.3 thymine glycols per 10^9 bases of DNA in cultured cells. At such low frequency, it would be unlikely for a lesion to occur within a gene or one of its regulatory regions, especially if it is realized that most of the lesions are not permanent but part of a steady state situation. That is, lesions are continuously removed through DNA repair pathways and re-introduced. Thus, it seems unlikely that under normal circumstances spontaneous DNA damage has any direct effect, in terms of impaired transcription. It could, however, increase the level of spontaneous mutagenesis through errors in DNA

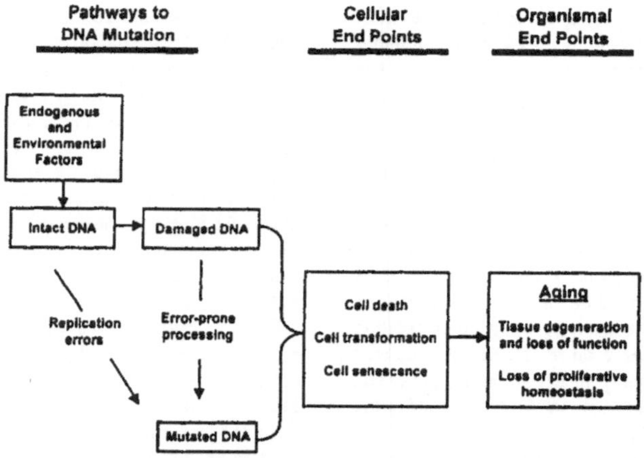

Fig. 1. Schematic representation of the hypothetical pathways that may lead to mutation accumulation as a possible proximate cause of aging. Spontaneous DNA damage is thought to be induced continually through endogenous mechanisms and from the environment, processed by the various DNA replication, repair and recombination systems to ultimately generate mutations as a molecular end point. Mutation accumulation may lead to cell death, cell transformation and cell senescence, which in turn could underlie the various symptoms of organismal aging, including organ dysfunction, tissue degeneration and a variety of pathological lesions.

damage processing. The resulting DNA sequence alterations can vary from large chromosomal alterations to point mutations. Such changes are irreversible except through cell elimination "repair". A hypothetical sequence of events, as to how DNA damage induction through DNA processing, mutation and its various cellular endpoints can lead to organismal aging is schematically depicted in Fig. 1. The different types of genomic targets, the percentage of the genome they comprise and their levels of instability are listed in Table 1. They will be briefly discussed below.

3.1. Cytogenetic changes

Due to a lack of methods to quantify and characterize mutation loads in vivo, i.e., in different organs and tissues, it has proved difficult to test the premise that mutations accumulate during aging of mammals in a sufficiently high number to cause some of the pathophysiological symptoms of aging. Curtis and Crowley [27] were the first to provide evidence for an increasing level of mutations during aging. They looked at mouse liver parenchymal cell metaphase plates after partial hepatectomy and found considerably higher numbers of cells with abnormal chromosomes in old as compared to young animals (i.e., from about 10% of the cells in 4- to 5-month-old mice to 75% in mice older than 12 months). Later, such large structural changes in DNA, i.e., aneuploidy, translocations, dicentrics, were observed to increase with donor age in white blood cells of human individuals, i.e., from about 2%–4% of the cells being chromosomally aberrated in young individuals to about six times higher in the elderly. The recent use of more advanced methods, such as chromosome painting have amply confirmed the increase in cytogenetic damage with age in both human [28] and mouse [29]. In both human and mouse the increase in chromosome aberrations appeared to be exponential.

3.2. Gene mutations

With the development of tests based on selectable endogenous target genes, it became possible to assess the mutant frequency at these loci among T-cells from human and animal donors. Using the hypoxanthine phosphoribosyl transferase (HPRT) locus test [30], different investigators have now shown that mutation frequencies at this locus increase with donor age. For example, results obtained with this assay suggest that mutant frequencies in humans increase with age from about 2×10^{-6} in young individuals to about 1×10^{-5} in middle aged and old individuals [31]. In mice mutant frequencies have been reported from about 5×10^{-6} in young animals to

Table 1
Instability of different genomic components

Sequence	Percentage of total (%)	Level of instability in vivo	Types of mutations	Increase with age	Reference
Whole chromosomes (e.g., translocations, aneuploidy)	100	2–10[a]	Deletions, copy loss/gain, transposition	Yes	[28,29]
Active genes (e.g., HPRT)	0.3	0.2×10^{-5} to 1×10^{-5} [b]	Deletions, point mutations	Yes	[30–33]
Non-coding DNA (e.g., lacZ reporter transgenic mice)	97	5×10^{-5} to 15×10^{-5} [b]	Deletions, point mutations	Yes	[42,43,45]
Retroposons (e.g., human L1, Alu)	20	?	Transposition	?	[49]
Tandem repeats (e.g., minisatellites)	15	10^{-5} to 10^{-5} [b]	Copy loss/gain Unequal cross-over	?	[50,51]
Telomeres	0.2	1–5 kb loss	Copy loss/gain	Yes	[58]
Mitochondrial genome	< 0.1–2	2%–100% of mutant copies in 0%–100% of all cells	Deletions, point mutations	Yes	[63,64]

[a]Percentage of cells with a chromosomal aberration.
[b]Mutant frequency per locus.

about 3×10^{-5} in middle aged mice [32]. However, in both mice and humans these values could be underestimates, due to the loss of HPRT mutants in vivo or in vitro. Indeed, results from Grist et al. [33], who assayed the HLA locus (using immunoselection for mutationally lost HLA antigen) in human lymphocytes, indicate 2–3 times higher mutant frequencies. Values higher than HPRT were also found with other assays involving selectable target genes and the discrepancy has been explained in terms of the inability of the HPRT test to detect mitotic recombination events (HPRT is X-linked) and a relatively strong in vivo selection against mutants [33]. Indeed, HPRT mutant frequencies have been found to decrease with time following exposure to mutagenic agents, such as ethyl nitrosourea and radiation, and has been considered as less suitable for long-term monitoring [34].

In mice subjected to caloric restriction, the only intervention demonstrated to increase life span [35], HPRT mutant frequencies were found to increase with age at a significantly slower rate than in the ad libitum fed animals [32]. These results were independently confirmed in rats by Aidoo et al. [36]. These results suggest that the level of accumulated somatic mutations reflects biological rather than chronological age. This conclusion was further strengthened by Odagiri et al. [37], who demonstrated accelerated accumulation of HPRT mutations in peripheral blood lymphocytes of so-called senescence-accelerated mice (SAM). Although the SAM model is not generally accepted as a mouse model of accelerated aging, these findings nevertheless demonstrate a link between somatic mutation rate and physiological decline.

Interestingly, the HPRT test has been used on tubular epithelial cells of kidney tissue from 2- to 94-year-old human donors. The mutation frequencies found were much higher than the above mentioned values for blood lymphocytes and were also increasing with age, from about 5×10^{-5} to about 2.5×10^{-4}; the increase with age appeared to be exponential rather than linear [38]. The higher mutant frequency in the kidney cells could reflect a relatively slow turnover as compared to T-cells.

With the development of transgenic mouse models harboring chromosomally integrated reporter genes, it became possible to directly test the hypothesis that somatic mutations in a neutral gene accumulate with age in different organs and tissues [39–41]. Using one of these models, harboring the *lacI* gene as a target, Lee et al. [42] were the first to demonstrate an age-related increase in mutant frequency in spleen, from about 3×10^{-5} in mice of a few weeks old to $1-2 \times 10^{-4}$ in 24-month-old animals. Subsequent results from other laboratories indicated age-related increases in mutant frequency in some, but not all organs. Dollé et al. [43], for example, demonstrated that mutant frequencies at a *lacZ* transgene increase with age in the liver, from about 4×10^{-5} in the young adults to about 15×10^{-5} in old animals (about 30 months), while such an increase was virtually absent in the brain. The increased susceptibility to spontaneous mutagenesis of liver versus vs. brain corresponds with observed higher frequency of focal pathological lesions in the mouse liver as compared to brain [44]. More recently, the pattern of organ-specificity in age-related mutation accumulation was expanded with the observation of age-related increases in mutation frequencies in spleen, heart and small intestine, but not in testes (Dollé et al., in preparation).

Essentially the same results as obtained by Dollé et al. [43] were found by Ono et al. [45]. These investigators used a mouse model of a different genotype (CD2 instead of C57Bl/6) harboring the same *lacZ* transgene [39]. They also observed an age-related increase in liver, heart and spleen, but virtually no increase in brain and testes. Dollé et al. also observed striking organ specificity with respect to the mutational spectra of the old animals. While in small intestine and brain almost only point mutations accumulated (the small increase in this latter organ was almost totally due to point mutations), in the liver and the heart large deletion mutations were a prominent part of the spectrum ([43]; Dollé et al., submitted).

In interpreting these data it should be noted that the observed increases were modest (varying from less than 2-fold, to more than 4-fold) and appeared to level off at middle age [42,43]. (Dollé et al. observed some evidence for an increased accumulation of especially large deletion mutations at very late age [43].) The relatively small age-related increase of mutant frequencies can be interpreted as evidence against a major role for somatic mutations in aging

[46]. However, it should be noted that although transgenic reporter genes do not suffer from a selection bias (as is the case with most selectable endogenous targets like the HPRT gene), it still provides an underestimate of the real mutation load and its adverse effects. Homologous (mitotic) recombination, for example, leading to deletion of entire reporter gene copies (loss of heterozygosity) is a frequent mutational event and goes undetected in the transgenic assays. Most of the transgenic models also do not account for mutational hotspots (see below) and such important functional end points as cell death are missed. Indeed, to put the results on mutant frequencies of different organs and tissues at various age levels into context, it will be necessary to also assess cell proliferation and cell death [46]. Most importantly, it will be necessary to determine at some point the critical level of cellular mutation loads in terms of physiological consequences.

3.3. Mutational hotspots

At some loci or regions of the genome the spontaneous mutant frequencies are higher than average. For example, Gossen et al. [47] studied a lacZ transgenic line with the reporter integrated near the pseudoautosomal region of the X-chromosome and found spontaneous somatic and germ line mutant frequencies that were up to hundred times higher than in all other lines tested.

Recognized mutation hot spots are the families of repeat elements that are a major component of the mammalian genome. Evidence is accumulating that rather than being "junk" DNA such repeat elements may have some functional relevance, for example, as transcription factor binding sites [48]. The two major families, termed L1 and Alu elements in humans, are considered to be retrotransposable. Although, the capability of such elements to retrotranspose has been demonstrated in human cell lines [49], there is no unambiguous evidence yet that they contribute much to somatic instability. For other repeat element loci, such as minisatellites, mutation frequencies as high as 10^{-2} to 10^{-1} have been found [50,51]. Related forms of genetic instability are triplet repeat expansion and telomere shortening (see below). Triplet repeat expansion is associated with human genetic diseases and increases with age [52].

A genetic instability hotspot that has recently received a lot of attention is the telomere. Telomere shortening, due to the loss of minisatellite repeat elements, has been observed to occur in normal somatic cells during aging in vivo and during repeated passaging of cultured cells. The latter results in so-called replicative senescence (see below) and crisis [53]. It is telomere shortening that provides the signals for entry into replicative senescence. Telomere shortening occurs because of the inability of the DNA synthesis machinery to replicate the very end of the lagging strand [54] and, possibly also, as a consequence of recombination. Since in most normal somatic cells telomerase, a ribonucleoprotein capable of adding TTAGGG repeats to chromosome ends, is suppressed, loss of telomere repeats can not be repaired. Eventually this results in chromosomal aberrations, e.g., end-to-end fusions. Reconstitution of telomerase prevents replicative senescence and appears to essentially immortalize normal human cells [55]. This strongly suggests that telomerase is a major and perhaps the most important genome stability system in actively proliferating human cells. Its potential importance for aging in vivo has been demonstrated by using mouse mutants with an inactivated telomerase RNA component (see below and Ref. [56]). Meanwhile, it has been demonstrated that areas of the endothelial lining of the blood vessels exposed to high hemodynamic stress undergo more rapid telomere loss with age [57]. This suggests some importance of focal replicative senescence in vivo in areas of high cell turnover. Recently, reductions in telomere length from the fetus to a 72-year-old male have been reported for 10 different tissues. In all tissues a decline with age was found and an increase in variation [58].

3.4. Mitochondrial DNA mutations

The 16.5 kbp mitochondrial genome has been implicated as a major target for somatic mutations during aging [59,60]. This seems not altogether justified in view of its high redundancy, which should compensate for accelerated mutation accumulation. Although the genes encoded in the mitochondrial genome, i.e., two rRNAs, 22 tRNAs and genes encoding 13 mitochondrial protein subunits, are necessary to produce essential enzyme complexes involved in oxidative phosphorylation, part of these are encoded in the nuclear genome. The importance assigned to the mtDNA as a target for age-related mutations is due to its close physical proximity to a major source of free radicals and its presumed lack of repair systems. The latter turned out not to be true; several repair pathways have been described for mtDNA, including double strand break repair [61,62]. However, there is evidence for a relatively high susceptibility of mtDNA to oxidative DNA damage [14].

Data on mutation accumulation in the mitochondrial genome have been difficult to interpret until assays were designed to analyze individual cells rather than homogenized tissue. A PCR-based method to scan whole mitochondrial genomes from single cells has recently been developed by Khrapko et al. [63]. These authors demonstrated that in human cardiac tissue, cardiomyocytes with a mtDNA deletion only occurred in old donors (centenarians). In these old individuals deletions were found at a frequency of up to one in seven cells. In such cells deleted mtDNA copies appeared to be present in up to 65% of the total number of copies. A similar approach was taken by Kopsidas et al. [64] for skeletal muscle. These authors detected much higher levels of deleted mtDNA copies, which could reflect tissue differences.

3.5. Mutations in germ cells

Other evidence for increased genome deterioration with age involves germ cells. It is generally assumed that an age-related increase in chromosomal abnormalities and other types of mutations in parental germ cells is responsible for increased risk of offspring with a genetic defect, such as Down syndrome [65]. Indeed, chromosomal abnormalities, including aneuploidy and chromosome structural rearrangements, occur in 10%–20% of spermatozoa and oocytes [66]. At least 15% of all recognized human pregnancies abort, most likely as a consequence of chromosomal mutations [67]. In fact, 50% of all spontaneous abortions contain a gross chromosomal mutation. The picture that emerges of the germ line, therefore, is that of an effective sieve that allows most mutations to be lost by negative selection, i.e., through the death of germ cells or early embryos. In most aged tissues such a sieve is absent. (Although recently evidence has emerged that stem cells are present in adult tissues, including neural tissue, suggesting some remaining regenerative capacity [68].) Interestingly, with the recently emerged possibility of cloning mammals by transferring nuclei from adult somatic cells into enucleated oocytes, genome deterioration with age can be directly investigated. Indeed, the success rate of cloning animals from somatic cells of old individuals should be markedly lower. Although the results thus far obtained with cloned sheep and mice suggest that cloning is less efficient from cells of adult animals as compared to fetal or embryonal cells, the numbers are still too low to draw conclusions [69,70].

4. Cellular end points of mutation accumulation and aging

To explain aging, at least in part, from the accumulation of somatic mutations, cellular end points common to mutation induction and aging need to be defined. In this respect three cellular end points, common in old individuals and known to be induced by mutations, are readily identified: cell death, cell transformation and cell senescence (Fig. 1).

4.1. Cell death

Cell death is a major component of the aging phenotype. It is supposed to contribute to a variety of functional deficits, including heart failure, kidney dysfunctioning, decline of immune function and, most notably neurological deterioration associated with memory loss and age-related diseases, such as Parkinson's disease, Alzheimer's disease and amyotrophic lateral sclerosis.

Both programmed cell death (apoptosis) and necrosis could occur as a result of DNA damage and mutation. It has been suggested that apoptosis is a defense mechanism against mutations and, more general, a mechanism for removing senescent and (pre) neoplastic cells during aging of multicellular organisms [71–73]. Such a role would be in keeping with the observed higher level of apoptosis in the livers of calorically restricted mice as compared to ad libitum fed animals [74,75]. In this view, loss of apoptosis control during aging would result in the accumulation of cells with damaged macromolecules, including DNA mutations. In general, it is conceivable that an accumulation of somatic mutations is primarily responsible for the loss of cells in various organs, perhaps primarily in renewable tissues, but possibly also in postmitotic tissues, such as the brain and the heart.

4.2. Cell transformation

Somatic mutations are generally considered as the major cause of cancer. This can be derived from various observations, including the actual presence of mutations in the tumor genome in genes thought to be critically involved in tumor initiation and/or progression, such as TP53, KRAS and RB1 [76]. Both exogenous [77] and endogenous [15] mutagenic mechanisms have been implicated in the induction of these mutations. Modeling of such gene defects, i.e., mutations in oncogenes and tumor suppressor genes, into the mouse genome often results in accelerated tumorigenesis [78], confirming the critical role of specific mutations as a cause of cancer. Other indirect evidence involve the general observation that most if not all mutagens are also carcinogens and that heritable mutations in genes controlling genome stability pathways often confer a high cancer susceptibility [78].

Somatic mutations are thought to activate and/or inactivate specific genes involved in such processes as cell cycle regulation and growth control, tissue invasion, neovascularization and DNA damage processing itself (i.e., to acquire a mutator phenotype; see below). The accumulation of the successive mutations that result in cancer must take place step-wise, with each mutation having a selective advantage [79]. However, thus far it is unclear how many different mutations are required for each particular tumor and the types of mutations (e.g., point mutations, genome rearrangements, aneuploidy) that are most important. A complicating factor in this respect is that most tumor genomes appear to contain numerous genomic alterations, of which only a part may be functionally important [80].

Based on the assumption of an in vivo mutation frequency not higher than 1.8×10^{-7} per cell generation per locus (HPRT) a mutator phenotype was considered necessary to account for the multistep nature of cancer pathogenesis in humans [81]. A mutator phenotype, i.e., an acquired somatic mutation rendering a tumor cell lineage more susceptible to spontaneous mutagenesis, would not only explain the multiple mutations necessary to give rise to a tumor, but also the large number of apparently random mutations so characteristic of the tumor phenotype. Although Tomlinson et al. [82] provided a model of tumorigenesis in which selection appeared to be more important than a high mutation rate, mutator phenotypes have actually been demonstrated in a number of cancers and cancer cell lines [83]. Indeed, in a subset of hereditary nonpolyposis colorectal cancer patients, a form of cancer caused by germline mutations in mismatch repair genes, a high level of somatic mutations (i.e., microsatellite instability) was found in their tumors and sometimes also in their non-neoplastic cells [84]. These results were confirmed by data from mice with a DNA mismatch repair deficiency, which had been cross-bred with a lacI-containing transgenic reporter mouse. Tumors of these hybrid mice exhibited greatly elevated lacI mutant frequencies, i.e., up to about 17-fold above the about 15-fold increased mutant frequencies of normal tissue in these MMR-deficient mice (Ref. [85]; see below).

However, mutation frequencies in tumors from both normally aged mice as well as p53-deficient mice were either in the normal range or were only moderately increased (Ref. [86]; H. Giese, unpublished). Mutator phenotypes in tumors are therefore not a general phenomenon (at least, not in mice), which leads to the conclusion that the background mutation rate of normal tissues should be able to explain tumor initiation and progression. Alternatively, it has been proposed that an age-related accumulation of somatic mutations could substitute for a mutator phenotype, at least for sporadic cancers that increase as a function of age [87].

Without neglecting the occurrence of mutator phenotypes in some tumors, it seems prudent to consider at this stage the possibility that the somatic mutation frequency in vivo in normal cells, which is much higher than the very low figures thus far adopted, is sufficient to cause most of the hyperplasia and neoplasia that is observed at old age. As mentioned above, the mutation rate at HPRT is likely to be higher than most data suggest, due to selection against cells with a mutated HPRT gene and the lack of somatic recombination of this X-linked gene. Results from Grist et al. [33], using the HLA-A locus as the selectable target, indicate mutation frequencies between 0.71×10^{-5} in neonates to 6.53×10^{-5} in elderly individuals. This is not dramatically different from results obtained with the lacZ and lacI reporter mice [42,43]. More recently, APRT heterozygous mice have been generated and at this non-X-linked locus mutation frequencies up to 12×10^{-5} have been reported for fibroblasts in vivo, 80% of which appeared to be due to mitotic recombination [88]. Interestingly, mutation frequencies in T lymphocytes from these mice were much lower, i.e., 3×10^{-5} (Jay Tischfield, personal communication).

Based on the increasing estimates of mutation loads in vivo, it seems very likely that without the need for mutator phenotypes most forms of cancer can be explained from mutation accumulation during aging. The question is if mutation accumulation also contributes to forms of age-related tissue degeneration and functional decline not due to cell death or cell transformation.

4.3. Cell senescence

Cell senescence was originally defined as the loss of replicative capacity of normal human fibroblasts in vitro [89]. More recently, replicative senescence has been considered in association with two potential physiologic consequences [90]. First, cell senescence could represent a tumor suppressive mechanism to prevent cells from acquiring the multiple mutations that are needed for malignant transformation. In this view, replicative senescence is part of the cellular defense network to maintain phenotypic fidelity, like apoptosis. Second, senescent cells (in vitro) appeared to have altered phenotypes, e.g., collagen overexpression, changes in cytokines, which may also occur in vivo and contribute to decreased tissue functioning. Thus, replicative senescence in vivo could contribute to the age-related degeneration of organs and tissues. Senescent cells may be more resistant to apoptosis than normal cells [91], in which case they can be expected to accumulate. An accurate assessment of the number of senescent cells is difficult to make without some phenotypic marker. Using a simple biomarker for replicative senescence, SA β-galactosidase, Dimri et al. [92] demonstrated the age-dependent accumulation of senescent cells in the human skin in vivo. This first direct link between replicative senescence and in vivo aging was followed by the demonstration that areas of the endothelial lining of the blood vessels exposed to high hemodynamic stress undergo more rapid telomere loss with age, suggesting focal replicative senescence in vivo [57].

While there seems little doubt that replicative senescence is a functional pathway per se, and probably one of the many aging-related programmatic responses induced by somatic damage, the definition of a senescent cell could be broadened to include every cell in an aged tissue with functional defects. Random somatic mutations, especially large genome rearrangements, could lead to impaired cell functioning rather than cell death or cell transformation. In this view, the accumulation of random mutations would result in a mosaic of cells at various levels of deficiency. It is conceivable that, especially large structural alterations at mutational hot spots, could gradually impair genome functioning. Indeed, rather than a catalog of useful genes interspersed with functionless DNA, each chromosome is now viewed as a complex information organelle with sophisticated maintenance and control systems. Destabilization of these structures by random DNA mutations may lead to changes in gene expression, for example, by influencing patterns of DNA methylation and conformation [93]. It should be noted that mutations do not necessarily depend on cell proliferation but rather on chronological time [94–96]. Indeed, mutations in postmitotic cells could be the result of errors during DNA repair processes (see, for example, Ref. [95]). In fact, it is conceivable that mutations accumulate more readily in postmitotic tissues than in actively proliferating ones, due to increased possibilities of negative selection in the latter.

5. Syndromes of accelerated aging and mutation accumulation

The potential impact of loss of genome integrity on the aging phenotype is underscored by the so-called segmental progeroid syndromes, which are characterized by the accelerated occurrence of certain aspects of the senescent phenotype [97]. (There are also unimodal progeroid syndromes, in which only one aspect of aging is accelerated, e.g., Alzheimer disease.) Most of these syndromes are caused by heritable mutations in genes involved in DNA processing, including DNA repair. While there is no single hereditary disorder that mimics all symptoms of human aging — which by itself is subject to individual variation — progeroid syndromes demonstrate enough aspects of accelerated senescence to be useful as models to study the genetic control of human aging. The Werner syndrome in particular has received ample attention after the relevant gene was identified as a member of the RecQ-like DNA helicases [98]. Other diseases based on a heritable mutation in a helicase gene are the Bloom syndrome and the Cockayne syndrome. The Bloom syndrome is a rare autosomal recessive disorder showing growth deficiency, immunodeficiency and a predisposition to cancer. An important feature, which is shared with the Werner syndrome is the high level of somatic mutations of all types in white blood cells of the Bloom patients, as well as a high rate of somatic recombination. This high level of genomic instability is caused by the loss of the DNA helicase activity [99].

The Werner syndrome is much richer than the Bloom syndrome in signs and symptoms of accelerated aging. That is, while Bloom syndrome is mainly associated with increased cancer risk, Werner patients show increased risk for atherosclerosis, cancer, osteoporosis and type 2 diabetes as well as early hair loss, graying of the hair, atrophy of the skin and cataracts. These symptoms might be caused by a defect in chromosome maintenance, although the exact mechanism of action of the WRN gene is as yet unknown. In cells from Werner syndrome patients, genomic instability is observed at the cytogenetic level in the form of chromosome breaks and translocations and at the molecular level by multiple large deletions [100]. A difference that might explain the difference in severity of aging symptoms between these two progeroid syndromes, i.e., the Bloom and Werner syndrome, is the existence of an exonuclease activity in the wrn protein. Purification of the recombinant wild type wrn protein revealed the existence of a $3' \rightarrow 5'$ exonuclease activity in the N-terminal region [101,102]. This exonuclease activity points towards a role of the WRN gene product in DNA repair, possibly in addition to other DNA transactions. At this point in time it is unclear if the loss of the helicase function alone is sufficient to cause Werner syndrome or whether loss of wrn exonuclease is also important. It should be noted that while in Bloom syndrome several basepair substitution mutations have been identified as the cause of the disease, Werner syndrome has been found exclusively associated with protein truncating mutations, which would be expected to inactivate both functions [103]. Hence, it is possible that inactivation of multiple DNA processing activities rather than only one lead to a more severe aging-related phenotype than single mutations.

The possible dual function in DNA transactions of the WRN gene product is interesting in view of the molecular basis of another segmental progeroid syndrome, i.e., the Cockayne syndrome (CS). The cause of this syndrome, which is also characterized by severe aspects of accelerated aging, including neurological abnormalities, involves a heritable mutation in either one of two genes, *CSA* and *CSB* [104]. These two genes play a role in both nucleotide excision repair (NER) and gene transcription. CS is apparently not associated with enhanced genetic instability [105], which is in contrast to Bloom syndrome and Werner syndrome. This lack of elevated mutation loads might explain why CS is not associated with predisposition to cancer. The reduced levels of RNA polymerase II transcription found in extracts of CS cells [106] support the notion that this progeroid syndrome is foremost a transcription-deficient disorder. It thus pleads against the idea that all syndromes that mimic accelerated aging are due to accelerated mutation accumulation.

5.1. Mouse models of genomic instability

The high level of genomic instability found associated with some premature aging syndromes supports the notion of genome stability systems as major longevity assurance systems. It is reasonable to assume that inactivation of one or more of such pathways accelerate in parallel both age-related deterioration and death and mutation accumulation. To test the hypothesis that accelerated mutation accumulation is always associated with the early appearance of some features of aging, genetic defects in genome stability systems can be modeled in the mouse, e.g., by mutational inactivation of the relevant gene in the germ line. The genes identified as involved in some human progeroid syndromes are a logical starting point for this line of research. For the most prominent accelerated aging syndrome, the Werner syndrome, this has thus far proved to be elusive. Mouse mutants with an inactivated WRN gene have been made, but a phenotype showing symptoms of accelerated aging comparable to Werner syndrome has not been observed. While displaying reduced embryonic survival, live-born WS mice otherwise appear normal during their first year of life [107].

More successful were attempts to model the Cockayne syndrome. For CS both CSB and CSA knockout mice have been made. In the CSB mouse mutant, initially a less severe phenotype as compared to the human disorder was found [108]. However, after crossing with another mouse mutant, i.e., an NER-deficient mouse harboring two inactivated copies of the XPA gene [109], a very severe CS phenotype was observed [110]. This underscores the notion that the inactivation of multiple functions in DNA transactions is more likely to cause severe senescence-related phenotypes than single defects. Indeed, the XPA-defective animals alone show normal development and in the absence of mutagenic treatments, virtually no symptoms of accelerated aging, at least until mid-age [111]. The latter suggests that the main problem in CS as a progeroid syndrome is not mutation avoidance but a transcription defect (see also above). This could of course also result in cell death and cell senescence (which would explain the symptoms of senescence), but probably not in cell transformation, which could account for the fact that CS patients are not cancer prone.

5.1.1. Mismatch repair mutants

Mice that are deficient in DNA mismatch repair genes have been generated at several laboratories [112]. Msh2-deficient mice succumb within 1 year to cancer with lymphomas, also a common cause of death in normally aged mice, observed in 80% of the cases [113]. Within the first year no difference was observed between wild type mice and the Msh2 heterozygotes. To investigate the spontaneous mutation frequencies in various organs and tissues in these animals Msh2- and Pms2-deficient mice were crossed with transgenic mutation reporter mice. The results indicated greatly elevated spontaneous mutation frequencies in several organs and tissues [114,115]. Virtually all of these mutations were point mutations, which could reflect the transgenic mutation systems used (based on bacteriophage lambda vectors, which lack the capability to recover large deletions) or the genome stability pathway involved (DNA mismatch repair deficiency appears associated with small insertions/deletions in small mononucleotide repeat runs.) Interestingly, these data immediately refute the possibility that high levels of random mutations (that is, point mutations) are incompatible with normal development. However, at this stage nothing is known about mutations accumulated at later ages since complete life span studies on these mice have not yet been performed. The possibility should be taken into account that smaller mutations, i.e., point mutations and small deletions or insertions, are less important in creating a general spectrum of age-related degeneration than large rearrangement mutations. Indeed, as suggested by the high level of DNA rearrangement mutations observed in lymphocytes of Werner syndrome patients (see above), large DNA rearrangements could be key DNA lesions in explaining adverse effects at old age [115].

5.1.2. NER mutants

Mice deficient in NER have been generated by the mutational inactivation of the gene XPA and appeared to mimic the phenotype of humans with xeroderma pigmentosum, that is, increased sensitivity to UVB-induced skin cancer [109]. However, at early age NER-deficient mice do not show spontaneous abnormalities. These mice develop normally, indistinguishable from wild type mice. At older age, i.e., from about 15 months onwards, $XPA^{-/-}$ mice show an increased frequency of hepatocellular adenomas [111], which in normal mice is only a common pathological lesion at old age [44]. The lack of spontaneous abnormalities in young mice might be due to the fact, that under normal conditions mice have only limited exposure to NER-mediated DNA damage. Hence, at early age NER appears to be dispensable. (This is in contrast to base excision repair, the inactivation of which is lethal: Ref. [116].) In order to test the hypothesis that loss of NER causes accelerated mutation accumulation, preceding the onset of accelerated tumor formation and aging, XPA-deficient mice were crossed with transgenic mutation reporter gene mice. In the liver, but not in the brain of the hybrids mutant frequencies showed an accelerated age-related increase from about 4 months onwards [117]. After 9–12 months, mutant frequencies were in the range of the maximum level reported earlier for 25- to 34-month-old lacZ control mice, with a similar high individual variation [43].

The absence of a severe spontaneous phenotype in the NER-deficient animals (at least until mid-age) is in striking contrast to the corresponding human syndrome, xeroderma pigmentosum. Humans with XP rarely survive beyond the third decade of life, a consequence of the dramatic increase in sunlight induced skin cancer [118]. Skin cancer does not occur in rodents, which have a fur that cannot be penetrated by UV and are also kept under conditions not permitting exposure to sunlight. This explains the lack of a phenotype of the XPA mutation at early ages. In the young animals NER could be essentially redundant, but become increasingly important at later ages. Indeed, it has been repeatedly argued that loss of redundancy in, e.g., cell number, gene copy number, functional pathways, could be responsible for the gradual loss of individual stability and increased incidence of disease associated with aging [119]. Complete life span studies of the NER-deficient animals will ultimately reveal if a late-age phenotype and shortened life span is associated with the observed accelerated mutation accumulation.

5.1.3. Telomerase mouse mutants

The potential importance of large genome rearrangements for the aging phenotype is illustrated by results obtained with telomerase null mice. Telomerase is a genome stability system, the absence of which has explicitly been implicated in replicative senescence (see above). Telomerase-deficient mice have been generated through disruption of the gene encoding the RNA component of the enzyme [56]. One of the surprising findings with respect to this mouse was their viability for at least six generations as well as the retained capacity of their cells to become established cell lines in vitro or form tumors. A decline in telomere length (which are much larger in mice than in humans) with each successive generation, however, was found in embryonic fibroblasts, as well as increased frequencies of aneuploid cells and end-to-end associations of chromosomes. The latter was expected, in view of the appearance of more and more cells without a detectable TTAGGG signal to stabilize their chromosome ends.

The late generation mice showed ample in vivo functional deficiencies. These include abnormalities in the hematopoietic system and the testes. Abnormalities in the hematopoietic system involved decreased proliferative capacity of splenocytes and bone marrow cells providing evidence that the progenitor-cell compartment and thereby the long-term renewal of haematopoietic stem cells was compromised. In addition, a dramatically high frequency (about 50%) of abnormal metaphases were present in cultured splenocytes. In the testes of late-generation animals severe atrophy was observed, due to checkpoint arrest and apoptosis. In the female mice uterine atrophy resulted in smaller ovaries and egg yield. Mice of generation 6 no longer produce offspring [120].

The importance of telomere maintenance for organismal aging, is exemplified by the recent demonstration of shortened life span, hematopoietic ablation, reduced capacity to respond to stresses, as well as increased incidence of spontaneous malignancies of third generation telomerase null mice [121]. At early age, these mice are phenotypically normal and adverse effects were only noted during a 2.5-year study of an aging cohort. These demonstrated effects of telomere instability and the accompanying genetic instability underscore that loss of genome integrity in vivo can lead to the loss of proliferative homeostasis as predicted by Martin [122].

The use of transgenic modeling as a tool for studying the relevance of genome stability systems in aging has just begun. Indeed, thus far very few if any life span studies on these models have been completed. Therefore, only the early effects of the introduced mutations and/or transgenes have been assessed. Moreover, very few combinations of genome stability mutations have been studied and information on the effects of (combinations of) less severe mutations is absent. The effects of genome stability mutants are likely to vary between different species and to interact (like the CSB with the XPA mutation). Some genome stability systems are essential for development and, therefore, less interesting to aging. (Although the possibility to make conditional transgenes or knockouts allows to study the effect of embryonically lethal mutations at later ages.) Others, including partially redundant systems, may become of major importance at the later stages of life. Even if loss of genome integrity would be the major cause of aging, then still it is important to realize that there is no single pathway that explains all the characteristics of aging. This explains why, for example, the Werner or Cockayne syndromes show some but not all characteristics of human aging.

6. Discussion and future prospects

Control of life span can be considered as a balance between somatic maintenance and reproductive effort [18]. In theory, therefore, aging can be explained in terms of an accumulation of somatic damage and in this paper the evidence is reevaluated that somatic mutations, i.e., irreversible alterations in DNA sequence, are a major component of the spectrum. There is currently abundant evidence that DNA mutations accumulate with age in most organs and tissues at a frequency that is much higher than originally estimated. Whether this is ultimately due to oxygen free radical damage or some other spontaneous form of DNA damage is currently unclear. For many environmental agents, such as UV, signature mutations have been established and demonstrated to occur in tumors. Although several potential signature mutations have been found for oxygen radicals [15,123], the myriad of possible adducts and other types of lesions (abasic sites, cross links, DNA breaks) severely hampers the use of mutation spectra as a guide to the original source of damage.

The main question that still remains to be addressed involves the functional impact of somatic mutation accumulation in vivo. This may be 3-fold. First, mutations, especially large chromosomal alterations (see above), should be considered capable of causing many of the observed age-related forms of cell and tissue degeneration. This is illustrated by the results obtained with the telomerase-deficient mice linking chromosomal instability to age-related decline of gastrointestinal, hematopoietic and cutaneous systems [120,121]. Second, the age-related increase in mutation frequency that has now been observed in most organs and tissues coincides with an increased incidence of neoplasms. Although in first instance this appears to conflict with the observed increase in cell loss that is also observed during aging, it is now well established that increased genomic instability, for example, due to the inactivation of a DNA repair gene (BRCA1; Ref. [124]), can lead to both increased cell death and cancer. Third, increased loss of genome integrity is likely to activate stress responses at the organ/tissue and organismal level. These may include error-prone repair systems, cell death systems and possibly many others, which in turn may have adverse effects on the organism on the long term.

While the above mentioned mechanisms may explain most aspects of aging, it still remains to be established if the actual mutation loads of cells in aged organisms is sufficient to cause normal aging. To test this, with the present relative luxury of an abundant supply of powerful molecular methods, at least three approaches can be distinguished. The first has already been referred to and involves the transgenic modelling of genome stability systems. A systematic search for the in vivo effects of complete or incomplete mutational inactivation of genes involved in genome stability pathways, e.g., DNA repair, replication, apoptosis, would likely reveal increased genetic instability accompanied by symptoms of accelerated aging. Such studies have already been initiated as described above, but are far from complete due to the lack of life span studies.

A second line of research that could be fruitful in testing the contribution of mutation accumulation to aging is to directly analyze human populations with varying aging rates for polymorphic variation in genes controlling genome stability. This molecular epidemiologic approach is greatly facilitated by the wealth of sequence information that is now becoming available through the human genome program. Unlike the situation in the known syndromes with symptoms of accelerated aging, normal aging is unlikely to be due to one or few completely inactivated genes. Instead, in normal human populations one would expect allelic variation at the many loci harboring genome stability genes to contribute to the observed variation in life span among individuals. As a step in providing a basis for molecular epidemiology studies to relate genetic variation at specific DNA repair genes to reduced DNA repair capacity and cancer susceptibility, Mohrenweiser and Jones [125] re-sequenced exons of nine genes involved in different repair pathways. They identified 15 different amino acid substitution variants in these nine genes and determined their frequencies in a small sample of individuals. Familial aggregation of cancer has been extensively described and there is increased interest in weakly predisposing alleles as risk factors [126]. Thus far, there has been no attempt to investigate subtle differences in genes specifying genome stability pathways between human populations differing greatly in aging rate. It is conceivable that polymorphic variation at multiple loci could have given rise to genotypes conferring extreme longevity, for example, by optimization of genome stability and other cellular defense systems. We have obtained evidence for a strong familial component of longevity in centenarians [127] and are currently comparing allelic variants of genes involved in recombinational repair in such long-lived individuals and control groups. Differences in frequency of the variant alleles would indicate a potential functional advantage of those alleles that are more frequent in the disease-resistant long-lived individuals. Protein bioinformatics methods can then help analyze the sequence and structural variation and provide insight into function.

Finally, DNA mutation accumulation in somatic cells, as far as this does not result in cell death or transformation, is expected to result in changes in gene expression. Age-related changes in gene expression can be a programmatic response to damage accumulation, such as various stress responses, including genotoxic stress. Such changes can be expected to be consistent, that is, involving all or most cells of a tissue. On the other hand, due to the stochastic nature of damage accumulation individual cells in a tissue may express unique gene expression defects, which can only be detected by assays that operate at the single cell level. Recently, there have been dramatic advances in the study of gene expression and some first steps have been taken towards comprehensive gene expression "profiling" of tissues [128] and even individual cells [129]. At the present time such assays lack the sensitivity to detect small increases in expression level. However, progress in the further development and optimization of assays is so rapid that it seems reasonable to assume that within the next decade complete expression profiles of aged tissues and individual cells will become available. Then the enormous task of linking such patterns of change to alterations in function must begin, which is the next great challenge and opportunity for physiology and medicine. Unraveling the different pathways of genome stabilization and a better understanding of their relative importance and long-term impact on health and disease at old age will provide new targets for prevention and therapeutic intervention based on reducing mutation loads [130].

Acknowledgements

I thank Drs. Huber Warner and Mitchell S. Turker for critically reading the manuscript and their useful comments, and Ms. Julia Perkins for her help with the preparation of the manuscript. This work was supported by NIH grants PO1 1801 AG10829-01, 1 P30 AG13319-05 and 1 RO1 ES/CA 08797-01.

References

[1] P.S. Henshaw, E.F. Riley, G.E. Stapleton, The biologic effects of pile radiations, Radiology 49 (1947) 349–364.
[2] G. Failla, The aging process and carcinogenesis, Ann. N. Y. Acad. Sci. 71 (1958) 1124–1135.
[3] L. Szilard, On the nature of the aging process, Proc. Natl. Acad. Sci. U.S.A. 45 (1959) 35–45.
[4] R.W. Hart, R.B. Setlow, Correlation between deoxyribonucleic acid excision-repair and life-span in a number of mammalian species, Proc. Natl. Acad. Sci. U.S.A. 71 (1974) 2169–2173.
[5] P. Alexander, D.I. Connell, Shortening of the life span of mice by irradiation with X-rays and treatment with radiomimetic chemicals, Radiat. Res. 12 (1960) 38–48.
[6] A.M. Clark, M.A. Rubin, The modification by X-irradiation of the life span of haploids and diploids of the wasp, Habrobracon SP., Radiat. Res. 15 (1961) 244–253.
[7] A.A. Morley, Is ageing the result of dominant and co-dominant mutations? J. Theor. Biol. 98 (1982) 469–474.

[8] J. Maynard Smith, A theory of ageing, Nature 184 (1959) 956–957.
[9] P.B. Medawar, An unsolved problem in biology, H.K. Lewis, London, 1952.
[10] B. Charlesworth, Evolution in Age-Structured Populations, Cambridge Univ. Press, Cambridge, 1980.
[11] N.J. Nusbaum, What good is it to get old? Med. Hypotheses 47 (1996) 77–79.
[12] G.M. Martin, S.N. Austad, T.E. Johnson, Genetic analysis of ageing: role of oxidative damage and environmental stresses, Nat. Genet. 13 (1996) 25–34.
[13] G.C. Williams, Pleiotropy, natural selection and the evolution of senescence, Evolution 11 (1957) 398–411.
[14] B.N. Ames, M.K. Shigenaga, T.M. Hagen, Oxidants, antioxidants, and the degenerative diseases of aging, Proc. Natl. Acad. Sci. U.S.A. 90 (1993) 7915–7922.
[15] D.I. Feig, T.M. Reid, L.L. Loeb, Reactive oxygen species in tumorigenesis, Cancer Res. 54 (1994) 1890s–1894s.
[16] A.R. Collins, Oxidative DNA damage, antioxidants, and cancer, BioEssays 21 (1999) 238–246.
[17] D.E.L. Promislow, M. Tatar, Mutation and sensescence: where genetics and demography meet, Genetica 102–103 (1998) 299–314.
[18] T.B.L Kirkwood, DNA, mutations and aging, Mutat. Res. 219 (1989) 1–7.
[19] H.J. Muller, The relation of recombination to mutational advance, Mutat. Res. 1 (1964) 2–9.
[20] T.T. Kibota, M. Lynch, Estimate of the genomic mutation rate deleterious to overall fitness in *E. coli.*, Nature 381 (1996) 694–696.
[21] S.F. Elena, R.E. Lenski, Test of synergistic interactions among deleterious mutations in bacteria, Nature 390 (1997) 395–398.
[22] D.I. Andersson, D. Hughes, Muller's ratchet decreases fitness of a DNA-based microbe, Proc. Natl. Acad. Sci. U.S.A. 93 (1996) 906–907.
[23] G. Bell, Sex and Death in Protozoa, Cambridge University Press, Cambridge, UK, 1988.
[24] T. Lindahl, Instability and decay of the primary structure of DNA, Nature 362 (1993) 709–715.
[25] D. Harman, Role of free radicals in aging and disease, Ann. N. Y. Acad. Sci. 673 (1992) 126–137.
[26] X.C. Le, J.Z. Xing, J. Lee, S.A. Leadon, M. Weinfeld, Inducible repair of thymine glycol detected by an ultrasensitive assay for DNA damage, Science 280 (1998) 1066–1069.
[27] H. Curtis, C. Crowley, Chromosome aberrations in liver cells in relation to the somatic mutation theory of aging, Radiat. Res. 19 (1963) 337–344.
[28] M.J. Ramsey, D.H. Moore II, J.F. Briner, D.A. Lee, L.A. Olsen, J.R. Senft, J.D. Tucker, The effects of age and lifestyle factors on the accumulation of cytogenetic damage as measured by chromosome painting, Mutation Res. 338 (1995) 95–106.
[29] J.D. Tucker, M.D. Spruill, M.J. Ramsey, A.D. Director, J. Nath, Frequency of spontaneous chromosome aberrations in mice: effects of age, Mutation Res. 425 (1999) 135–141.
[30] R.J. Albertini, J.A. Nicklas, J.C. Fuscoe, Th.R. Skopek, R.F. Branda, J.P. O'Neill, In vivo mutations in human blood cells: biomarkers for molecular epidemiology, Environ. Health Perspect. 99 (1993) 135–141.
[31] I.M. Jones, C.B. Thomas, B. Tucker, C.L. Thompson, P. Pleshanov, I. Vorobtsova, D.H. Moore II, Impact of age and environment on somatic mutation at the hprt gene of T lymphocytes in humans, Mutation Res. 338 (1995) 129–139.
[32] J.L. Dempsey, M. Pfeiffer, A.A. Morley, Effect of dietary restriction on in vivo somatic mutation in mice, Mutation Res. 291 (1993) 141–145.
[33] S.A. Grist, M. McCarron, A. Kutlaca, D.R. Turner, A.A. Morley, In vivo human somatic mutation: frequency and spectrum with age, Mutation Res. 266 (1992) 189–196.
[34] A.D. da Cruz, J. Curry, M.P. Curado, B. Glickman, Monitoring hprt mutant frequency over time in T-lymphocytes of people accidentally exposed to high doses of ionizing radiation, Environ. Mol. Mutagen. 27 (1996) 165–175.
[35] R. Weindruch, Caloric restriction and aging, Sci. Am. 274 (1996) 46–52.
[36] A. Aidoo, R.A. Mittelstaedt, L.E. Lyn-Cook, P.H. Duffy, R.H. Heflich, Effect of dietary restriction on lymphocyte HPRT mutant frequency in aging rats, Journal of Environ. Mole. Mutagen. 33 (1999) 5.
[37] Y. Odagiri, H. Uchida, M. Hosokawa, K. Takemoto, A.A. Morley, T. Takeda, Accelerated accumulation of somatic mutations in the senescence-accelerated mouse, Nat. Genet. 19 (2) (1998) 116–117.
[38] G.M. Martin, C.E. Ogburn, L.M. Colgin, A.M. Gown, S.D. Edland, R.J. Monnat Jr., Somatic mutations are frequent and increase with age in human kidney epithelial cells, Hum. Mol. Genet. 5 (1996) 215–221.
[39] J.A. Gossen, W.J.F. de Leeuw, C.H.T. Tan, P.H.M. Lohman, F. Berends, D.L. Knook, E.C. Zwarthoff, J. Vijg, Efficient rescue of integrated shuttle vectors from transgenic mice: a model for studying gene mutations in vivo, Proc. Natl. Acad. Sci. U.S.A. 86 (1989) 7971–7975.
[40] J. Vijg, J.A. Gossen, Somatic mutations and cellular aging, Comp. Biochem. Physiol. 104 B (1993) 429–437.
[41] H.-J. Martus, M.E.T. Dollé, J.A. Gossen, M.E.T.I. Boerrigter, J. Vijg, Use of transgenic mouse models for studying somatic mutations in aging, Mutation Res. 338 (1995) 203–213.
[42] A.T. Lee, C. DeSimone, A. Cerami, R. Bucala, Comparative analysis of DNA mutations in lacI transgenic mice with age, FASEB J. 8 (1994) 545–550.
[43] M.E.T. Dollé, H. Giese, C.L. Hopkins, H.J. Martus, J.M. Hausdorff, J. Vijg, Rapid accumulation of genome rearrangements in liver but not in brain of old mice, Nature Genet. 17 (1997) 431–434.
[44] R.T. Bronson, R.D. Lipman, Reduction in rate of occurrence of age-related lesions in dietary restricted laboratory mice, Growth, Development and Aging 55 (1991) 169–184.
[45] T. Ono, H. Ikehata, S. Nakamura, Y. Saito, Y. Hosoi, Y. Takai, S. Yamada, J. Onodera, K. Yamamoto., Age-associated increase of spontaneous mutant frequency and molecu-

lar nature of mutation in newborn and old lacZ-transgenic mouse, Mutation Res., in press.
[46] H.R. Warner, T.E. Johnson, Parsing age, mutations and time, Nat. Genet. 17 (1997) 368–370.
[47] J.A. Gossen, W.J.F. de Leeuw, A. Verwest, P.H.M. Lohman, J. Vijg, High somatic mutation frequencies in a *LacZ* transgene integrated on the mouse X-chromosome, Mutation Res. 250 (1991) 423–429.
[48] T.G. Krontiris, Minisatellites and human disease, Science 269 (1995) 1682–1683.
[49] M.L. Kimberland, V. Divoky, J. Prchal, U. Schwahn, W. Berger, H.H. Kazazian Jr., Full-length human L1 insertions retain the capacity for high freuquency retrotransposition in cultured cells, Hum. Mol. Genet. 8 (1999) 1557–1560.
[50] A.J. Jeffreys, R. Neumann, V. Wilson, Repeat unit sequence variation in minisatellites: a novel source of DNA polymorphism for studying variation and mutation by single molecule analysis, Cell 60 (1990) 476–485.
[51] Y.E. Dubrova, A.J. Jeffreys, A.M. Malashenko, Mouse minisatellite mutations induced by ionizing radiation, Nat. Genet. 5 (1993) 92–94.
[52] L.-J.C. Wong, T. Ashizawa, D.G. Monckton, C.T. Caskey, C.S. Richards, Somatic heterogeneity of the CTG repeat in myotonic dystrophy is age and size dependent, Am. J. Hum. Genet. 56 (1995) 114–122.
[53] W.E. Wright, J.W. Shay, Time, telomeres and tumours: is cellular senescence more than an anticancer mechanism?, Trends in Cell Biology 5 (1995) 293–297.
[54] A.M. Olovnikov, A theory of marginotomy. The incomplete copying of template margin in enzymic synthesis of polynucleotides and biological significance of the phenomenon, J. Theor. Biol. 41 (1973) 181–190.
[55] A.G. Bodnar, M. Ouellette, M. Frolkis, S.E. Holt, C.P. Chiu, G.B. Morin, C.B. Harley, J.W. Shay, S. Lichtsteiner, W.E. Wright, Extension of life-span by introduction of telomerase into normal human cells, Science 279 (1998) 349–352.
[56] M.A. Blasco, H.W. Lee, M.P. Hande, E. Samper, P.M. Lansdorp, R.A. DePinho, C.W. Greider, Telomere shortening and tumor formation by mouse cells lacking telomerase RNA, Cell; 91 (1) (1997) 25–34.
[57] E. Chang, C.B. Harley, Telomere length and replicative aging in human vascular tissues, Proc. Natl. Acad. Sci. U.S.A. 92 (1995) 11190–11194.
[58] M.G. Butler, J. Tilburt, A. DeVries, B. Muralidhar, G. Aue, L. Hedges, J. Atkins, H. Schwartz, Comparison of chromosome telomere integrity in multiple tissues from subjects at different ages, Cancer Genet. Cytogenet. 105 (1998) 138–144.
[59] G. Cortopassi, Y. Liu, Genotypic selection of mitochondrial and oncogenic mutations in human tissue suggests mechanisms of age-related pathophysiology, Mutation Res. 338 (1995) 151–159.
[60] D.C. Wallace, Mitochondrial DNA variation in human evolution, degenerative disease, and aging, Am. J. Hum. Genet. 57 (1995) 201–223.
[61] V.A. Bohr, G.L. Dianov, Oxidative DNA damage processing in nuclear and mitochondrial DNA, Biochimie 81 (1999) 155–160.
[62] U. Lakshmipathy, C. Campbell, Double strand break rejoining by mammalian mitochondrial extracts, Nucleic Acids Res. 27 (1999) 1198–1204.
[63] K. Khrapko, N. Bodyak, W.G. Thilly, N.J. van Orsouw, X. Zhang, H.A. Coller, T.T. Perls, M. Upton, J. Vijg, J.Y. Wei, Cell-by-Cell scanning of whole mitochondrial genomes in aged human heart reveals a significant fraction of myocytes with clonally expanded deletions, Nucleic Acids Res. 27 (1999) 2434–2441.
[64] G. Kopsidas, S.A. Kovalenko, J.M. Kelso, A.W. Linnane, An age-associated correlation between cellular bioenergy decline and mtDNA rearrangements in human skeletal muscle, Mutation Res. 421 (1998) 27–36.
[65] H.J. Evans, Mutation and mutagenesis in inherited and acquired human disease, Mutation Research Forum 1 (1996) 5–10.
[66] Y. Kamiguchi, H. Tateno, K. Mikamo, Chromosomally abnormal gametes as a cause of congenital anomalies in humans, Congenital Anomalies. 34 (1994) 1–12.
[67] M. Plachot, J. De Grouchi, A.-M. Junca, J. Mandelbaum, C. Turleau, P. Couillin, J. Cohen, J. Salat-Baroux, From oocyte to embryo: a model, deduced from in vitro fertilization, for natural selection against chromosome abnormalities, Ann. Génét. 30 (1987) 22–32.
[68] C.N. Svendsen, A.G. Smith, New prospects for human stem-cell therapy in the nervous system, Trends Neurosci. 22 (1999) 357–364.
[69] I. Wilmut, A.E. Schnieke, J. McWhir, A.J. Kind, K.H.S. Campbell, Viable offspring derived from fetal and adult mammalian cells, Nature 385 (1997) 810–813.
[70] T. Wakayama, A.C. Perry, M. Zuccotti, K.R. Johnson, R. Yanagimachi, Full-term development of mice from enucleated oocytes injected with cumulus cell nuclei, Nature 394 (1998) 369–374.
[71] C. Franceschi, Cell proliferation, cell death and aging, Aging 1 (1989) 3–15.
[72] P.J. Barr, L.D. Tomei, Apoptosis and its role in human disease, Biotechnology 12 (1994) 487–493.
[73] H.R. Warner, G. Fernandes, E. Wang, A unifying hypothesis to explain the retardation of aging and tumorigenesis by caloric restriction, J. Gerontol. 50A (1995) B107–B109.
[74] S.J. James, L. Muskhelishvili, Rates of apoptosis and proliferation vary with caloric intake and may influence incidence of spontaneous hepatoma in C57BL/6×C3H F1 mice, Cancer Res. 54 (1994) 5508–5510.
[75] L. Muskhelishvilli, R.W. Hart, A. Turturro, S.J .James, Age-related changes in the intrinsic rate of apoptosis in livers of diet-restricted and ad libitum-fed B6C3F1 mice, Am. J. Pathol. 147 (1995) 20–24.
[76] C. Lengauer, K.W. Kinzler, B. Vogelstein, Genetic instabilities in human cancers, Nature 396 (1998) 643–649.
[77] M.S. Greenblatt, W.P. Bennett, M. Hollstein, C.C. Harris, Mutations in the p53 tumor suppressor gene: clues to cancer etiology and molecular pathogenesis, Cancer Res. 54 (1994) 4855–4878.

[78] J. Vijg, H. van Steeg, Transgenic assays for mutations and cancer: current status and future perspectives, Mutation Res. 400 (1998) 337–354.
[79] E.R. Fearon, B. Vogelstein, A genetic model for colorectal tumorigenesis, Cell 61 (1990) 759–767.
[80] Y. Murakami, T. Sekiya, Accumulation of genetic alterations and their significance in each primary human cancer and cell line, Mutation Res. 400 (1998) 421–437.
[81] L.A. Loeb, Mutator phenotype may be required for multistage carcinogenesis, Cancer Res. 51 (1991) 3075–3079.
[82] I.P.M. Tomlinson, M.R. Novelli, W.F. Bodmer, The mutation rate and cancer, Proc. Natl. Acad. Sci. U.S.A. 93 (1996) 14800–14803.
[83] M. Perucho, Microsatellite instability: the mutator that mutates the other mutator, Nat. Med. 2 (1996) 676–681.
[84] R. Parsons, G.-M. Li, M. Longley, P. Modrich, B. Liu, T. Berk, S.R. Hamilton, K.W. Kinzler, B. Vogelstein, Mismatch repair deficiency in phenotypically normal human cells, Science 268 (1995) 738–740.
[85] A. Baross-Francis, S.E. Andrew, J.E. Penney, F.R. Jirik, Tumors of DNA mismatch repair-deficient hosts exhibit dramatic increases in genome stability, Proc. Natl. Acad. Sci. U.S.A. 95 (1998) 8739–8743.
[86] V.L. Buettner, K.A. Hill, H. Nishino, D.J. Schaid, C.S. Frisk, S.S. Sommer, Increased mutation frequency and altered spectrum in one of four thymic lymphomas derived from tumor prone p53/Big Blue double transgenic mice, Oncogene 13 (1996) 2407–2413.
[87] M.S. Turker, Estimation of mutation frequencies in normal mammalian cells and the development of cancer, Seminars in Cancer Biology 8 (1998) 407–419.
[88] C. Shao, L. Deng, O. Henegariu, L. Liang, N. Raikwar, A. Sahota, P.J. Stambrook, J.A. Tischfield, Mitotic recombination produces the majority of recessive fibroblast variants in heterozygous mice, Proc. Natl. Acad. Sci. U.S.A. 96 (1999) 9230–9235.
[89] L. Hayflick, P.S. Moorhead, The serial cultivation of human diploid cell strains, Exp. Cell Res. 25 (1961) 585–621.
[90] J. Campisi, Replicative senescence: an old lives' tale?, Cell 84 (1996) 497–500.
[91] E. Wang, Senescent human fibroblasts resist programmed cell death, and failure to suppress bcl2 is involved, Cancer Res. 55 (1995) 2284–2292.
[92] G.P. Dimri, X. Lee, G. Basile, M. Acosta, G. Scott, C. Roskelley, E.E. Medrano, M. Linskens, I. Rubelj, O. Pereira-Smith, M. Peacocke, J.A. Campisi, A biomarker that identifies senescent human cells in culture and in aging skin in vivo, Proc. Natl. Acad. Sci. U. S. A. 92 (1995) 9363–9367.
[93] K. Van Holde, J. Zlatanova, Unusual DNA structures, chromatin and transcription, Bioessays 16 (1993) 59–68.
[94] B.S. Strauss, The origin of point mutations in human tumor cells, Cancer Res. 52 (1992) 249–253.
[95] D.G. MacPhee, Mismatch repair, somatic mutations, and the origins of cancer, Cancer Res. 55 (1995) 5489–5492.
[96] A.A. Morley, The estimation of in vivo mutation rate and frequency from samples of human lymphocytes, Mutation Res. 357 (1996) 167–176.
[97] M.S. Turker, G.M. Martin, Genetics of human disease, longevity and aging, in: Principles of Geriatric Medicine and Gerontology 4th edition, W.R. Hazzard, J.P. Blass, W.H. Ettinger Jr., J.B. Halter, J.G. Outslander (Eds.), McGraw-Hill, 1999, pp. 21–44.
[98] C.E. Yu, J. Oshima, Y.-.H. Fu, E.M. Wijsman, F. Hisama, R. Alisch *et al.*, Positional cloning of the Werner's syndrome gene, Science 272 (1996) 258–262.
[99] N.F. Neff, N.A. Ellis, T.Z. Ye, J. Noonan, K. Huang, M. Sanz, M. Proytcheva, The DNA helicase activity of BLM is necessary for the correction of the genomic instability of bloom syndrome cells, Mol. Biol. Cell 10 (1999) 665–676.
[100] K. Fukuchi, G.M. Martin, R.J. Monnat Jr., Mutator phenotype of Werner syndrome is characterized by extensive deletions, Proc. Natl. Acad. Sci. U.S.A. 86 (1989) 5893–5897.
[101] S. Huang, B. Li, M.D. Gray, J. Oshima, I.S. Mian, J. Campisi, The premature ageing syndrome protein, WRN, is a 3′ → 5′ exonuclease, Nat. Genet. 20 (1998) 114–116.
[102] J.C. Shen, M.D. Gray, J. Oshima, A.S. Kamath-Loeb, M. Fry, L.A. Loeb, Werner syndrome protein.: I. DNA helicase and DNA exonuclease reside on the same polypeptide, J. Biol. Chem. 273 (1998) 34139–34144.
[103] E. Castro, C.E. Ogburn, K.E. Hunt, R. Tilvis, J. Louhija, R. Penttinen, R. Erkkola, A. Panduro, R. Riestra, C. Piussan, S.S. Deeb, L. Wang, S.D. Edland, G.M. Martin, J. Oshima, Polymorphisms at the Werner locus: I. Newly identified polymorphisms, ethnic variability of 1367Cy/Arg, and its stability in a population of Finnish centenarians, Am. J. Med. Genet. 82 (1999) 399–403.
[104] A.J. Van Gool, G.T.J. Van der Horst, E. Citterio, J.H.J. Hoeijmakers, Cockayne syndrome: defective repair of transcription?, EMBO J. 16 (1997) 4155–4162.
[105] Y.W. Lin, M. Kubota, H. Hirota, K. Furusho, K. Tomiwa, J. Ochi, Y. Kasahara, H. Sasaki, S. Ohta, Somatic cell mutation frequency at the HPRT, T-cell antigen receptor and glycophorin A loci in Cockayne syndrome, Mutation Res. 337 (1995) 49–55.
[106] G.L. Dianov, J.F. Houle, N. Iyer, V.A. Bohr, E.C. Friedberg, Reduced RNA polymerase II transcription in extracts of Cockayne syndrome and xeroderma pigmentosum/Cockayne syndrome cells, Nucleic Acids Res. 25 (1997) 3636–3642.
[107] M. Lebel, P.A. Leder, A deletion within the murine Werner syndrome helicase induces sensitivity to inhibitors of topoisomerase and loss of cellular proliferative capacity, Proc. Natl. Acad. Sci. U. S. A. 95 (1998) 13097–13102.
[108] G.T. van der Horst, H. van Steeg, R.J. Berg, A.J. van Gool, J. de Wit, G. Weeda, H. Morreau, R.B. Beems, C.F. van Kreijl, F.R. de Gruijl, D. Bootsma, J.H. Hoeijmakers, Defective transcription-coupled repair in Cockayne syndrome B mice is associated with skin cancer predisposition, Cell 89 (1997) 425–435.
[109] A. De Vries, F.M.A. C.Th.M. Van Oostrom, P.M. Hofhuis,

R.J.W. Dortant, F.R. Berg, P.W. de Gruijl, C.F. Wester, P.J.A. Van Kreijl, H. Capel, S.J. Van Steeg, Increased susceptibility to ultraviolet-B and carcinogens of mice lacking the DNA excision repair gene XPA, Nature 377 (1995) 169–173.

[110] J. de Boer, J.H. Hoeijmakers, Cancer from the outside, aging from the inside: mouse models to study the consequences of defective nucleotide excision repair, Biochimie 81 (1999) 127–137.

[111] A. De Vries, P.M. C.Th.M. Van Oostrom, R.B. Dortant, C.F. Beems, P.J.A. Van Kreijl, H. Capel, Spontaneous liver tumours and Benzo[a]pyrene-induced lymphomas in XPA-deficient mice, Mol. Carcinogen. 19 (1997) 46–53.

[112] N. de Wind, M. Dekker, A. van Rossum, M. van der Valk, H. te Riele, Mouse models for hereditary nonpolyposis colorectal cancer, Cancer Res. 58 (1998) 248–255.

[113] A.H. Reitmair, M. Redston, J.C. Cai, T.C. Chuang, M. Bjerknes, H. Cheng, K. Hay, S. Gallinger, B. Bapat, T.W. Mak, Spontaneous intestinal carcinomas and skin neoplasms in Msh2-deficient mice, Cancer Res. 56 (1996) 3842–3849.

[114] S.E. Andrew, A.H. Reitmair, J. Fox, L. Hsiao, A. Francis, M. McKinnon, T.W. Mak, F.R. Jirik, Base transitions dominate the mutational spectrum of a transgenic reporter gene in MSH2 deficient mice, Oncogene 15 (1997) 123–129.

[115] L. Narayanan, J.A. Fritzell, S.M. Baker, R.M. Liskay, P.M. Glazer, Elevated levels of mutation in multiple tissues of mice deficient in the DNA mismatch repair gene Pms2, Proc. Natl. Acad. Sci. U.S.A. 94 (1997) 3122–3127.

[116] H. Gu, J.D. Marth, P.C. Orban, H. Mossmann, K. Rajewsky, Deletion of a DNA polymerase beta gene segment in T cells using cell type-specific gene targeting, Science 265 (1994) 103–106.

[117] H. Giese, M.E.T. Dollé, A. Hezel,, H. van Steeg,, J. Vijg, Accelerated accumulation of somatic mutations in mice deficient in the nucleotide excision repair gene XPA, Oncogene 18 (1999) 1257–1260.

[118] J.E. Cleaver, K.H. Kraemer, in: The Metabolic and Molecular Bases of Inherited Disease III 7th edn., C.R. Scriver, A.L. Beaudet, W.S. Sly, D. Valle (Eds.), McGraw-Hill, New York, 1995, pp. 4393–4419.

[119] B.L. Strehler, M.R. Freeman, Randomness, redundancy and repair: roles and relevance to biological aging, Mech. Ageing Dev. 14 (1980) 15–38.

[120] H.W. Lee, M.A. Blasco, G.J. Gottlieb, J.W. Horner 2nd, C.W. Greider, R.A. DePinho, Essential role of mouse telomerase in highly proliferative organs, Nature 392 (1998) 569–574.

[121] K.L. Rudolph, S. Chang, H.W. Lee, M. Blasco, G.J. Gottlieb, C. Greider, R.A. DePinho, Longevity, stress response, and cancer in aging telomerase-deficient mice, Cell 96 (1999) 701–712.

[122] G.M. Martin, Genetic and environmental modulations of chromosomal stability: their roles in aging and oncogenesis, Ann. N. Y. Acad. Sci. 621 (1991) 401–417.

[123] M.S. Turker, B.M. Gage, J.A. Rose, D. Elroy, O.N. Ponomareva, P.J. Stambrook, J.A. Tischfield, A novel signature mutation for oxidative damage resembles a mutational pattern found commonly in human cancers, Cancer Res. 59 (1999) 1837–1839.

[124] X. Xu, K.U. Wagner, D. Larson, Z. Weaver, C. Li, T. Ried, L. Hennighausen, A. Wynshaw-Boris, C.X. Deng, Conditional mutation of Brca1 in mammary epithelial cells results in blunted ductal morphogenesis and tumour formation, Nat. Genet. 22 (1999) 37–43.

[125] H.W. Mohrenweiser, I.M. Jones, Variation in DNA repair is a factor in cancer susceptibility: a paradigm for the promises and perils of individual and population risk estimation?, Mutation Res. 400 (1998) 15–24.

[126] R.L. White, Excess risk of colon cancer associated with a polymorphism of the APC gene, Cancer Res. 58 (1998) 4038–4039.

[127] T.T. Perls, E. Bubrick, C.G. Wager, J. Vijg, L. Kruglyak, Siblings of centenarians live longer, Lancet 351 (1998) 1560.

[128] C.M. Perou, S.S. Jeffrey, M. van de Rijn, C.A. Rees, M.B. Eisen, D.T. Ross, A. Pergamenschikov, C.F. Williams, S.X. Zhu, J.C. Lee, D. Lashkari, D. Shalon, P.O. Brown, D. Botstein, Distinctive gene expression patterns in human mammary epithelial cells and breast cancers, Proc. Natl. Acad. Sci. U.S.A. 96 (16) (1999) 9212–9217.

[129] J.E. Kacharmina, P.B. Crino, J. Eberwine, Preparation of cDNA from single cells and subcellular regions, Methods Enzymol. 303 (1999) 3–18.

[130] J. Vijg, J.Y. Wei, Understanding the biology of aging: the key to prevention and therapy? J. Am. Geriatr. Soc. 43 (1995) 426–434.

Norman R. Klinman
Gregory H. Kline

The B-cell biology of aging

Authors' address
Norman R. Klinman, Gregory H. Kline,
Department of Immunology, The Scripps Research Institute, La Jolla, California, USA.

Correspondence to:
Norman R. Klinman
Department of Immunology, IMM-16
The Scripps Research Institute
10550 N. Torrey Pines Road
La Jolla CA 92037
USA
Fax: 1 619 784 2691
e-mail: nklinman@scripps.edu

Summary: Although both the number and responsiveness of peripheral B cells in aged mice remain relatively intact, there are dramatic changes in B-cell generation. Alterations in B-cell development include both a skewing of V-gene utilization, especially in cells responsive to phosphorylcholine, and a decrease in the generation of various developmental B-cell subsets. The altered representation of these subsets appears to be the consequence of two developmental blocks. The first developmental block occurs during the maturation of pro-B cells and is evidenced by a decrease in the number of pre-B cells. The second developmental block occurs at the earliest stage of sIg⁺-cell maturation ($sIgM^{very\ lo}$). Because of this block in B-cell maturation, in spite of a decrease in incoming pre-B cells, the number of $sIgM^{very\ lo}$ cells appears to increase in aged mice. Additionally, the time of residence of cells within this maturational stage increases dramatically, while the proportion of cells in more mature ($sIgM^{hi}$) stages of bone marrow development are decreased. In addition to the decreased number of maturing bone marrow B cells, the population of splenic B cells that represent recent bone marrow émigrés ($HSA^{very\ hi}$) is markedly decreased. In the face of this decrease in newly emerging cells from the bone marrow, the population of mature splenic B cells is maintained by their increased longevity.

Introduction

Humoral immune responsiveness and antibody-mediated defense mechanisms are markedly reduced with aging (1–3). Although the presumption has been that much of this decrease is due to decreased helper T-cell (T_H) function rather than an intrinsic primary B-cell deficit, in recent years aging-associated alterations in B-cell repertoire expression (4–8) and in the generation of both primary (9–14) and memory B cells (15–17) have been documented. Thus, the decreased humoral responsiveness of aged individuals is apparently due not only to alterations in the environment and ancillary stimulatory mechanisms, but also to alterations in B cells *per se*. Although this review will address the alterations in the stimulatory environment of B cells of aged mice, we will focus most of our attention on alterations with aging in the expression of the B-cell repertoire.

Immunological Reviews 1997
Vol. 160: 103–114
Printed in Denmark. All rights reserved

Immunological Reviews
ISSN 0105-2896

B-cell responsiveness in aged mice

In order to compare the antigen responsiveness of B cells obtained from aged vs young mice, this laboratory has used an *in vitro* fragment culture protocol wherein responses of limiting numbers of transferred B cells can be assessed at the single cell level (4, 5, 15, 18–25). Importantly, since the B cells obtained from both aged and young mice are transferred to identical carrier-primed, irradiated, young adult recipients prior to the establishment of fragment cultures and *in vitro* stimulation, the stimulatory environment for B cells of young and aged mice can be normalized and responses compared directly.

The impact of the aged environment

As detailed in other reviews in this issue, the response of B cells in aged individuals to T-cell dependent antigenic stimulation is, in part, diminished by a decrease in T_H and possibly accessory cell function (3). In addition, B-cell responsiveness is dampened by what appears to be an idiotype specific downregulation (26–30). While the physiological impact of "network" regulatory phenomena on the establishment and stimulation of the B cell repertoire of naive mice remains controversial (31–33), the generation of anti-idiotypic regulation following antigenic stimulation has been well documented (31–33, 34). Evidence for the acquisition of an antibody specific regulatory phenomenon following antigenic stimulation was obtained by analyses of the ability of transferred hapten-specific primary B cells to respond in fragment cultures derived from irradiated recipients that had been primed two months previously with either carrier protein alone or, alternatively, carrier as well as the entire hapten-carrier complex (26, 34). Responses of B cells in fragment cultures are greatly enhanced when spleen fragments are derived from irradiated recipients that previously have been carrier-primed, presumably as a result of the presence of carrier-specific T_H (18, 19). If, instead of using the carrier protein alone to immunize recipients for subsequent *in vitro* stimulation with hapten-carrier complexes, the hapten-carrier complex was also used to prime recipients, the responsiveness of hapten-specific B cells was markedly reduced (see *Table 1*) (34). Because this inhibition of responsiveness was obviated when B cells of a donor differing from the recipient in IgH (and therefore antibody V region) haplotype, the inhibition was assumed to be antibody or idiotype-specific. Importantly, such suppression was not present in young recipients prior to antigenic stimulation since priming with carrier alone optimized responses of both IgH-syngeneic and allogeneic B cells.

Reasoning that a lifetime of antigenic contacts may have engendered in aged individuals the accumulation of suppression of responses to a broad range of antigenic determinants, similar analyses were carried out using aged mice as carrier-primed recipients (26). Consistent with the spontaneous accumulation of antibody (idiotype) specific regulation with aging, responses of primary IgH-syngeneic but not allogeneic B cells were suppressed in aged recipients primed with carrier alone (*Table 1*). Indeed, this suppression could be transferred to young carrier-primed recipients by the transfer of spleen cells from aged to young mice.

Table 1. Antibody-specific immunoregulation develops spontaneously in aged mice

Donor* (B cells)	Recipient (BALB/c) Age (months)	Primed with Hy	DNP-Hy	Stimulatory† antigen	Clonal responses‡ per 10^6 B cells
BALB/c (Igh^a)	2	+		DNP-Hy	12
BALB/c (Igh^a)	2			DNP-Hy	4
B10.D2 (Igh^b)	2			DNP-Hy	13
B10.D2 (Igh^b)	2			DNP-Hy	14
BALB/c (Igh^a)	24			DNP-Hy	3
B10.D2 (Igh^b)	24			DNP-Hy	12
BALB/c (Igh^a)	2 (+ 4 x 10^7 spleen cells from 24-month old BALB/c)	+	-	DNP-Hy	4
B10.D2 (Igh^b)	2 (+ 4 x 10^7 spleen cells from 24-month old BALB/c)	+	-	DNP-Hy	12

*4 x 10^7 Igh^a or Igh^b spleen cells were transferred to each recipient. Each data point represents 10 or more recipients.

†All cultures were stimulated on days 0–1 with DNP-Hy at 10^{-8} M DNP.

‡Positive foci were detected by radioimmunoassay of culture fluids 1–2 weeks after stimulation.

Antigen responsiveness of primary B cells of aged mice

Although the total number of peripheral B cells is not reduced in aged mice, there is a reduction in the frequency of B cells responsive to some, but not all, antigens (15, 20–25). Thus, whereas the frequency of cells responsive to the haptenic determinants 2,4-dintrophenyl (DNP) and acetylnitrophenol (NP) is reduced 2–3-fold in aged mice (15, 20), the frequency of cells responsive to influenza virus antigens is unchanged (21) and those responsive to phosphorylcholine (PC) increases (4, 5) (*Table 2*).

Table 2. Antigen-specific precursor cell frequencies determined in fragment cultures

			Average frequency/10^6 injected cells			
			Spleen cells		sIg $^{lo-}$ bone marrow cells	
Frequency	Antigen	Mouse strain	Old[a]	Young[b]	Old[a]	Young[b]
Reduced	DNP	BALB/c	1.1	2.3	0.4	0.57
	NP	BALB/c	0.28	0.54	0.13	0.15
		B10.D2	0.39	0.75	-	-
	NNP	B10.D2	0.7	1.78	-	
Normal	PR8	BALB/c	0.57	0.56	-	
	PR8-HA	BALB/c	0.23	0.22	-	-
Elevated	PC	BALB/c	0.97	0.57	0.25	0.04

[a] Old mice were 24 months old.

[b] Young mice were 3 months old.

Source: abstracted from (4, 15, 20, 21, 22).

Table 3. Frequency of secondary B cells in aged and young mice

	Average frequency per 10^6 cells injected[a]		
	Old	Young	Young/old
DNP	7.6	22.1	2.9
NP	1.22	6.9	5.65

[a] The frequency of secondary splenic B cells was determined in fragment cultures 4 weeks after intraperitoneal injection of 100 µg of either DNP-BSA or NP-BSA in complete Fruend's adjuvant.

Source: abstracted from (15, 22).

Although the frequency of B cells responsive to various antigenic determinants differed between B cells obtained from aged *vs* young adult mice, the monoclonal responses obtained from responding cells was qualitatively similar (20). Most importantly, the amount of antibody produced by the resultant antibody forming cell (AFC) clones was comparable. This suggests that AFC clones derived from B cells of aged mice are deficient in neither their clonal expansion capacity nor in the amount of antibody produced by each AFC. Additionally, the clones of AFC derived from B cells of aged mice included antibodies of all tested isotypes. Thus, although switching to IgG_1 was somewhat less than for cells obtained from young mice, the clonal progeny of B cells of aged mice were competent to isotype switch.

In addition to the ability of the AFC progeny of B cells of aged mice to isotype switch and produce normal amounts of antibody, the affinity of the antibodies produced was indistinguishable from that produced from the AFC progeny of B cells of young mice (20). Consistent with findings that B cells of aged mice can make antibodies displaying a normal range of affinities, analyses of the repertoire of antibodies responsive to the influenza PR8 hemagglutinin (PR8-HA) showed extensive diversity indicating that antibody repertoire diversity was not demonstrably decreased in aged mice (21).

Memory B-cell generation in aged mice

It has long been known that aged mice respond poorly to immunization, generating both low antibody responses and little memory (1–3, 15–17). The decrease in memory B-cell generation in aged mice was quantified by the analysis of the frequency of cells responsive to the NP and DNP haptenic determinant after immunization. It can be seen in *Table 3* that aged mice generate far fewer memory B cells after immunization than do young adult mice (15, 22).

This decrease becomes even more dramatic in the light of subsequent findings indicating that the frequency of naive B cells that could give rise to memory B cells is at least as high in aged as in young adult mice (35). These findings were based on an analysis of splenic B cells obtained from young *vs* aged mice that were isolated on the basis of their low expression of cell surface heat-stable antigen (HSA^{lo}) as detected by the J11D monoclonal antibody (36). Previous studies from this laboratory had demonstrated that cells within the HSA^{lo} population are greatly enriched for cells that can give rise to memory B-cell responses both *in vivo* and *in vitro* (37–42). Consistent with this conclusion, the population of HSA^{lo} cells is also enriched for cells that can originate germinal centers (GC) in reconstituted SCID recipients (41) and can generate progeny that accumulate variable (V) region somatic mutations (37, 42). Conversely, splenic B-cell populations that have higher levels of HSA expression (HSA^{int-hi}) are enriched for cells that can give rise to primary AFC clones but are depleted of cells that can generate memory B cells, somatic mutations, or GC (37, 41, 42).

When DNP-specific HSA^{lo} cells were analyzed in fragment culture assays, wherein T-cell help is provided by carrier-primed T cells from young adults, the frequency of cells obtained from aged mice that could give rise to memory responses was at least as high as HSA^{lo} cells of young adult mice (35). Thus, in spite of an overall lower frequency of cells that could give rise to primary anti-DNP AFC responses, the population of naive progenitors of memory cells was intact.

Studies from other laboratories assessing the ability of cells of aged mice to generate GC responses *in vivo* have also concluded that aged mice are deficient in their capacity to generate memory B cells, especially those that are somatically mutated (16, 17). As anticipated, this defect was shown to be in part due to decreased T_H function. However, these authors also found that, even when B cells from aged mice were stimulated in the milieu of a young adult with young adult T_H, responses were also diminished (17). They concluded, therefore, that not only are T_H of aged mice deficient, but also that B cells of aged mice are intrinsically less able to generate somatically mutated memory B cells.

These findings would appear to contrast with findings from fragment culture assays wherein B cells from aged mice, stimulated in the presence of carrier primed T_H from young mice, mounted normal numbers of memory B-cell clones (35). However, although the frequency of responsive progenitors of memory B cells in aged mice may not be decreased, these cells may be deficient in their ability to accumulate somatic mutations. Recently, in studies of cells from young adult mice, this laboratory has been able to demonstrate that HSA^{lo} B cells stimulated *in vitro* generate clonal progeny that accumulate mutations at a rate comparable to that obtained *in vivo* (42). In order to determine if naive memory progenitors obtained from aged mice are comparable to those of young mice in their ability to accumulate somatic mutations when stimulated in fragment cultures derived from carrier primed and T_H reconstituted young adult mice, we have initiated an analysis of the rate of somatic mutation of HSA^{lo} cells from aged mice stimulated in fragment cultures. To date our findings indicate that the clonal progeny of naive memory progenitors obtained from aged mice can accumulate somatic mutations at a rate comparable to that of HSA^{lo} cells of young adults (N. Klinman, S. Zharavetiz, P. Linton, preliminary observations). However, our data does not address the possibility that newly generated memory B cells obtained from progenitors of aged mice may be deficient at the level of selection for high affinity.

B-cell development in aged mice

Although aged mice appear to have relatively normal numbers of mature peripheral B cells and normal responsiveness to some antigens, B-cell development is markedly altered in aged mice (4, 5, 9–14). These alterations in B-cell development include both alterations in the V-gene repertoire of newly generating B cells responsive to selected antigens (4, 5) and a dramatic difference in both the turnover kinetics and proportionate representation of various developmental and maturational B-cell subsets.

Altered repertoire development

The aforementioned studies of B-cell responses to the influenza viral antigens indicated that repertoire diversity was not measurably diminished in aged mice (21). Nonetheless, the possibility existed that, at least for some antigens, repertoire generation may be skewed. This possibility was suggested by analyses of responses of mature splenic B cells to the antigenic determinant PC (4). Unlike B cells responsive to the haptenic determinants DNP and NP, the frequency of B cells responsive to the PC was substantially increased in aged BALB/c mice. Since PC is present on bacteria and is a common environmental antigen, the possibility existed that this increased frequency in PC-responsive B cells could have been generated by antigenic stimulation and a resultant clonal expansion of peripheral PC-specific B cells. In order to investigate the origin of the increased frequency of B cells responsive to PC in aged mice, we assessed the frequency and clonotypic distribution of PC-specific precursors in newly generating bone marrow B cells of aged mice (4, 5). In Fig. 1, it can be seen that among newly generating bone marrow B cells, the overall frequency of PC-specific cells is markedly increased in aged *vs* young mice. This finding indicated that the alteration in expression of peripheral PC-specific B cells was due, at least in part, to an alteration at the level of B-cell generation in aged *vs* young mice.

Even more dramatic than the increase in the overall frequency of newly generated, anti-PC-specific, bone marrow B cells in aged mice was an alteration in the repertoire of antibody V-region utilization in B cells that were included in the response to PC in aged BALB/c mice. The response to PC has been well characterized in several murine strains (43–50). This response is unusual in that all tested inbred murine strains generate a highly restricted set of antibodies almost all of which utilize the V_HS107-V_1 H chain V-region gene segment in association with a small set of κ light (L) chains (44–47). In neonatal and adult BALB/c mice, the majority of all PC-responsive B cells express exactly the same clonotype which is identical to the TEPC-15 myeloma protein (T15) (43). This clonotype is comprised of the V_HS107-V_1 V_H-gene segment recombined with the DFL16.1 D_H-gene segment and J_{H1} at precise junctional residues with no N additions. The L chains incorporated in these antibodies utilize $V_\kappa T15$ of the $V_{\kappa 22}$-gene family recombined at a precise junctional residue with $J_{\kappa 5}$. Because anti-PC B cells are primarily found within the B-1 compartment (51, 52), the high frequency of T15 B cells may, in part, be accounted for by clonal expansion of mature B cells. However, analyses of newly generating bone marrow B cells have demonstrated that the T15 clonotype is generated recurrently in the marrow of numerous bones throughout the lifetime of individual mice (50).

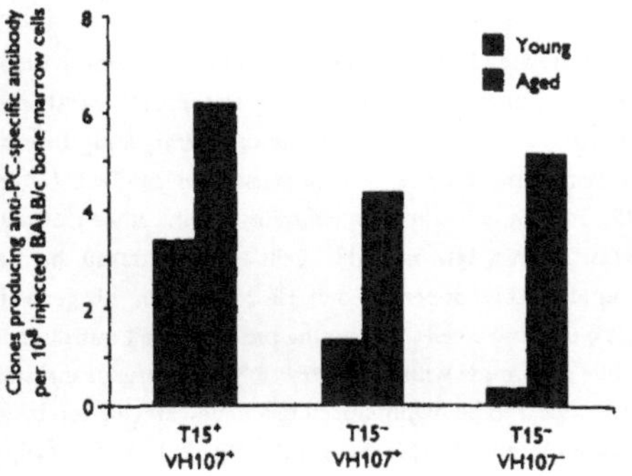

Fig. 1. Analysis of the frequency and Ig variable-region phenotype of anti-PC-responsive newly generating bone marrow cells (sIg$^{-\text{ to lo}}$) of aged vs young adult BALB/c mice. It can be seen that the majority of PC responsive cells in young adult mice are indistinguishable from T15 and over 90% are either T15 or share V_H with T15. In aged mice, the total number of PC-responsive cells increases greatly, including an increase in those that are indistinguishable from T15. In contrast to cells from young adult mice, PC-responsive cells of aged mice include a significant proportion of those whose H chains are encoded by a member of the V_HS107-gene segment family but whose antibody does not share idiotype with T15. Most dramatic is the increased frequency of anti-PC-specific B cells whose V_H genes are encoded by V_H-gene segments of families other than V_HS107, a phenotype that is very rare in young mice. Abstracted from (4, 5).

The predominance of the T15 clonotype among newly generating bone marrow B cells of young adult BALB/c mice can be seen in Fig. 1. Also shown in Fig. 1 is an analysis of newly generated bone marrow B cells of aged BALB/c mice (4, 5). It can be seen that the overall frequency of anti-PC B cells is increased in aged mice, and that this includes an increase in those B cells whose antibody product is identical to T15 in both V_H utilization and interaction with T15 specific anti-idiotypic antibodies. Additionally, however, there is a marked increase in those anti-PC B cells whose antibodies utilize a member of the V_HS107 V_H-gene segment family but do not bear T15 idiotypic determinants. Most dramatic is the marked increase in PC-responsive B cells whose antibodies utilize V_H-gene segment families other than V_HS107. Importantly, the binding of all of these antibodies could be inhibited with the PC-haptenic determinant indicating that these antibodies were truly PC specific.

Anti-PC antibodies that utilize non-V_HS107-encoded H chains are extremely rare in young adult BALB/c mice and mice of other strains as well. Thus the high frequency of such cells among newly generated B cells of aged mice represents a marked and novel aging-associated change in B-cell repertoire generation. This alteration has been verified by analyses of anti-PC hybridoma antibodies wherein anti-PC antibodies utilizing gene segments of the V_H J558, X-24 and 7183 gene segment families were obtained from B cells of aged but not young BALB/c mice (5). Furthermore, this finding has been confirmed by other investigators who have identified non-V_HS107-encoded antibodies in both aged BALB/c mice and aged mice of other murine strains (6, 7). Additionally, these investigators have shown that such antibodies are lower affinity and generally less protective than T15 antibodies. Thus, this aging-associated alteration in B-cell repertoire generation could have important consequences to the ability of aged mice to mount a protective immune response to certain bacterial pathogens. In addition to this increase and alteration in the repertoire of anti-PC-specific B cells in aged mice, others have found an alteration in B cells responsive to DNP (8). However, the molecular basis for these aging-associated alterations in B-cell specificity repertoire expression has yet to be identified.

The kinetics of B-cell development in young adult mice

In order to understand aging-specific alterations in B-cell development, it is important to review B-cell development in young adult mice. In fetal and early neonatal life, B cells are generated in the liver and spleen, but within the first week of neonatal life and throughout adulthood, B-cell generation shifts almost exclusively to the bone marrow (9–11, 53–55). For the first several weeks of post-natal development, there is vigorous B-cell neogenesis with most peripheral B cells representing relatively newly generated cells. However, with time, bone marrow B-cell neogenisis declines until, at 4 months of age, a steady state of 1–2 × 10^7 B cells per day leave the bone marrow. These cells, upon leaving the marrow, can be identified as HSA$^{\text{very hi}}$, IgM^{hi} and IgD$^{\text{neg-lo}}$ (56, 57). This newly generated subset represents only 5–10% of splenic B cells. Most B cells that are newly generated from the bone marrow appear to have a half-life of only 1–4 days, however, a small minority of these cells can enter the long-lived pool of splenic B cells (58). At the present time, the requisites for increased longevity are not known; however, some evidence exists that this selection for a long-lived vs short-lived status may be sIg-receptor mediated (59). Nonetheless, cells selected into the long-lived pool maintain a naive cell status since, unlike memory B cells that often express isotype-switched sIg receptors, their sIg receptors remain sIgM and sIgD, and the cells do not accumulate somatic mutations as would be the case for memory B cells. Recent analysis of 5-bromodeoxyuridine (BrdU)-labeled cells of mice fed BrdU for several days to several weeks indicate that the vast majority of the HSA$^{very\ hi}$ cells turn over with a half-life of several days, whereas the major population of splenic B cells that are HSA^{lo-int} turnover with a half-life of 4–6 weeks (58).

Bone marrow B-cell development in young adult mice has been extensively studied by numerous laboratories and has also been extensively reviewed (53–55, 60–63). Early commitment of cells to the B-cell lineage appears to be dependent on the expression of both transcription factors (including E2A and Pax5 (64–66)) and a set of intracellular proteins that include recombinase enzymes RAG1 and RAG2 (67–69); TdT (70); components of a pre-B cell receptor complex, including the transmembrane signaling components Igα and Igβ (71–73), and two polypeptides, VpreB and λ5, that co-combine with μ chains to form a L-chain-like complex (surrogate light chain (SL) (74–78)). Early pro-B cells are characterized by the expression of several cell-surface markers, including c kit, B220 and CD43 (60, 61). During this stage of pro-B cell development, H-chain gene segment rearrangement is initiated by the recombination of D_H and J_H gene segments on both chromosomes (60–63, 79). Ultimately, c kit expression decreases, HSA expression increases and cells begin to express BP-1 (60). During normal B-cell development, cells pass through these stages rapidly.

With the recombination of V_H-gene segments to D_H-J_H, H-chain expression is initiated, HSA and BP1 expression increase and these cμ+ pre-B cells initiate division (53, 60, 61). Whereas entrance and progression through the pro-B cell stage of development appears primarily dependent on the expression of appropriate transcription factors, the pro-B cell to pre-B cell transition is primarily dependent on V_H-D-J_H rearrangement leading to the expression of μ chain and, consequently, a pre-B-cell receptor complex (pre-BCR) comprised of μ chain plus SL, Igα and Igβ. Cells that do not express a functional pre-BCR fail to mature. Such failure can be due to: a) the absence of an H chain in cells wherein rearrangements on both chromosomes are non-productive; b) the H chain that is produced cannot interact with SL to form a pre-BCR (80–83), or c) a pre-BCR is formed but fails to mediate clonal maturation (84). Pre-B cells that express a functional pre-BCR mature clonally and undergo 5–6 divisions (80). Because of the death of cells that fail to mature and the clonal expansion of those that do mature, over 80% of B220$^+$, sIg$^-$ cells in the bone marrow are included within the pool of expanding or fully expanded pre-B-cell clones (80). Although maturing pre-B-cell clones undergo at least 5–6 divisions, these divisions appear to occur every 6–8 h so that clonal expansion is completed within 2 days. Subsequent to clonal expansion, resting cμ+ cells appear to express L chains and sIg only after a lag time of 1–2 days, further adding to the prevalence of cells of fully expanded pre-B-cell clones (61, 80, 85).

Following L-chain expression, cells remain non-divisional, begin to express sIg and rapidly progress through several poorly demarcated maturational stages prior to emigrating from the bone marrow, a process which *in toto* takes 24–36 h (60–63, 85). Initially, immature bone marrow B cells can be isolated as B220^{lo}, HSA^{int}, IgD$^-$ and IgM^{lo}. Cells appear to transit this stage within 4–8 h, representing only 10–20% of all newly generated bone marrow B cells. During the pre-B to B cell transition, or while cells reside within this very early sIg$^+$ maturational stage, cells appear to be highly susceptible to downregulation by tolerance or anti-idiotypic recognition (86–90). Additionally, at this stage cells may be induced to undergo receptor editing wherein sIg receptors are downregulated and RAG expression is increased and, as a result, cells express receptors with replaced L chains (91–93). Following this early stage of B-cell maturation, cells progressively increase the level of their sIgM, B220 and HSA. Ultimately, just prior to exiting the bone marrow, cells pass through a transitional stage wherein HSA expression begins to decrease and sIgD can begin to be expressed (94, 95). Some investigators have found that B cells may become tolerance susceptible at this stage as well (94, 95).

Aging-associated alterations in the kinetics of B-cell development

Although it is generally accepted that the rate of B-cell lymphopoiesis decreases substantially between the first and fourth months of post-natal development (9–14), there has been less agreement on whether there are decreases in development thereafter. This is largely because: a) the total number of mature peripheral (splenic) B cells remains relatively constant even into very old age; b) the majority of peripheral B cells after 4 months of age are long-lived (4–6 week half-life) (58, 96), and c) the population of newly generated bone marrow B cells does not decrease dramatically by 2 years of age (14). However, recent studies from three laboratories have demonstrated that between 12 and 24 months of age there is a dramatic decrease in the generation of pre-B cells (12–14). Most recently, Stephan et al. using the cell-surface markers established by Hardy (60), have carefully documented the representation of each developmental and maturational B-cell subset in the bone marrow and spleen of 1, 4, 12 and 24 month old mice (14). Their findings indicated that, whereas the splenic B-cell and bone marrow pro-B-cell populations remained intact in aged mice, the bone marrow pre-B-cell population, which is decreased by 2-fold between 1 and 4 months of age, is decreased another 2-fold between 12 and 24 months of age.

Surprisingly, in spite of the decrease in pre-B cells, in their analyses, the populations of neither newly generated bone marrow nor newly generated splenic B cells ($HSA^{very\ hi}$) were decreased in aged mice. In order to account for the apparent paradox that the population of newly generated B cells remains relatively intact even though fewer pre-B cells are being generated, these authors speculated that fewer newly maturing B cells may be eliminated by tolerance in aged mice.

In order to evaluate the cellular and molecular mechanisms that may underlie both the decrease in pre-B-cell generation and the paradoxical maintenance of both immature and mature B-cell populations, we have recently carried out an analysis of bone marrow and splenic B-lineage cells in 20–24 month old BALB/c mice similar to that carried out by Stephan et al. (14); however, we included an analysis of BrdU incorporation to assess the turnover kinetics of the various maturational subsets (58, 96). Additionally, we have evaluated the impact of negative selection on the various developmental subsets by an analysis of the ratio of productive to non-productive V_H-D-J_H rearrangements that utilize the V_H81X V_H-gene segment in the genomic DNA of various developmental subsets (80, 81, 88, 97–99). Previous analyses in this laboratory had shown that, relative to V_H-D-J_H rearrangements that utilize V_H-gene segments other than V_H81X, cells expressing H chains whose V regions are encoded by V_H81X are disfavored at several developmental stages (81, 88). First, there is a several-fold disfavoring of cells whose H-chains are encoded by V_H81X during the pro-B cell to pre-B-cell transition. This presumably occurs because many H chains encoded by V_H81X may not form a functional pre-BCR (81–84). There is a further 2-fold disfavoring of B cells whose H chains are encoded by V_H81X at the transition of pre-B cells to B cells (88), presumably in part due to the elimination of cells by tolerance to self-antigens, some of which may be recognized by V_H81X-encoded antibodies (100, 101). Finally, there is a further 2–3 fold disfavoring of cells whose H chains are encoded by V_H81X as cells emerge from the bone marrow and enter the spleen (88). Again, this may be due to the induction of tolerance of cells that recognize self-antigens in the periphery. Although the "self-antigens" relevant to the elimination of B cells whose H chains are encoded by V_H81X-D-J_H rearrangements are not known, the elimination of these cells as they pass through various developmental stages has been extremely useful in evaluating the developmental-stage dependence of negative selection. We are now applying this analysis to the B-cell subsets of aged mice and thus far have found the ratio of productive vs non-productive V_H81X-D-J_H rearrangements in both newly generated bone marrow and splenic B cells to be similar in cells of aged vs young adult mice (G. Kline, T. Hayden, N. Leaf, N. Klinman, preliminary observation). This implies that, in contrast to the speculation of Stephan et al. (14), there is no substantive difference in negative selection of newly generating B cells of aged vs young mice.

With respect to the representation of cells within the various maturational B-cell subsets in aged vs young adult mice, our findings are much like those published previously (12–14), (see Fig. 2). Similar to those studies, we have found a marked disparity in the number of pro-B cells vs pre-B cells, the latter being markedly decreased in aged mice. However, whether this block in B-cell development occurs at the pro-B-cell to pre-B-cell transition, as suggested by Stephan et al. (14), or whether it occurs at an earlier stage of pro-B-cell development, has yet to be determined.

With respect to the proportionate distribution of cells within various B-cell subsets as defined by cell-surface marker expression, our findings differ from earlier analyses in two respects (G. Kline, T. Hayden, N. Klinman, manuscript in preparation). First, while the total number of splenic B cells remains relatively unchanged, the proportion of those cells that express very high levels of HSA is as much as 2-fold less in aged compared with young adult mice. This would imply that fewer newly generated cells may be emigrating from the bone marrow to the spleen in aged vs young mice (see below). The second significant difference concerns the numbers and phenotype of newly generated bone marrow B cells (population "E" by the Hardy nomenclature) (Fig. 2). While we find that the total number of newly generated B cells is somewhat reduced in aged mice, this population varies greatly from aged mouse to aged mouse. Most significant, however, is that, whereas in young adult mice the subset of least mature (IgM^{lo}) B cells represents only 20% of all newly generated B cells, in aged mice this subset represents over 50%. Indeed, many cells of this population have such low expression of sIg that they may have been judged to be sIg negative in some earlier studies (15, 20). That these $sIgM^{very\ lo}$ cells are indeed sIg^- comes both from evidence obtained by FACS analysis for both sIgM and L-chain expression and from analyses of the ratio of productive vs non-productive V_H81X-D-J_H rearrangements in these cells, which indicates that many of these cells have been negatively selected in a manner similar to $sIgM^{lo}$ cells of young adults (88). The finding that, in spite of generating very few pre-B cells, aged mice have not only a disproportionate increase in the $sIgM^{lo}$ subset but also an absolute increase in the total number of such cells in comparison to young adults, suggests that, in addition to the developmental block at the pro-B-cell level in aged mice, there may be a second block in maturation, at a stage wherein newly generated B cells begin to upregulate sIgM expression (Fig. 2).

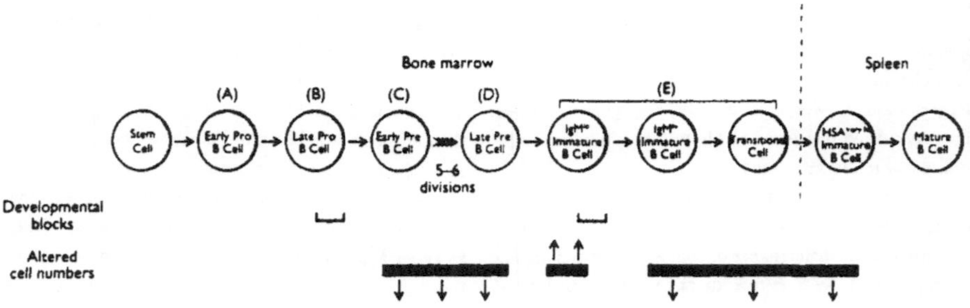

Fig. 2. A schematic representation of alterations in the course of B-cell maturation in aged vs young adult BALB/c mice. Because of a presumptive block in pro-B-cell maturation (see red bracket), pre-B cells are markedly decreased in numbers of aged mice (12–14). Although there are fewer maturing pre-B cells, the proportion of cells within the least mature population of B cells ($IgM^{very\ lo}$) is increased in aged mice. That this increase in immature B cells is the result of a second block in development at this stage (see red bracket) is evidenced by the accumulation of $sIgM^{very\ lo}$ cells, the slow rate of turnover of these cells, and the markedly diminished numbers of more mature stages ($IgM^{\pm}$) of bone marrow cells and least mature splenic B cells. In spite of the decrease in cells recently emigrated from the bone marrow, the population of mature splenic B cells remains intact due to an increase in their longevity.

In order to establish the kinetics of appearance of cells in each of the maturational subsets, we have conducted an analysis of BrdU incorporation in cells of mice fed BrdU for several days to several weeks (G. Kline, T. Hayden, N. Klinman, manuscript in preparation). The findings from these studies have not only provided a confirmation of the conclusions we have drawn from the phenotypic analyses, but have also suggested an internally consistent overview of the alterations of B-cell maturation in aged vs young adult mice (Fig. 2). Our initial analysis in this regard concerned the relative longevity of mature splenic B cells in aged vs young adult mice. Our findings with respect to young adults are consistent with those of Fulcher & Basten (58), who concluded that the average half-life of mature ($HSA^{int\text{-}lo}$) splenic B cells is 4–6 weeks. Since throughout the course of analysis (5 days to 4 weeks) a significantly smaller percentage of mature splenic B cells of aged vs young adults became BrdU-labeled, we have concluded that the average half-life of B cells in aged mice is substantially longer than in young adult mice. Greater B-cell longevity in aged mice could either reflect the gradual accumulation of the longest lived cells or an intrinsic difference in either the B cells or the mechanism that selects for B-cell longevity in aged mice. Experiments to address these alternatives have yet to be carried out.

The most dramatic difference in the rate of accumulation of BrdU-labeled cells in the spleens of aged vs young adult mice was seen in the population of $HSA^{very\ hi}$ cells, the majority of which, in young adults, represent the cells that most recently emigrated from the bone marrow (56–58). As mentioned above, the overall proportionate representation of $HSA^{very\ hi}$ cells is decreased approximately 2-fold in aged mice. Even more dramatic, however, is that, unlike young adults, wherein the vast majority (>70%) of these cells are BrdU-labeled within 5 days, even by day 8, only a minority of $HSA^{very\ hi}$ splenic B cells of aged mice are BrdU-labeled. Thus, overall, at least 5-fold fewer cells appear to be entering the spleen from the bone marrow in aged vs young adult mice. This slow rate of entry of newly generated B cells into the spleen of aged mice is consistent with the slow turnover of mature splenic B cells in these mice.

The finding that very few newly generated B cells are present in the spleen of aged mice differs from the conclusions reached by earlier studies (14) and thus requires a reassessment of the kinetics of B-cell generation in the bone marrow. Initially, we had investigated the kinetics of appearance of BrdU-labeled cells in subsets of bone marrow B-lineage cells of aged mice to determine if the decrease in pre-B cells was, at least in part, due to a more rapid progression of cells through the late pre-B-cell stage of development. In young adults, post-mitotic cells reside at this maturational stage for 24–36 h before expressing sIg (85). If cells of aged mice progressed through this stage more rapidly, it might have been expected that the population of newly generated bone marrow B cells in aged mice would accumulate BrdU more rapidly (in less than a day rather than 1–2 days). Surprisingly, the population of newly generated (IgM^{lo}, IgD^-) bone marrow B cells of aged mice was found to accumulate BrdU-labeled cells very slowly (G. Kline, T. Hayden, N. Klinman, manuscript in preparation). Whereas in young mice almost all $sIgM^+$, $sIgD^-$ cells were BrdU-labeled within 3 days, in aged mice less than half were labeled even after 5 days and the majority were not labeled until 8 days. Since the newly labeled cells ultimately accumulate in the $sIgM^{very\ lo}$ population of bone marrow B cells (which is expanded in aged *vs* young mice), these findings are consistent with the interpretation that B-cell maturation is blocked at the $sIgM^{very\ lo}$ stage in aged mice (Fig. 2). The conclusion that cells pass through this stage much more slowly in aged than in young adult mice would account for the maintenance of this population in spite of the very low production of pre-B cells and would rationalize, as well, the very low numbers of B cells that emigrate from the marrow to the spleen.

Concluding remarks

Given that the development, maturation and peripheralization of B cells is so markedly reduced in aged mice, it is remarkable that the numbers, cell surface phenotype and function of mature splenic B cells remain relatively unchanged even as late as 36 months of age. Thus, although the frequency of B cells responsive to certain haptenic determinants is decreased 2–3-fold in aged mice (15, 20), the frequency of naive B cells responsive to some antigens is maintained (4, 21), as is the frequency of naive progenitors of memory B cells (35). Most important, upon antigenic stimulation in splenic fragments derived from reconstituted young adults that provide young adult carrier-specific T_H, the B cells obtained from aged mice that do respond do so normally, making normal amounts of antibodies of diverse affinity and isotype. Thus, while humoral responses are decreased due to poor T_H function (3) and increased anti-idiotypic downregulation in aged mice (26–29), the B-cell component of the responses *per se* is not markedly affected. To the extent that the B-cell compartment is maintained in aged mice by increased B-cell longevity, the increased life-span of B cells does not appear to decrease their antigen responsiveness. In this regard, B cells appear to differ from T cells, wherein aging is accompanied by the accumulation of a major population of cells that have an altered phenotype and decreased responsiveness (3). A possible exception to the normal functionality of "aged" B cells may concern the generation of memory B cells (17). Whereas the frequency of cells that can give rise to memory B cells does not decrease (35), definitive experiments to test the suggestion that the intrinsic ability of these cells to generate high-affinity somatically mutated progeny is compromised have yet to be carried out.

As mice age beyond 20–24 months, bone marrow B-cell development is markedly decreased. This decrease is reflected in the frequency with which newly generated B cells enter the least mature population of splenic B cells. Concomitant with the diminished emigration of newly generated B cells from the bone marrow is the decrease in the development of both pre-B cells and in B cells at all stages of B-cell maturation beyond the developmental stage wherein cells first express sIgM ($sIgM^{very\ lo}$). At the present time, little is known concerning the mechanisms responsible for either the overall decrease in B-cell generation or the blocks that occur in B-cell development and maturation. Indeed, it has yet to be determined a) whether B-cell generation is downregulated in aged mice because the increased longevity and accumulation of long-lived cells diminishes the need for new cells and thus feeds back on the developing population or, alternatively, b) whether peripheral B cells become longer-lived because fewer new cells are being generated. Several studies have demonstrated that the longevity of mature B cells can be increased by a decrease in the generation of new cells (96, 102, 103). If this were the case, then the primary defect in aged mice would be the decrease in the maturation of pro-B cells. Studies to address this possibility are just being initiated and include an analysis of precisely which developmental stage of pro-B-cell development is aborted and whether, once generated, pre-B cells undergo a normal pattern of clonal maturation and expansion.

The alternative scenario, that a stable long-lived population of mature B cells feeds back on and inhibits the further generation of new B cells, is equally provocative. Does such a feedback mechanism exist and, if so, is it part of a normal physiological process responsible for maintaining B-cell homeostasis? Is the block in maturation of $sIgM^{very\ lo}$ cells also a consequence of homeostatic mechanisms and does this, in turn, mediate a feedback inhibition of pro-B-cell development? What mechanisms could account for a block in the further maturation of $IgM^{very\ lo}$ B cells? Cells at this maturational stage have been shown to accumulate as "anergic" cells in sIg transgenic mice in the presence of "self" antigens (90); not only self-antigens but abundant anti-idiotypic recognition would be present in aged mice (26). However, the sIgM^very lo cells that accumulate in aged mice differ from such anergic cells in that they express no sIgD and appear to remain highly antigen-responsive (4, 15, 20). Equally unknown would be the mechanism that could account for blocking further pro-B-cell maturation. One suggestion has been that, since pro-B cells from aged mice appear to proliferate normally under certain *in vitro* conditions (12, 14), either a stromal cell or IL-7 deficiency may be responsible (14). Alternatively, because the block in pro-B-cell maturation may result from feedback inhibition from either the increased pool of sIgM^very lo B-cell subset or the long-lived pool of splenic B cells, it is tempting to speculate that the downregulation of pro-B-cell development may be mediated by the regulation of the transcription factors, such as E2A (64, 65), which play so important a role in pro-B-cell development. In any case, while the phenotype of the aging associated alteration in B-cell development is now far more clearly defined, more questions have been identified than answered and the mechanisms that underlie altered B-cell expression in aged mice remain to be determined.

References

1. Makinodan T. Age related changes in antibody forming capacity. In: Sigal MM, Good RA, eds. Tolerance, autoimmunity and aging. Springfield IL: Charles C Thomas; 1970. p. 3–17.
2. Callard RE, Basten A, Waters LK. Immune function in aged mice. II. B cell function. Cell Immunol 1977,**31**:26–36
3. Miller RA. Aging and immune function. Int Rev Cytol 1991;**124**:187–215.
4. Zharhary D, Klinman NR. A selective increase in the generation of phosphorylcholine-specific B cells associated with aging. J Immunol 1986;**136**:368–370.
5. Riley SC, Froscher BG, Linton PJ, Zharhary D, Marcu K, Klinman NR. Altered V_H gene segment utilization in the response to phosphorylcholine of aged mice. J Immunol 1989;**143**:3798–3805.
6. Nicoletti C, Borghesi-Nicoletti C, Yang X, Schulze D, Cerny J. Repertoire diversity of antibody response to bacterial antigens in aged mice. II. Phosphorylcholine-antibody in young and aged mice differ in both V_H/V_L gene repertoire and in specificity. J Immunol 1991;**147**:2750–2755.
7. Yang X, Stedra J, Cerny J. Repertoire diversity of antibody response to bacterial antigens in aged mice. IV. Study of V_H-V_L gene utilization in splenic antibody foci by *in situ* hybridization. J Immunol 1994;**152**:2214–2221.
8. Goidl EA, Chen X, Schulze DH. Age-related changes in V_H gene use. J Cell Biochem 1993; Abstract Suppl 17B: F404.
9. Miller SC, Osmond DG. Lymphocyte populations in mouse bone marrow: quantitative kinetic studies in young, pubertal and adult C3H mice. Cell Tissue Kinet 1975;**8**:97–110.
10. Rolink A, Haasner D, Nishikawa S, Melchers F. Changes in frequencies of clonable pre B cells during life in different lymphoid organs of mice. Blood 1993;**81**:2290–2300.
11. Rolink A, Melchers F. Generation and regeneration of cells of the B-lymphocyte lineage. Curr Opin Immunol 1993;**5**:207–217.
12. Riley RL, Kruger MG, Elia J. B cell precursors are decreased in senescent BALB/c mice, but retain normal mitotic activity *in vivo* and *in vitro*. Clin Immunol Immunopathol 1991;**59**:301–313
13. Zharhary D. Age-related changes in the capability of the bone marrow to generate B cells. J Immunol 1988;**141**:1863–1869.
14. Stephan RP, Sanders VM, Witte PL. Stage specific alterations in murine B-lymphopoiesis with age. Int Immunol 1996;**8**:509–518
15. Zharhary D, Klinman NR. The frequency and fine specificity of B cells responsive to (4-hydroxy-3-nitrophenyl)acetyl in aged mice. Cell Immunol 1986;**100**:452–461.
16. Miller C, Kelsoe G. Ig V_H hypermutation is absent in the germinal centers of aged mice. J Immunol 1995;**155**:3377–3384.
17. Yang X, Stedra J, Cerny J. Relative contribution of T and B cells to hypermutation and selection of the antibody repertoire in germinal centers of aged mice. J Exp Med 1996;**183**:959–970.
18. Klinman NR. The mechanism of antigenic stimulation of primary and secondary clonal precursor cells. J Exp Med 1972;**136**:241–260.
19. Froscher BG, Klinman NR. The splenic focus assay. Methods Enzymol 1987;**150**:196–208
20. Zharhary D, Klinman NR. Antigen responsiveness of the mature and generative B cell populations of aged mice. J Exp Med 1983;**157**:1300–1308.
21. Zharhary D, Klinman NR. B cell repertoire diversity to PR8-influenza virus does not decrease with age. J Immunol 1984;**133**:2285–2287.
22. Zharhary D, Klinman NR. The effects of aging on murine B cell responsiveness. In: Goidl E, ed. Aging and the immune response. New York: Marcel Dekker; 1987. p. 81–93.
23. Zharhary D, Klinman NR. Responses of mature and newly generated B cells in aged mice. In: Facchini A, Haayman JJ, Labo G, eds. Immunoregulation in aging. Leiden: Eurage; 1986. p. 271–277.
24. Klinman NR. Aging and the expression of the B cell repertoire. Aging Immunol Infect Dis 1990;**2**:139–142
25. Klinman NR, Thorbecke GJ. Aging associated changes in the generation and development of B cell memory. Int Rev Immunol 1995;**12**:5–12.
26. Klinman NR. Antibody-specific immunoregulation and the immunodeficiency of aging. J Exp Med 1981;**154**:547–551.
27. Goidl EA, Innes JB, Weksler ME. Immunological studies of aging. II. Loss of IgG and high avidity plaque-forming cells and increased suppressor activity in aging mice. J Exp Med 1976;**144**:1037–1048
28. Szewczuk MR, Campbell JR. Loss of immune competence with age may be due to auto-anti-idiotypic antibody regulation. Nature 1980;**286**:164–166

29. Siskind GW, Goidl EA, Schrater AF, Thorbecke GJ, Weksler ME. The role of auto-anti-idiotype antibody in the regulation of the immune response. Cell Immunol 1982;**66**:34–42.
30. Flood PM, Urban JL, Kripke ML, Schreiber H. Loss of tumor-specific immunity with age. J Exp Med 1981;**154**:275–290.
31. Rodkey, LS. Studies of idiotypic antibodies: production and characterization of auto-anti-idiotypic antiserum. J Exp Med 1974;**139**:712–720.
32. Jerne N. Toward a network theory of the immune system. Ann Immunol (Paris) 1971;**125**:373–389.
33. Hart DA, Wang AL, Pawlak LL, Nisonoff A. Suppression of idiotypic specificities in adult mice by administration of antiidiotypic antibody. J Exp Med 1972;**135**:1293–1300.
34. Pierce SK, Klinman NR. Antibody specific immunoregulation. J Exp Med 1977;**146**:509–519.
35. Linton PJ. The status of progenitors of memory B cells in aged mice. Aging Immunol Infect Dis 1993;**4**:35–46.
36. Bruce J, Symington FW, McKearn TJ, Sprent J. A monoclonal antibody discriminating between subsets of T and B cells. J Immunol 1981;**127**:2496–2501.
37. Linton PJ, Decker DJ, Klinman NR. Primary antibody forming cells and secondary B cells are generated from separate precursor cell subpopulations. Cell 1989;**59**:1049–1059.
38. Linton PJ, Klinman NR. Self non-self discrimination by precursors of memory B cells. In: Melchers F et al. Progress in immunology, Vol. 7. Proceedings of the 7th International Congress of Immunology, Berlin, Germany. Heidelberg: Springer Verlag; 1989. p. 424–431.
39. Wu P, Ward RE. Ig repertoire expression of BALB/c primary and secondary B cell precursors specific for phosphorylcholine. J Immunol 1993;**150**:3862–3872.
40. Yin XM, Vitetta ES. The lineage relationship between virgin and memory B cells. Int Immunol 1992;**4**:691–698.
41. Linton PJ, Lo D, Lai L, Thorbecke GJ, Klinman NR. Among naive precursor subpopulations only progenitors of memory B cells originate germinal centers. Eur J Immunol 1992;**22**:1293–1297.
42. Decker DJ, Linton PJ, Jacobs SN, Biery M, Gingeras TR, Klinman NR. Defining subsets of naive and memory B cells based on their ability to somatically mutate in vitro. Immunity 1995;**2**:195–203.
43. Lieberman R, Potter M, Mushinski W, Humphrey W, Rudikoff S. Genetics of a new IgV_H (T15 idiotype) marker in the mouse regulating natural antibody to phosphorylcholine. J Exp Med 1974;**139**:983–1101.
44. Perlmutter RM, et al. The generation of diversity in phosphorylcholine-binding antibodies. Adv Immunol 1984;**35**:1–37.
45. Malipiero UV, Levy NS, Gearhart PJ. Somatic mutation in anti-phosphorylcholine antibodies. Immunol Rev 1987;**96**:59–74.
46. Crews J, Griffin H, Huang K, Calame K, Hood L. A single VH gene segment encodes the immune response to phosphorylcholine: somatic mutation is correlated with the class of the antibody. Cell 1981;**25**:59–66.
47. Gearhart PJ, Sigal NH, Klinman NR. Heterogeneity of the BALB/c anti-phosphorylcholine antibody response at the precursor cell level. J Exp Med 1975;**141**:56–71.
48. Sigal NH, Gearhart PJ, Klinman NR. The frequency of phosphorylcholine-specific B cells in conventional and germfree BALB/c mice. J Immunol 1975;**114**:1354–1358.
49. Sigal NH, Pickard AR, Metcalf ES, Gearhart PJ, Klinman NR. Expression of phosphorylcholine-specific B cells during murine development. J Exp Med 1977;**146**:933–948.
50. Klinman NR, Stone MR. Role of variable region gene expression and environmental selection in determining the anti-phosphorylcholine B cell repertoire. J Exp Med 1983; **158**:1948–1961.
51. Hardy RR. Variable gene usage, physiology and development of Ly-1$^+$ (CD5$^+$) B cells. Curr Opin Immunol 1993;**4**:181–185.
52. Kantor AB, Herzenberg LA. Origin of murine B cell lineages. Annu Rev Immunol 1993;**11**:501–538.
53. Park YH, Osmond DG. Phenotype and proliferation of early B lymphocyte precursor cells in mouse bone marrow. J Exp Med 1987;**165**:444–458.
54. Osmond DG. The turnover of B cell populations. Immunol Today 1993;**14**:34–37.
55. Opstelten D, Osmond DG. Pre B cells in mouse bone marrow: immunofluorescence stathmokinetic studies of the proliferation of cytoplasmic µ-chain bearing cells in normal mice. J Immunol 1983;**131**:2635–2640.
56. Allman DM, Gerguson SE, Cancro MP. Peripheral B cell maturation. I. Immature peripheral B cells in adults are heat-stable antigen^{hi} and exhibit unique signaling characteristics. J Immunol 1992; **149**:2533–2540.
57. Allman DM, Gerguson SE, Lentz VM, Cancro MP. Peripheral B cell maturation. II. Heat stable antigen^{hi} splenic B cells are an immature developmental intermediate in the production of long-lived marrow-derived B cells. J Immunol 1993;**151**:4431–4444.
58. Fulcher DA, Basten A. Influences on the lifespan of B cell subpopulations defined by different phenotypes. Eur J Immunol 1997;**27**:1188–1199.
59. Gu H, Tarlinton D, Muller W, Rajewsky K, Forster I. Most peripheral B cells in mice are ligand selected. J Exp Med 1991;**173**:1357–1371.
60. Hardy RR, Carmack CE, Shinton SA, Kemp JD, Hayakawa K. Resolution and characterization of pre-pro B cell stages in normal mouse bone marrow. J Exp Med 1991;**173**:1213–1225.
61. Melchers F, et al. Roles of IgH and L chains in the development of cells of the B lymphocyte lineage. Annu Rev Immunol 1994;**12**:209–225.
62. Ehlich A, Martin V, Muller W, Rajewsky K. Analysis of the B cell progenitor compartment at the level of single cells. Curr Biol 1994;**4**:573–583.
63. Löffert D, et al. Early B cell development in the mouse: insights from mutations introduced by gene targeting. Immunol Rev 1994;**137**:135–152.
64. Bain G, et al. E2A proteins are required for proper B cell development and initiation of immunoglobulin gene rearrangements. Cell 1994;**79**:885–892.
65. Zhuang Y, Soriano P, Weintraub H. The helix-loop-helix gene E2A is required for B cell formation. Cell 1994;**79**:875–884.
66. Urbanek P, Wang ZQ, Fetka I, Wagner EF, Busslinger M. Complete block of early B cell differentiation and altered patterning of the posterior midbrain in mice lacking Pax5/BSAP. Cell 1994;**79**:901–912.
67. Schatz DG, Oettinger MA, Baltimore D. The V(D)J recombination activating gene, RAG-1. Cell 1989;**59**:1035–1048.
68. Mombaerts P, Iacomini J, Johnson RS, Herrup K, Tonegawa S, Papaioannou VE. RAG-1 deficient mice have no mature B and T lymphocytes. Cell 1992;**68**:869–877.

69. Shinkai Y, et al. Rag-2-deficient mice lack mature lymphocytes owing to inability to initiate V(D)J rearrangement. Cell 1992;**68**:855–867.
70. Desiderio SV, et al. Insertion of N regions into heavy-chain genes is correlated with expression of terminal deoxytransferase in B cells. Nature 1984;**311**:752–755.
71. Hombach J, Leclerq L, Radbruch A, Rajewsky K, Reth M. A novel 34-kd protein co-isolated with IgM molecules in surface IgM-expressing cells. EMBO J 1988;**7**:3451–3456.
72. Papavasiliou F, Misulovin Z, Suh H, Nussenzweig MC. The role of Igβ in precursor B cell transition and allelic exclusion. Science 1995;**268**:408–411.
73. Hermanson GG, Eisenberg D, Kincade PW, Wall R. B29: a member of the immunoglobulin gene superfamily exclusively expressed on B lineage cells. Proc Nat Acad Sci USA 1988;**85**:6890–6894.
74. Pillai S, Baltimore D. Formation of disulphide linked μ2ω2 tetramers in pre B cells by the 18K ω-immunoglobulin light chain. Nature 1987;**329**:172–174.
75. Karasuyama H, Kudo A, Melchers F. The proteins encoded by the VpreB and λ5 pre B cell-specific genes can associate with each other and with μ heavy chains. J Exp Med 1990;**1172**:969–972.
76. Kitamura D, Kudo A, Schaal S, Muller W, Melchers F, Rajewsky K. A critical role of λ5 protein in B cell development. Cell 1992;**69**:823–831.
77. Löffert D, Ehlich A, Muller W, Rajewsky K. Surrogate light chain expression is required to establish immunoglobulin heavy chain allelic exclusion during early B cell development. Immunity 1996;**4**:133–144.
78. Lassoued K, et al. Expression of surrogate light chain receptors is restricted to a late stage in pre B cell differentiation. Cell 1993;**73**:73–86.
79. Alt FW, et al. Ordered rearrangement of immunoglobulin heavy-chain variable region segments. EMBO J 1984;**3**:1209–1219.
80. Decker DJ, Boyle NE, Koziol J, Klinman NR. The expression of the immunoglobulin heavy chain repertoire in developing bone marrow B lineage cells. J Immunol 1991;**146**:350–361.
81. Decker DJ, Boyle NE, Klinman NR. Predominance of non-productive rearrangements of V_H81X gene segments evidences a dependence of B cell clonal maturation on the structure of nascent H chains. J Immunol 1991;**147**:1406–1411.
82. Keyna U, Beck-Engesser GB, Jongstra J, Applequist SE, Jäck HM. Surrogate light chain-dependent selection of Ig heavy chain V regions. J Immunol 1995;**155**:5536–5541.
83. Klinman NR, Kline GH, Hartwell L, Beck-Engesser G, Keyna U, Jäck HM. Heavy chain assembly with surrogate light chain is required for allelic exclusion and pre B cell maturation. In: Keystone symposia on Molecular and Cellular Biology B lymphocytes in health and disease; 1997 Jan 10–16; Steamboat Springs (CO), USA. Silverthorne (CO): Keystone Symposia; 1997. p. 6 (Abstract).
84. Clarke SH, Ye J, Arnold LW. Pre B cell selection is dependent on the V_H structure of the μ/surrogate L chain receptor. In: Keystone Symposia on Molecular and Cellular Biology B lymphocytes in health and disease; 1997 Jan 10–16, Steamboat Springs (CO), USA. Silverthorne (CO): Keystone symposia; 1997. p.6 (Abstract).
85. Landreth KS, Rosse C, Clagett J. Myelogenous production and maturation of B lymphocytes in the mouse. J Immunol 1981;**127**:2027–2034.
86. Metcalf ES, Klinman NR. *In vitro* tolerance induction of bone marrow cells: a marker for B cell maturation. J Immunol 1977;**118**:2111–2116.
87. Metcalf ES, Schrater AF, Klinman NR. Murine models of tolerance induction in developing and mature B cells. Immunol Rev 1979;**43**:143–183.
88. Decker DJ, Kline GH, Hayden TA, Zaharevitz SN, Klinman NR. H chain V gene specific elimination of B cells during the pre B cell to B cell transition. J Immunol 1995;**154**:4925–4935.
89. Nemazee DA, Burki K. Clonal deletion of B lymphocytes in a transgenic mouse bearing anti-MHC class I antibody genes. Nature 1989;**337**:562–566.
90. Hartley SB, et al. Elimination of self-reactive B lymphocytes proceeds in two stages: arrested development and cell death. Cell 1993;**72**:325–335.
91. Tiegs SL, Russell DM, Nemazee D. Receptor editing in self-reactive bone marrow B cells. J Exp Med 1993;**177**:1009–1020.
92. Gay D, Saunders T, Camper S, Weigert M. Receptor editing: an approach by autoreactive B cells to escape tolerance. J Exp Med 1993;**177**:999–1008.
93. Prak EL, Trounstine M, Huszar D, Weigert M. Light chain editing in κ-deficient animals: a potential mechanism of B cell tolerance. J Exp Med 1994;**180**:1805–1815.
94. Carsetti R, Köhler G, Lamers MC. Transitional B cells are the target of negative selection in the B cell compartment. J Exp Med 1995;**181**:2129–2140.
95. Carsetti R, Köhler G, Lamers MC. A role for immunoglobulin D interference with tolerance induction. Eur J Immunol 1993;**23**:168–178.
96. Sprent J, Tough DF, Sun S. Factors controlling the turnover of T memory cells. Immunol Rev 1997;**156**:79–85.
97. Decker DJ, Klinman NR. Developmental selection plus evolutionary determination [contribute D_H reading frame bias]. Immunol Today 1991;**12**:421.
98. Carlsson L, Overmo C, Holmberg D. Developmentally controlled selection of antibody genes: characterization of individual V_H7183 genes and evidence for stage-specific somatic diversification. Eur J Immunol 1992;**22**:71–78.
99. Huetz F, Carlsson L, Tornberg UC, Holmberg D. V region directed selection in differentiating B lymphocytes. EMBO J 1993;**12**:1819–1826.
100. Countinho A, Freitas AA, Holmberg D, Grandien A. Expression and selection of murine antibody repertoire. Int Rev Immunol 1992;**8**:173–187.
101. Kearney JF, Bartels J, Hamilton AM, Lehuen A, Solvason N, Vakil M. Development and function of the early B cell repertoire. Int Rev Immunol 1992;**8**:247–257.
102. Sprent J, Schaefer M, Hurd M, Surh CD, Ron Y. Mature murine B and T cells transferred to SCID mice can survive indefinitely and many maintain a virgin phenotype. J Exp Med 1991;**174**:717–728.
103. Agenès F, Rosado MM, Freitas AA. Independent homeostatic regulation of B cell compartments. Eur J Immunol 1997;**27**:1801–1807.

Brain Research Bulletin, Vol. 44, No. 4, pp. 471–477, 1997

0361-9230/97 $17.00 + .00

ELSEVIER

PII S0361-9230(97)00228-1

Neuroendocrine Regulation of GnRH and Behavior During Aging in Birds

M. A. OTTINGER,*[1] N. THOMPSON,* C. VIGLIETTI-PANZICA† AND G. C. PANZICA†

*Department of Animal and Avian Science, University of Maryland, College Park, MD 20742, USA, and †Department of Anatomy, Pharmacology, and Forensic Medicine, University of Torino, Torino, Italy

ABSTRACT: Avian species exhibit a great variety of life-long patterns in reproduction. Japanese quail are relatively short lived and undergo an age-related loss of reproductive function, making this species an excellent model for the study of the basic biology of aging. Because individuals age at variable rates, sexual behavior has provided a useful index to assess reproductive status of individuals of the same chronological age. Further, exogenous testosterone restoras sexual behavior in reproductively senescent male quail, thereby providing evidence for a continued ability of the system to respond. In addition, we have been studying hypothalamic neuroendocrine systems that regulate the endocrine as well as behavioral components of reproduction. Overall, our findings point to the hypothalamic neuroendocrine systems as the site of initial age-related alterations that contribute to the reproductive deterioration. Specifically, we studied adrenergic, opioid peptide, vasotocin, and aromatase systems to understand their relationship to the cGnRH-I system and their potential role in the deterioration of the cGnRH-I system during aging. Our findings provide evidence for qualitative and quantitative alterations in the aromatase enzyme system, which can be partially restored with exogenous testosterone. In addition, other neuronal systems, including the vasotocin system, decline with the loss of gonadal steroids and are restimulated with exogenous testosterone. We will synthesize the data relative to these neuroendocrine systems with attention to the effects of gonadal steroids on these systems during aging. © 1997 Elsevier Science Inc.

KEY WORDS: Vasotocin, Aromatase, Japanese quail, Preoptic area, Septum, Neuropeptides, Neurotransmitters.

INTRODUCTION

Our view of the role of the gonadal steroids in endocrine and behavioral components of male reproduction has changed as we gain greater understanding of the reproductive system at all levels. Androgens and estrogens act at all stages of the life cycle, including sexual differentiation, maturation, adult function, and aging [43]. Often, changing patterns in production and plasma concentrations of these steroids have been the first clue for a role of steroids on specific target tissues. For example, increasing plasma androgen levels herald sexual maturation of the male and the initiation of sexual behavior. Testicular function accompanies this maturation-related process, thereby providing a synchronization of endocrine and behavioral components of reproduction by acting on specific brain circuitries. Several studies have demonstrated the presence of sexually dimorphic structures in cerebral areas involved in the control of behavioral aspects of reproduction and the importance of gonadal hormones in determining these differences during embryonic or early postnatal life (the so called organizational effects) or their later initiation and maintenance (activational effects; [2,43]). Often the relationship of these steroids and sexual behavior have been regarded as cause and effect in that courtship and mating behavior do not occur in the absence of some threshold level of gonadal steroid. There are many examples that support this relationship including studies in which the castrated male ceases sexual behavior and testosterone (T) replacement therapy restored sexual behavior (for a review see [43]). As mentioned above, this relationship also holds for sexual maturation in that plasma androgen levels increase with initiation of gonadal function to affect target areas in the brain, resulting in the stimulation of sexual behavior. This apparent cause–effect relationship breaks down during aging in mammals. In mammals, there appears to be a "disconnect" in that when males fully cease courtship and mating behavior, an external supply of T is not capable of restoring sexual behavior. In contrast to this, in quail the sexual behavior can be restored in senescent individuals by administration of exogenous T [37].

The focus of our present research is on how the process of aging is affecting neural circuitries (mainly hypothalamic) that control sexual behavior as well as neuroendocrine aspects of reproduction. This process ultimately leads to reproductive senescence (complete loss of both endocrine and behavioral components of reproductive function). The study of age-related changes in the gonadotropin releasing hormone (GnRH) system is central to this question, but being that the GnRH producing neurons are part of complex neural circuitries, it is equally important to consider possible interactions of neurotransmitter or neuropeptide systems and their response to gonadal steroids during the process of aging (Fig. 1). Our data have been collected in the male Japanese quail. However, there is a great deal of similarity between data from mammals and birds, and we will refer those data as appropriate.

TESTOSTERONE AND SEXUAL BEHAVIOR DURING AGING IN QUAIL

There is a great deal of variability in the sexual behavior of old male quail. Around 18 months of age, some individuals become spontaneously sexually inactive (senescent), whereas a large number of individuals still retain (frequently at a lower level than adult birds) sexual activity. Plasma levels of androgen decrease in all middle-age birds (18–30 months of age) and do not differ significantly between active and inactive males, even if the inactive

[1] To whom requests for reprints should be addressed.

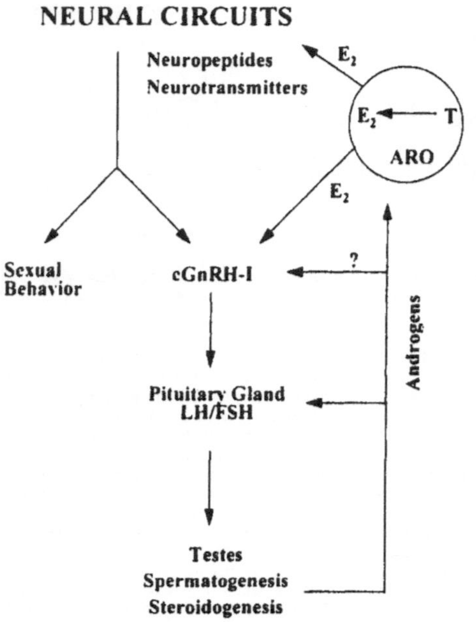

FIG. 1. Flow diagram illustrating the relationships among neural circuits, pituitary gland, and gonads that regulate endocrine and behavioral aspects of avian reproduction. During aging each of these levels can be affected.

males have normally lower circulating androgen levels. Once the animal has become completely senescent, the testes cease both spermatogenesis and steroid production [23,40]. Therefore, we typically observe senescence accompanied by regressed testes and decreased androgen levels primarily very late in aging (after 30 months or older). At this time, plasma androgen levels are in concert with reproductive status [36]. It appears, therefore, that the relationship between plasma gonadal steroids, production of hypothalamic hormones, and behavioral responses change during aging in comparison with what has been described in young and adult birds [39].

Moreover, it appears that the idea that the loss of gonadal steroids triggers the process of behavioral reproductive aging in the male Japanese quail is contradictory. Similar to data from mammals, plasma androgen levels do not significantly change during aging until the male has become reproductively senescent [36]. In both rats and quail, aging males that remain sexually active retain higher plasma androgen levels, and these levels are independent from the intensity of sexual activity. More intriguing is the fact that the sexual behavior can be restored in senescent male quail by administration of exogenous T, whereas mammals do not respond to this treatment [37]. If we assume that sexual behavior is primarily "written" in neuronal circuiteries, this means that compared to mammals, quail maintain a neuronal machinery that retains the capability to respond, even in aging. It has to be stressed however that the dose of T used to stimulate sexual behavior in senescent males was double the effective dose for young or in adult birds [37]. This suggests that part of the machinery does deteriorate during aging and the recovery of function may be based on a certain degree of plasticity still remaining in the old quail brain or to an overstimulation of the persisting system.

A further point that has to be considered is the fact that T is not the molecule that induces the cascade of neural events leading to the control of both endocrine and behavioral aspects of reproduction. The involved molecule is a T metabolite, estradiol (E_2), which is directly produced in the male brain by the action of the enzyme aromatase (ARO). Several studies demonstrated that E_2 stimulates male quail sexual behavior (for review, see [1]) and regulate the production of GnRH [31]. Therefore, studies on the location and response of the ARO-producing system are integral to understanding how the system works and what may become altered during aging.

HOW IS THE AROMATASE SYSTEM INFLUENCED IN SENESCENT MALES?

In young sexually mature birds, aromatase-containing cells are clustered within three nuclei: the medial preoptic nucleus (POM) in the preoptic area, the nucleus of the stria terminalis (nST) in the limbic complex, and the nucleus ventromedialis in the posterior hypothalamus [10]. ARO synthesis is stimulated by circulating T and the activity of this enzyme is strongly dimorphic [48]. In the POM, a loss of ARO-IR cells was observed in young castrated males coincident with the loss of male sexual behavior and presumably linked to the decreased levels of available T [3,20]. At the ultrastructural levels POM neurons in young male quail show changes in the extension of all protein synthesis-related organelles (rough endoplasmic reticulum, Golgi apparatus, vesicles) relative to variation in plasma T [44]. Subcutaneous administration of T by means of silastic implants restores aromatase staining and ultrastructure of POM neurons in castrate males. These data demonstrate that sexual behavior and ARO immunoreactivity show parallel changes in the young male quail. Finally, the distribution of ARO cells is unequal within the POM in that these cells are more numerous in the lateral than in the medial part of the nucleus (they represent about the 40% of the cells in the lateral POM, whereas they are about the 20% of the cells in the medial POM). The lateral population appears more sensitive to the decrease of circulating T: the number of IR cells decreases in castrated males to 10% of the initial number. On the contrary, the ARO-IR neurons in the medial part of the POM are reduced only by 30% of their initial number [3].

Age-related changes in this system were studied in three groups of male quail: 6-month-old adult sexually active, 36-month-old senescent males, and 36-month-old sexually active males. In old senescent males the decrease in the number of ARO cells was significant compared with the other two groups [17]. More detailed morphometrical study demonstrated that in contrast to young castrated males, the ARO-IR loss was more marked in the medial versus the lateral POM [41]. As in young male quail, the decrease in number was paralleled by a similar decrease in the cell size, indicating a loss in synthetic structures similar to young castrated males [44]. Spontaneously active old males had an intermediate number of ARO-IR cells [17]. These cells were more numerous in the lateral than in the medial POM and had a cell size that was significantly larger than in the adult sexually active males [41]. Similar increases in cell size have been already observed in other neurochemically identified systems of aging brain, and it is generally considered as a signal of overstimulation of the surviving portion of the system that is trying to compensate for the partial loss of cells [12]. Accordingly, the aging sexually active male quail may be able to compensate for loss of ARO-IR cells by

	AGE (weeks)		
Day post-implant	50	216 (active)	216 (inactive)
1	0	0	0
9	100	25	25
16*	100	50	40
20	70	50	60
21	70	80	90
23	100	100	80
28	100	100	90

FIG. 2. Percentage of photoregressed males showing reproductive behavior following implantation with testosterone (10 mm silastic implant). Males were aged (216-week-old) and categorized as behaviorally active or inactive prior to photoregression or males were adult (50-week-old) and behaviorally active prior to photoregression. After 14 days, all males received a second implant. Data from [37].

increased activity in a subset of cells, thereby maintaining sexual activity supported by sufficient production of E_2. Finally, a recent immunocytochemical study performed at ultrastructural level has demonstrated the presence of ARO immunoreactivity within synaptic terminals [34]. This means that the ARO neurons may convert T into E_2 not only at the level of the cell body (e.g., the POM), but also at the level of the axonal projections of these cells. In this respect it is possible that the communication between the POM and septal region (recently demonstrated by in vitro tracing studies [9]) where a large population of GnRH neurons is located (see below) would be facilitated via axonal projections.

HOW IS THE AVIAN GnRH SYSTEM AFFECTED DURING AGING?

Immunocytochemical studies have been conducted by many laboratories, including our own, on the distribution of chicken gonadotropin releasing hormones (cGnRH-I and II) in the avian brain (for a review see [56]) of sexually mature individuals. In particular, studies have concentrated on cGnRH-I because of evidence that implicates this form of GnRH in the regulation of avian reproduction [49]. According to immunohistochemical data in quail, cGnRH-I cell bodies are localized primarily in the rostral hypothalamus, in the preoptic area, as well as in the septal area (Fig. 2). GnRH-positive perikaria form two approximately continuous bilateral clusters that extend from the paraolfactory lobe and the preoptic area to end in the septal region [21,51]. This distribution is in agreement with older data showing that lesions in the preoptic area and anterior hypothalamus (probable destruction of GnRH producing elements) in the hen caused ovarian regression [47]. In addition, lesioning and deafferentation studies in the turkey and electrical stimulation and lesion studies in the quail are all consistent with the hypothesis that the GnRH system is located in the septopreoptic region [14,35]. Therefore, the search for neural factors that can influence the activity of the GnRH system should be focused on these regions.

In our initial studies, punches taken from the preoptic-septal area and from the median eminence showed that cGnRH-I concentrations decreased in both regions during aging to very low levels in the senescent male [38]. Moreover, the loss of cGnRH-I occurred first in the preoptic-septal region and then later in the median eminence, perhaps indicating the loss of synthetic capability and then later depletion of stored hormone. Using an in vitro perifusion system, we were able to assess cGnRH-I release from longitudinal hypothalamic slices in response to secretegogues and unmanipulated [30,31]. These data proved interesting in that slices from young reproductive males are strongly responsive to norepinephrine (NE) and epinephrine stimulation, whereas slices from old senescent males were significantly less responsive [38]. Further, there was a qualitative difference in cGnRH-I release in slices taken from young reproductive and old inactive males with reduced amplitude of release (see Fig. 3). In addition, multiple episodes of NE stimulation stimulated cGnRH-I release of similar amplitude from both young and old. However, cGnRH-I release by slices from old males was consistently lower than slices from young males. These data provide evidence that the cGnRH-I system is capable of response even in senescent males, but the qualitative response becomes altered during aging.

AGING AND GnRH: WHICH NEUROPEPTIDE SYSTEMS MODULATE RELEASE AND ARE LIKELY TO SIGNAL AGING?

It is likely that alterations in the regulatory neuropeptides, notably the opioid peptides, are the primary contributors to the age-related deterioration in hypothalamic response. A great deal of research has focused on CNS alterations during aging, especially those changes that relate to disease states such as Alzheimers disease or other dementia. Therefore, it is not surprising that neurotransmitter systems are considered fundamental to the changing function that occurs during aging. Many of these same systems have also been been implicated in reproductive failure. Accordingly, we have studied several neuronal systems that are known to have a close association with the GnRH system or are implicated in the regulation of the GnRH system. We will provide here a brief overview of these neurotransmitter and neuropeptide systems and provide greater detail on the systems that we have studied in the aging male quail.

Neurotransmitters/Neuropeptides Distribution in the Septopreoptic Area

Several peptidergic systems innervate the septo-preoptic region of Japanese quail (for a review see [42]). Moreover, the peptidergic innervation occurs in those regions that contain a number of gonadal hormone receptors [lateral septum (SL), POM, nST)], or where clusters of cGnRH-I positive neurons have been described [preoptic region (POA), SL, nucleus of the pallial commissure (nCPa)]. In general, a specific peptidergic pathway covers only part of a region; a noticeable exception to this pattern is the vasotocinergic innervation of the POM [54], which is found over the entire nucleus (but not limited to it). This area is critical in the control of male quail sexual behavior [46]. Interactions among peptidergic systems and the GnRH system may occur also at the level of the median eminence, in which the presence of several subcompartments may be observed in parasagittal and transverse sections [28,32]. Therefore, morphological data on histochemical or immunohistochemical distribution of neurotransmitters or neuropeptides in the quail septo-preoptic region provides evidence for a number of neurotransmitters or neuropeptides as putative regulators of gonadotropin release in the Japanese quail. Among the neuropeptides, neuropeptide Y (NPY), substance P, and vasoactive intestinal polypeptide are good candidates for the innervation of the preoptic components of the GnRH-I system, whereas VT is found in close correspondence of the larger septal clusters of the same system [42]. However, further information is needed to determine if this difference in the peptidergic supply to areas containing cGnRH-I positive cells corresponds to a functional role based on the anatomical differentiation of these systems.

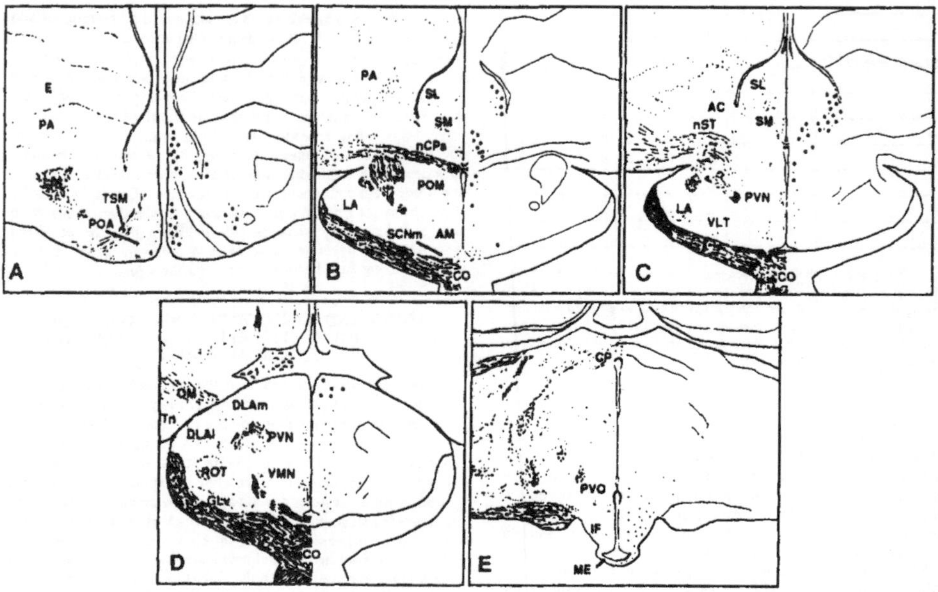

FIG. 3. Rostrocaudal (A–E) distribution of cGnRH-I immunoreactive cell bodies (closed circles) or fibers (crosses) within the septodiencephalic region of the Japanese quail brain. Drawings based on [21]. AC. nucleus accumbens; AM. nucleus anterior hypothalami; CO. optic chiasma; CP. pallial commissure; DLAl. nucleus dorsolateralis thalami. pars lateralis; DLAm. nucleus dorsolateralis thalami. pars medialis; E. ectostriatum; FPL. fasciculus prosencephalicus lateralis; GLv. nucleus geniculatus lateralis. pars ventralis; IF. infundibular region; LA. nucleus lateralis thalami; ME. median eminence; nCPa. nucleus of the pallial commissure; nST. nucleus of the stria terminalis; OM. tractus occipitomesencephalicus; PA. paleostriatum augmentatum; POA. nucleus preopticus anterior; POM. nucleus preopticus medialis; PVN. nucleus paraventricularis; PVO. paraventricular organ; ROT. nucleus rotundus; SCNm. nucleus suprachiasmaticus pars medialis; SL. septum lateralis; SM. septum medialis; Tn. nucleus taeniae; TSM. tractum septomesencephalicum; VLT. nucleus ventrolateralis thalami; VMN. nucleus ventromedialis.

ROLE OF SOME NEUROTRANSMITTERS AND NEUROPEPTIDES ON THE GnRH SYSTEM IN YOUNG AND OLD QUAIL

Catecholamines

The catecholamines have been of interest in the regulation of both endocrine and behavioral responses in many species for some time. There is clear evidence for noradrenergic stimulation of cGnRH release in vitro and in vivo [30,31,36]. In birds, catecholamine concentrations change with the ovulatory cycle [27]. Furthermore, catecholamine-containing neurons and high density of immunoreactive (IR) fibers are found in the preoptic area [5], which contains GnRH neurons [21,51] and regulates sexual behavior (for review see [46]). The distribution of adrenergic receptors has been documented in quail [6,8]. In vitro perifusion of longitudinal hypothalamic slices from Japanese quail has shown that both alpha and beta subtypes stimulate release of cGnRH-I [30]. Further, catecholamines are involved in the modulation of sexual behavior in quail (for reviews see [7,46]). These interactions become increasingly important as the animal ages [36] and are likely to underly age-related changes in episodic release of cGnRH-I (Fig. 4).

	young male quail (9 m)	old senescent male quail (3+ yrs)
Interpeak Interval (min±SE)	24±5	21±1.5
Pulse Duration (min±SE)	12±5	9±2
Pulse Frequency (pulses/hr±SE)	2.4±0.8	3.1±0.3
cGnRH-I release (pg/ml/5min±SE)	13±1.2	4.5±0.2*

FIG. 4. Parameters of cGnRH-I release from quail hypothalamic slices in vitro. The amplitude of the release is reduced in old senescent male quail (* $p < 0.05$).

Catecholamines are known to affect steroid receptor density [50], which then impacts GnRH release. As such, the declining levels of gonadal steroids during aging may then impact these adrenergic receptor systems, and age-related changes in adrenergic receptors have been found in the female rat [59]. Further, we found reduced NE and epinephrine (E) concentrations in median eminence and POA punches in aged quail, which also had reduced levels of cGnRH-I [38]. Collectively, these data show some evidence of declining function of the noradrenergic system, which in turn, results in a diminution of stimulatory input to the GnRH neurons (Fig. 4). However, it is not likely that declining noradrenergic function is the sole reason for age-related alterations in the cGnRH-I system for the following reasons: (1) noradrenergic systems appear to be modulated by opioid peptides in an inhibitory manner and it is not clear if this inhibition increases during aging; (2) the cGnRH-I neuron continues to respond to NE in vitro but at a reduced level; and (3) gonadal steroids modulate adrenergic as well as other neuropeptide systems, which, similar to the opioid peptides, are likely to regulate one another. Simply stated, the changes in the adrenergic systems are not sufficient to explain all the changes that are observed in neuroendocrine systems during aging.

Opioid Peptides

Opioid peptides inhibit sexual behavior and interact with adrenergic systems in regulating GnRH release in mammals. There is histological evidence for a close association of noradrenergic neurons and opioid peptides, dopamine, NPY, and GABA systems with GnRH neurons in mammals. As such, opioid peptides may inhibit GnRH release directly or indirectly via adrenergic systems [18,26]. Opiate receptor density changes with age and with steroid exposure [58,59]. Opioid peptides in the avian brain are similar to those found in mammals [29]: they have a widespread distribution within the galliform hypothalamus (for reviews see [28,32,56]). Neuronal projections containing enkephalin reach the median eminence and are in close association with cGnRH-I containing axons [33]. Further, enkephalin exposure reduced electrical activity of quail hypothalamic neurons in vitro [22]. Our research findings confirm that both β-endorphin and enkephalin inhibit cGnRH-I baseline and NE stimulated release in vitro from hypothalamic slices [19]. The activity and influence of the opioid peptides has not been investigated in aging birds. However, it is likely that this system continues to exert an "inhibitory tone" in the aging bird that may override declining stimulatory systems, such as the noradrenergic inputs that modulate cGnRH-I release in birds.

The Vasotocin System

The septal-preoptic region appears to be a phylogenetically stable target for the vasotocin (VT)/vasopressin (VP) family of peptides. VT/VP innervation has been observed in all vertebrates from fish to mammals, including man (for reviews see [11,16,25]). In several cases a sexually dimorphic distribution of the VT/VP innervation in the lateral septal region and the nST has been observed, with denser immunoreactivity in males than in females. Experimental treatments in which castration/photoregression reduced circulating T levels resulted in decreased VP/VT immunoreactivity and increased immunoreactivity after T replacement. Interestingly, castration decreased VP receptor binding in the hypothalamus of male hamsters [24]. These data suggest involvement of VT/VP in the control of male sexual behavior (for recent reviews see [11,15,16,25]). However, available data to this point are not conclusive and somewhat contradictory across the classes and species that have been studied (for birds see [13,57]). In addition, immunohistochemical studies demonstrated a close relationship among VT-IR fibers and aromatase-containing elements, suggesting the possibility of direct innervation of this behaviorally relevant neuronal system by the VT system [46,52].

Our recent studies demonstrated an age-related decrease in VT-immunoreactivity in areas (POM, SL, and nST) that are important for the regulation of both endocrine and behavioral aspects of reproduction. Other portions of the VT system do not show the same decrease in IR structures [45,55]. Further, the age-related VT decrease in SL, POM, and nST was restored to levels similar to those observed in adult male quail by T replacement therapy [45]. The same treatment also restored sexual behavior in old senescent males [37]. Again, we found a link between VT innervation of septo-preoptic region and sexual behavior in old male quail as was previously described for young castrated males [53,54]. From these data, it appears that the decline of sexual behavior induced by castration or photoregression in the young male or by physiological events in old senescent males is paralleled by a decline in VT immunoreactivity in specific cerebral areas. The VT system also retains a high degree of plasticity in the old male quail that is restimulated by exogenous T. Further studies should concentrate on the origin of these sexually dimorphic T-sensitive VT-IR fibers innervating the SL and POM. Very recent immunocytochemical and in situ hybridization studies have elucidated the presence of a sexually dimorphic population of weakly immunostained cells in a region with a high degree of homology with the mammalian bed nucleus of the stria terminalis [4,25], which is a major source of VP innervation of SL in the rat [15,16].

CONCLUSIONS

A great deal of information has been collected relative to aging in male quail. This now allows us to investigate the mechanisms involved in the age-related deterioration of hypothalamic function, particularly as it relates to cGnRH-I synthesis and release. Further, it is important to note that the deterioration of endocrine components of reproduction, as seen in the demise of the cGnRH-I system occur with loss of behavioral components of reproduction. This points to a fundamental synchrony of endocrine and behavioral aspects of reproduction and implies that there are common underlying neuroregulatory systems. Additional studies are needed in the quail model to obtain conclusive morphological, pharmacological, and endocrine data to understand these neurotransmitter/neuropeptide systems that regulate the cGnRH-I system in both young and old animals.

Steroids appear to be a critical element involved in the alterations that are observed during the process of aging. Several of the neuropeptide/neurotransmitter systems are located in regions in which the presence of steroid receptors has been detailed. Further, studies performed on young birds have established a gonadal hormone dependency of these systems. As estradiol is the centrally active metabolite of testosterone, changes in the aromatase system are also central to the understanding of morphological and functional alterations that occur during aging. In quail, there is a dramatic decrease in the aromatase system of old senescent males. This is likely to prevent production of sufficient E_2 to stimulate neuronal circuitries controlling sexual behavior as well as endocrine aspects of reproduction.

When testosterone is provided (endogenously or exogenously) to the senescent male, the results may be a stimulation of part of the surviving population of aromatase-producing cells. The increase in the cell size of the dorsolateral population suggests an overstimulation of the system to compensate the loss of cells.

In turn, the availability of T and of its metabolite E_2 can restimulate other systems. For example, the vasotocinergic system shows a very large decrease in immunostaining with aging. Exogeneous T restores a subpopulation of the fibers (lateral septum, POM) and in cells (nST) to a condition comparable to that observed in adult sexually mature birds. It is interesting to note that the vasotocinergic system has close association with the aromatase system. Further, the vasotocin innervation is found in the same regions as GnRH cells. It is a distinct possibility that changes in the cGnRH-I system during aging are linked to alterated function of neural circuitries that are dependent on suffient available T and its metabolite, E_2. As the animal becomes senescent, there is a decline in the availability of these steroids. Furthermore, data obtained comparing young and adult hypothalamic slices in vitro suggests that the GnRH system may also be altered itself.

Taken together, these data suggest that the total loss of sexual behavior in old senescent males might be linked to a fundamental alteration of the aromatase system that would limit the availability of locally synthesized E_2. The changes in the aromatase system are both qualitative and quantitative in nature, whereas changes in other systems appear to be directly be in response to the availability of gonadal steroids. Therefore, these other neuronal systems, such as the vasotocin system, shows changes that are due only to the decrease of circulating T and locally produced E_2; when these two conditions are restored, they can work and regulate the sexual behavior or the control the secretion of GnRH. Further, the aromatase system does exhibit the ability to be restimulated by exogeneous testosterone and, therefore, is not totally lost, but remains "plastid." This appears to be a fundamental attribute of the avian system that has been lost to mammals. It would be interesting to study other classes of vertebrates and determine if they maintain a similar neural plasticity.

ACKNOWLEDGEMENTS

This study was supported by the Maryland Agriculture Experimental Station, and by grants from MURST (40% to A. Fasolo, 60% to C.V.P. and G.C.P.), CNR (93.00372, 94.02462 to G.C.P.), NATO (CRG 92.1267 to G.C.P. and M.A.O.), EU (94-0472), USDA, and NRI

REFERENCES

1. Adkins Regan, E. K. Hormonal bases of sexual differentiation in birds. In: Balthazart, J., ed. Hormones, brain and behaviour in vertebrates, vol. 1. Sexual differentiation, neuroanatomical aspects, neuropeptides and neurotransmitters. Basel: Karger; 1990:1–14.
2. Arnold, A. P.; Breedlove, S. M. Organizational and activational effects of sex steroids on brain and behavior: A reanalysis. Horm. Behav. 19:469–498; 1985.
3. Aste, N.; Panzica, G. C.; Aimar, P.; Viglietti–Panzica, C.; Harada, N.; Foidart, A.; Balthazart, J. Morphometric studies demonstrate that aromatase-immunoreactive cells are the main target of androgens and estrogens in the quail medial preoptic nucleus. Exp. Brain Res. 101:241–252; 1994.
4. Aste, N.; Viglietti–Panzica, C.; Mühlbauer, E.; Grossmann, R.; Panzica, G. C. Sexual dimorphism of vasotocinergic structures in quail nucleus of stria terminalis. Ital. J. Anat. Embriol. 101(Suppl. 1):136–137; 1996.
5. Bailhache, T.; Balthazart, J. The catecholaminergic system of the quail brain: Immunocytochemical studies of dopamine β-hydroxylase and tyrosine hydroxylase. J. Comp. Neurol. 329:230–256; 1993.
6. Ball, G. F.; Nock, B.; McEwen, B. S.; Balthazart, J. Distribution of α_2-adrenergic receptors in the brain of the Japanese quail as determined by quantitative autoradiography: Implication for the control of sexually dimorphic reproductive processes. Brain Res. 491:68–79; 1989.
7. Balthazart, J.; Ball, G. F. Is dopamine interacting with aromatase to control sexual behavior in male quail? Poultr. Sci. Rev 4:217–233; 1992.
8. Balthazart, J.; Ball, G. F.; McEwen, B. S. An autoradiographic study of α_1-adrenergic receptors in the brain of the Japanese quail (*Coturnix coturnix japonica*). Cell Tissue Res. 258:563–568; 1989.
9. Balthazart, J.; Dupiereux, V.; Aste, N.; Viglietti–Panzica, C.; Barrese, M.; Panzica, G. C. Afferent and efferent connections of the sexually dimorphic medial preoptic nucleus of the male quail revealed by *in vitro* transport of DiI. Cell Tissue Res. 276:455–475; 1994.
10. Balthazart, J.; Foidart, A.; Surlemont, C.; Vockel, A.; Harada, N. Distribution of aromatase in the brain of the Japanese quail, ring dove, and zebra finch: An immunocytochemical study. J. Comp. Neurol. 301:276–288; 1990.
11. Boyd, S. K. Brain vasotocin pathways and the control of sexual behaviors in the bullfrog. Brain Res. Bull. 44:345–350; 1997.
12. Brody, H. The aging brain. Acta Neurol. Scand. 85:40–44; 1992.
13. Castagna, C.; Balthazart, J. Effects of vasotocin on sexual behavior and crowing in male Japanese quail. Ital. J. Anat. Embriol. 101(Suppl. 1):148–149; 1996.
14. Davies, D. T.; Follett, B. K. Neuroendocrine regulation of gonadotrophin-releasing hormone secretion in the Japanese quail. Gen. Comp. Endocrinol. 40:220–225; 1980.
15. De Vries, G. J. Sex differences in neurotransmitter systems. J. Neuroendocrinol. 2:1–13; 1990.
16. De Vries, G. J; Buijs, R. M. The origin of the vasopressinergic and oxytocinergic innervation of the rat brain with special reference to the lateral septum. Brain Res. 273:307–317; 1983.
17. Dellovade, T. L.; Rissman, E. F.; Thompson, N.; Harada, N.; Ottinger, M. A. Co-localization of aromatase enzyme and estrogen receptor immunoreactivity in the preoptic area during reproductive aging. Brain Res. 674:181–187. 1995.
18. Diez–Guerra, F. J.; Augood, S.; Emson, P. C.; Dyer, R. G. Opioid peptides inhibit the release of noradrenaline from slices of rat medial preoptic area. Exp. Brain Res. 66:378–384; 1987.
19. Fan, Y.; Ottinger, M. A. Inhibition of hypothalamic chcken gonadotropin releasing hormone (cGnRH-I) by opioid peptides in vitro. Soc. Neurosci. Abstr. 22:1996.
20. Foidart, A.; De Clerck, A.; Harada, N.; Balthazart, J. Aromatase-immunoreactive cells in the quail brain: Effects of testosterone and sex dimorphism. Physiol. Behav. 55:453–464; 1994.
21. Foster, R. G.; Panzica, G. C.; Parry, D. M.; Viglietti–Panzica, C. Immunocytochemical studies on the LHRH system of the Japanese quail: Influence by photoperiod and aspects of sexual differentiation. Cell Tissue Res. 253:327–335; 1988.
22. Furukawa, Y.; Kotegawa, T.; Tsutsui, K. Effects of opioid peptides on the electrical activity of preoptic and hypothalamic neurons in the quail brain. J. Exp. Zool. 273:96–103; 1995.
23. Gorham, S. L.; Ottinger, M. A. Sertoli cell tumors in Japanese quail. J. Avian Dis. 30:337–339; 1986.
24. Johnson, A. E.; Barberis, C.; Albers, H. E. Castration reduces vasopressin receptor binding in the hamster hypothalamus. Brain Res. 674:153–158; 1995.
25. Jurchevick, A.; Barth, S. W.; Aste, N.; Panzica, G. C.; Grossmann, R. Intracerebral sex differences in the vasotocin system in birds: Possible implication on behavioral and autonomic functions. Horm. Behav. 30:673–681; 1996.
26. Kalra, S. P. Neural circuits involved in the control of LHRH secretion: A model for estrous cycle regulation. J. Steroid Biochem. 23:733–742; 1985.
27. Knight, P. G.; Wilson, S. C.; Gladwell, R. T.; Cunningham, F. J. Hypothalamic contents of LHRH and catecholamines during the ovulatory cycle of the hen (*Gallus domesticus*). J. Reprod. Fertil. 71:289–295; 1984.
28. Korf, H. W.; Panzica, G. C.; Viglietti–Panzica, C.; Oksche, A. Pattern of peptidergic neurons in the avian brain: Clusters—Local circuitries—Projections. Basic Appl. Histochem. 32:55–75; 1988.
29. Kotegawa, T.; Takahashi, T.; Tsutsui, K.; Ikeda, T.; Minakata, H.; Nomoto, K. Isolation and characterization of opioid peptides in the avian brain. J. Exp. Zool 273:87–95; 1995.
30. Li, Q.; Paciotti, G. F.; Tamarkin, L.; Ottinger, M. A. LHRH-I release from quail hypothalamic slices measured by specific EIA. Gen. Comp. Endocrinol. 95:13–24; 1994.
31. Li, Q.; Tamarkin, L.; Levantine, P. L.; Ottinger, M. A. Estradiol and androgen modulate chicken luteinizing hormone-releasing hormone-I release *in vitro*. Biol. Reprod. 51:896–903; 1994.
32. Mikami, S. I. Immunocytochemistry of the avian hypothalamus and adenohypophysis. Int. Rev. Cytol. 103:189–248; 1986.
33. Millam, J. R.; Faris, P. L.; Cozzari, C. Immunohistochemical interactions between GnRH I or II with enkephalin in turkey hens. Soc. Neurosci. Abstr. 19:1993.
34. Naftolin, F.; Horvath, T. L.; Jakab, R. L.; Leranth, C.; Harada, N.; Balthazart, J. Aromatase immunoreactivity in axon terminals of the vertebrate brain: An immunocytochemical study on quail, rat, monkey and human tissues. Neuroendocrinology 63:149–155; 1996.
35. Opel, H. On hypothalamic control of ovulation in the turkey. Poult. Sci. 58:717–724; 1979

36. Ottinger, M. A. Altered neuroendocrine mechanisms during reproductive aging. Poult. Sci. Rev. 4:235–248; 1992.
37. Ottinger, M. A.; Balthazart, J. Altered endocrine and behavioral responses with reproductive aging in the male Japanese quail. Horm. Behav. 20:83–94; 1986.
38. Ottinger, M. A.; Li, Q.; Thompson, N.; Tamarkin, L. Catecholamines and LHRH during aging: Relationship to endocrinology and behavior. Soc. Neurosci. Abstr. 19:387; 1993.
39. Ottinger, M. A.; Nisbet, I. C. T.; Finch, C. E. Aging and reproduction: Comparative endocrinology of the common tern and Japanese quail. Am. Zool. 35:299–306; 1995.
40. Ottinger, M. A.; Thompson, N.; Bakst, M. R.; KubaKawa, K.; Kikuchi, M.; Ishii, S. Testicular morphology and gonadotropin receptors reflect age-related alterations in function. Am. Soc. Androl. 1996.
41. Panzica, G. C.; Aste, N.; Dellovade, T. L.; Rissman, E. F.; Foidart, A.; Balthazart, J.; Ottinger, M. A. Aromatase-containing cells of the medial preoptic nucleus respond differentially to testosterone in young and aged male quail. Ann. Endocrinol. 55:37; 1994.
42. Panzica, G. C.; Aste, N.; Viglietti-Panzica, C.; Fasolo, A. Neuronal circuits controlling quail sexual behavior. Chemical neuroanatomy of the septo-preoptic region. Poult. Sci. Rev. 4:249–259; 1992.
43. Panzica, G. C.; Aste, N.; Viglietti-Panzica, C.; Ottinger, M. A. Structural sex differences in the brain: Influence of gonadal steroids and behavioral correlates. J. Endocrinol. Invest. 18:232–252; 1995.
44. Panzica, G. C.; Castagna, C.; Aste, N.; Viglietti Panzica, C.; Balthazart, J. Testosterone effects on the neuronal ultrastructure in the medial preoptic nucleus of male Japanese quail. Brain Res. Bull. 39:281–292; 1996.
45. Panzica, G. C.; Garcia Ojeda, E.; Viglietti Panzica, C.; Thompson, N. E.; Ottinger, M. A. Testosterone effects on vasotocinergic innervation of sexually dimorphic medial preoptic nucleus and lateral septum during aging in male quail. Brain Res. 712:190–198; 1996.
46. Panzica, G. C.; Viglietti-Panzica, C.; Balthazart, J. The sexually dimorphic medial preoptic nucleus of quail: A key brain area mediating steroid action on male sexual behavior. Front. Neuroendocrinol. 17:1–75; 1996.
47. Ralph, C. L. Some effects of hypothalamic lesions on gonadotrophin release in the hen. Anat. Rec. 134:411–432; 1959.
48. Schumacher, M.; Balthazart, J. Sexual dimorphism in the hypothalamic metabolism of testosterone in the Japanese quail (*Coturnix coturnix japonica*). Prog. Brain Res. 305:51–59; 1984.
49. Sharp, P. J.; Talbot, R. T.; Main, G. M.; Dunn, I. C.; Fraser, H. M.; Huskisson, N. S. Physiological roles of chicken LHRH-I and -II in the control of gonadotrophin release in the domestic chicken. J. Endocrinol. 124:291–299; 1990.
50. Tetel, M. J.; Blaustein, J. D. Immunocytochemical evidence for noradrenergic regulation of estrogen receptor concentrations in the guinea pig hypothalamus. Brain Res. 565:321–329; 1991.
51. Van Gils, J.; Absil, P.; Grauwels, L.; Vandesande, F.; Balthazart, J. Distribution of luteinizing hormone-releasing hormones I and II (LHRH-I and II) in the quail and chicken brain: An immunocytochemical study using antibodies directed against synthetic peptides. J. Comp. Neurol. 334:304–323; 1993.
52. Viglietti-Panzica, C.; Absil, P.; Panzica, G. C.; Aste, N.; Balthazart, J. Immunohistochemical studies on vasotocin innervation of aromatase-containing regions in the male quail forebrain. Soc. Neurosci. Abstr. 20:1739; 1994.
53. Viglietti-Panzica, C.; Anselmetti, G. C.; Balthazart, J.; Aste, N.; Panzica, G. C. Vasotocinergic innervation of the septal region in the Japanese quail: Sexual differences and the influence of testosterone. Cell Tissue Res. 267:261–265; 1992.
54. Viglietti-Panzica, C.; Aste, N.; Balthazart, J.; Panzica, G. C. Vasotocinergic innervation of sexually dimorphic medial preoptic nucleus of the male Japanese quail: Influence of testosterone. Brain Res. 657:171–184; 1994.
55. Viglietti-Panzica, C.; García-Ojeda, E.; Aste, N.; Panzica, G. C.; Thompson, N.; Ottinger, M. A. The vasotocin system in the quail brain: Changes with age. Soc. Neurosci. Abstr. 22:1890; 1996.
56. Viglietti-Panzica, C.; Panzica, G. C. Peptidergic neurons in the avian brain. Ann. Sci. Nat. Zool. Paris 12:137–155; 1991.
57. Voorhuis, T. A. M.; De Kloet, E. R.; De Wied, D. Effect of a vasotocin analog on singing behavior in the canary. Horm. Behav. 25:549–559; 1991.
58. Weiland, N. G.; Wise, P. M. Estrogen and progesterone regulate opiate receptor densities in multiple brain regions. Endocrinology 126:804–808; 1990.
59. Wise, P. M.; Scarbrough, K.; Larson, G. H.; Lloyd, J. M.; Weiland, N. G.; Chiu, S. F. Neuroendocrine influences on aging of the female reproductive system. Front. Neuroendocrinol. 12:323–356; 1991.

LONGEVITY

Growing Old Together

Simple evolutionary arguments don't demand that organisms share mechanisms of aging. But new work is strengthening the case that common processes control the life-spans of biology's favorite model systems

At the Retirement Home for Aging Model Organisms, each creature might have rocked on the front porch complaining about its own species' particular way of growing old. Now a spate of new papers suggests that these organisms in fact break down in similar ways as they pass their primes. And it's seeming more likely that mammals, too, will pull up a chair and swap stories about similar failing molecular pathways.

In the past decade, researchers have discovered an array of genes that control aging, but it was questionable whether one organism's methods for dipping into the fountain of youth would apply to any other. Yet reports on pages 104 and 107 show that a signaling pathway that tunes the worm life-span also operates in flies. A third article, published online today by *Science* (www.sciencexpress.org), reports that a gene related to a member of this pathway likewise influences aging in yeast. Other recent work suggests that an aging gene discovered several years ago in yeast also controls life-span in worms. And researchers speculate that mammals share similar aging pathways.

"It's all coming together," says molecular geneticist Cynthia Kenyon of the University of California, San Francisco (UCSF). "This aging system that we know about in the worm is out there in other animals, regulating their aging. Maybe not in the same exact form, but it's there. It's unbelievable, simply unbelievable."

This congruence is surprising because natural selection acts on genes that influence reproductive success, so those that accelerate or decelerate aging might be expected to lie beyond the reach of evolution's mighty arm. As molecular geneticist Gordon Lithgow of the University of Manchester in the United Kingdom points out, "If aging is a collection of nonadaptive deleterious events, why on Earth should they be conserved across a wide variety of species?"

Aging gracefully without reproducing

No Hollywood star has grown old under as much scrutiny as the roundworm *Caenorhabditis elegans*. Extensive studies have revealed a detailed picture of a molecular signaling pathway that controls the worm's longevity. At the heart of the pathway lies the *daf-2* gene, which encodes a member of the insulin family of cell surface receptors. Upon binding an insulin-related hormone, daf-2 receptors trigger the activation and repression of a suite of other genes. Certain mutations in *daf-2* allow worms to wriggle into ripe old ages that greatly exceed those of their normal counterparts. Other genes in the pathway also affect longevity, and signals from both the reproductive and sensory systems feed into it as well. For example, genetically pinning a clothespin to a worm's nose by disrupting olfactory signaling appears to extend life via the *daf-2* pathway.

Now, two teams report that a related insulin-like pathway shapes life-span in flies as well. Female *Drosophila melanogaster* that carry only defective copies of the *daf-2* equivalent (*InR*) buzz around long after normal flies keel over, according to a group led by Marc Tatar, an evolutionary geneticist at Brown University in Providence, Rhode Island. Similarly, animals with inactive versions of another member of the pathway called *chico* (see diagram, p. 43) flit into extreme dotage, report geneticists Linda Partridge and David Gems of University College London and colleagues. In both cases, adding back an operational version of the appropriate gene returns life-span to almost normal. Together, these results suggest that functional *InR* and *chico* genes accelerate aging. The work is "tremendously exciting," says John Tower, a molecular biologist at the University of Southern California (USC) in Los Angeles. "To see [the aging pathway] conserved in two such different species makes you hypothesize that the pathway may be more general."

How these mutations confer longevity is a puzzle, however. Other work has shown that stalling or blocking reproduction in flies enhances longevity. Such findings bolster conventional wisdom about how animals allocate limited resources: If they make lots of babies, they burn through the energy that would otherwise preserve their own bodies. The newly described *chico* mutant females are almost sterile, so the Partridge and Gems team tested whether another mutation that prevents *Drosophila* from making eggs would also lengthen life to the same extent.

They found that sterility apparently isn't the only factor that sets back the clock in *chico* mutants: *chico* flies survived longer than the other sterile and long-lived mutant, called *ovo*. Furthermore, flies that carry one good and one bad copy of *chico* outlive *ovo* flies, even though these heterozygotes are more fertile than *ovo* mutants. The commonly observed trade-off between reproductive capacity and longevity might arise not simply because of energy investments, suggests Gems, but because a single system normally controls both processes. He cautions that the new results do not rule out a direct connection between fertility and life-span: The flies might be making the critical energy investment in reproductive activities before the mutation halts egg production.

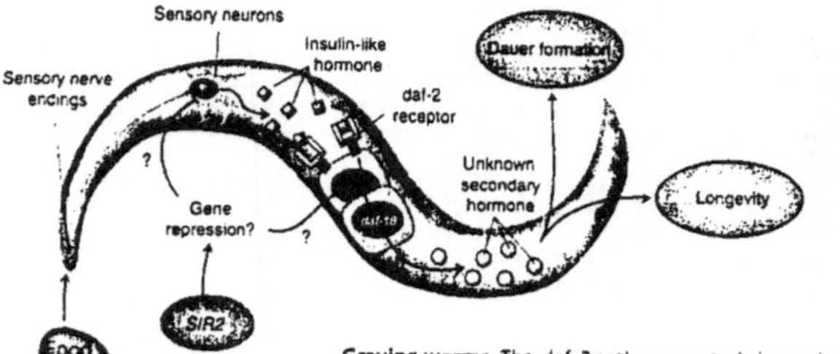

Graying worms. The *daf-2* pathway controls longevity and dauer formation in worms.

InR mutant flies, which are also sterile, provided another way to explore the relation between aging and reproduction—or at least fertility hormones. Normally, flies put reproduction on hold during the winter by tapering off levels of juvenile hormone (JH). It turns out that the long-lived *InR* mutants produce abnormally small amounts of this hormone, suggesting that low JH levels may mediate the mutants' life-span extension. To test this, the researchers exposed flies to a chemical that mimics JH.

The results indicate that JH might indeed be driving *InR* mutants' longevity, at least partially. The chemical didn't alter the life-spans of wild-type flies much, but it pushed mortality in *InR* flies back toward normal. This work "makes a leap from the genetic pathways to the hormonal signals that may be affecting life-span," says Lithgow, adding that the finding "begs the question of what endocrine signals in mammals might be involved in determining longevity."

Handling stress with maturity

Both fly papers examine another possible explanation for the staying power of *chico* and *InR* flies: resistance to stresses such as heat and oxidative damage. Long-lived mice, worms, and flies tolerate these and other physical and chemical insults better than do normal organisms. Furthermore, treatments that defend against damage from free radicals, such as overproduction of the enzyme superoxide dismutase or exposure to chemicals that perform a similar detoxifying reaction, confer long life on flies and worms. The researchers therefore wanted to know whether the mutants displayed unusual hardiness when subjected to harsh conditions. The results were inconclusive, but neither team conducted exhaustive tests. "These data don't move that argument either way," says Lithgow. "There's a lot of supportive evidence that there's a relationship between stress and aging."

Indeed, the work on yeast reported today advances the claim that stress resistance lengthens life. Valter Longo, a molecular geneticist at USC, and colleagues battered a population of yeast mutants with heat or paraquat, a chemical that creates reactive oxygen molecules. Among the survivors, they found two strains that endured both treatments. These mutants also lived longer than wild-type yeast and mutants that tolerated only heat or paraquat but not both. One of the mutants carries a defect in *Sch9*, which resembles *Akt1* and *Akt2*, genes that act in the *C. elegans* daf-2 pathway. When the researchers engineered a deletion of *Sch9*, they found that the resulting strain survived three times longer than did wild-type cells.

Both *Sch9* and the other yeast gene that emerged from both the heat and paraquat screens, adenylate cyclase (*Cyr1*), are know to participate in pathways that decrease stress resistance. Indeed, the two mutant strains tolerate not only heat and paraquat, but also hydrogen peroxide. And Longo's group found that genes that protect against heat shock increase survival in *Cyr1* mutants. He suggests that resistance to multiple stresses promotes extended longevity.

Other researchers previously implicated the adenylate cyclase pathway in yeast staying power. Several years ago, Leonard Guarente of the Massachusetts Institute of Technology and colleagues established that overproduction of the yeast protein Sir2 increases a different measure of longevity, the number of times a mother cell produces a daughter cell. Last September, his team showed that *SIR2* allows yeast to live longer via the adenylate cyclase pathway (*Science*, 22 September 2000, p. 2126). And Guarente recently extended the relevance of his work with *SIR2* by implicating a homologous gene in the process of worm aging. Overproducing the worm relative of Sir2 increases average life-span by up to 50%, he reported in the 8 March issue of *Nature*. Furthermore, he gathered genetic evidence that *Sir2* operates in the *daf-2* pathway.

Guarente suggests that *SIR2* and its kin mediate life-span by the same mechanism in many creatures: They remove an acetyl chemical group from DNA-binding proteins, allowing them to halt gene expression. But the targeted genes perform a variety of physiological roles. "Deacetylation of some protein is important for longevity, but how that plays out in different organisms is likely to be very, very different," Guarente says. His team found that nutrient deprivation spurs Sir2 to extend life-span, and this ability to respond surroundings, he says, "allows an organism to hang around longer when food is scarce. When things get better, you're still there and your neighbor's not."

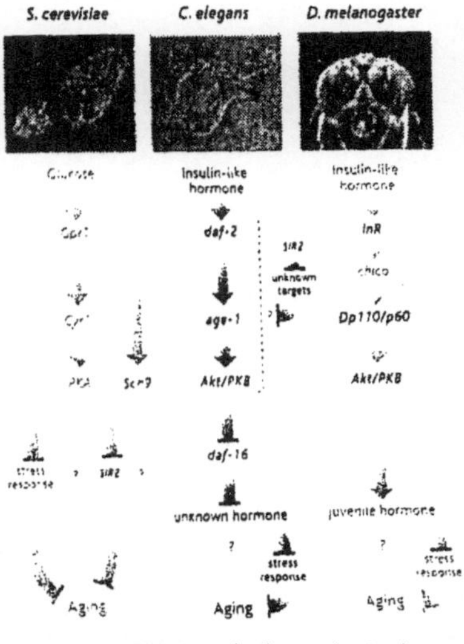

Copycats. Similar genes (red) control aging in yeast, worms, and flies. Arrows indicate gene activation (not necessarily direct); blocks indicate repression.

This link between environmental conditions and aging pathways also provides an explanation for how "aging" and "antiaging" genes might have evolved. In worms, the *daf-2* pathway controls life-span, fertility, and entry into a semidormant state that allows prepubescent worms to wait out rough times. Crowding and lack of food, for example, trigger this "dauer" state; the trade-off is that dauer worms can't reproduce. The life-span genes "all seem to be allowing these animals to survive periods of food deprivation and harsh conditions," says UCSF's Kenyon. "Now we're learning that by tinkering with this system, it's possible to change life-span in a healthy way. You don't have to be a dauer to do it."

Hungry but wizened

The connection between limited food and longevity is reminiscent of the one well-established treatment that stalls aging in mammals: calorie restriction. Calorie-restricted mammals have low levels of both insulin and insulin-like growth factor-1 (IGF-1). Several receptors respond to such hormones in mammals, including separate insulin and IGF-1 receptors, both of which resemble the single daf-2 receptor that underlies the aging pathway in *C. elegans*.

"You can make a very strong case that there is a link between IGF signaling and aging" in mammals, says Gems. Some of the best evidence comes from previous research on three strains of long-lived mice. They carry mutations in different genes that affect growth hormone signaling, and all have low levels of IGF-1 in the blood as well as alterations in the amounts of other hormones. Like many long-lived fly mutants, these mice "are small, have reduced reproductive function, mutations in what looks like a similar pathway, and more of a life-span effect on females than males," says Andrzej Bartke, a mammalian endocrinologist at Southern Illinois University in Carbondale. "To what extent the [fly and mouse] genes have corresponding functions, it's hard to know. But these papers are very exciting because they add an enormous amount of evidence that the parallelism does indeed exist."

The parallels among flies, worms, and yeast are falling into place, but connecting these pathways with mammalian aging will take more work. Other pathways are out of whack in the mouse mutants, warns William Sonntag, an endocrinologist at Wake Forest University School of Medicine in Winston-Salem, North Carolina, so it's too early to draw solid connections between IGF-1 and aging pathways in *C. elegans* and *Drosophila*. But he is cautiously optimistic. "The fly and nematode results make us more enthusiastic that similar mechanisms are working in vertebrates and mammals," he says. "If it transfers [from worms] to flies, there's the possibility that it will transfer to mammals."

–EVELYN STRAUSS

Review

Gerontology

Gerontology 2000;46:55–63

Ups and Downs of Aging Studies in vitro: The Crooked Path of Science

A. Macieira-Coelho

University of Paris VI, Versailles, France

Key Words

Cell senescence · Cell immortalization · Cell differentiation · Markers · Telomeres · Genetics · Energy transduction

Abstract

Different approaches using cell culture techniques to study the biology of aging are critically described. Most of the studies concerned the relationship between cell division potential and aging. The growth potential of cells is fundamental for aging of the organism, since it relates to phenomena such as the regeneration of tissues, wound healing, the immune response, and stem cell renewal. Unfortunately many experiments were misinterpreted disregarding the physiology of the mammalian organism. The terminal postmitotic cell, on which most research has been concentrated, seems irrelevant for aging of the organism. Nevertheless, some experiments yielded important contributions to the understanding of the biology of cell division. Future research should ascertain such interesting suggestions as the terminal differentiation hypothesis of the human fibroblast life cycle. It is important to elucidate the significance of the increased number of postmitotic cells in pathological processes. A neglected area should be further explored: the relationships between structural modifications of the cell, decreased probability of activating energy barriers, and decline of the division potential.

An extensive review of the approaches using cell culture techniques in gerontology research has been published [1]. In this paper, we will touch upon some of the highlights, mention some misinterpretations, raise some questions, and point to new research directions.

Growth Promoting Activity of Serum from Young and Old Donors

The first investigators to use cell culture in aging research were Carrel and Ebeling [2] when they compared the effects of serum from young and old chickens on the growth of cells in vitro. They found that the serum from young animals promoted the growth of homologous fibroblasts more than that from old animals. Kondo and Nomaguchi [3] obtained opposite results testing the effects of rabbit serum on homologous fibroblasts. Kondo et al. [4] also tested human serum on human fibroblasts and could not find a difference in the growth-promoting properties of sera from young and old donors. Sera from young and old rats had, in most cases, the same effect on proliferation of human fibroblasts [5]; however, a minor proportion of serum samples did contain inhibitory fractions. The authors attributed this to specific physiological or pathological conditions of the animals before death.

Wistrom et al. [6] also could not find any differences in the growth-promoting activities of sera from cows of widely different ages, as tested on human fibroblasts. Therefore, there seems to be enough evidence invalidating the results of Carrel and Ebeling [2].

KARGER

Fax + 41 61 306 12 34
E-Mail karger@karger.ch
www.karger.com

0304–324X/00/0462–0055$17.50/0

Accessible online at:
www.karger.com/journals/ger

A. Macieira-Coelho
INSERM, 73bis, rue Marechal-Foch
F–78000 Versailles (France)
Tel. +33 1 39 49 40 15, Fax +33 1 39 49 51 45
E-Mail a.macieira-coelho@universal.fr

Cell Migration

A slower migration of cells from explants originating from older chickens as compared with those of younger animals was observed early on in tissue culture studies. This finding proved to be longer lasting than the previous one. In fact, it has been confirmed by other investigators [7–9] who found that the latent period between explantation and cell migration increased in tissues from older animals.

A loss in the capacity to migrate was confirmed by other methods [10]. Monolayers of cultivated human fibroblasts from neonatal, adult, and aged donors were 'wounded', and the migration of cells from the edge of the wound was traced. The latent period prior to initiation of migration increased with donor's age.

Cell Division Potential

The use of cell cultures in experimental gerontology was greatly stimulated by the work of Hayflick and Moorhead [11] which dealt with the potential number of divisions of serially cultivated human fibroblasts. Based on the work of Swim and Parker [12], it was known that human fibroblasts stop growing in vitro after a period of rapid proliferation. Hayflick and Moorehead [11] proposed that the limited doubling potential of normal human fibroblasts is a characteristic of somatic cells and an expression of senescence at a cellular level. An experimental system thus became available to test Weissmann's speculation [13] that the limited proliferative potential of somatic cells is part of the mechanisms that lead to aging of the organism.

The issue of cell growth potential is obviously crucial for aging. It concerns phenomena such as the regeneration of tissues, wound healing, the immune response, and stem cell renewal. Nevertheless, many gerontologists were highly opposed to the use of cell culture to study aging. This was largely due to the lack of knowledge concerning the potential of cell culture in studying in vivo phenomena [reviewed in 1]. It was also the result of faulty interpretations by some investigators using the experimental system in vitro.

The first objection, and one which has been repeatedly raised, concerns the limited division potential of somatic cells. Many investigators have claimed that it is an in vitro artefact, the result of genetic damage caused by the artificial environment [14]. However, experiments maintaining human fibroblasts in quiescence [15] suggest that the nutrient media available are adequate for these cells. Moreover, contrary to other species, there is no description of a spontaneous, indefinitely dividing, normal somatic human cell population occurring in vitro.

Comparing the probability of immortalization (acquisition of an infinite division potential) of cells from different species, either spontaneous or induced by carcinogens, also supports the view that human fibroblasts are indeed endowed with a limited proliferative phenotype [16]. The relative probability of immortalization has a bearing on the physiopathology of the respective species. Furthermore, comparative biology suggests that the presence of a normal limited or unlimited proliferative phenotype varies along the evolutionary scale and also within an organism [17]. The presence of one or the other phenotype in somatic cells reflects attributes of the respective organisms.

In order to confirm his hypothesis, Hayflick [18] compared the long-term division potential of human embryonic fibroblasts with that of postnatal fibroblasts. He observed a shorter proliferative capacity of the latter cells. This work did not prove a relationship between the long-term division potential and aging, but demonstrated a decline in the capacity to proliferate during development. Other investigators compared the number of population doublings in vitro with the age of the respective donor and found an inverse correlation [19]. Identical results were obtained by different groups not only for fibroblasts but also for other types of cells [reviewed in 1].

Recent research [20] failed to reproduce these results. There are, however, several pitfalls in this publication. One concerns the cell bank used: indeed, our own experience with the same source of cells questions its validity with regard to the normalcy of the donors from which the cells were taken. For instance, we obtained results with cells from a supposedly normal donor, but these results suggested a certain pathologic condition. After much insistence, we finally obtained information about this donor who, in fact, had diabetes, hypertension and psoriasis. Another objection relates to the use of a mathematical formula to determine the number of population doublings which unfortunately is the method used by many laboratories. The kinetics of proliferation of these cell populations is very complex and thus does not fit into mathematical formulas. Indeed, it was shown [21–23] that after each subcultivation, some cells divide once, others twice or three times, and even others do not divide, but will do it after the following subcultivation. These kinetics of proliferation occur with increasing heterogeneity up to the end of the life span. Trying to express this chaotic behavior by a mathematical formula is unrealistic and produces fictitious numbers. A simpler, less sophisticated, but more reliable method to calculate the production of cells during their in vitro life span involves adding the increase in cell number after each passage [24]. On the other hand, the number of doublings is a very rough parameter which depends not only upon the pathophysiological conditions of the donor, but also on the in vitro maintenance of the cells. Furthermore, the significance that the limited proliferative life span of somatic cells has on aging of the organism is not the linear decrease of the cell dividing potential with age. The real significance, which has been diverted by wrong approaches, will be described below.

The correlation between cell population doublings in vitro and the maximum life span of the respective species was tested to confirm Hayflick's proposal. One study, which is rarely cited, did not find any correlation [25]. Another study that is repeatedly cited as proving the association between aging of the organism and potential doublings of the somatic cells did find a correlation [26]. Unfortunately, in both studies, sometimes embryonic cells were used, while sometimes postnatal cells were used, so we still have to wait for the right evidence. I would be surprised, however, if a correlation were to be eventually ascertained. Universal rules are rare or almost unknown in biology.

Implications of the Cell Growth Decline in Aging of the Organism

The first attempts to understand the dynamics of the changes progressing during the serial proliferation of cells with a limited division potential concluded that there is an entire spectrum of cells with different cycling probabilities which evolve with an increasing heterogeneity [21, 22, 27–30, 31]. These studies introduced the notion of uncertainty to the cycling probability, and to the nondividing state, and showed that it is irrelevant to determine the fraction of irreversibly nondividing cells, before the terminal stage is reached. According to these experiments, the connotation of this cell culture system for aging of the organism was that it revealed the drift created by replication well before the last mitoses. It is associated with a functional drift that creates new regulations and interactions with other cell compartments and is an important component of the permanent evolution taking place in a mammalian organism. It fits into what it is known of the physiology of the organism.

Other works [32, 33] concluded that the decline in growth potential during serial proliferation is due to the exponential increase of nondividing cells at each population doubling. The flaws in the methods used in these and subsequent studies reaching the same conclusions have been examined [23]. This Manichean interpretation was widely accepted by gerontologists using the cultivation of cells. It placed all the entire emphasis on the production of nondividers, and led to the use of the term 'cell senescence' to designate the terminal state of irreversible arrest of cell division which was considered the hallmark of cellular aging.

Most investigators were quick to jump to the assumption that there is a causality relationship between postmitotic cell and organism aging, despite the fact that there is no evidence showing that the human organism ages because somatic cells lose the potential to divide. In fact, the assumption goes against all the experiments that have tested the capacity of fibroblasts obtained from old donors to divide. Nevertheless, evaluating the number of nondividing cells almost became an obsession, and nowadays most research using fibroblasts in vitro to study cell aging concentrates on the final step leading to a postmitotic cell. This means, gerontologists using animal models rejected the cell culture model.

One marker has been proposed [34] and is being used by several laboratories to define the 'senescent cell'. It consists of the staining for a lysosomal enzyme: β-galactosidase. This enzyme was found to be expressed by several types of human cells reaching the terminal postmitotic stage, and it was claimed to increase with age in human skin cells.

It had previously been shown that lysosomal enzymes increase during prolonged resting phases in vitro [35]. These data showed that the accumulation of lysosomal enzymes occurred during a prolonged nonmitotic state, regardless of whether or not it was reversible. Therefore, although this marker expresses a long quiescence, it is not suitable for making a distinction between a reversible and a terminal resting stage.

Telomere length was used as a marker to detect nonreplicating cells in vivo [36]. Analysis of cells from donors aged 0–93 years showed a nonsignificant correlation between proliferative ability and donor age and between telomeric DNA and donor age.

Other attempts at the same goal [37] checked for the expression of genes involved in cell cycling, those that were reported to be repressed in the terminal fibroblast. The authors could not find any evidence for the presence of the latter cell in the skin from centenarians.

Modifications of the 30-nm chromatin fibers were also applied to ascertain the presence of the terminal nondividing human fibroblast: this was done using skin biopsy specimens from human donors of different ages [38]. The use of this marker showed that terminal postmitotic fibroblasts are not present in human fetal tissues. However, postnatal normal human skin contains a variable amount of cells unrelated to the donor's age, along with modifications in the nucleoplasm chromatin fibers characteristic of terminal cells, but without the nucleolar changes found in the postmitotic cells obtained in vitro [38]. An increase in the fraction of terminal fibroblasts identical to that observed in vitro (modifications in the nucleoplasm and nucleolus) was seen only in pathological states [38]. The telomere length in human vascular endothelial cells from sites submitted to higher hemodynamic stress [39] was consistent with the presence of terminal cells in cardiovascular diseases. These findings support other observations indicating the presence of an increased number of terminal postmitotic cells in arthritic joints [40, 41], in the fibrous plaques of arteries [42], and in the skin of patients with diabetes and kidney disease [43].

Since it seems possible that the terminal fibroblast is the result of a physiological cell differentiation cycle [44], it was proposed that an increased number of postmitotic cells may signal a disturbance of the homeostatic regulation of that cycle and be part of the pathological picture [45].

This leads us to a fundamental question at the core of gerontology research: is it possible to find a breaking point at the cellular level where one can distinguish an imbalance that signals a switch from physiological to pathological aging? An excess of terminal postmitotic cells could be an interesting biomarker to investigate that breaking point at the tissue level.

Genetics of a Limited Proliferative Phenotype

The two models described above concerning the kinetics of proliferation of mitotic cell aging imply opposite mechanisms. In the first model, division is accompanied by small, progressive changes in the cell, creating functional heterogeneity. The second model implies an all-or-none process where irreversible postmitotic cells are produced at each population doubling.

Experiments analyzing genomic modifications during serial divisions favor the first model [38, 46, 47]. It was shown that at each division, the distribution of DNA between daughter cells is asymmetric in a significant fraction of the cell population. Moreover, analysis of DNA at different levels of its structure revealed a progressive reorganization of the genome, suggesting an increased functional heterogeneity created through division. In the end, chaotic events occur in the genome that contribute to the postmitotic state [38, 45–47].

Cytogenetic data [reviewed in 1] also come out in favor of a genome reorganization through cell division and show that the capacity for chromosome recombination, in other words, the plasticity of the genome, declines during development and aging.

Those that adhered to the dualistic, Manichean model (dividers versus nondividers) studied the final step where the cell population enters the postmitotic stage. Although the postmitotic cell is of questionable significance for organism aging, many experiments along these lines produced results with important connotations for molecular and cell biology.

In general, investigators looked for the mechanism that could explain it all. Attempts to attribute limited proliferative phenotype to specific genes were made, hybridizing 'senescent' cells with immortal human cell lines [48]. The results were interpreted according to the rule of the excluded middle. Since the majority of clones had a finite life span, it was concluded that the 'senescent' phenotype is dominant, although a significant fraction of clones yielded immortal cell populations. This fraction was ignored and explained by stating that certain cell types are more unstable in hybrids and yield immortal variants more rapidly. Four complementation groups (A, B, C, and D) were assigned for indefinite division to explain the results obtained with crosses between immortal human cell lines [49]. It was suggested that there are at least four sets of genes or processes involved in the mechanisms leading to cellular 'senescence'. Further studies along these lines showed that four complementation groups were not enough to explain the data and that a subdivision of the immortal cell lines into multiple complementation groups A, B, C, D, E ... was necessary [50].

Inter alia, the hybridizations between mortal and immortal cells and between immortal cells have to be interpreted keeping in mind the data [51] showing that clones from the HeLa tumor derived cell line display a nonproliferative and a proliferative pool. The fastest growing clones and subclones segregate daughter cells with different growth potentials, showing a shift in the distribution toward lower growth rates and increases in the proportion of cells belonging to the nonproliferative pool. This phenomenon was called clonal attenuation. As such, in the fusions described above, some of the hybrids may have been obtained from cells committed to clonal attenuation. If HeLa cells display clonal attenuation, all other immortalized cell populations must have it to different degrees. It is possible that the different life spans obtained from the hybrids depend on the degree of clonal attenuation displayed by the parent cell lines.

On the other hand, the methods used for fusion may cause genetic changes which accentuate the phenomenon of clonal attenuation. Evidence was obtained showing that the use of drug selection to obtain the hybrids is deleterious and that the arrest of division may result from the cytotoxic effect of drug selection rather than from the expression of the mortal phenotype [52].

An additional flaw is found in the definition of the postmitotic state in the hybrids with a finite life span. The postmitotic terminal cell of a normal human fibroblast population is a well-defined state with identifiable markers, one in which the cell can remain metabolizing for an indeterminate period. The markers were rarely, if at all, used to ascertain their presence in the mortal hybrids. Arrested growth is not necessarily identical to the final physiologic postmitotic stage of the fibroblast proliferative life span.

Attempts to identify specific genes responsible for the mortal phenotype were also made by introducing human chromosomes into immortal cell lines. It soon became apparent that several chromosomes can suppress the immortal phenotype and that multiple genes are lost or mutated in immortal cell lines [53]. There were more chromosomes identified by monochromosomal transfer studies containing a putative gene responsible for the mortal phenotype than there were complementation groups.

There are caveats with experiments using chromosome transfers. Crosses between human cells and A9 mouse fibroblasts were undertaken to select the human chromosomes that will be transferred. It has been shown that cross-contamination of the two genomes occurs in hybrids of human and rodent cells [54]. Fusion products between human and mouse cells showed that species-related genetic determinants may interfere with the expression of certain phenotypes, making it impossible to interpret the results according to Mendelian genetics [55]. This interesting aspect was first reported in this study, but it has been subsequently overlooked when interpreting other experiments along these lines.

Repression of c-fos transcription was thought to be a main event in the terminal arrest of growth [56]. It had already been reported, however, that c-fos is not repressed and that the terminal cell still displays some early responses to growth factors [57].

Telomeres

It has been postulated that the inability of DNA polymerase to fully replicate the ends of the discontinuously synthesized DNA strand leads to a shortening of chromosome ends at each cell division cycle [58]. The loss to a critical level of TTAGGG repeats, that characterize mammalian chromosome ends, could cause the arrest of cells.

Hybridization of the terminal restriction fragments of DNA from serially dividing human fibroblasts with the appropriate probe, showed a shortening of the hybridization signal which was interpreted as a shortening of the telomeres as population doublings increased [59]. The extension of the human fibroblast life span accompanied by telomere elongation, after the introduction of telomerase by transfection of the cells with vectors encoding the human telomerase catalytic subunit, supports the relationship of causality between telomere shortening and the cell's finite division potential [60].

Since some immortal cell lines express the enzyme telomerase, and develop the capacity to reconstitute telomeres after each replication, the link between telomere integrity and replication potential seemed established [59].

These exciting results were obviously followed by feverish research along these lines. Screening of different kinds of tumors for the presence of the canonical human telomeric repeat $(TTAGGG)_n$ gave variable results. In some studies, 40% of the tumors analyzed revealed a reduction in telomere repeat arrays [61]. Moreover, it was reported that telomere fragment size does not limit the growth potential of leiomyomas [62].

Exceptions to the rule also occur in normal human cells: exhaustion of the growth potential of normal human oral keratinocytes is associated with a loss in telomerase activity without shortening of telomeres [63]. Results obtained with lymphocytes vary with the laboratory. Some investigators found that telomeric sequences are lost in proliferating lymphocytes [64], others could not find any alteration of chromosome ends [65, 66], while still others observed a decline early during the human life span, a plateau during young adulthood, and gradual attrition later in life [67].

Fibroblasts from patients with Werner's syndrome which have a shorter proliferative potential than those of normal age-matched control donors do not have shorter telomeres than control cells [68]. The mean terminal repeat fragments of cells from patients with Werner's syndrome that had ceased replication were significantly longer than those of the controls.

The situation also differs with the species. Mouse fibroblasts, the proliferative life span of which is shorter than that of human fibroblasts, have chromosomal terminal repeats that are many times longer [69]. Their size is also largely unchanged through somatic cell division. The mouse terminal repeat fragment is highly polymorphic, suggesting an unusually high mutation rate. Moreover, mice without telomerase, obtained with gene knockout techniques, reproduce normally for several generations, and their cells are as efficiently immortalized as those from telomerase-positive animals [70].

Understanding the connotations of changes at chromosome ends for the cell-proliferative life span depends upon the clarification of what really happens at the level of telomeres during replication. Approximately one quarter of in vitro immortalized cell lines have no detectable telomerase activity, but have long telomeres; therefore, they must have an alternative mechanism for chromosome lengthening [71]. Yeast apparently uses recombination, and Drosophila uses specific retrotransposons for this purpose [72].

The telomeres of human fibroblast chromosomes were analyzed with the canonical probe that detects all telomeres and with a TelBam11 probe that is specific for a subset of human telomeres [73]. A reduction in terminal repeat fragment size was found with the former probe; however, there was no evidence of a loss of TelBam11 homologous sequences even in cells at the end of their life span. The heterogeneity in fragment size was much greater when the canonical probe was used, relative to that seen using the TelBam11 probe. This could be attributable to a greater variability in the location of the restriction enzyme sites rather than to differences in the length of terminal repeats. Ben [73] concluded that variation in terminal repeat length my be related to how much telomeres participate in chromosome rearrangement. Telomere-promoted recombination can lead to degeneration of the telomeric sequence and subsequent loss of the hybridization ability [74]. Indeed, several examples of apparent terminal deletions are actually subtelomeric translocations [75]. So the exact nature of these modifications has yet to be ascertained. They could be part of the overall reorganization taking place in the genome through proliferation to which TTAGGG repeats seem particularly prone and which eventually contribute to a postmitotic state [76].

These are only a few questions raised concerning telomeres. A more extensive analysis of the work published is not within the scope of this short review.

Other Mechanisms That Have Been Implicated in the Growth Decline

The search for mechanisms implicated in the growth decline of cells with a limited proliferative life span went through several fads. Errors in proteins had its momentum: actually the desire to find errors was so great that any changes in proteins were taken as proof that errors were being made during serial proliferation [77]. Protein modifications, however, were found to be postsynthetic [78].

The role of free radicals was extensively studied [reviewed in 1]. Free radical metabolism has been proposed as the cornerstone, the 'theory of everything' in gerontology. However, it should not be forgotten that the production of free radicals is part of a vital cell process where electromagnetic energy (electron transport) is transformed into chemical energy (production of ATP). This is obviously a crucial phenomenon, and, as mentioned below, changes in energy transduction have been overlooked, but are a critical component of the cell growth decline. The real role of free radicals will probably only be known when the regulation of energy transduction in the cell is understood. This is a fundamental problem in cell biology that waits to be elucidated; for the moment we are a long way off.

Investigators have also searched for the growth inhibitor responsible for the growth decline and final arrest. The p21 kinase inhibitor coded by the SDI1 gene was thought to be a candidate [79]. It increases prior to the postmitotic state, but decreases when cells are finally arrested [80]. Since it is expressed in postmitotic cells at barely detectable levels, it is probably not required for the final growth arrest [81]. p53, a positive transactivator of p21 gene expression, must be one of the parameters responsible for the growth decline, since its activity increases in a stepwise fashion through the human fibroblast proliferative life span [82]. However, cells from individuals with Li-Fraumeni syndrome enter the terminal stage in spite of the absence of p53; therefore, this protein is probably not critical for final growth arrest [81]. In fact, several kinase regulators and glycopetide growth inhibitors play a role in the growth decline, in addition to the structural alterations that render more difficult the transduction of energy [reviewed in 83].

Conclusions

In our view, for the reasons mentioned above, and those reviewed previously [1], the limited in vitro proliferative life span of somatic cells is a valid experimental system to study aging, provided correct interpretation is made in accordance with the physiology of the mammalian organism. The relevance of this system used to study aging of the organism is that it demonstrates the gradual changes originated through serial divisions: they contribute to the permanent reorganization of the human organism which starts with the fusion of the gametes and proceeds until its extinction.

The growth decline and the irreversible postmitotic state are the result of the complex interplay between positive and negative growth regulators, modifications of surface and intracellular molecules, and reorganizations within the different hierarchical orders of the DNA structure.

There is, however, an aspect that has been neglected, and it concerns the problem of conformation biology [84]. Structure and function are coupled in a cell, the flow of information depending on structural flexibility. The structural reorganization at the supramolecular and cellular levels taking place through division decreases the conformational flexibility and lowers the probability of activating energy barriers [83]. This is a fundamental feature of the evolution of the fibroblast proliferative life span that is expressed by changes in cell morphology, volume [85], and contractility [86] and reaches its climax in the terminal stage. It is in itself responsible to a great extent for the deregulation of the different signalling pathways rendering the progression through the division cycle increasingly more difficult, becoming ultimately impossible with normal growth inducers. Therefore, the structure-dependent disturbances in energy transduction are at the core of the mechanisms underlying the growth decline. An increased focus on the study of this regulation is fundamental.

It is crucial to ascertain the interesting suggestion that the proliferative life span of the human fibroblast is a terminal differentiation cycle. It depends on establishing a role for the terminal cell in homeostasis. Another interesting and important point to ascertain is the role of the terminal cell in human pathology.

References

1 Macieira-Coelho A: Biology of Normal Proliferating Cells in vitro. Relevance for in vivo Aging. Interdiscipl Top Gerontol. Basel, Karger, vol 23, pp 1–218, 1988.

2 Carrel A, Ebeling AH: Age and multiplication of fibroblasts. J Exp Med 1921;34:599–623.

3 Kondo H, Nomaguchi TA: Effects of serum from rabbits of various ages on cell proliferation. Mech Ageing Dev 1985;32:1–10.

4 Kondo H, Nomaguchi TA, Sakurai Y, Yonezawa Y, Kaji K, Matsuo M, Okabe H: Effects of serum from human subjects of various ages on proliferation of human lung and skin fibroblasts. Exp Cell Res 1988;178:287–295.

5 Kondo H, Yonezawa Y, Nomaguchi TA: Effects of serum collected from rats of different ages on in vitro cell proliferation. Mech Ageing Dev 1988;42:159–172.

6 Wistrom C, Feng J, Villeponteau B: Proliferative capacity of human fibroblasts when cultured in serum from young or cold cows. J Gerontol. 1989;44:B160–B163.

7 Soukupova M, Holeckova E: The latent period of explanted organs of newborn, adult and senile rats. Exp Cell Res 1964;33:362–367.

8 Waters H, Walford RL: Latent periods for outgrowth of human skin explants as a function of age. J Gerontol 1970;25:381–383.

9 Bierman EL: The effect of donor age on the in vitro life span of cultured human arterial smooth-muscle cells. In Vitro 1978;14:951–955.

10 Muggleton-Harris AL, Reisert PS, Burghoff RL: In vitro characterization of response to stimulus (wounding) with regard to ageing in human skin fibroblasts. Mech Ageing Dev 1982;19:37–43.

11 Hayflick L, Moorhead PS: The serial cultivation of human diploid cell strains. Exp Cell Res 1961;25:585–621.

12 Swim HE, Parker RF: Culture characteristics of human fibroblasts propagated serially. Am J Hyg 1957;66:235–243.

13 Weissmann A: Essays upon Heredity and Kindred Biological Problems. Oxford, Clarendon Press, 1891.

14 Rubin H: Cell aging in vivo and in vitro. Mech Ageing Dev 1997;98:1–35.

15 Dell'Orco RT, Mertens JG, Kruse PF Jr: Doubling potential, calendar time, and senescence of human diploid cells in culture. Exp Cell Res 1973;77:356–362.

16 Macieira-Coelho A: Cancer and aging. Exp Gerontol 1986;21:483–495.

17 Macieira-Coelho A: Comparative biology of cell immortalization; in Macieira-Coelho A (ed): Progress in Molecular and Subcellular Biology: Cell Immortalization. Heidelberg, Springer, 1999, vol 24, pp 51–80.

18 Hayflick L: The limited in vitro lifetime of human diploid cell strains. Exp Cell Res 1965;37:614–636.

19 Martin GM, Sprague CA, Epstein CJ: Replicative life span of cultivated human cells: Effects of donor's age, tissue and genotype. Lab Invest 1970;23:86–92.

20 Cristofalo VJ, Allen RG, Pignolo RJ, Martin BG, Beck JC: Relationship between donor age and the replicative lifespan of human cells in culture: A reevaluation. Proc Natl Acad Sci USA 1998;95:10614–10619.

21 Macieira-Coelho A: Influence of cell density on growth inhibition of human fibroblasts in vitro. Proc Soc Exp Biol Med 1967;125:548–552.

22 Absher PM, Absher RG, Barnes WD: Genealogies of clones of diploid fibroblasts: Cinematographic observations of cell division patterns in relation to population age. Exp Cell Res 1974;88:95–104.

23 Macieira-Coelho A, Azzarone B: Aging of human fibroblasts is a succession of subtle changes in the cell cycle and has a final short stage with abrupt events. Exp Cell Res 1982;141:325–332.

24 Lima L, Macieira-Coelho A: Aging in vitro: Effect of low dose rate irradiation on the division potential of chick embryonic fibroblasts. Exp Cell Res 1972;73:345–350.

25 Stanley JF, Pye D, MacGregor A: Comparison of doubling numbers attained by cultured animal cells with life span of species. Nature 1975; 255:158–159.
26 Röhme D: Evidence for a relationship between longevity of mammalian species and life spans of normal fibroblasts in vitro and erythrocytes in vivo. Proc Natl Acad Sci USA 1981;78: 5009–5013.
27 Macieira-Coelho A, Pontén J, Philipson L: The division cycle and RNA synthesis in diploid human cells at different passage levels in vitro. Exp Cell Res 1966;42:673–684.
28 Macieira-Coelho A, Ponten J, Philipson L: Inhibition of the division cycle in confluent cultures of human fibroblasts in vitro. Exp Cell Res 1966;43:20–29.
29 Macieira-Coelho A, Berumen L: The cell cycle during growth inhibition of human embryonic fibroblasts in vitro. Proc Soc Exp Biol Med 1973;144:43–48.
30 Macieira-Coelho A: Are non-dividing cells present in ageing cell cultures? Nature 1974; 248:421–422.
31 Absher PM, Absher RG: Clonal variation and aging of diploid fibroblasts: Cinematographic studies of cell pedigrees. Exp Cell Res 1976; 103:247–255.
32 Merz GS, Ross JD: Viability of human diploid cells as a function of in vitro age. J Cell Physiol 1969;74:219–223.
33 Cristofalo VJ, Sharf BB: Cellular senescence and DNA synthesis. Thymidine incorporation as a measure of population age in human diploid cells. Exp Cell Res 1973;76:419–427.
34 Dimri G, Lee G, Basile M, Acosta M, Scott G, Roskelly C, Medranos E, Linskens M, Rubelj I, Pereira-Smith O, Peacocke M, Campisi J: A biomarker that identifies senescent human cells in culture and in aging skin in vivo. Proc Natl Acad Sci USA 1995;92:9363–9367.
35 Macieira-Coelho A, Garcia-Giralt E, Adrian M: Changes in lysosomal associated structures in human fibroblasts kept in resting stage. Proc Soc Exp Biol Med 1971;138:712–718.
36 Allsop RC, Vaziri H, Patterson C, Goldstein S, Younglai EV, Futcher AB, Greider CW, Harley CB: Telomere length predicts replicative capacity of human fibroblasts. Proc Natl Acad Sci USA 1992;89:10114–10118.
37 Grassilli E, Bellesia D, Salomoni P, Croce MA, Sikora E, Radiszewska E, Tesco G, Vergelli M, Latorraca S, Barbieri D, Fagiolo U, Santacaterina S, Amaducci L, Sorbi S, Franceschi C: C-fos/C-jun expression and AP-1 activation in skin fibroblasts from centenarians. Biochem Biophys Res Commun 1996;226:517–523.
38 Macieira-Coelho A: Chromatin reorganization during senescence of proliferating cells. Mutat Res 1991;256:81–104.
39 Chang E, Harley CB: Telomere length and replicative aging in human vascular tissues. Proc Natl Acad Sci USA 1995;92:11190–11194.
40 Silberberg R, Silberberg M, Feir D: Life cycle of articular cartilage cells: An electron microscopic study of the hip joint of the mouse. Am J Anat 1964;114:1459–1466.
41 Weiss C: Ultrastructural characteristics of osteoarthritis. Fed Proc 1975;32:1459–1466.
42 Ross R: The pathogenesis of atherosclerosis: An update. N Engl J Med 1986;314:488–500.
43 Morocutti A, Earle KA, Sethi M, Piras G, Richards D, Rodemann P, Viberti G: Premature senescence of skin fibroblasts from insulin-dependent diabetic patients with kidney disease. Kidney Int 1996;50:250–256.
44 Martin GM: Clonal attenuation: Causes and consequences. J Gerontol B Psychol Sci Soc Sci 1993;48:171–172.
45 Macieira-Coelho A: The last mitoses of the human fibroblast proliferative life span: Physiopathologic implications. Mech Ageing Dev 1995;82:91–104.
46 Macieira-Coelho A, Bengtsson A, van der Ploeg M: Distribution of DNA between sister cells during serial subcultivation of human fibroblasts. Histochemistry 1982;75:11–24.
47 Macieira-Coelho A: Reorganization in the different hierarchical structures of DNA during cell senescence; in Finch CE, Johnson TE (eds): Molecular Biology of Aging. UCLA Symp on Molecular Cell Biology. New York, Wiley-Liss 1990, vol 123, pp 351–364.
48 Pereira-Smith O, Smith J: Evidence for the recessive nature of cellular immortality. Science 1983;221:964–966.
49 Pereira-Smith O, Smith J: Genetic analysis of indefinite division in human cells: Identification of four complementation groups. Proc Natl Acad Sci USA 1988;85:6042–6046.
50 Moy EL, Duncan EL, Hukku B, Reddel RR: Reassessment of immortalization complementation group D. Exp Cell Res 1997;32:663–670.
51 Martinez AO, Norwood TH, Prothero JW, Martin GM: Evidence for clonal attenuation of growth potential in HeLa cells. In Vitro 1978; 14:996–971.
52 Ryan PA, Maher VM, McCormick JJ: Failure of infinite life span human cells from different immortality complementation groups to yield finite life span hybrids. J Cell Physiol 1994: 159:151–160.
53 Vojta PJ, Barrett JC: Genetic analysis of cellular senescence. Biochim Biophys Acta 1995: 1242:29–41.
54 Littlejohn MR, Camakaris J, Woodcock DM: Cross contaminaton of the genomes in human/hamster cell hybrids by multiple short recombination events. Somat Cell Mol Genet 1995;21: 385–398.
55 Kaouël LBC, Billard C, Macieira-Coelho: Growth characteristics in vitro of hybrids between normal and transformed cell lines. Int J Cancer 1978;21:338–347.
56 Seshadri T, Campisi J: Repression of c-fos transcription and an altered genetic program in senescent human fibroblasts. Science 1990: 247:205–209.
57 Paulsson Y, Bywater M, Pfeifer-Ohlsson S, Ohlsson R, Nilsson S, Heldin CH, Westermark B, Betsholtz C: Growth factors induce early pre-replicative changes in senescent human fibroblasts. EMBO J 1986;5:2157–2162.
58 Olovnikov AM: A theory of marginotomy: The incomplete copying of template margin in enzymic synthesis of polynucleotides and biological implication of the phenomena. J Theor Biol 1973;41:181–190.
59 Harley CB: Telomere loss: Mitotic clock or genetic time bomb. Mutat Res 1991;256:271–283.
60 Bodnar AG, Ouelette M, Frolkis M, Holt SE, Chiu CP, Morin GB, Harley CB, Shay JW, Lichtsteiner S, Wright WE: Extension of life-span by introduction of telomerase into normal human cells. Science 1998;279:349–352.
61 Schmitt H, Blin N, Zankl H, Schertan H: Telomere length variation in normal and malignant human tissues. Genes Chromosomes Cancer 1994;11:171–177.
62 Rogalla P, Rohen C, Henning Y, Deichert U, Bonk U, Bullerdiek J: Telomere repeat fragment sizes do not limit the growth potential of uterine leiomyomas. Biochem Biophys Res Commun 1995;211:175–182.
63 Kang MK, Guo W, Park NH: Replicative senescence of normal human oral keratinocytes is associated with the loss of telomerase activity without shortening of telomeres. Cell Growth Differ 1998;9:85–95.
64 Vaziri H, Schächter F, Uchida I, Wei L, Zhu X, Effros R, Cohen D, Harley CB: Loss of telomeric DNA during aging of normal and trisomic 21 human lymphocytes. Am J Hum Genet 1993; 52:661–667.
65 Luke S, Birnbaum R, Verma RS: Centromeric and telomeric repeats are stable in nonagenarians as revealed by the double hybridization fluorescent in situ technique. Genet Anal Tech Appl 1994;11:77–80.
66 Kormann-Bortolotto MH, Borsatto B, Smith MAC: Telomere shortening, ageing, and chromosome damage. Mech Ageing Dev 1996;89: 45–49.
67 Frenck RW Jr, Blackburn EH, Shannon KM: The rate of telomere sequence loss in human leukocytes varies with age. Proc Natl Acad Sci USA 1998;95:5607–5610.
68 Schultz VP, Zakian VA, Ogburn CE, McKay J, Jarzebowicz AA, Edland SD, Martin GM: Accelerated loss of telomeric repeats may not explain accelerated replicative decline of Werner syndrome cells. Hum Genet 1996;97:750–754.
69 Kipling D, Cook HJ: Hypervariable ultra-long telomeres in mice. Nature 1990;347:400–402.
70 Blasco MA, Lee HW, Hande MP, Smaper E, Lansdorp PM, DePinho RA, Greider CW: Telomere shortening and tumor formation by mouse cells lacking telomerase. Cell 1997;91: 25–34.
71 Bryan TM, Reddel RR: Telomere dynamics and telomerase activity in in vitro immortalized human cells. Eur J Cancer 1997;33:767–773.
72 Biessmann H, Mason JM: Telomere maintenance without telomerase. Chromosoma 1997; 106:63–69.
73 Ben P: Aging chromosome telomeres: Parallel studies with terminal repeat and telomere associated DNA probes. Mech Ageing Dev 1997; 99:153–166.

74 Ashley T, Ward DC: A 'hot spot' of recombination coincides with an interstitial telomeric sequence in the Armenian hamster. Cytogenet Cell Genet 1993;62:169–176.
75 Meltzer PS, Guan XY, Trent JM: Telomere capture stabilizes chromosome breakage. Nat Genet 1993;4:252–255.
76 Macieira-Coelho A: Reorganization of the genome during aging of proliferative cell compartments; in Macieira-Coelho A (ed): Molecular Basis of Aging. Boca Raton, CRC Press, 1995, pp 21–69.
77 Holliday R, Tarrant GM: Altered enzymes in ageing human fibroblasts. Nature 1972;238: 26–30.
78 Kahn A, Guillouzo A, Leibovitsch MP, Cottreau D, Bourel M, Dreyfus JC: Heat lability of glucose-6-phosphate dehydrogenase in some senescent human cultured cells: Evidence for its post-synthetic nature. Biochem Biophys Res Commun 1977;77:760–766.
79 Noda A, Ning Y, Venable SF, Pereira-Smith OM, Smith JR: Cloning of senescent cell-derived inhibitors of DNA synthesis using an expression screen. Exp Cell Res 1994;211:90–98.
80 Alcorta DA, Xiong Y, Phelps D, Hannon G, Beach D, Barrett JC: Involvement of the cyclin-dependent kinase inhibitor p16 (INK4a) in replicative senescence of normal human fibroblasts. Proc Natl Acad Sci USA 1996;93: 13742–13747.
81 Medcalf ASC, Klein-Szanto AJP, Cristofalo VJ: Expression of p21 is not required for senescence of human fibroblasts. Cancer Res 1996: 56:4582–4585.
82 Bond JA, Haughton M, Blaydes J, Gire V, Wynford-Thomas D, Wyllie FS: Mutant p53 rescues human diploid cells from senescence without the induction of SDI1/WAF1. Cancer Res 1996;13:2097–2104.
83 Macieira-Coelho A: Growth inhibition of human fibroblasts in vitro; in Macieira-Coelho A (ed): Progress in Molecular and Subcellular Biology: Inhibitors of Cell Growth. Heidelberg, Springer, 1998, vol 20, pp 249–270.
84 Ivanov VI, Minchenkova LE, Schyolkina AK, Poletayev AI: Different conformations of double-stranded nucleic acid in solution as revealed by circular dichroism. Biopolymers 1973;12:89–110.
85 Simons JWIM: The use of frequency distributions of cell diameters to characterize cell populations in tissue culture. Exp Cell Res 1967;45: 336–350.
86 Macieira-Coelho A, Azzarone B: Correlation between contractility and proliferation in human fibroblasts. J Cell Physiol 1990;142:610–614.

Oxidants, oxidative stress and the biology of ageing

Toren Finkel* **& Nikki J. Holbrook†**

*Laboratory of Molecular Biology, National Heart, Lung, and Blood Institute/National Institutes of Health, Building 10/6N-240, 10 Center Drive, Bethesda, Maryland 20892-1622, USA (e-mail: finkelt@nih.gov)
†Laboratory of Biological Chemistry, National Institute on Aging/National Institutes of Health, 5600 Nathan Shock Drive, Baltimore, Maryland 21224, USA (e-mail: nikki-holbrook@nih.gov)

Living in an oxygenated environment has required the evolution of effective cellular strategies to detect and detoxify metabolites of molecular oxygen known as reactive oxygen species. Here we review evidence that the appropriate and inappropriate production of oxidants, together with the ability of organisms to respond to oxidative stress, is intricately connected to ageing and life span.

Nearly a century ago it was noted that animals with higher metabolic rates often have shorter life spans. These observations led to the formulation of 'the rate-of-living hypothesis', which states that the metabolic rate of a species ultimately determines its life expectancy. Initially, the mechanistic link between metabolism and ageing was unknown. In the mid-1950s, Denham Harman articulated a 'free-radical theory' of ageing, speculating that endogenous oxygen radicals were generated in cells and resulted in a pattern of cumulative damage[1]. Although the concept of endogenous oxidants was at first controversial, the identification a decade later of superoxide dismutase (SOD)[2], an enzyme whose sole function seemed to be the removal of superoxide anions, provided mechanistic support for Harman's hypothesis. Given that the mitochondria produce most of the energy in the cell, and correspondingly consume the bulk of intracellular oxygen, the free-radical theory of ageing is now often thought of synonymously with the rate-of-living hypothesis; the higher the metabolic rate of an organism, the greater the production of reactive oxygen species (ROS) and hence the shorter the life span. However, in some species the strict correlation between metabolic rate and life span is not maintained. This is particularly true for birds and primates, which tend to live longer than would be predicted by their metabolic rates. Careful analysis of oxidant production demonstrated that at a given metabolic rate, mitochondria from these species tend to produce fewer ROS[3]. This indicates that ROS production rather then metabolic rate provides the strongest correlation with overall longevity.

The free-radical theory of ageing originally implied that the targets of ROS were random, indiscriminate and cumulative. However, although oxidants may certainly function stochastically, accumulating evidence implicates ROS as specific signalling molecules under both physiological and pathophysiological conditions. This more complex view of the importance of oxidants in biological processes is depicted in Fig. 1. As indicated, the generation of ROS, within certain boundaries, is essential to maintain homeostasis. For example, ROS generation by phagocytic cells constitutes an essential host defence mechanism necessary to combat infection. Likewise, cytosolic ROS produced in response to stimulation by growth factors are involved in regulating the proliferative response[4]. Under certain situations of metabolic stress, even mitochondrial-derived oxidants seem to function as signalling molecules[5,6]. Regardless of how or where they are generated, a rise in intracellular oxidant levels has two potentially important effects: damage to various cell components and triggering of the activation of specific signalling pathways. Both of these effects can influence numerous cellular processes linked to ageing and the development of age-related diseases. In this review, we address how ROS are generated, how the cell responds to oxidative stress and how these responses change with age. In addition, we describe the mounting genetic evidence that links oxidants and oxidative stress responsiveness to ageing and discuss the challenges associated with the potential development of anti-ageing therapies.

Oxidant production and antioxidant defences

For the purpose of this discussion, ROS encompass a variety of diverse chemical species including superoxide anions, hydroxyl radicals and hydrogen peroxide. Some of these species, such as superoxide or hydroxyl radicals, are extremely unstable, whereas others, like hydrogen peroxide, are freely diffusible and relatively long-lived. These various radical species can either be generated exogenously or produced intracellularly from several different sources. Cytosolic enzyme systems contributing to oxidative stress include, among others, the expanding family of NADPH oxidases, a superoxide-generating system that was first described in the neutrophil. The newly described non-phagocytic NADPH oxidases also generate superoxide anions, but depending on the specific NADPH oxidase expressed, can either trigger cellular transformation or replicative senescence[7,8]. The observation that different members of the NADPH oxidase family can have such widely differing biological outcomes reinforces the complexity in determining the cellular response to oxidants. Contributing factors may include the cell type, the absolute level and duration of oxidant production, the species of ROS generated, and the specific intracellular site of ROS production. Nonetheless, the NADPH oxidase family of enzymes, like the widely characterized nitric oxide synthase (NOS) family, illustrates the apparent purposeful and deliberate use of oxidant generation in normal cellular signalling and homeostasis.

Most estimates suggest that the majority of intracellular ROS production is derived from the mitochondria. The production of mitochondrial superoxide radicals occurs primarily at two discrete points in the electron transport chain, namely at complex I (NADH dehydrogenase) and at

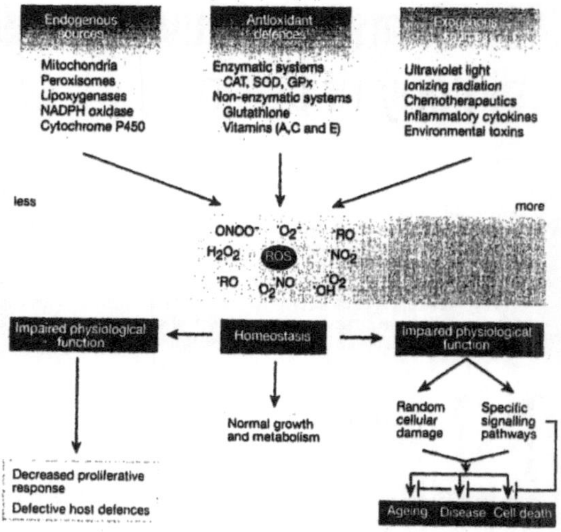

Figure 1 The sources and cellular responses to reactive oxygen species (ROS). Oxidants are generated as a result of normal intracellular metabolism in mitochondria and peroxisomes, as well as from a variety of cytosolic enzyme systems. In addition, a number of external agents can trigger ROS production. A sophisticated enzymatic and non-enzymatic antioxidant defence system including catalase (CAT), superoxide dismutase (SOD) and glutathione peroxidase (GPx) counteracts and regulates overall ROS levels to maintain physiological homeostasis. Lowering ROS levels below the homeostatic set point may interrupt the physiological role of oxidants in cellular proliferation and host defence. Similarly, increased ROS may also be detrimental and lead to cell death or to an acceleration in ageing and age-related diseases. Traditionally, the impairment caused by increased ROS is thought to result from random damage to proteins, lipids and DNA. In addition to these effects, a rise in ROS levels may also constitute a stress signal that activates specific redox-sensitive signalling pathways. Once activated, these diverse signalling pathways may have either damaging or potentially protective functions.

complex III (ubiquinone–cytochrome *c* reductase). Under normal metabolic conditions, complex III is the main site of ROS production[9]. With respect to human ageing, the Achilles' heel of this elegant system lies in the formation of the free radical semiquinone anion species (·Q−) that occurs as an intermediate in the regeneration of coenzyme Q (Fig. 2). Once formed, ·Q− can readily and non-enzymatically transfer electrons to molecular oxygen with the subsequent generation of a superoxide radical. The generation of ROS therefore becomes predominantly a function of metabolic rate and, as such, the rate of living can be indirectly translated to a corresponding rate of oxidative stress. In addition to generating oxidants, metabolism can produce a host of other by-products including glyoxal and methylglyoxal, both of which can contribute to advanced glycation end-product (AGE) formation that, in turn, seems to contribute to the ageing phenotype[10].

Evidence indicates that, *in vitro*, mitochondria convert 1–2% of the oxygen molecules consumed into superoxide anions[11]. Given that these initial estimates were made on isolated mitochondria in the presence of high, non-physiological concentrations of oxygen, the *in vivo* rate of mitochondrial superoxide production is undoubtedly considerably less. Whatever the absolute amount of mitochondrial ROS, given their potentially harmful effects, it is likely that numerous protective mechanisms have evolved to limit oxidant production and release. A further understanding of such regulatory mechanisms may reveal potential targets for therapeutic intervention. One postulated mechanism to reduce mitochondrial oxidant production is to increase the rate of metabolic uncoupling[12]. When oxygen consumption is uncoupled from ATP generation, heat is produced. This thermogenesis is mediated by an expanding family of uncoupling proteins (UCP-1, -2 and -3). However, the consumption of oxygen without ATP production would also reduce the level of free molecular oxygen potentially available for superoxide anion formation (Fig. 2). Consistent with the hypothesis that metabolic uncoupling might regulate ROS release is the recent evidence indicating that an increase in uncoupling reduces mitochondrial ROS release[3], whereas levels of mitochondrial oxidants rise in mice with a targeted deletion of UCP-3 (ref. 13).

The burden of ROS production is largely counteracted by an intricate antioxidant defence system that includes the enzymatic scavengers SOD, catalase and glutathione peroxidase. SOD speeds the conversion of superoxide to hydrogen peroxide, whereas catalase and glutathione peroxidase convert hydrogen peroxide to water. In addition to these well characterized antioxidant enzymes, at least five members of a new family of peroxide scavengers termed peroxiredoxins have recently been isolated[14]. A variety of other non-enzymatic, low molecular mass molecules are important in scavenging ROS. These include ascorbate, pyruvate, flavonoids, carotenoids and perhaps most importantly, glutathione, which is present in millimolar concentrations within cells.

The balance between ROS production and antioxidant defences determines the degree of oxidative stress. Consequences of this stress include modification to cellular proteins, lipids and DNA. The most widely studied oxidative stress-induced modification to proteins is the formation of carbonyl derivatives[15]. Carbonyl formation can occur through a variety of mechanisms including direct oxidation of certain amino-acid side chains and oxidation-induced peptide cleavage. Although all organs and all proteins can potentially be modified by oxidative stress, certain tissues and specific protein targets may be especially sensitive[16,17]. A recent report indicated that protein misfolding, independent of the cellular redox state, increases protein carbonylation[18]. As such, the notion that the rate of carbonyl formation is always directly proportional to the degree of oxidative stress may need to be re-examined.

Several studies have shown that ageing cells and organisms accumulate increased levels of oxidant-damaged nuclear DNA[19]. Perhaps because of its proximity to the main source of oxidant generation, or because of a limited DNA repair system, mitochondrial DNA is generally considered to be even more sensitive then nuclear DNA to oxidative damage. Two recent studies provided direct proof that oxidative stress can induce mitochondrial DNA damage. In these studies, oxidative stress was genetically engineered by targeted deletions in either Mn-SOD or the adenine nuclear transporter. These knockout mice had a respective defect in mitochondrial

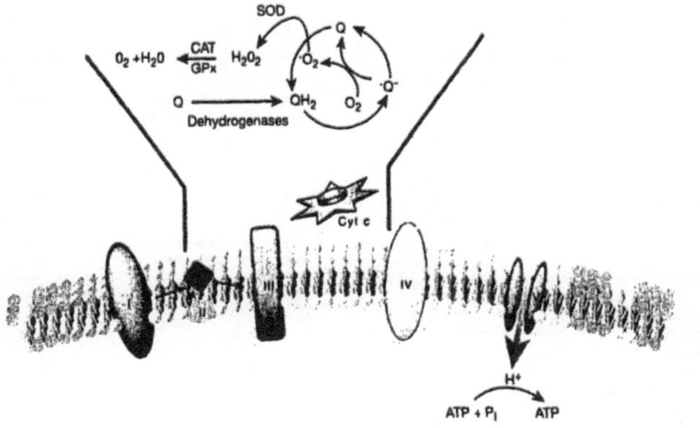

Figure 2 Complex III is the major source of mitochondrial ROS production. Electrons from complex I or II dehydrogenases are transfered to coenzyme Q (Q), also called ubiquinone. The resulting reduced form (QH_2) of coenzyme Q subsequently undergoes two sequential one-electron reductions (the Q cycle) using oxidized and reduced forms of cytochrome *b* and cytochrome *c* (Cyt *c*). The unstable intermediate in the Q cycle ($\cdot Q-$) can lead to superoxide formation by transferring electrons directly to molecular oxygen. Once generated, the superoxide can be enzymatically dismutated by SOD to form hydrogen peroxide that in turn is metabolized by enzymes such as catalase (CAT) and glutathione reductase (GPx) regenerating water and molecular oxygen. The generation of superoxide is non-enzymatic and hence the higher the rate of metabolism, the greater the production of ROS.

superoxide scavenging capacity or overall aerobic metabolism. Subsequent analysis of these animals showed significant increase in the level of mitochondrial DNA rearrangements[20,21]. Increasing damage to mitochondrial DNA inevitably leads to compromised mitochondrial function and integrity. Damaged mitochondria are thought to release more ROS and set in motion a vicious cycle of increasing DNA damage leading to increased ROS production that in turn leads to more DNA damage.

Cellular senescence and oxidative stress

After a finite number of divisions, primary cell cultures enter a state of replicative senescence in which they are growth-arrested and refractory to further mitogenic stimulation. Although the relevance of *in vitro* senescence to organismal ageing remains controversial, several studies indicate that oxidants are important in the development of the senescent phenotype. Early studies with human diploid fibroblasts revealed that cells grown in low oxygen tension exhibit a prolonged life span[22]. In contrast, cells grown in the presence of high oxygen concentrations have a reduced life span and show an accelerated rate of telomere shortening per population doubling[23]. Similarly, treatment of cultures of primary fibroblasts with moderate, non-lethal doses of exogenous hydrogen peroxide activates a rapid, senescence-like growth arrest[24].

The role of oxidants in cellular senescence was further underscored by recent observations that overexpression of an activated Ras gene can also induce a senescence-like state in human diploid fibroblasts[25]. Subsequent analysis demonstrated that expression of activated Ras in diploid fibroblasts resulted in an increase in oxidant levels[26]. In addition, although expression of activated Ras in human diploid fibroblasts produced growth arrest, this arrest could be reversed either by reducing ambient oxygen or by treatment with a cell-permeable antioxidant. As such, these results raise the possibility that a moderate, sustained rise in oxidants may function as a common trigger for activation of the senescence programme.

Oxidants in cellular signalling

Although the preceding discussions have focused mainly on the endogenous generation of ROS as a consequence of metabolic activities, many environmental stimuli including cytokines, ultraviolet (UV) radiation, chemotherapeutic agents, hyperthermia and even growth factors generate high levels of ROS that can perturb the normal redox balance and shift cells into a state of oxidative stress. When the stress is severe, survival is dependent on the ability of the cell to adapt to or resist the stress, and to repair or replace the damaged molecules. Alternatively, cells may respond to the insult by undergoing apoptosis, a process whereby severely damaged cells are removed from the multicellular host, and which, within limits, preserves the organism. A number of stress response mechanisms have evolved to help the cell and organism adapt to acute stress, and acting in either a cooperative or antagonistic fashion they serve to coordinate the acute cellular stress response and ultimately determine the outcome. Many of these pathways have been faithfully preserved throughout evolution. Thus, both the stresses and the response mechanisms represent players in an ancient battle. As will be discussed later, many mutations that prolong life seem to provide a global increase in stress resistance. Therefore, a more complete understanding of the cellular response to stress should provide significant insight into ageing.

Among the main stress signalling pathways and/or central mediators activated in response to oxidant injury are the extracellular signal-regulated kinase (ERK), c-Jun amino-terminal kinase (JNK) and p38 mitogen-activated protein kinase (MAPK) signalling cascades, the phosphoinositide 3-kinase (PI(3)K)/Akt pathway, the nuclear factor (NF)-κB signalling system, p53 activation, and the heat shock response (Fig. 3). Activation of these pathways is not unique to oxidative stress, as they are known to have central roles in regulating cellular responses to other stresses as well as regulating normal growth and metabolism. Indeed, in some situations the response to oxidants may involve overstimulation of normal ROS-regulated signalling pathways. In general, the heat shock response, ERK, PI(3)K/Akt and NF-κB signalling pathways exert a pro-survival influence during oxidant injury, whereas activation of p53, JNK and p38 are more commonly linked to apoptosis. However, numerous exceptions to these generalities can be found.

The initiating events leading to activation of pathways in response to oxidants are incompletely understood. In the case of p53 activation, oxidative stress may be sensed as a consequence of the DNA damage it causes. However, in some cells, elevated p53 expression results in increased oxidative stress[27,28], suggesting that an important consequence of oxidant-induced p53 activation is a further increase in the level of oxidative stress. This positive feedback loop may be important in committing to an apoptotic response.

Oxidants seem to activate the ERK and the PI(3)K/Akt pathways largely through stimulation of growth-factor receptors, mimicking

Table 1 **Mutations revealing links between longevity and stress resistance in multicellular organisms**

Species/mutations	Gene description	Phenotypic influence* Life span†	Stress resistance
C. elegans			
age-1	Human PI(3)K homologue	65% increase	Enhanced (UV, paraquat, heat)
daf-2	Human insulin-receptor homologue	100% increase	Enhanced (UV, paraquat, heat)
daf-16	Forkhead transcription factor	Suppresses longevity conferred by *age-1* and *daf-2* mutations	Suppresses stress resistance of *age-1* and *daf-2* mutants
clk-1	Homologue of yeast gene associated with coenzyme Q biosynthesis	40% increase	Enhanced increased (UVC)
spe-10	Unknown (sperm defective)	40% increase	Enhanced (UV, paraquat, but not heat)
spe-26	Unknown (sperm defective)	65% increase	Enhanced (UV, paraquat, heat)
old-1	Putative receptor tyrosine kinase	65% increase	Enhanced (UV, heat)
ctl-1	Cytosolic catalase	25% decrease; suppresses longevity conferred by *daf-2*, *age-1* and *clk-1*	Not determined
mev-1	Cytochrome *b* subunit of succinate dehydrogenase	37% decrease	Hypersensitive to oxygen
Drosophila			
mth	Putative G-protein-coupled receptor	35% increase	Enhanced (UV, paraquat, heat)
Mouse			
shc$^{-/-}$	Cytoplasmic signal-transduction adaptor protein	30% increase	Enhanced (UV, H_2O_2)

*See ref. 83 for original references describing the phenotypes of the *daf-2*, *daf-16*, *clk-1* and *spe-26* mutants. References for other mutants are as follows: *age-1* (refs 49, 50, 83); *spe-10* (ref. 84); *old-1* (ref. 85); *ctl-1* (ref. 56), *mev-1* (ref. 55); *mth* (ref. 60); and *shc*$^{-/-}$ (ref. 68).
†Numbers reflect changes in mean life span of the mutants relative to wild-type animals.

the actions of natural ligands. Many growth-factor receptors have been shown to undergo enhanced phosphorylation in response to direct treatment with oxidants, and agents or conditions that prevent receptor phosphorylation likewise inhibit the activation of ERK and Akt by oxidants[29–31]. One mechanism proposed to explain this effect is oxidant-mediated inactivation of critical phosphatases necessary for dephosphorylation (turning off) of the growth-factor receptors[31]. Support for such a mechanism has come from the finding that hydrogen peroxide, either derived exogenously or produced endogenously after growth-factor stimulation, can reversibly inactivate protein-tyrosine phosphatase 1B in cells[32]. The activation of growth-factor-receptor signalling pathways by oxidants is consistent with the demonstration that low concentrations of exogenous hydrogen peroxide are mitogenic[33].

Oxidative stress might induce activation of the JNK and p38 kinase pathways by an additional mechanism. The redox regulatory protein thioredoxin (Trx) has been shown to bind to apoptosis signal-regulating kinase (ASK1), an upstream activator of both JNK and p38, and under normal conditions inhibit its activity[34]. However, oxidative stress causes dissociation of the Trx–ASK1 complex and subsequent activation of the downstream JNK and p38 kinase[34]. Similarly, biochemical evidence indicates that under non-stressed conditions glutathione *S*-transferase binds to JNK to inhibit its activation, but that this interaction is also disrupted by oxidative stress[35]. These results show an intimate coupling between alterations in the intracellular redox state and the activity of downstream stress-activated pathways. The observation that multiple pathways are sensitive to a rise in ROS levels indicates that these pathways may have evolved, in part, to allow organisms to survive within an aerobic environment. In addition, it suggests that a rise in ROS might represent a common, if not universal, signal of cellular stress.

A common effect of the activation of these pathways is a change in pattern of gene expression mediated largely through modulation of the activities of transcription factors. Accordingly, a large number of oxidative stress-responsive transcription factors and genes have been identified[36] and some of these have been implicated in influencing ageing processes. The effect of ROS on expression and activity of transcription factors is complex and occurs at multiple levels, often in a seemingly antagonistic or paradoxical fashion. For example, although ROS generally cause an increase in AP-1 levels and increased nuclear translocation of NF-κB, oxidant stress can at the same time reduce the transcriptional activity of these molecules through the direct oxidation of critical cysteine residues contained within the DNA-binding domain[36].

Analysis of the integrated cellular response to oxidative stress is an area that is particularly suited for analysis using a genomic or proteomic approach. Initial characterization using the latter technique in peroxide-challenged yeast revealed over a 100 proteins whose levels changed after oxidative stress[37]. As might be expected, included among these were proteins involved in scavenging ROS, as well as heat shock proteins and chaperones. Nearly a quarter of the proteins identified were involved in carbohydrate metabolism. In general, oxidative stress repressed a number of proteins involved in glycolysis and the tricarboxylic acid cycle. This was interpreted as an attempt to restore levels of reducing equivalents such as NADPH at the expense of ATP generation. If these results are confirmed in higher eukaryotes it would imply that ROS may not only be by-products of metabolism but also be regulators of metabolic rates. Such a concept further blurs the distinction between the rate-of-living hypothesis and the free-radical theory of ageing and raises the possibility that the rate of living may determine the level of oxidant generation and that oxidant generation may in turn modulate the rate of living.

Several of the pathways activated by acute oxidative stress show diminished activity as a function of ageing. In a variety of model systems and stress paradigms including oxidative stress, the magnitude of induction of heat-shock proteins, and Hsp70 in particular, is attenuated with age[38]. The heat-shock protein family encompasses many chaperones involved in regulating the folding, transport and degradation of other cellular proteins. The relationship between ageing and a decline in the robustness of this stress response is unclear, but evidence indicates that elevating levels of Hsp70 enhances the survival of stressed cells and/or animals, whereas inhibiting this stress response reduces survival[39].

In contrast to the age-related attenuation in heat-shock protein induction in response to acute stress, two recent studies using complementary DNA microarray analysis to examine global age-associated changes in gene expression in mouse tissues provided evidence that basal expression (that seen in the absence of overt stress) of certain heat-shock proteins actually increases with ageing[40,41]. This elevated expression (which involves members other than Hsp70) was interpreted to occur as a response to the age-associated accumulation of oxidatively damaged proteins. Age-associated elevations in expression of heat-shock proteins have also been observed in *Drosophila*[42] and *Caenorhabditis elegans* (T. Johnson, personal communication). In some systems, stress-induced activation of the

ERK signalling pathway is also attenuated with ageing[43]. Like the heat-shock response, ERK activation exerts a pro-survival signal during oxidative stress[44]; thus, reduced ERK activity in aged cells may have a negative impact on survival. It is also worth noting that activation of ERK in response to mitogenic stimulation is also reduced as a function of ageing[45], again emphasizing the intimate link between proliferative and oxidative stress-response pathways. Basal DNA-binding activity of NF-κB has been shown to increase with age, which again has been suggested to reflect increased oxidative stress in aged cells and tissues[46,47]. However, as with heat-shock protein expression, evidence indicates that acute activation of this transcription factor by extracellular signals in T cells is diminished with ageing[48].

Genetics, oxidative phenomena and longevity

If oxidative stress and the ability to respond appropriately to it is important in ageing, then it follows that factors that increase resistance to stress should have anti-ageing benefits and lead to enhanced life span. In support of this claim, genetic links between stress responsiveness and longevity have been established in *C. elegans*, *Drosophila* and mice (Table 1). In *C. elegans*, such a relationship between life span and stress resistance was first demonstrated for *age-1* mutants which also displayed age-dependent elevations in CuZn-SOD and catalase activities[49,50]. A variety of other life-extending mutations that are likewise correlated with enhanced stress tolerance have since been described (see Table 1 for references).

To the extent that their functions are known, many of the mutated genes encode proteins involved in regulating energy use, and therefore potentially the level of ROS. For example, the *age-1*, *daf-2* and *daf-16* genes in *C. elegans* are associated with an insulin-like signalling pathway that regulates dauer larval formation allowing the worm to survive periods of food scarcity. *age-1* and *daf-2* seem to suppress the activity of the downstream target *daf-16*, a forkhead transcription factor[51]. Hence, loss of function of either of these upstream regulators enhances *daf-16* function and leads to increased life span. Importantly, loss-of-function mutations in *daf-16* not only prevent longevity conferred by the *age-1* and *daf-2* mutations, but also abolish stress resistance[52], thus strengthening the intimate link between longevity and stress responsiveness associated with this pathway. What remains to be determined is how the dauer mutations influence stress responsiveness and longevity. Identification of the transcriptional targets of daf-16 should provide insight into this process. Another *C. elegans* gene whose mutation confers both increased longevity and enhanced stress responsiveness is *clk-1* (ref. 53). *clk-1* encodes a mitochondrial protein homologous to a yeast protein involved in the synthesis of coenzyme Q, an electron carrier required for respiration (Fig. 2). Although its biological function in the nematode is unclear, *clk-1* mutants are believed to enhance life span by reducing the rate of metabolism, which in turn might lead to slower accumulation of damage resulting from metabolic by-products such as ROS. Consistent with this hypothesis, overexpression of *clk-1* leads to a reduction in life span[54].

Gene mutations resulting in reduced life span in *C. elegans* have also been described. Although a life-shortening phenotype is generally considered to be less reliable than a life-lengthening trait for the assessment of ageing-associated genes, two such mutations, *mev-1* and *ctl-1*, are worth special mention as they seem to be linked to oxidative stress. *mev-1* encodes a subunit of the enzyme succinate dehydrogenase cytochrome *b*, a component of complex II of the mitochondrial electron transport chain[55]. These animals have compromised mitochondrial function, reduced cytoplasmic SOD activity and show hypersensitivity to oxygen. *ctl-1* encodes a unique catalase localized in the cytosol[56]. It is likely to contribute to the cell's antioxidant defences, but *ctl-1* mutants have not been examined directly for responsiveness to oxidative stress. Interestingly, loss of *ctl-1* function also eliminates longevity conferred by both the dauer pathway mutants and *clk-1*. Additional evidence for a link between

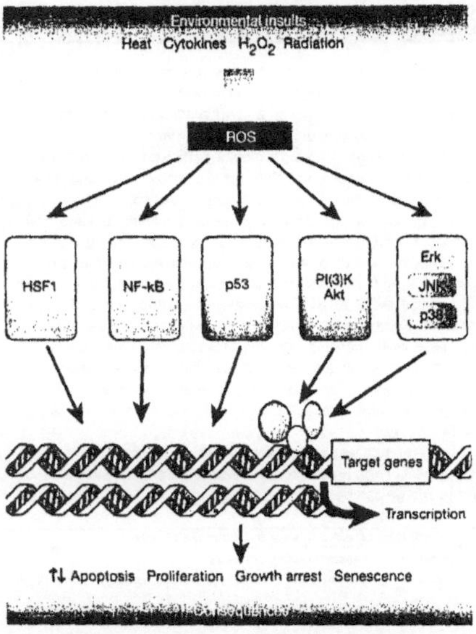

Figure 3 Major signalling pathways activated in response to oxidative stress. ROS can originate outside the cell, or may be generated intracellularly in response to external stimuli. Heat-shock transcription factor 1 (HSF1), NF-κB and p53 are themselves transcription factors, while the PI(3)K/Akt and MAPK pathways regulate transcription factors through phosphorylation. The degree to which a given pathway is activated is highly-dependent on the nature and duration of the stress, as well as the cell type. The consequences of the response vary widely, with the ultimate outcome being dependent on the balance between these stress-activated pathways. HSF1 is responsible for activation of the heat-shock response. Factors depicted in pink represent those pathways whose activities are altered with ageing.

longevity and oxidative stress resistance in the nematode has recently been obtained using a pharmacological approach to boost antioxidant defences[57]. Treatment of wild-type *C. elegans* with synthetic SOD/catalase mimetics was shown to extend the mean life span by 44%. Moreover, the agent was also effective in restoring normal life span to *mev-1* mutants.

A similar link between longevity and stress resistance also exists in *Drosophila*. Various strains of flies selected for extended life span display increased resistance to oxidative stress that in some cases is also correlated with enhanced activity of antioxidant enzymes[58,59]. In addition, at least one gene, *methuselah* (*mth*), has been identified whose mutation not only enhances longevity but also increases resistance to heat stress and paraquat (an intracellular ROS generator)[60]. Further support for the relationship between stress tolerance and life span in *Drosophila* has been obtained using a transgenic approach to overexpress antioxidant genes. Overexpression of glutathione reductase, an enzyme involved in the generation of reduced glutathione, increases the resistance of flies exposed to hyperoxic conditions and extends life span under such conditions, although it does not affect life span of flies maintained in normal air[61]. Several studies have shown that elevated expression of CuZn-SOD alone or in combination with catalase likewise results in enhanced stress tolerance and increased life span[62–64]. However, other studies have reported little or no extension in longevity or enhanced stress

Box 1
Oxidants and diseases of ageing

Although maximum life span is the most relevant and defined endpoint with regard to ageing, it seems likely that in large multicellular organisms ageing need not proceed uniformly. This concept of non-uniform or focal ageing, or what has been *termed* segmental progeria by others[illegible], may be particularly important in a variety of age-related human diseases. Certainly the incidence of a host of diseases, including both cardiovascular and neurodegenerative disorders, increases exponentially with age. The basis for this steep rise in disease incidence is unexplained, but one possibility is that such diseases may share common mechanisms with ageing. In this regard, increasing evidence indicates that reactive oxygen species (ROS) may participate in the pathogenesis of these diseases. In support of this notion is the experimental evidence that the vessel wall of patients with atherosclerotic risk factors, but no overt disease, is characterized by a significant increase in vascular ROS production[illegible]. In addition, many ophthalmological and neurodegenerative diseases seem to be mediated, at least in part, by oxidative stress.

One common feature to many of these conditions is the recruitment of inflammatory cells. These cells contribute to oxidative stress in large part because they contain the potent NADPH oxidase system. Once activated, the NADPH oxidase complex, which is present in neutrophils, macrophages, microglia and vascular cells, produces large amounts of superoxide. In both the brain and the vessel wall the interaction of superoxide with endogenously generated nitric oxide can lead to the formation of peroxynitrite and other damaging radical species. Although the relative contribution of these various radical species is unknown, the link between inflammation and ROS seems to provide a useful framework for understanding disease progression.

Although laboratory data clearly suggest a role for oxidants in the pathogenesis of many age-related diseases, the clinical use of antioxidant therapy has been at best equivocal. Indeed, most of the large randomized trials of antioxidant vitamins such as vitamin A and vitamin E have shown, at best, marginal results and occasionally, potentially harmful outcomes (see Table below). Nonetheless, antioxidant therapy is in its infancy and issues of dosing and duration remain largely unexplored. It therefore is interesting to speculate that, in contrast to the relatively rare progerias that are characterized by accelerated global ageing, the clinical manifestations of these more common diseases, namely an atherosclerotic plaque or an Alzheimer tangle, may represent the fingerprint of accelerated but focal ageing.

Box 1 Table **Oxidants, antioxidants and diseases of ageing**

Disease system	Laboratory/animal studies	Clinical data
Cardiovascular	Pre-atherosclerotic blood vessels have increased levels of ROS[illegible]	PHS I: no overall benefit of beta-carotene on CVD? Benefit in high-risk subgroup[illegible]
	Vitamin E protects against development of atherosclerosis[illegible]	CHAOS trail: vitamin E reduces rate of non-fatal myocardial infarct[illegible]
	Disruption of SOD leads to heart failure[illegible] and overexpression protects against injury[illegible]	ATBC study: no overall benefit on CVD rate with vitamin E or beta-carotene? Increase in CVD deaths with beta-carotene[illegible]
Ophthalmological	Offspring of pregnant mice depleted of glutathione develop cataracts[illegible]	PHS I: non-significant reduction in cataracts and macular degeneration with vitamin E and multivitamins[illegible]
	Retinal pigments produce ROS after light exposure[illegible]	NHS: carotenoids intake may decrease risk of cataracts[illegible]
	Retinal degeneration in primates with vitamin A or E deficiencies[illegible]	
Neurological	Mutations in SOD1 result in human ALS[illegible] and transgenic animal models rescued by antioxidants[illegible]	Vitamin E not protective in early Parkinson's disease[illegible]
	NMDA-receptor stimulation produces superoxide[illegible]	Vitamin E beneficial in Alzheimer's disease[illegible]
	Defects in the function of complex 1 seen in Parkinson's disease[illegible]	*N*-acetylcysteine does not effect survival in ALS[illegible]

The references cited above should be viewed as only representative examples derived from a much larger, relevant body of literature, which owing to space constraints cannot be fully presented. Acronyms and abbreviations: PHS I, Physicians' Health Study I; CHAOS, Cambridge Heart Antioxidant Study; ATBC, Alpha-Tocopherol, Beta-Carotene Cancer Prevention Study; NHS, Nurses' Health Study; CVD, cardiovascular disease; ALS, amyotrophic lateral sclerosis; NMDA, *N*-methyl-D-aspartate glutamate receptors.

resistance with overexpression of CuZn-SOD (refs 65, 66), and the basis for these discrepancies is presently unclear.

The effect of elevated SOD expression on stress resistance and life span has also been examined in mice. Although several reports have produced evidence to indicate that elevated CuZn-SOD levels afford protection against acute oxidative stress, a recent study showed that such acute protection did not correlate with increased longevity[67]. However, another study found an unexpected correlation between life span and control of the oxidative stress response: targeted mutation of the *shc* gene locus, which caused ablation of a splice variant of relative molecular mass 66,000 ($p66^{shc}$), resulted in selective resistance to oxidative stress and extended life span[68]. $p66^{shc}$ is one of three splice variants derived from the *shc* locus. $p52^{shc}$ and $p46^{shc}$ are important adaptor proteins that are regulated by tyrosine phosphorylation and participate in growth-factor- and stress-induced ERK activation. $p66^{shc}$ also undergoes tyrosine phosphorylation in response to extracellular signals, but its role in regulating ERK activation is less clear. However, the $p66^{shc}$ variant is unique in that it also becomes phosphorylated on serine residues in response to certain extracellular signals and, in the case of oxidative stress, this leads to apoptosis. Serine phosphorylation of $p66^{shc}$ is also stimulated by insulin[69]. This observation could provide another link to stress resistance and longevity, as an insulin-like pathway is clearly involved in regulating longevity and stress resistance in *C. elegans* (see above). Although better understanding of the mechanisms involved in $p66^{shc}$-mediated apoptosis and its importance to ageing await further investigation, these studies provide the first evidence for a genetic link between longevity and oxidative stress resistance in mammals.

Caloric restriction and oxidative stress

Limiting food intake, or caloric restriction, has been shown to extend life span in a wide range of species and in rodents it also slows the progression of a variety of age-associated diseases[70]. Although several theories have been advanced over the years to explain the anti-ageing effects of caloric restriction, one favoured hypothesis proposes that it acts by decreasing oxidative stress[71]. In support of this hypothesis, the rate of oxidant generation of mitochondria from calorically restricted mice is significantly lower than from their ad libitum-fed counterparts, and caloric restriction reduces the age-associated accumulation of oxidatively damaged proteins, lipids and DNA[72]. Caloric restriction also prevents many of the changes in gene expression and transcription-factor activity that normally occur with ageing, including basal elevations in expression of heat-shock proteins[40,41] and attenuation of stress-induced Hsp70 expression[72]. Finally, caloric restriction increases the ability of rodents to withstand a wide range of physiological stresses, improves thermotolerance and reduces heat-induced cellular damage in aged rats[73].

Therapeutic strategies against ageing

Understandably, there is tremendous public interest in the development of anti-ageing therapies. If one accepts the evidence that oxidative stress has a significant role in ageing processes, and that the ability to resist or prevent oxidative stress is a key determinant of longevity, then it is likely that strategies aimed either at reducing the oxidative burden or boosting host defence mechanisms involved in coping with the damage would have significant anti-ageing effects (see Box 1). Two key issues will then arise. First, can effective therapeutics be designed to combat oxidant damage? Second, would such approaches be effective in retarding ageing and/or preventing diseases associated with ageing? It is already known that caloric restriction fulfils both of these criteria in lower organisms, and experiments in non-human primates have produced encouraging results to suggest that reducing food intake might also delay ageing in humans[74]. However, the possible application of caloric restriction to the human population would surely face insurmountable ethical and practical difficulties that would prevent its use. On the other hand, better understanding of the specific mechanisms responsible for the anti-ageing influences of caloric restriction could lead to the development of pharmacological agents (that is, caloric restriction mimetics) that would be effective in slowing ageing or delaying the onset of age-associated diseases.

Enhancement of antioxidant defences through dietary supplementation would seem to provide a more reasonable and practical approach to reduce the level of oxidative stress and there is a wealth of evidence to support the effectiveness of such a strategy *in vitro*. However, although some success has been achieved using antioxidant therapy in mammalian models of disorders induced by oxidative stress, so far these strategies have had little or no effect in enhancing longevity[75,76]. Successful application of this approach will probably require much greater understanding of the pharmacological properties of many of the agents being used, including their rates of absorption, tissue distribution, metabolism and the microenvironment in which they must act. Indeed, the problem is clearly more complicated than simply adding a pharmacological agent, as most free-radical scavengers act in oxidation-reduction reactions that are reversible, and some, such as ascorbate, can act both as antioxidants and pro-oxidants, depending on the conditions[77]. In addition, given the role of ROS as mediators of normal signalling processes, determination of the optimal dosage of supplements may require fine tuning to avoid overshooting the desired effects to the point of perturbing the delicate redox balance required for the maintenance of normal cell functions (Fig. 1). However, the development of new synthetic compounds that act as mimetics of SOD and catalase has offered an alternative approach with some promise[78,79]. Use of such compounds in mouse models has been effective in attenuating oxidative stress-associated disease processes, and leads to extension of longevity in *C. elegans*[57].

Finally, consideration should be given to strategies to boost the host defence mechanisms that are known to be activated in response to oxidative stress. So far, the best mechanism for boosting such responses seems to be stress itself. That is, a sublethal or conditioning stress can lead to enhanced survival and reduced tissue damage following a subsequent, more severe stress. This concept, termed hormesis[19,70,71], is in many ways the physiological equivalent of the philosophical notion that 'what won't kill you, will make you strong'. Although the basis for this stress tolerance phenomenon is poorly understood, elevations in heat-shock proteins may be one important factor. Interestingly, such conditioning stress can also lead to life extension in both *C. elegans*[80,81] and *Drosophila*[82]. In humans as well as lower mammals, exercise is believed to have many anti-ageing benefits. Because even in conditioned athletes exercise results in acute stress, it is possible that some of its benefits are derived through such a stress-tolerance mechanism.

Conclusions

In summary, although ageing is likely to be a multifactorial process, there is now significant evidence implicating the generation of ROS and the corresponding response to oxidative stress as key factors in determining longevity. Much of the early evidence was correlative, such as the relationship between metabolic rate and life span, and the observed increase in oxidative damage as a function of age. However, the more recent identification of longevity-influencing genes, first in lower organisms and subsequently in mammals, significantly strengthens the mechanistic connection between oxidants, stress and ageing. Many questions remain including how these various longevity-associated genes actually function to influence stress resistance and ageing, and how applicable they are to human ageing. If oxidative stress does indeed contribute to ageing in the manner described, do ROS act purely as random, destructive agents or — as we have intimated in this review — as regulators of discrete pathways linked to stress responsiveness and ageing? Similarly, it remains unclear whether it is the absolute level of oxidative stress encountered or the subsequent appropriate or inappropriate response to oxidative stress, or perhaps some combination of both, that ultimately determines life expectancy. Finally, and very importantly, in what fashion does oxidative stress intersect, if at all, with other proposed determinants of ageing such as telomere length and telomerase activity, or gene products such as the Werner's helicase that are linked to accelerated ageing syndromes? Nearly a century after the rate-of-living hypothesis was first proposed, the answers to these and related questions remain largely unknown. However, given the rapid rate at which our knowledge in this area has increased over the past few years, it is likely that answers to many of these questions will be forthcoming within this first decade of the new century. These answers will undoubtedly help determine whether ROS are merely peripheral targets that correlate with longevity, or instead, are finally established as a central regulator of human ageing. □

1. Harman, D. Aging: a theory based on free radical and radiation chemistry. *J. Gerontol.* **2**, 298–300 (1957).
2. McCord, J. M. & Fridovich, I. Superoxide dismutase. An enzymatic function for erythrocuprein (hemocuprein). *J. Biol. Chem.* **244**, 6049–6055 (1969).
3. Ku, H. H., Brunk, U. T. & Sohal, R. S. Relationship between mitochondrial superoxide and hydrogen peroxide production and longevity of mammalian species. *Free Radical Biol. Med.* **15**, 621–627 (1993).
4. Finkel, T. Oxygen radicals and signaling. *Curr. Opin. Cell Biol.* **10**, 248–253 (1998).
5. Nishikawa, T. *et al.* Normalizing mitochondrial superoxide production blocks three pathways of hyperglycaemic damage. *Nature* **404**, 787–790 (2000).
6. Nemoto, S., Takeda, K., Yu, Z. X., Ferrans, V. J. & Finkel, T. A role for mitochondrial oxidants as regulators of cellular metabolism. *Mol. Cell Biol.* **20**, 7311–7318 (2000).
7. Suh, Y. A. *et al.* Cell transformation by the superoxide-generating oxidase Mox1. *Nature* **401**, 79–82 (1999).
8. Geiszt, M., Kopp, J. B., Varnai, P. & Leto, T. L. Identification of Renox, an NAD(P)H oxidase in kidney. *Proc. Natl Acad. Sci. USA* **97**, 8010–8014 (2000).
9. Turrens, J. F. Superoxide production by the mitochondrial respiratory chain. *Biosci. Rep.* **17**, 3–8 (1997).
10. Golubev, A. G. The other side of metabolism: a review. *Biochemistry* **61**, 1443–1460 (1996).
11. Boveris, A. & Chance, B. The mitochondrial generation of hydrogen peroxide. General properties and effect of hyperbaric oxygen. *Biochem. J.* **134**, 707–716 (1973).
12. Skulachev, V. P. Role of uncoupled and non-coupled oxidations in maintenance of safely low levels of oxygen and its one-electron reductants. *Q. Rev. Biophys.* **29**, 169–202 (1996).
13. Vidal-Puig, A. J. *et al.* Energy metabolism in uncoupling protein 3 gene knockout mice. *J. Biol. Chem.* **275**, 16258–16266 (2000).
14. Chae, H. Z., Kang, S. W. & Rhee, S. G. Isoforms of mammalian peroxiredoxin that reduce peroxides in presence of thioredoxin. *Methods Enzymol.* **300**, 219–226 (1999).
15. Stadtman, E. R. Protein oxidation and aging. *Science* **257**, 1220–1224 (1992).
16. Yan, L. J., Levine, R. L. & Sohal, R. S. Oxidative damage during aging targets mitochondrial aconitase. *Proc. Natl Acad. Sci. USA* **94**, 11168–11172 (1997).
17. Goto, S. *et al.* Carbonylated proteins in aging and exercise: immunoblot approaches. *Mech. Ageing Dev.* **107**, 245–253 (1999).
18. Dukan, S. *et al.* Protein oxidation in response to increased transcriptional or translational errors. *Proc. Natl Acad. Sci. USA* **97**, 5746–5749 (2000).
19. Beckman, K. B. & Ames, B. N. The free radical theory of aging matures. *Physiol. Rev.* **78**, 547–581 (1998).

20. Esposito, L. A., Melov, S., Panov, A., Cottrell, B. A. & Wallace, D. C. Mitochondrial disease in mouse results in increased oxidative stress. *Proc. Natl Acad. Sci. USA* **96**, 4820–4825 (1999).
21. Melov, S. *et al.* Mitochondrial disease in superoxide dismutase 2 mutant mice. *Proc. Natl Acad. Sci. USA* **96**, 846–851 (1999).
22. Packer, L. & Fuehr, K. Low oxygen concentration extends the lifespan of cultured human diploid cells. *Nature* **267**, 423–425 (1977).
23. von Zglinicki, T., Saretzki, G., Docke, W. & Lotze, C. Mild hyperoxia shortens telomeres and inhibits proliferation of fibroblasts: a model for senescence? *Exp. Cell Res.* **220**, 186–193 (1995).
24. Chen, Q. & Ames, B. N. Senescence-like growth arrest induced by hydrogen peroxide in human diploid fibroblast F65 cells. *Proc. Natl Acad. Sci. USA* **91**, 4130–4134 (1994).
25. Serrano, M., Lin, A. W., McCurrach, M. E., Beach, D. & Lowe, S. W. Oncogenic ras provokes premature cell senescence associated with accumulation of p53 and p16INK4a. *Cell* **88**, 593–602 (1997).
26. Lee, A. C. *et al.* Ras proteins induce senescence by altering the intracellular levels of reactive oxygen species. *J. Biol. Chem.* **274**, 7936–7940 (1999).
27. Johnson, T. M., Yu, Z. X., Ferrans, V. J., Lowenstein, R. A. & Finkel, T. Reactive oxygen species are downstream mediators of p53-dependent apoptosis. *Proc. Natl Acad. Sci. USA* **93**, 11848–11852 (1996).
28. Polyak, K., Xia, Y., Zweier, J. L., Kinzler, K. W. & Vogelstein, B. A model for p53-induced apoptosis. *Nature* **389**, 300–305 (1997).
29. Sachsenmaier, C. *et al.* Involvement of growth factor receptors in the mammalian UVC response. *Cell* **78**, 963–972 (1994).
30. Wang, X., McCullough, K. D., Franke, T. F. & Holbrook, N. J. Epidermal growth factor receptor-dependent Akt activation by oxidative stress enhances cell survival. *J. Biol. Chem.* **275**, 14624–14631 (2000).
31. Knebel, A., Rahmsdorf, H. J., Ullrich, A. & Herrlich, P. Dephosphorylation of receptor tyrosine kinases as target of regulation by radiation, oxidants or alkylating agents. *EMBO J.* **15**, 5314–5325 (1996).
32. Lee, S. R., Kwon, K. S., Kim, S. R. & Rhee, S. G. Reversible inactivation of protein-tyrosine phosphatase 1B in A431 cells stimulated with epidermal growth factor. *J. Biol. Chem.* **273**, 15366–15372 (1998).
33. Burdon, R. H. Superoxide and hydrogen peroxide in relation to mammalian cell proliferation. *Free Radical Biol. Med.* **18**, 775–794 (1995).
34. Saitoh, M. *et al.* Mammalian thioredoxin is a direct inhibitor of apoptosis signal-regulating kinase (ASK) 1. *EMBO J.* **17**, 2596–2606 (1998).
35. Adler, V. *et al.* Regulation of JNK signaling by GSTp. *EMBO J.* **18**, 1321–1334 (1999).
36. Allen, R. G. & Tresini, M. Oxidative stress and gene regulation. *Free Radical Biol. Med.* **28**, 463–499 (2000).
37. Godon, C. *et al.* The H₂O₂ stimulon in *Saccharomyces cerevisiae*. *J. Biol. Chem.* **273**, 22480–22489 (1998).
38. Richardson, A. & Holbrook, N. J. in *Cellular Aging and Cell Death* (eds Holbrook, N. J., Martin, G. R. & Lockshin, R. A.) 67–89 (Wiley, New York, 1996).
39. Morimoto, R. I. & Santoro, M. G. Stress-inducible responses and heat shock proteins: new pharmacologic targets for cytoprotection. *Nature Biotechnol.* **16**, 833–838 (1998).
40. Lee, C. K., Klopp, R. G., Weindruch, R. & Prolla, T. A. Gene expression profile of aging and its retardation by caloric restriction. *Science* **285**, 1390–1393 (1999).
41. Lee, C. K., Weindruch, R. & Prolla, T. A. Gene-expression profile of the ageing brain in mice. *Nature Genet.* **25**, 294–297 (2000).
42. Wheeler, J. C., Bieschke, E. T. & Tower, J. Muscle-specific expression of *Drosophila* hsp70 in response to aging and oxidative stress. *Proc. Natl Acad. Sci. USA* **92**, 10408–10412 (1995).
43. Guyton, K. Z. *et al.* Age-related changes in activation of mitogen-activated protein kinase cascades by oxidative stress. *J. Invest. Dermatol. Symp. Proc.* **3**, 23–27 (1998).
44. Guyton, K. Z. *et al.* Activation of mitogen-activated protein kinase by hydrogen peroxide: role in cell survival following oxidant injury. *J. Biol. Chem.* **271**, 4138–4142 (1996).
45. Liu *et al.* Age-related decline in MAP kinase activity in EGF-stimulated rat hepatocytes. *J. Biol. Chem.* **271**, 3604–3607 (1996).
46. Poynter, M. E. & Daynes, R. A. Peroxisome proliferator-activated receptor alpha activation modulates cellular redox status, represses nuclear factor-kappaB signaling, and reduces inflammatory cytokine production in aging. *J. Biol. Chem.* **273**, 32833–32841 (1998).
47. Supakar, P. C., Jung, M. H., Song, C. S., Chatterjee, B. & Roy, A. K. Nuclear factor kappa B functions as a negative regulator for the rat androgen receptor gene and NF-kappa B activity increases during the age-dependent desensitization of the liver. *J. Biol. Chem.* **270**, 837–842 (1995).
48. Ponnappan, U. Regulation of transcription factor NFkappa B in immune senescence. *Front. Biosci.* **3**, D152–D168 (1998).
49. Larsen, P. L. Aging and resistance to oxidative damage in *Caenorhabditis elegans*. *Proc. Natl Acad. Sci. USA* **90**, 8905–8909 (1993).
50. Vanfleteren, J. R. Oxidative stress and ageing in *Caenorhabditis elegans*. *Biochem. J.* **292**, 605–608 (1993).
51. Ogg, S. *et al.* The Fork head transcription factor DAF-16 transduces insulin-like metabolic and longevity signals in *C. elegans*. *Nature* **389**, 994–999 (1997).
52. Murakami, S. & Johnson, T. E. A genetic pathway conferring life extension and resistance to UV stress in *Caenorhabditis elegans*. *Genetics* **143**, 1207–1218 (1996).
53. Wong, A., Boutis, P. & Hekimi, S. Mutations in the clk-1 gene of *Caenorhabditis elegans* affect developmental and behavioral timing. *Genetics* **139**, 1247–1259 (1995).
54. Branicky, R., Benard, C. & Hekimi, S. clk-1, mitochondria, and physiological rates. *Bioessays* **22**, 48–56 (2000).
55. Ishii, N. *et al.* A mutation in succinate dehydrogenase cytochrome *b* causes oxidative stress and ageing in nematodes. *Nature* **394**, 694–697 (1998).
56. Taub, J. *et al.* A cytosolic catalase is needed to extend adult lifespan in *C. elegans daf-C* and *clk-1* mutants. *Nature* **399**, 162–166 (1999).
57. Melov, S. *et al.* Extension of life-span with superoxide dismutase/catalase mimetics. *Science* **289**, 1567–1569 (2000).
58. Dudas, S. *et al.* A coordinate upregulation of antioxidant gene activities is associated with the delayed onset of senescence in a long-lived strain of *Drosophila*. *J. Gerontol.* A **50**, B117–B127 (1995).
59. Harshman, L. G. & Haberer, B. A. Oxidative stress resistance: a robust correlated response to selection in extended longevity lines of *Drosophila melanogaster*. *J. Gerontol.* A **55**, B415–B417 (2000).
60. Lin, Y. J., Seroude, L. & Benzer, S. Extended life-span and stress resistance in the *Drosophila* mutant methuselah. *Science* **282**, 943–946 (1998).
61. Mockett, R. J., Sohal, R. S. & Orr, W. C. Overexpression of glutathione reductase extends survival in transgenic *Drosophila melanogaster* under hyperoxia but not normoxia. *FASEB J.* **13**, 1733–1742 (1999).
62. Orr, W. C. & Sohal, R. S. Extension of life-span by overexpression of superoxide dismutase and catalase in *Drosophila melanogaster*. *Science* **263**, 1128–1130 (1994).
63. Sun, J. & Tower, J. FLP recombinase-mediated induction of Cu/Zn-superoxide dismutase transgene expression can extend the life span of adult *Drosophila melanogaster* flies. *Mol. Cell Biol.* **19**, 216–228 (1999).
64. Parkes, T. L. *et al.* Extension of *Drosophila* lifespan by overexpression of human SOD1 in motorneurons. *Nature Genet.* **19**, 171–174 (1998).
65. Seto, N. O., Hayashi, S. & Tener, G. M. Overexpression of Cu-Zn superoxide dismutase in *Drosophila* does not affect life-span. *Proc. Natl Acad. Sci. USA* **87**, 4270–4274 (1990).
66. Orr, W. C. & Sohal, R. S. Effects of Cu-Zn superoxide dismutase overexpression on life span and resistance to oxidative stress in transgenic *Drosophila melanogaster*. *Arch. Biochem. Biophys.* **301**, 34–40 (1993).
67. Huang, T. *et al.* Ubiquitous overexpression of CuZn superoxide dismutase does not extend life span in mice. *J. Gerontol. Biol. Sci.* A **55**, B5–B9 (2000).
68. Migliaccio, E. *et al.* The p66^{shc} adaptor protein controls oxidative stress response and life span in mammals. *Nature* **402**, 309–313 (1999).
69. Kao, A. W. *et al.* Insulin stimulates the phosphorylation of the 66- and 52-kilodalton Shc isoforms by distinct pathways. *Endocrinology* **138**, 2474–2480 (1997).
70. Masoro, E. J. Caloric restriction and aging: an update. *Exp. Gerontol.* **35**, 299–305 (2000).
71. Sohal, R. S. & Weindruch, R. Oxidative stress, caloric restriction, and aging. *Science* **273**, 59–63 (1996).
72. Heydari, A. R., Wu, B., Takahashi, R., Strong, R. & Richardson, A. Expression of heat shock protein 70 is altered by age and diet at the level of transcription. *Mol. Cell Biol.* **13**, 2909–2918 (1993).
73. Hall, D. M. *et al.* Caloric restriction improves thermotolerance and reduces hyperthermia-induced cellular damage in old rats. *FASEB J.* **14**, 78–86 (2000).
74. Roth, G. S., Ingram, D. K. & Lane, M. A. Caloric restriction in primates: will it work and how will we know? *J. Am. Geriatr. Soc.* **47**, 896–903 (1999).
75. McCall, M. R. & Frei, B. Can antioxidant vitamins materially reduce oxidative damage in humans? *Free Radical Biol. Med.* **26**, 1034–1053 (1999).
76. Yu, B. P. Approaches to anti-aging intervention: the promises and the uncertainties. *Mech. Ageing Dev.* **111**, 73–87 (1999).
77. Podmore, I. D. *et al.* Vitamin C exhibits pro-oxidant properties. *Nature* **392**, 559–559 (1998).
78. Melov, S. *et al.* A novel neurological phenotype in mice lacking mitochondrial manganese superoxide dismutase. *Nature Genet.* **18**, 159–163 (1998).
79. Rong, Y. *et al.* EUK-134, a synthetic superoxide dismutase and catalase mimetic, prevents oxidative stress and attenuates kainate-induced neuropathology. *Proc. Natl Acad. Sci. USA* **96**, 9897–9902 (1999).
80. Lithgow, G. J. *et al.* Thermotolerance and extended life-span conferred by single-gene mutations and induced by thermal stress. *Proc. Natl Acad. Sci. USA* **92**, 7540–7544 (1995).
81. Johnson, T. E. & Hartman, P. S. Radiation effects on life span in *Caenorhabditis elegans*. *J. Gerontol. Biol. Sci.* **43**, B137–B141 (1988).
82. Tatar, M., Khazaeli, A. A. & Curtsinger, J. W. Chaperoning extended life. *Nature* **390**, 30 (1997).
83. Martin, G. M., Austad, S. N. & Johnson, T. E. Genetic analysis of ageing: role of oxidative damage and environmental stresses. *Nature Genet.* **13**, 25–34 (1996).
84. Cypser, J. R. & Johnson, T. E. The spe-10 mutant has longer life and increased stress resistance. *Neurobiol. Aging* **20**, 503–512 (1999).
85. Murakami, S. & Johnson, T. E. Life extension and stress resistance in *Caenorhabditis elegans* modulated by the tkr-1 gene. *Curr. Biol.* **8**, 1091–1094 (1998).
86. Abe, J. & Berk, B. C. Reactive oxygen species as mediators of signal transduction in cardiovascular disease. *Trends Cardiovasc. Med.* **8**, 59–64 (1998).
87. Wojcicki, J. *et al.* Effect of selenium and vitamin E on the development of experimental atherosclerosis in rabbits. *Atherosclerosis* **87**, 9–16 (1991).
88. Lebovitz, R. M. *et al.* Neurodegeneration, myocardial injury, and perinatal death in mitochondrial superoxide dismutase-deficient mice. *Proc. Natl Acad. Sci. USA* **93**, 9782–9787 (1996).
89. Li, Y. B. *et al.* Dilated cardiomyopathy and neonatal lethality in mutant mice lacking manganese superoxide dismutase. *Nature Genet.* **11**, 376–381 (1995).
90. Wang, P. H. *et al.* Overexpression of human copper, zinc-superoxide dismutase (SOD1) prevents postischemic injury. *Proc. Natl Acad. Sci. USA* **95**, 4556–4560 (1998).
91. Martensson, J., Steinherz, R., Jain, A. & Meister, A. Glutathione ester prevents buthionine sulfoximine-induced cataracts and lens epithelial cell damage. *Proc. Natl Acad. Sci. USA* **86**, 8727–8731 (1989).
92. Rozanowska, M. *et al.* Blue light-induced reactivity of retinal age pigment. *In vitro* generation of oxygen-reactive species. *J. Biol. Chem.* **270**, 18825–18830 (1995).
93. Hayes, K. C. Retinal degeneration in monkeys induced by deficiencies of vitamin E or A. *Invest. Ophthalmol.* **13**, 499–510 (1974).
94. Rosen, D. R. *et al.* Mutations in Cu/Zn superoxide dismutase gene are associated with familial amyotrophic lateral sclerosis. *Nature* **362**, 59–62 (1993).
95. Andreassen, O. A., Dedeoglu, A., Klivenyi, P., Beal, M. F. & Bush, A. I. N-acetyl-L-cysteine improves survival and preserves motor performance in an animal model of familial amyotrophic lateral sclerosis. *Neuroreport* **11**, 2491–2493 (2000).
96. Nagano, S., Ogawa, Y., Yanagihara, T. & Sakoda, S. Benefit of a combined treatment with trientine and ascorbate in familial amyotrophic lateral sclerosis model mice. *Neurosci. Lett.* **265**, 159–162 (1999).
97. Lafon-Cazal, M., Pietri, S., Culcasi, M. & Bockaert, J. NMDA-dependent superoxide production and neurotoxicity. *Nature* **364**, 535–537 (1993).
98. Parker, W. D. Jr, Boyson, S. J. & Parks, J. K. Abnormalities of the electron transport chain in idiopathic Parkinson's disease. *Ann. Neurol.* **26**, 719–723 (1989).
99. Hennekens, C. H. *et al.* Lack of effect of long-term supplementation with beta carotene on the incidence of malignant neoplasms and cardiovascular disease. *N. Engl. J. Med.* **334**, 1145–1149 (1996).

100. Stephens, N. G. *et al.* Randomised controlled trial of vitamin E in patients with coronary disease: Cambridge Heart Antioxidant Study (CHAOS). *Lancet* **347**, 781–786 (1996).
101. The Alpha-Tocopherol Beta Carotene Study Group. The effect of vitamin E and beta carotene on the incidence of lung cancer and other cancers in male smokers. *N. Engl. J. Med.* **330**, 1029–1035 (1994).
102. Seddon, J. M. *et al.* The use of vitamin supplements and the risk of cataract among US male physicians. *Am. J. Public Health* **84**, 788–792 (1994).
103. Hankinson, S. E. *et al.* Nutrient intake and cataract extraction in women: a prospective study. *Br. Med. J.* **305**, 335–339 (1992).
104. The Parkinson Study Group. Effects of tocopherol and deprenyl on the progression of disability in early Parkinson's disease. *N. Engl. J. Med.* **328**, 176–183 (1993).
105. Sano, M. *et al.* A controlled trial of selegiline, alpha-tocopherol, or both as treatment for Alzheimer's disease. The Alzheimer's Disease Cooperative Study. *N. Engl. J. Med.* **336**, 1216–1222 (1997).
106. Louwerse, E. S., Weverling, G. J., Bossuyt, P. M., Meyjes, F. E. & de Jong, J. M. Randomized, double-blind, controlled trial of acetylcysteine in amyotrophic lateral sclerosis. *Arch. Neurol.* **52**, 559–564 (1995).

Acknowledgements

We thank L. Rovira and R. Wange for their help in the design of illustrations, and T. Johnson, D. Longo, B. Howard, R. Levine, M. Gorospe and N. Epstein for thoughtful comments and discussions.

ANNALS OF CLINICAL AND LABORATORY SCIENCE, Vol. 26, No. 1

Laboratory Markers as an Index of Aging*

SHESHADRI NARAYANAN, Ph.D.

Department of Pathology,
New York Medical College,
Metropolitan Hospital Center
New York, NY 10029

ABSTRACT

At the cellular level, mutations in deoxyribonucleic acid (DNA) can lead to synthesis of altered proteins which are unable to sustain specific cell functions, eventually leading to its death. Veritably apoptosis, or programmed cell death, is a device to eliminate heavily mutated cells. Cell membranes with altered proteins can be recognized as foreign by the immune system, thus triggering autoimmunity. Molecular biology techniques allow us to examine changes that occur in DNA, reflected by polymorphisms and variable numbers of tandem repeats (VNTR). A general decline in organ function is associated with aging. However, these changes may also be precipitated by disease processes. Homeostatic control by the hypothalamus-pituitary-adrenal axis is also compromised with aging, leading to an increase in plasma adrenocorticotrophic hormone (ACTH) and corticosteroid levels. Derangement of the immune system with aging results in dysregulation of cytokine production. The ability of the cell to survive the onslaught of oxygen-free radicals with enzymatic and non-enzymatic antioxidants, and to repair DNA by activation of nuclear enzymes such as poly (ADP-ribose) polymerase (PAD-PRP), are some of determinants of aging.

Introduction

For a laboratory marker to be useful as an index of aging, it should reflect changes that are consonant with the mechanisms of aging. Veritably these changes occur at the cellular level, giving rise to a wide range of effects. Changes associated with aging are also evident at the neuroendocrine level, thus giving credence to neuro-endocrine markers as an index of aging.

Changes that occur at the cellular level are orchestrated by mutations in DNA.[1] These mutations in turn result in the production of altered proteins which disrupt the functions of the cell. The net result of the cell being unable to perform its specific functions is that the cell dies. Alternatively, the heavily mutated cells may program itself to self destruct, by a phenomenon called apoptosis.[2] On the other hand, the mutated DNA can be repaired

* Send reprint requests to Sheshadri Narayanan, Ph.D., Dept. of Pathology, New York Medical College, Metropolitan Hospital Center, New York, NY 10029.

by the activation of nuclear enzymes such as poly (ADP-ribose) polymerase (PADPRP).[2] A clue to aging may lie in the alteration of the activity of these nuclear enzymes. One other consequence of mutations occurring in DNA is the alteration of cell membrane proteins. These alterations are recognized as foreign by the immune system thus triggering autoimmune reactions associated with aging.

Laboratory markers associated with aging should be able to recognize the events that occur with aging. Thus, in general there is a decline in organ function which can be recognized by specific laboratory markers. Homeostatic control by the hypothalamus-pituitary-adrenal axis is compromised leading to an increase in the circulating level of hormones such as plasma adrenocorticotrophic hormone (ACTH) and corticosteroid levels.[3] Derangement of the immune system that occurs with aging is expressed by dysregulation of cytokine production. Decline of immune system with aging can also be due to the deficiency of zinc and other nutrients.[4] The ability to overcome the onslaught by oxygen-free radicals as we age is dependent on the level of enzymatic and nonenzymatic oxidants.

Renal function decreases with age leading to changes in creatinine clearance. Abnormal renal function in the elderly could also be due to renal vascular disease. The alteration of glucose tolerance with age leads to glycation of collagen, lens protein, and glomerular basement membrane. Decreased glucose tolerance in the elderly is usually related to genetics, obesity, and inactivity.[5] The accumulation of advanced glycation end products, such as pentosidine and carboxymethyllysine, represents not only a hallmark of aging but also to genetic susceptibility to diabetic complications.[6] While the laboratory markers associated with aging needs to be examined, attention should also be paid to changes occurring in syndromes of accelerated aging to focus on laboratory markers common to such syndromes.

Syndromes of Accelerated Aging

In addition to Down's syndrome, other syndromes of accelerated aging or progeroid syndromes that have received attention include Werner's syndrome. Cockayne's syndrome, and Hutchinson-Gilford syndrome.

In Down's syndrome, there is a derangement of the immune system. The absolute number of T lymphocytes is decreased. Thymic dysfunction, being a hallmark of Down's syndrome, is reflected in a preferential decrease in CD4+ cells (helper/inducer T-cell subset). While CD8+ cells (cytotoxic T-cell subset) do not decrease in children with Down's syndrome, in adults there is a marked increase in percentage of CD8+ cells leading to a decreased CD4/CD8 ratio.[7] In table I are summarized results reported in a study of lymphocyte subsets in children and adults with Down's syndrome.

TABLE I

Lymphocyte Subsets in Down's Syndrome*

	DSC	*Control*	*DSA*	*Control*
CD4 + %	33.4	41.3	29.5	46.2
n	743	1256	705	1008
CD8 + %	36.7	27.3	44.5	27.3
n	797	843	1126	595

**Blood* 1991;77:1263–70.

DSC = Down's syndrome children.

DSA = Down's syndrome adults.

Control are aged–matched healthy subjects (children and adults, respectively).

n = the absolute number of peripheral blood lymphocytes/mm³ positive to the monoclonal antibody used to measure specific subsets.

% = the percentage of subset population identified by specific monoclonal antibodies.

Other alterations in cytokines seen in Down's syndrome include an accelerated increase in percent of lymphocytes expressing markers of natural killer activity such as CD16, CD56, and CD57.[4,7] While the CD57+ subset is reported to increase during physiological aging, in Down's Syndrome the increase is accelerated.[7,8] While the percent of lymphocytes expressing natural killer activity increased, the natural killer activity, however, was markedly reduced in Down's syndrome compared to aged-matched healthy controls. The cytotoxic activity of these cells, however, were fully restored to levels seen in cells from healthy adults upon pre incubation with interleukin-2 (IL-2), interferon Beta (IFN-β), or interferon gamma (IFN-γ). The impairment of immune function in Down's syndrome may, perhaps, be related to defects of cell-adhesion molecules coded by genes on chromosome 21.[9]

Alteration in levels of immunoglobulin (IgG) subclasses in serum of older patients with Down's syndrome (mean age 55 ± 7 years) compared to healthy aged matched controls has been reported.[10] In the older Down's syndrome group, levels of IgG1 and IgG3 subclasses were higher when compared to aged matched healthy controls, and the increase was similar to those seen in subjects with autoimmune disease and viral infections. On the other hand, the level of IgG2 and IgG4 subclasses in the sera of this older Down's Syndrome group was lower than that found in healthy aged matched controls, and the decrease was similar to that seen in patients with recurrent infections. Thus, in Down's syndrome age-related impairment of the immune system is manifested by the decrease in the number of T lymphocytes (CD3+), B lymphocytes (CD19+), and reduced natural killer cell activity.[7]

Other progeroid syndromes are characterized by mutations. Thus, in Werner's syndrome there is a greater frequency (76 percent compared to 39 percent of controls) of deletions at the X chromosome-linked hypoxanthine phosphoribosyltransferase (HPRT) locus.[10] These deletions within the HPRT gene were uncovered by using standard molecular biology techniques.[11] Thus, the restriction fragments obtained from the X-chromosome-linked HPRT locus were studied with probes corresponding to the 5′ portion of the 1st HPRT intron and a 908-base pair complementary DNA (cDNA) probe spanning the exons 1, 3, 6, 7, and 8 or 9.[10] These studies showed that 58 percent of deletions involved the loss of a greater than 20 kilo base (kb) DNA from the HPRT gene.

In Hutchinson-Gilford syndrome, there is a reduced capacity to repair ultraviolet (UV) damaged DNA. This was demonstrated by using a skin biopsy specimen taken from the hypogastric region of the abdomen. Fibroblasts originating from the cultured skin of this progeria patient had a 22 to 41 percent reduction in unscheduled DNA synthesis after UV irradiation when compared to normal cells, thus demonstrating the reduced capacity to repair the damaged DNA.[12]

In yet another progeria syndrome, the Cockayne's syndrome, the preferential repair of active genes was demonstrated after treating cells with DNA damaging UV light.[13] Thus, the commonality in the progeria syndromes is the age-related changes seen at the cellular level.

Cytokines and Aging

The cytokine interleukin-6 (IL-6) has been aptly called a "cytokine for gerontologists."[14] Normally the levels of IL-6 in serum are low and are undetectable in absence of inflammation. With age IL-6 levels become measurable in serum even in absence of inflammation. Thus, during middle age IL-6 levels in serum are readily detectable. Clearly serum IL-6 levels are elevated in healthy elderly adults, apparently owing to a loss in the normal regulation of IL-6.

The lack of tight regulation of IL-6 that becomes noticeable in middle age is coincident with the development of autoimmunity and even the appearance of benign paraproteins. Increased levels of IL-6 in old age, accompanied by increased B-cell proliferation may be related to the development of lymphoma. Since IL-6 induces the synthesis of protease inhibitors such as α-1 antichymotrypsin and α-2 macroglobulin, the processing of amyloid precursor proteins may be affected. The IL-6 is reported to increase neuronal α-2 macroglobulin levels, thus resulting in the deposition of abnormal βA4 fragment in plaques associated with Alzheimer's disease. The negative influence of estrogen on IL-6 gene expression is blunted in menopause leading to increased IL-6 levels and osteoporosis. Thus, the loss of T-cell immunoregulation with age, as well as menopausal estrogen loss, is apparently responsible for the dysregulation of IL-6 synthesis and the appearance of diseases such as lymphoma, osteoporosis, and Alzheimers.[15]

The increase in IL-6 with age has been reported to be related to a decrease in the androgen, dehydroepiandrosterone sulfate (DHEA-SO_4), based on experiments on old mice in whom IL-6 levels were suppressed by administration of DHEA-SO_4.[16] High levels of IL-6 are reported to increase the percentage of cells bearing the receptor for transforming growth factor-beta (TGF-β), a cytokine which is a powerful immunosuppressive agent. Thus, the age-related immunosuppressive phenomena is perhaps related to an increase in IL-6 and the percentage of cells bearing TGF-β receptor.[17]

The decrease in immune function with age could also be secondary to protein malnutrition and vitamin and trace-element deficiencies.[18,19,20] Thus, dietary supplementation of antioxidant vitamins A, C, and E to a group of elderly subjects for 28 days improved cell-mediated immune function as indicated by a significant statistical increase in the absolute number of T-Lymphocytes, CD4 subsets, CD4 to CD8 ratio, and in the proliferation of lymphocytes on mitogen stimulation. In contrast, elderly subjects receiving placebo did not evidence changes in immune function.[18] The total number of elderly subjects included in the above study was 30, including those who received placebo.

Administration of vitamin E alone for 30 days in a placebo controlled trial utilizing 32 subjects demonstrated an improvement in immune function as judged by an increase in Interleukin-2 (IL-2) production and mitogenic response to concanavalin A.[19] Apparently vitamin E improves immune response by reducing prostaglandin E2 (PGE2) synthesis and the level of lipid peroxides.

Dietary supplementation with vitamins A, B6, B12, C, D, and E, minerals such as zinc, copper, selenium, calcium, magnesium, and iron, and other nutrients such as β-carotene, riboflavin, niacin, folate, and iodine improved immune response in healthy elderly subjects.[20] In this placebo controlled study, a total of 96 elderly subjects were followed for 12 months. The multi-nutrient supplementation in general corresponded to the recommended dietary allowances in the USA and Canada with the exception of vitamin E and β-carotene which were four times in excess of the upper quartile of customary intake. Subjects receiving the nutrient supplement had improved immune response measurable in terms of increased number of certain T-cell subsets and natural killer cells, increased IL-2 production, enhanced proliferative response to mitogen and increased antibody response. Infection-related illness was much less frequent in elderly subjects receiving the nutrient supplement as compared to subjects receiving placebo.

Apparently, decreased immune response associated with aging could be improved by dietary supplementation with physiological amounts of micronutrients. In this context, very large doses of micronutrients may actually impair immunity, since each micronutrient has minimum and maximum thresholds for optimum immune response.[20]

Role of DNA Repair Enzymes and Apoptosis

The enzyme poly (ADP-ribose) polymerase (ADPRP) presumably has a role in DNA repair and apoptosis.[21] The activity of ADPRP has been shown to correlate with age.[22] While small doses of mutagens have been shown to activate the enzyme ADPRP, large doses can actually be inhibitory unless zinc ions are present.[21] The enzyme ADPRP functions by utilizing nicotinamide adenine dinucleotide (NAD) as substrate for transferring ADP-ribose to acceptor nuclear proteins and to itself. However, excessive activation of the enzyme ADPRP results in the depletion of intracellular NAD and ATP pools, in turn triggering apoptosis, thus ascribing a role for this enzyme in the programmed cell death or apoptosis of the heavily mutated cell.

Apoptosis itself is promoted by the inhibition of protooncogene C-myc and by the expression of tumor suppressor gene p53. However, several growth factors such as Interleukins 3 and 6 (IL-3 and IL-6) and colony stimulating factor (CSF), on the contrary, favor cell survival by suppressing apoptosis.[23]

While a variety of laboratory procedures are available for the detection of apoptosis, one of the procedures using flow cytometry takes advantage of the fact that during apoptosis, a membrane phospholipid, phosphatidylserine, is exposed on the cell surface. Actually, the exposed phosphatidylserine molecule is recognized by macrophages which facilitates their removal of the apoptotic cells. Since the anticoagulant annexin V can bind to phosphatidyl serine, fluorescein isothiocyanate-labelled annexin V is used to bind and detect apoptotic cells by flow cytometry.[24]

Role of Zinc in Aging

Since zinc is needed for vital body functions such as DNA synthesis and cell-mediated immune functions, deficiency of this trace metal has an impact on aging. In a study of 180 elderly and healthy subjects, it was demonstrated that while plasma zinc levels were normal, zinc levels in granulocytes and lymphocytes were decreased compared to healthy younger control subjects. Indeed 36 out of 118 elderly subjects had deficient granulocyte and lymphocyte zinc levels.[25] The zinc deficiency was apparently related to the lower intake of this trace metal, since the mean dietary intake of zinc in the elderly population included in the study was 9.06 mg/day as compared to the recommended dietary allowance of 15 mg/day.[25] Since zinc has a reciprocal relationship to copper, zinc deficiency in this elderly population was accompanied with an increase in plasma copper level and a significant decrease in the production of the cytokine interleukin-1 (IL-1). However, zinc supplementation given to 13 elderly subjects deficient in zinc restored both zinc and plasma copper levels to normal. In addition, serum IL-1 level, and serum thymulin activity, and lymphocyte ecto-5^1-nucleotidase activity increased significantly in these zinc-deficiency corrected subjects.

Thus, it becomes apparent that zinc has a role in immune deficiency. Thus, in Down's syndrome, immune deficiency may be related to a defect in the bioavailability or utilization of zinc. In a study of 18 children with Down's syndrome with a mean age of seven years, the plasma level of thymic hormone, serum thymic factor (STF) was low.[26] These children also had a decrease in the absolute number of circulating T-lymphocytes and an increase in the level of inactive zinc-free STF molecules. After oral zinc supplementation for two months, there was a significant increase in the plasma level of STF accompanied by an almost total reduction in the number of inactive zinc-free STF molecules, and a marked increase in the absolute number of circulating T-lymphocytes. Apparently zinc supplementation given to these children with Down's Syndrome had an effect of reducing recurrent infections.

Oxygen Free Radicals and Role of Antioxidants in Aging

Oxygen free radicals (OFR), such as superoxide anions and hydroxyl radicals, can damage DNA. The sensitivity of peripheral blood mononuclear cells (PMNC) to OFR apparently increases with age. In the process of repairing DNA breaks caused by exposure to OFR, the enzyme poly (ADP-ribose) polymerase (PADPRP) is activated depleting the cell of adenosine triphosphate (ATP) and nicotinamide adenine dinucleotide (NAD) thus leading to an energy crisis in the cell threatening its survival.[27] However, this assault by OFR can be overcome by use of inhibitors of PADPRP, such as nicotinamide, which will prevent the cellular depletion of NAD. The use of antioxidants, such as carnitine, can exert a protective effect on the structure and function of mitochondria and generate ATP needed to repair promptly the DNA damage caused by OFR.[27]

Enzymatic antioxidants within the cell provide a defense against the attack by OFR. Thus copper-zinc superoxide dismutase catalyzes the dismutation of superoxide anions to hydrogen peroxide. The conversion of hydrogen peroxide to water and oxygen is accomplished by enzymes such as the selenium-dependent glutathione peroxidase and catalase.

In a study examining the age-related changes in the activity of some of the enzymatic antioxidants in human erythrocytes, it was found that the level of the enzyme copper-zinc superoxide dismutase was negatively correlated with age. The reduced activity is apparently due to a variety of reasons, including the inactivation of the enzyme by excess hydrogen peroxide generated during oxidative attack by superoxide anions, the glycation of the enzyme itself with age, or zinc or copper deficiency associated with aging.[28] The erythrocyte level of selenium-dependent glutathione peroxidase correlated positively with age apparently owing to the induction of the enzyme by increased levels of hydrogen peroxide generated as a result of increased oxidative attack with aging.

The activity of glutathione reductase, an enzyme in the erythrocytes, whose function is to regenerate reduced glutathione (GSH) needed for the activity of the selenium-dependent glutathione peroxidase was, however, significantly reduced with age. Thus, the increased selenium-dependent glutathione peroxidase activity accompanied by a decrease in the reduced glutathione reductase activity during aging may have a bearing on the decrease in the circulating levels of GSH associated with aging.[28]

Decline in Organ Function with Age

As we age, the efficiency of the organs in the body also decreases. Thus, renal function deteriorates with age as judged by the decrease in creatinine clearance values with advancing age.[29,30] In one study, 548 subjects who were considered normal were studied by 10-year age groups.[31] It was found that creatinine clearance progressively declined with age, decreasing from 140 ml/min/1.73 M^2 body surface area at age 30 to 97 ml/min/ 1.73M^2 at age 80. This decline was attributed to true renal aging since these subjects had neither specific diseases nor were they taking any medications that could alter the glomerular filtration rate. However, the presence of renal arteriosclerosis in elderly patients in varying degrees of severity must be kept in perspective, thus making it difficult to separate changes owing to disease from changes occurring as a result of aging. Even so, anatomical changes in the kidney, such as decrease in kidney size and volume after the age of 40, and electron microscopic changes observed in the basement membrane of the glomeruli and tubules are generally regarded as due to a consequence of aging.[32]

Glucose tolerance also decreases with age. In a study of 23 elderly subjects with a mean age of 79 years and 14 healthy young adults with a mean age of 28 years, it was noted that in 17 elderly patients the blood glucose level at 120 minutes after instituting the glucose tolerance test was greater than or equal to 140 mg/dL.[33] The pattern of insulin release during the glucose tolerance test in the elderly subjects was sluggish as compared to the young healthy adults. Injection of L-alanine to assess gluconeogenic capacity led to increases in both glucose and alanine, followed by a drop which occurred slowly, however, in the elderly. These data would suggest that there is an impaired transmembrane transport of both glucose and alanine in elderly people. The release of insulin owing to the increase of glucose resulting from the injection of L-alanine was lower in the elderly when compared to young adults. Plasma glucagon levels, however, increased to a greater extent in the elderly upon injection of L-alanine, when compared to young adults, thus offering an explanation for the diminished insulin sensitivity in the elderly.

The increase in the circulating levels of glucose with age favors the glycation of long-lived proteins, such as the lens crystallins of the eye and collagen. The fructosamine moiety resulting from the glycation of proteins undergoes further modification over a protracted period of time ranging from several months to many years. Essentially, the fructosamine moiety undergoes nonenzymatic changes characterized by the Maillard reaction, which includes dehydration and fragmentation to form pigmented, carboxyalkylated, and cross-linked proteins which are referred to as advanced glycation end products (AGE_s). The accumulation of AGE_s such as pentosidine and carboxymethyllysine have been reported to increase with age.[34] In addition to cross linking proteins such as collagen, and the lens crystallins protein, AGE_s can contribute to the aging process. The body's defense mechanism to degrade AGE_s is by way of specific receptors on macrophages which bind and process AGE_s. Therapeutic modalites to prevent formation of AGE_s is by administration of aminoguanidine which was successful in diabetic rats. While procedures for the determination of pentosidine in plasma by high performance liquid chromatography (HPLC) may be cumbersome, a simplified enzyme linked immunosorbent assay (ELISA) has been described for the measurement of pentosidine in plasma that correlates well with the HPLC procedure.[35]

However, it must be kept in perspective that the decrease in glucose tolerance in the elderly could be related to secondary conditions such as obseity, physical inactivity, and genetics. Thus, in one study comparing 10 healthy young subjects ranging in age from 23 to 29 years to 17 non hospitalized, non obese, elderly healthy subjects between the ages of 60 to 80, no significant differences in insulin sensitivity and beta-cell responsiveness were noted between the two groups.[5] It should also be noted that there is an inverse relationship between increased physical activity and the development of non insulin dependent diabetes mellitus (NIDDM).[36,37,38]

The question of whether or not genetics may play a role in the accumulation of AGE, is raised by the intriguing finding that while some poorly controlled diabetic patients escape complications, well-controlled patients do not. In the genesis of AGE$_s$, a highly reactive carbonyl compound such as 3-deoxyglucosone is produced, which in turn can react with free amino groups to produce a variety of intermediate and AGE$_s$. However, 3-deoxyglucosone could also be reduced by specific reductase enzyme to a relatively inactive compound such as 3-deoxyfructose.[6] Thus, the level and activity of the specific reductase enzyme could be a limiting factor in preventing the accumulation of AGE$_s$ and in turn the complications resulting from such accumulation.

The maintenance of homeostasis by the hypothalamic-pituitary-adrenal (HPA) axis is compromised during aging. This results in an increase in adrenocorticotropic hormone (ACTH) and corticosteroid levels.[3] Some of the changes observed during aging, such as immunosuppression, osteoporosis, hypertension, hyperglycemia and mood changes, may perhaps be related to the chronic hypersecretion of corticosteroids. It has been hypothesized that the imbalance of the mineralocorticoid and glucocorticoid receptors in the hippocampal neurons is responsible for the altered homeostatic control of the HPA axis.[3]

Changes in the circulating levels of arginine vasopressin (AVP) or antidiuretic hormone (ADH) released by the posterior pituitary have been related to aging. Thus, the plasma AVP level was found to be significantly higher (4.7 ± 0.6 ng/L) in 68 healthy subjects ranging in age from 53 to 87 years, compared to 2.1 ± 0.2 ng/L found in 45 subjects ranging in age from 21 to 51 years.[39]

Finally, thyroid homeostasis appears to be compromised during aging. In a longitudinal study of 74 healthy men in the age range of 30 to 96 years, serum thyroid stimulating hormone (TSH) levels were found to be 38 percent higher in the elderly group (mean age of 79.6) compared to the younger group (mean age 39.4). None of the elderly men included in the study had base line TSH levels greater than 7 μu/ml or a T4 level less than 5 μg/dl, thus ruling out the presence of primary hypothyroidism in these subjects.[40] Serum triiodothyronine (T_3) levels were 11 percent lower in the older age group (mean age 79.6) when compared to the younger group with a mean age of 39.4.

Incidentally, increased serum TSH levels were also noted in 25 children with Down's syndrome compared to 14 normal children. The increased TSH level in children with Down's syndrome was accompanied by low levels of plasma zinc and zinc dependant thymic hormone, thymulin. Zinc supplementation for four months normalized the TSH and thymulin levels in children with Down's syndrome.[41] The increased TSH levels noted in the healthy elderly population, could apparently be related to lower zinc levels reported in the elderly.[25]

Conclusions

Aging is marked by cellular and metabolic changes. While there may be an overlap in the changes in the levels of certain analytes between purely aged related changes and disease, changes in organ function owing to disease alone have also to be kept in perspective. Much remains to be learned about the mechanisms of aging. Hopefully, the progress made in molecular biology techniques will provide not only techniques that can be transferable to the laboratory, but also will allow the identification of sensitive laboratory markers based on the knowledge that is accumulating from the study of accelerated aging syndromes.

References

1. Giacomoni PU. Aging and cellular defense mechanisms. Ann NY Acad Sci 1992;663:1–3.
2. Franceschi C, Monti D, Scarfi MR, Zeni O, Temperani P, Emilia G, et al. Genomic instability and aging. Studies in centenarians (successful aging) and in patients with Down's syndrome (accelerated aging). Ann NY Acad Sci 1992;663:4–16.
3. De Kloet ER. Corticosteroids, stress and aging. Ann NY Acad Sci 1992;663:357–71.
4. Cossarizza A, Monti D, Montagnani G, Ortolani C, Masi M, Zannotti M, Franceschi C. Precocious ageing of the immune system in Down's Syndrome: Alteration of B-lymphocytes, T-lymphocytes and of cells with natural killer markers. Am J Med Genet Suppl 1990;7:213–8.
5. Pacini G, Valerio A, Beccaro F, Nosadini R, Cobelli C, Crepaldi G. Insulin sensitivity and beta-cell responsivity are not decreased in elderly subjects with normal OGTT. J Am Geriatr Soc 1988;36:317–23.
6. Brownlee M. Glycation and diabetic complications. Diabetes 1994;43:836–41.
7. Cossarizza A, Ortolani C, Forti E, Montagnani G, Paganelli R, Zannotti M, Marini M, Monti D, Franceschi C. Age-related expansion of functionally inefficient cells with markers of natural killer activity in Down's Syndrome. Blood 1991;77:1263–70.
8. Abo T, Cooper MD, Balch CM. Postnatal expansion of the natural killer and killer cell population in humans identified by the monoclonal HNK-1 antibody. J Exp Med 1982;155: 321–6.
9. Taylor GM, Williams A, D'Souza SW, Fergusson WD, Donnai D, Fennell J, Harris R. The expression of CD18 is increased on Trisomy 21 (Down Syndrome) lymphoblastoid cells. Clin Exp Immunol 1988;71:324–8.
10. Fukuchi K, Martin GM, Monnat RJ Jr. Mutator phenotype of Werner Syndrome is characterized by extensive deletions. Proc Natl Acad Sci USA 1989;86:5893–7.
11. Narayanan S. Overview of principles and current uses of DNA probes in clinical and laboratory medicine. Ann Clin Lab Sci 1992;22: 353–76.
12. Wang SM, Nishigori CK, Zhang JM, Takebe H. Reduced DNA-repair capacity in cells originating from a progeria patient. Mutat Res 1990; 237:253–7.
13. Hanawalt PC. Heterogeneity of DNA repair at the gene level. Mutat Res 1991;247:203–11.
14. Ershler WB. Interleukin-6: a cytokine for gerontologists. J Am Geriatr Soc 1993;41:176–81.
15. Ershler WB, Sun WH, Binkley N. Gravenstein S, Volk MJ, Kamoske G, Klopp RG, Roecker EB, Daynes RA, Weindruch R. Interleukin-6 and aging: blood levels and mononuclear cell production is modifiable by dietary restriction. Lymphokine Cytokine Res 1993;12:225–30.
16. Daynes RA, Araneo BA, Ershler WB, Maloney C, Li GZ, Ryu SY. Altered regulation of IL-6 production with normal aging. Possible linkage to the age-associated decline in dehydroepiandrosterone and its sulfated derivative. J Immunol 1993;150:5219–30.
17. Zhou D, Chrest FJ, Adler W, Munster A, Winchurch RA. Increased production of TGF-beta and IL-6 by aged spleen cells. Immunol Lett 1993;36:7–11.
18. Penn ND, Purkins L, Kelleher J, Heatley RV, Mascie-Taylor BH, Belfield PW. The effect of dietary supplementation with vitamins A, C and E on cell-mediated immune function in elderly long-stay patients: a randomized controlled trial. Age Ageing 1991;20:169–74.
19. Meydani SN, Barklund MP, Liu S, Meydani M, Miller RA, Cannon JG, Morrow FD, Rocklin R, Blumberg JB. Vitamin E Supplementation enhances cell-mediated immunity in healthy elderly subjects. Am J Clin Nutr 1990;52: 557–63.
20. Chandra RT. Effect of vitamin and trace-element. Supplementation on immune responses and infection in elderly subjects. Lancet 1992;340:1124–7.
21. Bertazzoni U, Scovassi AI. Correlation between poly(ADP-ribose) polymerase activity and cellular defense mechanisms. Ann NY Acad Sci 1992;663:215–7.
22. Grube K, Kupper JH, Burkle A. Direct stimulation of poly(ADP ribose) polymerase in permeabilized cells by double-stranded DNA oligomers. Anal Biochem 1991;193:236–9.
23. Ferrari S, Grande A, Manfredini R, Torelli U. Terminal differentiation. Ann NY Acad Sci 1992;663:180–6.
24. Koopman G, Reutelingsperger CPM, Kuijten GAM, Keehnen RMJ, Pals ST, Van Oers MHJ. Annexin V for flow cytometric detection of phosphatidylserine expression on B cells undergoing apoptosis. Blood 1994;84:1415–20.
25. Prasad AS, Fitzgerald JT, Hess JW, Kaplan J, Pelen F, Dardenne M. Zinc deficiency in elderly patients. Nutrition 1993;9:218–24.
26. Franceschi C, Chiricolo M, Licastro F, Zannotti M, Masi M, Mocchegiani E, Fabris N. Oral zinc supplementation in Down's Syndrome: restoration of thymic endocrine activity and of some immune defects. J Ment Defic Res 1988;32: 169–81.

27. Monti D, Troiano L, Grassilli E, Agnesini C, Tropea F, Barbieri D, et al. Cell proliferation and cell death in immunosenescence. Ann NY Acad Sci 1992;663:250–61.
28. Picot-Ceballos I, Trivier JM, Nicole A, Sinet PM, Thevenin M. Age-correlated modifications of copper-zinc superoxide dismutase and glutathione-related enzyme activities in human erythrocytes. Clin Chem 1992;38:66–70.
29. Narayanan S, Appleton HD. Creatinine: a review. Clin Chem 1980;26:1119–26.
30. Narayanan S. Renal Biochemistry and Physiology: pathophysiology and analytical perspectives. Adv Clin Chem 1992;29:121–59.
31. Rowe JW, Andres R, Tobin JD, Norris AH, Shock NW. The effect of age on creatinine clearance in men: a cross-sectional and longitudinal study. J Gerontol 1976;31:155–63.
32. Lubran MM. Renal function in the elderly. Ann Clin Lab Sci 1995;25:122–33.
33. Soerjodibroto WS, Heard CR, Exton-Smith AN. Glucose tolerance, plasma insulin levels and insulin sensitivity in elderly patients. Age Ageing 1979;8:65–74.
34. Lee AT, Cerami A. Role of glycation in aging. Ann NY Acad Sci 1992;663:63–70.
35. Taneda S, Monnier VM. ELISA of pentosidine, an advanced glycation end product, in biological specimens. Clin Chem 1994;40:1766–73.
36. Helmrich SP, Ragland DR, Leung RW, Paffenbarger RS. Physical activity and reduced occurence of non-insulin-dependent diabetes mellitus. N Engl J Med 1991;325:147–52.
37. Manson JE, Rimm EB, Stampfer MJ, Colditz GA, Willett WC, Krolewski AS, Rosner B, Hennekens CH, Speizer FE. Physical activity and incidence of non-insulin dependent diabetes mellitus in women. Lancet 1991;338:774–8.
38. Manson JE, Nathan DV, Krolewski AS, Stampter MJ, Willett WC, Hennekens CH. A prospective study of exercise and incidence of diabetes among US male physicians. J Am Med Assn 1992;268:63–7.
39. Crawford GA, Johnson AG, Gyory AZ, Kelly D. Change in arginine vasopressin concentrations with age. Clin Chem 1993;39:2023.
40. Harman SM, Wehmann RE, Blackman MR. Pituitary-thyroid hormone economy in healthy aging men: basal indices of thyroid function and thyrotropin responses to constant infusions of thyrotropin releasing hormone. J Clin Endocrinol Metab 1984;58:320–6.
41. Licastro F, Mocchegiani E, Zannotti M, Arena G, Masi M, Fabris N. Zinc affects the metabolism of thyroid hormones in children with Down's Syndrome: normalization of thyroid stimulating hormone and of reversal triiodothyronine plasmic levels by dietary zinc supplementation. Int J Neurosci 1992;65:259–68.

Cell, Vol. 96, 291–302, January 22, 1999, Copyright ©1999 by Cell Press

Molecular Biology of Aging

Review

F. Brad Johnson, David A. Sinclair,
and Leonard Guarente*
Department of Biology
Massachusetts Institute of Technology
Cambridge, Massachusetts 02139

Introduction

Is aging the final act in the script of developmental biology? The characteristic changes that are part and parcel of aging appear similar to developmentally regulated programs. But why would aging mechanisms have been evolutionarily selected as advantageous? Indeed, evolutionary biologists might argue that aging occurs by default due to the absence of selection in the postreproductive phase of life. By this view, the aging process is not programmed, but, rather, the detritus of the absence of selection for maintenance (Medawar, 1952; Kirkwood, 1977). However, it is quite reasonable that any mechanisms that sprang up to slow or regulate the pace of aging *would* be selected, because lucky individuals could potentially give rise to more progeny. Therefore, it is reasonable to suppose that life span *extending* processes have been selected and that these can be viewed as an elaboration of development itself. In principle, such extension mechanisms may act to slow or forestall deleterious changes in an organism that progressively lead to death. The life span of an organism, therefore, is the sum of deleterious changes and counteracting repair and maintenance mechanisms that respond to the damage (Figure 1).

A priori, one imagines such longevity mechanisms to be much less complex than those regulating embryonic development. The spatial and temporal constraints on embryonic development are many, while requirements for longevity mechanisms might be much more specific if there were a single process (or a few processes) whose breakdown is the limiting event in longevity (i.e., the Achilles heel).

Aging is defined when two criteria are met. First, the probability of death at any point in time increases with the age of the organism. This statistical definition applies from yeast to mammals and reflects the progressive nature of aging. Second, characteristic changes in phenotype occur in all individuals over time due to the limiting processes.

The phenotypic definition is equally general and is useful in distinguishing the aging process itself from diseases of aging, such as cancer and heart disease. Phenotypes of aging affect all of the individuals in a population, while diseases of aging affect only a subset. Both impact on life span, but in different ways. For example, the many advances in medicine and public health in this century have caused a large increase in the average life span of humans in developed countries. However, because these advances have not altered the aging process itself, there has been no substantial change in the maximum human life span.

Several avenues to studying aging have placed us on the threshold of understanding basic underlying mechanisms. These approaches include the identification of key genes and pathways important in aging; genetic studies of heritable diseases that cause the appearance of premature aging in affected people; physiological experiments that relate the pace of aging to caloric intake; and advances in human genetics, as well as cell and molecular biology leading to an understanding of the basis of many diseases of aging. Strikingly, single gene mutations have been found to significantly extend the life span in *C. elegans*, yeast, and, most recently, *Drosophila*, suggesting that aging may be relatively simple, at least in these organisms. Further, the limited replication potential of human cells in culture has been attributed to a specific mechanism (i.e., the shortening of telomeric ends of chromosomes). An important challenge is now to relate these recent findings to the more complex case of human aging.

In this review, we will discuss several important molecular models of aging that come from current research. These are damage by reactive oxygen species (ROS) generated by metabolism, genome instability, genetically programmed extension mechanisms, cell death, and systemic aging. Questions to be posed include the following. What evidence exists for and against these models? Can more than one of these models apply to aging of different tissues in humans—specifically do organs with continually dividing cells age by the same mechanism as organs that are postmitotic? Finally, is aging amenable to therapeutic intervention, and would such intervention be advisable?

Oxidative Damage

One theory of aging proposes that ROS which are generated by metabolism cause cumulative damage over a lifetime (Harman, 1981). Roughly two to three percent of oxygen taken up is chemically reduced by the addition of single electrons, which are sequentially converted into ROS, including the superoxide anion, hydrogen peroxide, and the hydroxyl radical. ROS have been shown to cause molecular damage relatively indiscriminately to proteins, lipids, and nucleic acids. In addition, specific damage has been observed in the mitochondrial DNA, which we consider below in Genome Instability.

What is the evidence that oxidative damage causes aging? One category of study that is supportive of this view involves animals transgenic for genes encoding antioxidants. Transgenic *Drosophila* overexpressing both Cu/Zn SOD and catalase live 34% longer than controls (Orr and Sohal, 1994). A more recent study shows that expression of human *SOD1* exclusively in *Drosophila* adult motor neurons leads to a 40% extension in life span (Parkes et al., 1998). Further experiments are necessary to clarify the nature of this primary role of motor neurons in life span. Conversely, mice knocked out for either *GPX1* (encoding glutathione peroxidase), *SOD1*,

*E-mail: leng@mit.edu.

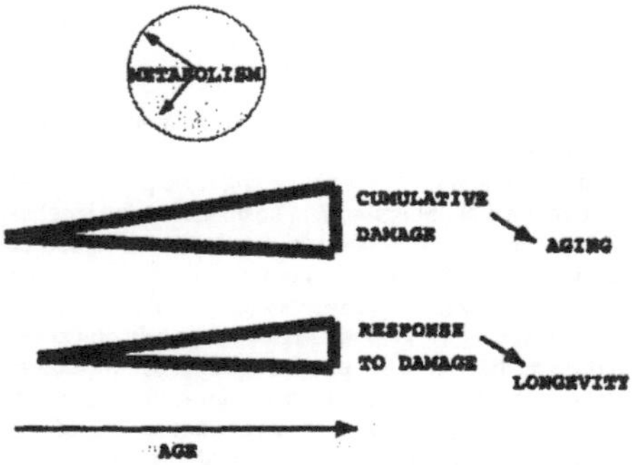

Figure 1. Life Span Is Determined by the Balance of Two Opposing Processes

Metabolism leads to the accumulation of damage (red), thus causing aging. Compensatory responses (green) limit or repair the damage, thus promoting longevity.

SOD2, or *SOD3* do not display a phenotype of rapid aging (Reaume et al., 1996; Ho et al., 1997; Melov et al., 1998); *GPX1*, *SOD1*, and *SOD3* knockout animals appear normal, whereas *SOD2* mice die of cardiac failure at day 8.

Another possible genetic link between oxidative damage and aging is provided by the long-lived *age-1* mutant of *C. elegans* (Klass, 1983; Friedman and Johnson, 1988; Larsen et al., 1995). *age-1* mutants live twice as long as the wild type, have increased levels of SOD and catalase (Larsen, 1993; Vanfleteren and De Vreese, 1996) and are more resistant to oxidative stresses (Larsen, 1993; Vanfleteren and De Vreese, 1996), as well as heat shock (Lithgow et al., 1994) and UV radiation (Murakami and Johnson, 1996). Similarly, the long-lived *Drosophila* mutant *methuselah* is resistant to oxidative stress and also high temperature and starvation (Lin et al., 1998). Mutation in another *C. elegans* gene, *clk-1*, slows development and rhythmic behavior and extends life span (Wong et al., 1995). Interestingly, *clk-1* is a homolog of yeast *CAT5* (Ewbank et al., 1997), a gene involved in synthesis of coenzyme Q, a component of the mitochondrial electron transport chain (Jonassen et al., 1998). Further, *clk-1* mutants contain reduced levels of coenzyme Q (Ewbank et al., 1997), suggesting that all the phenotypes of the *clk-1* mutant, including longevity, may be attributable to a slowing of metabolism.

It is interesting to note that these genetic links between aging and oxidative damage are found in animals in which most adult cells are postmitotic (i.e., *C. elegans* and *Drosophila*). Postmitotic cells may be more susceptible to the cumulative damage of ROS given their inability to replace themselves. In mammals, organs most vulnerable to this mechanism are the brain, heart, and skeletal muscle. The heart does, in fact, apparently experience a reduction in cell number over time, some of which may be due to damage-generated apoptosis (see Cell Death, below). However, the link between ROS and aging in organs consisting of dividing cells is even less clear.

Genome Instability

The importance of specific kinds of genome instability in aging is becoming increasingly apparent. The accumulation of genomic changes (i.e., point mutations [Szilard, 1959], loss of repeated DNA sequences such as ribosomal DNA [rDNA] [Strehler, 1986], rearrangements, and changes in chromosome number [Martin et al., 1985]) have long been proposed as causes of aging. The study of transgenic markers has revealed the accumulation of low levels of mutations with age in mice (Dolle et al., 1997). These changes were observed in the liver, which has mitotic capacity, but not in the brain, which is largely postmitotic. Age-dependent changes have also been observed in the *HPRT* and *HLA-A* genes of peripheral blood lymphocytes in humans (Grist et al., 1992; King et al., 1994). However, the low frequency of these genomic changes, even in old individuals, casts doubt on their importance in aging.

Ribosomal DNA

A strong causal link between genome instability and aging has been established in the model system *S. cerevisiae*. In this system, cell division is asymmetric, giving rise to a large mother cell and a smaller daughter cell. Aging is manifest both by the limited number of cell divisions and the changes in phenotype undergone by mother cells prior to senescence (Barton, 1950; Jazwinski, 1993).

The pace of aging in mother cells is dictated by changes occurring in the ribosomal DNA (rDNA) (Sinclair and Guarente, 1997), which is present in 100–200 tandem copies on chromosome XII. At some point in the middle of the life span of mother cells, a circular copy of rDNA pops out of the genomic array by homologous recombination (Figure 2). This extrachromosomal rDNA circle (ERC) replicates via an origin present in the rDNA sequence during subsequent cell cycles. Importantly, ERCs segregate in a highly biased fashion to mother cells. Thus, daughters are born youthful and ERC free, and mothers experience an exponential buildup in ERCs leading to about 1000 copies in senescent cells. The proliferation of ERCs elicits an apparent enlargement and fragmentation of the nucleolus (Sinclair et al., 1997), the site of rDNA transcription and ribosome assembly. The artificial creation of ERCs, by use of a site-specific recombinase, at an abnormally early age shortens life span, indicating that ERCs are a cause of aging. We imagine that the accumulated ERCs interfere with cell growth by titrating away critical components of the machinery for DNA replication and perhaps transcription.

The initial clue pointing to this mechanism was the genetic control of aging by *SIR4* (Kennedy et al., 1995). This gene, along with *SIR2* and *SIR3*, mediates genomic silencing at telomeres and the silent mating type loci *HML* and *HMR*. Loss-of-function mutations in these genes shorten life span, while a gain-of-function mutation in *SIR4* extends life span. This latter allele causes a relocalization of the SIR2/3/4p complex away from telomeres and *HM* loci to the nucleolus (Kennedy et al., 1997). This mutant constitutively mimics what happens with aging in wild-type cells. These findings suggest that the relocalization of the Sir complex to the nucleolus may be a genetic mechanism to extend life span in wild-type cells. A consequence of this relocalization is the

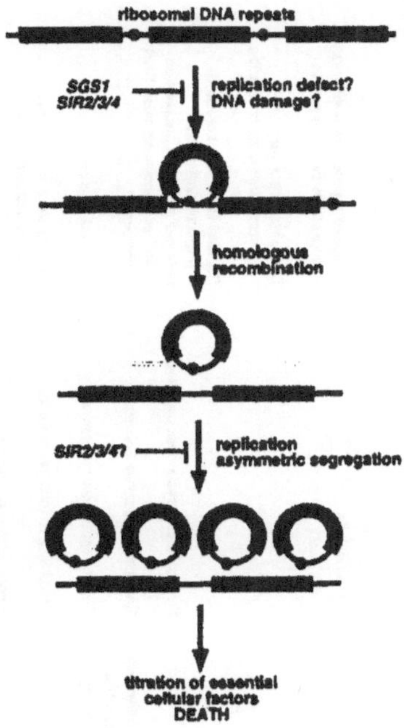

Figure 2. Essential Features of Aging in *Saccharomyces cerevisiae*

Between 150–200 copies of a 9.1 kb unit encoding rRNA is located on chromosome XII. The potentially unstable region is prone to homologous recombination between adjacent repeats at a DNA lesion, perhaps a stalled replication fork or DNA double-strand break. The extrachromosomal rDNA circle (ERC) replicates in ensuing cell cycles via the autonomously replicating sequences in the nontranscribed spacer. ERCs are preferentially held in the mother cell, causing them to accumulate exponentially. Mother cells grow large, and their cell cycle slows. When the ERC copy number reaches between 500 and 1000, the mother cells senesce, possibly due to titration of essential transcription or replication factors. The *SGS1* and *SIR* genes slow the rate of ERC accumulation; *SGS1* presumably limits damage or recombination, and the *SIR* genes act at undefined steps.

evacuation of *HM* loci leading to the expression of a and α mating type genes and the age-related phenotype of sterility (Smeal et al., 1996).

WRN DNA Helicase

Another link between aging and the nucleolus comes from studies of the disease Werner syndrome, which is due to loss-of-function mutations in a single gene, *WRN* (Epstein et al., 1966; Yu et al., 1996). This disease is one of a class of human progerias, which are characterized by shortened life span and the premature onset of phenotypic changes that resemble aging (Martin, 1978) (Table 1). For the model systems described above, the link between the genes identified and aging is clear, because mutations extending life span necessarily identify processes involved in aging. The connection between the mutated genes and normal aging is less certain in the case of the progerias. Figure 3 presents two alternative possibilities for the effects of the progeroid mutations. Consider normal aging as caused by a particular kind of cumulative damage that impinges on a key cellular process to debilitate its output. Normal longevity is promoted by the activity of Gene 1, which slows the accumulation of this damage. The study of a progeria due to mutation in Gene 1 will therefore directly illuminate a step that is the Achilles heel in normal aging. However, a mutation in a Gene 2 encoding a component of the process itself could, in principle, phenocopy rapid aging by limiting the output of the process. In this case, a study of the progeria will not immediately reveal the limiting step in longevity (i.e., the step in which the damage is produced) but may nonetheless identify a relevant process (i.e., the step impacted by the damage).

The *WRN* gene has been cloned and shown to encode a DNA helicase of the RecQ family (Yu et. al, 1996; Gray et al., 1997; Suzuki et al., 1997). Members of this family include *SGS1* of *S. cerevisiae*, *rqh1*$^{+}$ of *S. pombe*, the gene mutated in Bloom syndrome, *BLM*, and another human homolog *RECQL* (Gangloff et al., 1994; Puranam and Blackshear, 1994; Ellis et al., 1995; Stewart et al., 1997). Studies in yeast show that *SGS1* suppresses recombination in genomic DNA, including the rDNA (Gangloff et al., 1994; Watt et al., 1996). Further, null mutations in *SGS1* suppress the slow growth and rDNA hyperrecombination of *top3* (topoisomerase III) mutants (Gangloff et al., 1994) and decrease the fidelity of mitotic and meiotic chromosome segregation (Watt et al., 1995). Both Bloom and Werner syndromes are characterized by chromosome instability (reviewed in Ellis, 1997). In the case of Bloom syndrome, there are an elevated number of sister chromatid exchanges and varied rearrangements, and patients are prone to all cancers. Werner individuals display more subtle alterations in chromosome stability, but enhanced frequencies of deletions and rearrangements occur.

Strikingly, deletion of the *S. cerevisiae* homolog, *SGS1*, also causes premature aging in that organism (Sinclair et al., 1997). The mutants experience the characteristic fragmentation of the nucleolus and accumulation of ERCs in about half as many cell divisions as the wild type (Sinclair and Guarente, 1997) (Figure 2). Immunostaining reveals that both Sgs1p and the human WRN protein are concentrated in the nucleolus in yeast (Sinclair et al., 1997) and human cells (Gray et al., 1998; Marciniak et al., 1998), reinforcing the possibility that instability of the rDNA is a critical component of aging. In the absence of the Sgs1p helicase, ERCs form more readily, probably due to elevated intrachromosomal recombination.

What could be the biochemical defect in cells lacking *SGS1*-like DNA helicases? Recent findings suggest that a WRN-related protein is required for the formation of DNA replication foci in an vitro assay programmed by a *Xenopus* extract (Yan et al., 1998). A different study shows that the BLM helicase can unwind G quadruplexes in DNA (Sun et al., 1998). One possibility is that these helicases remove secondary structures from the template DNA during replication. In their absense, stalled replication forks may provoke the ensuing recombinational events.

The ERC mechanism of aging may be applicable to mammalian organs consisting of dividing cells. Circular

Table 1. Features of Selected Human Progeroid Syndromes

Progeroid Syndrome	Inheritance	Genetic Defect	Life Span (Approx. Maximum) in Years	Pathologies Possibly Related to Aging	Cellular Changes	Molecular Changes
Werner syndrome	Autosomal recessive	Loss of WRN, a RecQ family helicase.	60	Skin atrophy. Hair graying/loss. Arteriosclerosis. Osteoporosis. Muscle atrophy. Cataracts. Hyperlipidemia. Mild diabetes mellitus, type 2. Hypogonadism. Cancer (sarcomas).	Slow growth. Reduced replicative life span of cultured cells. Chromosome rearrangements. Sensitivity to 4-NQO and camptothecin (topoisomerase I poison)	Increased mutation rate, particularly DNA deletions. Rapid telomere shortening during cellular replicative life span. DNA replication defects (?).
Hutchinson-Gilford progeria	Autosomal dominant (?)	Unknown	25	Skin atrophy. Loss of subcutaneous fat. Hair loss. Arteriosclerosis. Bone loss. Inc. metabolic rate. Growth hormone insensitivity? Hypogonadism.		DNA repair defects?
Down syndrome	Sporadic	Trisomy 21; (sometimes trisomy only of critical regions of chromosome 21).	70	Hair graying/loss. Amyloidosis. Alzheimer-type CNS pathology. Cataracts. Vascular disease. Leukemia. Diabetes mellitus. Hypogonadism.		Elevated amyloid precursor protein (APP) synthesis. DNA repair defects?
Cockayne syndrome	Autosomal recessive	Mutation of CS-B, an ATPase of Swi2 family. Also mutation of CS-A and certain mutations in XP-B, XP-D (helicases in TFII-H), and XP-G.	40 (?)	Loss of subcutaneous fat. Skin photosensitivity. Neurodegeneration. Hypogonadism.	UV sensitivity	Impaired transcription-coupled nucleotide excision repair. Decreased recovery of transcription after irradiation. General RNA polymerase II transcription defect.
Ataxia telangiectasia	Autosomal recessive	Loss of ATM, a protein kinase of the PI-3 kinase family.	40–50	Neurodegeneration. Immunodeficiency. Cancer (leukemias and lymphomas). Occulocutaneous telangiectasias. Progeroid skin and hair changes. Hypogonadism.	Defective cell cycle checkpoint arrest. Reduced replicative life span. Chromosomal rearrangements. Sensitivity to ionizing radiation. Inappropriate apoptosis.	Delayed p53 phosphorylation and accumulation after DNA damage. Short telomeres. Defects in repair of DNA double-strand breaks.

Recent reviews: WS: Goto, 1997, Fry and Loeb, 1998; HG: Brown, 1992; DS: Hayes and Batshaw, 1993; CS: Bohr et al., 1998; AT: Rotman and Shilo, 1998.

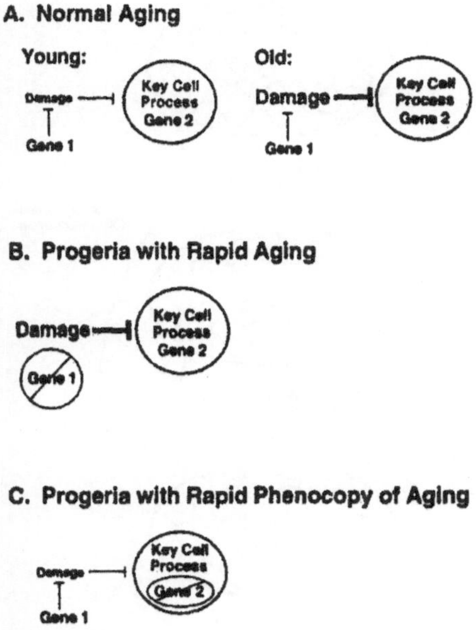

Figure 3. Models for Possible Relationships between Normal Aging and Progeroid Disorders

(A) Normal aging results from the accumulation of damage with age. The damage interferes with a key cellular process critical for longevity, of which Gene 2 is a component. Gene 1 slows the rate of damage accumulation, yielding a normal life span.

(B) A progeroid phenotype results when Gene 1 is inactivated, leading to rapid accumulation of damage and premature inhibition of the key cell process.

(C) Inactivation of Gene 2 leads to decreased activity of the key cell process and a premature aging phenotype. However, there is no effect on the accumulation of damage that normally limits longevity.

DNA has been observed in mammalian cells (reviewed in Gaubatz, 1990). However, although there are a few reports of changes in circular DNA with age in mammals, no dramatic accumulation of circular DNA with age has been observed, and mammalian circular DNA is not known to include rDNA. The strictest analog of yeast mother cells is probably stem cells of mitotically active organs. Although age-related changes have been noted in stem cell numbers and physiology in various studies, it is too early to discern a pattern of changes in rDNA that would be informative.

Telomeres

Telomeres, the repeated DNA sequences at the ends of linear chromosomes, are unable to be fully replicated by DNA polymerases. Thus, they will shorten with cell division unless maintained via telomerase, a ribonucleoprotein enzyme that can add telomeric repeat sequences to chromosome ends. Telomeres indeed shorten during replicative growth of many primary cultures of human cells, which like most somatic tissues lack detectable telomerase activity (Harley et al., 1990). It has thus been proposed that telomere shortening could be a molecular clock that signals the eventual growth arrest, termed replicative senescence, that is observed in all cultured primary human cells. Strong support for this notion has been recently provided by the demonstration that reactivation of telomerase in certain cultured human cells can extend their replicative life span beyond the normal limit (Bodnar et al., 1998; Vaziri and Benchimol, 1998; but see Kiyono et al., 1998). In addition, these apparently immortalized cells have properties indistinguishable from normal, untransformed cells by several phenotypic assays (Jiang et al., 1999; Morales et al., 1999).

There are several findings that relate telomere shortening to aging in vivo. There appears to be a rough correlation between telomere length and age in the human soma, for example in arteries at sites of elevated hemodynamic stress and presumably high cell turnover (Chang and Harley, 1995). It is possible that senescent cells accumulate focally in vivo and because of their altered physiology interfere with the tissues in which they reside. For example, elevated secretion of metalloproteases by senescent fibroblasts could degrade dermal collagen (reviewed in Campisi, 1998). Support for the importance of replicative senescence in human aging has included (1) a weak correlation between in vitro replicative life span and donor age (Martin et al., 1970), (2) correlation between in vitro life span and maximum life span in interspecies comparisons (Rohme, 1981), (3) diminished in vitro replicative life span of cells from people with premature aging syndromes (Faragher et al., 1993), and (4) evidence for the accumulation of senescent cells in elderly individuals (Dimri et al., 1995).

However, other findings and interpretations challenge the sweeping notion that telomere shortening causes aging. First, a recent study that used both cross-sectional and longitudinal samples obtained from the Baltimore Longitudinal Study on Aging indicated that if donor health status and biopsy conditions are controlled, there is no significant correlation between donor age and replicative life span of cultured fibroblasts (Cristofalo et al., 1998). Second, the interspecific correlation can be equally well explained as a correlation with animal size; larger animals generally live longer and might require cells with a greater replicative capacity to produce their larger bodies (Promislow, 1993). Third, the reduced replicative life span of some progeroid cells does not hold for all progerias and might reflect increased cell death or exit from the cell cycle for reasons unrelated to those of wild-type senescent cells. Indeed, WS cells exit the cell cycle before their telomeres are as short as wild-type senescent cells (Schulz et al., 1996) and there are differences in gene expression patterns between senescent wild-type and WS cells (Oshima et al., 1995; Toda et al., 1998). Fourth, the evidence for accumulation of senescent cells in vivo rests largely with the report of cells in elderly skin that express low pH β-galactosidase, an activity that is observed in senescent but not early-passage cultured cells (Dimri et al., 1995). However, the frequency of positive cells in vivo has not been determined. Moreover, there are clear differences between the morphology and gene expression profiles of cultured senescent fibroblasts and those in the skin of elderly individuals (Robbins et al., 1970; Takeda et al., 1992). On the other hand, the finding that $CD28^-$ T cells, which have shortened telomeres and decreased proliferative capacity in culture, accumulate with age is intriguing (Effros, 1998). More work needs to be done to examine the frequency of senescent cells in vivo, as well as whether sites of pathology are associated with them.

Also of note, telomeres in some mouse species tend to be very long throughout the lifetime of the animals. Telomerase-knockout mice have been constructed and have not yet displayed phenotypes in the first five generations (Blasco et al., 1997). Although highly mitotic tissues (testes, skin, and hematopoietic system) show defects in later generations that may result from chromosomal abnormalities, these changes have no clear relationship to normal aging phenotypes (Lee et al., 1998). It will be of interest to monitor any changes in the early generation animals as they grow older.

Mitochondrial DNA

Genetic instability outside of the nuclear genome might also contribute to aging (reviewed in Lee et al., 1997; Wallace et al., 1998). The mutation rate for mitochondrial DNA (mtDNA) is 10- to 20-fold greater than for nuclear DNA, and it is believed that mtDNA mutations may compromise mitochondrial functions in different ways (Figure 4). First, defects in electron transport and oxidative phosphorylation could lead to declines in ATP levels and the NAD:NADH ratio. Second, defective electron transport might increase free radical production by mitochondria—perhaps enabling a positive feedback loop to produce further mtDNA mutations. Third, and more speculatively, age-associated mitochondrial defects might induce cell death (see Cell Death, below).

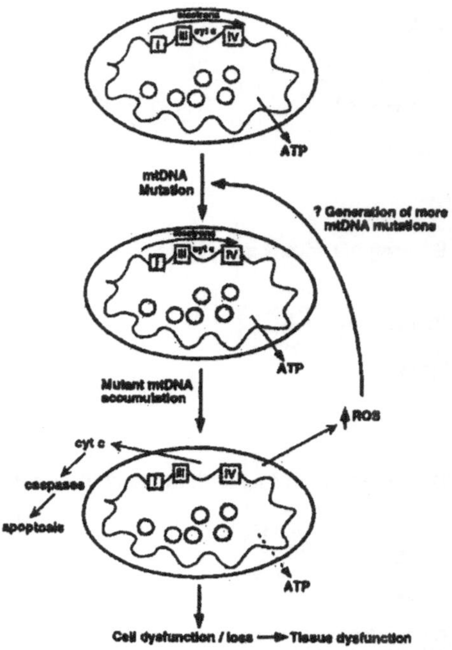

Figure 4. Mitochondrial Theory of Aging

A single mitochondrial genome (mtDNA) incurs a *loss-of-function* mutation. The mutant genome (red circle) accumulates perhaps due to a replicative advantage compared with wild type (green circle) to a point where mitochondrial and cell activities are compromised. Three types of interference with cell function could occur. First, a decline in electron transport leads to reduced ATP synthesis. The electron transport complexes containing mtDNA-encoded subunits are indicated by roman numerals; complexes I and IV appear most susceptible to age-associated declines in activity (red boxes). Second, increased generation of ROS leads to damage and possibly further mtDNA mutations. Third, cell death is triggered by necrosis or apoptosis, for example by release of cytochrome c. The resulting cell dysfunction or loss could lead to a decline in tissue function.

There is good evidence for a decline in the function of the mitochondrial electron transport chain with age in humans and other animals, particularly in postmitotic tissues (Boffoli et al., 1994). It is also clear that a variety of mtDNA alterations, including point mutations, deletions, and oxidative modifications increase with age, again preferentially in postmitotic tissues (reviewed in Lee et al., 1997). However, the connection between these two groups of findings is not as certain. Early studies suggested that although mtDNA alterations increased with age, they were quite rare, generally less than one percent of total mtDNA. Cells contain hundreds to thousands of mitochondrial genomes, and studies of hereditary diseases caused by mitochondrial mutations have shown that mutated genomes must accumulate to levels greater than 50%–80% to be apparent clinically (Sciacco et al., 1994).

Two recent findings suggest that age-associated mitochondrial mutations might nonetheless be significant. Using PCR amplification of the entire 16.5 kb mtDNA from skeletal muscle, numerous different deletion mutations are found to accumulate with age, and no one type of mutation predominates (Melov et al., 1995b). Thus, early studies that focused on only a few types of mutations may have underestimated the total number. Moreover, when PCR is performed on just a few or single muscle fibers (rather than on thousands as in earlier studies), deletions can account for a sizable fraction of total mtDNA, suggesting that individual mitochondria may incur unique mutations, which are selectively amplified over wild-type mtDNA (Schwarze et al., 1995; Brierley et al., 1998). The inability of postmitotic tissues to replace cells containing mutant mtDNA with healthy cells might explain a greater incidence of mtDNA mutations in postmitotic versus mitotic tissues. In skeletal muscle, fibers containing high levels of deletions are deficient for cytochrome c oxidase (COX) activity, indicating that the mutant mtDNA is interfering directly with electron transport (Brierley et al., 1998; Kopsidas et al., 1998). Indeed, in situ studies indicate that there are regional differences in COX activity along fibers and these correspond to regions containing deleted mtDNA (Muller-Hocker et al., 1993). Such COX-deficient fibers increase with age, but because they represent only a few percent of total cells it is still not clear that they contribute to the muscle weakening that occurs with normal aging. Another unexplored possibility is that certain mtDNA alterations could have a dominant inhibitory effect on mitochondrial function and thus might not need to accumulate to high levels to be deleterious.

Even if mtDNA mutations accumulate and contribute to aging, they might not be a primary cause. The *C. elegans age-1* gene encodes a presumptive cytosolic PI-3 kinase (see below), yet mutations in this gene that confer longevity cause mtDNA deletions to accumulate more slowly than wild type (Melov et al., 1995a). This finding suggests that mitochondrial deletions may occur downstream of other aging events that are more directly impacted by the *age-1* mutation.

Genetic Programs

As stated above, the universality of aging phenotypes within a species argues for an underlying genetic program. The redistribution of the Sir complex from telomeres to the nucleolus in yeast is a specific molecular example of how a genetically programmed response can extend life span and also lead to a gradual change in phenotype.

Another example of a genetic pathway that may be relevant in aging comes from *C. elegans*. After hatching, *C. elegans* can proceed through four larval stages directly to adulthood, or they can enter a dormant, juvenile stage known as dauer. Adult development can resume if environmental conditions improve. Which developmental pathway is taken depends primarily on the concentration of a pheromone that indicates environmental stresses, such as starvation, overcrowding, and high temperature. In the dauer state, worms can live up to 6 months, rather than the usual few weeks as adults.

Many genes have been isolated that are involved in this developmental decision. Mutants in these so-called dauer formation (*daf*) genes are either dauer defective (e.g., *daf-16*) or dauer constitutive (e.g., *daf-2* and *age-1*). Strikingly, certain mutations in two of these genes, *age-1* and *daf-2*, permit development to adulthood, but with life span extended 2- to 4-fold (Friedman and Johnson, 1988; Kenyon et al., 1993; Larsen et al., 1995; Tissenbaum and Ruvkun, 1998). This effect has been convincingly interpreted as evidence for a genetic program that influences aging. It would appear that a particular pattern of gene expression and physiology of dauer larvae has been activated in the adult organism. Mutation of *daf-16*, which suppresses dauer-constitutive mutations, also suppresses the life span extension in *daf-2* and *age-1* mutants, providing further evidence for this interpretation (Kenyon et al., 1993; Gottlieb and Ruvkun, 1994).

The cloning of these three *daf* genes has provided insights into the signaling pathway utilized by the worm to regulate dauer development (Figure 5). *daf-2* encodes a nematode member of the family of insulin-like receptors (Kimura et al., 1997), whereas *age-1* encodes a homolog of the p110 catalytic subunit of phosphatidylinositol 3-kinase (PI 3-kinase) that acts downstream of the *daf-2* receptor (Morris et al., 1996). Another gene, *daf-18*, mutations in which can reverse the longevity in certain *age-1* mutants, encodes a PI 3-phosphatase homolog (Ogg and Ruvkun, 1998).

daf-16 encodes a member of the hepatocyte nuclear factor 3/forkhead family of transcriptional activators that is a target of negative regulation by the PI 3-kinase-mediated signaling cascade (Lin et al., 1997; Ogg et al., 1997). In *age-1* and *daf-2* adults, this signal is partially on, allowing adult worms to enjoy an extended life span. The extension of life span may be in part due to the upregulation of a wide array of cellular defenses in the adult (as noted above), which combat oxidative and other damage that occur throughout the lifetime of the worm. The fact that developmental pathways can be manipulated to extend the life of adult worms suggests that genetic programs may regulate the pace of aging more generally.

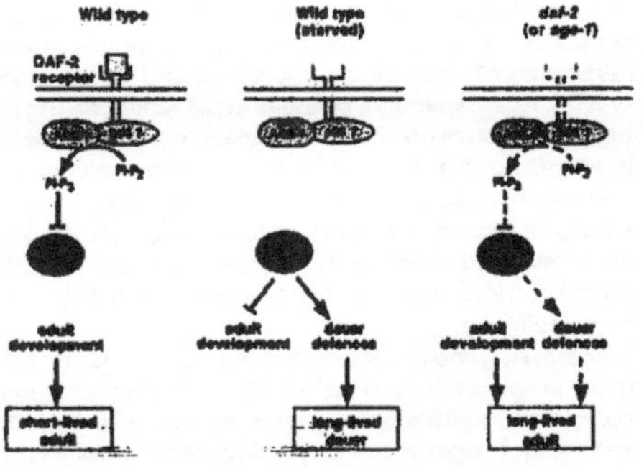

Figure 5. Proposed Mechanism of Life Span Extension by the Dauer Genetic Pathway in Adult *Caenorabditis elegans*

High levels of DAF-2 ligand signal adult development. Under conditions of starvation or overcrowding, increased activity of the DAF-16 forkhead transcription factor triggers dauer development. In certain *daf-2* or *age-1* mutants, a partially active DAF-2/AGE-1 signal results in simultaneous expression of genes involved in adult development and dauer-specific defenses that confer longevity.

Cell Death

Is there a connection between cell death and aging? Cells die via mechanisms that range from necrosis, which is a relatively passive outcome of cell injury, to apoptosis, or programmed cell death, which is an active and ordered process. The function of postmitotic tissues might especially be compromised by cell death, because the lost cells cannot be replaced (although the dogma that heart and brain are truly postmitotic has recently been challenged [Eriksson et al., 1998; Kajstura et al. 1998]). Apoptosis is triggered by a myriad of stimuli, such as steroid hormones, DNA damage, withdrawal of growth factors, or receptor-mediated apoptosis as in clonal deletion of immune cells. This pathway involves activation of a family of proteases, termed caspases, and nucleases, leading to the controlled degradation of cellular structures, followed by removal of the membrane-bound debris by phagocytic cells. There has been speculation that dysregulation of programmed cell death might contribute to aging (reviewed in Warner et al., 1997).

Both apoptotic and necrotic forms of cell death can be mediated by mitochondria (reviewed in Green and Reed, 1998). Given the evidence for changes in mitochondrial function during aging (see Figure 4 and above), one wonders whether these changes contribute to cell death. Mitochondrial events that can mediate cell death include (1) release of mitochondrial stores of caspase activators including cytochrome c, (2) disruption of the respiratory electron transport chain, and (3) production of ROS. Age-dependent changes in mitochondria might trigger the above mechanisms leading to cell loss. There is as yet no direct evidence for increased release of caspase activators from mitochondria with age, but the activity of electron transport complexes appears to decline (reviewed in Lee et al., 1997) and the rate of ROS production increases with age in many mammals (Sohal et al., 1994).

However, genetic data do not support a role for cell death in aging. For example, studies in *C. elegans* argue against any relationship between programmed cell death and aging. The genetic underpinnings of both processes have been extensively studied in this organism, and there is no apparent overlap between the genes that affect them (reviewed in Hengartner, 1997). In addition, knockout mice for the *bcl2* family proteins, which normally function to inhibit caspase activation and thus apoptosis, do not show clear aging phenotypes, although the *bcl2* knockout does develop some phenotypes that superficially resemble aging (Veis et al., 1993).

Nevertheless, there is correlative evidence in humans and other mammals that cell death is important in certain aspects of aging. One recent study reports that more than a third of ventricular cardiac myocytes are lost during normal aging in human males, although apparently not in females, and the loss of cells is accompanied by hypertrophy of the remaining cells (Olivetti et al., 1995). A similar loss has been observed in rats; in this case the majority of cells die via necrosis, and a substantial number are also lost via apoptosis (Kajstura et al., 1996). Neurodegenerative diseases, for example Alzheimers, are clearly accompanied by extensive loss of neurons. However, the loss of neurons with normal aging may not be as extensive as once thought (reviewed in Morrison and Hof, 1997). Recent studies indicate that the overall number of cortical neurons declines by only 10% with aging, and although particular brain regions do show higher levels of loss, it is not clear that they are responsible for the functional changes that accompany aging (Pakkenberg and Gundersen, 1997). It remains to be demonstrated whether genetic or pharmacological interventions that inhibit cell death can slow cardiac myocyte or neuronal loss and have beneficial effects on organ function.

On the other hand, it has been proposed that an *inability* of certain cells to undergo apoptosis might contribute to aging and diseases of aging. Cultured senescent fibroblasts are resistant to apoptosis, in part due to an inability to downregulate *bcl2*, and this could enable senescent cells to accumulate in vivo with deleterious consequences (Wang, 1995; see Genome Instability, above). There is also evidence from studies in the rat liver that apoptosis helps clear preneoplastic cells (Grasl-Kraupp et al., 1994). An inability of apoptosis to keep pace with increasing numbers of precancerous cells could thus contribute to age-associated elevations in cancer rates. Interestingly, calorically restricted rats increase their hepatocyte apoptosis rates with age more than rats fed ad libitum, with a consequent decline in the incidence of hepatocellular carcinoma (Muskhelishvili et al., 1995). Further studies are necessary before the contribution of cell death to aging can be correctly gauged.

Systemic Control of Aging

Many of the above mechanisms of aging are cell autonomous (i.e., are independent of external signals). However, it is possible, and, in fact, it has been suggested that systemic control of aging may occur (reviewed in Finch, 1990). By this view humoral factors (i.e., endocrine hormones) would coordinate the pace of aging in many different tissues and organs. Further, local mechanisms like cell–cell contact or paracrine hormones could coordinate aging in a proximate compartment of cells. Systemic and cell-autonomous processes are clearly not mutually exclusive; cell-autonomous and systemic aging may coexist or trigger each other.

A recent study in *C. elegans* has demonstrated that the longevity enjoyed by *daf-2* mutant animals is not cell autonomous (Apfeld and Kenyon, 1998). By constructing mosaic animals containing a mixture of *daf-2*(+) and *daf-2*(−) cells, the authors showed that a small number of mutant cells in any of several sublineages conferred a degree of longevity on the entire animal, which contained predominantly *daf-2*(+) cells. This was interpreted to mean that the *daf-2* gene controls the production of a secreted factor that dictates the pace of aging in the animal. In addition, the key determinant of life span was the ratio of *daf-2*(+) to *daf-2*(−) cells, and it did not depend on the presence of mutant cells in any particular lineage. This finding implies that there are not specific adult cells that dictate the rate of aging. Rather, the level of this secreted factor, averaged systemically, sets the life span. In *Drosophila*, a reduction in the activity of the gene *methuselah* extends life span. *methuselah* is homologous to genes encoding G protein–linked transmembrane receptors, raising the possibility that it might also play a role in transducing systemic aging signals (Lin et al., 1998).

Do systemic factors regulate aging in mammals? A genetic argument for a such a role might be provided by the mouse gene *Klotho*, which encodes a humoral factor (Kuro-o et al., 1998). *Klotho* mutants have a shortened life span and develop several pathologies that might have some relationship to aging. In normal human physiology, there are clear examples in which the levels of endocrine factors decline with age (reviewed in Lamberts et al., 1997). Estrogen levels decline sharply at menopause, and estrogen supplementation has been shown to slow several degenerative processes, including skin atrophy, osteoporosis, atherosclerosis, and cognitive decline. Moreover, a recent study suggests that the overall mortality rate declined in postmenopausal women receiving hormonal supplementaton (Grodstein et al., 1997). Other endocrine factors, including growth hormone, testosterone, and DHEA-S (dehydroepiandrosterone sulphate), decline gradually with age. Short-term supplementation with growth hormone has been shown to increase muscle mass, skin thickness, and bone density (Rudman et al., 1990).

While the effects of these hormones on specific organs are apparent, their relation to the aging process itself, if any, is not yet clear. To our knowledge, there has been no animal study in which hormone supplementation extended maximum life span. However, the recent findings in *C. elegans* provide a basis to believe that humoral factors may turn out to play an important role in at least some aspects of human aging.

Perspective

Recent advances in the study of aging indicate that this process is amenable to molecular analysis and may be relatively simple. The potential of single gene mutations to greatly extend life span in model systems suggests that relatively few limiting cellular or organismal processes control the rate of aging, at least in these species. In mammals, it is noteworthy that a simple dietary regimen can extend life span. Rodents that are calorically restricted live up to twice as long as controls that are fed ad libitum (reviewed in Weindruch, 1996). The restricted animals are about one-third smaller, have a lower body temperature than controls, and are relatively disease free (Weindruch and Sohal, 1997).

What new advances can we expect in the short term? One current limitation in studies of mammalian aging is the paucity of reliable and easily measured biological markers for the aging phenotype. We imagine that the DNA chip arrays that allow gene expression to be assessed across the entire genome will provide useful markers. If a characteristic pattern of changes is demonstrated over the lifetime of, for example, mice, the chip arrays will enable researchers to determine the rate of aging in genetically or physiologically altered animals.

Genome analysis may also identify human polymorphisms that are associated with longevity. So far, there is some evidence for such a link for alleles of *APO-E* (apolipoprotein-E), *ACE* (angiotensinogen-converting enzyme), *HLA-DR* (human leukocyte antigen-DR), and *PAI-1* (plasminogen activator inhibitor) (reviewed in Schachter, 1998). Interestingly, an allele of *WRN* may be associated with a reduced risk of myocardial infarction (Ye et al., 1997). As technologies for rapid assessment of polymorphisms develop, genomic analysis of centenarians may provide a comprehensive list of relevant polymorphic loci.

If only a few key processes are critical in human aging, there may be specific molecular targets that are amenable to therapeutic intervention. One can therefore imagine pharmacological means to slow or even stop aging in one or more organ systems. The fact that adult somatic cells have been used to clone whole animals suggests that aging-related nuclear changes may be reversible (Wilmut et al., 1997; Wakayama et al., 1998). However, it is not clear yet whether a full life span has been given to the clones. Moreover, the small fraction of somatic cells that yield clones may represent rare soma that have escaped aging-related cellular events.

Is it advisable to intervene in human aging? Over the past two centuries, human life expectancy has more than doubled in developed nations due to advances in medicine and public health. Assuming that future intervention will not affect causes of mortality that are independent of the aging process, average life span may be impacted, but probably to a smaller degree than what has occurred in the past 200 years. However, slowing the aging process may increase vitality and quality of life over the entire life span of individuals. In this regard, it is noteworthy that calorically restricted rodents have an extended life span that is relatively free of disease.

For society, the implications of slowing the aging process are more complex. Of course, in an increasingly overpopulated world, it would be important to offset any significant effects on longevity with a compensatory reduction in birth rates. In fact, in many industrialized countries, the current birth rate is sufficiently low to afford zero or negative growth. Most importantly, if the slowing of aging is associated with improved health and productivity of long-lived individuals, there may be a direct benefit to society. In an information-intensive era when societal progress increasingly depends on the knowledge and wisdom of individuals, this benefit should be at a premium.

Acknowledgments

We thank J. Wei, W. Thilly, and H. Tissenbaum, D. McNabb, S. Imai and R. Marciniak, and the other members of the Guarente lab for discussions and comments on the manuscript. Work from the Guarente laboratory is supported in part by National Institutes of Health grant R01 AG1119 from the National Institute on Aging. F. B. J. is supported by a Howard Hughes Medical Institute postdoctoral fellowship, and D. A. S. is supported by a Helen Hay Whitney postdoctoral fellowship.

References

Apfeld, J., and Kenyon, C. (1998). Cell nonautonomy of *C. elegans daf-2* function in the regulation of diapause and life span. Cell *95*, 199–210.

Barton, A.A. (1950). Some aspects of cell division in *Saccharomyces cerevisiae*. J. Gen. Microbiol. *4*, 84–86.

Blasco, M.A., Lee, H.W., Hande, M.P., Samper, E, Lansdorp, P.M., DePinho, R.A., and Grieder, C.W. (1997). Telomere shortening and tumor formation by mouse cells lacking telomerase RNA. Cell *91*, 25–34.

Bodnar, A., Ouellette, M., Frolkis, M., Holt, S., Chui, C.-P., Morin, G., Harley, C., Shay, J., Lichsteiner, S., and Wright, W. (1998). Extension of life-span by introduction of telomerase into normal human cells. Science *279*, 349–352.

Boffoli, D., Sacco, S.C., Vergari, R., Solarino, G., Santacroce, G., and Papa, S. (1994). Decline with age of the repiratory chain activity in human skeletal muscle. Biochim. Biophys. Acta *1226*, 73–82.

Bohr, V.A., Dianov, G., Balajee, A., May, A., and Orren, D.K. (1998). DNA repair and transcription in human premature aging disorders. J. Invest. Dermatol. Symp. Proc. *3*, 11–13.

Brierley, E.J., Johnson, M.A., Lightowlers, R.N., James, O.F.W., and Turnbull, D.M. (1998). Role of mitochondrial DNA mutations in human aging: implications for the central nervous system and muscle. Ann. Neurol. *43*, 217–223.

Brown, W.T. (1992). Progeria: a human-disease model of accelerated aging. Am. J. Clin. Nutr. *55*, 1222S-1224S.

Campisi, J. (1998). The role of cellular senescence in skin aging. J. Invest. Dermatol. Symp. Proc. *3*, 1–5.

Chang, E., and Harley, C.B. (1995). Telomere length and replicative aging in human vascular tissues. Proc. Natl. Acad. Sci. USA *92*, 11190–11194.

Cristofalo, V.J., Allen, R.G., Pignolo, R.J., Martin, B.G., and Beck, J.C. (1998). Relationship between donor age and the replicative life span of human cells in culture: a reevaluation. Proc. Natl. Acad. Sci. USA *95*, 10614–10619.

Dimri, G., Lee, X., Baile, G., Acosta, M., Scott, G., Roskelley, C.,

Medrano, E.E., Linskens, M., Rubelj, I., Pereira-Smith, O., et al. (1995). A biomarker that identifies senescent human cells in culture and in aging skin iv vivo. Proc. Natl. Acad. Sci. USA *92*, 9363–9367.

Dolle, M.E.T., Giese, H., Hopkins, C.L., Martus, H.J., Hausdorff, J.M., and Vijg, J. (1997). Rapid accumulation of genome rearrangements in liver but not in brain of old mice. Nat. Genet. *17*, 431–434.

Effros, R.B. (1998). Replicative senescence in the immune system: impact of the Hayflick limit on T-cell function in the elderly. Am. J. Hum. Genet. *62*, 1003–1007.

Ellis, N.A. (1997). DNA helicases in inherited human disorders. Curr. Opin. Genet. Dev. *7*, 354–363.

Ellis, N.A., Groden, J., Ye, T.Z., Straughen, J., Lennon, D.J., Ciocci, S., Proytcheva, M., and German, J. (1995). The Bloom's syndrome gene product is homologous to RecQ helicases. Cell *83*, 655–666.

Epstein, C.J., Martin, G.M., Schultz, A.L., and Motulsky, A.R. (1966). Werner syndrome: a review of its symptomatology, natural history, pathologic features, genetics and relationship to the natural aging process. Medicine *45*, 177–221.

Eriksson, P.S., Perfilieva, E., Bjork-Eriksson, T., Albom, A.M., Nordborg, C., Peterson, D.A., and Gage, F.H. (1998). Neurogenesis in the adult human hippocampus. Nat. Med. *4*, 1313–1317.

Ewbank, J.J., Barnes, T.M., Lakowski, B., Lussier, M., Bussey, H., and Hekimi, S. (1997). Structural and functional conservation of the *Caenorhabditis elegans* timing gene *clk-1*. Science *275*, 980–983.

Faragher, R.G., Kill, I.R., Hunter, J.A., Pope, F.M., Tannock, C., and Shall, S. (1993). The gene responsible for Werner syndrome may be a cell division "counting" gene. Proc. Natl. Acad. Sci. USA *90*, 12030–12034.

Finch, C. (1990). Longevity, Senescence, and the Genome. (Chicago: University of Chicago Press).

Friedman, D.B., and Johnson, T.E. (1988). A mutation in the *age-1* gene in *Caenorhabditis elegans* lengthens life and reduces hermaphrodite fertility. Genetics *118*, 75–86.

Fry, M., and Loeb, L.A. (1998). The three faces of the WS helicase. Nat. Genet. *19*, 308–309.

Gangloff, S., McDonald, J.P., Bendixen, C., Artur, L., and Rothstein, R. (1994). The yeast type 1 topoisomerase top3 interacts with Sgs1, a DNA helicase homologue: a potential eukaryotic reverse gyrase. Mol. Cell. Biol. *14*, 8391–8398.

Gaubatz, J.W. (1990). Extrachromosomal circular DNAs and genomic sequence plasticity in eukaryotic cells. Mutat. Res. *237*, 271–292.

Goto, M. (1997). Hierarchical deterioration of body systems in Werners syndrome: implications for normal ageing. Mech. Ageing Dev. *98*, 239–254.

Gottlieb, S., and Ruvkun, G. (1994). *daf-2, daf-16* and *daf-23*: genetically interacting genes controlling Dauer formation in *Caenorhabditis elegans*. Genetics *137*, 107–120.

Grasl-Kraupp, B., Bursch, W., Ruttkay-Nedecky, B., Wagner, A., Lauer, B., and Schulte-Hermann. (1994). Food restriction eliminates preneoplastic cells through apoptosis and antagonizes carcinogenesis in rat liver. Proc. Natl. Acad. Sci. USA *91*, 9995–9999.

Gray, M.D., Shen, J.C., Kamath-Loeb, A.S., Blank, A., Sopher, B.L., Martin, G.M., Oshima, J., and Loeb, L.A. (1997). The Werner syndrome protein is a DNA helicase. Nat. Genet. *17*, 100–103.

Gray, M.D., Wang, L., Youssoufian, H., Martin, G.M., and Oshima, J. (1998). Werner helicase is localized to transcriptionally active nucleoli of cycling cells. Exp. Cell Res. *242*, 487–494.

Green, D.R., and Reed, J.C. (1998). Mitochondria and apoptosis. Science *281*, 1309–1312.

Grist, S.A., McCarron, M., Kutlaca, A., Turner, D.R., and Morley, A.A. (1992). In vivo human somatic mutation: frequency and spectrum with age. Mutat. Res. *266*, 189–196.

Grodstein, F., Stampfer, M.J., Colditz, G.A., Willett, W.C., Manson, J.E., Joffe, M., Rosner, B., Fuchs, C., Hankinson, S.E., Hunter, D.J., Hennekens, C.H., and Speizer, F.E. (1997). Postmenopausal therapy and mortality. N. Engl. J. Med. *336*, 1769–1775.

Harley, C.B., Futcher, A.B., and Greider, C.W. (1990). Telomeres shorten during ageing of human fibroblasts. Nature *345*, 458–460.

Harman, D. (1981). The aging process. Proc. Natl. Acad. Sci. USA *78*, 7124–7128.

Hayes, A., and Batshaw, M.L. (1993). Down syndrome. Pediatr. Clin. North Am. *40*, 523–535.

Hengartner, M.O. (1997). Genetic control of programmed cell death and aging in the nematode *Caenorhabditis elegans*. Exp. Gerontol. *32*, 363–374.

Ho, Y.S., Magnenat, J.L., Bronson, R.T., Cao, J., Gargano, M., Sugawara, M., and Funk, C.D. (1997). Mice deficient in cellular glutathione peroxidase develop normally and show no increased sensitivity to hyperoxia. J. Biol. Chem. *272*, 16644–16651.

Jazwinski, S.M. (1993). The genetics of aging in the yeast *Saccharomyces cerevisiae*. Genetica *91*, 35–53.

Jiang, X., Jimenez, G., Chang, E., Frolkis, M., Kusler, B., Sage, M., Beeche, M., Bodnar, A.G., Wahl, G.M., Tlsty, T.D., and Chiu, C. (1999). Telomerase expression in human somatic cells does not induce changes associated with a transformed phenotype. Nat. Genet. *21*, 111–114.

Jonassen, T., Proft, M., Randez-Gil, F., Schultz, J.R., Marbois, B.N., Entian, K.D., and Clarke, C.F. (1998). Yeast Clk-1 homologue (Coq7/Cat5) is a mitochondrial protein in coenzyme Q synthesis. J. Biol. Chem. *6*, 3351–3357.

Kajstura, J., Cheng, W., Ranganathan, S., Li, P., Baoshen, L., Nitahara, J.A., Chapnick, S., Reiss, K., Olivetti, G., and Anversa, P. (1996). Necrotic and apoptotic myocyte cell death in the aging heart of Fischer 344 rats. Am. J. Physiol. *271*, H1215–H1228.

Kajstura, J., Leri, A., Finato, N., Loreto, C., Beltrami, C.A., and Anversa, P. (1998). Myocyte proliferation in end-stage cardiac failure in humans. Proc. Natl. Acad. Sci. USA. *95*, 8801–8805.

Kennedy, B.K., Austriaco, N.R., Jr., Zhang, J., and Guarente, L. (1995). Mutation in the silencing gene *SIR4* can delay aging in *S. cerevisiae*. Cell *80*, 485–496.

Kennedy, B.K., Gotta, M., Sinclair, D.A., Mills, K., McNabb, D.S., Murthy, M., Pak, S.M., Laroche, T., Gasser, S.M., and Guarente, L. (1997). Redistribution of silencing proteins from telomeres to the nucleolus is associated with extension of life span in *S. cerevisiae*. Cell *89*, 381–391.

Kenyon, C., Chang, J., Gensch, E., Rudner, A., and Tabtiang, R. (1993). A *C. elegans* mutant that lives twice as long as wild type. Nature *366*, 461–464.

Kimura, K.D., Tissenbaum, H.A., Liu, Y., and Ruvkun, G. (1997). *daf-2*, an insulin receptor-like gene that regulates longevity and diapause in *Caenorhabditis elegans*. Science *277*, 942–946.

King, C.M., Gillespie, E.S., McKenna, P.G., and Barnet, Y.A. (1994). An investigation of mutation as a function of age in humans. Mutat. Res. *6*, 79–90.

Kirkwood, T.B.L. (1977). Evolution of aging. Nature *270*, 301–304.

Kiyono, T., Foster, S.A., Koop, J.I., McDougall, J.K., Galloway, D.A., and Klingelhutz, A.J. (1998). Both Rb/$p16^{INK4a}$ inactivation and telomerase activity are required to immortalize human epithelial cells. Nature *396*, 84–88.

Klass, M.R. (1983). A method for the isolation of longevity mutants in the nematode *Caenorhabditis elegans* and initial results. Mech. Ageing Dev. *22*, 279–286.

Kopsidas, G., Kovalenko, S.A., Kelso, J.M., and Linnane, A.W. (1998). An age-associated correlation between cellular bioenergy decline and mtDNA rearrangements in human skeletal muscle. Mutat. Res. *421*, 27–36.

Kuro-o, M., Matsumura, Y., Aizawa, H., Kawaguchi, H., Suga, T., Utsugi, T., Ohyama, Y., Kurabayashi, M., Kaname, T., Kume, E., et al. (1998). Mutation of the mouse *klotho* gene leads to a syndrome resembling ageing. Nature *390*, 45–51.

Lamberts, S.W., van den Beld, A.W., and van der Lely, A.J. (1997). The endocrinology of aging. Science *278*, 419–424.

Larsen, P.L. (1993). Aging and resistance to oxidative damage in *Caenorhabditis elegans*. Proc. Natl. Acad. Sci. USA *90*, 8905–8909.

Larsen, P.L., Albert, P.S., and Riddle, D.L. (1995). Genes that regulate both development and longevity in *Caenorhabditis elegans*. Genetics *139*, 1567–1583.

Lee, C.M., Weindruch, R., and Aiken, J.M. (1997). Age-associated alteration of the mitochondrial genome. Free Radic. Biol. Med. *22*, 1259–1269.

Lee, H.W., Blasco, M.A., Gottlieb, G.J., Horner, J.W., Greider, C.W., and DePinho, R.A. (1998). Essential role of mouse telomerase in highly proliferative organs. Nature *392*, 569–574.

Lin, K., Dorman, J.B., Rodan, A., and Kenyon, C. (1997). *daf-16*: an HNF-3/forkhead family member that can function to double the life-span of *Caenorhabditis elegans*. Science *278*, 1319–1322.

Lin, Y.J., Seroude, L., and Benzer, S. (1998). Extended life-span and stress resistance in the *Drosophila* mutant *methuselah*. Science *282*, 943–946.

Lithgow, G.J., White, T.M., Hinerfeld, D.A., and Johnson, T.E. (1994). Thermotolerance of a long-lived mutant of *Caenorhabditis elegans*. J. Gerontol. *49*, B270–B276.

Marciniak, R.M., Lombard, D.B., Johnson, F.B., and Guarente, L. (1998). Nucleolar localization of the Werner syndrome protein in human cells. Proc. Natl. Acad. Sci. USA *95*, 6887–6892.

Martin, G.M. (1978). Genetic syndromes in man with potential relevance to the pathobiology of aging. Birth Defects Orig. Artic. Ser. *14*, 5–39.

Martin, G.M., Sprague, C.A., and Epstein, C.J. (1970). Replicative life-span of cultivated human cells. Effects of donor's age, tissue, and genotype. Lab. Invest. *23*, 86–92.

Martin, G.M., Smith, A.C., Ketterer, D.J., Ogburn, C.E., and Disteche, C.M. (1985). Increased chromosomal aberrations in first metaphases of cells isolated from the kidneys of aged mice. Isr. J. Med. Sci. *21*, 296–301.

Medawar, P.B. (1952). An Unsolved Problem in Biology (London: H.K. Lewis).

Melov, S., Lithgow, G.J., Fischer, D.R., Tedesco, P.M., and Johnson, T.E. (1995a). Increased frequency of deletions in the mitochondrial genome with age in *Caenorhabditis elegans*. Nucleic Acids Res. *23*, 1419–1425.

Melov, S., Shoffner, J.M., Kaufman, A., and Wallace, D.C. (1995b). Marked increase in the number and variety of mitchondrial DNA rearrangements in aging human skeletal muscle. Nucleic Acids Res. *23*, 4122–4126.

Melov, S., Schneider, J.A., Day, B.J., Hinerfeld, D., Coskunra, S.S., Crapo, J.D., and Wallace, D.C. (1998). A novel neurological phenotype in mice lacking mitochondrial manganese superoxide dismutase. Nat. Genet. *18*, 59–63.

Morales, C.P., Holt, S.E., Ouellette, M., Kaur, K.J., Yan, Y., Wilson, K.S., White, M.A., Wright, W.E., and Shay, J.W. (1999). Absence of cancer-associated changes in human fibroblasts immortalized with telomerase. Nat. Genet. *21*, 115–118.

Morris, J.Z., Tissenbaum, H.A., and Ruvkun, G. (1996). A phosphatidylinositol-3-OH kinase family member regulating longevity and diapause in *Caenorhabditis elegans*. Nature *382*, 536–539.

Morrison, J.H., and Hof, P.J. (1997). Life and death of neurons in the aging brain. Science *278*, 412–419.

Muller-Hocker, J., Seibel, P., Schneiderbanger, K., and Kadenbach, B. (1993). Different in situ hybridization patterns of mitochondrial DNA in cytochrome c oxidase-deficient extraocular muscle fibers in the elderly. Virchows Arch. A. Pathol. Anat. *422*, 7–15.

Murakami, S., and Johnson, T.E. (1996). A genetic pathway conferring life extension and resistance to UV stress in *Caenorhabditis elegans*. Genetics *143*, 1207–1218.

Muskhelishvili, L., Hart, R.W., Turturro, A., and James, S.J. (1995). Age-related changes in the intrinsic rate of apoptosis in livers of diet-restricted and ad libitum-fed B6C3F1 mice. Am. J. Pathol. *147*, 20–24.

Ogg, S., and Ruvkun, G. (1998). The *C. elegans* PTEN homolog, DAF-18, acts in the insulin receptor-like metabolic signaling pathway. Mol. Cell *2*, 887–893.

Ogg, S., Paradis, S., Gottlieb, S., Patterson, G.I., Lee, L., Tissenbaum, H.A., and Ruvkun, G. (1997). The Fork head transcription factor DAF-16 transduces insulin-like metabolic and longevity signals in *C. elegans*. Nature *389*, 994–999.

Olivetti, G., Giordana, M., Corradi, D., Melissari, M., Lagrasta, C., Gambert, S.R., and Anversa, P. (1995). Gender differences and aging: effects on the human heart. J. Am. Coll. Cardiol. *26*, 1068–1079.

Orr, W.C., and Sohal, R.S. (1994). Extension of life-span by overexpression of superoxide dismutase and catalase in *Drosophila melanogaster*. Science *263*, 1128–1130.

Oshima, J., Campisi, J., Tannock, T.C., and Martin, G.M. (1995). Regulation of c-fos expression in senescing Werner syndrome fibroblasts differs from that observed in senescing fibroblasts from normal donors. Cell. Physiol. *162*, 277–283.

Pakkenberg, B., and Gundersen, H.J.G. (1997). Neocortical neuron number in humans: effect of sex and age. J. Comp. Neurol. *384*, 312–320.

Parkes, T.L., Elia, A.J., Dickinson, D., Hilliker, A.J., Phillips, J.P., and Boulianne, G.L. (1998). Extension of *Drosophila* lifespan by overexpression of human SOD1 in motorneurons. Nat. Genet. *19*, 171–174.

Promislow, D.E.L. (1993). On size and survival: progress and pitfalls in the allometry of life span. J. Gerontol. *48*, B115–B123.

Puranam, K.L., and Blackshear, P.J. (1994). Cloning and characterization of RECQL, a potential human homologue of the *Escherischia coli* DNA helicase RecQ. J. Biol. Chem. *269*, 29838–29845.

Reaume, A.G., Elliott, J.L., Hoffman, E.K., Kowall, N.W., Ferrante, R.J., Siwek, D.F., Wilcox, H.M., Flood, D.G., Beal, M.F., Brown, R.H., Jr., et al. (1996). Motor neurons in Cu/Zn superoxide dismutase-deficient mice develop normally but exhibit enhanced cell death after axonal injury. Nat. Genet. *13*, 43–47.

Robbins, E., Levine, E.M., and Eagle, H. (1970). Morphologic changes accompanying senescence of cultured human diploid cells. J. Exp. Med. *131*, 1211–1222.

Rohme, D. (1981). Evidence for a relationship between longevity of mammalian species and life spans of normal fibroblasts in vitro and erythrocytes in vivo. Proc. Natl. Acad. Sci. USA *78*, 5009–5013.

Rotman, G., and Shilo, Y. (1998). ATM: from gene to function. Hum. Mol. Genet. *7*, 1555–1563.

Rudman, D., Feller, A.G., Nagraj, H.S., Gergans, G.A., Lalitha, P.Y., Goldberg, A.F., Schlenker, R.A., Cohn, L., Rudman, I.W., and Mattson, D.E. (1990). Effects of human growth hormone in men over 60 years old. N. Engl. J. Med. *323*, 1–6.

Schachter, F. (1998). Causes, effects, and constraints in the genetics of human longevity. Am. J. Hum. Genet. *62*, 1008–1014.

Schulz, V.P., Zakian, V.A., Ogburn, C.E., McKay, J., Jarzebowicz, A.A., Edland, S.D., and Martin, G.M. (1996). Accelerated loss of telomeric repeats may not explain accelerated replicative decline of Werner syndrome cells. Hum. Genet. *97*, 750–754.

Schwarze, S.R., Lee, C.M., Chung, S.S., Roecker, E.B., Weindruch, R., and Aiken, J.M. (1995). High levels of mitochondrial DNA deletions in skeletal muscle of old rhesus monkeys. Mech. Ageing Dev. *83*, 91–101.

Sciacco, M., Bonilla, E., Schon, E.A., DiMauro, S., and Moraes, C.T. (1994). Distribution of wild-type and common deletion forms of mtDNA in normal and respiration-deficient muscle fibers from patients with mitochondrial myopathy. Hum. Mol. Genet. *3*, 13–19.

Sinclair, D.A., and Guarente, L. (1997). Extrachromosomal rDNA circles—a cause of aging in yeast. Cell *91*, 1033–1042.

Sinclair, D.A., Mills, K., and Guarente, L. (1997). Accelerated aging and nucleolar fragmentation in yeast sgs1 mutants. Science *277*, 1313–1316.

Smeal, T., Claus, J., Kennedy, B., Cole, F., and Guarente, L. (1996). Loss of transcriptional silencing causes sterility in old mother cells of *S. cerevisiae*. Cell *84*, 633–642.

Sohal, R.S., Ku, H.H., Agarwal, S., Forster, M.J., and Lal, H. (1994). Oxidative damage, mitochondrial oxidant generation and antioxidant defenses during aging and in response to food restriction in the mouse. Mech. Ageing Dev. *74*, 121–133.

Stewart, E., Chapman, C.R., Al-Khodairy, F., Carr, A.M., and Enoch, T. (1997). *rqh1*$^+$, a fission yeast gene related to the Blooms and Werners syndrome genes, is required for reversible S phase arrest. EMBO J. *16*, 2682–2692.

Strehler, B.L. (1986). Genetic instability as the primary cause of human aging. Exp. Gerontol. *21*, 283–319.

Sun, H., Jarow, J.K., Hickson, I.D., and Mazels, N. (1998). The Blooms syndrome helicase unwinds G4 DNA. J. Biol. Chem. *273*, 27587–27592.

Suzuki, N., Shimamoto, A., Imamura, O., Kuromitsu, J., Kitao, S., Goto, M., and Furuichi, Y. (1997). DNA helicase activity in Werner's syndrome gene product synthesized in a baculovirus system. Nucleic Acids Res. *25*, 2973–2978.

Szilard, L. (1959). On the nature of the aging process. Proc. Natl. Acad. Sci. USA *45*, 30–45.

Takeda, K., Gosiewska, A., and Peterkofsky, B. (1992). Similar, but not identical, modulation of extracellular matrix components during in vitro and in vivo aging of human skin fibroblasts. J. Cell. Phys. *153*, 450–459.

Tissenbaum, H.A., and Ruvkun, G. (1998). An insulin-like signaling pathway affects both longevity and reproduction in *Caenorhabditis elegans*. Genetics *148*, 703–717.

Toda, T., Satoh, M., Sugimoto, M., Goto, M., Furuichi, Y., and Kimura, N. (1998). A comparative analysis of the proteins between the fibroblasts from Werner's syndrome patients and age-matched normal individuals using two-dimensional gel electrophoresis. Mech. Aging Dev. *100*, 133–143.

Vanfleteren, J.R., and De Vreese, A. (1996). Rate of aerobic metabolism and superoxide production rate potential in the nematode *Caenorhabditis elegans*. J. Exp. Zool. *274*, 93–100.

Vaziri, H., and Benchimol, S. (1998). Reconstitution of telomerase activity in normal human cells leads to elongation of telomeres and extended replicative life span. Curr. Biol. *26*, 279–282.

Veis, D.J., Sorenson, C.M., Shutter, J.R., and Korsmeyer, S.J. (1993). *bcl-2*-deficient mice demonstrate fulminant lymphoid apoptosis, polycystic kineys and hypopigmented hair. Cell *75*, 229–240.

Wakayama, T., Perry, A.C.F., Zuccotti, M., Johnson, K.R., and Yanagimachi, R. (1998). Full-term development of mice from enucleated oocytes injected with cumulus cell nuclei. Nature *394*, 369–374.

Wallace, D.C., Brown, M.D., Melov, S., Graham, B., and Lott, M. (1998). Mitochondrial biology, degenerative diseases and aging. Biofactors *7*, 187–190.

Wang, E. (1995). Senescent human fibroblasts resist programmed cell death, and failure to suppress *bcl2* is involved. Cancer Res. *55*, 2284–2292.

Watt, P.M., Louis, E.J., Borts, R.H., and Hickson, I.D. (1995). Sgs1: a eukaryotic homolog of *E. coli* RecQ that interacts with topoisomerase II in vivo and is required for faithful chromosome segregation. Cell *81*, 253–260.

Watt, P.M., Hickson, I.D., Borts, R.H., and Louis, E.J. (1996). *SGS1*, a homologue of the Blooms and Werners syndrome genes, is required for maintenance of genome stability in *Saccharomyces cerevisiae*. Genetics *144*, 935–945.

Warner, H.R., Hodes, R.J., and Pocinki, K. (1997). What does cell death have to do with aging? J. Am. Geriatr. Soc. *45*, 1140–1146.

Weindruch, R. (1996). The retardation of aging by caloric restriction: studies in rodents and primates. Toxicol. Pathol. *24*, 742–745.

Weindruch, R., and Sohal, R.S. (1997). Caloric intake and aging. N. Engl. J. Med. *337*, 986–994.

Wilmut, I., Schnieke, A.E., McWhir, J., Kind, A.J., and Campbell, K.H.S. (1997). Viable offspring derived from fetal and adult mamalian cells. Nature *385*, 810–813.

Wong, A., Boutis, P., and Hekimi, S. (1995). Mutations in the *clk-1* gene of *Caenorhabditis elegans* affect developmental and behavioral timing. Genetics *139*, 1247–1259.

Yan, H., Chen, C., Kobayashi, R., and Newport, J. (1998). Replication focus-forming activity 1 and the Werner syndrome gene product. Nat. Genet. *19*, 375–378.

Ye, L., Miki, T., Nakura, J., Oshima, J., Kamino, K., Rakugi, H., Ikegami, H., Higaki, J., Martin, G.M., and Ogihara, T. (1997). Association of a polymorphic variant of the Werner helicase gene with myocarial infarction in a Japanese population. Am. J. Med. Genet. *11*, 494–498.

Yu, C.E., Oshima, J., Fu, Y.H., Wijsman, E.M., Hisama, F., Alisch, R., Matthews, S., Nakura, J., Miki, T., Ouais, S., Martin, G.M., Mulligan, J., and Schellenberg, G.D. (1996). Positional cloning of the Werner's syndrome gene. Science *272*, 258–262.

Molecular biology's impact on our understanding of aging

David M A Mann

Department of Pathological Sciences, University of Manchester, Manchester M13 9PT
David M A Mann, reader

BMJ 1997;315:1078-81

Powerful molecular biological tools have begun to open up the very fabric of life—the human genome—and have allowed us to glimpse inside this Pandora's box. We now see that many common disorders of later life—for example, cancer, dementia, and vascular disease—are related to genetic variations that dictate an individual's likelihood of developing illnesses like these. These genetic variations differ from those that determine longevity, though both act synergistically to dictate how long and how well we might live. Control of gene expression will be needed to counteract the adverse actions of these to promote a healthy and productive old age.

Summary points

Aging and the disorders of later life are separate entities

Both are under genetic control

Aging involves defects in mitochondrial DNA which promote oxidative stress mediated cell damage

Age related disorders such as Alzheimer's disease are due to the effect of inherited genetic risk factors

Life span is determined by the effects of genetic risks associated with age and disease

Aging and disease—separate entities or continuum of change?

Common disorders such as cardiovascular disease, cancer, stroke, dementia, and diabetes become increasingly prevalent in later life. It is tempting to ascribe these to the body "wearing out"—a viewpoint supported by the frequent finding in many old, but otherwise healthy, people of low levels of the same kind of tissue changes generally associated with certain diseases when present in higher amounts. These diseases have been popularly equated with "normal aging," and the idea that in diseased individuals this normal process of aging may have become "exaggerated" or have "accelerated" out of control has often been put forward.

Health and disease might in this way be thought of as occupying a sliding scale of age determined tissue damage, with the one merging into the other at some point in life. Yet the argument is fallacious. Many disorders—for example, dementia—clearly become more common in later life; their incidence peaks in people in their eighties but then declines.[1] Furthermore, like humans, other animals (including the higher primates) age and die, yet they do not spontaneously develop these common disorders of humans. Hence, although growing older is a biological certitude, disease in old age may represent an additional burden of tissue damage superimposed on other ongoing alterations common perhaps to all cell types in the body and applicable, to a greater or lesser extent, to all individuals. These basic changes set the stage on which the disorders of old age can be played out. So what is aging, and what is disease? How important is each of these in determining how each of us will fare in later life? What causes aging and disease, and to what extent are they interdependent?

Molecular revolution

The recent molecular biological revolution has begun to make startling inroads into these areas of uncertainty, particularly in molecular genetics. Variations in the structure or stability of the human genome are now seen to be responsible for an ever growing list of common and not so common disorders by determining not only the probability of whether sufficient tissue damage causing disease will occur but also at what time in life this can be expected and for how long it will last. Moreover, these techniques are now providing the quantitative data necessary to clarify ideas on aging and disease that have been around for many years, and the boundaries are thus being established between which tissue damage might be thought of as being caused by the passage of time alone and which is a reflection of degenerative disease in later life.

This revolution was sparked by the advent of the polymerase chain reaction[2]—a means of replicating "in the test tube" a small number of copies, or even a single copy, of a segment of DNA (gene) to produce multiple, identical versions whose nucleotide sequence can be determined. This forms the base of many complex molecular analyses that, for example, assign biological characteristics, such as disease trait, to chromosomes or parts thereof (by linkage analysis) or permit the "narrowing down" of genes within that region to the one causing disease (by positional cloning). Once identified, screening techniques based on polymerase chain reaction can quickly and efficiently determine how many individuals with the disease share that particular genetic change, and DNA sequencing can show whether other disease-causing base changes or rearrangements of the DNA are present (fig 1).

In this way different types of mutation have been identified in all kinds of tissues and in many clinical situations (including aging), ranging from "simple" exchanges of nucleotide bases that alter the coding sequence for one or more amino acids (missense mutations) through to complex additions or expansions (insertions) or subtractions (deletions) of genetic material. Such changes, when transcribed into RNA and subsequently translated, produce a protein of abnormal sequence which creates structural or functional changes that render it defective at its normal task (dysfunctional) or confer a new function (gain of dysfunction) away from normality. These altered proteins may compromise the structural integrity or the metabolic or replicative capabilities of the cell in such a way as to trigger

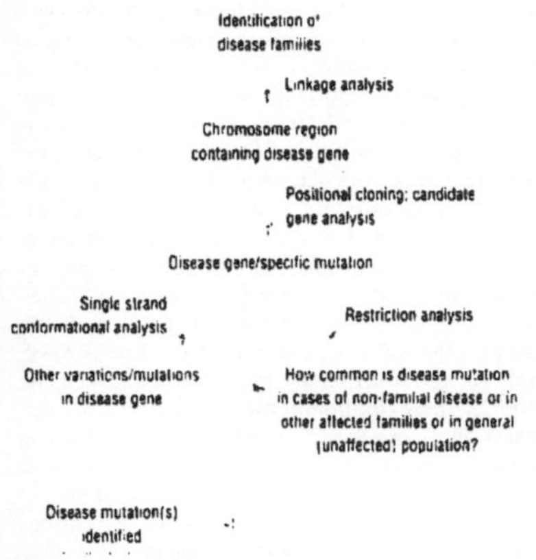

Fig 1 Molecular approaches to gene identification

allele of the apolipoprotein E gene, have been associated with Alzheimer's disease.[5] These polymorphic variations do not absolutely dictate whether disease will occur but cause a metabolic deficiency that increases the likelihood that the tissue damage of Alzheimer's disease will occur and thereby raise the level of susceptibility to clinical disease. Possession of one gene copy of the e4 allele doubles the risk of developing Alzheimer's disease: two copies increase the risk eight times (to about 90% certainty).[5] These same genetic factors are, however, associated with other disorders that share many of the tissue characteristics of Alzheimer's disease yet produce quite different and distinctive clinical syndromes. For example, in Down's syndrome the pathology is that of Alzheimer's disease,[6] but it is an extra (normal) copy of the amyloid precursor protein gene on the duplicated chromosome 21 that is causative.[7] In hereditary cerebral haemorrhage with amyloidosis other mutations in this gene are to blame.[8] Cortical Lewy body disease shares the same change in the apolipoprotein E e4 allele as Alzheimer's disease.[9] Hence, genetic changes like these confer metabolic alterations that in turn prescribe a particular pathology—β amyloid cascades that can eventually be recognised under the microscope as a tissue abnormality.

The nervous system as ideal model

The nervous system has been a traditional target for studies on aging as its principal cells, the neurones, do not divide nor are replaced routinely during adult life, as happens with cells of other tissues. They must thus endure a lifelong catalogue of insults to secure continued survival of both themselves and the individual. The nervous system therefore makes an ideal model in which to study aging and to differentiate from this the changes that relate to the common diseases of old age. Perhaps the best illustration of this conflict is Alzheimer's disease. For years the argument has raged as to whether Alzheimer's disease is a disease or represents the ultimate, or untimely, exaggerated effects of the process of aging.

Genetics of Alzheimer's disease

In the past decade the perception of Alzheimer's disease as a single entity with a characteristic clinical and pathological profile has changed into a realisation that this represents a broad neurodegenerative cascade to which entry can be gained through various routes (fig 2). Some cases of Alzheimer's disease involve the inheritance of mutations in particular genes (amyloid precursor protein gene[3] and the presenilin genes[4]) in a classic mendelian manner; these mutations are so lethal that disease will virtually always occur at some time, usually well before old age. Yet such mutational events account for only about 70% of all instances where inherited Alzheimer's disease occurs before the age of 60 and contribute nothing towards that greater mass of cases where onset is after this age or to those cases in earlier life where no clear pattern of inheritance is seen. Hence, other and more common genetic events known as polymorphisms or "natural" genetic variations, particularly the possession of the e4

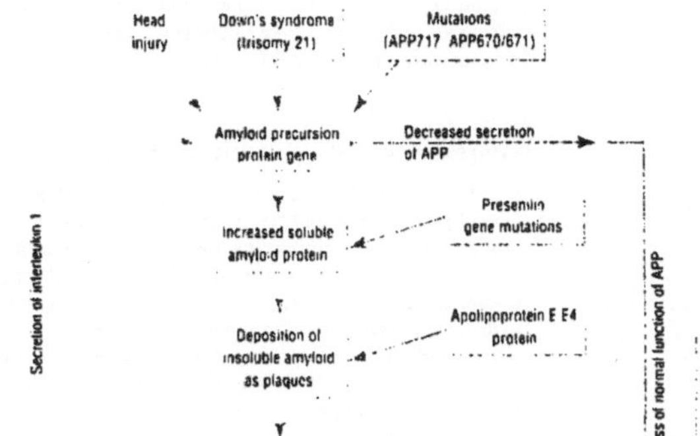

Fig 2 Pathogenesis of Alzheimer's disease. Mutations in amyloid precursor protein (APP) gene and presenilin genes increase production of soluble amyloid β protein. This deposits as insoluble plaques when concentration of amyloid in extracellular fluid is high. Apolipoprotein E E4 protein favours deposition by lowering the concentration threshold to plaque formation. Plaques are infiltrated by microglial cells, which secrete interleukin-1, increasing production of amyloid β protein and soluble amyloid and stimulating astrocytosis. Reaction oxygen species from microglia and astrocytes damage nerve endings, causing neuritic changes in plaques and the formation of neurofibrillary tangles in nerve cell cytoplasm. Damaged neurones die, causing clinical disability—dementia. Loss of normal secretion of APP (due to breakdown into amyloid) may also stimulate neuritic changes and hasten nerve cell death

deposition in this instance.[10] It is the extent and distribution of this pathology that will dictate the clinical profile that will ultimately emerge. Tissue changes associated with particular disorders are likely to be the biological products of adverse genetic variations. Modelling these genetic changes of Alzheimer's disease in animals (transgenic mice)[11-13] or in cell lines[14-16] has produced the same cellular metabolic changes and tissue alterations that characterise the human condition.

Thus Alzheimer's disease is a complex, multifactorial process that is essentially under genetic control and represents an interplay between various adverse genetic changes that have an impact on the pathological process in different ways. The nature and balance of these factors determine whether disease will occur, and if so, at what time of life and how rapidly progression will be. Most people will possess one or more of these factors and can thus expect to accumulate some of the pathology of Alzheimer's disease[17-19] in their brains given time—this risk is unlikely to be sufficient in most individuals to ensure progress into clinical disease, but it is none the less probable for many. Other pathologies seen in the brain in old age, such as Lewy bodies, can be equated with Parkinson's disease (clinical or subclinical)[20]; indeed the entire spectrum of neurodegenerative disorders is likely to be genetically determined. Pathologies such as plaques, tangles, and Lewy bodies should be taken as indicating nervous system disease whenever or wherever they might occur and are quite distinct from the changes of "normal aging."

Genetic changes in normal aging

So what is normal aging? Does this represent the good face of genes whose obverse side promotes pathology and disease in later life? For example, in contrast to the e4 allele, the apolipoprotein E e2 allele promotes longevity in both normal individuals[21] and those with Down's syndrome[22]; it delays the age of onset of Alzheimer's disease.[23 24] Although this is possible in some specific instances it is in general terms unlikely. Health promoting genes will fulfil housekeeping roles, maintaining the faithful production of proteins that sustain cell viability and resistance to damage; they may form part of a genetic cluster that specifies longevity in a programmed manner. Gradual failures in these cell maintenance systems may act as a springboard on which adverse genetic changes promoting pathology might gain momentum and take their toll in later life. But what are these basic changes that potentially affect all of us, whose actions progressively build up, thereby rendering the system so vulnerable in old age?

The core physiological process supporting the life of all cells is the oxidative metabolism of glucose in the mitochondrion to provide transducible energy. Unfortunately, undesirable byproducts (oxidants or reactive oxygen species) continuously result from this process, and these damage biomolecules (DNA, lipid, protein, carbohydrate) and impair their function (fig 3).[25] Not surprisingly therefore cells are equipped with antioxidant defences, which prevent or at least restrict these untoward effects.[26] It is a widely held view of cellular aging that it is this unchecked damage by reactive oxygen species that leads to an acquired decline in cellular function with time.[27]

Of particular importance is damage caused to the nuclear genes responsible for producing proteins vital for these defences and the mitochondrial genes critical for bioenergy maintenance. Mitochondrial DNA is at particular risk as it is located on the inner membrane, next to the sites of cellular respiration where reactive oxygen species are produced. Furthermore, it lacks the protective histone coat of nuclear DNA and is deficient in those repair enzymes that correct much of the nuclear damage. Moreover, this nuclear and mitochondrial genetic damage cannot be diluted out by cell division and selection in post-mitotically stable tissues, thereby leaving the cell poorly guarded to deal with further oxidative change. Changes in cytokines, growth factors, or hormonal regulators, which influence gene expression, may compound the situation at transcriptional level. A downward spiral thus sets in, leading eventually to metabolic collapse and apoptotic (programmed) cell death, particularly in times of physical, chemical, or biological stress (fig 4). Individuals who inherit a robust (mitochondrial) genotype may sustain adequate energy capacity into old age, despite these damages. However, other less genetically well blessed individuals may find their energy capacity eroded over time to produce a bioenergetic weakening that in some individuals may act synergistically with inherited gene defects to produce, at an early age, overt clinical disorders such as Parkinson's disease, type II diabetes, and mitochondrial myopathies.[27]

Hence, paradoxically, the very giver of life is also that which limits life span. The replicative potential (of dividing cells) may become progressively reduced through other DNA damage (telomere loss), and the stable daughter cells produced may, in the absence of their replacement by genetically more favoured cells, promulgate the chromosomal damage and dysfunction borne by their progenitors.[28]

Oxidative or other stochastic damage to DNA may also underpin certain cancers, resulting in the loss of

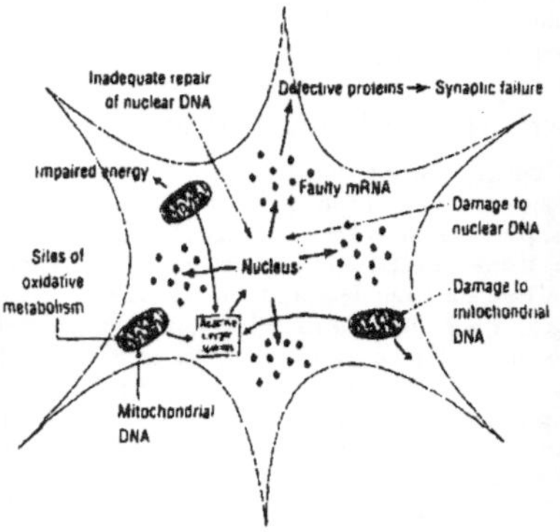

Fig 3 Reactive oxygen species damage nuclear and mitochondrial DNA. Unrepaired damage leads to faults in transcription and translation of proteins, compromising cell metabolism. Failures in DNA repair mechanisms permit a continuing cycle of oxidant damage, gradually weakening the function of the cell. Induction of cell death pathways (apoptosis) leads to loss of nerve cells

function of tumour suppressor genes and the activation of tumour promoting genes (oncogenes), with subsequent malignancy.[29] Oxidative damage to low density lipoproteins may contribute significantly to atherogenesis and cardiovascular disease.[30]

Conclusion

The recent molecular revolution argues that aging and the common diseases associated with age are fundamentally determined by an individual's own genetic make up, this being partly a function of the inherited genome and the modifications to this that occur over a lifetime. How well and how long a person lives depends on the net balance of this "genetic miasma"—what the Victorians used to call "constitution." None the less, lifestyle can have an important role.

Clearly, diet and hygiene, excessive alcohol consumption, cigarette smoking, drug misuse, sexual promiscuity, and occupational hazards can damage even the most perfect of cells and compromise life expectancy. At greater risk are cells and tissues already weakened by aging or disease. Avoiding these risks may increase an individual's likelihood of reaching old age but not necessarily lead to a high quality of life. The goal of future research will be to find how the effects of these adverse genetic changes can be minimised, by for example, gene therapy,[31] and the functioning of health promoting genes maximised, while social risks to health are avoided or controlled along the way.

This present review can only hint at the power that molecular biology has to explain why we age or become ill in later life. How we should respond to its future insights in terms of health care and preventive medicine in later life will be a major challenge for research and society alike.

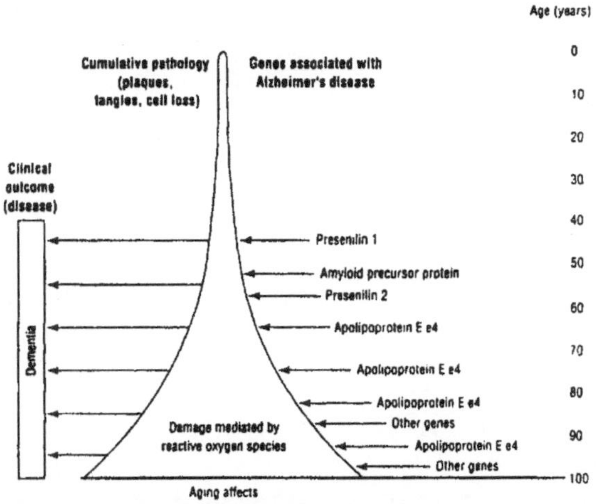

Fig 4 Relation between aging and dementia (Alzheimer's disease). Changes associated with aging (reactive oxygen species) increase as we get older. Genetic factors leading to the pathology of Alzheimer's disease can act at any time after age 40 but have a less powerful effect later in life (the larger the arrow the more powerful the effect). At age 50 dementia is due nearly entirely to the genetic influences of Alzheimer's disease. At 100 the development of dementia is uncommon—when it occurs it is mostly due to pathological changes of aging

1 Evans DA, Funkenstein H, Albert MS, Scherr PA, Cook NR, Chown MJ, et al. Prevalence of Alzheimer's disease in a community population of older persons. *JAMA* 1989;262:2551-6.
2 Saiki RK, Gelfand DH, Stoffel S, Scharf SJ, Higuchi R, Horn GT, et al. Primer-directed enzymatic amplification of DNA with a thermostable DNA polymerase. *Science* 1988;239:487-91.
3 Goate A, Chartier-Harlin MC, Mullan M, Brown J, Crawford F, Fidani L, et al. Segregation of a missense mutation in the amyloid precursor gene with familial Alzheimer's disease. *Nature* 1991;349:704-6.
4 Tanzi RE, Kovacs DM, Kim T-W, Moir RD, Guenette SW, Wasco W. The gene defects responsible for familial Alzheimer's disease. *Neurobiol Dis* 1996;3:159-68.
5 Roses AD. Apolipoprotein E alleles as risk factors in Alzheimer's disease. *Rev Med* 1996;47:387-400.
6 Mann DMA. The pathological association between Down syndrome and Alzheimer disease. *Mech Ageing Dev* 1988;43:99-136.
7 Neve RL, Finch EA, Dawes LR. Expression of the Alzheimer amyloid precursor gene transcripts in the human brain. *Neuron* 1988;1:669-77.
8 Levy E, Carman MD, Fernandez-Madrid IJ, Power MD, Lieberburg I, van Duinne SG, et al. Mutation of the Alzheimer's disease amyloid gene in hereditary cerebral haemorrhage, Dutch type. *Science* 1990;248:1124-6.
9 Pickering-Brown S, Mann DMA, Bourke JP, Roberts D, Balderson D, Burns A, et al. Apolipoprotein E4 and Alzheimer's disease pathology in Lewy body disease and in other β-amyloid forming diseases. *Lancet* 1994;343:1155.
10 Hardy J, Allsop D. Amyloid deposition as the central event in the aetiology of Alzheimer's disease. *Trends Pharmacol Sci* 1991;12:383-8.
11 Games D, Adams D, Alessandrini R, Barbour R, Berthelette P, Blackwell C, et al. Alzheimer-type neuropathology in transgenic mice overexpressing V717F β amyloid precursor protein. *Nature* 1995;373:523-7.
12 Duff K, Eckman C, Zehr C, Yu X, Prada C-M, Perez-Tur J, et al. Increased amyloid-β42(43) in brains of mice expressing mutant presenilin 1. *Nature* 1996;383:710-3.
13 Suzuki N, Cheung TT, Cai X-D, Odaka A, Otvos L, Eckman C, et al. An increased percentage of long amyloid β protein secreted by familial amyloid β protein precursor (βAPP717) mutants. *Science* 1994;264:1336-40.
14 Borchelt DR, Thinakaran G, Eckman CB, Lee MK, Davenport F, Ratovitsky T, et al. Familial Alzheimer's disease-linked presenilin 1 variants elevate Aβ1-42/1-40 ratio in vitro and in vivo. *Neuron* 1996;17:1005-13.
15 Scheuner D, Eckman C, Jensen M, Song X, Citron M, Suzuki N, et al. The amyloid β protein deposited in the senile plaques of Alzheimer's disease is increased *in vivo* by the presenilin 1 and 2 and APP mutations linked to familial Alzheimer's disease. *Nature Medicine* 1996;2:864-70.
16 Citron M, Westaway D, Xia W, Carlson G, Diehl T, Levesque G, et al. Mutant presenilins of Alzheimer's disease increase production of 42-residue amyloid β-protein in both transfected cells and transgenic mice. *Nature Med* 1997;3:67-71
17 Mann DMA, Brown AMT, Prinja D, Jones D, Davies CA. A morphological analysis of senile plaques in the brains of non-demented persons of different ages using silver, immunocytochemical and lectin histochemical staining techniques. *Neuropath Appl Neurobiol* 1990;16:17-25.
18 Morris JC, McKeel JDW, Storandt M, Rubin EH, Price JL, Grant EA, et al. Very mild Alzheimer's disease: informant-based clinical, psychometric, and pathologic distinction from normal aging. *Neurology* 1991;41:469-78.
19 Price JL, Davis PB, Morris JC, White DL. The distribution of plaques, tangles and related immunohistochemical markers in healthy ageing and Alzheimer's disease. *Neurobiol Ageing* 1991;12:295-312.
20 Gibb WRG. Idiopathic Parkinson's disease and the Lewy body disorders. *Neuropath Appl Neurobiol* 1986;12:223-34.
21 Schachter F, Faure-Delanef L, Guenot F, Rouger H, Froguel P, Lesueur-Ginot L, et al. Genetic associations with human longevity at the APOE and ACE loci. *Nature Genet* 1994;6:29-31.
22 Royston MC, Mann DMA, Pickering-Brown S, Perry RH, Raghavan R, Khin-Nu C, et al. ApoE2 allele promotes longevity and protects patients with Down's syndrome from the development of dementia. *NeuroReport* 1994;5:2583-5.
23 Talbot C, Lendon C, Craddock N, Shears S, Morris JC, Goate A. Protection against Alzheimer's disease with apoE ε2. *Lancet* 1994;343:1432-3.
24 West HL, Rebeck GW, Hyman BT. Frequency of the apolipoprotein E, E2 allele is diminished in sporadic Alzheimer disease. *Neurosci Lett* 1994;175:46-8.
25 Cohen G, Werner P. Free radicals, oxidative stress, and neurodegeneration. In: Calne DB, ed. *Neurodegenerative diseases*. Philadelphia: W B Saunders, 1994
26 Cohen G. Catalase, glutathione peroxidase, superoxide dismutase and cytochrome P-450 in the nervous system. In: Lajtha A, ed. *Handbook of neurochemistry*. New York: Plenum Publishing, 1983.
27 Wallace DC. Mitochondrial DNA mutations in human disease and aging. In: Esser K, Martin GM, eds. *Molecular aspects of aging*. Chichester: Wiley, 1995.
28 Osiewacz HD. Aging and genetic instabilities. In: Esser K, Martin GM, Eds. *Molecular aspects of aging*. Chichester: Wiley, 1995.
29 Miller RA. Gerontology: The study of aging as the study of cancer. In: Esser K, Martin GM, eds. *Molecular aspects of aging*. Chichester: Wiley, 1995.
30 Heinecke JW, Stocker R. Lipoprotein oxidation and atherosclerosis. In: Esser K, Martin GM, eds. *Molecular aspects of aging*. Chichester: Wiley, 1995.
31 Lowenstein PR. The use of adenovirus vectors to transfer genes to identified target brain cells in vitro. In: Lowenstein PR, Enquist LW, eds. *Protocols for gene transfer in neuroscience: towards gene therapy of neurological disorders*. Chichester: Wiley, 1996.

07019901

ELSEVIER

Maturitas 30 (1998) 105–111

MATURITAS
THE EUROPEAN MENOPAUSE JOURNAL

Ovarian ageing and the general biology of senescence

T.B.L. Kirkwood *

Biological Gerontology Group, Department of Geriatric Medicine and School of Biological Sciences, University of Manchester, Stopford Building, Oxford Road, Manchester M13 9PT, UK

Abstract

Ovarian ageing is not only of major importance in its own right but is also of interest for its relationship with the general biology of senescence. A key feature of ageing is the distinction in higher animals between the immortality of the germ-line and the mortality of somatic cells and tissues. The ovary contains the female germ cells, and it is through these cells that the female contribution to germ-line immortality is effected. It is abundantly clear that individual oocytes can and do age and that the ageing of the ovary plays a major role in initiating or accelerating a series of other senescent changes. To understand how ovarian ageing fits within the general biology of senescence, it is necessary to explain why ageing occurs at all, to examine the likely mechanisms of general ageing, and to ask whether there is anything special about ovarian ageing and its relationship with the human menopause. Research on ovarian ageing interacts with the our emerging understanding of the general biology of senescence at many levels, ranging from the evolution of the human life history to the biochemical and cellular mechanisms of ageing and longevity.

Keywords: Aging; Senescence; Menopause; Ovarian aging

1. Introduction

Ovarian ageing is not only of major importance in its own right but is also of interest for its relationship with the general biology of senescence. A key feature of ageing is the distinction in higher animals between the immortality of the germ-line and the mortality of somatic cells and tissues [1,2]. The ovary contains the female germ cells, and it is through these cells that the female contribution to germ-line immortality is effected, yet it is abundantly clear that individual oocytes can and do age. Furthermore, although the idea that general ageing is itself programmed is now regarded as untenable, it is clear that the ageing of the ovary plays a major role in initiating or accelerating a series of other senescent changes.

In order to understand how ovarian ageing fits within the general biology of senescence, it is necessary first to explain why ageing occurs at all, secondly to examine the likely mechanisms of

* Tel.: +44 161 2756851; fax: +44 161 2755654; e-mail: tom.kirkwood@man.ac.uk

0378-5122/98/$ - see front matter
PII S0378-5122(98)00065-6

general ageing, and thirdly to ask whether there is anything special about ovarian ageing and its relationship with the human menopause.

2. Why ageing occurs

One view of ageing is as a programmed process under active genetic control, for example as a means of population control. Despite the recurring popularity of this view, there are compelling reasons why it is unsound. Mortality in wild populations is usually so great that individuals rarely live long enough to show clear signs of senescence [3]. Furthermore, for ageing to have evolved in this way would have required that selection for advantage to the species or group was stronger than selection at the level of the individual for the advantages of a longer life. The conditions under which selection at the group level can outweigh selection at the level of the individual are stringent [4], and it is unlikely that they apply to the evolution of ageing [5].

If ageing is not directly programmed then its evolution needs to be understood in terms of the indirect action of natural selection. Two broad approaches to this problem can be distinguished. Firstly, we may note that the force of natural selection—that is, its ability to discriminate between alternative genotypes—progressively weakens with advancing age [6]. This happens simply because at older ages fewer individuals remain alive. The upshot is that there is only loose genetic control over the later stages of the life span. If germ-line mutations occur which give rise to deleterious effects late in life, there is little or no selection to eliminate the mutations from the population. Similarly, if genes have pleiotropic effects such that they produce favourable effects early in life, but have deleterious consequences later, they will be selected for on the basis of their early benefits with scant regard to their delayed side-effects. Either or both of these types of genes can accumulate within the genome and may cause ageing [3,7].

The second evolutionary approach is more specific about the nature of ageing processes. It examines the optimum investment in the maintenance of somatic parts of the organism and demonstrates that this optimum is less than is required for indefinite survival. Given the hazard of accidental death, to which no species is entirely immune, each individual has only a finite expectation of life, even in the absence of senescence. When the individual dies, the resources invested in the maintenance of its soma are lost. Too low an investment in the prevention or repair of somatic damage is obviously a mistake because then the individual may disintegrate too soon. However, too high an investment in maintenance is also wasteful because there is no advantage in maintaining the soma better than is necessary to survive the expected lifetime in the wild environment in reasonably sound condition, and excess investment in maintenance will reduce the resources available for growth and reproduction. Fitness is therefore maximized at a level of investment in somatic maintenance which is less than would be required for indefinite survival.

This conclusion is the basis for the disposable soma theory of ageing [6], named for its analogy with disposable goods that are manufactured with limited investment in durability on the principle that they have a short expected duration of use. The disposable soma theory has important implications concerning the mechanisms of ageing. It strongly supports the view that ageing results from the accumulation of random somatic damage.

3. Mechanisms of ageing

3.1. Biochemical mechanisms

A wide variety of biochemical mechanisms of ageing has been proposed involving various kinds of damage in cells and tissues. The major theories address the roles of oxidative damage, aberrant proteins, defective mitochondria, and DNA damage and mutations.

Oxidative damage is caused by reactive oxygen species, or 'free radicals', which are highly reactive molecules that are produced as by-products of oxidative metabolism within cells. Free radicals can damage almost any component in the cell, including proteins, nucleic acids, and membranes. Protection against the damaging actions of free radicals is provided by vitamins A, C and E and by enzymes such as superoxide dismutase, catalase, glutathione peroxidase, and glutathione reductase. The fact that all aerobic organisms have evolved defences against free radicals is evidence of the threat they pose. There is extensive evidence that oxidative damage accumulates with ageing and that long-lived species have better antioxidant defences than shortlived species [8,9].

Accumulation of aberrant proteins within cells and tissues is observed during normal ageing [10,11], and is specifically linked to the pathogenesis of age-associated conditions such as cataract and Alzheimer's disease. Aberrant proteins can arise as the results of errors in synthesis and/or abnormalities of post-translational modification such as misfolding of the peptide chain or abnormal phosphorylation. Protection against accumulation of aberrant proteins is mainly by means of selective proteolytic degradation of damaged, erroneous, or misfolded molecules. The proteins that accumulate during ageing tend to be variant forms that are resistant to proteolysis.

Eukaryotic cells rely for energy on oxidative phosphorylation occurring in their mitochondria, which are subcellular organelles containing their own DNA. Because free radicals are by-products of this process, the mitochondrial genome is particularly prone to mutational damage caused by free radicals. Damage to mitochondria is of great importance to the cell because it impairs energy production, but damaged mitochondria can still replicate because the proteins required for mitochondrial replication are encoded in the DNA of the cell nucleus. This arrangement means that defective mitochondria can accumulate in cells and may be an important cause of cellular ageing. There is growing evidence that defective mitochondria accumulate during ageing, especially in post-mitotic tissues, and that this is responsible for age-related declines in cell energy production [12,13].

The DNA in cells can be damaged in a great variety of ways and the importance of this threat is evident in the array of DNA repair mechanisms. Damage to DNA is a natural candidate for a molecular mechanism of ageing because once an alteration to the DNA sequence has occurred the altered sequence is preserved as faithfully as the original [14,15]. Thus, mutations can accumulate progressively during the lifetime and interrupt normal cell functions. Somatic mutations can also be a source of autoimmune reactions if they give rise to modified proteins that trigger an immune response [16]. Somatic mutations are known to be an important contributor to the age-associated development of many kinds of cancers, but current evidence suggests that the somatic mutation rate is too low to be a major primary cause of ageing. Nevertheless, the recent discovery that mutations in a gene coding for a DNA helicase, which has a role in DNA maintenance and repair, are responsible for a heritable progeroid syndrome (Werner syndrome) suggests that the role of DNA damage may be important at least in specific age-associated pathologies [17]. Furthermore, there is growing evidence that progressive telomere loss, due to the general inactivation of telomerase in somatic cells, may be a causal factor in the replicative senescence of dividing cells, which may in turn contribute to some aspects of the pathobiology of senescence.

An important prediction of the evolutionary theories of ageing is that there is not likely to be any single mechanism which is responsible for all of the age changes which occur. This is because each of the evolutionary theories proposes a class of gene actions which cause ageing. Thus, for example, the disposable soma theory suggests that multiple kinds of damage will accumulate because for each of the many somatic maintenance and repair systems, the evolved optimum level will be less than perfect. The upshot is that ageing is predicted to arise through the cumulative effects of multiple intrinsic and extrinsic stressors, and that longevity is regulated through the actions of a network of longevity assurance mechanisms [18,19].

3.2. Cellular mechanisms

Evidence of a cellular basis for ageing was first suggested by the discoveries that normal human fibroblasts have limited growth potential in vitro [20] and that the number of cell population doublings such cultures can achieve declines with increasing age of the cell donor [21]. Nevertheless, one of the least well understood aspects of the biology of ageing remains the precise nature of the relationship between cellular and organismic ageing [22]. The functional capacity of many organs within the body declines with age, probably due to the cumulative effects on cells of biochemical damage of the kinds described above.

Within multicellular organisms the different organs rely upon cell division and cell turnover to varying degrees. Neurons, for example, are postmitotic so age changes accumulating in these cells do so in the absence of cell division. Other tissues show high rates of cell turnover, and in such cases any accumulation of damage must be considered within the context of the cell replicative dynamics. The cells of the skin and intestinal epithelia and of the haemopoietic system are continuously replaced, and renewal of these cell types is from the division and clonal expansion of the progeny of small numbers of stem cells.

Age changes in stem cells are of great interest, because any deterioration with age in the numbers and functional integrity of stem cells may have profound consequences for the maintenance of function and proliferative homeostasis in the tissues that rely upon these cells. Individual stem cells in some tissues are pluripotent being capable of self maintenance and of differentiation to produce lineages of specialised cells which may undergo further differentiation as they too divide. Stem cells are long-lived with a large division potential although they undergo division at a lesser rate than their differentiated progeny in many tissues [23]. A number of studies involving serial transplantation of haemopoietic bone marrow to animals in which the resident bone marrow was destroyed by irradiation have shown that bone marrow cells, presumably including haemopoietic stem cells, have a large proliferative capacity but eventually die out [24–26]. These data suggest possible ageing of the stem cell population, although other explanations, such as the dilution and loss of, or damage to, the stem cells during serial transplantation, cannot be ruled out.

A stem cell system which offers many practical advantages for the investigation of age changes is found in the crypts of the mammalian small intestine. The small intestine is the most rapidly proliferating tissue in the mammalian organism, and the epithelium lining the intestinal mucosa comprises a structured hierarchical cell lineage. [27] examined mice aged between 5 and 32 months and found age-related alterations in the histology of the small intestine and in the apoptotic response of stem cells to low-dose irradiation. Animals aged 5, 15, 18 and 29 months were exposed to either 1 or 8 Gy γ-irradiation. A two-fold increase in the level of apoptosis was seen following 1 Gy γ-irradiation in the 29 month old animals, compared to the young and middle-age groups. After 8 Gy irradiation the level of apoptosis in all age groups was high and the age effect less pronounced. These data strongly suggest that stem cells do undergo some functional alteration with age, and further, unpublished data show that stem cells in old animals have impaired capacity for survival and tissue regeneration following high dose irradiation [27].

A longstanding debate in research on cell ageing has concerned the relative importance of programmed events versus the kinds of stochastic damage considered earlier. The absence of evolutionary support for the idea that ageing is programmed as an overall process to terminate the life of an organism weakens the case for programme theories at the mechanistic level. Nevertheless, it is known that programmed cell death (apoptosis) is essential in development and may serve as an important mechanism for tissue maintenance [28]. It also appears that active genetic controls on DNA synthesis and cell division play a part in the limited cell proliferation of fibroblast cultures that are commonly used as a model of cell ageing [29–31]. Although it is difficult to sustain the argument that apoptosis or programmed cell ageing are primary causes for the ageing of the organism, it is quite possible that they contribute, perhaps extensively, to senescence as secondary consequences of stochastic damage to cells.

4. Ovarian ageing

Ageing of the reproductive system and in particular the ageing of the human ovary can now be accommodated within the general biology of senescence. In one sense, reproductive senescence can be viewed simply as a part of the generalised ageing process. The cells of the reproductive system are subject to exactly the same kinds of biochemical stresses (oxidative damage, mutation, etc) which cause damage in other cells and may therefore accumulate damage with the passage of time, resulting in impaired function. Nevertheless, the ageing of the reproductive system is special for at least two reasons.

Firstly, although there is well-documented evidence of maternal and, to a lesser extent, paternal age effects in the frequency of genetic abnormalities in offspring, the germ-line must, in a sense, be immortal. That age-related changes occur in germ cells is not particularly surprising. The increase in abnormal progeny with parental age may reflect either less efficient screening for faults as a general consequence of ageing, and may be due to the weakness of natural selection for late reproductive viability. Over a timescale of generations, however, damage cannot be allowed to accumulate in the germ-line, and this may be prevented through several possible mechanisms [2,32]. Germ cells may be subject to unusually stringent 'quality assurance' mechanisms and/or special maintenance and repair. For example, the enzyme telomerase is expressed in germ cells in order to maintain constant telomere lengths, but switched off in the majority of somatic cells where gradual telomere loss does not matter. Mechanisms to ensure the long-term integrity of germ cells may be similar to those that act to maintain integrity of stem cell populations (see above), although the requirement for long-term effectiveness is even more exacting in germ cells than in stem cells. A second factor that may also be important in preventing accumulation of damage in the germ-line is selection. Selection may operate at all stages in the reproductive cycle, including the production of primordial germ cells in embryogenesis, gametogenesis, selective abortion of defective embryos, and finally phenotypic selection on the progeny.

A second reason why ovarian ageing may be special is that it appears to be programmed. We have seen that there is little or no support for the idea that ageing itself is programmed. Nevertheless, there is a clearly controlled shut-down of reproductive function that occurs in women at menopause, which happens around the age of 45–50 in all regions of the world [33]. The proximate cause of the menopause appears to be oocyte depletion, which triggers endocrine changes. True menopause is generally thought to be unique to humans, although there is some suggestion that a similar process may occur in female toothed whales [34]. Other species show a decline in fertility with age in females but this tends to be a more gradual process involving increasing irregularity of cycling, rather than a complete shutdown of reproductive function.

The paradox about menopause is why fertility should cease when a woman is only about halfway through her biological lifespan, and when the impact of senescence on most of her somatic functions is not yet very advanced. Assisted fertilisation techniques have recently demonstrated that post-menopausal women can successfully bear children without serious complications, and although the finite stock of oocytes places an intrinsic limit on fertility, there is no obvious reason why natural selection should not have increased the oocyte pool. It may of course be argued that the selection force for late fertility would have been weak, but this does not explain why the age of menopause appears to be rather precisely determined. The most plausible resolution of this paradox relates to factors that are believed to have been important in the evolution of human longevity [2]. Increased neonatal brain size, coupled with the constraint on the birth canal linked to bipedal gait, has made giving birth unusually difficult for human females, and the risks of child-bearing would presumably increase steeply if reproduction were to be continued during the later stages of the lifespan. It makes sense, therefore, to suppose that the menopause evolved as a means of removing older women from this risk and of preserving them for the important roles of rearing their later born children and possibly assisting with their grandchildren, as well as sharing their valuable knowledge and experience with their kin group. Seen in this way the menopause is not a primary feature of ageing but a secondary adaptation to lessen its deleterious effects [3,7,35–41].

In conclusion, it may be seen that research on ovarian ageing interacts with the our emerging understanding of the general biology of senescence at many levels, ranging from the evolution of the human life history to the biochemical and cellular mechanisms of ageing and longevity.

References

[1] Kirkwood TBL, Cremer T. Cytogerontology since 1881: a reappraisal of August Weismann and a review of modern progress. Hum Genet 1982;60:101–21.

[2] Kirkwood TBL. 1987. Immortality of the germ-line versus disposability of the soma. In: Woodhead AD, Thompson KH, editors. Evolution of Longevity in Animals. New York: Plenum Press, 1987:209–218.

[3] Medawar PB. An Unsolved Problem of Biology. London: Lewis, 1952.

[4] Maynard Smith J. Group selection. Q Rev Biol 1976;51:277–83.

[5] Kirkwood TBL. Comparative and evolutionary aspects of longevity. In: Finch CE, Schneider EL, editors. Handbook of the Biology of Aging. New York: Van Nostrand Rheinhold, 1985:27–44.

[6] Kirkwood TBL, Rose MR. Evolution of senescence: late survival sacrificed for reproduction. Philos Trans R Soc Lond 1991;B332:15–24.

[7] Williams GC. Pleiotropy, natural selection and the evolution of senescence. Evolution 1957;11:398–411.

[8] Sohal RS. The free radical hypothesis of aging: an appraisal of the current status. Aging Clin Exp Res 1993;5:3–17.

[9] Martin GM, Austad SN, Johnson TE. Genetic analysis of aging: role of oxidative damage and environmental stresses. Nat Genet 1996;13:25–34.

[10] Adelman RC, Roth GS. Altered Proteins and Aging. Boca Raton: CRC Press, 1985.

[11] Rosenberger RF. Senescence and the accumulation of abnormal proteins. Mutat Res 1991;256:255–62.

[12] Linnane AW, Marzuki S, Ozawa T, Tanaka M. Mitochondrial DNA mutations as an important contributor to ageing and degenerative diseases. Lancet 1989;25:642–5.

[13] Kadenbach B, Munscher C, Frank V, Muller-Hocker J, Napiwotski J. Human aging is associated with stochastic somatic mutations of mitochondrial DNA. Mutat Res 1995;338:161–72.

[14] Hanawalt PC. On the role of DNA damage and repair processes in aging: evidence for and against. In: Warner HR, Butler RN, Sprott RL, Schneider EL, editors. Modern Biological Theories of Aging. New York: Raven Press, 1987:183–198.

[15] Vijg J. DNA sequence changes in aging: how frequent?, how important? Aging Clin Exp Res 1990;2:105–23.

[16] Burnet FM. Intrinsic Mutagenesis: A Genetic Approach to Ageing. New York: Wiley, 1974.

[17] Yu CE, Oshima J, Fu YH, Wijsman EM, Hisama F, Alisch R, Matthews S, Nakura J, Miki T, Ouasis S, Martin GM, Mulligan J, Schellenberg GD. Positional cloning of the Werner's syndrome gene. Science 1996;272:193–4.

[18] Kirkwood TBL, Franceschi C. Is aging as complex as it would appear? Ann NY Acad Sci 1992;663:412–7.

[19] Kirkwood TBL, Kowald A. Network theory of ageing. Exp Gerontol 1997;32:395–9.

[20] Hayflick L. The limited in vitro lifetime of human diploid cell strains. Exp Cell Res 1965;37:614–36.

[21] Martin GA, Sprague CA, Epstein CJ. Replicative life span of cultivated human cells: effect of donor's age, tissue and genotype. Lab Invest 1970;23:86–92.

[22] Kirkwood TBL. Towards a unified theory of cellular ageing. Monogr Dev Biol 1984;17:9–20.

[23] Potten CS, Loeffler M. Stem cells: attributes, cycles, spirals, pitfalls and uncertainties. Development 1990;110:1001–20.

[24] Williamson AR, Askonas BA. Senescence of an antibody forming cell clone. Nature 1972;238:337–8.

[25] Ogden DA, Micklem HS. The fate of serially transplanted bone marrow cell populations from young and old donors. Transplantation 1976;22:287–93.

[26] Harrison DE. Do haematopoietic stem cells age? Monogr Dev Biol 1984;27:21–41.

[27] Martin K, Kirkwood TBL, Potten CS. Age changes in stem cells of murine small intestinal crypts. Exp Cell Res 1998;241:316–323.

[28] Potten CS. The significance of spontaneous and induced apoptosis in the gastrointestinal track of mice. Cancer Methods Rev. 1992;11:179–95.

[29] Smith JR. DNA synthesis inhibitors in cellular senescence. J Gerontol 1990;45:B32–5.

[30] Ning Y, Pereira-Smith OM. Molecular genetic approaches to the study of cellular senescence. Mutat Res 1991;256:303–10.

[31] Smith JR, Pereira-Smith OM. Replicative senescence: implications for in vivo aging and tumor suppression. Science 1996;273:63–7.

[32] Medvedev ZA. On the immortality of the germ line: genetic and biochemical mechanisms. A review. Mech Ageing Dev 1981;17:331–59.

[33] Gosden RG. Biology of Menopause. The Causes and Consequences of Ovarian Ageing. London: Academic Press, 1985.

[34] Marsh H, Kasuya T. Changes in the ovaries of the short-finned pilot whale, *Globicephala macrorhynchus*, with age and reproductive activity. In: Perrin WF, Brownell RL, DeMaster DP, editors. Reproduction in

Whales, Dolphins and Porpoises. Cambridge, MA: Report of the International Whaling Commision (special issue 6), 1986:311–335.

[35] Hamilton WD. The moulding of senescence by natural selection. J Theor Biol 1966;12:12–45.

[36] Kirkwood TBL, Holliday R. Ageing as a consequence of natural selection. In: Collins AJ, Bittles AH, editors. The Biology of Human Ageing. Cambridge: Cambridge University Press, 1986:1–16.

[37] Hill K, Hurtado AM. The evolution of premature reproductive senescence and menopause in human females: an evaluation of the 'grandmother hypothesis'. Hum Nat 1991;2:313–50.

[38] Hill K, Hurtado AM. Ache Life History: The Ecology and Demography of a Foraging People. New York: Aldine de Gruyter, 1996.

[39] Rogers AR. Why menopause? Evol Ecol 1993;7:406–20.

[40] Austad SN. Menopause: an evolutionary perspective. Exp Gerontol 1994;29:255–63.

[41] Peccei JS. The origin and evolution of menopause: the altriciality-lifespan hypothesis. Ethol Sociobiol 1995;16:425–49.

American Journal of Pathology, Vol. 147, No. 1, July 1995

Review

Human Replicative Senescence

A Molecular Study

Patrick J Hensler* and
Olivia M. Pereira-Smith*†

From the Roy M. and Phyllis Gough Huffington Center on Aging, The Cell and Molecular Biology Graduate Program,* Division of Molecular Virology,* and Departments of Cell Biology and Medicine† Baylor College of Medicine, Houston, Texas*

As our understanding of cellular proliferation and growth control has progressed, it has become clear that there are numerous positive and negative regulatory mechanisms that maintain proliferative homeostasis. Therefore, any alterations of these processes could have deleterious consequences at the level of both the cell and the organism. Changes can manifest as either loss of proliferation or as uncontrolled proliferation such as hyperplastic or neoplastic growth. They encompass many examples of how alterations in normal regulatory processes participate in the development and progression of disease, as well as the normal aging process. Normal animal cells in culture exhibit tightly regulated growth control and undergo a finite number of divisions after which they cease proliferating. This phenomenon has been termed cellular, or more correctly replicative, senescence. Some populations of cancer cells differ from normal cells in that they do not undergo replicative senescence and have an indefinite lifespan in culture. This finding indicates that changes in processes or genes related to replicative senescence are involved in the progression to tumorigenicity and neoplasia. Numerous studies have attempted to understand the mechanisms controlling replicative senescence and how these are undermined to allow for the evolution of tumor cells. Evidence has accumulated to support a genetic basis for replicative senescence and it has become clear that an understanding of the genetic and molecular controls of that aspect of cellular senescence will provide insight into not only aspects of aging but also cancer.

The Model System of Replicative Senescence

The complex processes of aging involves numerous changes in various organ systems. The study of human aging has been difficult not only because of this complexity but also because of the extensive genetic heterogeneity associated with different individuals. In addition, it is difficult to distinguish between true age-related changes and disease-induced phenomena. It is for this reason that simpler model systems have been utilized to analyze certain aging processes.

It has been well documented that normal human cells have a limited division potential *in vitro*.[1] At the end of their *in vitro* life span, after the achievement of a variable number of population doublings (PDs), these cultured cells exhibit morphological changes and cease proliferating and are referred to as senescent. The observation of the phenomenon of cellular senescence led to the proposal that the limited division potential of these cells *in vitro* is "an expression of aging or senescence at the cellular level."[2] A variety of human cell types have been found to undergo senescence, including epidermal keratinocytes,[3] smooth muscle cells,[4] lens epithelial cells,[5] glial cells,[6] endothelial cells,[7] T lymphocytes,[8,9] and adrenocortical cells.[10] It is therefore clear that changes

Supported by National Institutes of Health Grants R37-AG05333, T32-AG00183, PO1 AG07123, and RO1 AG11066.

Accepted for publication March 13, 1995.

Address reprint requests to Dr. Patrick J. Hensler, Division of Molecular Virology, One Baylor Plaza, T-236, Baylor College of Medicine, Houston, TX 77030.

in proliferative homeostasis involving any of these cell types could affect healthy aging and be associated with age-related pathologies. In fact, it has been found that the decline in immune system function that occurs with age is the result of a loss of or decreased proliferative response by T cells to antigen. Increased proliferation is a component of the increasing tumor incidence observed with advancing age. Several other lines of evidence suggest that the aging of cells in culture may be related to aging of the organism as a whole. Cells derived from shorter-lived animal species have fewer PDs *in vitro* than cells derived from longer-lived species.[11-13] *In vitro* replicative potential is inversely related to the age of the tissue donor.[4-16] Fibroblasts obtained from patients with genetic disorders that mimic premature aging, such as Hutchinson-Gilford and Werner's syndromes, undergo fewer PDs than cells derived from normal, age-matched controls.[17-20] Although these observations do not prove the validity of *in vitro* cellular senescence as a model system for the study of *in vivo* cellular aging, they provide compelling support for the hypothesis that aging of cells is related to the aging process of the organism.

Cellular senescence is not a process of programmed cell death as these cells remain viable in culture for long periods ranging from several months up to 3 years[21] (O. Pereira-Smith, unpublished observations). However, the distinguishing feature of replicative senescence is an irreversible block in the cell cycle. These cells are unable to synthesize DNA[21] yet remain metabolically active.[22] As a population of cells approaches replicative senescence, the time required to transit the cell cycle grows longer, primarily because of increases in the time it takes to complete the G1 phase of the cell cycle.[23] Once they reach senescence, the cells typically arrest in the cell cycle with a G1 DNA content.[24-26] Actively proliferating, early passage fibroblasts can be induced to enter a quiescent (GO) state by the removal of mitogenic stimuli; this state is reversible, however, when appropriate growth factors are reintroduced.[27] In contrast, no known combination of mitogens can stimulate DNA synthesis and/or cell division in senescent cells.[28] The inability of mitogens to induce DNA synthesis in senescent cells does not appear to be a result of a lack of DNA synthesis machinery, as these cells can be induced to undergo semi-conservative synthesis of DNA by infection with SV40[29] or cytomegalovirus,[30] although this appears to be part of an abortive cell cycle. These data suggest that senescence is the result of an active, genetically controlled process.

Approaches to the Study of Replicative Senescence

Comparison of Young and Senescent Cells

To gain a better understanding of the mechanisms responsible for replicative senescence, one can make comparisons between young and senescent cells. These studies have resulted in the accumulation of a large amount of data regarding morphological, biochemical, and molecular changes that occur during the life span of the cell. However, the significance of these changes is not fully understood as it is difficult to determine which changes are causal to, or the result of, senescence. From these studies it has become clear that the program of cellular senescence results in the inability of metabolically active cells to initiate DNA synthesis and progress through the cell cycle. Therefore, it has been hypothesized that the end point of the senescence program is the build-up or specific production of an inhibitor that has the function of blocking the cell cycle machinery from initiating DNA synthesis. Inhibitors of DNA synthesis are thought to be involved in normal tissue homeostasis, and several inhibitors have been isolated from primary tissues and established cell lines.[31,32] The identification of a senescence-specific inhibitor of DNA synthesis has been a major goal of investigators characterizing the *in vitro* model of cellular senescence.

The hypothesis that senescent human diploid fibroblasts express a DNA synthesis inhibitor is supported by several lines of evidence. Norwood et al[33] and Yanishevsky and Stein[34] demonstrated that heterokaryons formed when senescent human fibroblasts were fused with young, proliferation-competent cells resulted in an inability of either nucleus to initiate DNA synthesis. This phenomenon was not observed in fusions between young cells. These data suggest that senescent cells produce an inhibitor of DNA synthesis that can diffuse through the common cytoplasm of a heterodikaryon to affect a young nucleus. Additional work by Stein and Yanichevsky[35] demonstrated that senescent cells could inhibit initiation of DNA synthesis when fused with various immortal cell lines. However, the inhibition did not occur in fusions with cell lines that were immortalized by DNA tumor viruses,[36,37] presumably because of the counteraction by specific DNA tumor viral genes (ie, SV40 T antigen and papillomavirus E6, E7). Carcinogen-derived immortal cell lines were unable to overcome the senescent cell inhibitor, suggesting that elements encoded by DNA tumor viruses might interact with the senescent cell inhibitory mechanism.

Drescher-Lincoln and Smith[38] and Burmer et al[39] performed experiments in which senescent cytoplasts were fused with young proliferating cells and found that the senescent cell-derived cytoplast retained the inhibitory activity. If RNA or protein synthesis were blocked in the cytoplasts before fusion, the inhibitory activity was lost. These experiments indicated that the senescent cytoplasm contains an inhibitor that is actively produced and that the inhibition is not the result of mere dilution of positive growth factors present in young cells that are missing in senescent cells.[40]

Additional evidence for the DNA synthesis inhibitor was obtained from studies with senescent cell mRNA. Microinjection of mRNA isolated from senescent cells into young cells resulted in inhibition of initiation of DNA synthesis[41]; however, this activity was not present after RNAse treatment or in the non-polyA$^+$ fraction of the senescent RNA. The presence of an inhibitory mRNA expressed by senescent cells was used to devise an expression screen of a senescent cell cDNA library. This led to the identification and isolation of several senescent cell-derived inhibitors of DNA synthesis (SDI-1 to -3). These inhibitory transcripts have been partially characterized for their ability to inhibit DNA synthesis in young cells after transfection and microinjection.[42] Only SDI-1 mRNA levels were up-regulated in nondividing cells (senescent and quiescent). Although full characterization of this protein is ongoing, results demonstrate that SDI-1 has several functional characteristics that are similar to the proposed inhibitors identified by whole cell fusion studies. In particular, co-transfection of SDI-1 and SV40 T antigen into young cells revealed that SV40 T antigen antagonizes the ability of SDI-1 to inhibit initiation of DNA synthesis.[43] Other groups have also identified the SDI-1 gene products as a Cdk-interacting protein (CIP1/p21) and a p53-transactivated gene (WAF1).[43,44] The inhibitory activity is thought to be the result of interaction with and inhibition of Cdk-cyclin kinase activity that may interfere with phosphorylation of Rb.[43,45] In addition, the gene is regulated by p53-dependent and -independent pathway(s).[44,46] Johnson et al[46] have provided evidence for the p53-independent pathway for regulation by showing that SDI-1 expression can be up-regulated upon addition of DNA-damaging agents that cause growth arrest independently of p53. More direct evidence was obtained by treating immortal human cells that did not express wild-type p53 with some of these agents. Agents such as hydrogen peroxide and hydroxyurea caused an elevation in RNA levels of SDI-1 and growth arrest in the absence of p53.

The importance of SDI-1 in cell survival is indicated by the fact that no mutations or deletions in this gene have been detected in a large number of immortal human cell lines. Putative mutations were found to be polymorphisms when normal human cell lines were examined.[47] This is in contrast to the p53 gene, in which many mutations are observed in immortalized cells. Collectively, these data suggest that SDI-1 is required for growth arrest both in the normal cell cycle and at the end of the *in vitro* lifespan, and it also controls cell growth to allow for DNA repair. The regulation and production of SDI-1 is of paramount interest to investigators in the fields of cellular senescence and cancer biology, as loss of this function would result in genetic catastrophe.

Somatic Cell Hybrid Studies

The second major approach to the study of cellular senescence is to examine the mechanisms by which cells escape senescence to become immortal. By understanding which genetic changes have occurred to allow for indefinite division, one can then identify important growth regulatory genes and pathways that could play a role in tumor formation. The frequency with which cells escape senescence to become immortal varies between species. Spontaneous and induced immortalization occurs much more frequently for rodent cells than for human cells. As human cells are resistant to spontaneous immortalization, and induced immortalization after the introduction of various viral DNAs or treatment with chemical carcinogens occurs rarely (approximately 1 in 10^7 cells), these are preferred for studies of cellular aging.

Fusion of Normal Cells with Immortal Cells

Fusions of normal cells with immortal cell lines has provided additional evidence for a dominant genetic basis of senescence. If random damage resulted in senescence, one would assume that immortal cells would have acquired mechanisms to cope with such damage. These new mechanisms would be present in a hybrid resulting from fusion of a normal with an immortal human cell; therefore these hybrids would have an extended life span or an immortal phenotype. This has been demonstrated to not be the case as fusions utilizing clonal populations of various immortal and normal human cells yielded hybrid cells with a limited doubling potential.[48–51] In early studies of tumorigenicity, normal and immortal human cells were fused and the resulting hybrids analyzed for growth properties.[52,53] On the basis of these studies, the phenotype of unlimited growth potential (immortal) was thought to be dominant. However, later careful studies demonstrated that fusion of human fibroblasts with human cervical carcinoma-derived cells, HeLa, resulted in some hybrids with a limited division potential.[48] These results were questioned as the genetic stability of the parental HeLa cell line was suspect. The limited life span of the hybrids might have resulted from the loss of immortalizing HeLa DNA, however, this seemed unlikely as Bunn and Tarrant[48] observed that nondividing hybrids could be maintained in culture and that, over time, some of the nondividing cultures gave rise to foci of dividing cells at a low frequency (1×10^5 to 2×10^5 cells). These dividing cells had regained the immortal phenotype and could be cultured indefinitely. It was most likely such immortal variants that were analyzed in the earlier tumorigenicity studies.

Muggleton-Harris and De Simone[49] fused normal human cells with SV40-transformed (immortal) cells and reported that the large majority of the hybrid clones had very limited division potential (<8 PDs). They did not further study the 1% large hybrid clones obtained. Pereira-Smith and Smith[50] extended these observations and, after fusion of normal human fibroblasts with SV40-transformed immortal cells, reported limited proliferation (<7 PDs) in 70% of the hybrids. However, approximately 30% of the hybrids continued proliferating and eventually achieved between 16 and 52 PDs after which division ceased. Expression of SV40 large T antigen was assessed and found to be present in all hybrids which had stopped dividing. As was reported by Bunn and Tarrant,[48] immortal variants did arise from nondividing populations at a low frequency.

Fusion of different immortal cell lines with highly differentiated cell types such as normal human T cells and endothelial cells also yielded hybrids with limited division potential.[9] Collectively, these data support a genetic basis for replicative aging and clearly indicate that cells escape replicative senescence to become immortal because of recessive changes such as loss/inactivation of both copies of growth regulatory gene(s). They also suggest that the dominance of senescence is governed by common mechanism(s) in very different cell types.

Fusion of Immortal Cells with Immortal Cells

To determine the number of genes involved in cellular senescence, a series of experiments was performed in which different immortal human cell lines were fused to each other. These experiments were based on the idea that, if the immortal parental lines have the same recessive changes, the resulting hybrids would be immortal and the cell lines would assign to the same complementation group. On the other hand, fusion of immortal cells with different recessive changes should result in complementation of these defects as a normal copy of a growth regulatory gene(s) would still be present in the parental lines. The resulting hybrids would have a limited *in vitro* life span, and these cell lines would assign to different complementation groups. With this approach, more than 40 different immortal human cell lines have assigned to four complementation groups for indefinite division.[51,54–57] In earlier studies,[57] complementation group assignment did not show a correlation with cell type, tumor type, embryonal layer of origin, expression of an activated oncogene, or the mode of transformation whether by viral or chemical means. However, a recent study revealed that seven independently derived immortal T and B cell lines all assigned to the same complementation group for indefinite division.[56] These data suggest a common mechanism for immortalization of human lymphocytes.

The identification of only four complementation groups for indefinite division indicates that a limited number of genes or gene pathways are altered to produce immortal cells and that a small number of primary genes are involved in the control of senescence. These results are consistent with the fact that spontaneous immortalization of human cells from normal individuals has not been reported and that the frequency of induced immortalization of human cells is low. The complementation group studies greatly facilitate a focused approach to the identification of genes involved in group-specific growth control in an attempt to understand senescence and the changes responsible for immortalization.

Microcell-Mediated Chromosome Transfer Studies

Microcell-mediated chromosome transfer has been used in an attempt to identify chromosomes encoding senescence-related genes specifically altered in cell lines assigning to a given complementation group. With this technique, single normal human chromosomes can be introduced into immortal human cell lines assigned to the four complementation groups for indefinite division.[58-61] Each individual human chromosome is carried in a mouse or hamster cell line background and is tagged with a selectable biochemical marker that allows for single human chromosome transfer. The chromosome can be introduced into each cell line of interest, and clones that have incorporated the chromosome can be selected. By analysis of the growth potential of these microcell hybrids, the chromosome can be assessed for its ability to complement the immortal defect in the parental cell line and induce senescence in a complementation group-specific manner. As cellular senescence has been proposed to be an anti-tumor mechanism,[62-64] chromosomes implicated in tumor suppression have been examined for the ability to restore finite proliferation to immortal cell lines. Human chromosome 11 was reported to suppress tumorigenicity in several tumor cell lines, and a region of chromosome 11 was implicated in Wilms' tumor.[65-68] This chromosome was assessed for the ability to restore finite proliferation in cell lines assigning to the various complementation groups but was found to have no effect on cell lines representative of all four complementation groups.[58] Therefore, there is no cell senescence-related gene on human chromosome 11.

Whole cell fusion experiments had demonstrated that the tumorigenic phenotype of HT-1080, a human fibrosarcoma cell line, was suppressed when fused with normal human diploid fibroblasts. Cytogenetic analysis of nontumorigenic hybrids and hybrids that regained the tumorigenic phenotype (tumorigenic segregates) showed a correlation between loss of chromosomes 1 and 4 and acquisition of tumorigenic potential[69]; therefore, chromosome 4 was introduced into cell lines assigning to the complementation groups. These studies demonstrated that chromosome 4 could reverse the immortal phenotype and induce cellular senescence in three tumor lines assigned to complementation group B but had no effect on the proliferation potential of cell lines representative of the other complementation groups. This result suggests that a gene, altered in immortal cell lines assigned to complementation group B, resides on human chromosome 4 and that the normal counterpart of this gene is required for expression of the senescent phenotype.[59]

Human chromosome 1 has also been implicated in growth control. Sugawara et al[70] reported that human chromosome 1 was able to induce senescence in an immortal hamster cell line. In these experiments, interspecies hamster-human fusions between normal human fibroblasts and an immortal Syrian hamster cell line produced hybrids that exhibited limited life spans similar to that which has been observed in human cell fusions. In hybrids that failed to undergo senescence, karyotypic analysis revealed that all of the clones had lost both copies of human chromosome 1, whereas all other human chromosomes were observed in at least some of the immortal hybrids. With the hypothesis that human chromosome 1 is important in the control of cellular senescence, microcell-mediated chromosome transfer was used to introduce a normal human chromosome 1 into various immortal human cell lines representing the complementation groups for indefinite division. It was found that human chromosome 1 is able to selectively reverse the immortal phenotype of three cell lines that assign to complementation group C.[61] Therefore, human chromosome 1 contains a gene or set of genes, important in the control of cellular senescence, that are defective in group C cell lines. Again, this effect was group specific in that introduction of human chromosome 1 into cell lines assigning to the remaining complementation groups (A, B, or D) had no effect on proliferation. This activity maps to the long arm of chromosome 1 (1q), and efforts are in progress to identify and clone this senescence-related gene(s).

Work by several other laboratories have implicated other chromosomes in normal growth control or senescence. For example, Klein et al[71] reported a loss of cellular proliferation after the introduction of a normal human chromosome X into a nickel-transformed rodent cell line, and Ogata et al[60] reported that human chromosome 7 suppresses proliferation of the cell lines SUSM-1 and KMST1, 4NQO-transformed human fibroblast cell lines. It is possible that these chromosomes do carry genes involved in cell senescence. However, as the chromosomes were transferred into cell lines chosen somewhat randomly, it is difficult to judge the specificity of the effect and determine whether the loss of proliferation was caused by aberrations in the particular cell line used or induction of true senescence. The strength of using cell lines already assigned to complementation groups is that it allows one to test whether multiple cell lines within a group respond by ceasing proliferation after the introduction of a particular chromosome. It is also necessary to demonstrate that cell lines assigned to the other groups are not affected by the transfer of this same chromosome.

Summary

Replicative senescence is an inability of serially cultured cells to synthesize DNA and divide, resulting in a terminal loss of proliferation despite the maintenance of basic metabolic processes. This system has been proposed as a model for the study of aspects of aging at the cellular level. The basis for this model system and its features have been summarized. Although strong experimental evidence exists to support the hypothesis that replicative senescence is a dominant active process, the mechanisms responsible for this phenomenon remain a mystery.

Investigators have taken several approaches to gain a better understanding of replicative senescence. Several groups have documented the differences between young and senescent cells whereas others have identified changes that occur during the course of a cell's *in vitro* life span. With molecular and biochemical approaches, important changes in gene expression and function of cell cycle-associated products have been identified. The active production of an inhibitor of DNA synthesis has been demonstrated, may represent the final step in a cascade of events governing replicative senescence.

The study of immortal cells that have escaped senescence has also provided useful information particularly with regard to the genes governing the senescence program. These studies have identified four complementation groups for indefinite division, which suggests that there are at least four genes or gene pathways in the senescence program. Through the use of microcell-mediated chromosome transfer, chromosomes encoding senescence genes have been identified, and efforts to clone these genes are ongoing. The isolation and characterization of these genes and their protein products will be an important step in understanding replicative senescence.

Collectively, these data demonstrate that disruption of specific gene products can alter the proliferative capacity of a population of cells and may result in immortalization. Although conclusive evidence has not been accumulated to distinguish between the role of these gene products as an immediate cause of replicative senescence, the identification, isolation, and characterization of the genes governing the senescence program will provide a greater understanding of the changes that occur during normal aging. As these genes clearly play regulatory roles, understanding their normal function is also likely to provide insight into the mechanisms underlying a number of diseases. This information could prove useful in the design of therapies for the treatment of disorders that result from altered growth regulation. For example, further characterization of the inhibitor of DNA synthesis may make this protein a target for manipulation in diseases such that the inhibitor could be introduced/up-regulated in diseases of uncontrolled proliferation (ie, tumors) or down-regulated when proliferation is needed (ie, augmentation of wound healing). Clearly, an understanding of the mechanisms responsible for the phenomenon of replicative senescence will contribute not only to our understanding of aging but also of general growth regulatory processes.

Acknowledgments

We thank George M. Martin, MD, for his critical and valuable evaluation of this manuscript.

References

1. Hayflick L, Moorhead PS: The serial cultivation of human diploid cell strains. Exp Cell Res 1961, 25:585–621
2. Hayflick L: The limited *in vitro* lifetime of human diploid cell strains. Exp Cell Res 1965, 37:614–636
3. Rheinwald JG, Green M: Serial cultivation of strains of human epidermal keratinocytes: the formation of keratinizing colonies from single cells. Cell 1975, 6:331–334
4. Bierman EL: The effect of donor age on the *in vitro* life span of cultured human arterial smooth muscle cells. In Vitro 1978, 14:951–955
5. Tassin J, Malaise E, Courtois Y: Human lens cells have an *in vitro* proliferative capacity inversely proportional to the age. Exp Cell Res 1979, 123:388–392
6. Blomquist E, Westermark B, Ponten J: Aging of human glial cells in culture: increase in the fraction of nondividers as demonstrated by a mini cloning technique. Mech Ageing Dev 1980, 12:173–182
7. Thornton SC, Mueller SN, Levine EM: Human endothelial cells: use of heparin in cloning and long-term serial cultivation. Science 1983, 222:623–625
8. Effros RB, Walford RL: T cell cultures and the Hayflick limit. Hum Immunol 1984, 9:49–65
9. Pereira-Smith OM, Robetorye S, Ning Y, Orson FM: Hybrids form fusion of normal human T lymphocytes with immortal human cells exhibit limited life span. J Cell Physiol 1990, 144:546–549
10. McAllister JM, Hornsby PJ: Improved clonal and nonclonal growth of human, rat, and bovine adrenocortical cells in culture. In Vitro Cell Dev Biol 1987, 23:677–685
11. LeGuilly Y, Simon M, Lenoir P, Bourel M: Long term culture of human adult liver cells: morphological changes related to *in vitro* senescence and effect of donors age on growth potential. Gerontologia 1973, 19:303–313

12. Goldstein S: Aging *in vitro*: growth of cultured cells from the Galapagos tortoise. Exp Cell Res 1974, 83: 297–302
13. Rohme D: Evidence for a relationship between longevity of mammalian species and life spans of normal fibroblasts *in vitro* and erythrocytes *in vivo*. Proc Natl Acad Sci USA 1981, 78:5009–5013
14. Martin GM, Sprague CA, Epstein CJ: Replicative life span of cultivated human cells: effects of donor's age, tissue, and genotype. Lab Invest 1970, 23:86–92
15. Schneider EL, Mitsui Y: The relationship between *in vitro* cellular aging and *in vivo* human age. Proc Natl Acad Sci USA 1976, 73:3584–3588
16. Goldstein S, Moerman EL, Soeldner JS, Gleason RE, Barnett DM: Chronologic and physiologic age affect replicative life span of fibroblasts from diabetic, prediabetic, and normal donors. Science 1978, 199:781–782
17. Goldstein S: Life span of cultured cells in progeria. Lancet 1969, 1:424
18. Martin GM, Gartler SM, Epstein CJ, Molulsky AG: Diminished lifespan of cultured cells in Werner's syndrome. Fed-Proc 1965, 24:678–682
19. Goldstein S: Studies on age-related diseases in cultured skin fibroblasts. J Invest Dermatol 1979, 73: 19–23
20. Norwood TH, Hoehn H, Salk D, Martin GM: Cellular aging in Werner syndrome: a unique phenotype? J Invest Dermatol 1979, 73:92–96
21. Matsumura T, Pfendt EA, Hayflick L: DNA synthesis in the human diploid cell strain WI-38 during *in vitro* aging: an autoradiographic study. J Gerontol 1979, 34: 328–334
22. Matsumura T, Zerrudo Z, Hayflick L: Senescent human diploid cells in culture: survival, DNA synthesis and morphology. J Gerontol 1979, 34:328–334
23. Grove GL, Kress ED, Cristofalo VJ: The cell cycle and thymidine incorporation during aging *in vitro*. J Cell Biol 1976, 70:133a
24. Yanishevsky R, Mendelsohn ML, Mayall BH, Cristofalo VJ: Proliferative capacity and DNA content of aging human diploid cells in culture: a cytophotometric and autoradiographic analysis. J Cell Physiol 1974, 84: 165–170
25. Yanishevsky R, Carrano AV: Prematurely condensed chromosomes of dividing and non-dividing cells in aging human cell cultures. Exp Cell Res 1975, 90:169–174
26. Schneider EL, Fowlkes BJ: Measurement of DNA content and cell volume in senescent human fibroblasts utilizing flow multiparameter single cell analysis. Exp Cell Res 1976, 98:298–302
27. Pardee AB, Dubrow R, Hamlin JL, Kletzien RF: Animal cell cycle. Annu Rev Biochem 1978, 47:715–750
28. Cristofalo VJ, Phillips PD, Sorger T, Gerhard G: Alterations in the responsiveness of senescent cells to growth factors. J Gerontol 1989, 44:55–62
29. Gorman SD, Cristofalo VJ: Reinitiation of cellular DNA synthesis in BrdU-selected nondividing senescent WI-38 cells by simian virus 40 infection. J Cell Physiol 1985, 125:122–126
30. Ide T, Tsuji Y, Ishibashi S, Mitsui Y, Toba M: Induction of host DNA synthesis in senescent human diploid fibroblasts by infection with human cytomegalo-virus. Mech Ageing Dev 1984, 25:227–235
31. Nakamura T, Yoshimoto K, Nakayama Y, Tomita Y, Ichihara A: Reciprocal modulation of growth and differentiated functions of mature rat hepatocytes in primary culture by cell-cell contact and cell membranes. Proc Natl Acad Sci USA 1983, 80:729–7233
32. Iype PT, McMahan JB: Hepatic proliferation inhibitor. Mol Cell Biochem 1984, 59(1–2):57–80
33. Norwood TH, Pendergrass WR, Sprague CA, Martin GM: Dominance of the senescent phenotype in heterokaryons between replicative and post-replicative human fibroblast-like cells. Proc Natl Acad Sci USA 1974, 71:2231–2235
34. Yanishevsky RM, Stein GH: Ongoing DNA synthesis continues in young human diploid cells (HDC) fused to senescent HDC, but entry into S phase is inhibited. Exp Cell Res 1980, 126:469–472
35. Stein GH, Yanishevsky RM: Entry into S phase is inhibited in two immortal cell lines fused to senescent human diploid cells. Exp Cell Res 1979, 120:155–165
36. Norwood TH, Pendergrass WR, Martin GM: Reinitiation of DNA synthesis in senescent human fibroblasts upon fusion with cells of unlimited growth potential. J Cell Biol 1975, 64:551–556
37. Stein GH, Yanishevsky RM, Gordon L, Beeson M: Carcinogen-transformed human cells are inhibited from entry into S phase by fusion to senescent cells but cells transformed by DNA tumor viruses overcome the inhibition. Proc Natl Acad Sci USA 1982, 79:5287–5291
38. Drescher-Lincoln CK, Smith JR: Inhibition of DNA synthesis in proliferating human diploid fibroblasts by fusion with senescent cytoplasts. Exp Cell Res 1983, 144:455–462
39. Burmer GC, Motulsky H, Zeigler CJ, Norwood TH: Inhibition of DNA synthesis in young cycling human diploid fibroblast cells upon fusion to enucleate cytoplasts from senescent cells. Exp Cell Res 1983, 15: 79–84
40. Pendergrass WR, Norwood TH: Cell size and the regulation of DNA polymerase α in senescent human diploid fibroblast-like cultures. Molecular Biology of Aging: Gene Stability and Gene Expression. Edited by RS Sohal, LS Birnbaum, RG Cutler. New York, Raven Press, 1985, pp 195–201
41. Lumpkin CK, McClung JK, Pereira-Smith OM, Smith JR: Existence of high abundance antiproliferative messenger RNA in senescent human diploid fibroblasts. Science 1986, 232:393–395
42. Noda A, Venable SF, Pereira-Smith OM, Smith JR: Cloning of senescent cell derived inhibitors of DNA synthesis using an expression screen. Exp Cell Res

American Journal of Epidemiology
Vol. 150, No. 9
Printed in U.S.A.

Longitudinal Change in Height of Men and Women: Implications for Interpretation of the Body Mass Index

The Baltimore Longitudinal Study of Aging

John D. Sorkin,[1,2] Denis C. Muller,[1] and Reubin Andres[1]

Age differences in height derived from cross-sectional studies can be the result of differential secular influences among the age cohorts. To determine the magnitude of height loss that accompanies aging, longitudinal studies are required. The authors studied 2,084 men and women aged 17–94 years enrolled from 1958 to 1993 in the Baltimore Longitudinal Study of Aging, Baltimore, Maryland. On average, men's height was measured nine times during 15 years and women's height five times during 9 years. The rate of decrease in height was greater for women than for men. For both sexes, height loss began at about age 30 years and accelerated with increasing age. Cumulative height loss from age 30 to 70 years averaged about 3 cm for men and 5 cm for women; by age 80 years, it increased to 5 cm for men and 8 cm for women. This degree of height loss would account for an "artifactual" increase in body mass index of approximately 0.7 kg/m² for men and 1.6 kg/m² for women by age 70 years that increases to 1.4 and 2.6 kg/m², respectively, by age 80 years. True height loss with aging must be taken into account when height (or indexes based on height) is used in physiologic or clinical studies. *Am J Epidemiol* 1999;150:969–77.

age factors; body height; body mass index; longitudinal studies

Height is a basic biologic characteristic known to change with aging. A large number of cross-sectional studies of age differences in height have been reported. For example, in 1950 Büchi (1) catalogued some 50 reports published between 1829 and 1947. While a number of studies have described the longitudinal rate of loss of height with aging (2), only seven covered the entire adult life span and in only three of these were there more than two height measurements per subject. A collation of 17 studies on longitudinal change in height with aging will be reported separately (2).

Any age-related change in height has implications beyond descriptive anthropometry. If height changes with age, indexes of obesity such as the body mass index (BMI; weight in kilograms divided by height in meters squared) would change with age independent of any change in obesity. Because BMI is related to the square of the height, even a small change in height may have a large effect on the BMI. Thus, a given BMI could have different meanings for adults at different ages, even if no other changes in body composition occurred with aging. Because height is lost with normal aging, it is important to quantify the effect the height "artifact" has on the computed BMI. Recent proposals of best weight for height suggest that a modest increase in BMI with age is related to optimum survival (3, 4). It is possible that the suggested magnitude of the BMI increase is no more than a result of the height artifact on BMI. To ascertain whether BMI should increase beyond the magnitude of the effect due to the age-related height artifact, the magnitude of the artifact must be determined.

To ascertain the pattern of adult height change as a function of age, to compare the patterns for men and women, and to examine the effect of height change with age on BMI, we analyzed data from the Baltimore Longitudinal Study of Aging (BLSA) (Baltimore, Maryland). This paper reports the results of cross-sectional and longitudinal analyses carried out for men and women across the entire adult span of years.

Received for publication October 9, 1997, and accepted for publication March 4, 1999.

Abbreviations: BLSA, Baltimore Longitudinal Study of Aging; BMI, body mass index.

[1] Metabolism Section, Laboratory of Clinical Investigation, Intramural Research Program, Gerontology Research Center, National Institute on Aging, Baltimore, MD.

[2] Division of Gerontology, Baltimore Veterans Affairs Medical Center, University of Maryland School of Medicine, Baltimore, MD.

Reprint requests to Dr. John D. Sorkin, Intramural Research Program, National Institute on Aging, 5600 Nathan Shock Drive, Baltimore, MD 21224.

MATERIALS AND METHODS

Subjects

The subjects for this study were men and women aged 17–94 years when they joined the BLSA, an open cohort study of normative aging (5). Men have been studied since 1958, women since 1978. Initially, men aged 70 years or younger were examined every 18 months, those over age 70 years annually. In 1970, the examination schedule was changed to every 2 years for men aged 20–59 years, every 18 months for those aged 60–69 years, and annually for those aged 70 years or older. Since 1978, all subjects have been examined every 2 years. BLSA subjects generally are well educated, middle to upper-middle class, community dwelling, and in good health. The present study reports on longitudinal analysis of height change with age for 1,068 men and 390 women and on cross-sectional height differences obtained for 1,430 men and 654 women. Because only 6 percent of BLSA participants are members of minority groups, and because the change in height with age may not be the same for different racial groups, these analyses were limited to White subjects. Throughout the course of the study, height was measured by using a movable anthropometer with subjects in their stocking feet. Subjects were asked to stand tall with heels together, head in a "normal" position, and eyes looking straight ahead. Data for this report were collected from 1958 through 1993.

Statistical methods

Cross-sectional data. The sex-specific cross-sectional age differences in height were quantified by regressing height on age using both linear and quadratic models. In these regressions, each subject was characterized by his or her mean height and mean age during participation in the study. In addition to these regressions based on data from all of the subjects, we tabulated age decade-specific mean heights.

Longitudinal data. Longitudinal analyses of the change in height with age included only those subjects whose height was measured three or more separate times. A two-stage random effects model was used to analyze the data. The first stage consisted of computing each subject's rate of height loss ($\beta_{Subject}$) by regressing the subject's heights on the ages at which the heights were measured, as follows:

$$\text{Height} = \beta_0 + \beta_{Subject}\text{Age}$$

The second stage began by characterizing each subject by $\beta_{Subject}$ and age at enrollment. The subjects were then assigned to one of nine age groups on the basis of their entry age: 17–19, 20–29, 30–39, . . . , and 90–94 years. Within each age group, the subject-specific slopes and entry ages were averaged to produce a decade-specific mean slope and a decade-specific entry age, as well as their associated standard errors. A two-stage random effects model allows for unbalanced and incomplete data. Each subject may have had his or her height measured a different number of times, resulting in unbalanced data. In the data series for any subject, data for one or more of the "scheduled" collections times may not have been obtained, resulting in incomplete data.

To quantify the rate at which the slope of height on age changes with age, sex-specific regressions were performed in which the subject-specific slopes of height on age ($\beta_{Subject}$ from the linear model above) were regressed on entry age in a quadratic model. The following quadratic equation was used:

$$\beta_{Subject} = \beta_0 + \beta_1(\text{Age} - \overline{\text{Age}}) + \beta_2(\text{Age} - \overline{\text{Age}})^2$$

where

$\beta_{Subject}$ = the slope of height on age for a specific subject

Age = the subject's age at entry to the study

$\overline{\text{Age}}$ = the sex-specific mean age of the subjects included in this study

This analysis quantifies the rate at which the slope of height on age (i.e., the velocity of height change, $\beta_{Subject}$) changes with age; it gives (through the coefficients β_1 and β_2) the acceleration or deceleration in the rate of height loss that occurs with aging.

The effect of initial height on the rate of height loss ($\beta_{Subject}$) was assessed by adding entry height to the sex-specific quadratic equations relating age to the velocity of height loss:

$$\beta_{Subject} = \beta_0 + \beta_1(\text{Age} - \overline{\text{Age}}) + \beta_2(\text{Age} - \overline{\text{Age}})^2 + \text{Height}_{\text{At entry}}$$

To enable the statistical significance of the coefficients of the quadratic models (both cross-sectional and longitudinal) to be interpretable, that is, to eliminate the colinearity of age and age^2, the regressions were performed by using "centered" terms for age and age^2 (6). Centering was accomplished by subtracting the appropriate sex-specific mean age from each subject's entry age prior to entering age or age^2 into the quadratic models.

The cumulative longitudinal changes in height of men and women that occur over the adult life span, from young adult life (e.g., age 20 years) to late life (e.g., age 80 years), were calculated by integrating, with respect to age, the equations relating the longitudinal slope of height to age:

$$\text{Cumulative height change} = \int \beta d\text{Age} = \int \beta_0 + \beta_1(\text{Age} - \overline{\text{Age}}) + \beta_2(\text{Age} - \overline{\text{Age}})^2 d\text{Age}$$

where dAge indicates that the integration is performed with respect to age.

The cumulative change in height was used to determine the change in BMI associated with loss of height, assuming weight remains constant:

$$\text{Change in BMI} = \text{weight}_{\text{initial}} / \left(\text{height}_{\text{initial}} - \int (\beta_0 + \beta_1(\text{Age} - \overline{\text{Age}}) + \beta_2(\text{Age} - \overline{\text{Age}})^2)d\text{Age} \right)^2$$

where

$\text{weight}_{\text{initial}}$ = the weight at the youngest age
$\text{height}_{\text{initial}}$ = the height at the youngest age

RESULTS

Cross-sectional

Among both men and women, cross-sectional height decreased with age (figure 1, table 1). The relation between age and height was curvilinear: a quadratic equation described the relation better than a linear equation did (for both men and women, partial F test $p < 0.02$). The curvilinear relation shows that height was lost at an increasing rate with increasing age. For men, age explained 14 percent of the variance in height (equation 1, appendix). For women, age explained 22 percent of the variance (equation 2, appendix). The coefficients for both men and women were significant at $p < 0.001$.

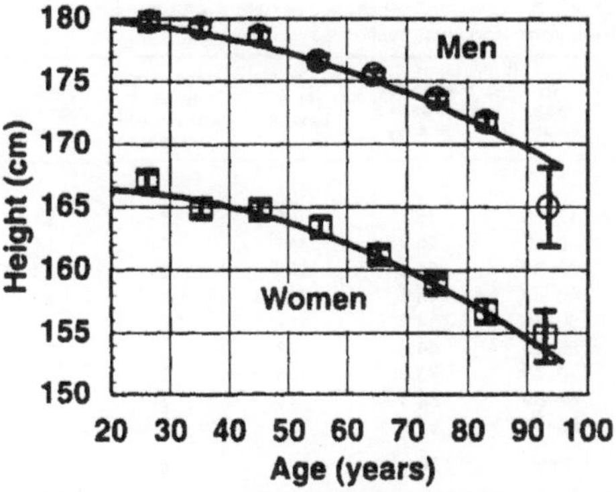

FIGURE 1. Cross-sectional differences in height with age among men and women from the Baltimore Longitudinal Study of Aging, Baltimore, Maryland, 1958–1993. The curves represent sex-specific cross-sectional equations derived by regressing data from 1,430 men and 654 women on age. The eight points plotted for each sex are the decade-specific cross-sectional means; the error bars represent plus or minus one standard error.

Longitudinal

A total of 1,068 men, aged 17–94 years (mean, 50.6) when they entered the BLSA, were followed for an average of 15 years (table 2). A total of 390 women, aged 17–93 years (mean, 53.5) at entry, were followed for an average of 9 years. The follow-up period ranged from 1.9 to 33.7 years. Subjects' height was measured three or more times. During follow-up, men had their height measured an average of nine times, women five times. Among both men and women, the longitudinal slope of height on age became increasingly negative with increasing age (table 2, figure 2), and the rate of loss of height increased with increasing age. In each age group except 90–94 years, where the sample size

TABLE 1. Cross-sectional relation between age and height among men and women in the Baltimore Longitudinal Study of Aging,* Baltimore, Maryland, 1958–1993

Age group (years)	Men: Mean age (years)	Men: No. of subjects	Men: Mean no. of observations per subject	Men: Mean height (SE†) (cm)	Women: Mean age (years)	Women: No. of subjects	Women: Mean no. of observations per subject	Women: Mean height (SE) (cm)
17–19	17.7	1	1	184.8	19.6	1	2	151.3
20–29	26.5	89	2	179.8 (0.66)	26.1	83	2	167.1 (0.66)
30–39	35.2	212	6	179.3 (0.45)	35.2	105	3	164.9 (0.61)
40–49	45.2	180	8	178.6 (0.51)	45.2	66	4	164.8 (0.66)
50–59	55.2	239	9	176.6 (0.43)	55.5	86	4	163.4 (0.74)
60–69	64.6	291	8	175.5 (0.35)	65.3	114	4	161.2 (0.58)
70–79	75.0	293	7	173.6 (0.38)	74.7	127	4	158.9 (0.56)
80–89	83.2	117	5	171.8 (0.49)	83.0	64	3	156.7 (0.86)
90–94	93.6	8	4	165.0 (3.09)	92.9	8	3	154.7 (2.06)
Total	57.6	1,430	7	176.1 (0.18)	56.0	654	4	162.2 (0.28)

* Subjects were assigned to an age group on the basis of their mean age during their participation in the study and were characterized by their mean height.

† SE, standard error.

TABLE 2. Longitudinal change in height with age among men and women in the Baltimore Longitudinal Study of Aging,* Baltimore, Maryland, 1958–1993

Age group (years)	Mean age (years) at entry	No. of subjects	Mean no. of height measurements per subject	Mean follow-up (years)	Mean height (cm)	Mean slope (SE†) (cm/year)	*p* value	Slope for women-men (SE difference) (cm/year)	*p* value
					Men				
17–19	18.4	6	5	13	180.8	0.106 (0.080)	0.25		
20–29	26.9	141	7	16	179.2	0.013 (0.006)	0.04	−0.018 (0.023)	0.43
30–39	34.8	202	9	18	178.9	−0.024 (0.006)	0.00	−0.002 (0.022)	0.93
40–49	45.3	206	12	21	176.7	−0.063 (0.005)	0.00	−0.047 (0.024)	0.06
50–59	54.8	167	10	17	175.1	−0.102 (0.009)	0.00	−0.063 (0.018)	0.00
60–69	64.9	147	9	12	174.2	−0.143 (0.011)	0.00	−0.080 (0.019)	0.00
70–79	73.9	155	7	8	172.8	−0.192 (0.013)	0.00	−0.098 (0.033)	0.00
80–89	82.3	41	5	5	171.1	−0.308 (0.038)	0.00	−0.164 (0.119)	0.17
90–94	93.4	3	5	5	167.6	−0.575			
Total	50.6	1,068	9	15	176.1	−0.091 (0.004)	0.00		
					Women				
20–29	26.4	46	4	8	166.7	−0.005 (0.022)	0.81		
30–39	35.1	70	5	9	164.0	−0.026 (0.021)	0.21		
40–49	44.4	46	6	10	165.1	−0.110 (0.024)	0.00		
50–59	55.0	63	6	11	160.7	−0.165 (0.015)	0.00		
60–69	65.0	71	6	10	160.6	−0.223 (0.016)	0.00		
70–79	74.2	78	4	7	157.5	−0.290 (0.030)	0.00		
80–89	82.8	12	4	6	154.3	−0.472 (0.113)	0.00		
90–93	91.7	3	3	4	157.7	−0.336	0.00		
Total	53.5	390	5	9	161.6	−0.161 (0.011)	0.00		

* Subjects were assigned to age groups on the basis of their age at first visit.
† SE, standard error.

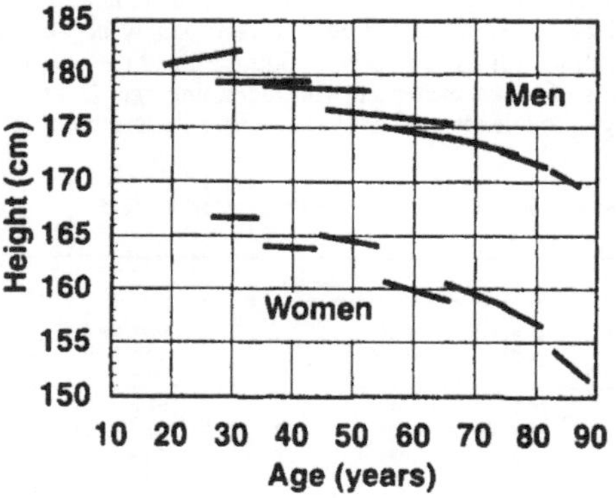

FIGURE 2. Longitudinal change in height, by decade of age, among men and women from the Baltimore Longitudinal Study of Aging, Baltimore, Maryland, 1958–1993. Each line segment starts at the point corresponding to subjects' mean entry age and height. The projection of each line segment on the abscissa represents the average follow-up period; the slope of each segment is the slope of height on age for each age group. Data for the oldest age group, 90–94 years, were not included because of the small sample sizes (table 2).

was very small, the slope was more negative for women than it was for men (table 2, figure 3), indicating that at all ages women lose height more rapidly than men do. The rate of loss was statistically significantly more rapid among women than among men in the age groups 50–59, 60–69, and 70–79 years. (In the age group 40–49 years, the rate of loss was more rapid among women than among men, $p < 0.06$.) The rate at which women lost height increased more rapidly with age than it did for men. Thus, the difference between the sex-specific slopes increased progressively from age 30–39 through 80–89 years (table 2). For both men and women, the slopes were significantly negative in the five decades of 40–49, 50–59, 60–69, 70–79, and 80–89 years.

Age explained 33 percent of the variance in the rate at which men's height changed (equation 3, appendix). For women, age explained 28 percent of the variance (equation 4, appendix). The coefficients of both equations were significant at $p < 0.001$. For men, the relation between age and the rate of change in height was curvilinear. An equation containing age and age^2, that is, a quadratic equation (equation 3, appendix), fit the data better than an equation linear in age (i.e., contain-

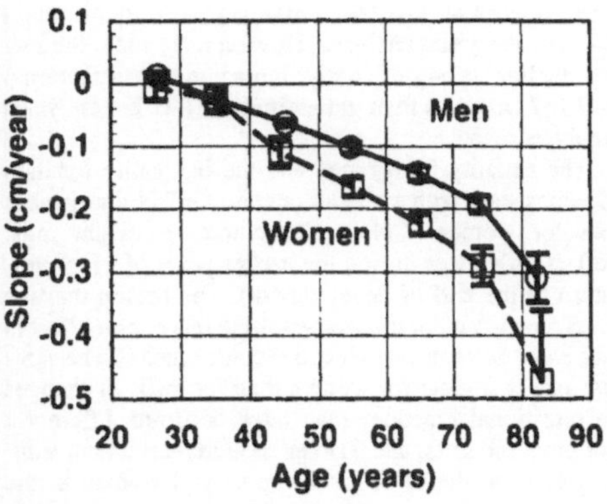

FIGURE 3. Comparison of the decade-specific slopes of change in height for men and women from the Baltimore Longitudinal Study of Aging, Baltimore, Maryland, 1958–1993. The error bars represent one standard error. The oldest age group of men and women, 90–94 years, and the youngest age group of men, 17–19 years, were not included because of small numbers of subjects observed (table 2).

ing age but not age^2; equation 5, appendix), $p < 0.001$. For women, there was a suggestion that the quadratic equation (equation 4, appendix) fit the data better than the linear equation did (equation 6, appendix), $p < 0.16$. The slope from the linear equation (equation 6, appendix) for women, −0.00654 (standard error, 0.00054) cm/year per year, was more negative than the slope from the linear equation (equation 5, appendix) for men, −0.00478 (standard error, 0.00021) cm/year per year, $p < 0.003$.

Rate of loss of height as a function of initial height

The rate of loss of height was a function of initial height for men but not for women; taller men lost height faster than shorter men did (equation 7, appendix). For men, the intercept and the coefficients of the linear and quadratic age terms in the equation (equation 7, appendix) were significant at $p < 0.001$, and the coefficient of the height term was significant at $p = 0.04$. The combination of age, age^2, and initial height explained 33 percent of the variance in the rate of loss of height (equation 7, appendix), the same proportion that was explained by using a model that contained only age and age^2 (equation 3, appendix). In an equivalent analysis for women, the coefficient for the height term was not significant at $p = 0.60$. Although initial height was a statistically significant predictor of the rate of height loss for men, the effect was relatively small compared with the effect of aging. The effect of initial height across the entire range of heights of BLSA males was to increase the rate of height loss 0.054 cm/year for the tallest compared with the shortest men. For men, the addition of initial height to the equation giving slope as a function of age made only minor changes to the value of the intercept and the coefficients of age and age^2. Thus, the effect of age on the rate of height loss was largely independent of the effect of initial height.

Cumulative loss of height over the adult life span

The cumulative longitudinal changes in height of men and women that occur over the adult life span, from young adult life (e.g., age 20 years) to late life (e.g., age 80 years), can be calculated by integrating, with respect to age, the equations that relate longitudinal slope of height to age (men: equation 8, appendix; women: equation 9, appendix). Evaluation of the sex-specific integrals for any two ages (e.g., 20 and 80 years) produces the difference between the heights at the two ages, the change in height that occurs over the 60-year period from age 20 to 80 years. The average aggregate loss of height over the adult life span is greater for women than it is for men (figure 4).

Change in BMI associated with loss of height

The changes in height that occur with aging can be used to determine the cumulative change in BMI with age due solely to the change in height, that is, under

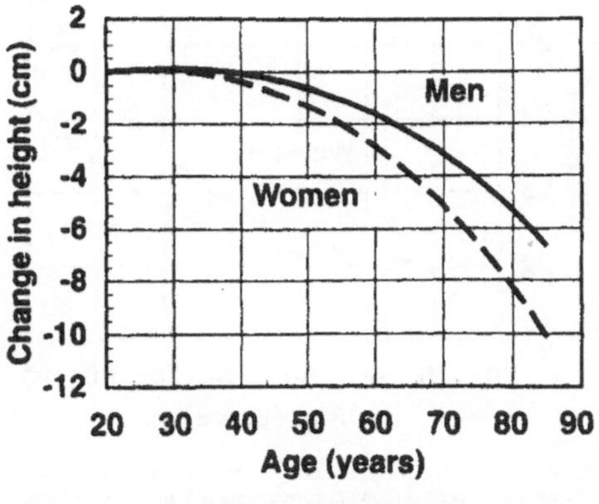

FIGURE 4. Average cumulative loss of height with aging for men and women from the Baltimore Longitudinal Study of Aging, Baltimore, Maryland, 1958–1993. The curves were created by integrating, with respect to age, the equations giving the sex-specific slopes of height as a function of age (described in Implications of loss of height for interpretation of the BMI) and evaluating the integrals from age 20 years to the ages shown along the abscissa.

the assumption that weight remains constant with age. The change in height with age had a substantial and increasing effect on BMI beyond middle age (figure 5). The results of our study demonstrated that in the 60-year period from age 20 to 80 years, BMI increased by an average of 1.5 kg/m^2 for men and 2.5 kg/m^2 for women (figure 5), independent of any change in weight.

DISCUSSION

Our results show that in the BLSA population, men and women lose height with age (cross-sectional analysis) and with aging (longitudinal analysis). The relation between height and aging is curvilinear for men (both the cross-sectional and longitudinal associations were statistically significantly quadratic) and somewhat curvilinear for women (the cross-sectional association was statistically significantly quadratic, and the longitudinal association was suggestive at $p < 0.16$), indicating that the rate of height loss increases with increasing age (figures 1 and 2). The increasing rate of height loss with aging is evident in the decade-specific longitudinal slopes of height on age (table 2), which become increasingly negative with increasing age. Women lose height more rapidly than men do; the sex difference is not significant cross-sectionally but is highly significant longitudinally. Interestingly, for women, the decline in height is almost identical in cross-sectional (–0.167 cm/year) and longitudinal (–0.161 cm/year) analyses. However, for men, the rate of decline is significantly larger in cross-sectional (–0.147 cm/year) than in longitudinal (–0.091 cm/year) analysis.

The equality for women and the inequality for men are consistent with a secular or cohort effect for men but not for women. Cohort differences in height must reflect influences during the earlier years of life (intrautero to the end of development). The reason that the BLSA men but not the women show this cohort effect is not clear. Several factors could contribute: 1) The sample size is smaller for women than for men; 2) changes in nutritional practices may have occurred differently for boys and girls; and 3) even though recruitment techniques were the same for the men and women in the BLSA study, it is possible that there were differences in the ethnic or cultural makeup of the men and women recruited over the years, but we are not aware of any such differences.

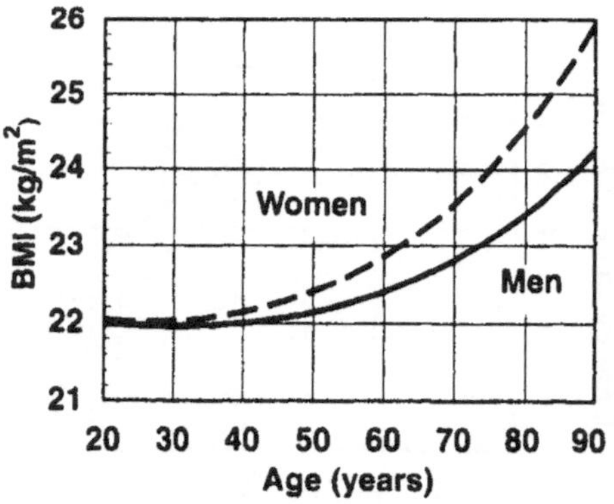

FIGURE 5. Average change in body mass index (BMI) associated with change in height (assuming constant weight) during adult life for men and women in the Baltimore Longitudinal Study of Aging, Baltimore, Maryland, 1958–1993. Loss of height associated with aging was computed by integrating the sex-specific longitudinal equations relating slope to age, resulting in equations that related height to age. The graphs were constructed by assuming a height of 175.3 cm and a weight of 67.6 kg for men and a height of 162.6 cm and a weight of 58.1 kg for women at age 20 years.

Implications of loss of height for interpretation of the BMI

The change in height associated with aging has implications for interpretation of the BMI, a commonly used index of adiposity. The rationale for "correcting" weight for height is obvious. BMI is generally assumed to be an age-invariant measure of obesity, that is, a given BMI at age 20 years is tacitly assumed to measure the same degree of obesity at ages 20, 30. . . . , and 90 years. Because BMI is inversely proportional to the square of height, even a small change in height might have a large effect on BMI. The change in height with age has a substantial and increasing effect on BMI beyond middle age (figure 5). In the 60-year period from age 20 to 80 years, BMI on average will increase by 1.5 kg/m^2 for men and 2.5 kg/m^2 for women (figure 5), independent of any change in weight. Therefore, the tacit assumption that BMI measures the same degree of obesity at all ages is incorrect. BMI values increase with age as height decreases, which could lead to an interpretation that obesity has increased. To our knowledge, the values shown in figure 5 represent the first attempt to quantify this artifact.

Although loss of height with aging has a substantial effect on BMI, the relatively minor quantitative effect on BMI of loss of height with aging on the interpretation of guidelines for best weight for height can be appreciated by comparing the height artifact with several published recommendations for optimal BMI at increasing ages. Table 3 presents the optimal BMI at specific ages and the BMI changes due solely to height loss for men and for women. From age 22 to 70 years, the optimum increase in BMI according to the

TABLE 3. Optimal BMI* values from several sources compared with BMI changes due solely to loss of height with aging, the Baltimore Longitudinal Study of Aging, Baltimore, Maryland, 1958–1993

	Age (years)					
	22	30	40	50	60	70
	Recommended BMI					
Source of recommendation (reference)†						
Gerontology Research Center (5)	20.1	21.3	22.9	24.5	26.1	27.7
National Research Council (6)	21.5	22.5	23.5	24.5	25.5	26.5
Dietary Guidelines for Americans, 1990 (7)	22.0	22.0	24.0	24.0	24.0	24.0
Dietary Guidelines for Americans, 1995 (9)	22.0	22.0	22.0	22.0	22.0	22.0
Metropolitan Life Insurance Company (8)		22.4	22.4	22.4		
	BMI changes due to height change					
Men	22.0	22.0	22.0	22.2	22.4	22.7
Women	22.0	22.0	22.2	22.4	22.9	23.6

* BMI, body mass index (kg/m²).

† The Gerontology Research Center, National Research Council, and Dietary Guidelines for Americans tables are not sex specific, that is, separate tables for men and women were not deemed necessary. In the Metropolitan Life Insurance tables, the BMIs at the midpoint for a medium frame are the same for men and women at their sex-specific average heights: heights were corrected for shoe height (1 inch for men and women; 1 inch = 2.54 cm), and weights were corrected for weight of clothes (5 pounds for men, 3 pounds for women; 1 pound = 0.45 kg).

Gerontology Research Center table (7) is 7.6 kg/m², while there is an artifactual increase in BMI of 0.7 kg/m² for men and 1.6 kg/m² for women due to a decrease in height (table 3). Similar computations show that the National Research Council (8) recommends an increase of 5.0 kg/m². The 1990 Dietary Guidelines for Americans (9) recommend an increase of 2.0 kg/m² at age 35 years, while the 1983 Metropolitan Life Insurance Company table (10) indicates that BMI should be constant from ages 25 to 59 years. Similarly, the 1995 Dietary Guidelines for Americans (11) do not recommend an increase in BMI with aging. The increase in weight recommended by both the National Research Council and the Gerontology Research Center is significantly larger than that attributable to the age-related loss of height alone.

If the denominator in the equation used to calculate BMI (weight/height²) is decreased, the age-associated loss of height artifactually increases the BMI without increasing true adiposity. The artifact becomes important for men after age 50 years and for women 10 years earlier (figure 5). To our knowledge, this artifact has not been quantified previously. However, height is not the only aspect of body composition that changes with age. There is also loss of lean body mass (primarily muscle) and usually an increase in adipose tissue mass. The effect of these changes is to increase true adiposity (percent body fat). The effects of the loss of height and of lean body mass with aging are therefore opposite: height loss increases apparent obesity through an artifactual increase in BMI, while loss of muscle mass (with its consequent decrease in weight) makes older subjects appear less obese than they truly are. Thus, interpretation of any given BMI across the age span as it pertains to partitioning the body into fat and fat-free mass is complex, and a simple adjustment of the observed BMI for height loss will not make BMI an age-invariant measure.

Suggestions for future studies

Our study population was limited to White men and women, and we are unaware of any study of longitudinal change in height in non-White populations. Studies of non-White populations are clearly needed. During each BLSA visit, each subject's height was measured only once. Therefore, we have no data on the reliability of repeated height measurements, a limitation of our study that should be avoided in future studies. Of note, we recently completed a review of the world's literature describing longitudinal change in height (2). None of the papers we found provided data describing the reliability of repeated height measurements.

REFERENCES

1. Büchi EC. Änderungen der Körperform beim erwachsenen Menschen. Eine Untersuchung nach der Individual-methode. (In German). Horn-Wien, Germany: Verlag Ferdinand Berger, 1950.

2. Sorkin JD. Muller DC. Andres R. Longitudinal changes in the height of men and women. Consequential effects on body mass index. Epidemiol Rev (in press).
3. Andres R. Elahi D. Tobin JD. et al. Impact of age on weight goals. Ann Intern Med 1985:103:1030–3.
4. Troiano RP. Frongillo EA. Sobal J. et al. The relation between body weight and mortality: a quantitative analysis of combined information from existing studies. Int J Obes Relat Metab Disord 1996:20:63–75.
5. Shock NW. Greulich RC. Andres R. et al. Normal human aging: the Baltimore Longitudinal Study of Aging. Washington. DC: US Government Printing Office. 1984. (NIH publication no. 84-2450).
6. Kleinbaum DG. Kupper LL. Miller KE. Applied regression analysis and other multivariable methods. Boston. MA: PWS-Kent Publishing Company. 1988.
7. Andres R. Mortality and obesity: the rationale for age-specific height-weight tables. In: Andres R. Bierman EL. Hazzard WR. eds. Principles of geriatric medicine. New York. NY: McGraw-Hill. 1985:311–18.
8. National Research Council Committee on Diet and Health. Diet and health: implications for reducing chronic disease risk. Washington. DC: National Academy Press. 1989:563–5.
9. US Department of Agriculture. US Department of Health and Human Services. Nutrition and your health: dietary guidelines for Americans. 3rd ed. Washington. DC: US Government Printing Office. 1990. (US GPO publication no. 1990-273-930) (Home and garden bulletin no. 232).
10. Metropolitan Life Insurance Company. 1983 Metropolitan height and weight tables. Statistical bulletin of the Metropolitan Life Insurance Company. 1983;64(January–June):2.
11. US Department of Agriculture. US Department of Health and Human Services. Nutrition and your health: dietary guidelines for Americans. 4th ed. Washington. DC: US Government Printing Office. 1995. (US GPO publication no. 1996-402-519) (Home and garden bulletin no. 232).

APPENDIX

Equations. obtained from cross-sectional analyses, relating height to age:

Men:

$$\text{Height} = 178 - 0.1316(\text{Age} - 50.6) - 0.00151(\text{Age} - 50.6)^2;\ r^2 = 0.14, p < 0.001 \quad (1)$$

Women:

$$\text{Height} = 163 - 0.1635(\text{Age} - 53.5) - 0.00191(\text{Age} - 53.5)^2;\ r^2 = 0.22, p < 0.001 \quad (2)$$

Equations. obtained from longitudinal analyses. describing the relation between the rate at which height is gained or lost (i.e., the slope of height ($\beta_{Subject}$)) and age:

Men (modeled as a quadratic relation):

$$\beta_{Subject} = -0.0784 - 0.00463(\text{Age} - 50.6) - 0.0000439(\text{Age} - 50.6)^2,\ r^2 = 0.33 \quad (3)$$

Women (modeled as a quadratic relation):

$$\beta_{Subject} = -0.1455 - 0.00660(\text{Age} - 53.5) - 0.0000476(\text{Age} - 53.5)^2,\ r^2 = 0.28 \quad (4)$$

Men (modeled as a linear relation):

$$\beta_{Subject} = -0.0912 - 0.00478(\text{Age} - 50.6),\ r^2 = 0.32 \quad (5)$$

Women (modeled as a linear relation):

$$\beta_{Subject} = -0.1605 - 0.00654(\text{Age} - 53.5),\ r^2 = 0.28 \quad (6)$$

Equation, obtained from longitudinal analyses, describing the relation between the rate at which height is gained or lost (i.e., the slope of height ($\beta_{Subject}$)) and age; the equation is adjusted for the initial height:

Men:

$$\beta_{Subject} = 0.13537 - 0.00479(\text{Age} - 50.6) - 0.0000443(\text{Age} - 50.6)^2 - 0.00121\text{Height}, \quad (7)$$

$r^2 = 0.33$, $p < 0.001$ for all coefficients except height, coefficient for height $= p < 0.04$,
where Height is the subject's initial height in centimeters.

Equations, obtained from longitudinal analyses, describing the cumulative loss in height with aging:

Men:

$$\text{Cumulative height change} = \int -0.0784 - 0.00463(\text{Age} - 50.6) - 0.0000439(\text{Age} - 50.6)^2 \, d\text{Age} \quad (8)$$

$$\text{Cumulative height change} = 0.0435\text{Age} - 0.00009\text{Age}^2 - 0.000015\text{Age}^3$$

Women:

$$\text{Cumulative height change} = \int -0.1455 - 0.00660(\text{Age} - 53.5) - 0.0000476(\text{Age} - 53.5)^2 \, d\text{Age} \quad (9)$$

$$\text{Cumulative height change} = 0.0714\text{Age} - 0.00075\text{Age}^2 - 0.000016\text{Age}^3$$

0021-972X/01/$03.00/0
The Journal of Clinical Endocrinology & Metabolism

Vol. 86, No. 2
Printed in U.S.A.

Longitudinal Effects of Aging on Serum Total and Free Testosterone Levels in Healthy Men

S. MITCHELL HARMAN, E. JEFFREY METTER, JORDAN D. TOBIN, JAY PEARSON, AND MARC R. BLACKMAN

The Intramural Research Program, National Institute on Aging, National Institutes of Health (S.M.H., E.J.M., J.D.T., J.P.) and Department of Medicine (M.R.B.), The Johns Hopkins University School of Medicine, Baltimore, Maryland 21224

ABSTRACT

Many studies have shown cross-sectional (and two small studies, longitudinal) declines in total and/or free testosterone (T) levels, with age, in men. The extent to which decline in T is the result of the aging process *per se*, as opposed to chronic illness, medication use, and other age-related factors, remains controversial. The frequency with which aging leads to T levels consistent with hypogonadism has also not been defined. These issues bear on the potential use of T replacement in aging men, because aging and hypogonadism have, in common, reduced bone and lean body mass and muscle strength and increased total and abdominal fat. We measured T and sex hormone-binding globulin (SHBG), by RIA, in stored samples from 890 men in the Baltimore Longitudinal Study on Aging. Using a mixed-effects model, we found independent effects of age and date of sampling to reduce T levels. After compensating for date effects, which investigation suggested was artifactual, we observed significant, independent, age-invariant, longitudinal effects of age on both T and free T index (free T index = T/SHBG), with an average change of −0.124 nmol/L·yr and −0.0049 nmol T/nmol SHBG·yr. T, but not free T index, also decreased with increasing body mass index. Use of β−blocking drugs was associated with higher T and higher free T index levels. Using total T criteria, incidence of hypogonadal T levels increased to about 20% of men over 60, 30% over 70 and 50% over 80 yr of age, and even greater percentages when free T index criteria were employed. Our observations of health factor independent, age-related longitudinal decreases in T and free T, resulting in a high frequency of hypogonadal values, suggest that further investigation of T replacement in aged men, perhaps targeted to those with the lowest serum T concentrations, are justified. (*J Clin Endocrinol Metab* **86**: 724–731, 2001)

NUMEROUS CROSS-SECTIONAL INVESTIGATIONS have demonstrated lower concentrations of circulating testosterone (T) and/or free T in older men (1–9). However, the finding, in some studies, that T levels did not fall significantly with age in exceptionally healthy men (10–12) raised the question of the relative roles of chronic age-related illness *vs.* aging *per se* in producing the observed decreases. Studies of this issue have suggested that age, illness, and smoking all exert independent effects on T and free T levels during aging (8, 13, 14). In the absence of studies of longitudinal change in T with age, it is uncertain to what extent confounds such as secular or cohort effects (15) may have influenced the above results. Two small-scale longitudinal investigations have observed decreases, with aging, in total T (16, 17); but no large longitudinal series has been reported to date. The above studies suggest that T levels decline at a more or less constant rate, with age, in men, with no period of accelerated decline. However, no quantitative evaluation of the longitudinal influence of aging on the incidence of hypogonadism has yet been reported.

These issues are of more than theoretical concern because aging in men is associated with decreases in bone mineral density (BMD) (18, 19), lean body and muscle mass (20, 21), strength (22, 23) and aerobic capacity (24), as well as with increases in total and abdominal body fat, low-density lipoprotein cholesterol, and/or low-density lipoprotein/high-density lipoprotein cholesterol ratios (25–28), all of which also occur in nonelderly hypogonadal men (29, 30). These changes in body composition and metabolism predispose to musculoskeletal frailty, to osteoporotic fractures [a major health problem in elderly men (31)], and to cardiovascular disease.

Because of the potential for using T replacement to prevent or ameliorate age-related osteoporosis, sarcopenia, and the abdominal obesity/glucose intolerance/hyperlipidemia syndrome (syndrome X) in men, a more thorough understanding of the rate and extent of longitudinal changes in T with age and how such changes are related to other variables (such as disease and smoking) is needed. In the current study, we measured concentrations of T and sex hormone binding globulin (SHBG), and we calculated free T indices in sequential serum samples obtained from a large group of men well-characterized with regard to health-related variables, in the Baltimore Longitudinal Study of Aging (BLSA).

Materials and Methods

Subjects and samples

All men were participants in the BLSA, a largely middle class, 87% Caucasian population, whose characteristics have been described (32). The BLSA, an open registration study of the physiology of aging, has, for more than 40 yr, accumulated data on men studied, at approximately 2-yr intervals, during visits to the NIA's Gerontology Research Center in Baltimore. The BLSA investigative protocol is approved by the combined Institutional Review Board of the Johns Hopkins Bayview Medical Center and the Gerontology Research Center. All subjects studied signed

Received May 8, 2000. Revision received October 4, 2000. Accepted October 13, 2000.

Address all correspondence and requests for reprints to: S. Mitchell Harman, M.D., Ph.D., Director, Kronos Research Foundation, 4455 East Camelback Road, Suite B135, Phoenix, Arizona 85018. E-mail: harman@kronofoundation.org.

IRB-approved informed-consent documents. Each man receives an extensive interim medical and psychological history and physical examination at each visit, and serum samples are banked for future investigation. Blood samples were obtained from BLSA subjects in the morning between 0700 and 0930 h, after an overnight fast. Before 1992, samples were stored at −20 C. Samples collected after 1992 were kept at −80 C. All men (n = 890) from whom serum samples of adequate volume were available were included in the study. The BLSA men studied were somewhat, but not significantly, younger (mean age, 58.8 ± 15.8 vs. 61.8 ± 17.1) than the remaining 821 BLSA men who were not studied, and were seen, on average, 8 yr more recently than the BLSA men who were not included in the study. However, there is no reason to believe that the men who were studied differed in any substantive way from those who were not. Samples assayed in this study were selected from the frozen serum bank as follows: During a 6-month period in 1995, sera from each subject's most recent and previous 3 visits and from visits closes: to 10, 15, 20, 25, and 30 yr from the most recent one were retrieved and sent to Covance Laboratories, Inc. (Vienna, VA) for assay of T and SHBG. The number of samples assayed was 3763, with 3565 producing acceptable assay results for T and 3537 for both T and SHBG. The number of samples per subject ranged from 1–10 (mean 4 ± 1.9), with samples stored from less than 1, up to 33, yr (mean 10 ± 8 yr).

Assays

T levels were determined, in duplicate, on aliquots of 100 μL serum, using ^{125}I, double antibody RIA kits obtained from Diagnostic Systems Laboratories, Inc. (Webster, TX). Minimum detectable T levels averaged 0.42 nmol/L, with intra- and interassay coefficients of variance, respectively, of 4.8% and 5.7% at concentrations of 7.74 and 7.29 nmol/L, and 3.3% and 6.4% at concentrations of 44.7 and 42.9 nmol/L. SHBG concentrations were measured in 50-μL aliquots using RIA kits purchased from Radim (Liege, Belgium) which employ ^{125}I labeled SHBG and PEG-complexed second antibody. The sensitivity of the SHBG assay was approximately 10 nmol/L. The CV at 5 nmol/L was 22%; and at 25 nmol/L, 5%; with intra- and interassay coefficients of variance, respectively, of 3.4 and 10.8% at concentrations of 22 and 19 nmol/L, and 1.8% and 7.7% at concentrations of 117 and 118 nmol/L.

Confounds and validation

Preliminary analysis of data from 3565 samples, stored between 1961 and early 1995, revealed a significant increase in T level with length of storage, independent of age, so that mean levels of T in samples taken from 1961 to 1975 were, on average, about 40% higher than levels in samples stored between 1985 and 1995. To investigate whether this finding was an artifact or a true secular trend in the population, we first assayed an additional set of 120 serum samples from later visits (1995–1998) for T and SHBG, by the same methods used for the original samples. Some of these samples were from men whose samples were included in the first set of assays; others were not. We detected no apparent discontinuity between the more recent set of samples and those immediately preceding them, for any of the above measures (Fig. 1). Next, we reassayed 221 of the previously examined samples, evenly distributed across the entire date range, using a different RIA method at Endocrine Sciences, Inc. Laboratories (Calabasas Hills, CA). This latter assay employed ^{125}I-labeled T and second antibody-coated tubes after extraction and column chromatography of samples (33). The minimum detectable level of T was 1.04 nmol/L; and intraassay CV's were, respectively, 9.4, 5.3, and 7.1% at concentrations of 3.61, 9.65, and 17.7

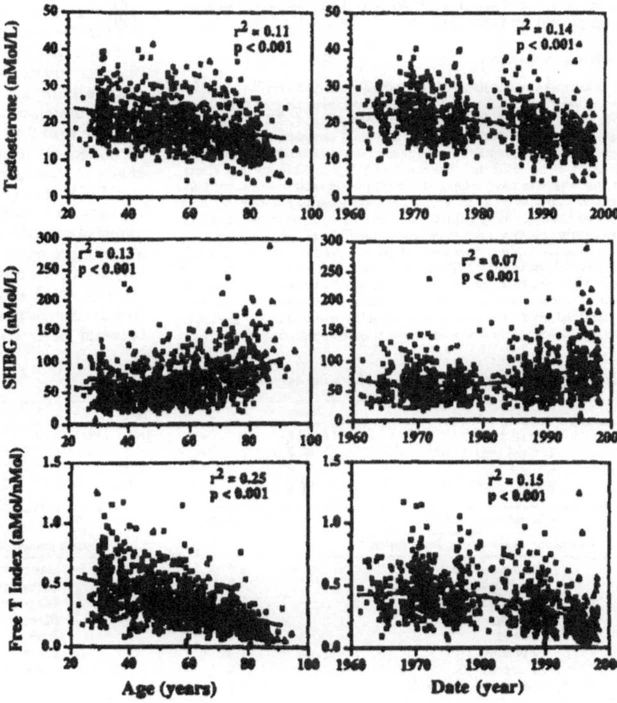

FIG. 1. Effects of age (*left panels*) and date (*right panels*) on serum T, SHBG, and free T index. Individual data points for T (*upper panel*), SHBG (*middle panel*), and free T index (*lower panel*) from the earliest available (usually first) visits during 1963–1994 are shown as *closed squares*, and the additional set of 120 serum samples from later visits (1995–1998) are plotted as *closed triangles*, against age (*left panels*) and sample date (*right panels*). Best-fit regression lines, r^2 and P values for the 1963–1994 data are shown. Total T concentrations and free T index values decreased linearly with increasing age, whereas SHBG exhibited a curvilinear increase with age, rising at a slightly greater rate in the older, than in younger, men. With progressively more recent sample date, total T concentrations fell, SHBG level was higher, and free T index decreased, all with significant curvilinear relationships. There were no apparent discontinuities between the original samples and the post-1994 samples assayed later.

ng/dL, with interassay CV's of 5.7, 2.8, and 7.0% at concentrations of 3.12, 8.84, and 16.1 nmol/L. By the new assay, T showed no significant correlation with visit date ($r = -0.088$, $P > 0.15$), whereas the T values derived from the original assays on the same 221 samples were inversely related to visit date ($r = -0.199$, $P < 0.01$). Using date intervals that divided the samples into equal numbers (1963.9–1986.8; 1986.8–1988.3; and 1988.3–1994.6), the mean T values from the repeat assay were significantly higher in each time period, compared with T values in the original assay [being, respectively, 18.5 ± 7.1 vs. 15.5 ± 5.7 ($P < 0.01$); 19.1 ± 6.2 vs. 14.1 ± 4.3 ($P < 0.001$); and 17.4 ± 6.1 vs. 13.0 ± 4.3 ($P < 0.001$); with a mean difference of 4.2 nmol/L, even after adjustment for date effects]. To detect evaporation of samples as a possible cause of higher T concentrations in the older samples, we determined serum Na^+ levels, by a routine laboratory method (automated spectrophotometry at Endocrine Sciences, Inc. Laboratories) in 191 of these same 221 samples. Sodium concentrations were elevated (>145 mmol/L) in only 12 samples distributed fairly evenly over the full range of dates, and low (<135 mmol/L) in 18 samples, with 11 of 18 values occurring in the oldest samples. There was no significant relationship between either sample date ($r^2 = 2 \times 10^{-4}$, $P > 0.8$) or serum T level ($r^2 = 0.003$, $P > 0.4$) and Na^+ concentration, by linear regression analysis. Finally, we added either pure T in charcoal-extracted female serum, in amounts sufficient to increase the T concentrations by about 6.9 nmol/L, or an equal amount of blank serum, to duplicate aliquots of 24 samples stored between 1968 and 1972 and to 24 samples from visits between 1988 and 1992. These matched aliquots were then coded and assayed blindly at Covance Laboratories, Inc. using the original reagents and method. This assay read an average increase of 9.0 ± 1.0 nmol/L in the older samples and 6.5 ± 1.2 nmol/L in the more recent samples, a 29% difference ($P < 0.001$). This discrepancy is of an order of magnitude similar to the difference in estimated T levels by date in the original assays and, along with the above findings, strongly suggests the presence of a date (*i.e.* storage time)-related assay artifact.

Statistical analysis

The sample size consisted of 3651 samples analyzed for T and SHBG, after inclusion of the 120 newer samples used in assessing the confounds. Thirty of the FTI estimates were unreasonably high (>3.3 SD above the mean) and were excluded, producing a final sample size of 3621 FTI measurements. Descriptive statistics used to characterize the data were calculated using SPSS, Inc. Version 9 (SPSS, Inc., Chicago, IL). Cross-sectional analyses used simple and multiple regression to examine the effects of age and date. Best-fit regression lines are shown for each relationship (either simple linear plots or, where higher order equations significantly improved the r^2 values, curvilinear plots)

The longitudinal analyses were based on mixed effects models using the statistical package MLWIN (34) to examine the effect of age on T and free T index. We first corrected all values of T by adding 4.2 nmol/L to adjust for the constant and systematic underestimates, compared with the standard (extraction) assay. The model considered initial age, year, and elapsed time from first measurement in the prediction. The form of the equation was: T = (b0+b0i)+b1 × first-age+(b2+b2i) × time+(b3+b3I) × time^2+b4 × first-date+b5 × first-date^2+b6 × time × first-age+bj × covariatej+error (where b0i, b2i, and b3i are random effects that reflect individual variation from the mean effect). To account for the above described effect of date on apparent T concentrations, date-adjusted estimates were developed by setting the date to 1995, the most recent date from which the original samples assayed were obtained. A similar approach was used to estimate free T index. The date effect we observed was systematic, so that adjustment of total T levels to a common date of 1995 resulted in a distribution in which mean T for men 20–45 was 17.6 nmol/L (508 ng/dL) and the 2.5th percentile (used by convention to define hypogonadism) was at 11.3 nmol/L (325 ng/dL). These values resemble those generally accepted for normal young adults (35, 36), suggesting that the adjustment was appropriate.

Based on the models, individual regression equations were developed for all men with two or more determinations ($n = 782$), and the equations were used to establish grouped average effects based on the age decade at first evaluation. Initial age, slopes, and time were calculated for each age decade group of men and were subsequently plotted. For analyses that examined the impact of covariates, the additional terms were added to the equations as fixed effects.

Results

Characterization of study population

Table 1 shows the means and ranges for the age, weight, height, and body mass index (BMI = kg/m^2) for the men at date of entry into this study. To investigate whether effects of medications, use of tobacco and alcohol, or chronic illness interacted with the those of aging *per se*, we classified men, with regard to each of these variables, by review of computerized visit records. At each visit, subjects were identified as to whether they were taking glucocorticoids, β-blockers, or psychotropic medications, all of which can alter reproductive hormone dynamics. Men were also classified as never, former, or current smokers and as users of less than or more than 2 oz of alcohol daily, and their illnesses were identified by review of diagnostic classifications in visit charts. Percentages of men in each classification are depicted in Table 1. The ages ranged from the third through the ninth decades. Men varied in body fatness from very lean to severely obese, with a mean BMI value in the overweight zone according to the new NHLBI and WHO classification scheme (37, 38). Of the medications classified, only β-blockers were used by a high-enough proportion of the population (~15%) to exert a significant effect. Men were about evenly divided between never and former smokers. The former smokers

TABLE 1. Cross-sectional description of men in sample at entry

Continuous variables	Mean	SD	Minimum	Maximum
Age (yr)	53.8	16.0	22.5	91.3
Weight (kg)	80.2	12.3	53.3	152.1
Height (cm)	176.8	6.8	156.1	197.6
BMI (kg/m^2)	25.6	3.2	18.3	43.0

Classification variables	Percentage		Percentage
Medication usage during study period		Diseases present during study period	
Glucocorticoids	4.4	Cancer	9.4
Beta blockers	14.7	Coronary disease	20.3
Psychiatric	5.6	Diabetes	16.3
Smoking		Stroke	4.9
Never	40.8	Renal	0.2
Former	41.3	Liver	0.4
Current	17.9	Thyroid	4.3
Alcohol (>2 gm/day)	6.1	COPD	12.8

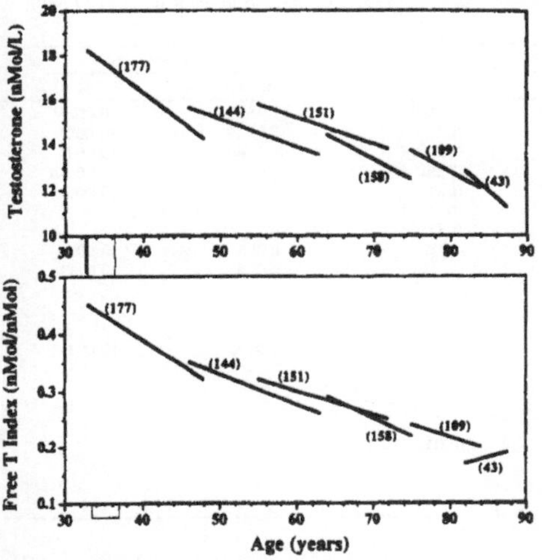

FIG. 2. Longitudinal effects of aging on date-adjusted T and free T index. Linear segment plots for total T and free T index *vs.* age are shown for men with T and SHBG values on at least two visits. Each linear segment has a slope equal to the mean of the individual longitudinal slopes in each decade, and is centered on the median age, for each cohort of men from the second to the ninth decade. *Numbers in parentheses* represent the number of men in each cohort. With the exception of free T index in the ninth decade, segments show significant downward progression at every age, with no significant change in slopes for T or free T index over the entire age range.

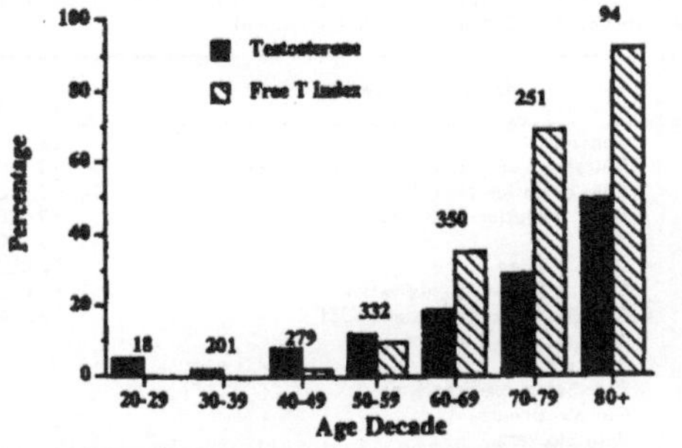

FIG. 3. Hypogonadism in aging men. *Bar height* indicates the percent of men in each 10-yr interval, from the third to the ninth decades, with at least one T value in the hypogonadal range, by the criteria of total T < 11.3 nmol/L (325 ng/dL) (*shaded bars*), or T/SHBG (free T index) < 0.153 nmol/nmol (*striped bars*). *Numbers above each pair of bars* indicate the number of men studied in the corresponding decade. The fraction of men who are hypogonadal increases progressively after age 50 by either criterion. More men are hypogonadal by free T index than by total T after age 50, and there seems to be a progressively greater difference, with increasing age, between the two criteria.

tended to be older than the other two smoking groups (56.0 yr for former smokers, 53.6 for nonsmokers, and 48.0 for current smokers). Relatively few men used excessive amounts (>2 ounces/day) of alcohol. The most common chronic diseases were coronary artery disease, diagnosed as either history of a clinical event (myocardial infarction, angina, heart failure) or definitely abnormal treadmill exercise testing, and diabetes mellitus, which is probably present at a relatively high rate in the BLSA, compared with the general population, because all BLSA volunteers undergo glucose tolerance testing, and thus the detection rate can be assumed to be nearly 100%. Cancer (other than nonserious skin cancers) and chronic obstructive pulmonary disease (COPD) were present in significant numbers of subjects as well. Each of the other chronic illnesses classified occurred in less than 5% of the population studied.

Cross-sectional analysis of T, SHBG, and free T index

Figure 1 illustrates values, from the earliest available (usually first) visit for each man, for the total T concentrations, SHBG levels, and free T indices (calculated as total T/SHBG) plotted against age of the subjects and against the date on which the sample was obtained. Total T concentrations and free T index values decreased linearly with age, whereas SHBG exhibited a curvilinear relationship with age, increasing at a slightly greater rate in the older than in younger men. Total T, SHBG, and free T index all showed significant curvilinear relationships with sample date as well.

Longitudinal analysis

Figure 2 illustrates linear segment plots for date-adjusted total T and free T index *vs.* age, based on samples from all men with sera available for at least two visits ($n = 782$). Using mixed effects models to predict individual equations, we plotted segments that represent the longitudinal trend (mean slope) for cohorts of men, in each decade from the 30's to the 80's. These segments show significant downward progression at every age, with no significant differences among slopes for T or the free T index. The apparent exception was the slope of free T index for the relatively small number ($n = 43$) of men in their 80's, which was not significantly negative or positive and had a large SE. The earliest point of each segment was within acceptable limits, relative to the point at the corresponding age for men in the previous decade, to exclude significant discontinuities Thus, there was no evidence of secular effects on these date-adjusted measures.

Multivariate analysis

A mixed-effects model analysis (Table 2) confirmed independent longitudinal effects of age on both T and free T index, with respective coefficients of −0.110 nmol/L·yr and −0.005 nmol T/nmol SHBG·yr ($Z = 14.5$ and 14.2, $P < 0001$). Date and date^2 were also significant contributors to estimates of T, requiring adjustment of T values to a common date, as noted above. In this analysis, T decreased (−0.350 nmol/L·kg·m^2) with increasing BMI, independent of age, whereas there was a small (0.003 nmol/L·kg·m^2), but significant, increase in the free T index with increasing BMI. In this population, smoking did not contribute significantly to either T levels or the free T index, whereas alcohol consumption decreased the free T index. Of the medications studied, only

TABLE 2. Coefficients and their SD's for mixed effects models with testosterone or free testosterone index as dependent variables (only significant, $P < 0.05$, coefficients are shown)

Independent Variable	Testosterone		Free testosterone index	
	Coefficient	SD	Coefficient	SD
Constant	15.8	(0.26)[a]	0.1890	(0.026)[a]
Initial age deviation from mean initial age	−0.110	(0.01)	−0.0050	(0.000)
Date deviation from 1995	−0.510	(0.03)	−0.0180	(0.001)
Date² deviation from 1995	−0.010	(0.001)	−0.0004	(0.0000)
Time	−0.170	(0.03)[a]	−0.0050	(0.000)[a]
Time × Time	0.006	(0.001)[a]		
Time × Initial Age deviation	−0.002	(0.001)		
BMI deviation from mean BMI	−0.350	(0.03)	0.003	(0.001)
Alcohol			−0.023	(0.011)
Smoking				
Psychiatric medications				
Glucocorticoids				
Beta blockers			0.0290	(0.001)
Renal disease				
Liver disease				
Thyroid disease				
COPD				
Cancer	−0.0920	(0.01)		
Diabetes				
Stroke				
Coronary heart disease				

[a] Significant random effect.

β-blockers had an apparent effect, with users having higher free T indices than nonusers. Of the disease categories examined, only cancer showed an influence, lowering total T but not the free T index.

Occurrence of hypogonadism

We used date-adjusted T and free T index values to calculate the percentages of men in each decade who were hypogonadal, defined as having at least one visit in that age decade at which T was less than 11.3 nmol/L (325 ng/dL) or the free T index was less than 0.153 nmol/nmol (the 2.5th percentile values, defined as: the mean values − 1.96 SD, for men 21–45 yr of age in our study). As shown in Fig. 3, there were progressive increases in the prevalence of hypogonadism, defined by either set of criteria, from relatively low levels for men less than 49 yr of age to 12%, 19%, 28%, and 49% (by total T) or to 9%, 34%, 68%, and 91% (by free T index) in men in their 50's, 60's, 70's, and 80's, respectively. Using the free T index, the percentage of men classified as hypogonadal tended to be lower under age 50 and higher above age 50 than by the total T criterion. Further analysis showed that 78% of the men identified as hypogonadal by a single total T determination, who had subsequent samples evaluated, had low total T levels in all subsequent (mean, 2.6) visits. This percentage was even higher (97%) for men identified as hypogonadal by free T index.

Discussion

Most (1, 5–9), but not all (10–12), cross-sectional studies have demonstrated a decrease, with age, in total T in men. To date, there have been only two longitudinal studies reported, each of which showed decreases in total T in, respectively, 66 men, 41–61 yr old, followed for 13 yr (16) and 77 men, 61–87 yr old, followed for 15 yr (17). In the present study in 890 generally healthy, middle class, American men in the BLSA, we found that both T and free T index (a calculated value related to free or bioavailable T) decreased progressively at a rate that did not vary significantly with age, from the third to the ninth decades. The magnitude of the decrease in total T averaged 0.110 nmol/L (3.2 ng/dL) per year, was similar whether examined cross-sectionally or longitudinally, and was comparable to the 0.121 nmol/L (3.5 ng/dL) per year reported by Zmuda *et al.* in 66 men, 41–61 yr old, followed for 13 yr (16) but less than the rate of 0.382 nmol/L (11 ng/dL) per year observed in 77 men, 61–87 yr old, followed for 15 yr by Morley *et al.* (17). The decrease in free T index was somewhat steeper than that of total T, owing to a trend for an increase in SHBG with age. These latter results are also consistent with findings in earlier investigations (2–4, 9, 10).

Cohort and secular effects are two potential confounds of cross-sectional studies of aging (15). A cohort effect is one that occurs when individuals of similar age vary significantly from older and/or younger groups, in one or more measured parameters, because of a historical factor common to their generation. A secular effect is one that results when there is progressive alteration of a critical condition in the environment, producing the appearance of a change with age. Our study contained an apparent secular effect that was probably factitious, *i.e.* systematically higher T levels at every age in samples obtained less recently, leading to an apparent decrease in serum T of about 40% from the 1970's to the 1990's. Such an effect could have occurred because of a change in environmental exposure of our study cohort, over time, such as an increase in estrogenic substances in food and/or water, which has been suspected to cause a decrease in sperm counts and a higher incidence of male infertility (39, 40). However, careful evaluation suggested a progressive alteration in the frozen samples, with time in storage, leading to a systematic upward variation in assay estimates of total T.

The latter change may have been attributable to the observed decrease, with increasing storage time, in levels of SHBG. This protein is stable in short-term frozen storage (41) but was progressively depleted in specimens stored for years in our study. If the antibody we employed in the T RIA competed incompletely with SHBG for binding of T, samples with higher SHBG levels would have lower apparent total T estimates. Our finding, that known additions of pure T to older samples assayed as about 30% greater, compared with identical additions to more recent samples, is consistent with such a mechanism.

Another potential confound, common to both cross-sectional and longitudinal studies, is a progressive effect of age-related variables, such as chronic illness or cumulative exposure to alcohol or tobacco, which are independent of the aging process itself (8, 14, 39). In our study, adjustment for illness, medications, and smoking produced little change in the overall effect observed because of aging *per se*. Of the chronic illnesses identified in our study population, only cancer, present (at any time during the period of follow-up) in 9%, was associated with a greater decrease in T levels than occurred with aging alone. This is consistent with our prior observations in men with cancer (14) and accords with other studies suggesting that serious chronic illness is associated with decreased T secretion in men (8, 13). The present finding that coronary artery disease had little or no impact on serum T levels is consistent with our previous prospective study in BLSA men (42), but it contrasts with other studies in which coronary heart disease and/or coronary risk factors have been associated with diminished T levels (43) or a greater rate of decline in serum T over time (16).

We found no significant effects of current smoking on T and free T index. In a previous large cross-sectional study, smokers were found to have somewhat higher levels of total T (8). Moreover, excessive alcohol consumption is a well-known cause of male hypogonadism (12). The apparent lack of effect of smoking and the relatively small alcohol effect on free T index were probably attributable to the relatively small number of smokers and, possibly, to the lighter use of both tobacco and alcohol in this health-conscious population. The use of β-blocking agents was associated with a trend to somewhat higher levels of T and free T index. Whether this association was caused by an effect of the drugs themselves, or some characteristic of men more likely to be prescribed these drugs, is currently unknown. Hypertensive men treated with β-blocking agents often experience sexual dysfunction and have mild reductions in serum T levels (44, 45). Although the mechanism(s) by which β-blocking agents might increase serum T is unknown, propranolol or nadolol have been reported to prevent melatonin-induced short photoperiod gonadal regression in male hamsters (46).

Our finding that total T, but not free T index, tended to decrease with greater BMI is consistent with prior studies showing that obesity is associated with decreases in both SHBG and total T, with an unchanged T-to-SHBG ratio (47), but contrasts with other studies showing diminished free, as well as total, T in with increasing total (48) or abdominal (49) obesity in men.

Our analysis of date-adjusted T and free T index levels, by decade, showed that relatively high numbers of older men in this generally healthy population had at least one hypogonadal value (defined as below the 2.5th percentile for young men). By the free T index criterion, this fraction rose from nearly 35% of men in their 60's to approximately 90% of men in their 80's. Using a total T criterion, the rates of hypogonadism were still significant but somewhat less dramatic. The issue of how properly to define hypogonadism, or indeed any hormone deficiency, remains problematic. The conventional definition for T levels is statistical (values more than 2 SD below the mean), rather than functional. Such a definition does not reflect clinical realities, such as the existence of characteristic individual set points for circulating hormone levels, below which one, but not another, individual may develop metabolic changes of hormone deficiency; nor does it address the concept of reserve capacity, the possibility that persons with hormone levels 2 SD below the population mean still may have adequate hormone concentrations to meet their metabolic needs. Using a dependent indicator hormone (*e.g.* LH for gonadal function) is helpful, to some extent, but is also potentially circular (*i.e.* the issue becomes that of determining what is the upper limit of normal for an LH level). It would clearly be better to define the lower limit of normal for a hormone as: the blood level at which metabolic and/or clinical sequelae of hormone deficiency begin to appear, or the level below which definite benefits can be demonstrated for hormone supplementation for a significant proportion of the population. However, insufficient data exist to do this with a high degree of confidence for T. The recent observation that elderly men with T levels less than 200 ng/dL at baseline had significant improvements in BMD after 3 yr of T treatment, whereas men with higher T levels did not (50), is a good beginning, but more information is needed. Higher bioavailable (but not total) T levels have been associated with greater BMD (51, 52) in men and are inversely associated with fat mass (53) in older men. Free or total T level is also a significant independent predictor of lean body mass and muscle strength (53, 54). These findings highlight the importance of determining the extent to which the age-related decrease in T is associated with changes common to pathological hypogonadism (29, 30) and the so-called aging phenotype, including increased total and central body fat (25–28), decreased lean body and skeletal muscle mass and strength (20–23), bone calcium depletion (18, 19, 55), and the metabolic and clinical consequences of these alterations, as well as the increase in erectile dysfunction and loss of libido (56).

Given the likelihood that age-related sarcopenia contributes importantly to frailty, (57) and the importance of osteoporotic fractures as a cause of morbidity and mortality in elderly men (58–60), T replacement is a potentially useful strategy for reducing age-associated disabilities in some aged men (61). Three months of T enanthate injections increased lean body mass in men over 60 yr of age (62); T treatment improved hamstring and quadricep muscle strength, after 4 weeks, in healthy older men (63); and administration of T for 2 yr produced a gain in bilateral grip strength in elderly hypogonadal men (64). In contrast, 3 yr of T replacement, using transdermal patches, failed to improve muscle strength in men more than 65 yr of age, despite significant increases in lean body mass (65).

To our knowledge, the current study is the largest longitudinal evaluation of the effects of normal aging on male gonadal hormone function reported to date. Results strongly support the concept of an effect of aging to lower both total and bioavailable circulating T levels at a relatively constant rate, independent of obesity, illness, medications, cigarette smoking, or alcohol intake. In addition, our findings suggest that a significant proportion of men over 60 yr of age have circulating T concentrations in the range conventionally considered to be hypogonadal. Whether such men may benefit from T replacement therapy deserves further investigation.

Acknowledgments

We gratefully acknowledge the expert nursing and administrative assistance of the BLSA staff, especially Mr. Howard Baldwin; Mr. Norman Beaudry of Covance Laboratories, Inc. for performance of T and SHBG RIAs; and Drs. Reubin Andres and Adrian Dobs for constructive and critical reviews of this manuscript.

References

1. **Vermeulen A, Rubens R, Verdonck L.** 1972 Testosterone secretion and metabolism in male senescence. J Clin Endocrinol Metab. 34:730–735.
2. **Rubens R, Dhont M, Vermeulen A.** 1974 Further studies on Leydig cell function in old age. J Clin Endocrinol Metab. 39:40–45.
3. **Baker HW, Burger HG, de Kretser DM, et al.** 1976 Changes in the pituitary-testicular system with age. Clin Endocrinol (Oxf). 5:349–372.
4. **Pirke KM, Doerr P.** 1975 Age related changes in free plasma testosterone, dihydrotestosterone and oestradiol. Acta Endocrinol (Copenh). 80:171–178.
5. **Purifoy FE, Koopmans LH, Mayes DM.** 1981 Age differences in serum androgen levels in normal adult males. Hum Biol. 53:499–511.
6. **Bremner WJ, Prinz PN.** 1983 A loss of circadian rhythmicity in blood testosterone levels with aging in normal men. J Clin Endocrinol Metab. 56:1278–1281.
7. **Tenover JS, Matsumoto AM, Plymate SR, Bremner WJ.** 1987 The effects of aging in normal men on bioavailable testosterone and luteinizing hormone secretion: Response to clomiphene citrate. J Clin Endocrinol Metab. 65:1118–1126.
8. **Gray A, Berlin JA, McKinlay JB, Longcope C.** 1991 An examination of research design effects on the association of testosterone and male aging: results of a meta-analysis. J Clin Epidemiol. 44:671–684.
9. **Ferrini RL, Barrett-Connor E.** 1998 Sex hormones and age: a cross-sectional study of testosterone and estradiol and their bioavailable fractions in community-dwelling men. Am J Epidemiol. 147:750–754.
10. **Harman SM, Tsitouras PD.** 1980 Reproductive hormones in aging men. I. Measurement of sex steroids, basal luteinizing hormone and Leydig cell response to human chorionic gonadotropin. J Clin Endocrinol Metab. 51:35–41.
11. **Nieschlag E, Lammer U, Freischem CW, Langer K, Wickings EJ.** 1982 Reproductive function in young fathers and grandfathers. J Clin Endocrinol Metab. 51:675–681.
12. **Sparrow D, Bosse R, Rowe JW.** 1980 The influence of age, alcohol consumption, and body build on gonadal function in men. J Clin Endocrinol Metab. 51:508–512.
13. **Deslypere JP, Vermeulen A.** 1984 Leydig cell function in normal men: effect of age, life-style, residence, diet, and activity. J Clin Endocrinol Metab. 59:955–962.
14. **Blackman MR, Weintraub BD, Rosen SW, Harman SM.** 1988 Comparison of the effects of lung cancer, benign lung disease, and normal aging on pituitary-gonadal function in men. J Clin Endocrinol Metab. 66:88–95.
15. **Sorkin JD, Katzel LI.** 1998 Methodological problems in research on aging. In: Tallis R, Fillit H, Brocklehurst JC, eds. Textbook of geriatric medicine and gerontology. 5th ed. London: Churchill Livingstone; 179–192.
16. **Zmuda JM, Cauley JA, Kriska A, et al.** 1997 Longitudinal relation between endogenous testosterone and cardiovascular disease risk factors in middle-aged men. A 13-year follow-up of former Multiple Risk Factor Intervention Trial participants. Am J Epidemiol. 146:609–617.
17. **Morley JE, Kaiser FE, Perry 3rd HM, et al.** 1997 Longitudinal changes in testosterone, luteinizing hormone, and follicle-stimulating hormone in healthy older men. Metabolism. 46:410–413.
18. **Orwoll ES, Oviatt SK, McClung MR, Deftos LJ, Sexton G.** 1990 The rate of bone mineral loss in normal men and the effects of calcium and cholecalciferol supplementation. Ann Intern Med. 112:29–34.
19. **Marcus R.** 1991 Skeletal aging. Understanding the functional and structural basis of osteoporosis. Trends Endocrinol Metab. 2:53–58.
20. **Forbes GB, Reina JC.** 1970 Adult lean body mass declines with age: some longitudinal observations. Metabolism. 19:653–663.
21. **Gallagher D, Visser M, De Meersman RE, et al.** 1997 Appendicular skeletal muscle mass: effects of age, gender, and ethnicity. J Appl Physiol. 83:229–239
22. **Kallman DA, Plato C, Tobin J.** 1990 The role of muscle loss in age-related decline in grip strength: A cross-sectional and longitudinal analysis. J Gerontol. 45:M82–M88.
23. **Bruce SA, Newton D, Woledge RC.** 1989 Effects of age on voluntary force and cross-sectional area of human adductor policis muscle. Q J Exp Physiol. 74:359–362.
24. **Fleg JL, Lakatta EG.** 1990 Role of muscle loss in the age-associated reduction in VO2max in older men. J Appl Physiol. 68:329–333.
25. **Horber FF, Gruber B, Thomi F, Jensen EX, Jaeger P.** 1997 Effect of sex and age on bone mass, body composition and fuel metabolism in humans [see Comments]. Nutrition. 13:524–534.
26. **Siervogel RM, Wisemandle W, Maynard LM, et al.** 1998 Serial changes in body composition throughout adulthood and their relationships to changes in lipid and lipoprotein levels. The Fels Longitudinal Study. Arterioscler Thromb Vasc Biol. 18:1759–1764.
27. **Kreisberg RA, Kasim S.** 1987 Cholesterol metabolism and aging. Am J Med. 82:54–60.
28. **Couillard C, Lemieux S, Moorjani S, et al.** 1996 Associations between 12 year changes in body fatness and lipoprotein- lipid levels in men and women of the Quebec Family Study. Int J Obes Relat Metab Disord. 20:1081–1088.
29. **Mauras N, Hayes V, Welch S, et al.** 1998 Testosterone deficiency in young men: marked alterations in whole body protein kinetics, strength, and adiposity. J Clin Endocrinol Metab. 83:1886–1892.
30. **Katznelson L, Rosenthal DI, Rosol MS, et al.** 1998 Using quantitative CT to assess adipose distribution in adult men with acquired hypogonadism. AJR Am J Roentgenol. 170:423–427.
31. **Kannus P, Parkkari J, Sievanen H, et al.** 1996 Epidemiology of hip fractures. Bone. [1 Suppl] 18:57S–63S.
32. **Shock NW, Gruelich RC, Andres RA, et al., eds.** 1984 Design and operation of the Baltimore Longitudinal Study of Aging. Washington, DC: U.S. Government Printing Office. NIH Publication 84–2450.
33. **Furuyama S, Mayes DM, Nugent CA.** 1970 A radioimmunoassay for plasma testosterone. Steroids. 16:415–428.
34. **Goldstein H, Rasbash J, Plewi I, et al.** 1998 A User's guide to MLWIN. London: University of London.
35. **Boyle WJ, Compton AA, Rigg G, Menon KM, McCann DS.** 1984 Serum testosterone measurements. Am J Clin Pathol. 81:754–761.
36. **Wang C, Baker HW, Burger HG, De Kretser DM, Hudson B.** 1975 Hormonal studies in Klinefelter's syndrome. Clin Endocrinol (Oxf). 4:399–411.
37. **NHLBI Obesity Education Initiative Expert.** 1998 Panel on the identification and treatment of overweight and obesity in adults. Clinical guidelines on the identification, evaluation, and treatment of overweight and obesity in adults Bethesda, MD: National Heart, Lung, and Blood Institute.
38. **World Health Organization.** 1998 Obesity: preventing and managing the global epidemic: report of a WHO consultation on obesity. Geneva: WHO.
39. **Vermeulen A.** 1993 Environment, human reproduction, menopause, and andropause. Environ Health Perspect. [101 Suppl] 2:91–100.
40. **Jensen TK, Toppari J, Keiding N, Skakkebaek NE.** 1995 Do environmental estrogens contribute to the decline in male reproductive health? Clin Chem. 41:1896–1901.
41. **Kley HK, Rick W.** 1984 The effect of storage and temperature on the analysis of steroids in plasma and blood. J Clin Chem Clin Biochem. 22:371–378.
42. **Contoreggi CS, Blackman MR, Andres RA, et al.** 1990 Plasma levels of estradiol, testosterone and DHEAS do not predict risk of coronary artery disease in men. J Androl. 11:460–470.
43. **Phillips GB, Pinkernell BH, Jing TY.** 1994 The association of hypotestosteronemia with coronary artery disease in men. Arterioscler Thromb. 14:701–706.
44. **Suzuki H, Tominaga T, Kumagai H, Saruta T.** 1988 Effects of first-line antihypertensive agents on sexual function and sex hormones. J Hypertens Suppl. 6:S649–S651.
45. **Rosen RC, Kostis JB, Jekelis AW.** 1988 Beta-blocker effects on sexual function in normal males. Arch Sex Behav. 17:241–255.
46. **Champney TH.** 1989 β-Adrenergic blockers prevent short photoperiod-induced gonadal regression, but not melatonin-induced regression in male Syrian hamsters. J Exp Zool. 249:221–228.
47. **Amatruda JM, Harman SM, Pourmotabbed G, Lockwood DH.** 1978 Depressed plasma testosterone and fractional binding of testosterone in obese males. J Clin Endocrinol Metab. 47:268–271.
48. **Zumoff B, Strain GW, Miller LK, et al.** 1990 Plasma free and non-sex-hormone-binding-globulin-bound testosterone are decreased in obese men in proportion to their degree of obesity. J Clin Endocrinol Metab. 71:929–931
49. **Haffner SM, Valdez RA, Stern MP, Katz MS.** 1993 Obesity, body fat distribution and sex hormones in men. Int J Obes Relat Metab Disord. 17:643–649.
50. **Snyder PJ, Peachey H, Hannoush P, et al.** 1999 Effect of testosterone treatment on bone mineral density in men over 65 years of age [In Process Citation]. J Clin Endocrinol Metab. 84:1966–1972.
51. **Murphy S, Khaw KT, Cassidy A, Compston JE.** 1993 Sex hormones and bone mineral density in elderly men. Bone Miner. 20:133–140.

52. Greendale GA, Edelstein S, Barrett-Connor E. 1997 Endogenous sex steroids and bone mineral density in older women and men: the Rancho Bernardo Study. J Bone Miner Res. 12:1833–1843.
53. Abbasi AA, Mattson DE, Duthie Jr EH, et al. 1998 Predictors of lean body mass and total adipose mass in community-dwelling elderly men and women. Am J Med Sci. 315:188–193.
54. Baumgartner RN, Waters DL, Gallagher D, Morley JE, Garry PJ. 1999 Predictors of skeletal muscle mass in elderly men and women. Mech Ageing Dev. 107:123–136.
55. Ongphiphadhanakul B, Rajatanavin R, Chailurkit L, et al. 1995 Serum testosterone and its relation to bone mineral density and body composition in normal males. Clin Endocrinol (Oxf). 43:727–733.
56. Korenman SG, Morley JE, Mooradian AD, et al. 1990 Secondary hypogonadism in older men: its relation to impotence. J Clin Endocrinol Metab. 71:963–969.
57. Evans W. 1997 Functional and metabolic consequences of sarcopenia. J Nutr. [5 Suppl] 127:998S–1003S.
58. Orwoll ES, Klein RF. 1995 Osteoporosis in men. Endocr Rev. 16:87–116.
59. Jackson HA, Kleerkoper M. 1990 Osteoporosis in men: diagnosis, pathophysiology, and prevention. Medicine. 69:137–152.
60. Francis RM, Peachck M, Marshall DH, Horsman A, Aaron JE. 1989 Spinal osteoporosis in men. Bone Miner. 5:347–357.
61. Morley JE, Kaiser FE, Sih R, Hajjar R, Perry 3rd HM. 1997 Testosterone and frailty. Clin Geriatr Med. 13:685–695.
62. Tenover JS. 1992 Effects of testosterone supplementation in the aging male. J Clin Endocrinol Metab. 75:1092–1098.
63. Urban RJ, Bodenburg YH, Gilkison C, et al. 1995 Testosterone administration to elderly men increases skeletal muscle strength and protein synthesis. Am J Physiol. 269:E820–E826.
64. Sih R, Morley JE, Kaiser FE, et al. 1997 Testosterone replacement in older hypogonadal men: a 12-month randomized controlled trial [see Comments]. J Clin Endocrinol Metab. 82:1661–1667.
65. Snyder PJ, Peachey H, Hannoush P, et al. 1999 Effect of testosterone treatment on body composition and muscle strength in men over 65 years of age. J Clin Endocrinol Metab. 84:2647–2653.

FOCUS

SKIN AGING: OBSERVATIONS AT THE CELLULAR AND MOLECULAR LEVELS

MINA YAAR

Department of Dermatology, Boston University School of Medicine, Boston, Massachusetts, USA

Isr J Med Sci 1996;**32**:1053–1058

Keywords: skin aging; apoptosis; growth factors; signal transduction; receptors

Understanding the aging process is of great relevance to health care providers in developed countries because of a marked increase in the absolute number and the proportion of the population who are elderly (1). The elderly, those 65 years of age and older, visit physicians and use hospitals and nursing care facilities considerably more than those less than 65 years old. As well, the current leading causes of morbidity and mortality are related to deterioration of individual organ systems and to cancer, two conditions whose incidence is highest in the elderly population.

Aging can best be defined as a progressive deterioration in the ability of an organism to respond adaptively to environmental changes. Such deterioration leads to increased vulnerability and ultimately death. Since most of the processes in the body begin at the cellular level, it is important to understand how the aging process affects the individual cells in our body.

The skin offers an excellent model system in which to study aging, because of its easy accessibility, wide variety of known functions, and well-characterized cell types including keratinocytes, fibroblasts and melanocytes. Recently, sophisticated tissue culture systems have been developed. They enable us to isolate cutaneous cells from newborns, young donors (age 20–25) and old donors (over 65 years). Cells isolated from these skin specimens are maintained in culture under identical conditions and their behavior can be compared.

The age-associated cellular and molecular changes that will be discussed in this review are only those observed in cells isolated from sun-protected skin, and therefore can be considered as representing intrinsic aging. Changes that occur in habitually sun-exposed older skin are termed photoaging. Photoaging is a different process from intrinsic aging because it includes decrements that are the result of ultraviolet irradiation and not only changes that occur with chronologic aging.

In vitro studies from our laboratory have shown for both keratinocytes and fibroblasts that there is a significant age-associated decrease in cellular proliferative capacity (Fig. 1). This decreased proliferation is most pronounced between newborn and adult donors, but can also be detected between young adult and old adult donors (2). The decreased mitogenic response is observed when cells are stimulated with serum that contains several mitogens. It is also observed in response to specific mitogens like insulin, epidermal growth factor, or platelet-derived growth factor. Therefore, the behavior of cultured skin-derived cells reflects the chronologic age of the tissue donor.

Address for correspondence: Dr. M. Yaar, Department of Dermatology, Boston University School of Medicine, 80 East Concord Street, Boston, MA 02118-2394, USA.

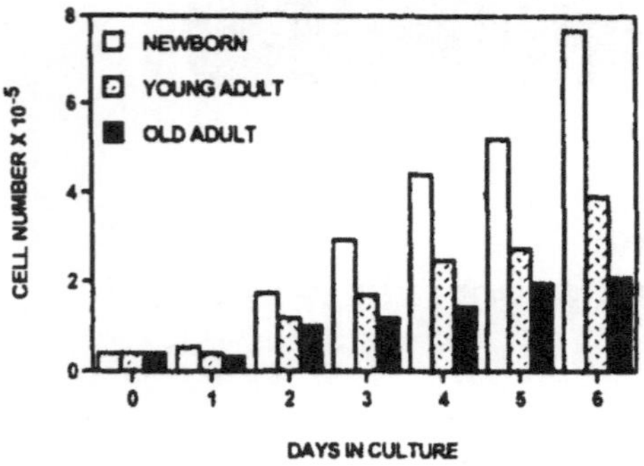

Fig. 1. Effect of donor age on mitogen-stimulated fibroblast proliferation.
Fibroblasts from newborn (white bar), young adult (24–32 years) (stippled bar) and old adult (67–82 years) (black bar) donors were provided with 10% calf serum, a known mitogen for fibroblasts. Over the course of 6 days, newborn cells proliferated significantly more than adult fibroblasts, and young adult fibroblasts proliferated more than old adult cells. This experiment shows an age-associated decrease in the proliferative capacity in response to mitogenic stimulation.

Since the early 1960's, much emphasis has been placed on identifying the homeostatic pathways that fail with aging and lead to the observed age-associated decreased proliferative capacity. Chromosomal alterations — including aneuploidy (extra or missing chromosomes), structural abnormalities (breaks or amplifications), single-strand breaks, and DNA cross-linking — are reported to increase with age (3–5). Decreased DNA methylation, a mechanism employed by vertebrate cells to regulate transcription, is also noted with aging (6). Failure of this mechanism may interfere with the proper control of cellular proliferation.

It was also reported that telomeres become shorter with aging (7). Telomeres are protein-DNA structures, at the end of eukaryotic chromosomes, whose proposed function is to protect chromosomes from degradation. They are also required for proper gene replication (8). A significant correlation is found between telomere length and the replicative capacity of the cell (7). Therefore, as telomere length becomes shorter with aging, the replicative capacity of the cell decreases.

Studies in our laboratory have focused on identifying mechanisms upstream of gene replication that compromise the ability of old adult cells to proliferate in response to mitogenic stimuli. Cellular proliferation requires the presence of extracellular signaling molecules, and a set of receptors that enable the cell to bind and respond to these molecules (Fig. 2). In the absence of these mitogenic signals, the cell is under the influence of specific inhibitory mechanisms. However, once a mitogenic signal is provided to the cell, it undergoes a series of events that lead to cellular proliferation (9).

The mitogenic signal is provided by proteins called growth factors. They are effective in very low concentrations, about 10^{-9} to 10^{-11} M. For each type of growth factor there is a specific receptor or set of receptors. Cells respond to a given growth factor only if they display the appropriate receptor. Binding of the growth factor to the receptor transiently activates the receptor, which in turn activates several intracellular proteins. Through a cascade of activated cytosolic proteins, the signal is transduced from the cell surface to the nucleus (10) (Fig. 2).

Because there is an age-associated decrease in cellular proliferation it appears plausible that aging may affect the level or function of growth factors and/or molecules that are part of the signal transduction pathway. *In vivo* studies suggest that there is a gradual age-associated decrease in the levels of at least some growth factors and hormones. These include insulin-like growth factors I and II (11,12), transforming growth factor-beta (12), nerve growth factor (13), growth hormone and calciotropic hormone (14). Therefore, the age-associated decrease in cellular proliferation, resulting in impaired wound healing in the elderly, may be due, at least in part, to insufficient amounts of growth factors in the wound bed.

The first step in growth factor biosynthesis is the production of mRNA transcripts from specific genes. Few studies have measured the level of mRNA species that encode growth factors. The data from these studies show that although the effect of age on growth factor mRNA levels varies, the levels of many transcripts decline with age. These include insulin-like growth factor I and II (15), nerve growth factor (16), growth hormone (17), interleukins 2 and 3 (18,19), and epidermal growth factor (20). However, the level of some mRNA species does not change significantly with age (21) and, in a few cases, levels have been reported to increase with age (22,23). These discrepancies appear to reflect in part differences in mRNA levels among different tissues and may also reflect differences in the culture conditions or the cell type used. Nonetheless, it appears that age-associated decrements in transcription and/or mRNA stability of some genes that encode growth factors may account for decreased amounts of specific factors that are crucial for an adequate proliferative response. However, *in vitro* studies from our and other laboratories suggest that cells derived from old donors, even when

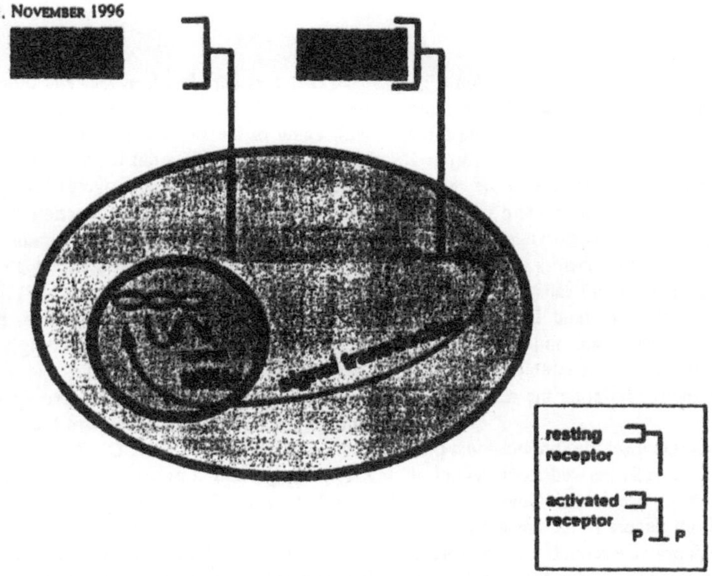

Fig. 2. Growth factor-mediated signal transduction pathway.
The previously resting receptor becomes phosphorylated on its intracellular portion upon growth factor binding (as indicated by the letter P). The phosphorylated (activated) receptor activates intracellular and intranuclear signal transduction pathways that eventually lead to cellular proliferation.

provided with optimal concentrations of growth factors, proliferate less than cells derived from young donors (Fig. 1). Therefore, other age-associated decrements, either in growth factor receptor function, in the intracellular and intranuclear signal transduction pathways, or in both, may also play a role in the aging process.

Indeed, studies from our laboratory, using dermal fibroblasts and epidermal growth factor as a prototype mitogen, showed a significant age-associated decrease in epidermal growth factor binding to its receptor (24). Compared to newborn cells, young adult fibroblasts were found to have half the number of functional receptors, and old adult fibroblasts had less than 20%. Further investigations revealed that there was also decreased affinity of the existing receptors for their ligand (24). Further studies from our laboratory displayed that epidermal growth factor-receptor mRNA level decreased with age (25). mRNA levels for the androgen receptor (26), glucocorticoid receptor II (27), and interleukin II receptor (28,29) were also found by others to decrease with age. These results suggest that the decreased cellular proliferation observed with aging is not due solely to decreased amounts of growth factors but may also reflect impairment in the level of some growth factor receptors and/or decrements in growth factor–growth factor receptor interaction. Further studies are required to determine whether these changes constitute a general mechanism of cellular aging, affecting a wide range of growth factor and cytokine pathways, or whether they are limited to selected growth factors and their receptors.

A critical role in growth factor-mediated signal transduction is played by transcription factors. This diverse group of proteins acts within the nucleus to form a complex with RNA polymerase and to initiate the synthesis of RNA. Most of the studies that investigate transcription factors have focused on the fos-jun family, which encodes proteins that bind the AP-1 element in the DNA (Fig. 3) (30). It was shown that even in the presence of optimal concentrations of growth factors and functioning receptors, the induction of c-fos mRNA was blocked in late-passage fibroblasts. This suggests that senescent cells are unable to proliferate because of selective repression of c-fos (31). c-fos appears to initiate gene transcription in response to a wide range of environmental stimuli. Its induction in tissues such as vascular endothelium after hemodynamic stress (32), and brain after electroconvulsive shock (33) was also reported to decrease with age. These results suggest that with aging there is a diminished capacity of the cell to upregulate c-fos in response to various exogenous stimuli.

Other proteins that can control cellular proliferation have also been identified. For example, the retinoblastoma protein binds and sequesters specific transcription factors that are crucial for activation of genes that are required for cellular proliferation (Fig. 4) (34). In the absence of mitogenic stimulation, these regulatory proteins are bound to the retinoblastoma protein and cellular proliferation cannot be initiated. However, the binding of these factors to the retinoblastoma protein depends on this protein's state of phosphorylation. In its dephosphorylated state, it strongly binds the transcription factors. Phosphorylation of the retinoblastoma protein causes it to release the proteins, allowing them to bind DNA and start cellular proliferation. When the mitogenic signal reaches the nucleus it relieves the inhibition on cellular proliferation by phosphorylating the retinoblastoma protein.

We have analyzed the level of retinoblastoma protein phosphorylation in cells derived from donors of different ages after mitogenic stimulation. Before stimulation, cells from donors in all age groups displayed only unphosphorylated retinoblastoma protein, as expected in resting cells. However, within 16 hours after mitogenic stimulation of newborn cells the majority of the retinoblastoma protein became phosphorylated. In contrast, with increasing donor age a decrease in the level of retinoblastoma protein phosphorylation was observed (35). These combined results suggest that intranuclear events downstream of the receptor–ligand interaction are also affected with aging.

Taken together, data from our and other laboratories imply that aging at the cellular level is consistent with aging of the whole organism. We observed an age-associated deterioration in the ability of the cell to respond to environmental signals like growth factors, leading to compromised function of individual organ systems, as well as the entire organism.

Recently, a family of genes called growth arrest and DNA damage inducible genes, or GADD, has been cloned. The function of the protein products of these genes is still unknown. Interestingly, however, these genes are induced *in vivo* and *in vitro* by stresses such as growth factor deprivation or DNA damage (36). We examined *in vitro* the expression of one GADD gene, GADD 153, in newborn keratinocytes, and compared its expression in adult keratinocytes (2). When cells were not provided fresh growth factors, GADD mRNA levels were high in both newborn and adult cells. However, within 4 hours of growth factor provision, GADD expression was nondetectable in newborn keratinocytes. In contrast, a low level of GADD expression persisted in adult cells, and its level was strongly induced in these cells sooner than in newborn cells, as early as 14 hours after provision of growth factors. In newborn keratinocytes, GADD was induced only 24 hours after growth factor provision. This experiment suggests that adult keratinocytes never fully respond

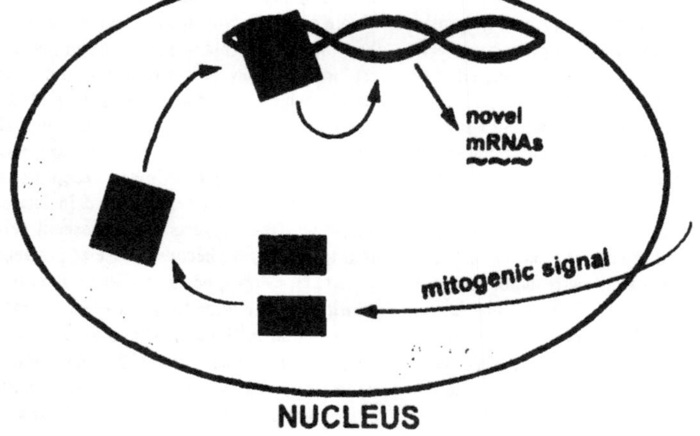

Fig. 3. Fos/jun mediated RNA transcription.
The nuclear transcription factors fos and jun are transiently induced by mitogenic stimulation. They bind each other, and the complex then binds nuclear DNA to induce gene transcription and the synthesis of proteins required for cellular proliferation.

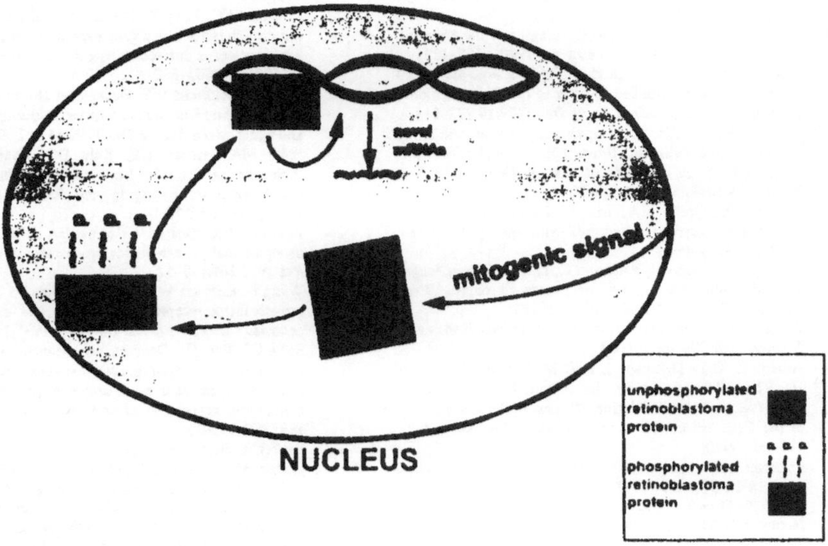

Fig. 4. Phosphorylation of the retinoblastoma protein.
The retinoblastoma protein (Rb) binds and sequesters specific transcription factors required for activation of genes that participate in cellular proliferation. Upon mitogenic stimulation, the retinoblastoma protein becomes phosphorylated by specific kinases (as indicated by the letter p, on its side). Retinoblastoma phosphorylation results in the dissociation of transcription factors from the retinoblastoma protein and these transcription factors bind nuclear DNA and induce gene transcription and the synthesis of novel proteins required for cellular proliferation.

to the presence of the signalling growth factors, and they perceive lack of growth factors earlier than newborn keratinocytes, consistent with a loss of sensitivity or awareness to environmental signals.

Interestingly, it was recently suggested that the GADD genes may be involved in the process of programmed cell death or apoptosis. Apoptosis is an active process in which cells execute their own death when necessary (37). Apoptosis occurs normally during embryonic development, as well as in the adult organism. One of the well-characterized signals for the initiation of apoptosis in cells is growth factor deprivation. In the absence of growth factors the apoptotic program is activated and cells die. It is therefore intriguing to speculate, based on our observations, that those aged cells in our body that are incapable of perceiving the presence of growth factors, as manifested by continuous expression of the GADD gene, are, in a sense, growth factor deprived and therefore nonfunctional, and are eliminated by the process of apoptosis.

In summary, work on skin aging provides insights into the aging process in general. It suggests that aging is accompanied by a loss of response to growth factors and consequently a loss of proliferative ability. Furthermore, recent reports also raise the possibility that aging is an apoptotic process aimed at eliminating dysfunctional cells. However, whether aging occurs as an inherent genetic program, or whether it is the result of random environmental damage, remains to be elucidated. The advent of sophisticated molecular biology techniques makes it possible to understand the aging process at the cellular and molecular levels. Let us hope that with this understanding, novel approaches will be designed to overcome the debilitating effects of the aging process.

REFERENCES

1. Johnson M-LT. Aging of the United States population: The dermatologic implications. In: Peterson BL, Gilchrest BA, eds. Clinics in Geriatric Medicine. Philadelphia: WB Saunders Co.,1989:41–51.
2. Yaar M. Molecular mechanisms of skin aging. In: Dahl

MB, ed. Advances in Dermatology, vol. 10. Chicago: Mosby-Year Book, Inc., 1995:63–75.
3. Vijg J. DNA sequence changes in aging: How frequent, how important? Aging: *Clin Exp Res* 1990;**2**:105–123.
4. Nakanishi K, Shima A, Fukuda J, Fujita S. Age-associated increase of single-stranded regions in the DNA of mouse brain and liver cells. *Mech Aging Dev* 1979;**10**:273–281.
5. Tice RR, Setlow RB. DNA repair and replication in aging organisms and cells. In: Finch CE, Schneider EL, eds. Handbook of the Biology of Aging. New York: Van Nostrand Reinhold, 1985:173–224.
6. Wilson VL, Smith RA, Ma S, Cutler RG. Genomic 5-methyldeoxycytidine decreases with age. *J Biol Chem* 1987;**262**:9948–9951.
7. Allsopp RC, Vaziri H, Patterson C, et al. Telomere length predicts replicative capacity of human fibroblasts. *Proc Natl Acad Sci USA* 1992;**89**:10114–10118.
8. Zakian VA. Telomeres: Beginning to understand the end. *Science* 1995;**270**:1601–1607.
9. Alberts B, Bray D, Lewis J, Raff M, Roberts K, Watson JD. The cell division cycle. In: Alberts B, Brey D, Lewis J, Raff M, Roberts K, Watson JD, eds. Molecular Biology of the Cell, 3rd ed. New York: Garland Publishing Inc., 1994:863–910.
10. Feldman SR, Yaar M. Oncogenes: The growth control genes. *Arch Dermatol* 1991;**127**:707–711.
11. Cohen P, Ocrant I, Fielder PJ, et al. Insulin-like growth factors (IGFs): Implications for aging. *Psychoneuroendocrinology* 1992;**17**:335–342.
12. Nicolas V, Prewett A, Bettica P, et al. Age-related decreases in insulin-like growth factor-I and transforming growth factor-beta in femoral cortical bone from both men and women: Implications for bone loss with aging. *J Clin Endocrinol Metab* 1994;**78**:1011–1016.
13. Nishizuka M, Katoh-Semba R, Eto K, Arai Y, Iizuka R, Kato K. Age- and sex-related differences in the nerve growth factor distribution in the rat brain. *Brain Res Bull* 1991;**27**:685–688.
14. Rosen CJ. Growth hormone, insulin-like growth factors, and the senescent skeleton: Ponce de Leon's Fountain revisited? *J Cell Biochem* 1994;**56**:348–356.
15. Park GH, Buetow DE. Genes for insulin-like growth factors I and II are expressed in senescent rat tissues. *Gerontology* 1991;**37**:310–316.
16. Larkfors L, Ebendal T, Whittemore SR, Persson H, Hoffer B, Olson L. Decreased level of nerve growth factor (NGF) and its messenger RNA in the aged rat brain. *Mol Brain Res* 1987;**3**:55–60.
17. Martinoli MG, Ouellet J, Rhéaume E, Pelletier G. Growth hormone and somatostatin gene expression in adult and aging rats as measured by quantitative *in situ* hybridization. *Neuroendocrinology* 1991;**54**:607–615.
18. Wu WT, Pahlavani MA, Cheung HT, Richardson A. The effect of aging on the expression of interleukin 2 messenger ribonucleic acid. *Cell Immunol* 1986;**100**:224–231.
19. Li DD, Chien YK, Gu MZ, Richardson A, Cheung HT. The age-related decline in interleukin-3 expression in mice. *Life Sci* 1988;**43**:1215–1222.
20. Gresik EW, Wenk-Salamone K, Onetti-Muda A, Gubits RM, Shaw PA. Effect of advanced age on the induction by androgen or thyroid hormone of epidermal growth factor and epidermal growth factor mRNA in the submandibular glands of C57BL/6 male mice. *Mech Ageing Dev* 1986;**34**: 175–189.
21. Compton CC, Tong Y, Trookman N, Zhao H. Transforming growth factor alpha gene expression in cultured human keratinocytes is unaffected by cellular aging. *Arch Dermatol* 1995;**131**:683–690.
22. Yang F, Freidrichs WE, Buchannan JM, et al. Tissue specific expression of mouse transferrin during development and aging. *Mech Ageing Dev* 1990;**56**:187–197.
23. Salih MA, Herbert DC, Kalu DN. Evaluation of the molecular and cellular basis for modulation of thyroid c-cell hormones by aging and food restriction. *Mech Ageing Dev* 1993;**70**:1–21.
24. Reenstra WR, Yaar M, Gilchrest BA. Effect of donor age on epidermal growth factor processing in man. *Exp Cell Res* 1993;**209**:118–122.
25. Wang B, Reenstra WR, Yaar M, Gilchrest BA. Epidermal growth factor receptor transcription and mRNA stability decrease with age. *J Invest Dermatol* 1995;**104**:674.
26. Song CS, Rao TR, Demyan WF, Mancini MA, Chatterjee B, Roy AK. Androgen receptor messenger ribonucleic acid (mRNA) in the rat liver: Changes in mRNA levels during maturation, aging, and caloric restriction. *Endocrinology* 1991;**128**:349–356.
27. Peiffer A, Barden N, Meaney MJ. Age-related changes in glucocorticoid receptor binding and mRNA levels in the rat brain and pituitary. *Neurobiol Aging* 1991;**12**:475–479.
28. Nagel JE, Chopra RK, Chrest FJ, et al. Decreased proliferation, interleukin 2 synthesis, and interleukin 2 receptor expression are accompanied by decreased mRNA expression in phytohemagglutinin-stimulated cells from elderly donors. *J Clin Invest* 1988;**81**:1096–1102.
29. Travali S, Camazzo G, Distefano AM, et al. Expression of cell cycle-dependent genes and proliferative state of lymphocytes in aging. *Arch Gerontol Geriatr* 1990;**11**: 133–139.
30. Lamb P, McKnight S. Diversity and specificity in transcriptional regulation: the benefits of heterotypic dimerization. *TIBS* 1991;**16**:417–422.
31. Seshadri T, Campisi J. Repression of c-fos transcription and an altered genetic program in senescent human fibroblasts. *Science* 1990;**247**:205–209.
32. Takahashi R, Schunkert H, Isoyama S, et al. Age-related differences in the expression of proto-oncogene and contractile protein genes in response to pressure overload in the rat myocardium. *J Clin Invest* 1992;**89**:939–946.
33. Rogue PJ, Ritz MF, Malviya AN. Impaired gene transcription and nuclear protein kinase C activation in the brain and liver of aged rats. *FEBS Lett* 1993;**334**:351–354.
34. Kaelin WG Jr, Krek W, Sellers WR, et al. Expression and cloning of a cDNA encoding a retinoblastoma-binding protein with E2F-like properties. *Cell* 1992;**24**:351–364.
35. Yaar M, DiBenedetto P, Garmyn M, Gilchrest BA. Retinoblastoma protein phosphorylation is defective during aging and enhanced after malignant transformation of keratinocytes. *Clin Res* 1992;**40**:522A.
36. Fornace AJ, Nebert DW, Hollander MC, et al. Mammalian genes coordinately regulated by growth arrest signals and DNA-damaging agents. *Mol Cell Biol* 1989;**9**:4196–4203.
37. Kerr JFR. Definition of apoptosis and overview of its incidence. In: Lavin M, Watters D, eds. Programmed Cell Death. Chur, Switzerland: Harwood Academic Publishers GmbH, 1993:1–15.

Horm Res 1995;43:89–92

Jean-Hilaire Saurat

Department of Dermatology, University
Canton Hospital, Geneva, Switzerland

Retinoids and Ageing

Key Words
Ageing
Photo ageing
Retinoic acid

Abstract
The observation that topical retinoic acid could attenuate wrinkles and other features of skin photoageing has projected this molecule into the media spotlight. Behind this effect are in fact major basic biological issues related to the role of retinoids in tissue maintenance and their possible interference with the molecular biology of cellular ageing.

Retinoids

Retinoids such as retinol (ROL), retinaldehyde (RAL) and retinoic acid (RA) play a fundamental role in embryogenesis, reproduction, vision, and control of cell growth and differentiation of many adult tissues. Most, if not all, the activities of retinoids occur through nuclear receptors that belong to the superfamily of steroid, thyroid hormones and vitamin D; two families of receptors are involved in the gene regulation induced by retinoids: the retinoic acid receptors (RARs) whose natural ligand is RA and the retinoid X receptor family (RXR) whose ligand is 9-*cis* RA. These biologically active ligands are thought to be generated within the target cell, a process designated as an 'intracrine' system [1]. Moreover, low-molecular-weight binding proteins that bind either precursors (ROL) or the active ligand RA are present in target cells and should be potential regulators of the intracrine process. Therefore three steps are currently actively investigated: (1) the production of active ligands from inactive precursors, (2) the role of cellular ROL and RA binding proteins and (3) the nuclear receptors.

Transformation of ROL into Active Ligands

The process comprises two steps; the first one, oxydation of ROL into RAL, is rate-limiting and involves a ROL deshydrogenase related to the family of alcohol deshydrogenases: the second step involves a retinal dehydrogenase that transforms RAL into RA. Cultured keratinocytes transform ROL into RA [2]; this indicates that epidermis is able to produce the active ligand from stored precursors. Interestingly only differentiating keratinocytes showed significant enzyme activity; this links the generation of RA to terminal differentiation of keratinocytes; RAL was a key metabolite in this process, which suggests that it might be a candidate for use as a proretinoid [3]. It is presently not known if there are age-related changes in these processes.

Cytosolic ROL and RA Binding Proteins

These proteins are members of the superfamily of fatty acid binding proteins. Cytosolic retinol binding protein 1 (CRBP1) appears to be involved in the uptake of ROL by the cell, as well as in the enzymic process of ROL esterification. The human epidermis expresses low amounts of CRBP1; these amounts are not dramatically modified upon exposure to exogenous (topical or systemic) retinoids. It has recently been shown that there are at least

Jean-Hilaire Saurat
Department of Dermatology
University Canton Hospital
CH-1211 Geneva 14 (Switzerland)

0301–0163/95/0433–0089
$8.00/0

two cytosolic RA binding proteins CRABP1 and CRABP2. They influence the synthesis and metabolism of retinoids and therefore the steady state concentration of free RA [4]. CRABP1 and CRABP2 differ in their tissue distribution, affinity for RA (CRABP1 7–16 n*M*; CRABP2 50 n*M*), and regulation by RA (CRABP2 is upmodulated by RA). The human epidermis appears to express low or undetectable mRNA for CRABP1 and 2 as shown by both Northern blot and in situ hybridization, whereas low levels of CRABP1 and 2 proteins can be identified by appropriate techniques [5]. To date there is no indication that ageing induces an alteration in the expression of these proteins in vivo; however, exposure to UVB, a known inducer of skin ageing, increases CRABP2 [6].

Nuclear Receptors: RAR and RXR

The RAR subfamily comprises three types of RAR, the RAR-α, -β and -γ. Analysis of the mRNA transcripted from these three genes has shown many isoforms (up to 6–8 for each) that are generated by alternative splicing with separate promoter regions for alternative A domains that are coded on separate exons; there may be differences at the 3′ region as well. However, it is not known whether all these RAR isoforms, as defined at the nucleic acid level, are indeed translated into functional RAR proteins [7]. There is therefore an extreme complexity in this field, which functional relevance may take time to be explained.

The natural ligand for RXRs has recently been shown to be an isomer of RA, 9-*cis* RA [8]. The RXR subfamily also appears to comprise three genes encoding for three distinct receptors RXR-α, -β and -γ; they are closely related to each other but quite divergent from the RAR subfamily in both structure and ligand specificity [9]. Altogether these ligand-dependent transcription factors constitute a family of retinoid nuclear receptors. Upon ligand binding they bind to short, specific DNA sequences, so-called response elements, in the vicinity of target genes and stimulate, or in some cases repress, transcription. Like all nuclear receptors, RAR and RXR consist of three main domains: an amino-terminal domain (responsible in part for transactivation, and which is not conserved within the superfamily), a highly conserved DNA-binding domain and a moderately conserved carboxy-terminal, ligand-binding domain. The DNA-binding domain consists of two zinc fingers, whose fine structure determines the subgroup of the nuclear receptor superfamily; RARs seem to fall into the subgroup that also comprises the receptors for vitamin D and T3. Following the current model, the members of this subgroup function most likely as heterodimers with the RXR. This indicates that retinoids, via the isomer 9-*cis*, influence the regulation of many genes, including vitamin D- and thyroid hormone-responsive genes [10, 11]. Such promiscuity considerably increases the number of potential retinoid-responsive genes. So far a limited number of genes that have nucleotide sequences capable of binding RAR and RXR proteins and, therefore, designated as RA response element (RARE) or retinoid X response element has been identified. These include for RARs: growth hormone, laminin B1, osteocalcin, alcohol deshydrogenase 3, phenol-enol-pyruvate-carboxykinase ocytocin, RAR-β, RAR-α_2, CRBP1 and CRABP2 genes; sequences for RXR binding have been identified in the CRBP2, CRABP2 and the apolipoprotein A_1 genes. The role of these genes in the mediation of the biological activities of retinoids is not evident. So far the genes of either important structural skin proteins, such as keratins, or key regulatory factors, such as TGF-β (that are modulated by RA), have not been shown to have either RARE or retinoid X response element. It might be that the regulatory effect of RA in these systems occurs through binding of RA-RAR not to RARE but rather to transcription factors; thus RARs are potent inhibitors of AP-1 (*jun/fos* dimer) in the presence of RA [12]. The profound effect of such a mechanism in cell differentiation has recently been shown in the model of acute promyelocytic leukemia [13]. In this respect, it is interesting that the human collagenase gene, an important factor in the ageing process (see below), is a target for transcriptional activation by c-*jun* and c-*fos*, and that this activation is repressed by RA.

Retinoids and Ageing

That retinoids may interfere with the ageing process is not unexpected since senescent arrest resembles a process of terminal differentiation involving repression of proliferation-promoting genes with reciprocal new expression of antiproliferative genes [14]. Retinoids modulate many genes; in the epidermis in vivo, they induce proliferation and inhibit terminal differentiation of keratinocytes. However, the notion that retinoids may indeed have antiageing properties came from observations made almost by serendipity during clinical studies conducted in humans by dermatologists.

Topical RA (tretinoin) was first shown to be useful in treating disorders of epidermal differentiation by the German dermatologist Stüttgen in the early 1960s; Kligman later showed its effect in acne vulgaris. By the late 1970s, oral retinoids were introduced, etretinate for psoriasis and ichthyoses by Bollag and 13-*cis* RA for severe forms of acne. That a single class of compounds might affect so many seemingly distinct conditions was linked to the effects of retinoids in modulating cell differentiation, and notably epidermal differentiation. This further led to the trial of RA in premalignant lesions of the skin known as actinic keratoses. Since these lesions develop in skin that has been chronically exposed to sunlight, other signs of chronic sun damage, such as elastosis and wrinkling (photoageing), were also treated with topical RA. Although the effect on actinic keratoses was not dramatic, a significant improvement in the cosmetic appearance of the subjects was observed. Cordero [15] first used topical RA for the treatment of human photoageing. Kligman then developed a mouse model for testing the hypothesis that topical RA could be useful in treating UV-induced dermal ageing (elastosis). Using the hairless mouse he showed that UVB irradiation reproduced the changes found in human photodamaged skin. In mice treated topically with RA, new collagen was formed and numerous fibroblasts appeared in the dermis, thus suggesting reversal of photodamage. In open clinical studies in humans, he obtained convincing evidence that topical RA was useful in moderating many of the changes associated with photoageing. Light and electron microscopy studies of treated skin showed that atrophic epidermis became hyperplastic, and that new collagen deposition took place in the dermis [for a review, see 16]. These observations were confirmed in a double-blind trial [17] and in many subsequent reports.

These observations should be considered in the perspective of the differences between *intrinsic* – i.e. chronologic skin ageing – and *photoageing*. The former is observed in its purest form only in sun-protected skin, whereas the latter occurs in chronically sun-exposed skin. The most dramatic differences between intrinsic ageing and photoageing are seen in the dermis [18]; in intrinsic skin ageing there is hypocellularity and hypoplasia like in other organs whereas in photoaged skin there are some signs of inflammation (heliodermatitis) which may be responsible for dermal component degradation due to elastases and metalloproteinases; collagen is reduced and elastotic material is increased and fragmented. These observations suggest that the action of topical RA on photoaged skin might not be extrapolated to intrinsic ageing, and thus that RA might not be active in the general process of ageing. However, it has recently been shown that the effects of topical RA are not limited to photoaged skin: although the drug modestly enhanced clinical appearance in intrinsically aged skin, there was histologic improvement of the involutional structural changes [19].

The mechanism of action of RA in skin ageing is not completely understood. It has been demonstrated that RA penetrates into the skin and is converted into many metabolites including 9-*cis* RA [20]: epidermal keratinocytes as well as dermal fibroblasts express RARs and RXRs as well as CRABPs. Topical application of RA induces CRABP2, which is considered as a specific retinoid response [21, 22]. Therefore it is likely that the effects of RA in skin ageing are mediated through the retinoid nuclear receptor pathways. Of the many biological effects of topical RA, two appear to be pertinent in terms of reversing the signs of ageing. One is the in vitro observation that RA downregulates collagenase expression in fibroblasts [23] and upregulates TIMP-1 messenger RNA [24]; senescent fibroblasts are thought to be locked into a constitutively activated state in which there is upregulation of secreted proteases [25]. This might be inhibited by RA. Another is the in vivo increased production of collagen I (which accounts for 85% of total collagen in human dermis) upon topical RA in humans; this has been shown by immunohistochemistry with antibody to the aminopropeptide portion [26]. Senescent fibroblasts display a marked decrease in the production of collagen [27]. Since TGF-β_1, a stimulator of fibroblast production of collagen, is induced by RA [28], this growth factor may be involved in this effect of RA.

To what extent the observations made in human skin with RA may apply to the general biology of ageing remains to be demonstrated. It should be understood that the doses of RA applied to the skin for obtaining these effects are very high; it is unlikely that such doses of RA might be delivered systemically for acting in senescent fibroblasts in other tissues without inducing intolerable side effects. The possibility still exists that synthetic receptor-specific retinoids may fill this gap in the future.

Mechanisms of Bone Loss

TIMOTHY M. MURRAY

ABSTRACT. We have seen rapid and important advances in the understanding of bone resorption processes at the cellular and molecular levels. Indeed, the field of molecular and cellular biology of bone is proliferating at a virtually exponential rate. Osteoclastic bone resorption, the major process involved in Type 1 osteoporosis, is the target of several pharmacological agents for the treatment of osteoporosis, and is the potential target of several agents that are proposed for the treatment of osteoporosis. This review will focus on (1) the factors known or postulated to affect bone loss; (2) the function of the differentiated osteoclast and the pharmacological effects on osteoclast function; (3) the processes involved in the generation and differentiation of mature osteoblasts and, again, the pharmacological implications of these processes. (*J Rheumatol 1996*;**(suppl 45)23:6–10)**

Key Indexing Terms:
OSTEOPOROSIS — BONE MINERAL DENSITY — BONE RESORPTION
POSTMENOPAUSAL — VITAMIN D — PARATHYROID HORMONE

PATHOGENESIS OF OSTEOPOROSIS

The main pathogenetic factor in osteoporosis is reduction in bone mass. It is generally accepted that low bone mass is the major risk factor for osteoporotic fracture, and measurement of bone mass is a most useful clinical guide for estimating fracture risk. Bone quality (as opposed to quantity) may also be important in pathogenesis, but this has not been well studied. Aspects of bone quality include reduced connectivity, alteration in collagen or other protein constituents, and osteonecrosis.

The amount of bone (bone mass) results from the effects of 2 dynamically coordinated cellular processes, namely bone formation and bone resorption. In the normal bone remodeling cycle, osteoclasts migrate onto a quiescent bone surface and form resorption lacunae. Osteoblasts subsequently occupy these lacunae, where they deposit bone matrix, which is subsequently mineralized. On balance, bone formation equals bone resorption. When the rate of resorption exceeds that of formation, the result is osteoporotic bone loss, which can occur in either of 2 ways. In the first, bone resorption is increased and bone formation fails to keep pace. Clinical markers of bone turnover and resorption, such as urinary hydroxyproline and pyridinolines, are increased, indicating a high rate of cellular activity. Second, bone turnover may be decreased such that there is a reduction in bone formation in the presence of a normal rate of resorption. In this situation, cellular activity is less than normal and biochemical indices of bone turnover and resorption are decreased. Type 1 osteoporosis is generally considered to be due to the former osteoclastic mechanism[1].

From the University of Toronto, St. Michael's Hospital, Toronto, Canada.
T.M. Murray, MD, Professor of Medicine, University of Toronto, Director, Metabolic Bone Clinic, St. Michael's Hospital.
Address reprint requests to Dr. T.M. Murray, St. Michael's Hospital, 233–38 Shuter St., Toronto, Ontario M5B 1A6.

CHANGES IN BONE WITH AGE

The period of skeletal development culminates when peak bone mass is reached at about the age of 20 years. Bone mass declines gradually thereafter in both men and women, with more rapid loss in women at the time of menopause[1]. The age at which an individual reaches the fracture threshold is dependent on 2 factors: (1) the height of peak bone mass; and (2) the rate of subsequent bone loss. Factors known to influence the attainment of peak bone mass include dietary calcium and vitamin D, hormonal factors (including delayed puberty or hypogonadism), physical factors, heredity, and intercurrent illness. Studies by Danish investigators indicate that the measurement of bone mass and the determination of biochemical indices of bone turnover can be used to predict the amount of bone loss over a 10 to 12 year period[2]. There is no firm evidence, however, that bone loss is a bimodal process (i.e., that there are fast losers and slow losers). Thus, although diagnostic tests may be applied to establish a general index of future bone loss, the pathogenic reasons cannot be stated with certainty in any given patient.

Evaluation of spine bone mineral density (BMD) with the dual photon technique can document the sharp bone loss following menopause. In addition to the decrease in bone mass over the 10 year period after menopause, bone loss can be demonstrated by chemical indices of bone turnover and resorption, such as increased serum concentrations of osteocalcin and increased urinary excretion of hydroxyproline and pyridinoline. The average woman loses between 20 and 25% of bone mass during the 10 years following menopause. A prospective study from Denmark examined bone loss at 3 different skeletal sites and found an overall decrease of 20 to 25% in bone mass at all 3 sites 16 years after menopause, although the rate of loss varied at the different sites[3].

Histological study of bone from women with surgically

induced menopause shows an early increase in bone resorption followed by a compensatory increase in bone formation, with subsequent attainment of a new steady state. Resorption is evidently the primary event, and the loss is much greater in cancellous than in cortical bone (Figure 1)[4].

There is further evidence that increased bone resorption is the primary mechanism in early menopausal bone loss. A group of women in early menopause received a course of hormone replacement therapy (HRT), which was withdrawn after 2 months. Bone resorption was determined by urinary excretion of Type I collagen and N-terminal extension peptide, a new and specific marker for bone resorption. Bone resorption decreased during HRT and increased when HRT was discontinued[5].

It has traditionally been thought that bone loss tapers with age. In fact, research by Nguyen, *et al* shows that between the ages of 60 and 80 years, the rate of bone loss (measured longitudinally) doubles in both sexes[6]. The skeletal changes that take place between 70 and 80 years of age are among the least studied and least understood. The emphasis on treating only perimenopausal women may be misguided: rapid bone loss occurs in older men and women, and treatment during that period of life may be even more important than treatment for women at menopause.

Earlier observations (by spinal bone densitometry) that vertebral bone loss tends to level off with age in older women should not lead to complacency with regard to the management of osteoporosis. A recent prospective study by Greenspan in women older than 65 years found evidence of continued loss of roughly 1% per year, measured by BMD at the hip[7]. It has not yet been determined whether the pathogenic factors involved in bone loss at the hip in older women are the same as the factors already described. Whatever the cause, it is clear that women continue to lose bone mass, if at a somewhat slower rate, even after the rapid decrease that occurs in the first decade of menopause.

Another study found a significant relationship between urinary hydroxyproline and BMD in 3 skeletal sites in 196 healthy women more than 10 years after menopause[8]. The result suggests that, contrary to general belief, bone resorption continues to play a determinant role in the production of bone mass. Moreover, it suggests important clinical implications and provides a strong rationale for the use of antiresorptive therapy in menopausal women.

RISK FACTORS FOR BONE LOSS

Epidemiological studies to identify risk factors for osteoporosis generally employ cross sectional retrospective analyses, which fail to distinguish between factors that relate to bone gain and those that concern bone loss. A prospective study by Reid, *et al* confirmed that most of the major risk factors for osteoporosis relate to bone loss[9]. The reason for the influence of initial BMD on bone loss is not altogether clear, but it suggests a longstanding influence, not only on bone density, but also on continuing bone loss and possibly a genetic factor in bone loss. Body weight and body

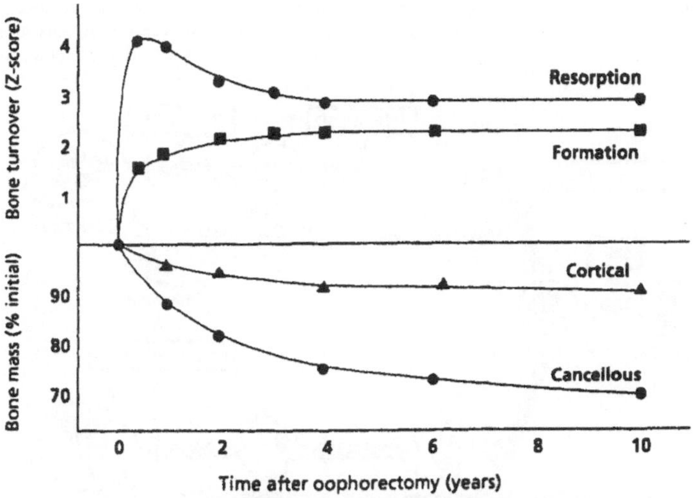

Figure 1. Resorption appears to be the primary event in bone mineral density loss following surgical menopause. An early increase in bone resorption is followed by a compensatory increase in bone formation and, subsequently, attainment of a new steady state. The loss is greater in cancellous than in cortical bone. (With permission, from Kanis, 1994[4].)

fat mass are important determinants, with an observable difference in bone density in women on weight reduction diets. These investigators also found that the tubular maximum concentration for calcium reabsorption was important, suggesting that renal factors in calcium conservation could determine rates of bone loss. Methodological flaws in this study may account for the absence of a correlation between osteoclastic bone loss and dietary calcium, sodium, caffeine, and protein, as well as concentrations of sex hormones, parathyroid hormone (PTH) and 25-OH-D, and exercise and muscle strength.

Hormonal factors and dietary factors should be included in the list of factors implicated in osteoclastic bone loss. There is evidence that ethanol can cause bone loss directly by influencing osteoclastic bone resorption[1], and further evidence indicates that bone loss may occur in alcoholics without liver disease. A striking interaction has been shown between cigarette smoking and body weight. A Norwegian study showed a more marked effect of smoking in lean females, judged by fracture incidence[10]. The results suggested that a lean smoker adds 10 years to her skeletal age in terms of fracture incidence.

Many studies have shown that early menopause is a risk factor for osteoporosis. However, a study by French investigators examined BMD as a function of age in patients with normal as opposed to early menopause (before age 40)[11]. After correction for chronological age, an effect was evident not only on bone mass but also on the rate of bone loss. These results again suggest the influence of a factor that has a long lived effect on bone. Patients were losing more bone with age in consequence of an event that had taken place many years earlier.

A pivotal study of identical twins showed abnormalities in the vitamin D receptor alleles[12]. Specific polymorphisms strongly correlated with BMD. The variance of bone mass was in the range of 10 to 15%, which is highly significant in clinical terms. The investigators postulated that the difference could account for a 70% genetic component of bone mass in the twin subjects. Although these findings were initially considered highly controversial, acceptance is steadily growing. It has yet to be determined whether the abnormality is important in bone formation during childhood and development, or plays a role in subsequent bone loss. It may well prove to be important in both types of effects on bone mass. Furthermore, the abnormality described may not be a primary risk factor, but rather a closely linked gene may be of primary importance.

DIFFERENTIATION OF OSTEOCLASTS

Osteoclasts (bone resorbing cells) and osteoblasts (bone forming cells) are derived from separate precursors in bone marrow. The osteoblast is derived from the mesenchymal stem cell CFU-F, whereas the osteoclast is derived from a hematopoietic stem cell precursor, CFU-GM (granulocyte macrophage precursor). The differentiation of the mature, functional osteoclast from the precursor is shown in detail in Figure 2, with regulatory mechanisms at the various steps[13].

The physiological regulators of osteoclast function, as well as drugs, can influence the differentiated osteoclast. However, the effects of PTH and 1,25-dihydroxyvitamin D occur at an earlier stage of the differentiation process. Several important control points on mature osteoclasts have been discovered in recent years. Attachment of the cell to the bone surface seals off a resorption zone with specific

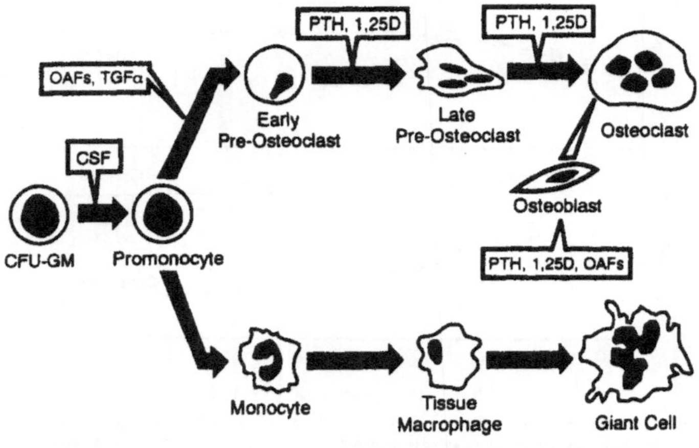

Figure 2. Differentiation of mature, functional osteoclast from precursors. Physiological regulators of osteoclast function, as well as drugs, influence the differentiated osteoclast. (With permission, from Favus, 1993[13].)

integrin molecules. After attachment, the cell pumps hydrogen ions into the resorption cavity by means of a specific hydrogen adenosine triphosphatase (H-ATPase). A carbonic anhydrase II enzyme generates the protons to be pumped via the ATPase. The osteoclast maintains electrical neutrality through the actions of a chloride pump and a chloride bicarbonate exchanger. The calcium channel is important in the function of the osteoclast. Studies of isolated osteoclasts have shown that blockade of the calcium channel can block bone resorption by fully differentiated osteoclasts. The cell contains numerous other important channels, as well as a lysosomal apparatus that produces lysosomal enzymes to degrade the organic components of the bone.

BIOCHEMICAL PROPERTIES OF OSTEOCLASTS

Several biochemical properties could provide targets for therapeutic agents. In particular, targeting the $\alpha_v\beta_3$ integrin (RGD protein) could prevent the attachment of osteoclasts. Echostatin, a compound under development that is derived from snake venom, contains RGD protein and could be used to prevent osteoclast attachment. Other agents could be developed to block the effects of H-ATPase. Carbonic anhydrase inhibitors are unlikely to be sufficiently specific, because carbonic anhydrase II is also present in the distal renal tubule. There is also interest in the development of osteoclastic calcium channel blockers and tyrosine kinase inhibitors, both of which have the potential to block osteoclast activity.

FACTORS IN OSTEOCLAST MEDIATED BONE RESORPTION

It has been demonstrated through animal experiments that the regulation of osteoclast differentiation is an indirect effect[14]. When animals were treated with PTH, a bone resorbing agent, it was shown that PTH was not acting primarily and directly on osteoclasts. Most PTH receptors are, in fact, detected on osteoblastic cells. Further studies have shown that the majority of bone resorbing agents act via osteoblastic cells in the bone marrow, which release intermediates to stimulate osteoclasts. These have, at times, been referred to as "osteoclast activating factors." Several studies have identified these factors as peptides released by osteoblastic cells, prostaglandins, colony stimulating factors, and other growth factors, including interleukins and cytokines[14].

The most interesting recent data concern bone resorptive cytokines, of which there are many. Those that can be most strongly implicated in osteoporotic bone loss are interleukin 1 (IL-1), IL-6, IL-11, granulocyte macrophage colony stimulating factor (GMCSF), MCSF, and tumor necrosis factor (TNF). They are present in other tissues, but are strongly linked to the osteoporotic process in bone. IL-1, IL-6, IL-11, and TNF have not only been implicated in bone loss but also been shown to be blocked by specific antibodies or, in the case of IL-1, to be blocked by an IL-1 receptor antagonist that is naturally occurring and is now available in recombinant form. Cytokines that are increased during menopause are IL-1, IL-6, and TNF.

Experimental and clinical studies have shown that many of these factors decrease during estrogen therapy. Some of them are stimulated by PTH. Ralston compared gene expression for these factors in patients with osteoporotic fracture and age matched healthy patients without osteoporotic fractures. The results suggested there was increased expression of IL-1 and IL-6 in the bone cells of patients with osteoporosis[15].

IL-1 was the first cytokine to be implicated in the osteoporotic process. IL-1 production in circulating monocytes was shown to be increased in patients with the disease. It was subsequently shown that IL-6 mRNA was more highly expressed in the bone cells of osteoporotic women. It has been shown in the mouse model that IL-1 stimulates bone resorption *in vitro* and will induce hypercalcemia when injected into normal mice[16]. IL-1 also inhibits bone formation.

IL-6 has been shown to stimulate osteoclast formation in murine bone cultures and human marrow cultures. It has also been strongly implicated in the regulation of bone loss by estrogen. Estrogen not only diminishes the production of IL-6 but also interferes with the response to IL-6 by affecting GAP-130, a transduction protein linked to the IL-6 receptor.

GMCSF and MCSF are also regulated by estrogen. Although there is still considerable disagreement about the relative importance of the various cytokines, it is evident that they are all important and that all are acting in concert in a complicated and as yet poorly understood relationship.

These factors have other important clinical implications. For example, bisphosphonates, such as etidronate and alendronate, have been shown to diminish IL-6 production. 1,25-dihydroxyvitamin D in doses that cause bone resorption can increase IL-6 as well. IL-6 has been implicated in leukemia and Paget's disease of bone, while IL-1, TNF, and IL-6 have all been implicated in multiple myeloma.

CONCLUSION

The apparent complexity of the bone remodeling process may prove to be a help and not a hindrance to the development of drugs for osteoporosis. Agents could be designed to take advantage of the multiple agonists, receptors, cell types, and control points to block osteoclastic bone resorption at virtually any step. Drugs could be combined to ensure the greatest blockade of resorption with the least risk of side effects.

REFERENCES

1. Mundy GR: Osteopenia. In: *Disease-a-Month 1987*;33:537–600.
2. Hansen MA, Overgaard K, Riis BJ, Christiansen C: Potential risk factors for development of postmenopausal osteoporosis — examined over a 12-year period. *Osteoporosis Int 1991*;1:95–102.

3. Hansen MA, Overgaard K, Christiansen C: Spontaneous postmenopausal bone loss in different skeletal areas — followed up for 15 years. *J Bone Min Res 1995*;10:205-10.
4. Kanis JA: *Osteoporosis*. Oxford: Blackwell, 1994:46.
5. Gertz BJ, Shao P, Hanson DA, *et al:* Monitoring bone resorption in early postmenopausal women by an immunoassay for cross-linked collagen peptides in urine. *J Bone Min Res 1994*;9:135–42.
6. Nguyen TV, Kelly PJ, Sambrook PN, Gilbert C, Pocock NA, Eisman JA: Lifestyle factors and bone density in the elderly: Implications for osteoporosis prevention. *J Bone Min Res 1994*;9:1339–46.
7. Greenspan SL, Maitland LA, Myers IA, *et al:* Femoral bone loss progresses with age: A longitudinal study in women over age 65. *J Bone Min Res 1994*;9:1959–65.
8. Prince R, Dick I, Devine A, *et al:* Importance of bone resorption in the determination of bone density in women more than 10 years past the menopause. *J Bone Min Res 1993*:1273–9.
9. Reid IR, Ames RW, Evans MG, *et al:* Determinants of the rate of bone loss in normal postmenopausal women. *J Clin Endocrinol Metabol 1994*;79:950–4.
10. Forsen L, Bjorndal A, Bjartveit K, *et al:* Interaction between current smoking, leanness and physical inactivity in the prediction of hip fracture. *J Bone Min Res 1994*;9:1671–8.
11. Pouilles JM, Tremollieres F, Bonneu M, Ribot C: Influence of early age at menopause on vertebral bone mass. *J Bone Min Res 1994*;9:311–5.
12. Morrison NA, Qi JC, Tokita A, *et al:* Prediction of bone density from vitamin D receptor alleles. *Nature 1994*;367:284–7.
13. Mundy GR: Bone resorbing cells. In: Favus MJ, ed. *Primer on the Metabolic Bone Diseases*. 2nd ed. Philadelphia: Lippincott-Raven, 1993:28.
14. Suda T, Takahashi N, Martin TJ: Modulation of osteoclast differentiation. *Endocr Rev 1992*;13:66–80.
15. Ralston SH: Analysis of gene expression in human bone biopsies by polymerase chain reaction: Evidence for enhanced cytokine expression in postmenopausal osteoporosis. *J Bone Min Res 1994*;9:883–90.
16. Sabatini M, Boyce B, Aufdemorte T, Bonewald L, Mundy GR: Infusions of recombinant human interleukin 1 alpha and beta cause hypercalcemia in normal mice. *Proc Natl Acad Sci USA 1988*;85:5235–9.

Am J Physiol Cell Physiol
279: C611–C618, 2000.

Skeletal muscle fiber quality in older men and women

WALTER R. FRONTERA,[1,2] DONGWON SUH,[2] LISA S. KRIVICKAS,[2] VIRGINIA A. HUGHES,[1] RICHARD GOLDSTEIN,[2] AND RONENN ROUBENOFF[1]
[1]*Nutrition, Exercise Physiology, and Sarcopenia Laboratory, Jean Mayer United States Department of Agriculture Human Nutrition Research Center on Aging at Tufts University, Boston 02111; and* [2]*Department of Physical Medicine and Rehabilitation, Harvard Medical School and Spaulding Rehabilitation Hospital, Boston, Massachusetts 02114*

Received 6 October 1999; accepted in final form 5 April 2000

Frontera, Walter R., Dongwon Suh, Lisa S. Krivickas, Virginia A. Hughes, Richard Goldstein, and Ronenn Roubenoff. Skeletal muscle fiber quality in older men and women. *Am J Physiol Cell Physiol* 279: C611–C618, 2000.—Whole muscle strength and cross-sectional area (WMCSA), and contractile properties of chemically skinned segments from single fibers of the quadriceps were studied in 7 young men (YM, 36.5 ± 3.0 yr), 12 older men (OM, 74.4 ± 5.9 yr), and 12 older women (OW, 72.1 ± 4.3 yr). WMCSA was smaller in OM compared with YM (56.1 ± 10.1 vs. 79.7 ± 13.1 cm^2; $P = 0.031$) and in OW (44.9 ± 7.5; $P < 0.003$) compared with OM. Age-related, but not sex-related, differences in strength were eliminated after adjusting for WMCSA. Maximal force was measured in 552 type I and 230 type IIA fibers. Fibers from YM (type I = 725 ± 221; type IIA = 792 ± 271 μN) were stronger ($P < 0.001$) than fibers from OM (I = 505 ± 179; IIA = 577 ± 262 μN) even after correcting for size. Type IIA fibers were stronger ($P < 0.005$) than type I fibers in YM and OM but not in OW (I = 472 ± 154; IIA = 422 ± 97 μN). Sex-related differences in type I and IIA fibers were dependent on fiber size. In conclusion, differences in WMCSA explain age-related differences in strength. An intrinsic defect in contractile proteins could explain weakness in single fibers from OM. Sex-related differences exist at the whole muscle and single fiber levels.

sarcopenia; chemically skinned segments of single fibers; aging; specific force

THE CAPACITY OF SKELETAL MUSCLE to generate force declines with age (17). A loss of muscle mass can explain part of this decline, but it is possible that alterations in muscle quality also occur (26). The extent and nature of qualitative changes in muscle and their contribution to sarcopenia and muscle dysfunction in old age are poorly understood. Since impairment of skeletal muscle function leads to disability and loss of independence (36), it is important to understand the basic cellular mechanism underlying muscle dysfunction in the elderly. This knowledge is essential to optimize rehabilitation and preventive strategies for this population.

Whole muscle quality has been estimated by normalizing strength to muscle cross-sectional area (specific force, SF). In humans, some reports show a significant reduction in SF with age, whereas other reports show no change (17, 26; for review see Ref. 5). Significant methodological limitations in the estimates of muscle mass or size in humans may explain these conflicting results. For example, imaging techniques produce measurements of whole muscle cross-sectional areas (WMCSA), but in pennate muscles (like m. quadriceps femoris), where muscle fibers attach to tendons at an angle, WMCSA underestimates the amount of muscle generating force; i.e., the physiological cross-sectional area (PCSA). Even improved measurements of PCSA suggest that factors other than WMCSA account for part of the force produced by a muscle (12).

The ability of skeletal muscle to generate force and produce movement may be compromised by an incomplete activation of motor units (central motor drive), peripheral nerve dysfunction, loss of hormonal influences, changes in the excitation-contraction coupling mechanisms, or by alterations in the contractile elements of muscle cells. Since elderly men and women can fully activate their muscles (34, 38) and show minimal changes in nerve conduction velocity (7), it appears likely that one site of possible qualitative alterations is in the contractile elements of the muscle cell. The skinned muscle fiber preparation allows the direct investigation of the function of muscle proteins in segments of a cell with an intact myofilament lattice, but without the confounding effects of neural influences, the variations in fiber architecture and the spatial orientation of fibers, the effects of intercellular connective tissue, or of the protein heterogeneity between cells found in multicellular preparations (22, 46). Finally, it has been suggested that changes in the levels of hormones such as estrogen and its effects on the number and function of cross-bridges could result in different rates of loss of strength between the sexes (35, 44). Direct measurements of muscle fiber quality could be used to test this hypothesis.

The purpose of the present investigation was to compare the SF of m. vastus lateralis in vivo and the SF of

Address for reprint requests and other correspondence: W. R. Frontera, Dept. of Physical Medicine and Rehabilitation, Spaulding Rehabilitation Hospital, 125 Nashua St., Boston, MA 02114 (E-mail: frontera.walter@mgh.harvard.edu).

The costs of publication of this article were defrayed in part by the payment of page charges. The article must therefore be hereby marked "*advertisement*" in accordance with 18 U.S.C. Section 1734 solely to indicate this fact.

segments of single muscle fibers from young and old men and old women. Our hypotheses were: *1*) there is a difference in muscle strength and WMCSA between young and old men but no difference in SF at the whole muscle level; *2*) compared with fiber segments from younger subjects expressing the same myosin heavy chain (MyHC) isoform, type I and IIA fibers from older subjects show lower SF; and *3*) there are no sex-related differences in SF in whole muscle, type I, or IIA single fiber segments.

MATERIALS AND METHODS

Subjects. A total of 31 healthy subjects not involved in strength training volunteered for the study. The general characteristics of the subjects are presented in Table 1. The volunteers received a complete explanation of the purposes and procedures and gave their written consent. A comprehensive medical evaluation including a medical history, physical examination, and routine blood and urine tests was performed prior to their participation in the study. A resting electrocardiogram was also done in older volunteers. The number of self-reported diagnoses was 1.3 ± 1.2 in women (2 reported no diagnoses) and 1.8 ± 1.0 in men (2 reported no diagnoses). The number of medications taken by the subjects was 0.9 ± 1.4 in women (6 were not taking medications) and 1.5 ± 1.0 in men (2 were not taking medications). Subjects with conditions that could interfere with neuromuscular function were excluded.

The level of recreational activity and sports participation over the last year was estimated using a questionnaire (25) (Table 1). The study was approved by the Human Investigation Review Committees of Tufts University-New England Medical Center and the Spaulding Rehabilitation Hospital.

Muscle strength measurements. An isokinetic dynamometer (Cybex II, Medway, MA) was used to measure strength (Newton-meters) of the left knee extensors at 1.04 rad/s as previously reported (11). The subjects performed five maximal voluntary contractions, and the highest or peak torque was recorded.

Computerized tomography. A computerized tomography (CT) scan of the thigh was performed between 38 and 62% (mean ± SD = 50 ± 4%) of the length of the femur measured from the distal end of the bone. The scan was obtained using a GE Highspeed Advantage CT scanner (Milwaukee, WI) operating at 100 kV and 170 mA. A slice width of 10 mm and a scanning time of 1 s were used. From the CT image, the WMCSA occupied by m. quadriceps femoris was measured. The image was analyzed according to optical density on a computer (Macintosh Power PC 8500; Apple Computer, Cupertino, CA) by a single investigator in a blinded fashion with IMAGE software (developed at the National Institutes of Health and available on the Internet) modified by G. Solares of our group for quantification of cross-sectional areas of fat, muscle, and bone to the nearest 0.1 cm^2 (32). The coefficient of variation (CV) for repeated measurements of muscle area in a single scan by the same observer was <0.25%. The CV for measurements done 3 wk apart in two subjects was <0.8%.

Table 1. *General characteristics of subjects*

	Younger Men	Older Men	Older Women
n	7	12	12
Age, yr	37.3 (3.5)	74.4 (6.1)	72.1 (4.0)
Height, cm	178.6 (8.3)	174.7 (6.2)	159.5 (5.1)
Weight, kg	81.9 (10.5)	76.3 (9.4)	66.6 (6.2)
BMI, kg/m²	25.6 (1.5)	25.0 (2.6)	26.3 (3.6)
Physical activity, kcal/wk	657 (747)	1,431 (1,445)	820 (693)

Values are means, with SD in parentheses; *n* = no. of subjects. BMI, body mass index.

Muscle biopsies and permeabilization of fibers. Biopsy specimens were taken from m. vastus lateralis under local anesthesia using biopsy needles (4) and suction (8). The specimens were placed in relaxing solution (see below) at 4°C. Bundles of ~30 fiber segments were dissected free from the samples and then tied with surgical silk to glass capillary tubes at slightly stretched lengths. The fiber segments were chemically skinned for 24 h in relaxing solution containing 50% (vol/vol) glycerol at 4°C and were subsequently stored at −20°C for up to 4 wk before use.

Experimental procedure. A detailed explanation of the general methods used in this study has been published by others (22, 30). Briefly, on the day of an experiment, fiber segments were placed for 30 min in relaxing solution (see below) containing 0.5% Brij-58 (polyoxyethylene 20 cetyl ether; Sigma Chemical) prior to mounting in an experimental apparatus, similar to the one described previously by Moss (30). A fiber segment length of 1 to 2 mm was left exposed to the solution between connectors leading to a force transducer (model 400A, Cambridge Technology) and a lever arm system (model 308B, Cambridge Technology). The apparatus was mounted on the stage of an inverted microscope (model IX70; Olympus, Tokyo, Japan). While the fiber segments were in relaxing solution, sarcomere length (SL) was set to 2.75–2.85 μm by adjusting the overall segment length. The segments were observed through the microscope at a magnification of ×320.

The sarcomere length, the segment diameter, and the length of segment between the connectors were measured with an image analysis system (Image-Pro Plus; Media Cybernetics, Silver Spring, MD). Fiber depth was measured by recording the vertical displacement of the microscope nosepiece while focusing on the top and bottom surfaces of the fiber. The focusing control of the microscope was used as a micrometer. Fiber cross-sectional area (FCSA) was calculated from the diameter and depth, assuming an elliptical circumference, and was corrected for the 20% swelling that is known to occur during skinning (14, 30). Maximum force (Po) was adjusted for FCSA.

Relaxing and activating solutions contained (in mM) 4 Mg-ATP, 1 free Mg^{2+}, 20 imidazole, 7 EGTA, 14.5 creatine phosphate, and KCl to adjust the ionic strength to 180 mM. The pH was adjusted to 7.0. The concentrations of free Ca^{2+} were 10^{-9} M (relaxing solution) and $10^{-4.5}$ M (maximum activating solution) and are expressed as pCa (i.e., $-\log[Ca^{2+}]$). Apparent stability constants for Ca^{2+}-EGTA were corrected for temperature (15°C) and ionic strength (180 mM) (9). The computer program of Fabiato (9) was used to calculate the concentrations of each metal, ligand, and metal-ligand complex.

Immediately preceding each activation, the fiber was immersed for 10–20 s in a solution with a reduced Ca^{2+}-EGTA buffering capacity. This solution was identical to the relaxing solution except that EGTA was reduced to 0.5 mM, which resulted in a faster attainment of steady tension during subsequent activation. Maximum active force (Po) was calculated as the difference between the total force in activating solution (pCa 4.5) and the resting tension measured in the same segment while in the relaxing solution. All contractile measurements were carried out at 15°C.

MyHC composition. After mechanical measurements, each fiber was placed in SDS sample buffer in a plastic microcen-

Table 2. *Multiple regression analysis of whole muscle strength and CSA of the knee extensors*

	Coefficient	SE	t	P
Age comparison				
Constant	11.1	69.78	0.16	0.876
Age	−0.36	0.53	−0.68	0.506
CSA	2.62	0.64	4.07	0.001
Sex comparison				
Constant	28.8	26	1.11	0.28
Sex	−24.08	9.23	−2.61	0.016
CSA	1.81	0.45	3.98	0.001

CSA, cross-sectional area; SE, standard error.

trifuge tube and stored at −20°C for up to 1 wk or at −80°C if the gels were to be run later. The MyHC composition of single fibers was determined by SDS-PAGE (21). The acrylamide concentration was 4% (wt/vol) in the stacking gel and 6% in the running gel, and the gel matrix included 30% glycerol. Sample loads were kept small (equivalent to ~0.05 mm of fiber segment) to improve the resolution of the MyHC bands (types I, IIA, and IIB). Proteins were identified using a combination of human myosins from vastus lateralis muscles and from reports in the literature (see Ref. 22).

Statistical analysis. Multiple regression analysis was used to test all three hypotheses. For the first hypothesis, a regression using only males was used to test for age differences, whereas a regression using only older subjects, but including both sexes, was used to test for differences between men and women.

The second and third hypotheses were tested using a regression with standard errors corrected for the lack of independence (clustering effect) caused by using multiple fibers from the same person. In Tables 4 and 5 the corrected standard errors are referred to as "robust standard errors." This approach allowed both the inclusion of standard covariates [age, gender, fiber type (I and IIA), fiber size] and the inclusion of the interaction terms to correctly model a ratio (SF). The use of ratios in correlation and regression analysis in medicine and physiology has been discussed by Tanner in 1949 (41) and Kronmal in 1993 (20). The residuals from all regressions were examined for possible violations of underlying assumptions.

The CV for three measurements done by the same observer in four fibers was 0.5% for diameter and 3.7% for depth. We thought that the error in the depth measurement could increase the variability in CSA. Therefore, a sensitivity analysis was used to determine whether the amount of error in the measure might affect any results. For each of the two regressions, four sets of 10,000 new sets of depth measurements and CSA were simulated; for each Monte Carlo simulation, the regression was re-estimated (thus 80,000 regressions were estimated). The new depth measures were simulated from a uniform distribution (first, ±10% error; and second, ±15% error), from a normal distribution, and from a log-normal distribution. The results of all simulations showed that the error in the measurement of depth did not alter the results.

The relationship between the SF at the whole muscle and single fiber levels was examined using a simplified multilevel analysis. We calculated the mean SF across individual fibers for each subject in the study and correlated this with the whole muscle SF. Weighted correlations for number of fibers per person and for SD of the SF among fibers for each person were tried with the same results. Due to the nonlinear nature of the association, we also calculated Kendall's tau rank correlation coefficient.

Sample size was determined using the PinT software for two level (number of subjects and fibers) analysis (40). At a power of about 0.80, a sample size of 24 older persons and 20 fibers per person were needed to detect a significant difference. Analyses were performed using Stata, version 5 (Stata, College Station, TX), and Systat Mac version 5.2.1.

RESULTS

In vivo measurements of muscle function and size. The results of the multiple regression analysis of isokinetic peak torque and WMCSA are presented in Table 2. There was no interaction between age or sex and WMCSA. Older men (56.1 ± 10.1 cm^2) showed smaller WMCSA compared with younger men (79.7 ± 13.1 cm^2) and larger WMCSA compared with older women (44.9 ± 7.5 cm^2). Significant differences in strength between young (209 ± 50 Nm) and older (130 ± 32 Nm) men were eliminated after adjusting for WMCSA. On the other hand, differences in strength between older men and women (86 ± 14 Nm) persisted even after adjusting for WMCSA.

Contractile properties of single fibers. The results for the maximal force (Po) and fiber diameter, depth, CSA, and SF of type I and IIA single muscle fibers are presented in Table 3. A total of 552 type I and 230 type IIA fibers were studied (an average per subject of 17 and 8 type I and IIA fibers, respectively).

Age comparisons. The results of the multiple regression analysis comparing single fiber maximal force in young and old males are presented in Table 4. Fibers (I and IIA) from older men had lower maximal force compared with fibers from younger men. The differences in force between young and older men were significant even after adjusting for differences in fiber size (diameter, depth). There was no interaction between age and fiber type. Therefore, in both age groups, type IIA fibers were significantly stronger than type I fibers.

Table 3. *Maximal force, diameter, depth, CSA, and SF of type I and IIA single muscle fibers from m. vastus lateralis*

	Younger Men (N = 7) (n: I = 91, IIA = 92)	Older Men (N = 12) (n: I = 221, IIA = 82)	Older Women (N = 12) (n: I = 240, IIA = 56)
Po, μN			
Type I	725 (221)	505 (179)	472 (154)
Type IIA	792 (271)	577 (262)	422 (97)
Diameter, μm			
Type I	100 (18)	94 (17)	91 (20)
Type IIA	96 (15)	90 (20)	80 (13)
Depth, μm			
Type I	66 (13)	67 (12)	69 (14)
Type IIA	67 (12)	67 (14)	64 (16)
CSA, μm²			
Type I	5,270 (1,580)	4,960 (1,380)	4,990 (1,620)
Type IIA	4,890 (1,230)	5,110 (2,000)	4,140 (1,080)
SF, N/cm²			
Type I	20.8 (6.7)	14.9 (4.4)	14.2 (4.4)
Type IIA	23.8 (6.8)	16.4 (5.3)	15.7 (5.4)

Values are means, with SD in parentheses; N = number of subjects; n = number of fibers. Po, maximal force; SF, specific force.

Table 4. *Multiple regression analysis of maximal force comparing younger and older males*

	Coefficient	Robust SE	*t*	*P*	95% CI
Constant	−203	71	−2.88	0.01	(−)352–(−)55
Age (old)	−211	35	−6.086	<0.001	(−)283–(−)138
Fiber type (IIA)	92	29	3.155	0.005	30–152
Diameter	6	0.5	10.075	<0.001	4.4–6.7
Depth	7	0.5	11.06	<0.001	4.9–7.1

Analysis of maximal force (Po) comparing younger (n = 7) and older (n = 12) males. The coefficient (*column 1*) represents the mean difference in Po between age groups (*row 2*) and fiber types (*row 3*) and the mean change in Po for a one unit change in the size variables (diameter and depth) holding all other variables constant. The robust standard error (*column 2*) is the SE corrected for the lack of independence or clustering effect caused by using multiple fibers from the same person. Statistical significance is reflected in columns t and P and the 95% confidence interval (CI) of the differences in Po is presented in the last column to the *right*.

Table 5. *Multiple regression analysis of maximal force comparing older men and women*

	Coefficient	Robust SE	*t*	*P*	95% CI
Constant	−163	109	−1.49	0.15	(−)388–63
Sex (women)	363	99	3.68	0.001	159–567
Fiber type (IIA)	68	30	2.28	0.033	6–131
Interaction: fiber type × sex	−79	33	−2.39	0.025	(−)146–(−)10
Diameter	2.8	1	2.65	0.014	0.6–4.9
Interaction: diameter× sex	−1.4	0.8	−1.74	0.094	(−)3.0–0.3
Depth	2.8	1.2	2.28	0.032	0.26–5.26
Interaction: depth × sex	−3.9	0.9	−4.49	<0.001	(−)5.6–(−)2.1
CSA	0.05	0.01	3.36	0.003	0.02–0.07

Analysis of maximal force (Po) comparing older men (n = 12) and women (n = 12). For explanation of rows and columns see Table 4.

There was no interaction between age and any of the fiber size variables (diameter, depth). When FCSA was added to the regression model (in addition to age, fiber type, diameter, and depth), no significant influence was seen. The relationship between FCSA and maximal force in men for both age groups and fiber types is presented in Fig. 1. SF was lower in older men than in younger men. In both age groups, type I fibers showed a lower SF compared with IIA fibers.

Sex comparisons. The results of the multiple regression analysis comparing single fiber maximal force in older men and women are presented in Table 5. The relationship between FCSA (in percentiles) and maximal force for both fiber types in older men and women is presented in Fig. 2.

There was a significant interaction between sex and fiber type. Thus the differences in maximal force between sexes depended on the fiber type. In Fig. 2, the sex and fiber type interaction is shown by the different amount of vertical separation between curves for men and women.

Differences between men and women in fibers expressing the same MyHC isoform were influenced by fiber size. For example, the difference in force between men's and women's type IIA fibers increased with fiber size (see Fig. 2). Smaller type I fibers (i.e., fibers in the lower range of the size continuum) from women were stronger than similar fibers from men. In contrast, larger type I fibers from men were stronger than those from women.

In men, type IIA fibers were stronger than type I fibers even after adjusting for differences in fiber size (i.e., IIA fibers had a higher SF). In women, on the other hand, there was no difference in maximal force between fiber types (i.e., similar SF).

Whole muscle and single fiber relation. The association between the whole muscle and single fiber SF for all subjects is shown in Fig. 3. The unweighted correlation between the two variables was significant (r = 0.37; P = 0.042). Kendall's tau rank correlation coefficient (used because of the nonlinear nature of the association) was 0.25 (P = 0.054).

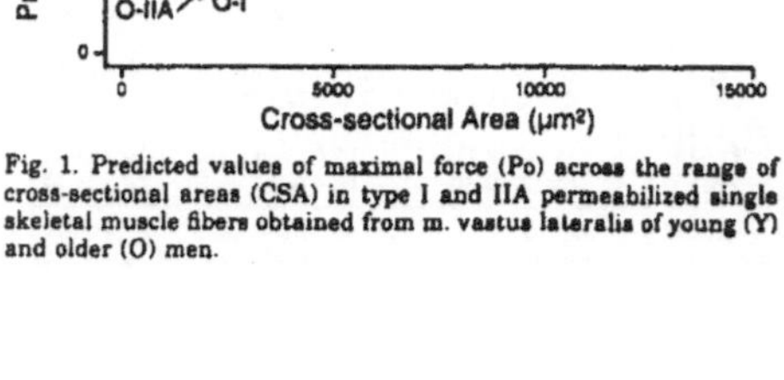

Fig. 1. Predicted values of maximal force (Po) across the range of cross-sectional areas (CSA) in type I and IIA permeabilized single skeletal muscle fibers obtained from m. vastus lateralis of young (Y) and older (O) men.

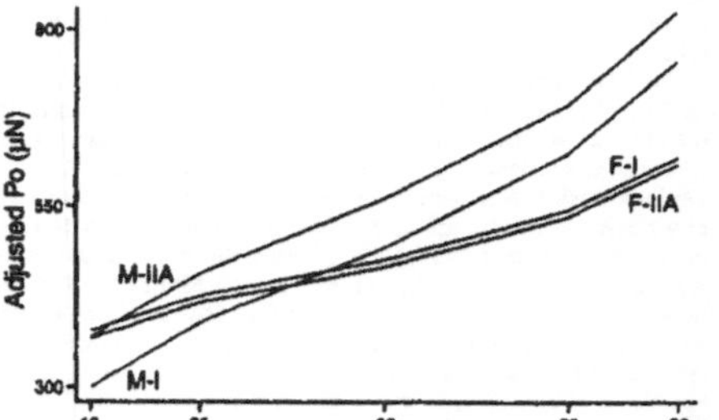

Fig. 2. Relationship between depth/diameter/CSA and maximal force (Po) in type I and IIA fibers from old men (M) and women (F). The 10th percentile is that value that is greater than 10% of all values and less than 90% of all values. For this graph, depth, diameter, and CSA were each set at the 10th percentile of their distribution. The predicted value of Po was then calculated for each of the four groups. This was repeated for the 25th, 50th (median), 75th, and 90th percentiles.

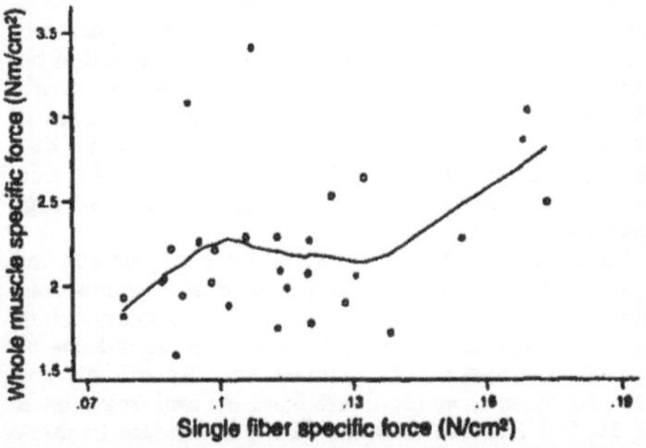

Fig. 3. Relationship between specific force (SF) at the whole muscle and single fiber levels (n = 31). The curve is the locally weighted scatter plot smoother (lowess), using a bandwidth of 0.65 (the curve is very similar for any bandwidth from about 0.25 up). This curve clearly shows the nonlinear nature of the relationship. Kendall's rank correlation coefficient (tau) is 0.25 (P = 0.054).

DISCUSSION

To our knowledge, this is the first study to investigate the relationship between strength and size at the whole muscle and single fiber levels in the same cohort of young men and older men and women. Our main results were: *1*) strength and WMCSA of the knee extensors were significantly higher in younger men compared with older subjects and in older men compared with older women; *2*) the age-related, but not the sex-related, differences in whole muscle strength were eliminated after adjusting for WMCSA; *3*) fibers from younger men were stronger than fibers from older men expressing the same MyHC isoform even after adjusting for size; *4*) in general, type I and IIA fibers from older men were stronger than similar fibers from older women even after adjusting for size; and *5*) in men, but not in women, type IIA fibers were stronger than type I fibers.

Whole muscle. Many studies have evaluated the association between age and changes in whole muscle SF or quality (force/size) with variable results (see Refs. 5 and 27). In the present study, adjusting for WMCSA eliminated differences in strength between young and older men. This is similar to our previous report (11), in which differences in isokinetic strength of the knee extensors tested at 1.04 rad/s among men and women (age range = 45–78 yr) were significantly reduced or eliminated after adjusting for whole body muscle mass estimated using urinary creatinine. Häkkinen et al. (15) and Kent-Braun and Ng (19) also reported no age-related differences in isometric strength of the knee extensors and ankle dorsiflexors, respectively, after adjusting for muscle CSA.

Other studies have reported significantly higher isokinetic (26, 33) and isometric (48) strength-to-WMCSA ratios in younger men compared with older men. Lynch et al. (26) recently reported a reduction in muscle quality with age in both men and women (range 19–93 yr). However, they estimated muscle mass from appendicular fat-free mass determined by dual-energy X-ray absorptiometry and measured isokinetic strength at an angular velocity of 0.52 rads/s. Metter et al. (27) estimated muscle quality in cross-sectional and longitudinal analyses of men and women (range 18–93 yr) and concluded that the relationship between muscle quality and age is dependent, among other factors, on how muscle mass is estimated. For example, they reported no significant age effect in strength expressed per kilogram of creatinine but a linear decline when expressed per CSA.

In the present study we also report sex-related differences in isokinetic muscle strength that persisted after adjusting for size. Our data are similar to that reported by Häkkinen et al. (15) who tested isometric strength in 70-yr-old men and women. However, Miller et al. (29) found no differences in the isometric strength-to-WMCSA ratio between younger men and women (mean age = 23–25 yr).

Discrepancies among studies could be explained by differences in: *1*) the architecture of the muscle groups tested; *2*) techniques used to estimate muscle CSA, not all of which are precise in vivo; *3*) measurements of strength (static vs. dynamic) (33); and *4*) statistical techniques (ratio vs. allometric) to assess the relationship between muscle strength and size (31).

Several experimental interventions have been shown to alter whole muscle quality, including compensatory hypertrophy after tenotomy (18), bed rest (23), hind limb suspension (13), and strength training (10). These studies suggest that the relationship between strength and size is not constant. The fact that in several of the experimental models the measurements are made after isolating the muscle from neural influences suggests that the intrinsic properties of muscle fibers may be affected. To test this hypothesis, we studied permeabilized single muscle fibers.

Age comparisons at the single fiber level. Our single fiber SF values (Table 3) for younger men of 20.8 (type I) and 23.8 N/cm^2 (type IIA) are in the same range as those published by others. For type I fibers, published values for SF measured under similar conditions are 19 (24), 21 (22), and 20.5 N/cm^2 (16). For type IIA fibers, published values are 25 (24), 20 (22), and 22.3 N/cm^2 (16).

In the present study we found significantly lower maximal force in both type I and type IIA fibers from older men compared with similar fibers from younger men. These differences were significant even after adjusting for fiber size and confirm previous observations by Larsson et al. (24). That study, however, was limited by a small sample size (4 young and 4 older men) and the absence of WMCSA comparisons.

The results of the present study suggest that the intrinsic ability (quality or SF) of muscle fibers to generate force is lower in older subjects. Conceptually, the events leading to the generation of force could be classified into three domains: *1*) excitation-contraction coupling, *2*) fuel metabolism and energy supply, and *3*) cross-bridge formation and myofilament sliding. It is possible that with advanced adult age, one or more of these domains is altered, resulting in muscle weakness and disability.

The elderly maintain the capacity to fully activate all motor units (34), and mean motor unit firing rates at given force levels are similar in young and older men (38). However, the effectiveness of the neural impulse in generating force may be limited by the excitation-contraction events in the muscle cell. This hypothesis is supported by the work of Delbono et al. (6) showing that aging interferes with the linking of the dihydropyridine receptors in the transverse tubules and the ryanodine receptors in the sarcoplasmic reticulum, resulting in uncoupling of calcium release and reduced availability of calcium to initiate contraction. In the present study it is unlikely that the above alterations could explain the differences between young and older men, since the permeabilization of fibers and the use of Brij solution removes the sarcoplasmic reticulum. Furthermore, only concentrations of calcium known to produce maximal activation were used. Another important difference between our study and that of Delbono et al. (6) is that we found lower SF values in both type I and IIA fibers, whereas they reported changes that were limited to fast-twitch fibers. It appears that under the conditions of the present study, weakness resulted from alterations beyond the excitation-contraction coupling steps.

Energy supply could potentially limit force generation. In this regard, Taylor et al. (42) used nuclear magnetic resonance to study muscle energetics in young and old volunteers. Under resting conditions, no differences were seen in intracellular pH or levels of ATP and phosphocreatine (PCr). Similar reductions in pH and PCr were seen during hand-grip exercise (7.5 min of fatiguing exercise), and resynthesis of ATP during recovery (12 min) was similar among age groups. The authors concluded that the aging process does not alter the energetics of human skeletal muscle. Furthermore, since energy supply is carefully controlled in the present experiments and the composition of the solutions standardized, alterations in this domain cannot explain the observed differences in force production.

Abnormalities in the quantity and/or function of the contractile or regulatory proteins (including troponin and tropomyosin) could explain the reduction in the intrinsic capacity of muscle cells to generate force. In younger subjects (mean age = 28.3 yr), a 6-wk bed rest period resulted in a 40% reduction in single fiber SF with a decline in myofibrillar protein content (23). The mean fractional rate of myofibrillar protein synthesis is slower in older muscles, probably due to a slower elongation of the peptide chain (45). Also, an age-related decline of MyHC synthesis rate (but not of sarcoplasmic protein) has been observed (2). Since, in that study, the MyHC synthesis rate correlated with in vivo measurements of muscle strength, the reported decline could contribute to weakness in the elderly. Although there is no direct evidence that with age newly synthesized proteins are abnormal, a slow synthesis rate could delay the replacement of dysfunctional proteins. Thompson and Brown (43) reported in single fibers from rat soleus muscle a dissociation between the loss of FCSA and force with age and attributed this to changes in the molecular dynamics of myosin. Brooks and Faulkner (5) have concluded that in mammals the deficit in SF may result from a reduction in the number of cross-bridges or in the average force per cross-bridge.

Alterations in the internal architecture of the cell could potentially limit force generation. For example, the amount of force generated by each cross-bridge may be reduced if the space between myofilaments increases. Evidence to support this hypothesis has been reported in single fibers from rat and frog muscle (1, 14, 28). Studies in single fibers from older humans testing this hypothesis are needed. Another possible explanation is a loss of thin myofilaments as reported in human soleus after bed rest (37). However, in the latter study, SF was only slightly decreased.

The reported reduction in force with age after adjusting for size suggest that atrophy of muscle fibers in itself offers only a partial explanation for the muscle weakness of old age. Atrophy can result in the loss of muscle mass without appreciable changes in the physiological properties of the cells (39). Degeneration and cell death, on the other hand, involve changes in both structure and function. It remains to be seen whether the strength training that reverses muscle atrophy in the elderly (10) can also reverse the age-related alterations in the intrinsic quality of muscle fibers.

Sex comparisons at the single fiber level. To our knowledge, this is the first report to compare single human muscle fibers from both sexes. Previous studies have not included women (24) or have combined the results from both sexes (22). Our results at the single fiber level demonstrated significant sex-related differences in maximal force that were dependant on fiber type and not explained by differences in fiber size.

Other authors have noted the influence of sex on the age-related changes in skeletal muscle. Young et al. reported that the age-related reduction in strength was proportional to the reduction in WMCSA in women (47) but not in men (48). These differences in younger subjects have been attributed to greater fiber area, greater fiber number, and/or higher voluntary motor unit activation in men (3). Since our experimental preparation isolated single fibers from the influence of the nervous system and we adjusted for the differences in size, those hypotheses cannot explain our observations and suggest that differences may exist in the contractile elements of the muscle cell. The fact that the sex-related differences were fiber type dependent supports this hypothesis.

The reduction in estrogen levels after menopause could explain the different rates of loss of strength between sexes (35). This hypothesis is supported by the observation that 50-yr-old men and women have similar strength-to-WMCSA ratio but 70-yr-old women show lower values than 70-yr-old men (15). Balagopal and colleagues (2) observed that the correlation between muscle strength and MyHC synthesis rate is stronger in women and suggested that protein synthesis in sarcopenia could be regulated by different parameters in men and women. The mechanism by which estrogen may exert this influence remains to be determined, but it may be due to an influence on the number of cross-bridges or the force generated per cross-bridge (44) or the sensitivity of the cross-bridges to metabolites, such as hydrogen ions or inorganic phosphate, that reduce force (35). It is also possible that immunologic, neurological, or genetic influences play a role.

Experiments with whole muscle and with single fibers represent different levels of study. Extrapolation of observations made in experiments with single fiber segments should be made with caution. The whole muscle condition represents a summation of various influences and intermediate steps eliminated in the single fiber preparation. It would be of scientific interest and practical significance to understand the nature of the association between the single fiber level and determinations of whole muscle function. Figure 3 illustrates the complex and nonlinear nature of the relationship between the two levels. We have initiated a more in-depth analysis, including random effects modeling.

In conclusion, the present study demonstrates an age-related reduction in the intrinsic ability of single muscle fibers expressing type I and IIA MyHC isoforms to generate force. It is likely that alterations at the level of cross-bridge formation and interaction play a significant role. More research is needed to understand the basic mechanisms underlying muscle dysfunction in old age.

This material is based upon work supported by the US Department of Agriculture, under agreement 58-1950-9-001 and contract 53-K06-1.

Any opinions, findings, conclusions, or recommendations expressed in this publication are those of the author(s) and do not necessarily reflect the view of the US Department of Agriculture.

REFERENCES

1. **Bagni MA, Cecchi G, and Colomo F.** Myofilament spacing and force generation in intact frog muscle fibers. *J Physiol (Lond)* 430: 61–75, 1990.
2. **Balagopal P, Rooyackers OE, Adey DB, Ades PA, and Nair KS.** Effects of aging on in vivo synthesis of skeletal muscle myosin heavy-chain and sarcoplasmic proteins in humans. *Am J Physiol Endocrinol Metab* 273: E790–E800, 1997.
3. **Behm DG and Sale DG.** Voluntary and evoked muscle contractile characteristics in active men and women. *Can J Appl Physiol* 19: 253–265, 1994.
4. **Bergström J.** Muscle electrolytes in man. *Scand J Clin Lab Invest* 14 *Suppl* 68: 1962.
5. **Brooks SV and Faulkner JA.** Skeletal muscle weakness in old age: underlying mechanisms. *Med Sci Sports Exerc* 26: 432–439, 1994.
6. **Delbono O, O'Rourke KS, and Ettinger WH.** Excitation-calcium release uncoupling in aged single human skeletal muscle fibers. *J Membr Biol* 148: 211–222, 1995.
7. **Dumitru D** *Electrodiagnostic Medicine*. Philadelphia, PA: Hanley and Belfus, 1995, p. 139.
8. **Evans WJ, Phinney SD, and Young VR.** Suction applied to a muscle biopsy maximizes sample size. *Med Sci Sports Exerc* 14: 101–102, 1982.
9. **Fabiato A.** Computer programs for calculating total from specified free or free from specified total ionic concentrations in aqueous solutions containing multiple metals and ligands. *Methods Enzymol* 157: 378–417, 1988.
10. **Frontera WR, Meredith CN, O'Reilly KP, Knuttgen HG, and Evans WJ.** Strength conditioning in older men: skeletal muscle hypertrophy and improved function. *J Appl Physiol* 64: 1038–1044, 1988.
11. **Frontera WR, Hughes VA, Lutz KJ, and Evans WJ.** A cross-sectional study of muscle strength and mass in 45- to 78-yr-old men and women. *J Appl Physiol* 71: 644–650, 1991.
12. **Fukunaga T, Roy RR, Shellock FG, Hodgson JA, and Edgerton VR.** Specific tension of human plantar flexors and dorsiflexors. *J Appl Physiol* 80: 158–165, 1996.
13. **Gardetto PR, Schlutter JM, and Fitts RH.** Contractile function of single muscle fibers after hindlimb suspension. *J Appl Physiol* 66: 2739–2749, 1989.
14. **Godt RE and Maughan DW.** Swelling of skinned muscle fibers of the frog. *Biophys J* 19: 103–116, 1977.
15. **Häkkinen K, Kraemer WJ, Kallinen M, Linnamo V, Pastinen UM, and Newton RU.** Bilateral and unilateral neuromuscular function and muscle cross-sectional area in middle-aged and elderly men and women. *J. Gerontol.* 51A: B21–B29, 1996.
16. **Harridge SDR, Botinelli R, Canepari M, Pellegrino MA, Reggiani C, Esbjörnsson M, and Saltin B.** Whole-muscle and single-fibre contractile properties and myosin heavy chain isoforms in humans. *Pflügers Arch* 432: 913–920, 1996.
17. **Hurley BF.** Age, gender, and muscular strength. *J Gerontol* 50A: 41–44, 1995.
18. **Kandarian SC and White TP.** Mechanical deficit persists during long-term muscle hypertrophy. *J Appl Physiol* 69: 861–867, 1990.
19. **Kent-Braun JA and Ng AV.** Specific strength and voluntary muscle activation in young and elderly men and women. *J Appl Physiol* 87: 22–29, 1999.
20. **Kronmal RA.** Spurious correlation and the fallacy of the ratio standard revisited. *J Royal Statis Soc (series A)* 156: 379–392, 1993.
21. **Laemmli UK.** Cleavage of structural proteins during the assembly of the head of bacteriophage T4. *Nature* 227: 680–685, 1970.
22. **Larsson L and Moss RL.** Maximum velocity of shortening in relation to myosin isoform composition in single fibres from human skeletal muscles. *J Physiol (Lond)* 472: 595–614, 1993.
23. **Larsson Li LX, Berg HE, and Frontera WR.** Effects of removal of weight-bearing function on contractility and myosin isoform composition of single human skeletal muscle fibers. *Pflügers Arch* 432: 320–328, 1996.
24. **Larsson L, Li X, and Frontera WR.** Effects of aging on shortening velocity and myosin isoform composition in single human skeletal muscle cells. *Am J Physiol Cell Physiol* 272: C638–C649, 1997.
25. **Lee IM, Paffenbarger RS, and Hsieh CC.** Time trends in physical activity among college alumni, 1962–1988. *Am J Epidemiol* 135 915–925, 1992.
26. **Lynch NA, Metter EJ, Lindle RS, Fozard JL, Tobin JD, Roy TA, Fleg JL, and Hurley BF.** Muscle quality. I. Age-associated differences between arm and leg muscle groups. *J Appl Physiol* 86: 188–194, 1999.
27. **Metter EJ, Lynch NA, Conwit R, Lindle R, Tobin J, and Hurley BF.** Muscle quality an age: cross-sectional and longitudinal comparisons. *J. Gerontol. Biol. Sci.* 54A: B207–B218, 1999.
28. **Metzger JM and Moss RL.** Shortening velocity in skinned single muscle fibers: influence of filament lattice spacing. *Biophys J* 52: 127–131, 1987.
29. **Miller AEJ, MacDougall JD, Tarnopolsky MA, and Sale DG.** Gender differences in strength and muscle fiber characteristics. *Eur J Appl Physiol* 66: 254–262, 1993.
30. **Moss RL.** Sarcomere length-tension relations in frog skinned muscle fibers during calcium activation at short lengths. *J Physiol (Lond)* 292: 177–192, 1979.
31. **Neder JA, Nery LE, Silva AC, Andreoni S, and Whipp BJ.** Maximal aerobic power and leg muscle mass and strength related to age in non-athletic males and females. *Eur J Appl Physiol* 79: 522–530, 1999.

Allostatic load as a marker of cumulative biological risk: MacArthur studies of successful aging

Teresa E. Seeman*†, Bruce S. McEwen‡, John W. Rowe§, and Burton H. Singer¶

*Division of Geriatrics, University of California School of Medicine, 10945 Le Conte Avenue, Suite 2339, Los Angeles, CA 90095-1687; ‡Laboratory for Neuroendocrinology, The Rockefeller University, New York, NY 10021; §Aetna, Inc, Hartford, CT 06156; and ¶Office of Population Research, Princeton University, Princeton, NJ 08544

Contributed by Burton H. Singer, February 13, 2001

Allostatic load (AL) has been proposed as a new conceptualization of cumulative biological burden exacted on the body through attempts to adapt to life's demands. Using a multisystem summary measure of AL, we evaluated its capacity to predict four categories of health outcomes, 7 years after a baseline survey of 1,189 men and women age 70–79. Higher baseline AL scores were associated with significantly increased risk for 7-year mortality as well as declines in cognitive and physical functioning and were marginally associated with incident cardiovascular disease events, independent of standard socio-demographic characteristics and baseline health status. The summary AL measure was based on 10 parameters of biological functioning, four of which are primary mediators in the cascade from perceived challenges to downstream health outcomes. Six of the components are secondary mediators reflecting primarily components of the metabolic syndrome (syndrome X). AL was a better predictor of mortality and decline in physical functioning than either the syndrome X or primary mediator components alone. The findings support the concept of AL as a measure of cumulative biological burden.

In elderly populations, comorbidity in the form of multiple co-occurring chronic conditions is the norm rather than the exception. For example, in the U.S. 61% of women and 47% of men age 70–79 report two or more chronic conditions. These figures rise to 70% of women and 53% of men age 80–89 with 2+ chronic conditions (1). No single form of comorbidity occurs with high frequency, but rather a multiplicity of diverse combinations are observed (e.g., osteoarthritis and diabetes, colon cancer, coronary heart disease, depression, and hypertension). This diversity underscores the need for an early warning system of biomarkers that can signal early signs of dysregulation across multiple physiological systems.

One response to this challenge was the introduction of the concept of allostatic load (AL) (2–4) as a measure of the cumulative physiological burden exacted on the body through attempts to adapt to life's demands. The ability to successfully adapt to challenges has been referred to by Sterling and Eyer (5) as allostasis. This notion emphasizes the physiological imperative that, to survive, "an organism must vary parameters of its internal milieu and match them appropriately to environmental demands" (5). When the adaptive responses to challenge lie chronically outside of normal operating ranges, wear and tear on regulatory systems occurs and AL accumulates.

An initial operationalization of the notion of AL (6) used assessments of 10 biological parameters reflecting functioning of the hypothalamic-pituitary-adrenal (HPA) axis, sympathetic nervous system, cardiovascular system, and metabolic processes. Higher scores on a summary numerical measure of AL were shown to predict four major health outcomes: incident cardiovascular disease (CVD), decline in physical functioning, decline in cognitive functioning, and mortality (6). Although the summary AL measure was a significant predictor of outcomes, none of the individual components exhibited strong predictive capacity. This finding suggests that risk for the above outcomes is related to the overall impact of dysregulation across multiple regulatory systems. In addition, examination of the most commonly occurring pairs and triples of factors contributing to overall AL scores clearly indicated that there were multiple pathways of relatively comparable frequency through which individuals achieved higher AL scores. This finding is consistent with the idea that a cumulative measure of "biological dysregulation" across multiple systems may act as an early warning indicator of multiple health outcomes. It also emphasizes the point that there are multiple pathways to comorbidity and mortality.

Despite these positive indications, some important caveats are associated with previous analyses. First, the predictive ability of AL was assessed over a relatively short time period; i.e., 2.5 years. Second, six of the 10 components of the measure of AL are standard indicators of syndrome X, the metabolic syndrome that has been related to increased risk of CVD (7). These include levels of seated systolic blood pressure (SBP) and diastolic blood pressure (DBP), high density lipoproteins (HDL), and total cholesterol, glycosylated hemoglobin, and abdominal obesity [i.e., high waist-hip ratio (WHR)]. Thus, the question arises as to whether the extant operationalization of AL is something more than an alternative measure of syndrome X. Third, there is a cascade of events through which AL develops, with chemical messengers such as cortisol (C), epinephrine (EPI), and norepinephrine (NE) playing the role of primary mediators (4). Previous studies give no indication of the potential predictive value of such primary mediators, nor do they clarify how they contribute to a summary measure of AL beyond the syndrome X components.

The purposes of this paper are to investigate: (*i*) the predictive value of a summary measure of AL over a period of 7 years beyond baseline survey of an elderly population; (*ii*) the predictive value of syndrome X parameters, which are all secondary mediators in the cascade that leads to cumulating AL, for outcomes other than CVD, such as decline in physical and cognitive functioning, and (*iii*) the kinds of outcomes for which a set of primary mediators are good predictors in their own right and those for which they contribute, above and beyond syndrome X parameters, to the predictive capacity of a summary measure of AL.

Methods

Data for these analyses come from the MacArthur Successful Aging Study, a longitudinal study of relatively high functioning men and women aged 70–79. As described in greater detail elsewhere (8), subjects were subsampled on the basis of age and both physical and cognitive functioning from three community-

Abbreviations: AL, allostatic load; CVD, cardiovascular disease; CI, confidence interval; C, cortisol; NE, norepinephrine; EPI, epinephrine; DBP, diastolic blood pressure; SBP, systolic blood pressure; HDL, high density lipoproteins; HPA, hypothalamic-pituitary-adrenal; WHR, waist-hip ratio; DHEA-S, dihydroepiandrosterone sulfate; OR, odds ratio.

†To whom reprint requests should be addressed. E-mail: tseeman@mednet.ucla.edu.

The publication costs of this article were defrayed in part by page charge payment. This article must therefore be hereby marked "*advertisement*" in accordance with 18 U.S.C. §1734 solely to indicate this fact.

based cohorts in Durham, NC, East Boston, MA, and New Haven, CT that were part of the Established Populations for Epidemiological Studies of the Elderly. Age was restricted to 70–79 years, and age-eligible men and women were screened on the basis of four criteria of physical functioning and two criteria of cognitive functioning to identify those functioning in the top third of the age group.

Of the 4,030 age-eligible men and women, a cohort of 1,313 subjects met all screening criteria; 1,189 (90.6%) agreed to participate and provided informed consent. Baseline data collection was completed between May 1988 and December 1989 and included a 90-min, face-to-face interview as well as blood samples and 12-h overnight urine; 80.3% agreed to provide blood samples and 85.8% consented to provide urine samples.

The cohort was reinterviewed in 1991 and again in 1996; with reassessments of all measures included in the baseline interview. Outcomes examined here are based on the 1996 follow-up (data collection from October 1995 to February 1997). The mean interval from the first follow-up was 85 months (i.e., 7 years; SD = 7 months). Of the 1,118 individuals alive at the end of data collection in 1991, 202 had died by the time of the 1996 follow-up (18%); of the remaining 916 subjects, 722 completed face-to-face interviews (79%), 107 (11.7%) had proxy-partial interviews, 44 (4.8%) refused, and 43 (4.7%) were alive but couldn't be contacted.

AL. Our measure of AL, described previously (6), is designed to summarize levels of physiological activity across a range of regulatory systems pertinent to disease risks (2–4). These parameters included: SBP and DBP (indices of cardiovascular activity); WHR (an index of more chronic levels of metabolism and adipose tissue deposition, thought to be influenced by increased glucocorticoid activity) (9); serum HDL and total cholesterol (markers whose levels are known to influence the development of atherosclerosis, increased risk being seen with higher levels in the case of total cholesterol and lower levels in the case of HDL); blood plasma levels of glycosylated hemoglobin (Hb_{A1c}; an integrated measure of glucose metabolism over several months time); serum dihydroepiandrosterone sulfate (DHEA-S; a functional HPA axis antagonist) (10) ; 12-h urinary C excretion (an integrated measure of 12-h HPA axis activity); and 12-h urinary NE and EPI excretion levels (integrated indices of 12-h sympathetic nervous system activity).

SBP and DBP were calculated as the average of the second and third of three seated blood pressure readings (11). WHR was calculated based on waist circumference (measured at its narrowest point between the ribs and iliac crest) and hip circumference (measured at the maximal buttocks) (12). Blood samples for assays of HDL cholesterol, total cholesterol, glycosylated hemoglobin, and DHEA-S were obtained by phlebotomists who visited subjects' homes the morning after their home interview to collect the blood sample and an overnight urine sample (see below). Although subjects were not required to fast, most blood samples were taken early in the morning before subjects had eaten. A total of 9 cm^3 of blood was drawn in serum-separator tubes, and another 10 cm^3 was drawn into heparinized tubes. Samples were processed and frozen within 4 h of blood drawing. Blood in the serum-separator tubes was allowed to clot, and the tube was centrifuged at 15 relative centrifugal force for 10 min in a refrigerated (4°C) centrifuge. The sera were sent to Nichols Laboratories (San Juan Capistrano, CA) for measurements of HDL and total cholesterol, and DHEA-S. A total of 2 cm^3 of blood from a heparinized tube was removed after mixing for assays of glycosylated hemoglobin (Hb_{A1c}), assayed by affinity chromatography methods (13).

Subjects also completed an overnight urine collection from 8 p.m. on the evening after their home interview to 8 a.m. the next morning. The overnight collection protocol serves to minimize the potential confounding effects of physical activity as subjects generally spent this time at home (and much of that time in bed). Thus, individual differences in overnight excretion of C as well as NE and EPI provide integrated measures of differences in more "steady-state" operating levels of the HPA axis and sympathetic nervous system (i.e., estimates of "basal," nonstimulated levels of activity). All samples were sent to Nichols Institute immediately after collection for assays of NE, EPI, and C. Determinations were made by HPLC (14, 15). Interassay variation was 8–9% for catecholamines regardless of range; interassay variation for C was 1% in low range (<50 μg/dl) and 8% in higher range. Results for each of the three outcomes are reported as μg (NE, EPI, or C)/g creatinine to adjust for body size.

Table 1. AL: Criteria for contributions from individual biological parameters

Highest quartile	
SBP (> = 148 mm Hg)	
DBP (> = 83 mm Hg)	
WHR (> = 0.94)	
Ratio total cholesterol/HDL (> = 5.9)	
Glycosylated hemoglobin (> = 7.1%)	
Urinary CI (> = 25.7 ug/g creatinine)	
Urinary NE (> = 48 ug/g creatinine)	
Urinary EPIN (> = 5 ug/g creatinine)	
Lowest quartile	
HDL cholesterol (< = 37 mg/dl)	
DHEA-S (< = 350 ng/ml)	

For each of the 10 indicators of AL, subjects were classified into quartiles based on the distribution of scores in the baseline cohort. AL was measured by summing the number of parameters for which the subject fell into the highest risk quartile (i.e., top quartile for all parameters except HDL cholesterol and DHEA-S for which membership in the lowest quartile corresponds to highest risk). Use of the top/bottom quartile criterion to define contributions to higher AL from these various biological parameters reflects a data-driven partitioning of the sample. Membership in the upper/lower quartile represents a quantitative way of classifying those exposed to more extreme levels of system activity relative to the rest of the population and thus potentially at greater risk of disease pathology. Table 1 indicates the actual criterion cutoff for each of the 10 indices. Other methods of summarizing the data, including averaging z scores and use of other criterion cut points also were examined in earlier analyses and yielded comparable results (6).

Contributions of syndrome X and non-X (i.e., primary mediator) components of AL were examined by dividing the total AL score into two component parts: one reflecting the sum of the syndrome X components (i.e., SBP, DBP, HDL and total cholesterol, Hb_{A1c}, WHR) and the other reflecting the sum of the non-syndrome X components (i.e., urinary C, NE, EPI, and DHEA-S).

Health Outcomes. *All-cause mortality.* Deaths were identified through contact with next of kin at the time of subsequent interviews (1991 and 1996), ongoing local monitoring of obituary notices, and National Death Index searches.

Incident CVD (1988–1996). Incident CVD events were assessed through self-reports of new, doctor-diagnosed myocardial infarction, stroke, hypertension, or diabetes between 1988 and 1996. Reports of such newly diagnosed conditions at the time of the 1991 and 1996 follow-up interviews were compared with annual interview data from Established Populations for the Epidemiologic Study of the Elderly interviews for the previous 5 years (for Durham subjects) and 10 years (for New Haven and

East Boston subjects) to confirm that the condition had not been previously reported. Although medical records were not available for further confirmation, self-reports of such conditions have been found to agree with medical records for these conditions with percent agreement of 0.86 or higher and kappas of 0.70 or higher (16). Because of the small number of cases for individual diseases, we used a summary measure reflecting the occurrence of any incident CVD (i.e., doctor-diagnosed myocardial infarction, hypertension, diabetes, or stroke). Of the total 161 incident CVD events reported between 1988 and 1995, there were 10 incident reports of a diagnosis of hypertension in subjects who also had high blood pressure at baseline as a contributor to their AL score. To reduce potential confounding between baseline AL and subsequent reports of incident hypertension, these 10 cases were deleted from the count of incident CVD events.

Cognitive and physical performance. Detailed, performance-based assessment of both cognitive and physical performance were available at both baseline (1988) and the final follow-up (1996), permitting examination of both cross-sectional and longitudinal associations with AL.

Cognition. Measures of cognition were designed to assess those aspects of higher cortical function that are necessary to perform complex cognitive activities. Measures included six tasks: language, abstraction, spatial ability (copying), delayed spatial recognition, delayed recall of confrontation naming items, and delayed recall of a story. A measure of overall cognitive performance was created by summing scores on the six tasks (17).

Physical performance. A summary measure of physical performance, based on five separate tests of physical ability, was used. The five abilities assessed included timed measures of balance, gait, chair stands, foot taps, and manual ability. Scores for each of the subscales were scaled to indicate the proportion of the best possible score that a subject achieved, i.e., rescaled scores range from 0 (worst possible performance) to 1 (best possible performance), and a summary measure of performance was created by adding the scaled subscale scores for the respective years (range each year = 0 to 5) (18).

Covariates. Multivariate analyses included age, sex, race, income, and education along with summary measures of baseline health status. Age was measured as of the date of the baseline interview (range = 70 to 79). Race was coded African-American versus Caucasian. Education was measured as less than high school vs. high school or higher. Annual household income was measured in $10,000 increments to $30,000+; final analyses use a dichotomous classification: <$10,000 versus $10,000 or more. Summary measures of baseline prevalence of CVD (i.e., myocardial infarction, stroke, diabetes, high blood pressure) and of other conditions (cancer, broken hip, or other bones) were developed based on self-reports of doctor-diagnosed conditions. These measures were included in analyses of cognitive and physical functioning and of mortality.

Analyses. Logistic regression models were used to test for associations between AL and incident CVD and mortality. For the continuous outcome measures of cognitive and physical functioning, linear regression models were used to assess associations between AL and changes in performance. Comparison of the 720 individuals with complete baseline data to those missing either AL or covariate data indicated that those with complete data did not differ in terms of their age, education, income, burden of chronic conditions, or incidence of CVD; mortality was only marginally higher among those excluded from the analyses (25% vs. 21%, $P = 0.08$). Those with complete data were more likely to be male and to exhibit somewhat better cognitive and physical performance abilities at baseline, although the actual differences in performance were relatively minor (data not shown).

To assess associations between AL and changes in cognitive and physical functioning, 1996 function was regressed on AL scores, controlling for baseline function. With inclusion of baseline performance scores, the analysis is modeling residualized change, i.e., change adjusted for baseline level of physical performance (19) and coefficients for other variables provide estimates of the impact of these factors on change in performance from baseline. Models were examined controlling for possible confounding by socio-demographic and/or health status characteristics. Similar multivariable logistic (rather than proportional hazards) regression models were examined with respect to incident and recurrent CVD and mortality outcomes because information on time to event was not available. Possible nonlinear risk relationships for AL were examined by including a series of four dummy variables reflecting increasing levels of AL (levels 1–2, 3–4, 5–6, 7+) with scores of zero as the reference group.

To assess the question of whether AL is something more than simply an alternative measure of syndrome X, models were run, replacing the summary AL score with the two separate subscales measuring syndrome X components and non-X components. SAS version 6.12 was used for all analyses (20).

Results

The sample was 49% male, 19% non-white and showed considerable socio-economic diversity. Forty percent of the subjects reported annual household incomes of less than $10,000/year whereas 14% reported annual incomes of $30,000 or more. Educational attainment also varied from less than eighth grade (17%), to high school (25.6%), to some college or more (22%). At baseline, subjects reported on average slightly more than one chronic condition; 56% reported prevalent CVD and 26% reported no prevalent chronic conditions.

Over the following 7.5 years, there were 153 deaths among the 720 subjects with complete baseline data and a total of 151 incident CVD events among the 582 subjects free of baseline CVD. Average declines in physical performance as measured by our five-point scale were equivalent to approximately 1 SD ($\bar{x} = -0.49$), whereas average declines in cognitive performance based on the 90-point scale were equivalent to approximately half of a SD ($\bar{x} = -3.7$). The actual range of change scores for both performance domains was quite wide: changes in physical performance ranged from −3.35 to +0.62 (maximum possible range: −5 to +5), changes in cognitive performance ranged from −49 to +21 (maximum possible range = −89 to +89).

AL and 7-Year Outcomes. Logistic regression analyses revealed significantly increased risk of mortality with higher AL scores, independent of age, sex, ethnicity, education, income, and baseline morbidity [odds ratio (OR) = 1.23; 95% confidence interval (CI) = 1.09, 1.39]. Use of dummy-coded indicator variables for categories of increasing AL revealed an approximately monotonic increase in relative risk with increasing AL. As shown in Fig. 1*a*, those with scores of 1–2 experienced 1.67 greater mortality compared with those scoring zero on AL (95% CI = 0.71, 4.59); those scoring 3–4 on AL experienced 2.45 times greater mortality (95% CI = 1.04, 6.77) whereas those at the highest levels of AL (i.e., 7+) experience the greatest relative risk for mortality (OR = 6.42; 95% CI = 1.36, 32.12).

Logistic regression models for incident CVD revealed a marginally significant association between baseline AL and subsequent incident CVD events (OR = 1.12; 95% CI = 0.99, 1.27; $P = 0.06$). Examination of the individual dummy-coded indicators for increasing levels of AL revealed a generally increasing pattern of risk for CVD with increasing AL although none of the individual comparisons reached statistical significance (see Fig. 1*b*).

Linear regression analyses of changes in physical functioning also revealed that higher AL scores were associated with greater

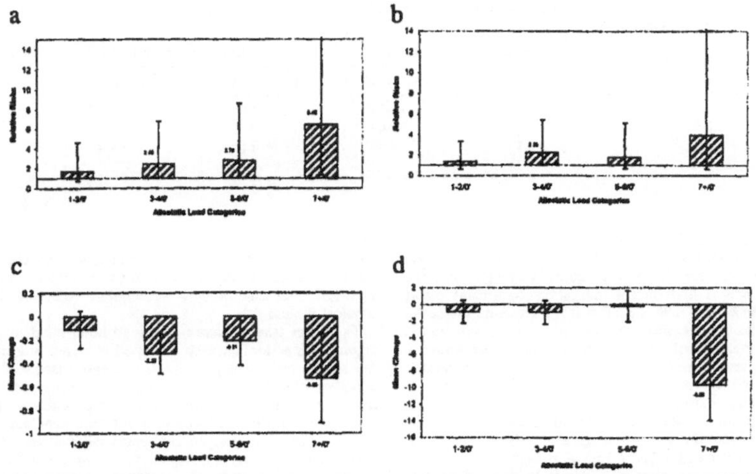

Fig. 1. Associations between baseline AL and 7-year health outcomes. Effects reaching statistical significance ($P <= 0.05$) are indicated by numbers at top of bars. Indicated are 95% CIs for each estimated relative risk. (*a*) Mortality risks. (*b*) Incident CVD risks. (*c*) Changes in physical functioning. (*d*) Changes in cognitive functioning.

average decline ($b = -0.06, P = 0.02$). As shown in Fig. 1*c*, those with highest AL scores (i.e., those with scores of 7+) experienced the greatest declines and, although not strictly monotonic, there was a general trend toward smaller declines moving from higher to lower AL groups.

For cognitive declines, linear regression models again revealed a significant negative effect of AL ($b = -0.58, P = 0.03$). As shown in Fig. 1*d*, this effect appears to reflect the significantly greater declines among those with the highest AL scores.

Cumulative AL versus Syndrome X Risk. We next considered the question of whether these observed associations reflect the cumulative impact of the 10 components or were simply reflections of the effects of a specific subset of measures such as those associated with syndrome X. To test this hypothesis, models were examined that included two separate scores: one score reflected the number of syndrome X characteristics that were present (i.e., high SBP and DBP, high ratio of total cholesterol to HDL, high WHR, and high Hb_{A1c}), and the other score reflected the number of non-X characteristics that were present (i.e., high urinary C, NE, EPI, and low DHEA-S).

The possibility that each of these subscales might contribute independently to health risk was supported by evidence indicating considerable nonoverlap between these two scores. For example, among those with a 0 score on the syndrome X subscale, over 70% had scores of 1 or more on the non-X subscale; likewise, among those with 0 scores on the non-X subscale, 80% had a score of 1 or more on the syndrome X subscale. Analyses including the two subscale scores provide an indication of the independent contribution of each subset of biological measures (i.e., an assessment of the extent to which the non-X components of AL are contributing independently to the prediction of our four health outcomes).

Analyses of mortality revealed significant effects for both syndrome X and non-X risk scores (for syndrome X: OR = 1.18, $P = 0.03$, 95% CI = 1.02, 1.36; and for non-X: OR = 1.36, $P = 0.002$, 95% CI = 1.12, 1.66). The pattern of mortality risks across the levels of syndrome X and non-X scores based on dummy-coded indicators for each of the levels of syndrome X and non-X is shown in Table 2. For non-X scores, there is a pattern of increasing mortality as one moves from lower to higher non-X scores; as also indicated in Table 2, each of these groups experiences significantly increased mortality as compared with those scoring 0 on the non-X risk factors. By contrast, syndrome X scores of 1, 2, and 3 are associated with similar and nonsignificant increases in mortality risk (OR ~ 1.7) when compared with those with scores of 0. Scores of 4 or more on syndrome X, however, are associated with significantly greater risk (OR = 2.47; 95% CI = 1.28, 2.66)

Table 2. Mortality and cardiovascular risks associated with syndrome X and non-X factors

Syndrome X & non-X scores	Mortality		Incident CVD	
	Syndrome X risks OR	Non-X risks OR	Syndrome X risks OR	Non-X risks OR
1/0	1.68 (0.96, 3.01)	1.9* (1.21, 3.02)	1.59 (0.93, 2.75)	1.55* (1.0, 2.44)
2/0	1.65 (0.91, 3.06)	2.01* (1.14, 3.51)	1.64 (0.92, 2.96)	1.21 (0.68, 2.14)
3/0	1.54 (0.77, 3.07)	2.69* (1.30, 5.45)	1.65 (0.85, 3.2)	1.34 (0.64, 2.72)
4+/0	2.47* (1.19, 5.13)		1.71 (0.78, 3.68)	

AL = 0 is reference group in all analyses. *, $P < 0.05$.

Table 3. Changes in physical and cognitive performance associated with syndrome X and non-X factors

Syndrome X & non-X scores	Change in physical function		Change in cognitive function	
	Syndrome X risks *b*	Non-X risks *b*	Syndrome X risks *b*	Non-X risks *b*
1/0	−0.09	−0.11*	1.03*	1.22*
2/0	−0.11	−0.03	−0.17	−1.43*
3/0	−0.4*	−0.34*	−2.69*	−0.58
4+/	−0.15		−0.15	

*, $P < 0.05$.

Analyses of incident CVD events revealed a stronger, positive association between the syndrome X scores and incident CVD than was the case for the non-X scores, although even the syndrome X scores had only a marginally significant effect (OR = 1.14; 95% CI = 0.98, 1.32; $P = 0.08$). Examination of the patterns of risk for specific levels of syndrome X and non-X revealed an unexpectedly flat pattern of risk for syndrome X scores of 1 or more (see Table 2); in no case did the estimated relative odds of incident CVD achieve statistical significance. A somewhat similar pattern was seen for the non-X scores, although in the latter case, the estimated relative odds of an incident CVD event did achieve marginal significance for those with scores of 1 versus 0 (OR = 1.55; $P = 0.06$).

Linear regression modeling of declines in physical performance revealed only a marginal association for syndrome X ($b = -0.062$, $P = 0.07$) and no significant association for non-X ($b = -0.066$, $P = 0.13$). However, as shown in Table 3, analyses based on the dummy-coded indicators for levels of these subscales suggested that these linear estimates may be inaccurate as there appear to be threshold effects for both syndrome X and non-X scores—the largest declines in performance being evident among those with highest or next highest scores on syndrome X and/or non-X scales. This pattern is consistent with the previously observed risks for decline in physical performance associated with the highest AL scores (i.e., scores of 7+). Indeed, the stronger overall effect of AL appears to reflect more modest effects of both syndrome X and non-X factors.

Linear regression models for cognitive performance revealed a significant, negative effect for syndrome X risk scores ($b = -0.61$, $P = 0.04$), with a nonsignificant effect for non-X risk scores ($b = -0.46$, $P = 0.25$). Analyses based on the dummy-coded indicators did not reveal a consistent pattern of effects (see Table 3).

Discussion

A primary goal of these analyses was to provide a more thorough examination of the concept of AL as a measure of cumulative biological burden. A first set of analyses confirmed that overall AL was a significant predictor of 7.5-year health outcomes in this cohort of older adults as it had been in earlier analyses of 2.5-year outcomes (6). Higher baseline AL scores were associated with significantly greater risk for mortality as well as for declines in cognitive and physical functioning; and a marginally significant effect was also seen for incident CVD. The risk relationship was generally linear for mortality but appeared to reflect more of a threshold pattern with respect to risk for decline in cognitive and physical functioning, where only those at the highest levels of AL experienced significantly greater declines in functioning.

A second set of analyses, focusing on the contributions of the syndrome X and non-X components of AL, indicated that for two of our outcomes, mortality and physical functioning, both subscales appear to contribute to subsequent risks. In the case of mortality, both the syndrome X and non-X components of AL contributed significantly to risk. In the case of physical functioning, although neither of the two subscales was significantly related to patterns of change in its own right, each was marginally and negatively related to risk for decline, and the summary measure of AL, combining these subscale effects, did exhibit a significant association with subsequent risks for decline in physical functioning.

The other two outcomes examined here, incident CVD and cognitive functioning, each provided evidence suggesting that the syndrome X components are largely responsible for the observed effects of AL—a finding that is not surprising in the case of incident CVD but more unexpected with respect to risks for cognitive decline. This pattern of results provides somewhat mixed support for the general hypothesis that the more comprehensive summary measure of AL would offer the best prediction of outcomes. The evidence for this is strongest with respect to mortality and physical functioning but less so for incident CVD and cognition.

Limitations of the data presented here are important to consider in evaluating the findings. Two features of the MacArthur Study of Successful Aging database are worth noting. First, the analyses use secondary data (a convenience sample) and are thus constrained by the available biological data (a set of measures that were not collected specifically to assess AL). Longitudinal data collection explicitly designed to assess AL lies in the future. Thus, our inability to consistently demonstrate significant contributions from non-syndrome X factors may be due, at least in part, to the fact that the available data did not provide as comprehensive coverage of these latter factors as was true for the syndrome X characteristics. One might speculate, for example, that stronger evidence for AL as a predictor of both cognition and CVD, and indeed as a predictor of all of the outcomes, would have been found had our measure of AL included a more comprehensive assessment of parameters of dysregulation in all important regulatory systems. Thus, for example, inclusion of information on missing elements such as measures of hemostatic or immune/inflammatory processes, which have been linked to CVD (21, 22) and/or cognition (J. Weaver, M. Albert, T. Harris, J.W.R., and T.E.S., unpublished work), might improve the ability of our measure of AL to predict outcomes.

Clearly, the measure of AL used in these analyses should be seen as an initial, and only partial, assessment of the concept. AL has been conceptualized as a marker of cumulative biological dysregulation across all major regulatory systems; contributing factors include dysregulation in both system state (or usual) levels of activity and in system dynamics. Although our measure of AL included information on metabolism, cardiovascular activity and both HPA and sympathetic nervous system activity, it does not include information on such things as inflammatory markers, clotting factors, or immune function. Component measures are also snapshots of levels of activity; there is no information on system dynamics.

A second potential limitation of the MacArthur study is the fact that our cohort was selected at baseline to represent the top third of those aged 70–79 in terms of their physical and cognitive functioning. As we have demonstrated in previous analyses (6), the resulting cohort has a more restricted range of AL scores than do groups representing those individuals who did not meet criteria for enrollment into the cohort (i.e., those with greater impairments). Thus, although our cohort does exhibit a range of scores on AL that are associated with significant differences in risk for major health outcomes, it is possible that because of our relatively more restricted range of baseline AL scores we have underestimated the impact of AL on health risk. Notably, despite these limitations, our initial summary measure does provide evidence to support the value of a more cumulative approach to assessment of biological burden. This evidence is particularly strong for two of our outcomes, namely, mortality and declines in physical performance.

Further evaluation of the concept of AL will require attention to a number of issues related to measurement. First, what would be the *a priori* set of component measures needed to comprehensively assess AL (i.e., which regulatory systems and which parameters of those regulatory systems, including both system dynamics and system state/level assessments)? As noted above, our existing measure does not include assessments of hemostatic or inflammatory parameters, or assessment of parasympathetic activity; nor does it encompass assessments of system dynamics. A second issue concerns the procedures for developing optimal operational measures of AL from these component measures. In this vein, alternatives to our initial, summated measure where each component had equal weight and "higher risk" was uniformly designated by values in the uppermost or lowermost quartile need to be explored. Examples of alternatives include use of recursive partitioning (23) to identify nonlinear patterns of association between the various components of AL that are associated with greater versus lesser health risk. In addition, parameters associated with different biological systems should be weighted differentially according to the health outcomes to be predicted. Construction of such weighting systems based on canonical correlation analyses has been done (A. Karlamangla, T.E.S., B.H.S., J.W.R., and B.S.M., unpublished work).

Finally, the concept of AL postulates that a more comprehensive, cumulative measure of biological wear and tear will provide a better index of health risk over the life course, better that is, in comparison to more traditional approaches that assess biological risks in terms of individual parameters. Further development of this concept, however, requires clearer delineation of the cascade of events (both within and across systems) through which AL develops, including specification of the relationships among regulatory systems and the pattern of cumulating dysregulation (e.g., what are most frequently the initial markers of dysregulation and how does this ripple outward to other systems and parameters?). Along these lines, a reformulation of AL that more explicitly addresses the cascade of events by delineating three levels of events that are seen as contributing to AL (24) has been developed. At the first level are primary mediators, e.g., chemical messengers such as C, NE, and EPI. A second level of events is represented by what we refer to as primary effects, e.g., cellular events, like enzymes, receptors, ion channels, or structural proteins induced genomically or phosphorylated via second messenger systems, that are regulated as part of allostasis by the primary mediators. A third level is represented by secondary mediators, e.g., more integrated processes such as blood pressure, heart rate, and metabolic profiles that are the outcome of effects of one or more primary mediators. The final outcome of this cascade of effects would be the actual disease outcomes.

The analyses completed to date suggest that the concept of AL offers considerable insight into the cumulative risks to health from biological dysregulation across multiple regulatory systems. Further specification of the roles of primary and secondary mediators is likely to offer greater insight into the cascade of events leading from allostasis to AL. This should allow us to more sharply delineate the health consequences of cumulating AL.

Work on this article was supported by National Institute on Aging Grants AG-00586 and AG 17056 (to T.E.S.) and by the MacArthur Research Network on Successful Aging and the MacArthur Research Network on Socioeconomic Status and Health through grants from the John D. and Catherine T. MacArthur Foundation.

1. Jaur, L. & Stoddard, S. (1999) *Chartbook on Women and Disability in the U.S.* (U. S. National Institute on Disability and Rehabilitation Research, Washington, DC).
2. McEwen, B. S. & Stellar, E. (1992) *Arch. Int. Med.* **153**, 2093–2101.
3. McEwen, B. S. (1998) *N. Eng. J. Med.* **338**, 171–179.
4. McEwen, B. S. & Seeman, T. E. (1999) *Ann. N.Y. Acad. Sci.* **896**, 30–47.
5. Sterling, P. & Eyer, J. (1988) in *Handbook of Life Stress, Cognition, and Health*, eds. Fisher, S. & Reason, J. (Wiley, New York), pp. 631–651.
6. Seeman, T. E., Singer, B., Rowe, J. W., Horwitz, R. & McEwen, B. S. (1997) *Arch. Int. Med.* **157**, 2259–2268.
7. Reaven, G. M. (1988) *Diabetes* **27**, 1595–1607.
8. Berkman, L. F., Seeman, T. E., Albert, M., Blazer, D., Kahn, R., Mohs, R., Finch, C., Schneider, E., Cotman, C., McClearn, G., *et al.* (1993) *J. Clin. Epidemiol.* **46**, 1129–1140.
9. Bjorntorp, P. (1987) *Acta Med. Scand.* **723**, 121–134.
10. Svec, S. & Lopez. S. A. (1989) *Lancet* **ii**, 1335–1336.
11. Hypertension Detection and Follow-up Program Cooperative Group (1978) *J. Chronic Dis.* **31**, 651–667.
12. Lohman, T. G., Roche, A. F. & Martorell, R. (1988) *Anthropometric Standardization* (Human Kinetics Books, Champaign, IL).
13. Little, R. R., England, J. D., Wiedmyer, H. M. & Coldstein, D. E. (1983) *Clin. Chem.* **29**, 1080–1082.
14. Canalis, E., Reardon, G. & Caldarella, A. (1982) *Clin. Chem.* **28**, 2418–2420.
15. Krstulovic, A. M. (1983) *J. Chromatogr.* **22**, 1–34.
16. Bush, T. L., Miller, S. R., Golden, A. L. & Hale, W. E. (1989) *Am. J. Pub. Health* **79**, 1554–1556.
17. Inouye, S., Albert, M., Mohs, R. C., Sun, K. & Berkman, L. F. (1993) *J. Gerontol.* **48**, M146–M151.
18. Seeman, T. E., Charpentier, P. & Berkman, L. F. (1994) *J. Gerontol.* **49**, M97–M108.
19. Kendall, M. G. & Stuart, A. (1986) *The Advanced Theory of Statistics* (Charles Griffin, London), Vol. 2.
20. SAS Institute (1996) *SAS/STAT Guide for Personal Computers* (SAS Institute, Cary, NC), Version 6.
21. Tracy, R. (1995) *Arterioscler. Thromb. Vasc. Biol.* **15**, 1269–1279.
22. Ridker, P., Cushman, M., Stampfer, M., Tracy, R. & Hennekens, C. (1997) *N. Engl. J. Med.* **336**, 973–979.
23. Zhang, H. & Singer, B. (1999) *Recursive Partitioning in the Health Sciences* (Springer, New York).
24. McEwen, B. S. & Seeman, T. E. (1999) *Ann. N.Y. Acad. Sci.* **896**, 30–47.

Molecular and Cellular Biology of the Senescent Hypertrophied and Failing Heart

Bernard Swynghedauw, PhD, MD, Sophie Besse, PhD, Patrick Assayag, MD, François Carré, PhD, MD, Brigitte Chevalier, PhD, Daniéle Charlemagne, PhD, Claude Delcayre, PhD, Sandrine Hardouin, Christophe Heymes, and Jean-Marie Moalic, PhD

During aging, experimental studies have revealed various cellular changes, principal among which is myocyte hypertrophy, which compensates for the loss of myocytes and is associated with fibrosis. The expression of α-myosin heavy chain is replaced by that of the isogene β-myosin, which leads to decreased myosin adenosine triphosphatase (ATPase) activity. In consequence, contraction is slower and more energetically economical. The Ca^{2+}-ATPase of the sarcoplasmic reticulum and Na^+/Ca^{2+} exchange activity are decreased, which probably explains the reduced velocity of relaxation. Membrane receptors are also modified, since the density of both the total β-adrenergic and muscarinic receptors is decreased. The senescent heart is able to hypertrophy in response to overload and to adapt to the new requirements. Similar alterations are observed both in the senescent heart and in the overloaded heart, in clinical as well as in experimental studies; however, differences do exist, especially in terms of fibrosis and arrhythmias.

(Am J Cardiol 1995; 76:2D–7D)

Epidemiologic studies have shown that cardiology is the main branch of geriatrics and that diseases of the heart are one of the major cause of mortality among the overall population and the main cause in persons >65 years of age.[1] Therefore, it is of major importance to determine the heart morphology in healthy aged persons without heart disease and in experimental animal models. From a purely economical point of view, it is important to know whether the aged myocardium is enough like the myocardium of adult persons that we can prescribe the same drugs to both age groups and also whether even the first screening of new therapies needs to be done in senescent animal models. As explained later, the aged heart is not only enlarged, but also different in terms of receptor densities, a major target for drug design.

This review will focus on the capacity of the senescent heart to hypertrophy and on a few aspects of the function of the normal senescent heart, namely, systolic and diastolic function and the renin–angiotensin and autonomous nervous systems.

The pioneer work in this field was from the group of E. Lakatta in Baltimore, who recently published an important review focused mostly on the physiologic aspects of the problem.[2] Other review articles were published from our laboratory[3] and also by Folkow and Svanborg.[4] The more general problem of the biochemistry of senescence was reviewed in 1992 by Kirkland.[5]

CAPACITY OF THE SENESCENT HEART TO HYPERTROPHY

Most, if not all, of the experimental studies on myocardial senescence have been performed on rats. In rats, the 50% survival age is around 30 months, but usually 24-month-old rats are used and considered as equivalent to a 75-year-old human both in terms of survival rate and molecular/cellular cardiac modifications.[3]

Cardiac hypertrophy is usually defined by the heart weight to body weight ratio. This ratio may vary in senescent rats, and the variations do not depend on the strain of rat. This ratio may also vary because obviously animals >27 months, like humans >85 years, do not feed normally and can be overfed, or starved, depending on various environmental conditions. Finally, the development of ventricular hypertrophy is more pronounced on the left side than on the right one,[6,7] which may reflect the role of aortic impedance as a determinant of senescent cardiac hypertrophy.

The modifications in both protein synthesis and degradation and mRNA content, which occurs during aging, would suggest that the senescent heart may also be unable to adapt to an increased load and would be unable to hypertrophy further in response to mechanical stress. This question has more than an academic interest, since both arterial hypertension and coronary insufficiency, the 2 main causes of cardiac hypertrophy and failure, are indeed much more frequent after 65 years than before.[1]

From INSERM, Hôpital Lariboisière, Paris, and the Département de Physiologie, Faculté de Médecine, Renens (F.C.), France.

Address for reprints: Bernard Swynghedauw, PhD, MD, U127-INSERM, 41 Boulevard de la Chapelle, 75010, Paris, France.

Several authors, using different models of overload, found that the response of the senescent myocardium in terms of compensatory hypertrophy was significantly attenuated (reviewed by Besse et al[7]). The hearts of 18- and 22-month-old Fischer 344 rats were unable to hypertrophy in response to known and calibrated arterial hypertension induced by 4 weeks of aortic constriction. Nevertheless, the same constriction induced a 23% left ventricular hypertrophy in young rats and this in spite of the same degree of hypertension in the different groups of animals. Others found that the senescent myocardium has a diminished hypertrophic response to overload. In the same strain, 4 weeks of aortic constriction in 25–28-month-old rats induced a significant but modest hypertrophic response. Seven months of chronic atrioventricular block caused a smaller degree of cardiac hypertrophy in the 24-month group compared with the 12-month group.[8] The same results have been reported by Capasso et al[9] in a model of renovascular hypertension in which the hypertrophic response in young rats was greater than in old rats (50% vs 10%).

However, more recently, different results were obtained by others and also in our laboratory and showed a normal response to mechanical stress during senescence. The response of the senescent myocardium in a model of renovascular hypertension in terms of both hypertrophic response and changes in genetic expression and protein synthesis was similar to that seen in younger animals. A study from this laboratory in senescent Wistar rats subjected to deoxycorticosterone acetate (DOCA)–salt hypertension showed the same degree of hypertrophy, the same changes in isomyosins, and no effect on both collagen and its mRNA, compared with young animals.[7] As already explained, it is not easy to appreciate the degree of cardiac hypertrophy during senescence, mainly because the nutritional status can vary enormously, independently of the genetic strain. Also because interventions used to create overload, such as uninephrectomy, also modify feeding conditions. These considerations are likely to explain these discrepancies.

CELLULAR BIOLOGY OF THE UNDISEASED SENESCENT HEART

Myocyte size and number: Experimental studies had revealed that in the aged heart a loss of myocytes and fibrosis is associated with myocyte hypertrophy, which is more significant in the left ventricle (60%) than in the right ventricle (10–20%). It was commonly admitted that, after birth, the cardiac cells lose their capacity to undergo DNA synthesis, cytokinesis, and mitosis.

A study by Anversa et al[10] using 3- and 21-month-old male Sprague-Dawley rats showed a significant degree of cardiac *hypo*trophy, since the heart weight to body weight ratio decreased by 23% with aging. This hypotrophy was a consequence of a pronounced loss in the total number of cardiocytes, located mostly in the endocardium of the left ventricle. Such a loss was only partly compensated for by an enhanced myocyte volume (from 15 to $31 \cdot 10^{-3}$ mm^3). This cardiocyte hypertrophy is a constant finding during senescence and is much more pronounced in the left than the right ventricle.

In most of the mammalian species, including rat, rabbit, guinea pig, and cat, most (47–81%) cardiac myocytes are binucleated. In humans binucleation is observed in only 10% of the cardiocytes. In fact, human cardiocytes became progressively polyploid with aging, with a progressive increase in both the degree of polyploid cells and number of highly (>16 n) polyploid nuclei, and number of nucleoli. The process is accelerated by addition of mechanical overload, which can result in an increased number of octoploid nuclei up to 70%, instead of 15% in control subjects.

Extracellular matrix: Earlier pathologic studies in humans have revealed myocardial fibrosis and an increased collagen concentration in aged hearts. In addition, the collagen structure is also modified with an increased collagen crosslinking. Similar results were reported in rat hearts. Histochemical studies have shown an abnormal accumulation of fibrillar collagen, both around the vessels and within the interstitium. The concentration of the 2 main components of fibrosis, namely, collagen I and fibronectin, increases during aging.[11] By contrast, both type I and type III procollagen mRNA levels were reduced in the senescent rat heart (−63% and −51%, 3-month-old animals compared with 16- and 24-month-old animals, respectively) showing that changes in myocardial collagen mRNA and protein were not synchronous and suggesting that during senescence collagen concentration is not transcriptionally regulated.[11] Collagen concentration is a major determinant of both the diastolic compliance and arrhythmias.

SYSTOLIC FUNCTION

Physiologic studies: The Baltimore study,[2] which used only subjects with normal electrocardiograms during exercise, has clearly demonstrated that ejection fraction and cardiac output at rest are not

TABLE I Papillary Muscle Energetics During Senescence in Rat Heart*

	3 Months Old	22–24 Months Old
Isotonic contraction and relaxation		
Maximum shortening velocity (V_{cmax}, in L_{max}/sec)	2.38 ± 0.06	1.99 ± 0.08†
Maximum unloaded shortening velocity (V_{max}, in L_{max}/sec)	3.20 ± 0.06	2.81 ± 0 10†
Maximum lengthening velocity (V_{rmax}, in L_{max}/sec)	3.25 ± 0.1	2.75 ± 0.09†
Isometric contraction		
Active force per surface area (mN/mm^2)	55 ± 2	45 ± 2‡
Energetics		
Curvature of the Hill's hyperbola (G)	1.6 ± 0.07	1.93 ± 0.13‡
Peak power output (E_{max}, in $mN \cdot L_{max} \cdot sec^{-1} \cdot mm^{-2}$)	30 ± 1	20 ± 1*

*Recalculated from Besse et al.[7]
†$p < 0.05$.
‡$p < 0.01$.

altered with aging. Aortic impedance is greatly increased with aging because of an increased aortic stiffness and diameter.[12] During senescence, the enhanced characteristic impedance of the aorta is responsible for a left ventricular overload, which in turn activates the expression of several specific genes, as detailed later.

During aging, the myocyte action potential is prolonged without any modification of the amplitude, resting potential, or overshoot.[2] The myoplasmic calcium transient, which immediately follows sarcolemmal depolarization, is also prolonged.[13] Studies on contractility on papillary muscle were complicated by the fact that the senescent papillary muscle is both thicker and stiffer than young muscles. Nevertheless, isometric and isotonic contraction is prolonged in every species studied and both the maximum velocity of shortening and the maximum unloaded shortening velocity were depressed (Table I).

Relaxation is also impaired and both the relaxation velocity and the peak of negative force derivative normalized per cross-sectional area were depressed. The diminution in the active force, even normalized per cross-sectional area, is less informative because of the stiffness of this type of preparation. Nevertheless, active force per surface area correlates with the biochemical parameters and more particularly with the isomyosin shift, which suggests that this parameter reflects in part the state of contractility. Moreover, the skinned myofilament response to calcium was unaltered, suggesting that the changes in the velocity of contraction that have been observed depend on contractile proteins and on the different determinants of intracellular calcium movements.[7]

The slowing of the contraction allows the senescent muscle to maintain the active tension within normal limits to the detriment of the velocity at which this tension is developed. Both physiologic and biochemical changes have, therefore, a beneficial effect in terms of muscle energetics. Contraction energetics of papillary muscle were calculated from the mechanical parameters of force and velocity.[7] For every muscle, the force/velocity curve had an accurate fit with the hyperbolic model; the same correlation was found equally in young and senescent animals, and in both groups the curvature of Hill's hyperbola ("G") was positive. During senescence, the peak power output was significantly depressed and the curvature of the force/velocity curve was increased, which indicates that each contraction was more economical (Table I). In other words, the changes in physiologic and biochemical properties are clearly adaptational.

Biochemical and molecular biological studies: From a biologic point of view, a decrease in maximum velocity may result from a slowing of the intracellular calcium transient as a consequence of a change in the membrane proteins responsible for the calcium movements, or a modification in the contractile protein apparatus.

Although the V1 isoform is predominant in the young adult rat ventricle, the isomyosin content shifts progressively with age from V1 to V3, so that by 24 months the rat ventricle contains 80% V3, the ββ isomyosin heavy chain (MHC). Rat is a particularly suitable species to perform such a study because the ventricles of a 2-month-old rat contain 100% V1; in fact, the content of V1 can almost be used as a means to determine age. This shift of the isomyosins, which also occurs during cardiac overload, results in a decreased adenosine triphosphatase (ATPase) activity that parallels the decline in the unloaded shortening velocity and, at least in rats, is likely to constitute the main determinant of the adaptational process.[7,9,11,14–17]

A study from this laboratory[14] has reported that skeletal α-actin, which represents 40% of the mRNA coding for sarcomeric actin at birth, is almost absent at 2 months and does not reappear in the senescent heart. This is in opposition to the MHC expression where the ββ isoform, which is the main isoform present in the newborn rat, disappears in the adult but is reexpressed in senescent rat heart, which demonstrates that the regulation of the sarcomeric actins and MHCs is independent.

The modifications of the membrane proteins occur at several different levels, including the

Ca^{2+}-ATPase of sarcoplasmic reticulum (SR)[7,17] and also adrenergic and vagal receptors,[18,19] calcium channels,[13] the Na^+/Ca^{2+} exchange,[16] and the Na^+,K^+)-ATPase, the so-called sodium pump. This phenoconversion of the membrane protein is directly responsible for the slowing of the intracellular calcium transient as measured with a fluorescent probe specific for calcium[13] and can take into account the decrease in maximum velocity. In addition, we know that the duration of the action potential is prolonged.[13]

Correlation analysis demonstrates a strong positive correlation between shortening velocity and the content of both slow isomyosin and Ca^{2+}-ATPase of the sarcoplasmic reticulum; partial correlation analysis showed that the relaxation parameters depended more on the isoforms of myosin than on the density in Ca^{2+}-ATPase of the sarcoplasmic reticulum, which outlines the strong dependency of active relaxation relative to contraction.[7]

DIASTOLIC FUNCTION

Physiologic studies: During cardiac hypertrophy, active relaxation is impaired in parallel with the shortening velocity. During senescence, the global diastolic function remains unchanged despite a pronounced impairment in the early ventricular filling because the contribution of atrial systole to ventricular filling is increased. The impairment in early ventricular filling and the diminution of the E wave shown by Doppler echocardiography studies result from both a prolonged active isovolumic relaxation and an increased myocardial stiffness. Thus, a 50% reduction in the rate of left ventricular filling during early diastole is observed from 30 to 80 years of age. The enhanced myocardial stiffness is a major point and had been evident in rats but had in fact never been directly measured in normal senescent humans[7] (reviewed by Lakatta [2]).

Basic determinants of diastolic function: Cardiac overload is associated with a quantitative change in the Ca^{2+}-ATPase of the sarcoplasmic reticulum, the enzyme that is responsible for the reuptake of calcium by the sarcoplasmic reticulum.[17] In this condition the calcium uptake, an essential property of the sarcoplasmic reticulum, is impaired, and the number of the functionally active Ca^{2+}-ATPase molecules is decreased; this correlates negatively with the degree of cardiac hypertrophy, suggesting that the gene encoding this particular enzyme is simply not activated by the process of hypertrophy. Consequently, the products of the gene, namely, the mRNA and the protein, are diluted in a larger heart. This gene belongs to a family of genes whose expression is not sensitive to mechanical stress. Similar results were obtained in humans. The diminished concentration of Ca^{2+}-ATPase correlates with a decrease in the relaxation velocity.[7]

Calcium extrusion from the sarcoplasm also depends on the activity of the Na^+/Ca^{2+} exchange. It has been shown by Heyliger et al[16] that the activity of this exchanger is attenuated in the aged myocardium as in cardiac hypertrophy of mechanical origin. In vitro, the activity of the Na^+/Ca^{2+} exchanger is progressively reduced from 4 to 27 months in the senescent rat ventricle as the capacity of the sarcoplasmic reticulum to pump calcium.

In general, a good correlation has been reported between myocardial stiffness and fibrosis. Myocardial fibrosis did not directly depend on mechanical overload and is observed in 3 conditions: during senescence, after prolonged ischemia as a reparative process, or under certain hormonal conditions. Aldosterone, angiotensin II, or catecholamines can induce ventricular fibrosis, independently and through different mechanisms. During senescence, myocardial fibrosis may result from ischemia—especially in humans, but not in rats, which do not usually have atherosclerosis—but not from hormonal status, since the plasma aldosterone level is unchanged, the renin–angiotensin sytem is strongly depressed, and the plasma catecholamine level is nearly normal. The most likely hypothesis is that fibrosis replaces myocyte loss, but this important finding is far from being demonstrated.

Atrial contraction is an active participant in ventricular filling. During senescence, atrial hypertrophy and a slowing of atrial contraction compensates for the impairment in the early diastolic filling. Compared with the ventricle, filling in the normal atria from young animals, including rats, is more rapid and the maximum shortening velocity for an unloaded muscle is 2–3 times faster in the atria than in the ventricle. In parallel, most mammalian atria contain almost 100% V1, the fast isoform of myosin; during mechanical overload the isomyosin composition of human atria is modified and there is a shift from V1 to V3. This shift correlates with the degree of atrial hypertrophy as quantitated by echocardiography and can be considered as a biologic mechanism that compensates for the increased atrial contraction. Although, for ethical reasons, this particular point has never been explored in aged persons, it is easy to hypothesize

TABLE II β-Adrenergic/Muscarinic System of the Senescent Rat Heart*

	Young Adults	Senescence
β_1-Adrenergic system		
Receptors:		
B_{max} (fmol/mg)	31 ± 2	23 ± 2*
mRNA (pg mRNA/μg RNA)	2.6 ± 0.4	1.4 ± 0.25†
$G_{\alpha s}$ mRNA (pg mRNA/μg RNA)	18.5 ± 1.6	13.7 ± 0.85‡
Muscarinic system		
Receptors:		
B_{max} (fmol/mg)	104 ± 7	54 ± 3§
mRNA (pg mRNA/μg RNA)	1.2 ± 0.11	2.3 ± 0.40‡
$G\alpha i_2$ mRNA (pg mRNA/μg RNA)	8.2 ± 1	9.7 ± 1
β_1-Adrenergic/muscarinic receptor ratios		
B_{max}	0.29 ± 0.1	0.42 ± 0.2‡
mRNA	2.1 ± 0.3	0.56 ± 0.1†

*Molecular biologic data, recalculated from Chevalier et al[18] and Hardouin et al.[19]
†$p < 0.05$.
‡$p < 0.01$.
§$p < 0.001$.

TABLE III Renin–Angiotensin–Aldosterone System and the Atrial Natriuretic Factor in Senescence in Rats*†

	Young	Senescence
Circulating RAAS and plasma hormone concentrations		
Renin activity (ng ang I/ml/h)	38 ± 5	6 ± 1‡
Ang I (ng/ml)	6 ± 1	1.7 ± 0.2‡
Aldosterone (fmol/ml)	844 ± 99	715 ± 85
Cortisol (nmol/ml)	42 ± 3	68 ± 9§
Free T_4, (pg/ml)	29 ± 2	12 ± 2‡
Free T_3, (pg/ml)	3.1 ± 0.2	2.6 ± 0.2
Plasma ANF (pmol/l)[75]	21 ± 6	52 ± 8§
Ventricular content in mRNAs of RAAS and ANF		
Left ventricle		
Angiotensinogen (fg mRNA/μgRNA)	18 ± 2	92 ± 29‖
Ang I (fg mRNA/μgRNA)	50 ± 5	120 ± 21‖
ANF (densitometric score)	2 ± 0.5	23 ± 5‡
Right ventricle		
Angiotensinogen, (fg mRNA/μgRNA)	215 ± 53	181 ± 41
Ang I (fg mRNA/μgRNA)	1.9 ± 0.2	2.0 ± 0.2
ANF (densitometric score)	2.5 ± 0.7	2.0 ± 0.4

*Reprinted with permission from Circulation[20] and Hypertension.[21]
†Ang I = angiotensin I; ANF = atrial natriuretic factor; RAAS = renin–angiotensin–aldosterone system; T_3 = triiodothyronine; T_4 = thyroxine.

that such an adaptational mechanism exists during senescence in the human as well.

CARDIOVASCULAR RESPONSE TO EXERCISE

Physiological studies: The exercise-induced increase in cardiac output in normal young adults is mainly due to an increased heart rate. In elderly persons during exercise, there is a normal enhancement of the cardiac output, which is mainly due to an increased ejection fraction with only slight tachycardia, or no augmentation of the heart rate at all. There are presently several clinical and experimental data that the responsiveness to catecholamines is attenuated with aging.[2]

Molecular studies on the cardiac autonomous system during senescence: In the senescent Wistar rat heart, the density of the total β-adrenergic receptors and the corresponding mRNA, and the density of muscarinic receptors but not of their corresponding mRNAs, are diminished. The drop in muscarinic receptor density is greater than that in total β adrenergic receptors (Table II).[18,19] It is interesting that the important decrease in the number of muscarinic receptors, which has been observed in the senescent myocardium, has also been reported in the cerebral cortex, striatum, and hippocampus, suggesting that, for unknown reasons, aging has a rather specific and pronounced effect on the muscarinic system. In contrast, during compensatory cardiac hypertrophy, the cellular density in both the β_1-adrenergic and muscarinic receptors drops in parallel and their ratio remains constant.

Several findings suggest that, in fact, the receptors were not the only component of the myocardial autonomous system to be modified during aging. A diminution of the $G_{\alpha s}$ mRNA was also observed and the adenylate cyclase is also modified. Aging, like mechanical overload, is accompanied by several modifications of the autonomous myocardial system, which are likely to be located "down the road," as initially suggested by Lakatta.[2]

RENIN–ANGIOTENSIN SYSTEM

Plasma levels of atrial natriuretic factor (ANF) are elevated in healthy elderly people compared with young individuals, but the factors responsible for these differences are still poorly known. Aged rats also had high plasma ANF levels. In vitro the ANF secretory rate was also greater in response to phenylephrine, but not to stretch. The relative levels of mRNA coding for ANF showed a strong activation in the left, but not in the right, ventricle (Table III).[20,21]

In contrast, aging in humans as well as in the rat (Table III) is associated with low plasma levels of angiotensin I, renin activity, and angiotensinogen. The latter is likely to reflect a diminution of the hepatic level of angiotensinogen mRNA. In contrast, the myocardial mRNA levels for both angiotensinogen and angiotensin-converting enzyme were both up-regulated in the left ventricle. During senescence, the specific left ventricular activation of normally poorly expressed genes as the genes coding for ANF, angiotensinogen, and angiotensin-converting enzyme (Table III) suggests that the trigger for activation is common to the 2 systems and is, in fact, aortic impedance.[12,20]

Cardiac failure is mainly a disease of the elderly, and the average age of most of the clinical trials, including the Cooperative North Scandinavian Enalapril Survival Study on angiotensin-converting enzyme inhibitor devoted to this disease, is around 65 years. Obviously, most of the patients have a very low level of plasma angiotensin I and the question is, how do angiotensin-converting enzyme inhibitors work in the elderly? The response is likely to be found in the various tissues of the renin–angiotensin system, at least the one located in the myocardium.

CONCLUSION

Several, but not all, alterations observed in the hearts of the elderly result from the initiation of the general process of adaptation of any muscle to a mechanical overload. The senescent heart is an overloaded heart and reflects the vascular effects of aging.

Nevertheless, several changes occur in the aged myocardium that cannot be explained by a simple mechanical overload, including myocyte loss, a pronounced fibrosis and its clinical consequences, permanent or nearly permanent arrhythmias, and a pronounced down-regulation of the muscarinic system.

The senescent heart is potentially a diseased heart. The myocardial function of the elderly is still normal because of several cellular or tissular adaptational mechanisms, including a slowing of calcium transient and action potential, Starling's law, and an increased contribution of atria to ventricular filling.

Acknowledgments: This study was supported in part by a grant from Fondation de la Recherche Médicale de France and a grant from Fondation de France.

1. Brock DW, Guralnik JM, Brody JA. Demography and epidemiology of aging in the United States. In: Schneider EL, Rowe JW, eds. Handbook of the Biology of Aging. San Diego, CA: Academic Press, 1990:3–23.

2. Lakatta E. Cardiovascular regulatory mechanisms in advanced age. *Physiol Rev* 1993;73:413–467.

3. Assayag P. Besse S. Delcayre C. Hardouin S. Carré F. Moalic JM. Swynghedauw B. Biological and molecular characteristics of the senescent heart. In: Beregi E. Gergely IA. Rajczi K. eds. Recent Advances in Aging Science. Bologna: Monduzzi Editore, 1993:341–346.

4. Folkow B. Svanborg A. Physiology of cardiovascular aging. *Physiol Rev* 1993;73:725–764.

5. Kirkland JL. The biochemistry of mammalian senescence. *Clin Biochem* 199[illegible] 5[illegible]–66.

6. Yin FCP. Spurgeon HA. Weisfeldt ML. Lakatta EG. Mechanical properties of myocardium from hypertrophied rat hearts. A comparison between hypertrophy induced by senescence and by aortic bandin. *Circ Res* 1980:46:292–300.

7. Besse S. Assayag P. Delcayre C. Carré F. Cheav SL. Lecarpentier Y. Swynghedauw B. Normal and hypertrophied senescent rat heart. Mechanical and molecular characteristics. *Am J Physiol* 1993:265(suppl):H183–H190.

8. Watford GD. Spurgeon HA. Lakatta EG. Diminished cardiac hypertrophy and muscle performance in older compared with younger adult rats with chronic atrio ventricular bloc. *Circ Res* 1988;63:502–511.

9. Capasso JM. Malhotra A. Scheuer J. Sonnenblick EH. Myocardial. biochemical. contractile and electrical performance following imposition of hypertension in young and old rats. *Circ Res* 1986:58:445–460.

10. Anversa P. Palackal T. Sonnenblick EH. Olivetti G. Meggs LG. Capasso JM. Myocyte cell loss and myocyte cellular hyperplasia in the hypertrophied aging rat heart. *Circ Res* 1990:67:871–885.

11. Besse S. Robert V. Assayag P. Delcayre C. Swynghedauw B. Non-synchronous changes in myocardial collagen mRNA and protein during aging. Effect of DOCA-salt hypertension. *Am J Physiol* 1994:267(suppl):H2237–H2244.

12. Safar M. Ageing and its effects on the cardiovascular system. *Drugs* 1990: 39(supp 1):1–8.

13. Orchard CH. Lakatta EG. Intracellular calcium transients and developed tension in rat heart muscle. *J Gen Physiol* 1985;86:637–651.

14. Carrier L. Boheler KR. Chassagne C. de la Bastie D. Wisnewsky C. Lakatta E. Schwartz K. Expression of the sarcomeric actin isogenes in the rat heart with development and senescence. *Circ Res* 1992;70:1–18.

15. Maciel LMZ. Polikas R. Rohrer D. Popovich BK. Dillmann WH. Age-induced decreases in the messenger RNA coding for the sarcoplasmic reticulum Ca^{2+} ATPase of the rat heart. *Circ Res* 1990:67:230–234.

16. Heyliger CE. Prakash AR. McNeill JH. Alterations in membrane Na^{+}–Ca^{2+} exchange in the aging myocardium. *Age* 1988;11:1–6.

17. Lompré AM. Lambert F. Lakatta EG. Schwartz K. Expression of sarcoplasmic reticulum Ca^{2+}ATPase and calsequestrine genes in rat heart during ontogenic development and aging. *Circ Res* 1991;69:1380–1388.

18. Chevalier B, Mansier P, Teiger E, Callens-El Amrani F, Swynghedauw B. Alterations in β adrenergic and muscarinic receptors in aged rat heart. Effects of chronic administration of propranolol and atropine. *Mech Ageing Dev* 1991:60: 215–224.

19. Hardouin S, Bourgeois F, Besse S, Machida CA, Swynghedauw B, Moalic JM. Decreased accumulation of β1 adrenergic receptors, Gαs and total myosin heavy chain messenger RNAs in the left ventricle of senescent rat heart. *Mech Ageing Dev* 1993;71:169–188.

20. Heymes C, Swynghedauw B, Chevalier B. Activation of angiotensinogen and angiotensin converting enzyme gene expression in the left ventricle of senescent rats. *Circulation* 1994;90:1328–1333.

21. Tummala PE, Dananberg J, Grekin RJ. Alterations on the secretion of arial natriuretic factor in atria from aged rats. *Hypertension* 1992;20:85–88.

0733–8619/98 $8.00 + .00

AGE-RELATED BIOLOGY AND DISEASES OF MUSCLE AND NERVE

Kevin M. Flanigan, MD, Guiseppe Lauria, MD, John W. Griffin, MD, and Ralph W. Kuncl, MD, PhD

SKELETAL MUSCLE

The effects of aging on skeletal muscle can be recognized in three ways. First, muscle mass and strength both decline generally. Second, alterations in muscle histopathology—mainly denervation or mitochondrial abnormalities—reflect underlying physiologic changes that occur with aging. Third, and perhaps reflecting the influence of the two previous points, susceptibility to certain neuromuscular diseases is age related.

Declining Muscle Strength and Mass

Loss of muscle mass and function recently has gained the descriptive label of "sarcopenia," although this term of the moment may yet go the way of the even more colorful french phrase "myosclerosis of the aged."[76] Such a loss is clearly a function of aging: whereas studies in working men demonstrate that a decline in strength does begin after attainment of peak strength in the early 20s or around the age of 30,[12,46] other studies point out the general stability of strength until a substantial decline begins in the fifth or sixth decade of life.[52,64] These alterations in strength coincide with a decrease in speed of muscle movement, measured both by muscle twitch contraction velocities[14] and knee extension velocities.[52]

This decline in strength may be due to attrition of functioning motor units. Electrophysiologic estimating methods demonstrate a decreased number of motor

From the University of Utah, Salt Lake City, Utah (KMF), the Institute of Clinical Neurology, University of Ferrara, Ferrara, Italy (GL), and the Department of Neurology, Johns Hopkins University School of Medicine, Baltimore, Maryland (JWG, RWK)

units in elderly humans.[14,21] This correlates with the declining counts of anterior horn cells in the spinal cord of elderly individuals at autopsy.[75] That decrease might reflect age-related changes in the levels of, or response to, trophic factors that have an effect on motor neuron survival, such as brain-derived neurotrophic factor, neurotrophin-3, neurotrophin-4, ciliary neurotrophic factor, or insulin-like growth factor-1 (IGF-1), among others.[54]

The tempo, mechanism, and compensations for sarcopenia are easier to analyze in aging rodents. There, aging is evidenced by weakness, slowing of muscle twitch, axonal loss, and neurogenic atrophy of muscle. As in humans, these are explained largely by drop-out of entire motor units.[13] In earlier stages of aging, compensatory reinnervation is indicated by fiber type grouping; the declining number of functioning motor units is accompanied by an increase in the innervation ratio—representing the number of myofibers innervated by a given motor neuron—in the remaining motor units.[13,45] As a result, muscle fiber type, which is determined by innervation and activity, converts from fast twitch to slow twitch as the relative proportion of slow oxidative myofibers increases and fast oxidative/glycolytic myofibers decline.[13] But eventually, in late aging, reinnervation is not sustained.[13]

The neuromuscular junction may also be a site of vulnerability in aging. The rate by which motor nerve terminals regenerate is slowed in aged rats, and nerve terminals become concomitantly smaller and less complex than in younger adult rats.[63] Age-related changes at the neuromuscular junction resemble those seen during early developmental innervation when synapses are eliminated[2]; it remains unclear whether reduced reinnervation at the motor nerve terminal level is caused by presynaptic or postsynaptic factors.

Whatever the mechanism, human and rodent studies together suggest that age-related muscle weakness is caused, at least in part, by a decline in the number of functional motor units followed by inefficient reinnervation by surviving motor neurons.

An alternative explanation for age-related weakness is related to inefficient excitation-contraction coupling. The dihydropyridine receptor (DHPR), central to excitation-contraction coupling, is found in decreased numbers in the sarcoplasmic reticulum of aged muscle fibers that show concomitant uncoupling of sarcolemmal excitation from sarcoplasmic reticulum Ca^{2+} release.[19] This may be related to IGF-1, as activation of the IGF-1 receptor leads to phosphorylation of DHPR, modulating the DHPR function; this signaling pathway is altered in aged muscles.[19]

The metabolic mechanisms of the loss of muscle mass are not clear. Resistance training in elderly individuals results in an increase in strength and endurance.[27,48] This generally occurs without a large proportionate increase in muscle mass, but increases in strength are specific to muscle length and type of contraction (i.e., isometric, shortening, or lengthening); that specificity argues that the gains in strength are due to changes in neural drive, presumably from the central nervous system, rather than alterations in intrinsic muscle factors such as an increase in contractile proteins.[26,48] Muscle wasting, in general, is mediated through the ubiquitin-proteasome pathway, which can be stimulated by glucocorticoids or by cytokines such as tumor necrosis factor (and hence by disease states such as sepsis or neoplasia).[60] Although aging may not increase muscle protein degradation, muscle protein synthesis declines with age (see below), and the balance may tip in favor of degradation with increasing age.

Histopathology of Aged Muscle

Much of the histopathologic change reported in normal aged muscle supports the hypothesis that the major component of weakness in old age is neurogenic atrophy. In studies based on autopsy or biopsy samples from elderly individuals without neuromuscular disease, type II atrophy has been a common feature, but other clearly neurogenic changes have been noted, including groups of angular atrophic fibers, frequent target or "targetoid" fibers, and fiber type grouping.[41,53,61,76] These changes generally are more severe in distal muscles (they begin as early as the second decade in the extensor digitorum brevis[42]), but they are also common in proximal muscles (such as the deltoid and biceps).[41,76]

In addition to neurogenic changes, the frequency of other cytoarchitectural alterations increases with age. These are generally absent until after age 70 and include necrosis, cytoplasmic bodies, ring-fibers, and increased central nuclei.[40,41] The increased incidence of these architectural abnormalities could relate to inefficient protein synthesis with age, as elderly people have decreased synthesis rates of both the myofibrillar protein fraction[79] and total muscle proteins[80]; at least in part, this may be due to the general decrease in mRNA synthesis in aged muscle.[10]

One histochemical alteration, the ragged-red fiber, holds special significance, as it is considered a hallmark of mitochondrial dysfunction. Ragged-red fibers contain pleomorphic and proliferative mitochondria, which are considered compensatory changes, and also abnormal inclusions and paracrystalline arrays in inner mitochondrial membranes, which probably represent abnormal protein assembly and interactions. In muscle, they are commonly seen in disorders related to mutations of the mitochondrial DNA (mtDNA), but they also occur with increased frequency with increasing age.[67] Similarly, in aged muscle there is an increased frequency of fibers that do not stain at all for cytochrome c oxidase (COX).[63] Both ragged-red fibers and COX-negative fibers correlate with an increase in multiple mtDNA deletions with age.[3,17] Several studies have reported a decline in respiratory chain function with age,[16,77] but the reported decline in measured function (up to 50%) seems excessive in light of the low levels of mutated mtDNA (which can be less than 1%); more recent evidence suggests that these reported decrements in respiratory chain function are due to altered physical activity rather than age, as spectrophotometric comparisons reveal no differences in respiratory chain function in muscle from young and elderly conditioned athletes.[8] Different mtDNA mutations are present in different COX-negative fibers, and nuclear genes also encode COX subunits, explaining why there is a low overall incidence of an individual mtDNA mutation in aged subjects.[9]

Age-associated Neuromuscular Diseases

Several neuromuscular disorders with muscle weakness have an age-related incidence. The increased susceptibility of aged muscle to disease probably represents a "double hit" in which an inciting risk factor (e.g., gene or a toxin) acts synergistically either with a mechanism present in normal aging (e.g., motor neuron loss or multiple mtDNA deletions) or with another age-related disease.

Amyotrophic lateral sclerosis (ALS) has an overall median age of onset close to 60 years in population-based studies,[1] but the age-specific incidence steadily increases over all decades.[44] The age-related attrition of motor neurons and motor units has been postulated to contribute significantly to the age-specific presentation of ALS.[58] It is said that patients and doctors do not perceive weakness until half the complement of motor units is lost. One can easily appreciate, therefore, how even a fairly abrupt but subclinical toxin-induced loss of motor neurons could appear to present later on as an age-related degeneration when the "normal," age-related attrition of motor units is added as the "double hit." The neurogenic amyotrophy and less-efficient reinnervation of aging may also explain why increased age is an independent prognostic risk factor for shorter survival in ALS.[50] In familial ALS caused by SOD-1 mutations, the resultant toxic gain of function may be to generate free radical peroxidations of proteins and lipids that slowly accumulate with age. The SOD-1 gene mutation may synergize with age-related mtDNA mutations, which lessen the cell's ability to withstand oxidative stress.[4] Whatever the primary insult is in ALS, the pathophysiology must be at least accentuated due to age-related changes. Another age issue with ALS is diagnostic confusion with combined cervical and lumbar spondylotic amyotrophy, which is also more common in the aged.

Myasthenia gravis has two age-related incidence peaks. Women are more affected than men (2:1) at an early peak in the second to third decade; this reflects the increased autoimmune vulnerability of younger women in general. Men are more commonly affected at a peak in the sixth to seventh decade, correlating with a somewhat increased incidence of thymoma.[72] This second peak may also represent susceptibility of a less efficient, altered neuromuscular junction due to the aforementioned age-related changes. Why men predominate in the later ages is unknown. Older myasthenic patients tend to respond more easily than younger patients to immunosuppressive therapies.

Drug-induced myopathies are more common in older adults. First, common myotoxic drugs are often prescribed for disorders that themselves have an increased incidence with age (e.g., lovastatin for hypercholesterolemia or colchicine for gout). Second, drugs by themselves are rarely myotoxic but require the superimposition of a second risk factor, some of which are age related. Renal insufficiency is the risk factor for myopathy from colchicine or clofibrate.[49,78] The declining creatinine clearance with age therefore helps explain the age-related incidence of those toxicities. It is also conceivable that the alterations in muscle protein synthesis could accentuate the effects of drugs that affect structural muscle proteins.

Late-onset mitochondrial myopathy is characterized by progressive proximal weakness and histopathologic changes suggestive of a mitochondrial myopathy.[43] These changes, specifically COX-negative fibers, are present in a greater number than is to be expected for the patient's age. This is a convincing clinical syndrome of muscle weakness associated with an increased proportion of deleted mtDNA. Somehow, muscle dysfunction must reflect accumulation of mitochondrial dysfunction with age, via either age-related respiratory chain deficits in deconditioned muscle[8] or slowly accumulated oxidative injury.[59]

Other myopathies in the elderly do not fall neatly into a relationship to the age-related functional or structural changes just described. Inflammatory myopathies remain the most common acquired myopathies in the elderly. This may seem surprising in view of the youthful presentation of most autoimmune diseases. Dermatomyositis and, to a lesser extent, polymyositis, however, are associated with neoplasm and hence also increase with age.[70] Inclusion body myositis is most common in men older than 50. Given the presence of ubiquitin within the inclusions, one is tempted to postulate that a decline in function of the ubiquitin-proteasome pathway is somehow responsible for the disease; however, the pathogenesis of inclusion body myositis is unknown. Another vacuolar myopathy is oculopharyngeal muscular dystrophy, which is the only dystrophy that presents more commonly in the elderly (usually with ptosis and dysphagia).

Evaluation of the elderly patient with muscle weakness must take into account the increased incidence of these syndromes in old age, although evaluation can sometimes yield surprises. In a retrospective study of 77 weak patients aged 65 or older who had biopsies as part of an evaluation for myopathy, the accuracy of the clinical diagnosis of myopathy (as correlated to histopathologic diagnosis) approximated that in a younger control group of patients.[51] Either electromyography or serum creatine kinase were abnormal in only 68% of the elderly patients who had histopathologic evidence of a myopathy, which is typical of the low sensitivity of those diagnostic tests. Histopathologic diagnoses unexpected from clinical data were found in 6, including 3 cases of unexpected inflammatory myopathies, 2 cases of oculopharyngeal dystrophy, and 1 case of adult rod-body (nemaline) myopathy.

PERIPHERAL NERVOUS SYSTEM

Clinical Manifestations

Every neurologist has encountered the clinical correlates of age-related changes in the peripheral nervous system (PNS) manifested most universally by elevated vibratory thresholds. Quantitative sensory testing has abundantly confirmed the loss of vibratory, tactile, and thermal sensibility. Between childhood and the ninth decade the displacement of vibratory stimulus necessary for detection on the foot increases several fold.[23,74] Clinical experience indicates that loss of vibratory sensibility is greatest in the toes and feet[18,37,71] as compared to the hands, and this pattern has been confirmed in most[74] but not all[23] quantitative studies. Similarly, touch pressure thresholds increase with age, as tested by von Frey hairs[74] or by computer-driven devices.[25] Small fiber functions also decline with age.[65] Thermal sensibility, as measured by warm-cool differences or by heat pain, is as severely affected as large fiber functions.[23,74] In all of these modalities the falloff in sensibility appears to begin in youth and progress in a steady fashion. For example, with the Case IV device the average vibratory and thermal sensibility thresholds appear to decrease linearly.[23] That such population data apply to the individual is the experience of all neurologists who compare their own vibratory thresholds to those of patients, then realize that the threshold of "abnormal" has to be adjusted each decade they practice. At some point the young patient's threshold in the toe is superior to that of the examiner in the test finger.

Electrophysiologic Correlates

Most electrophysiologic measures of nerve function show a similar decade-by-decade decline. For example, in one study the mean sensory nerve action potential amplitude in the sural nerve decreased from 24 to 11 μV between ages 30 and 70,[38,74] and sensory nerve conduction velocities declined as well. Similarly, the median and the peroneal motor conduction velocities decrease. Most data suggest that the compound motor action potential amplitudes also decrease, but this measure is more variable and, therefore, is less robust for repeated comparisons.

Autonomic tests also show age-related decrements in function:[74] (1) relative orthostatic hypotension[5,7,11]; (2) decreased R-R variability in the respiration-induced sinus arrhythmia[5,47]; (3) reduced sudomotor responses; and (4) reduced pupillary responses.

With all of the physiologic measures, one of the striking features is that the decrease begins in early adulthood and progresses in a relatively steady fashion. Theories of pathogenesis must take into account the lack of an abrupt falloff with age.

Structural Changes in the PNS with Aging

Age-related changes in the neurons of the PNS include loss of cells and axons, decreases in the size of neurons and their axons, and predegenerative changes including the accumulation of lipofuscin in the perikarya and of glycogen in axons.[34] Less often considered is the importance of changes in the Schwann cells and their myelin sheaths.

An appreciation of the changes produced by aging depends on understanding some features of normal PNS development. First, in normal development most of the growth of peripheral axons occurs segment by segment along their length, rather than at the "growth cone." This assertion may at first appear anomalous to neurologists and neurobiologists schooled in studies of nerve regeneration or in growth of neurites in tissue culture. In these settings, outgrowth unmistakably occurs primarily at the growing tip.[66] By contrast, in vivo most of the developmental elongation of the axons innervating, for example, a leg occurs during the developmental lengthening of the leg. Each Schwann cell initially establishes a site of ensheathment and myelination at a point along the axon. During subsequent growth, that myelin internode elongates in direct proportion to the lengthening of the extremity,[28,29] so that in a sciatic nerve, for example, internodes may lengthen from an initial 100 µm to 1500 µm. In each individual fiber, the internode length is relatively constant.[22,28,29] The internodes of a lumbar root are substantially longer than that of a cervical root, reflecting the greater growth during development.[29] Throughout all or most of normal adult life, these internodes remain of the same length that was achieved when the adult limb length was obtained. The implication of these morphometric observations is that the Schwann cells of myelinated fibers survive without renewal for the life of the organism; one can conclude this because a degenerated myelin sheath would be replaced by multiple short intercalated internodes. Thus, the myelin forming Schwann cell undergoes a terminal differentiation with its initial commitment to myelinate a segment of axon and survives in that same relative position along the fiber. Unlike neurons, Schwann cells are capable of dedifferentiating and reentering the cell cycle if they lose their myelin sheaths.

This morphologic review summarizes first the changes in axons, then changes in Schwann cells and myelin sheaths.

Age-Related Alterations in Axons

The number of myelinated axons in long nerves of the PNS decreases with age.[39] Detailed studies of the spatial distribution of fiber loss have not been undertaken, but the data available suggest that the degree of loss is greatest in the distal regions of long nerves. There are also changes in the transducers. The best documented example is the 16-fold reduction in the density of Meissner's corpuscles in between the ages 4 and 76.[6]

The number of unmyelinated axons in the sural nerve decreases somewhat, and denervated Schwann cell bands, characteristic of unmyelinated fibers, are more frequently encountered.[39] These bands can be assigned to previous myelinated fibers on the basis of the characteristic stacked lamellae of Schwann cell process, a feature not encountered in degenerating myelinated fibers.[39,68,69]

Low and colleagues[55,56] documented the reduction in the number of preganglionic neurons and their myelinated axons.

The recent advent of skin biopsies has provided the opportunity for systematic analysis of epidermal innervation and whether this changes with age. McArthur et al[57] have examined the density of nerve fibers within the epidermis in 90 normal individuals between 20 and 80 years of age. The utility of this measure was established by comparison to stereologic analysis.[73] Throughout the age range studied, the normal spatial gradient in epidermal nerve fiber density persists unchanged. For example, the fiber density in the L5 distribution is greatest in the paraspinous region and decreases sequentially in the proximal thigh, the distal thigh above the knee, and in the leg below the knee. Epidermal fiber density, however, does not significantly decrease with age in either the distal leg or distal thigh.[37] Similarly, there are no significant abnormalities of fiber morphology, such as the axonal swellings seen in many types of neuropathy.[35,36]

Changes in Axonal Size and Morphology

The size of the large myelinated fibers substantially decreases with age. This is due in part to atrophy of the large fibers, as documented by the increase in relative thickness of the myelin sheaths (g ratio)[39] and by the deviation from roundness of the fibers (shape factor increasingly less than 1.0). This atrophy of the large axons undoubtedly contributes to the reduction in conduction velocities with age.[39]

Changes in Schwann Cells and Myelin Sheaths

Age-related remodeling of myelin is best indicated by studies of teased nerve fibers. The proportion of fibers showing internodal demyelination or remyelination is increased in patients older than 45 years of age. The increased number of fibers with irregular myelin reflecting axonal atrophy and the decreased caliber of the largest myelinated fibers in older age groups[39] were already noted. This suggests that the age-related demyelination may be "secondary demyelination." That term implies demyelination consequent to a primary axon change and here correlates with the presence of axonal atrophy.[24]

Another striking feature of Schwann cell changes with age is the increased prominence of π granules of Reich. These are lamellated electron-dense profiles located in the Schwann cell perikaryon. Their composition and derivation are not entirely certain, but they contain sulfated glycoconjugates and stain metachromatically.

A dramatic feature of aging in nerves of many species, including rodents, is vacuolar degeneration of myelin.[5,33,34] In rats, intramyelinic vacuoles are detectable in some fibers within the ventral root by 9 months of age, but their number in both dorsal and ventral spinal roots increases markedly after 18 months of age. The basis for these structures is not known, but similar profiles can be brought on earlier by experimental enlargement of the axon followed by a collapse in its size. This has been accomplished with the neurotoxin 3,3'-iminodipropionitrile. The toxin is known to retard axonal transport of neurofilament proteins,[31] with an associated accumulation of neurofilaments in the most proximal region of the ventral roots and the most distal region of the dorsal roots (that is, regions nearest the nerve cell bodies).[15] When administration of the agent is withdrawn, neurofilamentous masses begin to migrate from proximal to distal down the axon,[30] leaving in their wake a decrease in axonal caliber and vacuolar myelinopathy.[32] Thus, the pathogenesis of vacuolar degeneration of myelin in the spinal roots may involve, in part, age-related decrease in axonal caliber.[24]

SUMMARY

Age-related biologic changes in neurons and skeletal muscle commonly affect neuromuscular function and strongly influence the expression of neuromuscular disease. Of primary importance is the attrition of entire motor units, with resultant neurogenic atrophy of skeletal muscle. Other age-related processes are sensory neuronal loss, distal axonal degeneration, axonal atrophy, accumulation of multiple mitochondrial DNA mutations in muscle, and physical inactivity and deconditioning. The decline for most of these begins in early life and proceeds steadily; the curious lack of an abrupt falloff with age is not yet accounted for by any theory of pathogenesis. Fortunately, some of the motor loss in the elderly can be overcome in part by exercise training programs.[27]

ACKNOWLEDGMENTS

This work was supported by the Jay Slotkin Fund for Neuromuscular Research, the Cal Ripken/Lou Gehrig Fund for Neuromuscular Research, and NIH grant 1-K08-NS01912.

References

1. Armon C, Kurland LT: Classic and western Pacific amyotrophic lateral sclerosis: Epidemiologic comparisons. *In* Hudson AJ (ed): Amyotrophic Lateral Sclerosis: Concepts in Pathogenesis and Etiology. Toronto, University of Toronto Press, 1990, pp 144–165
2. Balice-Gordon RJ: Age-related changes in neuromuscular innervation. Muscle Nerve (supplement 5): S83, 1997
3. Baumer A, Zhang C, Linnane AW, et al: Age-related human mtDNA deletions: A heterogeneous set of deletions arising at a single pair of directly repeated sequences. Am J Hum Genet 54:618–630, 1994
4. Beal MF, Howell N, Bodis-Wollner I (eds): Mitochondria and Free Radicals in Neurodegenerative Diseases. New York, Wiley-Liss, 1997
5. Berg BN, Wolf A, Simms HS: Degenerative lesions of spinal roots and peripheral nerves in aging rats. Gerontologia 6:72–80, 1962
6. Bolton CF, Winkleman RK, Dyck PJ: A quantitative study of Meissner's corpuscles in man. Neurology 16:1, 1966
7. Brannon W, McCormick W, Lampert P: Axonal dystrophy in the gracile nucleus of man. Acta Neuropathol 9:1–6, 1967
8. Brierley EJ, Johnson MA, Bowman A, et al: Mitochondrial function in muscle from elderly athletes. Ann Neurol 41:114–116, 1997
9. Brierley EJ, Johnson MA, Lightowlers RN, et al: Role of mitochondrial DNA mutations in human aging: Implications for central nervous system and muscle. Ann Neurol 43:217–223, 1998
10. Britton VJ, Sherman FG, Florini JR: Effect of age on RNA synthesis by nuclei and soluble RNA polymerases from liver and muscle of C57BL/6J mice. J Gerontol 27:188, 1972
11. Brown WF: A method for estimating the number of motor units in thenar muscles and the changes in motor unit count with aging. J Neurol Neurosurg Psychiatry 35:845, 1972
12. Burke WE, Tuttle WW, Thompson CW: The relation of grip strength and grip-strength endurance to age. J Appl Phys 5:628, 1953
13. Caccia MR, Harris JB, Johnson MA: Morphology and physiology of skeletal muscle in aging rodents. Muscle Nerve 2:202, 1979
14. Campbell MJ, McComas AJ, Petito F: Physiological changes in ageing muscles. J Neurol, Neurosurg Psychiatr 36:174, 1973
15. Clark AW, Griffin JW, and Price, DL: The axonal pathology in chronic IDPN intoxication. J Neuropathol Exp Neurol 39:42, 1980
16. Cooper JM, Mann VM, Schapira AH: Analyses of mitochondrial respiratory chain function and mitochondrial DNA deletion in human skeletal muscle: Effect of ageing. J Neurol Sci 133:91, 1992
17. Cortopassi GA, Arnheim N: Detection of a specific mitochondrial DNA deletion in tissues of older humans. Nucleic Acids Res 18:6927, 1990
18. Critchley M: The neurology of old age. Lancet 1:1119, 1931
19. Delbono O, Renganathan M, Messi ML: Excitation-Ca^{2+} release-contraction coupling in single aged human skeletal muscle fiber. Muscle Nerve (supplement 5):S88, 1997

20. Delbono O, Sonntag WE: Influence of insulin-like growth factor (IGF-1) on dihydropyridine receptor (DHPR) function and SR calcium release on rat skeletal muscle fibers at different ages. Biophys J 70:A388, 1996
21. Doherty TJ, Brown WF: The estimated numbers and relative sizes of thenar motor units as selected by multiple point stimulation in young and older adults. Muscle Nerve 16:355, 1993
22. Duncan D: A relation between axon diameter and myelination determined by measurements of myelinated spinal root fibres. J Comp Neurol 60:437, 1934
23. Dyck PJ, Karnes J, O'Brien PC, et al: Detection thresholds of cutaneous sensation in humans. In Peripheral Neuropathy. Philadelphia, WB Saunders, 1993, p 706
24. Dyck PJ, Lais AC, Karnes JL, et al: Permanent axotomy, a model of axonal atrophy and secondary segmental demyelination and remyelination. Ann Neurol 9:575, 1981
25. Dyck PJ, Schultz PW, and O'Brien PC: Quantitation of touch-pressure sensation. Arch Neurol 26:465, 1972
26. Enoka RM: Neural strategies in the control of muscle force. Muscle Nerve (Supplement 5): S66, 1997
27. Fiatarone MA, O'Neill EF, Ryan ND, et al: Exercise training and nutritional supplementation for physical frailty in very elderly people. New Engl J Med 330:1769–1775, 1994
28. Friede R, Meier T, Diem M: How is the exact length of an internode determined? J Neurol Sci 50:217, 1981
29. Friede RL, Bischhausen R: The precise geometry of large internodes. J Neurol Sci 48:367, 1980
30. Griffin JW, Drucker N, Gold BG, et al: Schwann cell proliferation and migration during paranodal demyelination. J Neurosci 7:682, 1987
31. Griffin JW, Hoffman PN, Clark AW, et al: Slow axonal transport of neurofilament proteins: Impairment by 3,3'-iminodipropionitrile administration. Science 202:633, 1978
32. Griffin JW, Price DL: Demyelination in experimental IDPN and hexacarbon neuropathies: Evidence for an axonal influence. Lab Invest 45:130, 1981
33. Griffiths IR, Duncan ID: Age changes in the dorsal and ventral lumbar nerve roots of dogs. Acta Neuropathol 32:75, 1975
34. Grover-Johnson N, Spencer PS: Peripheral nerve abnormalities in aging rats. J Neuropathol Exp Neurol 38:316, 1979
35. Harvey GL, Gold R, Hartung HP, et al: Nonneural-specific T lymphocytes can orchestrate inflammatory peripheral neuropathy. Brain 118:1263, 1995
36. Holland NR, Griffin JW, Cornblath DR, et al: Idiopathic pure small fiber sensory neuropathy: Clinical course and intraepidermal nerve fiber densities. Ann Neurol 44: in press, 1998
37. Howel TH: Senile deterioration of the central nervous system. BMJ 1:56, 1949
38. Huang CY: Peripheral neuropathy in the elderly: A clinical and electrophysiological study. J Am Geriatr Soc 29:49–54, 1981
39. Jacob JM, Love S: Qualitative and quantitative morphology of human sural nerve at different ages. Brain 108:897, 1985
40. Jennekens FGI: Disuse, cachexia, and ageing. In Mastaglia FL, Walton J (eds): Skeletal Muscle Pathology. Edinburgh, Churchill Livingstone, 1982, p 605
41. Jennekens FGI, Tomlinson BE, Walton JN: Histochemical aspects of five limb muscles in old age. J Neurol Sci 14:259, 1971
42. Jennekens FGI, Tomlinson BE, Walton JN: The extensor digitorum brevis: Histological and histochemical aspects. J Neurol Neurosurg Psychiatr 35:124, 1972
43. Johnston W, Karpati G, Carpenter S, et al: Late-onset mitochondrial myopathy. Ann Neurol 37:16–23, 1995
44. Juergens SM, Kurland Lt, Okazaki H, et al: ALS in Rochester, Minnesota, 1925–1977. Neurology 30:463–470, 1980
45. Kadhiresan VA, Hassett CA, Faulkner JA: Properties of single motor units in medial gastrocnemius muscles of adult and old rats. J Physiol 493:543, 1996
46. Kamon E, Goldfuss AJ: In-plant evaluation of the muscle strength of workers. Am Indust Hygeine Assoc J 39:801, 1978
47. Katzman R, Terry R: The Neurology of Aging. Philadelphia, FA Davis, 1983
48. Keen DA, Yue GH, Enoka RM: Training-related enhancement in the control of motor unit output in elderly humans. J Appl Physiol 77:2648, 1994

49. Kuncl RW, Wiggins WW: Toxic myopathies. Neurol Clin 6:593, 1988
50. Lacomblez L, Bensimon G, Leigh PN, et al: Dose-ranging study of riluzole in amyotrophic lateral sclerosis. Lancet 347:1425–1431, 1996
51. Lacomis D, Chad DA, Smith TW: Myopathy in the elderly: Evaluation of the histopathologic spectrum and the accuracy of clinical diagnosis. Neurology 43:825, 1993
52. Larsson L, Grimby G, Karlsson J: Muscle strength and speed of movement in relation to age and muscle morphology. J Appl Physiol 46:451, 1979
53. Larsson L, Sjodin B, Karlsson J: Histochemical and biochemical changes in human skeletal muscle with age in sedentary males, age 22–65 years. Acta Physiol Scan 103:31, 1978
54. Lindsay RM, Wiegand SJ, Altar CA, et al: Neurotrophic factors: From molecule to man. Trends in Neurosci 17:182, 1994
55. Low PA, Dyck PJ: Splanchnic preganglionic neurons in man. II. Morphometry of myelinated fibers of T7 ventral spinal root. Acta Neuropathol 40:219, 1977
56. Low PA, Okazaki H, Dyck PJ: Splanchnic preganglionic neurons in man. I. Morphometry of preganglionic cytons. Acta Neuropathol 40:55, 1977
57. McArthur JC, Stocks A, Hauer P, et al: Epidermal nerve fiber density: Normative reference range and diagnostic efficiency. Arch Neurol. In press, 1998
58. McComas AJ, Upton AR, Sica REP: Motoneurone disease and ageing. Lancet 2(844):1477, 1973
59. Mendell JR: Mitochondrial myopathy in the elderly: Exaggerated aging in the pathogenesis of disease. Ann Neurol 37:16, 1995
60. Mitch WE, Goldberg AL: Mechanisms of muscle wasting. N Engl J Med 355:1897, 1996
61. Moore MJ, Rebeiz JJ, Holden M, et al: Biometric analysis of normal skeletal muscle. Acta Neuropathologica (Berlin) 19:51–69, 1971
62. Müller-Hocker J: Cytochrome c oxidase deficient fibres in the limb muscle and diaphragm of a man without muscular disease: An age-related alteration. J Neurol Sci 100:14, 1990
63. Pestronk A, Drachman DB, Griffin JW. Effects of aging on nerve sprouting and regeneration. Exp Neurol 70:65, 1980
64. Petrofsky JS, Lind AR: Aging, isometric strength and endurance, and cardiovascular responses to static effort. J Appl Physiol 38:91, 1975
65. Procacci BG, Bozza G, Buzzelli G, et al: The cutaneous pricking pain threshold in old age. Gerontologia Clinica 12:213, 1970
66. Ramon y Cajal S: Degeneration and regeneration of the nervous system (RM May, Trans). London, Oxford University Press, 1928
67. Rifai Z, Welle S, Kamp C, et al: Ragged red fibers in normal aging and inflammatory myopathy. Ann Neurol 37:24, 1995
68. Schaumburg HH, Spencer PS, Ochoa J: The aging human peripheral nervous system. *In* The Neurology of Aging. Philadelphia, FA Davis, 1983, p 111
69. Sharma AK, Thomas PK: Quantitative studies on age changes in unmyelinated nerve fibers in the vagus nerve of man. *In* Studies on Neuromuscular Diseases. Basel, S Karger, 1975, p 211
70. Sirurgeirson B, Lindelof B, Edhag O, et al: Risk of cancer in patients with dermatomyositis or polymyositis: A population-based study. N Engl J Med 326:363–367, 1992
71. Skre H: Neurological signs in a normal population. Acta Neurol Scand 48:575, 1972
72. Sommer N, Willcox N, Harcourt GC, et al: Myasthenic thymus and thymoma are selectively enriched in acetylcholine receptor-reactive T cells. Ann Neurol 28:312–319, 1990
73. Stocks EA, Mouton P, Hauer P, et al: Stereological estimation of epidermal nerve fibers in normal human skin. Society for Neuroscience Abstracts 20:701, 1994
74. Thomas PK: Peripheral nerve changes in the elderly. *In* New Issues in Neurosciences. London, John Wiley, 1988
75. Tomlinson BE, Irving D: The numbers of limb motor neurons in the human lumbosacral cord throughout life. J Neurol Sci 34:213, 1977
76. Tomlinson BE, Walton JN, Rebeiz JJ: The effects of ageing and of cachexia upon skeletal muscle. J Neurol Sci 9:321, 1969
77. Trounce I, Byrne E, Marzuki S. Decline in skeletal muscle mitochondrial respiratory chain function: Possible factor in aging. Lancet 8639:637, 1989
78. Wallace SL, Singer JZ, Duncan GJ, et al: Renal function predicts colchicine toxicity: Guidelines for the prophylactic use of colchicine in gout. J Rheumatol 18:264–269, 1991

79. Welle S, Thornton C, Jozefowicz R, et al: Myofibrillar protein synthesis in young and old men. Am J Physiol 264:E693, 1993
80. Yarasheski KE, Zachwieja JJ, Bier DM: Acute effects of resistance exercise on muscle protein synthesis rate in young and elderly men and women. Am J Physiol 265:E210, 1993

Address reprint requests to

Ralph W. Kuncl, MD, PhD
Johns Hopkins University School of Medicine
600 N. Wolfe Street, Meyer 5-119
Baltimore, MD 21287-7519

Journal of Neuroscience Research 52:1–6 (1998)

Mini-Review

Neurodegeneration and Aging: Role of the Second Genome

Manuel B. Graeber,[1]* Eva Grasbon-Frodl,[1,2] Ulrich V. Eitzen,[2] and Siegfried Kösel[2]
[1]Molecular Neuropathology Laboratory, Department of Neuromorphology, Max-Planck-Institute of Neurobiology (formerly Max-Planck-Institute of Psychiatry, Basic Sciences Institute), Martinsried, Germany
[2]Molecular Neuropathology Laboratory, Institute of Neuropathology, Ludwig-Maximilians-University, Munich, Germany

The latest Health Report of the World Health Organization predicts a significant increase in the age of human populations over the next two decades. In the developed world, at least 20% of the population will be older than 65 years. This development together with the as yet unknown etiology of many neurodegenerative disorders has caused an increased interest in the biology and pathophysiology of mitochondria. Dysfunction of mitochondria has been linked to both normal aging and neurodegenerative disorders, with the latter occurring much more frequently at higher age. Specifically, genetic defects in mitochondria have been shown to accumulate during life, and certain mutations of mitochondrial genes have been implicated in the etiology of Parkinson's and Alzheimer's diseases. In addition, a large number of new mitochondrial diseases have been identified following the first description of mitochondrial mutations 10 years ago. While there can be little doubt that DNA defects of mitochondria play a role in aging, specific mutations of mitochondrial genes underlying Parkinson's or Alzheimer's diseases remain to be identified. There is evidence, however, that mutations of the mitochondrial genome may increase the susceptibility to neurodegeneration. J. Neurosci. Res. 52:1–6, 1998.

Key words: neurodegenerative disorders; mitochondria; Parkinson's disease; Alzheimer's disease

INTRODUCTION

The field of mitochondrial molecular genetics did not exist a decade ago. Leber's hereditary optic neuropathy (LHON) was the first disease shown to be due to a point mutation of mitochondrial DNA (Wallace et al., 1988), and at about the same time the first disease-associated deletions of the mitochondrial chromosome were described (Holt et al., 1988). Since then, the field of mitochondrial disorders has become a focus not only within neurology but also of various other disciplines (Moraes, 1996; Ozawa, 1997). In few areas of medicine has progress been more spectacular (DiMauro, 1996). Primary defects in mitochondrial function are now implicated in over 100 diseases, and the list continues to grow (Luft, 1994). Yet, the importance of mitochondrial dysfunction is not confined to established pathological conditions but seems of very general relevance as mitochondrial damage has been associated with normal aging (Richter, 1995; Sastre et al., 1996; Gabbita et al., 1997). Consistent with this notion, mitochondrial diseases and aging share a decline in mitochondrial respiratory function (Linnane et al., 1995). The nervous system is one of the organs most dependent on mitochondrial respiration. Thus, observations that mitochondrial mutations accumulate in post-mitotic tissue such as the nervous system and muscle (Cortopassi and Liu, 1995; Liu et al., 1997) and that neurodegenerative disorders are generally more common at higher age make these organelles an attractive target for neuroscientists.

MITOCHONDRIA CONTAIN THEIR OWN DNA

Mitochondria are known as the "powerhouses" of the cell where oxygen is metabolised, resulting in the production of the cellular energy carrier, ATP. In addition, mitochondria have a number of other biosynthetic functions (Schatz, 1994). A very peculiar feature of mitochondria is their content of DNA, which is due to their

Contract grant sponsor: German Research Foundation; Contract grant sponsor: Max Planck Society.

*Correspondence to: Dr. Manuel B. Graeber, Molecular Neuropathology Laboratory, Department of Neuromorphology, Max-Planck-Institute of Neurobiology, Am Klopferspitz 18a, 82152 Martinsried, Germany. E-mail: neuropat@neuro.mpg.de

Received 10 December 1997; Accepted 16 December 1997

TABLE I. Specifics of Mitochondrial Genetics

- Mitochondria are semiautonomous organelles; possess their own replication-, transcription- and translation system
- Exclusively maternal inheritance of mitochondrial DNA
- Mitotic segregation of mitochondrial DNA can lead to heteroplasmy, i.e., the proportion of genetically different populations of mitochondria differs between generations of mitotically active cells
- Approximately tenfold higher mutation rate compared with nuclear DNA: reasons include lack of protective histones, absence of introns, lack of sufficient DNA repair machinery, physical proximity of the mitochondrial chromosome to the respiratory chain where DNA-damaging oxygen radicals are produced

evolutionary origin (Hill and Singh, 1997) and which makes mitochondria unique among the cellular organelles. The mitochondrial genome is a double-stranded, 16569-base pair circular DNA molecule which is present in several copies per organelle. Since each cell contains hundreds or thousands of mitochondria, there are multiple copies of each mitochondrial chromosome in individual cells. Mitochondrial genes code for a total of 13 polypeptides of the respiratory chain as well as two rRNAs and 22 tRNAs which are required for the expression of this genome. All 13 peptides form components of the respiratory chain/oxidative phosphorylation (OXPHOS) system which is located in the inner mitochondrial membrane. The OXPHOS peptides include subunits of complex I or NADH ubiquinone oxidoreductase (ND1, ND2, ND3, ND4, ND4L, ND5, ND6), complex III or ubiquinol-cytochrome c oxidoreductase (cytb), complex IV or cytochrome c oxidase (CO1, CO2, CO3), and complex V or ATP synthase (ATPase 6 and 8). The remaining subunits of the OXPHOS system are encoded by nuclear genes.

DIFFERENCES BETWEEN MITOCHONDRIAL AND NUCLEAR GENETICS

A realistic assessment of the relevance of mitochondrial DNA sequence variation seems impossible without an understanding of some important differences between nuclear and mitochondrial genetics (Table I). Mitochondrial DNA replicates autonomously and is inherited via the cytoplasm of the parent cell with the individual mitochondrion being the segregating unit (Attardi et al., 1995). Thus, in the case of mitochondrial mutations both mutated as well as normal mitochondria may be present within the same cell. This situation has been termed "heteroplasmy" and can principally occur at the organelle, cellular, and tissue level. Although a particular mitochondrial DNA mutation may be present in all mitochondrial DNA molecules of an organism ("homoplasmy"), many disease-causing mutations of mitochondrial DNA are heteroplasmic (vide infra). Therefore, mitochondrial genetics is inherently quantitative. In addition, unlike blood (Thart et al., 1996), post-mitotic tissues such as the nervous system and muscle tend to accumulate defunct mitochondrial DNA. At the same time, nervous tissue and muscle are characterized by an especially high energy demand which renders these tissues prime targets for mitochondrial disease. However, it is important to note that less than 1% of all mitochondrial proteins are encoded by mitochondrial genes. This implies a necessity for cooperation between the nuclear and mitochondrial genomes which is as yet incompletely understood (Poyton and McEwen, 1996).

ESTABLISHED MITOCHONDRIAL DISEASES

Inheritance of classical mitochondrial diseases is exclusively maternal, with both males and females being affected (for review, see DiMauro and Wallace, 1993; Graeber and Müller, 1998). The most important mitochondrial diseases are mitochondrial encephalomyopathy with lactic acidosis and stroke-like episodes (MELAS), myoclonic epilepsy with ragged red fibres (MERRF), Leber's hereditary optic neuropathy (LHON), and Kearns-Sayre syndrome (KSS). MELAS is typically due to single base exchanges that lie within the tRNA-Leu^{UUR} gene (np 3243, 3271, and 3291). Point mutations are also the underlying cause of the MERRF syndrome, but here mutations usually affect the tRNA-Lys gene (np 8344). MELAS and MERRF mutations are heteroplasmic, and both diseases show molecular as well as clinical overlap ("overlap syndromes"). In contrast, both homoplasmic as well as heteroplasmic point mutations have been identified in LHON (np 3460, 11778, 14484), which is an exception in that males are more frequently affected than females. Neuropathy, ataxia, and retinitis pigmentosa (NARP) as well as Leigh's syndrome are two rare diseases caused by point mutations within the ATPase6 gene (np 8993). Leigh syndrome occurs if the mutation is present in the majority of mitochondria, i.e., when the degree of heteroplasmy is very high (>90–95%). Kearns-Sayre syndrome (KSS) is a well-known mitochondrial disease caused by DNA deletions (Shoffner et al., 1989).

MITOCHONDRIA AND AGING

Our knowledge of the biological basis of cellular aging is still fragmentary. Studying synaptic and cytoplasmic mitochondria in the cerebellum, Fattoretti and colleagues concluded that the number of fully functional mitochondria decreases with advancing age (Bertoni-Freddari et al., 1993; Fattoretti et al., 1996). It is unclear, however, at which cellular level the underlying defects occur (Craig and Hood, 1997). One possibility is that there are age-related alterations in the mitochondrial membrane potential, the driving force for ATP synthesis, which results in a significantly increased production of oxidants (Hagen et al., 1997). The damaging effect of such reactive intermediate metabolites is well established. Thus, at least some degenerative cellular changes occurring during aging may be explained by the effect of oxygen free radicals (Cortopassi and Wang, 1995; Kalous and Drahota, 1996). This view is supported by the established relationship between dietary restriction, retarded age-associated increases in mitochondrial free-radical production, and reduced oxidative damage within cells (Wachsmann, 1996). The accumulation of lipofuscin, which is typical of aged human nerve cells, has been associated with an age-related decrease in mitochondrial "detoxification" mechanisms (Chen and Yu, 1996). Wei et al. (1996a) could show that the amount of lipid peroxides measured as malondialdehyde and the activity of manganese-superoxide dismutase in mitochondria exhibit an age-dependent increase in different human tissues. In addition, it has been suggested that tandem duplications and large-scale deletions of mtDNA are early molecular events of the human aging process (Corral-Debrinski et al., 1992; Soong et al., 1992; Wei et al., 1996b), and it has been demonstrated that certain point mutations of mitochondrial DNA accumulate in the aging human brain (Zhang et al., 1993; Liu et al., 1997). However, the functional implications of these findings are controversial (Hayashi et al., 1994). To complicate the matter further, Takai and co-workers discuss the possibility that the common age-associated changes in human and mouse mitochondria are limited to a decrease in mitochondrial translational activity (Takai et al., 1995). Thus, there are clear species differences (Calleja et al., 1993; Eimon et al., 1996). In addition, differences in the susceptibility of different organs to mitochondrial damage have to be envisioned.

An explanation for the biological fitness and survival of genetically and/or functionally compromised mitochondria has been provided on theoretical grounds. It has been suggested that mitochondria with reduced respiratory function, due to a mutation or deletion affecting the respiratory chain, suffer less frequent lysosomal degradation because they inflict free radical damage more slowly on their own membranes (de Grey, 1997). Therefore, once such a mutation occurs in a mitochondrion of a non-dividing cell, e.g., a neuron, mitochondria carrying the defect will rapidly populate that cell and cause the destruction of its respiratory capability. Damaged mitochondria are still able to replicate because the enzymes for mitochondrial replication are encoded entirely in the cell nucleus (Kowald and Kirkwood, 1993).

DO MITOCHONDRIAL DEFECTS HAVE A PRIMARY ROLE IN NEURODEGENERATION?

An involvement of mitochondrial genetic defects in the pathogenesis of neurodegenerative disorders has been extensively discussed in the context of Parkinson's and Alzheimer's diseases. Parkinson's disease is a very common disorder of movement characterized by neuronal cell death which mainly affects dopaminergic neurons of the substantia nigra. In 1983, the observation was made that a neurotoxin, MPTP (1-Methyl-4-phenyl-1,2,3,6-tetrahydropyridine), causes a Parkinson's disease-like syndrome in humans (Langston et al., 1983). It is now known that this toxin inhibits complex I of the mitochondrial respiratory chain (McNaught et al., 1996). These findings are of great interest, as several groups have independently reported on a brain-region as well as disease-specific reduction in the activity of mitochondrial complex I in the substantia nigra of patients who died with Parkinson's disease (Schapira et al., 1989, 1990; Mizuno et al., 1989; Janetzky et al., 1994). Several nuclear genetic defects have been shown to be associated with parkinsonism (Graeber et al., 1992; Furukawa et al., 1996; Polymeropoulos et al., 1997). However, the vast majority of classical Parkinson's cases lack a family history, which makes the molecular genetic analysis of this disorder difficult. Due to its strong representation in the mitochondrial genome, complex I of the respiratory chain is the most likely target of mitochondrial DNA mutations. Since its complete DNA sequence is known (Anderson et al., 1981), several groups (Ikebe et al., 1995; Kapsa et al., 1996; Kösel et al., 1997) have conducted systematic sequencing studies of mitochondrial complex I genes. As a result, a role of mitochondrial complex I mutations has been excluded in more than 90% of the sporadic Parkinson's cases (Kösel et al., 1997). However, mitochondrial DNA sequence variation could well contribute to Parkinson's disease pathogenesis in a subset of patients. This view is in line with studies of Swerdlow et al. (1996) who created an in vitro system in which the potential contributions of environmental toxins, complex I nuclear DNA mutations, and mitochondrial DNA mutations could be analysed. These authors concluded that the complex I defect found in Parkinson's disease is genetic and may arise from mitochondrial DNA (Swerdlow et al., 1996). Direct evidence that a single point mutation of complex I can, in fact, cause basal ganglia dysfunction comes from findings in a rare form of familial dystonia (Jun et al., 1994).

The identification of several different genetic defects which cause Alzheimer's disease illustrates the genetic heterogeneity of this disorder (see Roses, 1996, for review). Alzheimer's disease is also the best-studied example of a disease the onset of which is heavily influenced by a susceptibility gene, the apolipoprotein E epsilon 4 allele (Roses, 1997). Furthermore, a reduction in the activity of key mitochondrial enzymes has been described in Alzheimer's disease, and oxidative damage is increased in those brain areas most severely affected by the disease (Sims, 1996). There is evidence for a defect of mitochondrial complex IV, cytochrome c oxidase, at the enzymatic as well as at the messenger RNA and DNA level in the Alzheimer's brain (Parker et al., 1994; Chandrasekaran et al., 1994; Davis et al., 1997). From a pathophysiological point of view, mitochondrial dysfunction in Alzheimer's disease could lead to both reduced ATP synthesis and an increased production of reactive oxygen radicals.

The risk of developing both Parkinson's and Alzheimer's diseases appears to be modified by a mitochondrial tRNAGln mutation (np 4336; Shoffner et al., 1993; Wallace, 1994). This finding has been confirmed by some (Hutchin and Cortopassi, 1995; Egensperger et al., 1997) but not all groups (Wragg et al., 1995; Tysoe et al., 1996). Yet, the latter did not study histopathologically confirmed cases.

CONCLUSIONS AND OUTLOOK

The role of mitochondria in aging and neurodegeneration is beginning to be elucidated. Energy depletion and tissue degeneration are the consequences of mitochondrial dysfunction and are also found in association with normal aging (Finkel, 1997). Defects of mitochondrial DNA can both be the result and the cause of aging. Whether specific mutations of the mitochondrial genome underlie distinct neurodegenerative diseases or simply "negatively" complement cellular dysfunction already in place awaits clarification. Sequencing studies focusing on neuropathologically confirmed disease cases from diverse ethnic backgrounds (Kösel et al., 1998) should help clarify whether certain mitochondrial haplotypes predispose to Parkinson's and/or Alzheimer's diseases (Hofmann et al., 1997). The next decade of mitochondrial genetics is likely to provide crucial answers to these questions.

ACKNOWLEDGMENTS

We thank U. Müller and C. Kloss for reading of the manuscript.

REFERENCES

Anderson S, Bankier AT, Barrell BG, De Bruijn MHL, Coulson AR, Drouin J, Eperon IC, Nierlich DP, Roe BA, Sanger F, Schreier PH, Smith AJH, Staden R, Young IG (1981): Sequence and organization of the human mitochondrial genome. Nature 290:457–470.

Attardi G, Yoneda M, Chomyn A (1995): Complementation and segregation behavior of disease-causing mitochondrial DNA mutations in cellular model systems. Biochim Biophys Acta 1271:241–248.

Bertoni-Freddari C, Fattoretti P, Casoli T, Spagna C, Meier-Ruge W, Ulrich J (1993): Morphological plasticity of synaptic mitochondria during aging. Brain Res 628:193–200.

Calleja M, Pena P, Ugalde C, Ferreiro C, Marco R, Garesse R (1993): Mitochondrial DNA remains intact during Drosophila aging, but the levels of mitochondrial transcripts are significantly reduced. J Biol Chem 268:18891–18897.

Chandrasekaran K, Giordano T, Brady DR, Stoll J, Martin LJ, Rapoport SI (1994): Impairment in mitochondrial cytochrome oxidase gene expression in Alzheimer disease. Mol Brain Res 24:336–340.

Chen JJ, Yu BP (1996): Detoxification of reactive aldehydes in mitochondria: effects of age and dietary restriction. Aging 8:334–340.

Corral-Debrinski M, Horton T, Lott MT, Shoffner JM, Beal MF, Wallace DC (1992): Mitochondrial DNA deletions in human brain: regional variability and increase with advanced age. Nature 2:324–329.

Cortopassi G, Liu Y (1995): Genotypic selection of mitochondrial and oncogenic mutations in human tissue suggests mechanisms of age-related pathophysiology. Mutat Res 338:151–159.

Cortopassi G, Wang E (1995): Modelling the effects of age-related mtDNA mutation accumulation; complex I deficiency, superoxide and cell death. Biochim Biophys Acta 1271:171–176.

Craig EE, Hood DA (1997): Influence of aging on protein import into cardiac mitochondria. Am J Physiol 41:H2983–H2988.

Davis RE, Miller S, Herrnstadt C, Ghosh SS, Fahy E, Shinobu LA, Galasko D, Thal LJ, Beal MF, Howell N, Parker WD (1997): Mutations in mitochondrial cytochrome c oxidase genes segregate with late-onset Alzheimer disease. Proc Natl Acad Sci USA 94:4526–4531.

de Grey, AD (1997): A proposed refinement of the mitochondrial free radical theory of aging. Bioessays 19:161–166.

DiMauro S (1996): Mitochondrial encephalomyopathies: what next? J Inherit Metab Dis 19:489–503.

DiMauro S, Wallace DC (eds) (1993): "Mitochondrial DNA in Human Pathology." New York: Raven Press.

Egensperger R, Kösel S, Schnopp NM, Mehraein P, Graeber MB (1997): Association of the mitochondrial tRNA A4336G mutation with Alzheimer's and Parkinson's diseases. Neuropathol Appl Neurobiol 23:315–321.

Eimon PM, Chung SS, Lee CM, Weindruch R, Aiken JM (1996): Age-associated mitochondrial DNA deletions in mouse skeletal muscle: comparison of different regions of the mitochondrial genome. Dev Genet 18:107–113.

Fattoretti P, Bertoni-Freddari C, Caselli U, Paoloni R, Meier-Ruge W (1996): Morphologic changes in cerebellar mitochondria during aging. Anal Quant Cytol Histol 18:205–208.

Finkel E (1997): Piecing together the puzzle of ageing. Lancet 350:1150.

Furukawa Y, Shimadzu M, Rajput AH, Shimizu Y, Tagawa T, Mori H, Yokochi M, Narabayashi H, Hornykiewicz O, Mizuno Y, Kish SJ (1996): GTP-cyclohydrolase I gene mutations in hereditary progressive and dopa-responsive dystonia. Ann Neurol 39:609–617.

Gabbita SP, Butterfield DA, Hensley K, Shaw W, Carney JM (1997): Aging and caloric restriction affect mitochondrial respiration and lipid membrane status—an electron paramagnetic resonance investigation. Free Rad Biol Med 23:191–201.

Graeber MB, Müller U (1998): Recent developments in the molecular genetics of mitochondrial disorders. J Neurol Sci (in press).

Graeber MB, Kupke KG, Muller U (1992): Delineation of the dystonia-parkinsonism syndrome locus in Xq13. Proc Natl Acad Sci USA 89:8245–8248.

Hagen TM, Yowe DL, Bartholomew JC, Wehr CM, Do KL, Park JY, Ames BN (1997): Mitochondrial decay in hepatocytes from old rats: membrane potential declines, heterogeneity and oxidants increase. Proc Natl Acad Sci USA 94:3064–3069.

Hayashi J. Ohta S. Kagawa Y. Kondo H. Kaneda H. Yonekawa H. Takai D. Miyabayashi S (1994): Nuclear but not mitochondrial genome involvement in human age-related mitochondrial dysfunction. Functional integrity of mitochondrial DNA from aged subjects. J Biol Chem 269:6878–6883.

Hill KA. Singh SM (1997): The evolution of species-type specificity in the global DNA sequence organization of mitochondrial genomes. Genome 40:342–356.

Hofmann S. Jaksch M. Bezold R. Mertens S. Aholt S. Paprotta A. Gerbitz KD (1997): Population genetics and disease susceptibility: characterization of central European haplogroups by mtDNA gene mutations, correlation with D loop variants and association with disease. Hum Mol Genet 6:1835–1846.

Holt IJ. Harding AE. Morgan-Hughes JA (1988): Deletions of muscle mitochondrial DNA in patients with mitochondrial myopathies. Nature 331:717–719.

Hutchin T. Cortopassi G (1995): A mitochondrial DNA clone is associated with increased risk for Alzheimer disease. Proc Natl Acad Sci USA 92:6892–6895.

Ikebe S. Tanaka M. Ozawa T (1995): Point mutations of mitochondrial genome in Parkinson's disease. Mol Brain Res 28:281–295.

Janetzky B. Hauck S. Youdim MBH. Riederer P. Jellinger K. Pantucek F. Zochling R. Boissl KW. Reichmann H (1994): Unaltered aconitase activity, but decreased complex I activity in Substantia nigra pars compacta of patients with Parkinson's disease. Neurosci Lett 169:126–128.

Jun AS. Brown MD. Wallace DC (1994): A mitochondrial DNA mutation at nucleotide pair 14459 of the NADH dehydrogenase subunit 6 gene associated with maternally inherited Leber hereditary optic neuropathy and dystonia. Proc Natl Acad Sci USA 91:6206–6210.

Kalous M. Drahota Z (1996): The role of mitochondria in aging. Physiol Res 45:351–359.

Kapsa RM. Jean-Francois MJ. Lertrit P. Weng S. Siregar N. Ojaimi J. Donnan G. Masters C. Byrne E (1996): Mitochondrial DNA polymorphism in substantia nigra. J Neurol Sci 144:204–211.

Kösel S. Grasbon-Frodl EM. Mautsch U. Egensperger R. von Eitzen U. Frishman D. Hofmann S. Gerbitz KD. Mehraein P. Graeber MB (1997): Novel mutations of mitochondrial complex I in pathologically proven Parkinson disease. Neurogenet 1: (in press).

Kösel S. Grasbon-Frodl EM. Graeber MB (1998): A new approach to the genetic analysis of nervous system diseases: retrospective genotyping of archival brains. Prog Brain Res (in press).

Kowald A. Kirkwood TB (1993): Mitochondrial mutations, cellular instability and ageing: modelling the population dynamics of mitochondria. Mutat Res 295:93–103.

Langston JW. Ballard P. Tetrud JW. Irwin I (1983): Chronic parkinsonism in humans due to a product of meperidine-analog synthesis. Science 219:979–980.

Linnane AW. Degli EM. Generowicz M. Luff AR. Nagley P (1995): The universality of bioenergetic disease and amelioration with redox therapy. Biochim Biophys Acta 1271:191–194.

Liu VW. Zhang C. Linnane AW. Nagley P (1997): Quantitative allele-specific PCR: demonstration of age-associated accumulation in human tissues of the A → G mutation at nucleotide 3243 in mitochondrial DNA. Hum Mutat 9:265–271.

Luft R (1994): The development of mitochondrial medicine. Proc Natl Acad Sci USA 91:8731–8738.

McNaught KS. Thull U. Carrupt PA. Altomare C. Cellamare S. Carotti A. Testa B. Jenner P. Marsden CD (1996): Effects of isoquinoline derivatives structurally related to 1-methyl-4-phenyl-1,2,3,6-tetrahydropyridine (MPTP) on mitochondrial respiration. Biochem Pharmacol 51:1503–1511.

Mizuno Y. Ohta S. Tanaka M. Takamiya S. Suzuki K. Sato T. Oya H. Ozawa T. Kagawa Y (1989): Deficiencies in complex I subunits of the respiratory chain in Parkinson's disease. Biochem Biophys Res Commun 163:1450–1455.

Moraes CT (1996): Mitochondrial disorders. Curr Opin Neurol 9:369–374.

Ozawa T (1997): Genetic and functional changes in mitochondria associated with aging. Physiol Rev 77:425–464.

Parker WD, Jr, Parks J. Filley CM. Kleinschmidt-DeMasters BK (1994): Electron transport chain defects in Alzheimer's disease brain. Neurology 44:1090–1096.

Polymeropoulos MH. Lavedan C. Leroy E. Ide SE. Dehejia A. Dutra A. Pike B. Root H. Rubenstein J. Boyer R. Stenroos ES. Chandrasekharappa S. Athanassiadou A. Papapetropoulos T. Johnson WG. Lazzarini AM. Duvoisin RC. Di Iorio G. Golbe LI. Nussbaum RL (1997): Mutation in the alpha-synuclein gene identified in families with Parkinson's disease. Science 276: 2045–2047.

Poyton RO. McEwen JE (1996): Crosstalk between nuclear and mitochondrial genomes. Annu Rev Biochem 65:563–607.

Richter C (1995): Oxidative damage to mitochondrial DNA and its relationship to ageing. Int J Biochem Cell Biol 27:647–653.

Roses AD (1996): The Alzheimer diseases. Curr Opin Neurobiol 6:644–650.

Roses AD (1997): A model for susceptibility polymorphisms for complex diseases: apolipoprotein E and Alzheimer disease. Neurogenetics 1:3–11.

Sastre J. Pallardo FV. Pla R. Pellin A. Juan G. O'Connor JE. Estrela JM. Miquel J. Vina J (1996): Aging of the liver: age-associated mitochondrial damage in intact hepatocytes. Hepatology 24: 1199–1205.

Schapira AHV. Cooper JM. Dexter D. Jenner P. Clark JB. Marsden CD (1989): Mitochondrial complex I deficiency in Parkinson's disease. Lancet i:1269.

Schapira AHV. Mann VM. Cooper JM. Dexter D. Daniel SE. Jenner P. Clark JB. Marsden CD (1990): Anatomic and disease specificity of NADH CoQ1 reductase (complex I) deficiency in Parkinson's disease. J Neurochem 55:2142–2145.

Schatz G (1994): Mitochondria—the next 100 years. KargerGazette 58:1–2.

Shoffner JM. Lott MT. Voljavec AS. Soueidan SA. Costigan DA. Wallace DC (1989): Spontaneous Kearns-Sayre/chronic external ophthalmoplegia plus syndrome associated with a mitochondrial DNA deletion: a slip-replication model and metabolic therapy. Proc Natl Acad Sci USA 86:7952–7956.

Shoffner JM. Brown MD. Torroni A. Lott MT. Cabell MF. Mirra SS. Beal MF. Yang CC. Gearing M. Salvo R. Watts RL. Juncos JL. Hansen LA. Crain BJ. Fayad M. Reckord CL. Wallace DC (1993): Mitochondrial DNA variants observed in Alzheimer disease and Parkinson disease patients. Genomics 17:171–184.

Sims NR (1996): Energy metabolism, oxidative stress and neuronal degeneration in Alzheimer's disease. Neurodegeneration 5:435–440.

Soong NW. Hinton DR. Cortopassi G. Arnheim N (1992): Mosaicism for a specific somatic mitochondrial DNA mutation in adult human brain. Nature 2:318–323.

Swerdlow RH. Parks JK. Miller SW. Tuttle JB. Trimmer PA. Sheehan JP. Bennett JPJ. Davis RE. Parker WDJ (1996): Origin and functional consequences of the complex I defect in Parkinson's disease. Ann Neurol 40:663–671.

Takai D. Inoue K. Shisa H. Kagawa Y. Hayashi J (1995): Age-associated changes of mitochondrial translation and respiratory function in mouse brain. Biochem Biophys Res Commun 217:668–674.

Thart LM. Jansen JJ. Lemkes HHPJ. Deknijff P. Maassen JA (1996): Heteroplasmy levels of a mitochondrial gene mutation associated with diabetes mellitus decrease in leucocyte DNA upon aging. Hum Mutat 7:193–197.

Tysoe C, Robinson D, Brayne C, Dening T, Paykel ES, Huppert FA, Rubinsztein DC (1996): The tRNA(Gln) 4336 mitochondrial DNA variant is not a high penetrance mutation which predisposes to dementia before the age of 75 years. J Med Genet 33:1002–1006.

Wachsman JT (1996): The beneficial effects of dietary restriction: reduced oxidative damage and enhanced apoptosis. Mutat Res 350:25–34.

Wallace DC (1994): Mitochondrial DNA sequence variation in human evolution and disease. Proc Natl Acad Sci USA 91:8739–8746.

Wallace DC, Singh G, Lott MT, Hodge JA, Schurr TG, Lezza AMS, Elsas LJ, Nikoskelainen EK (1988): Mitochondrial DNA mutation associated with Leber's hereditary optic neuropathy. Science 242:1427–1430.

Wei YH, Kao SH, Lee HC (1996a): Simultaneous increase of mitochondrial DNA deletions and lipid peroxidation in human aging. Ann NY Acad Sci 786:24–43.

Wei YH, Pang CY, You BJ, Lee HC (1996b): Tandem duplications and large-scale deletions of mitochondrial DNA are early molecular events of human aging process. Ann NY Acad Sci 786:82–101.

Wragg MA, Talbot CJ, Morris JC, Lendon CL, Goate AM (1995): No association found between Alzheimer's disease and a mitochondrial tRNA glutamine gene variant. Neurosci Lett 201:107–110.

Zhang C, Linnane AW, Nagley P (1993): Occurrence of a particular base substitution (3243 A to G) in mitochondrial DNA of tissues of ageing humans. Biochem Biophys Res Commun 195:1104–1110.

Aging and the Brain: A New Frontier

David A. Drachman, MD

With aging, both "normal" senescent age-related changes (ARCs) and late-onset diseases affect the brain, producing declines in performance. The brain as a postmitotic structure is particularly vulnerable to ARCs, and senescence is by far the most powerful risk factor for neurological diseases of the elderly such as sporadic Alzheimer's disease. The concept of senescence as an immutable result of the passage of time is yielding to understanding of the biology of ARCs. Both individual and species differences in longevity illustrate the variable effects of time. Whereas human life *expectancy* has been extended by prevention and treatment of specific diseases, life *span* can be altered by modifying the processes producing ARCs. Models of prolonged life span (eg, modifications of *Caenorhabditis elegans* longevity genes, restricted caloric intake) demonstrate the feasibility of extending longevity throughout the phylogenetic spectrum. Both programmed and variable factors produce ARCs. Cell survival depends on a balance of opposing factors—oncogene and anti-oncogene products, cyclins, growth factors, and so on; apoptotic death results when the balance shifts. Variable factors, including accumulation of oxygen free radicals, protein conformational changes, decline in chaperone functions, and secondary loss of mitochondrial energy production, can also result in neuronal degeneration. To prevent the increased neuronal vulnerability of senescence, ARCs must be modified. The "new frontier" in neurology is the challenge of understanding the changes of aging, both to determine their impact on disease and to prevent their consequences.

Drachman DA. Aging and the brain: a new frontier. Ann Neurol 1997;42:819–828

Aging and its effects on the brain are considered to be virtually inevitable. Twenty years ago, Bernard Strehler [1] noted that, with regard to aging of the brain, "deterioration in function will occur in any semiclosed system . . . [and is] simply one effect of the general law of entropy increase," a statement suggesting that a physical law predicts and governs the disorganization of brain structure and function over time. The brain (along with the heart and striated muscle) is the oldest part of the mammalian body, since neurons are postmitotic and those still present at the age of 85 years are remnants of the ones with which we were born. The teleological—perhaps survival—advantage of this neuronal persistence is the ability of the brain to preserve its memories and other learned functions; the disadvantage is that the brain and nervous system cannot be renewed, like the gut or skin. With the passage of time, and the varied intrinsic and extrinsic events that affect the brain, entropy increases—a fact that may explain why neurologic disorders are so important in the elderly.

Aging and *senescence* are often used as synonymous terms, but *aging* suggests a passive, essentially chronological process, whereas *senescence* refers to the age-related changes (ARCs) that occur in mature creatures over time, resulting in increased vulnerability to impairment and disease and in decreased survival [2]. Advances in understanding of the mechanisms of senescence now make it clear that these changes are the result of specific biological mechanisms, which are neither passive nor immutable.

Human survival curves and life-tables show that with advancing age there is an exponential increase in the death rate [3, 4]. While this is true for most living creatures, it is not true for all, as discussed later. It is also evident that ARCs—if not the direct cause of many of the neurological disorders of the elderly—are important (often the *most* important) predisposing factors contributing to these conditions. For these reasons, the process of senescence, its effects on the nervous system, its biological mechanisms, the means of modification, and strategies for mitigating its impact are crucial areas for neurologists to understand and investigate, but they represent an as yet little-explored frontier of clinical neuroscience.

Measures of Aging and Longevity

Life expectancy and life span are often confused, leading to a misconception of how each could be improved or lengthened. *Life expectancy* [5] is the average number of years of life remaining for a population of a given age, and up to the present time has been increased primarily by the treatment or elimination of diseases. In the United States at birth life expectancy is about 75 years—just under 73 years for men and more than 79 years for women. At age 65 life expectancy is 17 years,

From the Department of Neurology, University of Massachusetts Medical Center, Worcester, MA.

Received Apr 25, 1997, and in revised form Aug 20. Accepted for publication Aug 22, 1997.

Address correspondence to Dr Drachman, Department of Neurology, University of Massachusetts Medical Center, 55 Lake Avenue North, Worcester, MA 01655.

and then decreases sharply with each passing quinquennium. *Life span* is the genetically endowed usual limit to life for an individual with no special risk factors for disease. The human life span is about 85 years; to increase it, the biological mechanisms underlying senescent ARCs would have to be modified. The ***maximum life span potential***, however, is considerably longer and is determined as the age of the verified longest-lived individual—122 years for humans. Finally, ***active life expectancy*** [4] defines the number of years of independent life remaining at any age, assuming the usual age-related risks of disability. As life expectancy increases, it approaches the life span, and with this "squaring of the survival curve" disability and disease increase. At the age of 65 years, two thirds of a person's remaining years will be active, but at 85 years of age, only about a third of the remaining time, on average, will be independent [4]. A remarkable proportion of the disability of the elderly is the result of neurological problems: half of all the disabled elderly have a neurological cause for their disability, and of those who are totally incapacitated, over 90% are limited by neurological disease [6, 7].

Neurological Changes in the Elderly

Both the structure and the function of the brain are affected in the elderly. By the age of 80 years, the brain loses about 10% of its weight and volume [8], with shrinkage evident on imaging studies. Neurons are lost selectively—nearly 50% from the substantia nigra [9] and perhaps 25% of those in the mesial temporal region [10] (although the extent of losses in the hippocampus is disputed [11, 12], whereas neurons in many of the brainstem nuclei are preserved to advanced old age. Large cortical neurons shrink in size as well as in numbers. Synapses decrease slightly but not as much in normal aging as in Alzheimer's disease [13, 14]. Neuritic plaques and neurofibrillary tangles increase in number with advancing age [15], and some appear in the temporal lobes and other cortical areas in more than half of cognitively normal people over the age of 65, with large numbers of plaques, tangles, or both seen at autopsy in more than 35% [16]. It is unclear whether these changes would have led to Alzheimer's disease had the subjects survived a few more years [17].

Clinically, two changes are nearly universal in the elderly: a decline in the ability to learn new information ("memory storage") [18–20], and a decline in reaction time [21, 22]. The terms *benign senescent forgetfulness* [23] and *age-associated memory impairment* [24] have been applied to the decrease in memory, often manifested by frequent complaints about difficulty remembering names and misplacing of common objects. Although immediate memory span (what can be remembered after single hearing) is unchanged, an extensive literature documents the decreased capacity of the elderly to store information that exceeds immediate memory span [19, 20]. Similarly, simple reaction time (eg, finger tapping, following a pursuit rotor) is minimally decreased, and complex reaction time [25] (handwriting, video games) shows consistent and significant decline with age. Performance on intelligence tests, such as the Wechsler Adult Intelligence Scale declines significantly with age, but IQ scores are age corrected to compensate for these changes. To achieve an IQ of 100 at age 75, for example, one need get only half as many items correct as at age 21 [26]. Not all aspects of intelligence decline equally, however; "crystallized intelligence" (eg, vocabulary) may remain unchanged into advanced age, whereas "fluid intelligence" (eg, digit-symbol substitution) undergoes significant decline [27]. Despite these age-related changes in function, age accounts for only about one-fourth of the measurable variability in cognitive testing among groups of individuals, and many intelligent old people perform better than less well-endowed younger ones. Studies of creativity and achievements show that in most fields peak productivity occurs in the 30's [28] despite the occasional Picasso or Grandma Moses who paints into the tenth decade of life. Although social or motivational theories are thought by some to explain the decrease in productivity in the elderly, there is biological evidence that, with advancing age, activity levels decline in many creatures other than humans [29].

Other neurosensory and neuromuscular changes occurring in aged people include visual decline with presbyopia and decreased contrast sensitivity, hearing loss, diminished sense of smell and taste, decline of muscle mass, a stooped, slowed near-parkinsonian gait [30]; diminished ankle jerks, and decreased peripheral sensation. The amplitudes of sensory nerve action potentials and, to a lesser degree, compound muscle action potentials decrease with age, and the electroencephalographic background alpha rhythm slows. Neuroimaging studies reveal a moderate degree of brain shrinkage; quantitative microscopy suggests that the brain volume lost in normal aging is proportionally greater than the loss of neuronal cell bodies (Lippa C, Drachman D, Ragland R, unpublished observations, 1997). Finally, most of the supporting systems, on which the brain depends, also decline, including cardiac index, vital capacity, renal function, and immune competence [31].

The neurological changes that occur with age have been ascribed to alterations occurring at several different levels: *"system" impairments* (eg, hippocampal complexes, basal ganglia, cholinergic system); *neuronal impairments* (eg, synaptic losses, slowed axoplasmic flow, decreased plasticity); and *cellular/molecular mechanisms* (eg, apoptosis, oxygen free radical damage, protein conformational changes). Specific clinical manifestations, such as impaired memory or a gait disorder, are often most directly explained at a system level and result from identifiable neuronal losses in an anatomical system. But the ultimate causes of the neuronal losses underlying these neurological changes occur at the level of cellular and molecular mechanisms; these mechanisms can be modified, and it is here that recent progress promises powerful new tools for the future.

Senescence as a Risk Factor for Neurological Disease

The changes described earlier—in memory, vision, gait, and so forth—although nearly universal, are not the most important neurological problems of aging. Instead, the fact that the *ARCs of senescence are the most powerful risk factors for many degenerative neurological diseases* far outweighs the significance of the relatively modest functional losses occurring with advancing age. Several examples will help to illustrate this point:

Alzheimer's disease. The discovery by Roses and his colleagues that the ε4 allele of the apolipoprotein E (ApoE) gene is a major risk factor for Alzheimer's disease may provide an important insight into the disease mechanism and eventually perhaps a tool to delay or prevent this condition [32]. Having an ApoE ε4 allele, rather than an ApoE ε2 or ApoE ε3, increases the lifetime risk of Alzheimer's disease by about four-fold. By contrast, surviving to 85 years of age, rather than 65, increases the lifetime risk of Alzheimer's disease by 20- to 30-fold [33]. Nor is this merely a cumulative risk: at age 85 the *annual* risk of development of Alzheimer's disease is 20 times greater than at age 65.

Head trauma, aneurysmal subarachnoid hemorrhage. With advancing age, the survival and neurological recovery from serious head trauma dramatically declines. Jennett and Teasdale noted that "age is the most important factor influencing recovery after severe head injury" [34]. Whereas loss of consciousness for less than an hour, without focal neurological deficit, is usually benign in young adults, it is likely to result in permanent cognitive impairment in those over 60 years of age. Jane and colleagues [35] have similarly shown that both the immediate clinical effects and the outcome of aneurysmal subarachnoid hemorrhage worsen with age, with greater impairment of consciousness, size of clots, occurrence of hydrocephalus and probability of rebleeding, and worse overall outcome on the Glasgow Outcome Scale [35].

Brain tumors. The clinical effects of brain tumors are highly dependent on age. Of patients with dominant hemisphere primary brain tumors, over 75% of those more than 50 years of age presented with aphasia, whereas less than 10% of patients under 50 were aphasic—independent of tumor histology, precise location or size [36].

These examples—only three of many—illustrate the importance of senescence in the increased vulnerability of the brain both to degenerative disease and to intercurrent insults. To understand *why* the aging brain is more vulnerable we must examine the nature of the aging process itself.

The Variability of Life Expectancy and Life Span

Life expectancy and life span vary considerably both within a species, and to a greater extent, between species.

Within a given species there is approximately a 50% variation in longevity among individuals who live out their "natural" life expectancy, an observation illustrated daily by a glance at a newspaper's obituary page. This difference is not purely chance, since women live, on average, nearly 10% longer than men. This robust sex difference is found in many cultures throughout the world, and similar sex differences in longevity are found in many species [3]. The reason for female longevity is unknown but may be X-linked or related to hormonal effects or other factors. In the population over the age of 65 years, there are 1.5 times as many women as men, and the predominance of women increases further with advancing age. This increased survival of women explains why there are many more widows than widowers and is reflected in nursing homes, where there are twice as many women as men.

The *interspecies* difference in life span is even more striking but is a fact of nature that we often take for granted [37]. At 85 years, humans have one of the longest life spans, while the seagull has a span of 50 years, the domestic mouse less than 4 years [38], and the fruit fly about 6 weeks. These remarkable differences in life span have been variably attributed directly to body size, brain size, and logarithm of brain weight and inversely to heart rate and metabolic rate, but there are many exceptions to these relatively simple rules. The tiny 20-g brown bat has a life span of 20 years—considerably longer than that of the mouse or the larger rat. Some creatures are *semelparous*, indicating that they die rapidly after spawning; the best known of these are the Pacific salmon, who die after their second year of life if they go upstream to spawn or, if they wait another season, after their third year [3]. The immediate mechanism resulting in the salmons' death is believed to be a tremendous release of glucocorticoids, although the biological explanations for the lack of a homeostatic mechanism to prevent their abrupt demise remains obscure.

There is no creditable evidence that natural selection plays a direct role in choosing members of a species that die at an early age, especially since selection and mating take place before individual longevity is determined (ie, by the time of death) [39]. The theory of "group selection" for short life spans, suggesting that the group does better if the elderly members die off earlier, is universally rejected by evolutionary biologists. It is possible, however, that in some species ***antagonistic pleiotropy*** favors selection of those genes that promote survival and vigor in young creatures but that inadvertently decrease longevity in the elderly [40].

A few creatures do not undergo senescence over time. The giant sequoias and bristlecone pine trees live for hundreds or thousands of years but are inanimate. Some animals show similar characteristics in that while they are not immortal, they have indeterminate longevity [3] and do not show the usual senescent changes. In these creatures the likelihood of dying does not increase with each passing year, and the females continue to ovulate and reproduce indefinitely in some species [3]. The Pacific rockfish, sturgeon, and some giant tortoises are creatures of this sort.

The remarkable interspecies differences in life spans across the animal kingdom, and the fact that some creatures do not undergo senescence, suggest that aging is not an inevitable wearing out that must be expected over time.

One keystone of aging research has been *Gompertz's law*, named for an English actuary who first proposed it over 170 years ago, which states that *mortality rates increase exponentially with age*. Until recently, this law had seemed unassailable, although it had not been tested with very old creatures. Recently, Vaupel and colleagues [41] examined the validity of Gompertz's law by studying a million medflies. As predicted, the mortality rate increased exponentially for the first 7 to 8 weeks and peaked with 15% of the remaining population dying each day. Beyond 8 weeks, however, the mortality rate *fell*, dropping to 4% per day. This strikingly demonstrated that certain of the oldest flies were, in fact, more fit to survive than the younger ones. Besides showing that Gompertz's law is not invariable, the observation suggested the need to elucidate the biological basis of the differences that permitted the cohort of older flies to survive not only longer but with a different distribution of mortality.

Prolongation of Life Span

An increasing number of studies now demonstrate the possibility of prolonging life span. One of the best-explained studies of extended life span was performed with aged tomatoes at the University of California at Berkeley [42]. Senescence in tomatoes is characterized by ripening and turning red, a process requiring exposure to ethylene, a chemical intrinsically produced in aging tomatoes, catalyzed by the enzyme ACC synthase. Genetic modification of tomato plants by introducing antisense RNA to ACC synthase produced tomatoes that remained "youthful"—green and unripened—indefinitely. When these tomatoes were exposed to extrinsic ethylene in the atmosphere they ripened (senesced) normally. Although this is a model remote from human senescence, it provides a useful insight into aging mechanisms that exist in nature.

Life extension in the roundworm *Caenorhabditis elegans* has been extensively studied with transgenic mutants of several groups of genes but especially alleles of the *daf*, *clk*, and *age* groups. *Daf* genes are related to the stress-induced "dauer" larval form; *clk* genes are thought to regulate metabolic rate; and *age-1* may increase resistance to free oxygen radical species and expression of heat shock proteins [43–45]. With various transgenic combinations of these alleles, *C. elegans* adults can live up to five times longer than the wild-type life span of about 9 days. These studies provide further insights into the mechanisms of aging and the life span, pointing to the effects of metabolic rate, accumulation of oxygen free radicals, mitochondrial damage, and conformational change in proteins as key—and modifiable—factors in controlling life span.

A number of studies of higher animals have demonstrated the ability to prolong not only life expectancy but also life span. The prolongation of life *expectancy* can, of course, be accomplished by the successful prevention or treatment of diseases and the elimination of predators. The remarkable extension of human life expectancy during the 20th century is largely the result of public health measures and the successful treatment of many infectious diseases, but life span has not been appreciably increased.

In animal studies, restricted feeding has been the most reliable means of extending life span [46, 47]. Laboratory rats fed a diet restricted to 60% of that eaten ad lib by controls live 40 to 50% longer and suffer fewer malignancies or other diseases [48]. Fibroblasts from calorie-restricted old rats, when grown in tissue culture, divide, synthesize protein [49], and are able to respond to thermal stress (heat shock) like cells from young animals.

Low-dose ionizing radiation has also been shown to prolong life span—a phenomenon called *hormesis* [50]. A similar effect has been shown in humans living on the outskirts of Nagasaki: men exposed to low doses of radiation had lower mortality rates for noncancer diseases than did unexposed controls [51].

These observations, demonstrating both the variability and the modifiability of longevity in nature, provide evidence that senescence is neither a fixed nor a passive process.

Causes of Senescence

There are two groups of theories about the basis of aging and senescence—those that propose a specific program for aging and death, and those that regard senescence to be the result of multiple random events.

Programmed Aging and Death

Until 35 years ago, it was generally believed that cells in tissue culture were immortal. In the early 1960s, Hayflick and Moorhead [52] carried out rigorous studies demonstrating that, to the contrary, cultured cells could undergo only a limited number of divisions; they then underwent cell growth arrest at the G_1–S boundary, became dormant, and eventually died [53–55]. Hayflick and his colleagues observed that human fetal fibroblasts could divide about 50 times, whereas adult-derived cells could divide only 20 to 30 times, a number that decreased with advancing age. Cells from long-lived animals were able to divide more times than those from short-lived animals. The capacity to divide, they found, resides in the cellular nucleus rather than the cytoplasm: when the nuclei of young cells were transplanted into old cell bodies, they could divide many times, whereas old nuclei transplanted into young cell bodies had a limited capacity to divide. These landmark cell-cycle studies proved that cells undergo senescence in culture and provided a valuable experimental system with which to measure the cellular and molecular events in aging and to evaluate factors affecting survival.

Hayflick's studies raised the question of how cells "keep count" of the number of divisions they have undergone. An important mechanism is related to the telomeres, or chromosomal tails, that are found on each end of every chromosome and that consist of hundreds of TTAGGG repeat sequences [56, 57]. Each time a cell divides, it loses part of its telomere tails; when the telomeres become too short, the cell stops dividing. Telomerase is an enzyme that, although present in normal cells, is inactive except in germinal cells; it allows cells to regenerate their telomeres. Cells that are "immortalized" in culture and cancer cells also have activated telomerase, enabling them to replace their telomeres after division and to continue dividing. The understanding of telomeric shortening helps to explain the mechanism of the "Hayflick phenomenon" of limited cellular division in culture.

The third programmed mechanism of aging—*apoptosis*—is the one most likely relevant to neuronal aging and death [58–62]. This is an intrinsically (ie, genetically) coded and regulated process that leads to cell death by means of a predetermined degenerative pathway, in contrast to cellular *necrosis*. It is characterized by a lack of inflammatory changes, bullae appearing in apoptotic cells, and eventually segmentation of the nuclear DNA into small pieces to produce a "laddered" appearance on electrophoretic gels. The programmed cellular degeneration can be triggered by many different mechanisms, such as radiation, withdrawal of necessary growth factors, or aging itself. Apoptotic cell death occurs when the balance of opposing factors that promote or prevent cell survival or division tilts sufficiently to invoke the cell death sequence. In the regulation of cell survival and division, oncogene proteins (eg, Bcl-2, p35) promote survival and division, and anti-oncogene proteins (eg, Rb, p53, ICE) oppose cell division; the different cyclins vary in their roles [59]. Certain viruses (eg, SV-40, adenovirus type 5) are able to "immortalize" cells in culture, so they may continue to divide without limit: these viruses produce immortality by binding the p53 and Rb gene products, preventing the initiation of the apoptotic program and, in addition, activating telomerase.

A vast array of opposing factors regulates the survival, growth, and division of cells [61–64]. In young cells, these opposing factors appear to be strongly arrayed in a stable detente enabling the cells to survive, neither on the brink of death nor en route to the uncontrolled division of malignant change. As cells age, however, the regulatory control is weakened: the cells become metastable and may tip more easily toward either growth or death [60, 65]. Although neurons rarely develop malignant change, control of their survival, and support of their vital functions, is believed to be regulated in a manner similar to that of dividing cells.

Extrinsic Causes of Cell Death

Random events are also thought to play an important role in the aging process. In the 1950s, Leo Szilard first suggested the ***somatic mutation*** theory of aging, in which random mutations of DNA accumulate over time and eventually cause the degenerative changes of aging [66]. Later, Orgel [67] proposed his ***error catastrophe*** theory, suggesting that errors occurring in critical proteins, such as key enzymes, might start a cascade of increasingly disruptive biochemical mistakes, resulting in senescence and eventually death. Neither form of this random mutation theory has found experimental support, however [68, 69], and chemically abnormal proteins are found only occasionally in aging cell cultures and rarely in vivo.

Although genetic mutations producing chemical changes in important proteins are not consequential in aging, ***conformational changes*** may be [68, 70–72]. These posttranslational third-order structural changes are due to misfolding of proteins or alteration of their normal, functional, shapes [71, 73, 74]. In support of this idea, it is known that about 0.1% of the body's proteins undergo dextrorotation each year. These right-handed conformations are largely nonfunctional. The conformation of other proteins may be altered by a wide variety of processes, including initial misfolding, covalent changes, phosphorylation, glycation and cross-linking [70, 74]. The chaperone systems (eg, the heat shock proteins [HSPs] of the HSP-70 group) normally aid in the initial folding of proteins and recognition of conformationally altered proteins and may refold or aid in the removal of abnormal proteins. The function of HSPs declines significantly with age [75], related in part to the energy requirements of the process. At present, the only neurological disorder believed to be directly caused by conformational abnormalities is Creutzfeld-Jakob disease, in which the normal C form of the prion protein is conformationally (but not chemically) altered to the pathological Sc form [76, 77]. At present, the precise role of conformational protein changes in senescence is uncertain, but the possibility that other degenerative neurological disorders such as Alzheimer's disease may be the result of post-translational conformational changes in neural proteins remains.

Another important mechanism that may explain the ARCs of senescence is the damage induced by *oxygen free radicals* [39, 78–82]. As a byproduct of oxidative metabolism and normal energy production, oxygen alone or in combined forms (hydroxyl radicals, peroxide radicals) may be generated with an unpaired extra electron. These oxygen free radicals seek opportunities to donate the unpaired electron, damaging recipient DNA, proteins, or lipids. The mitochondrial DNA is especially vulnerable to free radical damage, since it both is proximate to the site of oxidative energy production and lacks the repair capabilities of nuclear DNA [78, 83, 84]. This initiates a vicious cycle: impaired mitochondrial function results in decreased energy production with its consequences—calcium influx into cells, excitotoxic damage, failure of chaperone repair mechanisms, and so forth [78, 79, 83, 85–87]. It is not clear whether oxygen free radicals are responsible for neural senescence or whether they play a direct role in causing parkinsonism, Alzheimer's disease, amyotrophic lateral sclerosis, and other disorders of the central nervous system that develop with advancing years [78, 80, 87–91].

In the experimental setting, the brains of aged gerbils are found to contain twice the amount of oxidized protein as those of young gerbils as well as diminished amounts of two enzymes (neutral protease and glutamine synthetase) helpful in disposing of oxidized products [92]. Older gerbils make many more errors than young adult gerbils in running radial arm mazes, but both the behavioral impairment and the biochemical abnormalities are reversed by *N*-tert-butyl-α-phenylnitrone (PBN), a powerful "spin-trapping" compound that reacts with and removes free radicals [90, 92].

Many other mechanisms have been proposed to explain how aging affects the brain—some of the ARC mechanisms may initiate neural degeneration; some amplify degeneration; some permit it to occur. Glycation [39] or cross-linking of proteins and lack of nerve growth factors [93] may initiate degenerative changes; excitotoxicity may amplify degeneration [78, 87]; and hormonal lack [94] (eg, dihydroepiandrosterone, human growth hormone), immune failure [95], and impairment of support systems [18] may permit degeneration to occur.

Strategies to Alter Brain Senescence

Methods of preventing the onset of brain degenerative changes that occur with age—or of slowing their evolution—are a crucial concern of neurologists and neuroscientists. Some strategies are already available, whereas others remain possibilities for the future.

Current Methods of Delaying Brain Aging

The influence of genetics both on longevity and brain degeneration is well established. To some extent, longevity can be inherited [55], although in humans no single gene, group of genes, or mode of inheritance has been identified. The Total Immediate Ancestral Longevity (TIAL) index, calculated by adding the ages at death of both parents and all four grandparents, provides some prediction of longevity [96]. Ninety- to 100-year-olds have a TIAL index that is 16% above the mean, and seven of eight 90-year-olds have at least one parent who lived past 70. The role of genetic factors (eg, ApoE ε4 [32, 97], amyloid precursor protein gene mutations, presenilin 1 and 2 [98, 99]) in causing or accelerating the occurrence of Alzheimer's disease changes has provided important clues to the nature of that disease. It has been facetiously suggested that "choosing one's parents wisely" is necessary to ensure a long life with successful aging.

Experimental evidence has firmly established the dramatic extension of life in animals maintained on a restricted-calorie diet [100, 101]. Rats, mice, fruitflies, and spiders all evidence the prolongation of longevity already described. Even in vitro, cells from underfed old animals function like cells from young animals, with greater doubling ability, energy production, heat shock response, and lower production of free radicals. A longitudinal study of nurses showed that thin women lived longer and had fewer diseases than heavy ones [102]. The mechanism of this phenomenon is not yet clear, although it may be the result of reduced free radicals. It is not known, however, whether restricted-calorie diets delay senescence of the brain or alter brain function either positively or negatively [103].

There is extensive documentation of declines in cognitive functions with age, including memory and learning, complex reaction time, channel capacity, and other functions [19, 22, 24, 103–107]. Aerobic exercise has been shown to produce modest improvements of performance on a number of cognitive tests, whereas strengthening exercises (eg, weight lifting) and stretching exercises have little or no effect [108]. Similarly, consistent exercise produces a modest increase in longevity. The mechanism of these beneficial effects of physical exercise is unclear.

Antioxidants such as vitamins E and C are already widely used, although the evidence that these readily available dietary supplements modify brain aging is not available. Far more powerful antioxidants, such as the spin traps (eg, PBN), have been studied in experimental animals [90, 92, 109]. Mice given PBN survive about one-third longer than controls, and old gerbils given PBN improve performance on a radial arm maze to a level comparable with young gerbils [92]. In tissue culture, cells grown with PBN are able to divide more times than untreated cells. No human studies of PBN have yet been published, however.

Future Possibilities for Modifying Brain Aging

This review points up the importance of aging and senescence in both the decline of cognitive function in the elderly and the vulnerability of the aging nervous system to degenerative diseases. It also suggests numerous possibilities for the modification of brain senescence and the prevention of age-associated brain disorders. The possibilities include repairing conformational alterations in crucial brain proteins, resetting regulatory control of cell-cycle factors, preventing apoptosis by providing critical growth factors or influencing oncogenes and antioncogenes, using highly effective antioxidants, transplanting brain tissue [110, 111], developing new neurons from neural stem cells [112], substituting for endocrine changes [94, 113], and directly treating Alzheimer's disease through growing knowledge of the roles of amyloid precursor protein, ApoE ε4, presenilins, and so on.

Conclusions

Neurologists and neuroscientists have focused attention and research on degenerative neurological *diseases*, which have a predilection for the aging nervous system. Important advances have been made in the understanding of the specific genetic, molecular, biochemical, pathological, physiological, and clinical changes that occur in Alzheimer's disease, Parkinson's disease, amyotrophic lateral sclerosis, and other disorders. This disease-specific information will certainly identify important steps in these conditions that may be amenable to therapies.

Less attention has been focused on the implications of the dramatic acceleration in the occurrence of these conditions with advancing age and the ARCs of senescence. Fortunately, we are now at a point where the scientific bases of these aging changes—the causes of increasing neuronal vulnerability—have become approachable. The various intrinsic and extrinsic aging events that accumulate over time likely interact to produce or contribute to the familiar patterns of "normal aging," dementia, movement disorders, sensory impairments, falls, and other neurological disorders of the elderly. Neuronal apoptosis, protein conformational changes, the insults of reactive oxygen species, loss of mitochondrial energy production, accumulation of β-amyloid and cerebrovascular lipids, and the array of genetically exaggerated vulnerabilities of aging can be analyzed and modified, singly and conjointly.

The opportunity to "square the survival curve" for the aging brain is an important new frontier in neurological science. The search for the unique causes of specific degenerative diseases has been successful to a point. Strategies must now shift to protecting and defending the postmitotic and vulnerable aging brain from the varied insults and changes that accumulate over time, resulting in the degenerative brain disorders of aging.

This study was supported in part by NIA grant #P50 AG05134, the Sterling Morton Neurological Research Fund, and the Stanley and Harriett Friedman Research Fund.

Presented in part as the Presidential Address at the 120th Annual Meeting of the American Neurological Association, Washington, DC, October 23, 1995.

References

1. Strehler B. Introduction: aging and the human brain. In: Terry R, Gershon S, eds. Neurobiology of aging. New York: Raven Press, 1976:1–22
2. Masoro E. Aging: current concepts. In: Masoro E, ed. Handbook of physiology: aging. New York: Oxford University Press, 1995:3–24
3. Finch C. Longevity, senescence and the genome. Chicago: University of Chicago Press, 1990
4. Cassel C, Brody J. Demography, epidemiology and aging. In: Cassel C, Riesenberg D, Sorensen L, Walsh J, eds. Geriatric medicine. 2nd ed. New York: Springer-Verlag, 1990:16–27
5. Olshansky S, Carnes B, Cassel C. In search of Methuselah: estimating the upper limits to human longevity. Science 1990; 250:634–640
6. Akhtar A, Broe D, Crombie A, et al. Disability and dependence in the elderly at home. Age Aging 1973;2:102–110
7. Broe G, Akhtar A, Andrews G, et al. Neurologic disorders in the elderly at home. J Neurol Neurosurg Psychiat 1976;39: 362–366
8. Tomlinson B, Corsellis J. Ageing and the dementias. In: Adams J, Corsellis J, Duchen L, eds. Greenfield's neuropathology. 4th ed. New York: John Wiley & Sons, 1984:951–1025
9. DeKeyser J, Ebinger G, Vauquelin G. Age-related changes in the human nigrostriatal dopaminergic system. Ann Neurol 1990;27:157–161
10. Coleman P, Flood D. Neuron numbers and dendritic extent in normal aging and Alzheimer's disease. Neurobiol Aging 1987;8:521–545
11. West MJ. Regionally specific loss of neurons in the aging human hippocampus. Neurobiol Aging 1993;14:287–293
12. Wickelgren I. Is hippocampal cell death a myth? Science 1996;271:1229–1230
13. Hamos J, DeGennaro L, Drachman D. Synaptic loss in Alzheimer's disease and other dementias. Neurology 1989;39: 58–61

14. Lippa CF, Hamos JE, Pulaski-Salo D, et al. Alzheimer's disease and aging: effects on perforant pathway perikarya and synapses. Neurobiol Aging 1992;13:405–411
15. Katzman R, Terry R, DeTeresa R. Clinical, pathological, and neurochemical changes in dementia: a subgroup with preserved mental status and numerous neocortical plaques. Ann Neurol 1988;23:138–144
16. Miller F, Hicks S, D'Amato C, Landis J. A descriptive study of neuritic plaques and neurofibrillary tangles in an autopsy population. Am J Epidemiol 1984;120:331–341
17. Ulrich J. Alzheimer changes in nondemented patients younger than sixty-five: possible early stages of Alzheimer's disease and senile dementia of the Alzheimer type. Ann Neurol 1985;17:273–277
18. Drachman D, Long R, Swearer J. Neurological examination of the elderly patient. In: Albert M, Knoefel J, eds. Clinical neurology of aging. 2nd ed. New York: Oxford University Press, 1994:159–180
19. Drachman D. Memory and cognitive function in normal aging. Dev Neuropsychol 1986;2:277–285
20. Smith A. Memory. In: Birren J, Schaie K, eds. Handbook of the psychology of aging. 4th ed. San Diego: Academic Press, 1996:236–250
21. Salthouse T. Speed of behavior and its implications for cognition. In: Birren J, Schaie K, eds. Handbook of the psychology of aging. 2nd ed. New York: Van Nostrand Reinhart, 1985:400–426
22. Salthouse T. The nature of the influence of speed on adult age differences in cognition. Dev Psychol 1994;30:240–259
23. Branconnier RJ, Cole JO, Spera KF, DeVitt DR. Does "benign senescent forgetfulness" exist? Recall and recognition as diagnostic indices of malignant memory loss in senile dementia: a Bayesian analysis. Clin Geriatr Med 1988;4:897–916
24. Larrabee GJ, McEntee WJ. Prevalence of age-associated memory impairment in a randomly selected population from eastern Finland. Age-associated memory impairment: sorting out the controversies. Neurology 1995;45:741–747 (Editorial)
25. Salthouse TA. Processing speed as a mental capacity: influence of processing speed on adult age differences in working memory. Acta Psychol 1994;86:199–225
26. Wechsler D. The Wechsler Adult Intelligence Scale—revised manual. New York: Psychological Corporation, 1981
27. Horn JL. Psychometric studies of aging and intelligence. In: Gershon S, Raskin A, eds. Aging. New York: Raven Press, 1975:19–43
28. Lehman H. Age and achievement. Princeton: Princeton University Press, 1953
29. Gutmann E, Hanzlikova V. Age changes in the neuromuscular system. Bristol: Scientechnica, 1972
30. Larish D, Martin P, Mungiole E. Characteristic patterns of gait in the healthy old. Ann NY Acad Sci 1988;515:18–32
31. Shock N. The science of gerontology. In: Jeffers E, ed. Proceedings of seminars. Durham: Duke University Press, 1962
32. Saunders A, Strittmatter W, Schmechel D, et al. Association of apolipoprotein E allele $\varepsilon 4$ with late-onset familial and sporadic Alzheimer's disease. Neurology 1993;43:1467–1472
33. Seshadri S, Drachman D, Lippa C. Apolipoprotein E $\varepsilon 4$ allele and the lifetime risk of Alzheimer's disease: what physicians know, and what they should know. Arch Neurol 1995;52:1074–1080
34. Jennett B, Teasdale G. Management of head injuries. Philadelphia: FA Davis, 1981:321–322
35. Lanzino G, Kassell N, Germanson T, et al. Age and outcome after aneurysmal subarachnoid hemorrhage: why do older patients fare worse? J Neurosurg 1996;85:410–418
36. Recht LD, McCarthy K, O'Donnell BF, et al. Tumor-associated aphasia in left hemisphere primary brain tumors: the importance of age and tumor grade. Neurology 1989;39:48–50
37. Finch C. New models for new perspectives in the biology of senescence. Neurobiol Aging 1991;12:625–634
38. Finch C, Pike M, Witten M. Slow mortality rate accelerations during aging in some animals approximate that of humans. Science 1990;249:902–905
39. Masoro E. Biology of aging: facts, thoughts, and experimental approaches. Lab Invest 1991;65:500–510
40. Rose M. Evolutionary biology of aging. New York: Oxford University Press, 1991
41. Carey J, Liedo P, Orozco E, Vaupel J. Slowing of mortality rates at older ages in large medfly cohorts. Science 1992;258:457–461
42. Oeller P, Min-Wong L, Taylor L, et al. Reversible inhibition of tomato fruit senescence by antisense RNA. Science 1991;254:437–439
43. Larsen P, Albert P, Riddle D. Genes that regulate both development and longevity in *Caenorhabditis elegans*. Genetics 1995;139:1567–1583
44. Lakowski B, Hekimi S. Determination of life-span in *Caenorhabditis elegans* by four clock genes. Science 1996;272:1010–1013
45. Lithgow G. Molecular genetics of *Caenorhabditis elegans* aging. In: Schneider E, Rowe J, eds. Handbook of the biology of aging. 4th ed. San Diego: Academic Press, 1995:55–73
46. McCay C, Crowell M, Maynard L. The effects of retarded growth upon the length of life span and upon the ultimate body size. J Nutr 1935;10:63–79
47. Sprott R, Austad S. Animal models for aging research. In: Schneider E, Rowe J, eds. Handbook of the biology of aging. 4th ed. San Diego: Academic Press, 1996:3–23
48. Weindruch R, Walford R, Fliegel S, Guthrie D. The retardation of aging in mice by dietary restriction: longevity, cancer, immunity and lifetime energy intake. J Nutr 1986;116:641–654
49. VanRemmen H, Ward W, Sabia R, Richardson A. Gene expression and protein degradation. In: Masoro E, ed. Aging. New York: Oxford University Press, 1995:171–234
50. Johnson T, Hartman P. Radiation effects on life span in *Caenorhabditis elegans*. J Gerontol Biol Sci 1988;43:B137–141
51. Mine M, Okumura Y, Ichimaru M, et al. Apparently beneficial effect of low to intermediate doses of A-bomb radiation on human lifespan. Int J Radiat Biol 1990;58:1035–1043
52. Hayflick L, Moorhead P. The serial cultivation of human diploid cell strains. Exp Cell Res 1961;25:585–621
53. Hayflick L. The limited in vitro lifetime of human diploid cell strains. Exp Cell Res 1965;37:614–636
54. Hayflick L. The cell biology of human aging. Sci Am 1980;242:58–65
55. Hayflick L. How and why we age. New York: Ballantine Books, 1994
56. Blasco M, Funk W, Villeponteau B, Greider C. Functional characterization and developmental regulation of mouse telomerase RNA. Science 1995;269:1267–1270
57. Smith J, Pereira-Smith O. Replicative senescence: implications for in vivo aging and tumor suppression. Science 1996;273:63–67
58. Majno G, Joris I. Apoptosis, oncosis, and necrosis: an overview of cell death. Am J Pathol 1995;146:3–15
59. Monti D, Grassilli E, Troiano L, et al. Senescence, immortalization, and apoptosis. In: Fabris N, Harman D, Knook D, et al, eds. Physiopathological processes of aging. New York: New York Academy of Sciences, 1992:70–82
60. Adams J, Mukherjee S, Klaidman L, et al. Apoptosis and oxidative stress in the aging brain. In: Kitani K, Aoba A, Goto S,

eds. Pharmacological intervention in aging and age-associated disorders. New York: New York Academy of Sciences. 1996: 135–151

61. Steller H. Mechanisms and genes of cellular suicide. Science 1995;267:1445–1462
62. Rubin L. Gatchalian C. Rimon G. Brooks S. The molecular mechanisms of neuronal apoptosis. Curr Opin Neurobiol 1994;4:696–702
63. Bump N. Hackett M. Hugunin M. et al. Inhibition of ICE family proteases by baculovirus antiapoptotic protein p35. Science 1995;269:1885–1888
64. Davies A. The Bcl-2 family of proteins, and the regulation of neuronal survival. Trends Neurosci 1993:18:355–358
65. Smale G. Nichols NR. Brady DR. et al. Evidence for apoptotic cell death in Alzheimer's disease. Exp Neurol 1995;133: 225–230
66. Szilard L. On the nature of the aging process. Proc Natl Acad Sci USA 1959:45:30–45
67. Orgel L. The maintenance of the accuracy of protein synthesis and its relevance to ageing. Proc Natl Acad Sci USA 1963:49: 517–521
68. Danner D. Holbrook N. Alterations in gene expression with aging. In: Schneider E. Rowe J. eds. Handbook of the biology of aging. 3rd ed. San Diego: Academic Press. 1990:97–115
69. Norwood T. Smith J. Stein G. Aging at the cellular level: the human fibroblastlike cell model. In: Schneider E. Rowe J. eds. Handbook of the Biology of aging. 3rd ed. San Diego: Academic Press. 1990:131–154
70. Gafni A. Structural modifications of proteins during aging. J Am Geriatr Soc 1997;45:871–880
71. Taubes G. Misfolding the way to disease. Science 1996;271: 1493–1495
72. Johnson T, Lithgow G. Murakami S. et al. Genetics of aging and longevity in lower organisms. In: Holbrook N, Martin G. Lockshin R, eds. Cellular aging and cell death. New York: Wiley-Liss, 1996:1–17
73. Levine R, Stadtman E. Protein modifications with aging. In: Schneider E, Rowe J. eds. Handbook of the biology of aging. 4th ed. San Diego: Academic Press. 1996:184–197
74. Richardson A, Holbrook N. Aging and the cellular response to stress: reduction in the heat shock response. In: Holbrook N, Martin G, Lockshin R. eds. Cellular aging and cell death. New York: Wiley-Liss, 1996:67–80
75. Wu B, Gu M, Heydari A. Richardson A. The effect of age on the synthesis of two heat shock proteins in the HSP70 family. J Gerontol Biol Sci 1993:48:B50–B56
76. Beyreuther K, Masters C. Catching the culprit prion. Nature 1994;370:419–420
77. Prusiner S, Hsiao K. Human prion diseases. Ann Neurol 1994;35:385–395
78. Beal M. Aging, energy, and oxidative stress in neurodegenerative diseases. Ann Neurol 1995;38:357–366
79. Harman D. Role of free radicals in aging and disease. In: Fabris N, Harman D, Knook D, et al, eds. Physiopathological processes of aging. New York: New York Academy of Sciences, 1992:126–141
80. Harman D. Aging and disease. In: Kitani K, Aoba A, Goto S. eds. Pharmacological intervention in aging and age-associated disorders. New York: New York Academy of Sciences, 1996: 321–336
81. Richter C. Oxidative damage to mitochondrial DNA and its relationship to ageing. Int J Biochem Cell Biol 1995;27:647–653
82. Yu BP, Yang R. Critical evaluation of the free radical theory of aging. In: Kitani K, Aoba A. Goto S. eds. Pharmacological intervention in aging and age-associated disorders. New York: New York Academy of Sciences, 1996:1–11
83. Nagley P. Mackay I. Baumer A. et al. Mitochondrial DNA mutation associated with aging and degenerative disease. In: Fabris N. Harman D. Knook D. et al. eds. Physiopathological processes of aging. New York: New York Academy of Sciences. 1992:92–102
84. Wei Y-H. Kao S-H. Lee H-C. Simultaneous increase of mitochondrial DNA deletions and lipid peroxidation in human aging. In: Kitani K. Aoba A. Goto S. eds. Pharmacological intervention in aging and age-associated disorders. New York: New York Academy of Sciences. 1996:24–43
85. Julius M. Lang C. Gleiberman L. et al. Glutathione and morbidity in a community-based sample of elderly. J Clin Epidemiol 1994:47:1021–1026
86. Fletcher R. Fletcher S. Glutathione and ageing: ideas and evidence. Lancet 1994;344:1379–1380
87. Mattson MP. Calcium and neuronal injury in Alzheimer's disease: contributions of beta-amyloid precursor protein mismetabolism. free radicals, and metabolic compromise. Ann NY Acad Sci 1994:747:50–76
88. Knight J. Diseases related to oxygen-derived free radicals. Ann Clin Lab Sci 1995;25:111–121
89. Chiueh C. Mohanakumar K. Sternberger L. et al. In vivo generation of hydroxyl radicals and MPTP-induced dopaminergic toxicity in the basal ganglia. Ann NY Acad Sci 1994:738: 25–36
90. Floyd R. Oxidative damage to behavior during aging. Science 1991;254:1597
91. Wolozin B, Luo Y, Wood K. Neuronal loss in aging and disease. In: Holbrook N, Martin G, Lockshin R, eds. Cellular aging and cell death. New York: Wiley-Liss, 1996:283–302
92. Carney JM. Starke-Reed PE, Oliver CN, et al. Reversal of age-related increase in brain protein oxidation, decrease in enzyme activity, and loss in temporal and spatial memory by chronic administration of the spin-trapping compound N-tert-butyl-alpha-phenylnitrone. Proc Natl Acad Sci USA 1991;88: 3633–3636
93. Tolkovsky A. Buckmaster E, Edwards S, et al. Signaling for survival: The biochemistry of NGF dependence in primary sympathetic neurons. In: Holbrook N, Martin G, Lockshin R. eds. Cellular aging and cell death. New York: Wiley-Liss, 1996:267–282
94. Birkenhager-Gillesse E, Derksen J, Lagaay A. Dehydroepiandrosterone sulphate (DHEAS) in the oldest old, aged 85 and over. Ann NY Acad Sci 1994:719:543–552
95. Miller R. Aging and the immune response. In: Schneider E, Rowe J, eds. Handbook of the biology of aging. 4th ed. San Diego: Academic Press, 1996:355–392
96. Pearl R, Pearl R. The ancestry of the long-lived. Baltimore: Johns Hopkins University Press, 1934
97. Strittmatter W, Saunders A, Schmechel D, et al. Apolipoprotein E: high-avidity binding to B-amyloid and increased frequency of type 4 allele in late-onset familial Alzheimer disease. Proc Natl Acad Sci USA 1993:90:1977–1981
98. Levy-Lahad E, Wijsman E, Nemens E, et al. A familial Alzheimer's disease locus on chromosome 1. Science 1995;269: 970–977
99. Sherrington R, Rogaev E, Liang Y, et al. Cloning of a gene bearing missense mutations in early-onset familial Alzheimer's disease. Nature 1995;375:754–760
100. Masoro E. Retardation of aging processes by food restriction: an experimental tool. Am J Clin Nutr 1992:55:1250S–1252S
101. Holloszy J. Exercise and food restriction in rats. J Nutr 1992; 122:774–777
102. Manson J, Willett W, Stampfer M, et al. Body weight and mortality among women. N Engl J Med 1995;333:677–685
103. Castro C, Rudy J. Impaired short-term memory resulting

from postnatal undernutrition is attenuated by physostigmine. Behav Neural Biol 1993;60:282–285
104. Drachman D. If we live long enough, will we all be demented? Neurology 1994:44:1563–1565
105. Barker A, Jones R, Jennison C, et al. A prevalence study of age-associated memory impairment. Br J Psych 1995;167: 642–648
106. Hanninen T, Hallikainen M, Koivisto K, et al. Cognitive problems in the elderly. Curr Opin Neurol 1995;8:256–261
107. Salthouse TA. Selective influences of age and speed on associative memory. Am J Psychol 1995;108:381–396
108. Dustman R, Ruhling R, Russell E, et al. Aerobic exercise training and improved neuropsychological function of older individuals. Neurobiol Aging 1984;5:35–42
109. Joseph J, Cao G, Cutler R. In vivo or in vitro administration of the nitrone spin-trapping compound n-tert-butyl-alpha-phenylnitrone (PBN) reduces age-related deficits in striatal muscarinic receptor sensitivity. Brain Res 1995;671:73–77
110. Kordower J, Freeman T, Snow B, et al. Neuropathological evidence of graft survival and striatal reinnervation after the transplantation of fetal mesencephalic tissue in a patient with Parkinson's disease. N Engl J Med 1995;332:1118–1124
111. Olanow C, Kordower J, Freeman T. Fetal nigral transplantation as a therapy for Parkinson's disease. Trends Neurosci 1996;19:102–109
112. Reynolds B, Weiss S. Generation of neurons and astrocytes from isolated cells of the adult mammalian central nervous system. Science 1992;255:1707–1710
113. Herbert J. The age of dehydroepiandrosterone. Lancet 1995; 345:1193–1194

Oxidative stress and Alzheimer disease[1,2]

Yves Christen

ABSTRACT Research in the field of molecular biology has helped to provide a better understanding of both the cascade of biochemical events that occurs with Alzheimer disease (AD) and the heterogeneous nature of the disease. One hypothesis that accounts for both the heterogeneous nature of AD and the fact that aging is the most obvious risk factor is that free radicals are involved. The probability of this involvement is supported by the fact that neurons are extremely sensitive to attacks by destructive free radicals. Furthermore, lesions are present in the brains of AD patients that are typically associated with attacks by free radicals (eg, damage to DNA, protein oxidation, lipid peroxidation, and advanced glycosylation end products), and metals (eg, iron, copper, zinc, and aluminum) are present that have catalytic activity that produce free radicals. β-Amyloid is aggregated and produces more free radicals in the presence of free radicals; β-amyloid toxicity is eliminated by free radical scavengers. Apolipoprotein E is subject to attacks by free radicals, and apolipoprotein E peroxidation has been correlated with AD. In contrast, apolipoprotein E can act as a free radical scavenger and this behavior is isoform dependent. AD has been linked to mitochondrial anomalies affecting cytochrome-*c* oxidase, and these anomalies may contribute to the abnormal production of free radicals. Finally, many free radical scavengers (eg, vitamin E, selegeline, and *Ginkgo biloba* extract EGb 761) have produced promising results in relation to AD, as has desferrioxamine—an iron-chelating agent—and antiinflammatory drugs and estrogens, which also have an antioxidant effect. *Am J Clin Nutr* 2000;71(suppl):621S–9S.

KEY WORDS Alzheimer disease, free radicals, antioxidants, β-amyloid, oxidative stress, metal, humans

INTRODUCTION

Alzheimer disease (AD) is no longer the condition described in texts written some 10 y ago, ie, a pathology caused by quite mysterious phenomena. Research in the fields of genetics and molecular biology has produced many findings about this common neurodegenerative disease. Many genes involved with the disease have been identified. Some of these genes are involved in early-onset forms of the disease and have a direct causal effect: the amyloid precursor protein (APP) gene located on chromosome 21 and presenilin genes 1 and 2 located on chromosomes 14 and 1, respectively. The apolipoprotein E (apo E) gene (located on chromosome 19), the α_2-macroglobulin gene located on chromosome 12, and other unidentified genes may determine susceptibility in late-onset forms and sporadic cases.

Genetic research along these lines, together with research in molecular biology investigating the main components of the 2 principal hallmarks of AD, ie, senile plaques (typically bearing β-amyloid) and neurofibrillary tangles (mainly composed of tau protein), have helped produce hypotheses about the pathogenesis of AD that no doubt come close to reflecting the real situation. In particular, it appears that a cascade of events may occur that leads to amyloidogenesis and, more specifically, to the formation and deposition of a long β-amyloid peptide (of 42 or 43 amino acids; the shorter form of 40 amino acids has no pathologic effect). Various mutations that have been identified on the *APP*, *PS1*, or *PS2* gene all feature increased production of this peptide (1). Various complementary factors, including cytokines, transforming growth factor β1, and interleukin 1, seem to be involved in triggering the process of amyloidogenesis (2, 3).

Although this amyloid hypothesis is feasible, it is still patently incomplete, particularly because there is still no proper understanding of the relations between amyloidogenesis and the development of the neurofibrillary tangles. An effort must also be made to produce a proper account of the actual neurodegenerative process itself and of neuronal death. Free radicals probably play a role in these processes. The free radical hypothesis of aging, which was proposed many years ago, posits that the age-related accumulation of reactive oxygen species (ROS) results in damage to major components of cells: nucleus, mitochondrial DNA, membranes, and cytoplasmic proteins. The imbalance between the generation of free radicals and ROS may be involved in the pathogenesis of most of the neurodegenerative disorders, including AD, as suggested by many authors for many years (4–6).

Free radicals are in fact potent deleterious agents causing cell death or other forms of irreversible damage, eg, free radicals appear to modify ≈10000 DNA base pairs every day (7). Neurons appear to be particularly vulnerable to attack by free radicals for the following reasons: *1*) their glutathione content, an important natural antioxidant, is low (8); *2*) their membranes contain a high proportion of polyunsaturated fatty acids (9); and *3*) brain metabolism requires substantial quantities of oxygen (10). The fact that age is a key risk factor in AD provides considerable support for the free radical hypothesis because effects of the attacks by free radicals, particularly those produced by ROS, can accumulate over the years (11). Other identified risk

[1] From the Fondation Ipsen, Paris.

[2] Address reprint requests to Y Christen, Fondation Ipsen, 24 rue Erlanger, 75016 Paris, France. E-mail: yves.christen@beaufour-ipsen.com.

factors, such as brain trauma, are also likely to be involved in the production of free radicals. The free radical hypothesis can account for the vastly heterogeneous nature of AD and the fact that both genetic and nongenetic causes are involved. Such general considerations suggest that free radicals are involved in many age-related pathologies, specifically in AD and all neurodegenerative diseases (12).

THE ROLE OF METALS IN ALZHEIMER DISEASE

Metals play a major catalytic role in the production of free radicals, and attention has centered on the role of many metals in AD, including iron, aluminum, mercury, copper, and zinc. Iron is involved in the formation of the free hydroxyl radical, which has recognized deleterious effects, as described in Fenton's and Haber-Weiss's classical reactions. Many observations provide proof that the metabolism of iron is involved in AD. The concentration of iron in the brains of AD patients is elevated. Iron, transferrin, and ferritin have been found in senile plaques (13, 14). Smith et al (15) studied the distribution of iron in the brains of AD patients using various histochemical methods and observed that the iron distribution matched the distribution of senile plaques and neurofibrillary tangles, the 2 key features of AD. The iron related to the lesions could be part of an oxidative process in situ. The characteristics of the iron-binding sites can be determined by the histidine residues and protein conformation. Indeed, when the iron chelating agent desferrioxamine is used, the histochemical labeling disappears. It is therefore reasonable to infer that an alteration in the homeostasis of iron could be involved in AD. One study showed elevated blood serum and cerebrospinal fluid concentrations of the iron-binding protein p97 in AD patients (16). The authors suggested that p97 concentrations could be used as a marker of the disease. They also suggested that p97 concentrations are useful in identifying AD patients, for following the course of the disease, and for assessing the effect of possible therapeutic approaches.

Aluminum has been suggested as a causal factor in AD, in part because of reports showing the toxicity of aluminum, the elevation of aluminum concentrations in the brains of patients with AD, and an association between aluminum concentrations in water and the prevalence of AD (17). However, some studies clearly showed that the aluminum content is not elevated in the brains regions of AD patients that are selectively vulnerable to the neuropathologic changes associated with the disease (18). The possibility of copper's involvement in AD is supported by the fact that copper can act as a catalyst in the production of ROS and by data showing that the APP molecule contains a copper-binding site (19–21). The binding of Cu(II) leads to the modification of APP via the oxidation of cysteines 144 and 158, which leads to the formation of cystine and Cu(I). In this respect, APP serves in the electron transfer to Cu(II), at least in vitro. The oxidation of the 2 cysteines to cystine results in the production of 2 electrons, yet only 1 electron is required for the reduction of Cu(II) to Cu(I). The remaining electron can be involved in the production of hydroxyl radicals (21).

The possible involvement of copper in neurodegeneration is also suggested by the fact that this metal is essential for many enzyme activities, including cytochrome-*c* oxidase and Cu/Zn superoxide dismutase (22). In a recent study, Deibel et al (23) showed lower concentrations of copper in 5 zones of the brains of AD patients, particularly in the hippocampus.

The latest metal cited as a possible factor in AD is zinc. Zinc induces a rapid tinctorial amyloid formation in humans but not in rats, which perhaps explains the scarcity of cerebral β-amyloid in these animals (24). APP binds Zn(II), and this binding modulates the functional properties of APP (eg, inhibits the cleavage of APP by α-secretase and increases binding to heparin). In addition to these data, which are directly related to AD, there is increasing evidence suggesting that zinc accumulation can mediate neuronal death associated with other brain injuries, including ischemia (25).

OXIDATIVE STRESS PRODUCTS AND ALZHEIMER DISEASE

Although data may not always be consistent, many studies have provided evidence for the deleterious consequences of oxidative stress products on certain cellular targets in AD. The oxidation of mitochondrial DNA and, to a lesser extent, of nuclear DNA has been observed in the parietal cortex of AD patients (26) as well as in elderly patients without AD. Protein oxidation has also been observed in elderly individuals with and without AD, but appears to be more marked in AD patients in the regions presenting the most severe histopathologic alterations (27). Many studies have shown increased lipid peroxidation in the brains of AD patients, particularly in the temporal lobe, where histopathologic alterations are very noticeable (28–30). These observations, however, were not corroborated by other studies (31, 32), which failed to find any difference in the basal level of peroxidation. Ramassamy et al (33) suggested that these inconsistent results are related to the apo E genotype. Those with the *E4* allele are likely to be more susceptible to peroxidation than are those without this allele.

Additionally, several studies have identified within the brains of AD patients, particularly in the neurofibrillary tangles, the end products of peroxidation: malondialdehyde (34), peroxynitrite (35, 36), carbonyls (10, 37), advanced glycosylation end products (AGEs; 34, 38), superoxide dismutase-1 (39), and heme oxygenase-1 (40, 41). Heme oxygenase-1 is a cellular enzyme that is up-regulated in the brain and in other tissues in response to an oxidative challenge or other noxious stimuli.

The lipoperoxidation phenomena could have a major, or even causal, influence on the pathogenesis of the disease. Busciglio and Yankner (42), in particular, showed that in Down syndrome, which involves a neurodegenerative component similar to or even identical to that of AD, neuronal death occurs according to a process of apoptosis that is related to an increase in lipid peroxidation and can be stopped by catalase and free radical scavengers.

The effect of ROS on membrane phospholipids could prove to be important because some alterations to membrane phospholipids may be specific to the pathogenesis of AD (43). Markesbery (14) showed that lipid peroxidation is a major cause of depletion of membrane phospholipids in AD. One of the products of lipid peroxidation, 4-hydroxynonenal, which is found in high concentrations in AD patients (44), proved to be toxic to hippocampal cells in culture (45). This highly reactive α,β-aldehyde is thought to cause neuronal death by altering the ATPases involved in ionic transfers and calcium homeostasis (14). The increased calcium concentration could itself cause a cascade of intracellular events, resulting in increased ROS and cellular death (46). Supporting evidence for this link between calcium homeostasis and free radicals derives from data showing that the glutamate-dependent flux of calcium is associated with the production of free radicals by the mitochondria (47, 48). Nitric acid and peroxynitrite appear to play a crucial role in the excitotoxicity related to *N*-methyl-D-aspartate glutamate-receptor activation. Experiments in mice have shown that cells from neuronal nitric oxide synthase–deficient animals are resistant to toxicity induced by *N*-methyl-D-aspartate (49).

Recently, Montine et al (50) found that cerebrospinal fluid F_2-isoprostane concentrations are elevated in AD patients. These compounds are produced by free radical–catalyzed peroxidation of arachidonic acid, independent of the cyclooxygenase enzyme. This discovery is important because it confirms that lipid peroxidation is elevated in AD, but also because it suggests the possible use of the quantification of cerebrospinal fluid F_2-isoprostane concentrations as a biomarker of this disease.

PROTEIN GLYCATION AND ALZHEIMER DISEASE

Glycation phenomena have been held responsible for many age-related pathologies. Glycation of proteins starts as a nonenzymatic process with the spontaneous condensation of ketone or aldehyde groups of a sugar with a free amino acid group to form a labile Schiff base, consistent with the classical reaction first described by Maillard in 1912 (38). A cascade of reactions then results in the formation of AGEs, which are composed of irreversibly cross-linked heterogeneous protein aggregates.

Some studies showed the presence of AGEs in association with the 2 major proteins of AD, β-amyloid (51) and tau (34, 38). This observation supports the argument that AGEs are involved in the pathogenesis of AD (52, 53). Free radicals are involved in glycation processes and clearly can foster the formation of cross-linkings of β-amyloid (54). However, it has not been clearly established that these cross-linking phenomena play a causal role in AD. Mattson et al (54) propose that β-amyloid glycation is a late event in the evolution of AD and is the result of free radical generation by β-amyloid itself. On the basis of this hypothesis, glycation would not play a role in the disease. To quote Mattson et al (54), it would be "a tombstone, but not a cause."

Yan et al (55) showed that the AGE receptor, known as RAGE, is also a β-amyloid receptor. This discovery supports the idea of a relation between AGE and AD as well as between the production of free radicals and these 2 elements. The combination of AGE and RAGE can cause oxidative stress, as shown in the production of thiobarbituric acid–reactive substances, heme oxygenase-1, and the activation of nuclear transcription factor κB (NF-κB). β-Amyloid binding with RAGE also elicits the macrophage colony stimulating factor (56). This introduces an inflammatory pathway, according to a procedure linking oxidative processes and inflammation in AD. This observation could provide a partial explanation for the effect of nonsteroidal antiinflammatory drugs: their NF-κB–blocking action. However, many other studies showed that activation (not inhibition) of NF-κB can protect against neurodegeneration (57–60).

PRODUCTION OF ROS AND β-AMYLOID PROTEIN

Antioxidants providing protection from β-amyloid toxicity

Many studies have observed a direct toxic effect of β-amyloid on cultures of neurons or cell lines. If this reflects the situation in vivo, the phenomenon may offer an explanation for the role of amyloidogenesis in the pathogenesis of AD. Behl et al (61) showed that this β-amyloid toxicity on PC12 cell lines is prevented by vitamin E and by antioxidants in general. A second study by Behl et al (62) showed that hydrogen peroxide mediates this β-amyloid toxicity, which explains why catalase, which degrades hydrogen peroxide, protects the cells from β-amyloid toxicity. The same research team selected β-amyloid toxicity–resistant PC12 cell line clones and showed that they contained high concentrations of the antioxidant enzymes catalase and glutathione peroxidase (63). They also showed that the different amyloid peptides involved in various forms of amyloidosis in humans (amylin, calcitonin, and atrial natriuretic peptide) have a toxic effect similar to that of β-amyloid (64). In fact, experiments with synthetic peptides suggest that a common oxidative mechanism is found in all cases and that the amphiphilic form of the peptides rather than the structure in β layers may be the cause. The authors therefore concluded that β-amyloid toxicity could not be caused solely by an interaction through specific membrane receptors.

These pioneering studies provide clear confirmation of the involvement of the free radical process in β-amyloid toxicity. They are supported by studies of other free radical scavengers, such as *Ginkgo biloba* extract EGb 761 (65), melatonin (66), and EUK-8 (67). These scavengers also afford protection to hippocampal cells subjected to free radical stress. EGb 761 can also protect hippocampal cells from β-amyloid toxicity 4 h posttreatment (68).

β-Amyloid aggregation and production of free radicals

A dual relation exists between β-amyloid and the production of free radicals. Not only can the oxidative processes transform nonaggregated β-amyloid into aggregated β-amyloid in vitro (69), but β-amyloid itself is a source of free radicals. β-Amyloid interacts with vascular endothelial cells, producing a surfeit of free superoxide radicals that can scavenge the endothelium-derived relaxing factor and produce oxidizing agents causing lipid peroxidation (70). Even though this concerns the vascular endothelium and not the neurons, these data support the hypothesis that β-amyloid acts via the production of free radicals and plays a role in neurodegenerative processes.

Mass spectrometric and electron paramagnetic resonance spin trapping indicate that β-amyloid, in aqueous solution, fragments and generates free radical peptides (71). Some of these free radical peptides (the β-amyloid 25–35 fragment) have a potent lipoperoxidizing effect on the synaptosomal membranes in the neocortex (72). The many recent studies that focused on the involvement of the microglia as affected by β-amyloid, and therefore of inflammatory-type processes in the brains of AD patients, implicitly lead to the conclusion that free radicals are being produced because it is a well-known phenomenon associated with inflammation. McDonald et al (73) showed that the β-amyloid fibrils activate protein-tyrosine kinase–dependent signaling and superoxide production in the microglia. This research confirms the findings of other studies showing that β-amyloid stimulates ROS production in the microglia. Meda et al (74) showed that β-amyloid triggered the release of the NO_2^- radical by the microglia in rodents.

NF-κB and β-amyloid

The increase in the activity of NF-κB protects neurons against β-amyloid toxicity (75). Many new data suggest that NF-κB plays an important role in neurodegenerative disorders.

Goodman and Mattson (57) showed that activation of NF-κB may constitute a cytoprotective signaling pathway that induces expression of protective gene products such as calbindin and antioxidant enzymes. In fact, exposure of cells to oxidative stress results in activation of NF-κB (58, 76), and high constitutive NF-κB activity mediates resistance to oxidative stress in neuronal cells (77).

These data contradict those of Grilli et al (78) concerning the neuroprotective role of aspirin, which would be mediated through inhibition of NF-κB. According to Lipton (79), "it remains for future studies to determine whether we can explain the Janus faces of NF-κB activation that cause excitotoxic neurodestruction in Grilli's study but neuroprotection in the hands of Mattson." Barger et al (59) provide an explanation for these contradictory data: NF-κB activation of glial cells is probably neurotoxic but its involvement in neurons is probably neuroprotective.

It is also important to understand that antioxidants block the activation of NF-κB. However, this fact is quite logical: the activation of NF-κB is related to the importance of the free radical pool in the cells. When antioxidants decrease the quantity of free radicals, the cells no longer need to activate the protective mechanism of activation of NF-κB.

APOLIPOPROTEIN E AND OXIDATIVE STRESS

The *E4* allele of apo E has been linked to both late-onset family forms and sporadic forms of AD, whereas the *E2* allele appears to offer protection. This discovery inspired many new research projects and working hypotheses suggesting that oxidative processes play a role in AD. Apo E helps transport cholesterol through both the vascular system and the neurons. In the brain, apo E is the crucial cholesterol carrier, which explains the role it plays in neuroplasticity-related phenomena because these require changes to membrane lipids and, therefore, a cholesterol-dependent function. Poirier (80) maintains that apo E has a beneficial effect for neuronal protection but that the apo E4 isoform is less effective than are the apo E2 and E3 isoforms. (In other words, apo E4 may not be toxic, but simply capable of producing a less-favorable effect.) Miyata and Smith (81) showed that apo E has a beneficial effect against free radicals. Apo E reduced neuronal death caused by hydrogen peroxide and β-amyloid through antioxidant activity. This effect is clearly perceptible with the E2 isoform, but less so with E3 and even less so with E4.

However, Leininger-Muller et al (82) showed that apo E is sensitive to attacks by free radicals, with the responses being isoform dependent. The E4 isoform is more sensitive than is the E3 isoform, which in turn is more sensitive than is the E2 isoform. An antioxidant such as EGb 761 protects apo E from oxidation in a similarly isoform-dependent way; the effect is seen most clearly with apo E4 (82).

Ramassamy et al (33) established that another relation existed between the apo E genotype and lipoperoxidation in AD. They showed that the level of peroxidation in the brains of AD patients depended on the apo E genotype and was higher when the *E4* allele was present. This discovery is important because it explains why previous studies sometimes found an increase and sometimes found no change in lipoperoxidation in the brains of AD patients. Ramassamy et al also showed that the level of peroxidation in the brains of AD patients is inversely proportional to the concentration of apo E, which confirms the hypothesis that apo E has a beneficial effect against lipoperoxidation and that the effect is more pronounced when the patient has no *E4* allele. Ramassamy et al proved, in vitro, that EGb 761 effectively reduces the level of lipoperoxidation in the brains of AD patients.

MITOCHONDRIAL ANOMALIES AND ALZHEIMER DISEASE

Defects in the electron transport chain within the mitochondria are major factors contributing to the production of free radicals. Many studies have shown a low level of oxidative phosphorylation in AD, this being expressed both by energy deficits and the potentially toxic production of free radicals. Note that the electron transport chain that results in the formation of ATP via the reduction of oxygen to water is a complex enzymatic system consisting of 5 distinct phases: complex 1 (NADH dehydrogenase), complex 2 (succinate dehydrogenase), complex 3 (ubiquinol–cytochrome-*c* reductase), complex 4 (cytochrome-*c* oxidase), and complex 5 (ATP synthase).

Many observations have confirmed alterations in mitochondrial function in the course of aging and neurodegenerative diseases such as AD, Huntington disease, and Parkinson disease, and, for AD in particular, effects on cytochrome-*c* oxidase in complex 4 (83). Mutisaya et al (84) showed that postmortem cytochrome-*c* oxidase activity was ≈25–30% lower than normal in the cerebral cortex (frontal, parietal, temporal, and occipital) and in the platelets of AD patients. Simonian and Hyman (85) used histochemical methods to show the lower than normal cytochrome-*c* oxidase activity in the dentate gyrus and the CA4, CA3, and CA1 zones of the hippocampus. A decline in cytochrome-*c* oxidase activity is associated with decreased expression of messenger RNA (mRNA) molecules, which is lower in the midtemporal region of the brains of AD patients. Chandrasekaran et al (86) showed lower than normal mRNA concentrations in subunits 1 and 3 of cytochrome-*c* oxidase. In fact, normal amounts of cytochrome-*c* oxidase are found in the brains of AD patients; it is only the enzyme activity that is affected (87). Lakis et al (88) transferred this cytochrome-*c* oxidase defect to cybrid lines. The cybrid technique, which appears to hold great promise for the study of neurodegenerative diseases (89), uses cell lines stripped of their mitochondria and then transfers to them mitochondria from platelets (or other tissues) taken from patients with the disease to be studied. Lakis et al (88) observed increased production of free radicals by cybrids, which confirms that the cytochrome-*c* oxidase defect can produce damaging amounts of free radicals. The same reconstitution with mitochondria from AD platelets resulted in reduced cytochrome-*c* oxidase activity (90). Any possible relation between these mitochondrial anomalies and other processes involved in AD, amyloidogenesis in particular, is as yet unproved. However, Askanas et al (91), using an adenovirus as vector, showed that the transfer of the *APP* gene caused mitochondrial anomalies in a culture of muscular cells. This phenomenon could provide the explanation for inclusion-body myositis, a muscular condition similar to AD, but, on a broader scale, it could also establish a link between amyloid metabolism and mitochondrial anomalies.

Recently, the hypothesis of a link between mitochondrial function and cytochrome-*c* oxidase received considerable support with the discovery of mutations in cytochrome-*c* oxidase genes that segregate with late-onset AD (92). Cytochrome-*c* oxidase is produced by the combined effect of mitochondrial and nuclear genes. However, these catalytic centers are encoded exclusively by 2 mitochondrial genes, *CO1* and *CO2*, which encode for subunits I and II, respectively. The gene encoding subunit III is *CO3*. Davis et al (92) described several DNA polymorphisms in *CO1* and *CO2* (but not in *CO3*) that segregate with AD. However, subsequent work by other groups concluded that these polymorphisms are not present in the mitochondrial genome but are rather nuclear pseudogenes (93, 94). Davis et al (95) recognized that the presence of these nuclear mitochondria-like DNA sequences is unlikely to cause the cytochrome-*c* oxidase defect in humans. However, they believe that an elevation in the ratio of this pseudogene relative to authentic mitochondrial DNA in the blood of AD patients still holds true.

How can mitochondrial defects have an influence on the development of AD? In theory, they can trigger 2 harmful events: the production of destructive free radicals and a reduction in energy resources. The slowing down of energy metabolism in AD has been extensively reported in pyruvate dehydrogenase (lipoamide), oxoglutarate dehydrogenase (lipoamide), and oxidative phosphorylation enzymes (96, 97). A reduction to ≈50% of normal activity of oxoglutarate dehydrogenase (lipoamide) was found in cultured AD fibroblasts (98). Observations in AD patients were also made in vivo, in which positron emission tomography was used to study glucose metabolism in patients. A reduction in the regional cerebral metabolic glucose concentration was observed at rest. This was seen throughout the neocortex and was found to correlate with the severity of dementia (99). Despite this reduced metabolism at rest, often accompanied by left-right asymmetry, some cortical regions of the brains of AD patients can still be activated during certain cognitive tasks (100). Rapoport et al (101) interpreted these findings to suggest the existence of a down-regulation process in AD that, in the early stages, is reversible. The study of gene expression supports this hypothesis. A drop in mRNA can be detected in the mitochondrial genes of the cytochrome-*c* oxidase complex and also in the mRNA of the cytochrome-*c* oxidase complex 4 subunit encoded by a nuclear gene. Rapoport et al (101, 102) maintained that, because of the lower neuronal energy requirement caused by synaptic damage in AD, there may be a down-regulation of cytochrome-*c* oxidase subunit gene expression. Reduced brain glucose utilization in the early stages of AD, as measured with positron emission tomography, would reflect this down-regulation. If this were the case, the free radicals would not be the cause of the process, but simply a consequence.

WHICH FREE RADICALS ARE INVOLVED IN ALZHEIMER DISEASE?

Many free radicals are likely involved in AD. There are excellent reasons to suspect the free hydroxyl radical as a factor in AD (15) because of its toxicity and its role in various chemical reactions, such as Fenton's, involving iron. Other suspects include the superoxide radical hydrogen peroxide, which is implicated in amyloid neurotoxicity, and peroxynitrite, which can be formed by combining superoxide and nitric oxide. Peroxynitrite can react with carbon dioxide to form unstable transmitters that act as free hydroxyl radicals. Many recent studies have confirmed the toxicity of peroxynitrite to neurons and the involvement of nitric oxide in neurologic pathologies. Smith et al (36) observed traces of oxidative modifications caused by peroxynitrite in the brains of AD patients, and this protein nitration included, among others, neurofibrillary tangles. This was not, however, observed in brains that were not affected by AD.

THERAPEUTIC USE OF ANTIOXIDANTS

All observations suggesting the involvement of ROS in the pathogenesis of AD raise the possibility of the therapeutic use of free radical scavengers and antioxidants. The hypothesis is particularly attractive because many free radical scavengers are known and many (eg, vitamins E and C, *Ginkgo biloba* extract EGb 761, melatonin, flavonoids, and carotenoids) have no major side effects. This hypothesis has been tested with a reasonable degree of success under both experimental and clinical conditions. Many experimental studies showed that free radical–scavenging substances inhibit the toxic effect of β-amyloid or hydrogen superoxide on cell cultures and organotypic hippocampal cultures (66–68).

Many preliminary clinical studies have been completed and published. Crapper McLachlan et al (103) showed that administration over a 2-y period of an iron-chelating agent, desferrioxamine, slowed the clinical development of the disease. It was hoped that the chelating effect would make it possible to inhibit iron-dependent lipid peroxidation, which, as has been seen, can play a role in the etiology of AD.

Three free radical–scavenging drugs used for therapeutic purposes in different fields were also scrutinized in clinical studies of AD and produced beneficial results: vitamin E (α-tocopherol), selegiline (also a monoamine oxidase B inhibitor), and *Ginkgo biloba* extract EGb 761(Tanakan).

The first 2 products were investigated as part of the Alzheimer's Disease Cooperative Study in 324 patients with moderately severe AD (104). The authors did not observe any significant improvement on cognitive tests (unlike observations of certain cholinesterase inhibitors, tetrahydroaminoacridine and E2020) but did observe significant delays in the time of the occurrence of the following outcomes: death, institutionalization, loss of the ability to perform basic activities of daily living, and severe dementia. These findings provided supporting evidence for the hypothesis that antioxidants may be capable of slowing the pathogenic process.

EGb 761 has been studied in clinical investigations both in Germany (105) and in the United States (106). The studies concluded that it had a positive effect on cognitive indexes, similar to the results obtained with tetrahydroaminoacridine. A formal review of >30 articles pointed to a small but significant effect of this drug on cognitive function in AD (107). Interestingly, this result differs from the findings for α-tocopherol and selegiline because neither of these 2 drugs had any significant effect on cognition. Note, however, that EGb 761 has many other effects in addition to its free radical–scavenging activity, including a protective effect on neuroreceptors in aging subjects, a monoamine oxidase–inhibiting effect (less of an effect than selegiline, but no doubt an indirect effect), and a clear effect on cognitive indexes in both animals and humans (108, 109).

Nonsteroidal antiinflammatory drugs have also been shown to have a useful effect in countering AD (110, 111). These drugs operate principally by acting on cyclooxygenases and inhibiting prostaglandin synthesis, or by decreasing glutamate-related excitotoxicity and reducing the production of ROS. Lastly, the beneficial effect of estrogen in AD may be partly attributable to its antioxidant activity. Behl et al (112) showed that 17β-estradiol protects neuroblastoma cells against oxidative stress, and Goodman et al (113) showed that 17β-estradiol and estriol suppress membrane oxidation in hippocampal neurons induced by β-amyloid.

NUTRITION AND ALZHEIMER DISEASE

Although free radical scavengers are known to have a beneficial effect when taken in small doses as a preventive treatment, the influence of free radical scavengers in food is still difficult to prove. Many free radical scavengers are present in food, particularly in fruit and vegetables (eg, as carotenoids and flavonoids). Regular consumption of these nutritive substances may have beneficial effects. Some preliminary results have shown that diets high in antioxidant activity (strawberry extracts, spinach, and blueberry extracts) prevent the deleterious effects of oxidative stress on signal transduction and nerve growth factor in rats (114). There was also a trend toward the prevention of deficits in motor behavior. However, these experiments in animals are not directly related to AD.

Regular consumption of antioxidants in the diet may have a beneficial effect in humans. Cognitive impairment has been associated with lower vitamin C intakes (115). Fruit and vegetables could also have protective effects against stroke and vascular dementia (116). Epidemiologic studies to confirm the beneficial effects of antioxidants in AD have not yet been conducted.

CONCLUSION

The idea that free oxygen radicals might be involved in AD was originally based on general considerations focusing on free radical processes involved in aging and many medical conditions, particularly in cerebral pathologies (4, 6). This idea then inspired many research projects, which were often unable to reach any firm conclusions. Data on activities of antioxidant enzymes in AD are contradictory: some researchers have detected alterations to glutathione peroxidase and superoxide dismutase (Mn or Cu/Zn), whereas others have not (14).

Still, the involvement of free radicals in the pathogenesis of AD is, however, now widely accepted for the following reasons:

1) Neurons are particularly sensitive to free radicals.

2) Aging is the principal AD risk factor and is itself related to accumulated free radical attacks.

3) Examinations of the brains of AD patients show many signs of free radical attacks, ie, damage to mitochondrial and nuclear DNA, protein oxidation, lipid peroxidation, and AGEs.

4) Traces of substances have been found in the brains of AD patients that indicate the presence of metals (iron, copper, zinc, and aluminum) capable of catalyzing reactions that produce free radicals.

5) Free radical scavengers reduce the toxicity of β-amyloid.

6) β-Amyloid is sensitive to the action of free radicals, contributing to aggregation and itself producing peptides in free radical form.

7) Apo E is subject to free radical attacks and a correlation exists between apo E peroxidation and AD; it can also act as an isoform-dependent free radical scavenger. The oxidative status of the brain is related to the apo E genotype.

8) AD is related to mitochondrial anomalies, particularly for cytochrome-*c* oxidase, and these anomalies may explain the abnormal production of free radicals.

9) The use of many free radical scavengers (vitamin E, selegiline, and *Ginkgo biloba* extract EGb 761) has produced positive therapeutic results, as has the use of antiinflammatory drugs, estrogens, and the iron-chelating agent desferrioxamine, which inhibits the catalytic action of iron.

This evidence clearly supports the involvement of free radicals and ROS in AD. However, it is unknown whether free radicals are really one of the basic causes of the pathogenesis and neuronal damage in AD. Free radicals may act indirectly as a result of other pathologic processes. Their involvement does not seem specific to any particular disease and evidence for their role can be challenged for this reason. Metals, for example, are present in the brains of all people and not just in those with AD. It is plausible that we can describe neuropathologic disorders (eg, AD, but also Parkinson disease, Huntington disease, and amyotrophic lateral sclerosis) as a combination of 2 kinds of events: a genetic and specific defect related to a particular molecule (eg, presenilin or APP in AD, parkin in Parkinson disease, and tau in frontotemporal dementia with Parkinsonism) and the age-related abnormal production of free radicals. The development of these diseases requires these 2 events to occur. In AD, the occurrence of these 2 events is clear: recent data from rhesus monkeys show that the microinjection of β-amyloid results in profound neuronal loss, tau phosphorylation, and microglial proliferation only in aged animals (117). Even if they are not the specific cause of the disease, free radicals are likely an essential factor in the pathogenesis of AD and, consequently, are a potential target for therapy.

REFERENCES

1. Younkin SM, Tanzi RE, Christen Y, eds. Presenilins and Alzheimer's disease. Heidelberg, Germany: Springer-Verlag, 1998.
2. Wyss-Coray T, Masliah E, Mallory M, et al. Amyloidogenic role of cytokine TGF-β1 in transgenic mice and in Alzheimer's disease. Nature 1997;389:603–6.
3. Patterson P, Kordon C, Christen Y, eds. Neuro-immune interactions and neurological and psychiatric disorders. Heidelberg, Germany: Springer-Verlag, 1999.
4. Hartman D. Free radical theory of aging: Alzheimer's disease pathogenesis. Age 1995;18:97–119.
5. Halliwell B, Gutteridge JMC. Free radicals in biology and medicine. Oxford, United Kingdom: Oxford University Press, 1989.
6. Packer L, Prilipko L, Christen Y, eds. Free radicals in the brain. Heidelberg, Germany: Springer-Verlag, 1992.
7. Ames B, Shingenaga M, Park EM. Oxidation damage and repair: chemical, biological and medical aspects. Elmsford, United Kingdom: Pergamon Press, 1991.
8. Cooper AJL. Glutathione in the brain: disorders of glutathione metabolism. In: Rosenberg RN, Prusiner SB, DiMauro S, Barchi RL, Klunk LM, eds. The molecular and genetic basis of neurological disease. Boston: Butterworth-Heinemann, 1997:1242–5.
9. Hazel JR, Williams EE. The role of alterations in membrane lipid composition in enabling physiological adaptation of organisms to their physical environment. Prog Lipid Res 1990;29:167–227.
10. Smith MA, Rudnicka-Nawrot M, Richey PL, et al. Carbonyl-related posttranslational modification of neurofilament protein in the neurofibrillary pathology of Alzheimer's disease. J Neurochem 1995;64:2660–6.
11. Benzi G, Moretti A. Are reactive oxygen species involved in Alzheimer's disease? Neurobiol Aging 1995;16:661–74.
12. Beal MF, Howell N, Bodis-Wollner IB, eds. Mitochondria and free radicals in neurodegenerative diseases. New York: Wiley-Liss, 1997.
13. Loeffler DA, Connor JR, Juneau PL, et al. Transferrin and iron in normal, Alzheimer's disease, and Parkinson's disease brain regions. J Neurochem 1995;65:710–6.

14. Markesbery WR. Oxidative stress hypothesis in Alzheimer's disease. Free Radic Biol Med 1997;23:134–47.
15. Smith MA, Richey Harris PL, Sayre L, Perry G. Iron accumulation in Alzheimer disease is a source of redox-generated free radicals. Proc Natl Acad Sci U S A 1997;94:9866–8.
16. Kennard ML, Feldman H, Yamada T, Jefferies WA. Serum levels of the iron binding protein p97 are elevated in Alzheimer's disease. Nat Med 1996;2:1230–5.
17. Crapper DR, Quittkat S, Krishnan SS, Dalton AJ, DeBoni U. Intranuclear aluminum content in Alzheimer's disease, dialysis encephalopathy, and experimental aluminum encephalopathy. Acta Neuropathol (Berl) 1980;50:19–24.
18. Bjertness E, Candy JM, Torvik A, et al. Content of brain aluminum is not elevated in Alzheimer disease. Alzheimer Dis Assoc Disord 1996;10:171–4.
19. Multhaup G, Schlicksupp A, Hesse L, et al. The amyloid precursor protein of Alzheimer's disease in the reduction of copper (II) to copper (I). Science 1996;271:1406–9.
20. Multhaup G. Amyloid precursor protein, copper and Alzheimer's disease. Biomed Pharmacother 1997;51:105–11.
21. Multhaup G, Masters CL, Beyreuther K. Oxidative stress in Alzheimer's disease. Alzheimer Rep 1998;1:147–54.
22. Linder MC, Hazegh-Azam M. Copper biochemistry and molecular biology. Am J Clin Nutr 1996;63(suppl):797S–811S.
23. Deibel MA, Ehmann WD, Markesbery WR. Copper, iron, and zinc imbalances in severely degenerated brain regions in Alzheimer's disease: possible relation to oxidative stress. J Neurol Sci 1997;143:137–42.
24. Bush AI, Pettingell WH, Multhaup G, et al. Rapid induction of Alzheimer Aβ amyloid formation by zinc. Science 1994;265:1464–7.
25. Choi DW, Koh JY. Zinc and brain injury. Annu Rev Neurosci 1998;21:347–75.
26. Mecocci P, MacGarvey U, Beal MF. Oxidative damage to mitochondrial DNA is increased in Alzheimer's disease. Ann Neurol 1994;36:747–51.
27. Hensey K, Hall N, Subramaniam R, et al. Brain regional correspondence between Alzheimer's disease histopathology and biomarkers of protein oxidation. J Neurochem 1995;65:2146–56.
28. Palmer AM, Burns MA. Selective increase in lipid peroxidation in the inferior temporal cortex in Alzheimer's disease. Brain Res 1994;645:338–42.
29. Lovell MA, Ehmann WD, Butler SM, Markesbery WR. Elevated thiobarbituric acid-reactive substances and antioxidant enzyme activity in the brain in Alzheimer's disease. Neurology 1995;45:1594–601.
30. Marcus DL, Thomas C, Rodriguez C, et al. Increased peroxidation and reduced antioxidant enzyme activity in Alzheimer's disease. Exp Neurol 1998;150:40–4.
31. Hayn M, Kremser K, Singewald N, et al. Evidence against the involvement of reactive oxygen species in the pathogenesis of neuronal death in Down's syndrome and Alzheimer's disease. Life Sci 1996;59:537–44.
32. Lyras L, Cairns NJ, Jenner P, Halliwell B. An assessment of oxidative damage to proteins, lipids and DNA in the brain from patients with Alzheimer's disease. J Neurochem 1997;68:2061–9.
33. Ramassamy C, Krzywokowski P, Bastianetto S, et al. Apolipoprotein E, oxidative stress and EGb 761 in Alzheimer's disease brain. In: Packer L, Christen Y, eds. *Ginkgo biloba* extract (EGb 761) study: lesson from cell biology. Paris: Elsevier, 1998:69–83.
34. Yan SD, Chen X, Schmidt AM, et al. Glycated tau protein in Alzheimer disease: a mechanism for induction of oxidant stress. Proc Natl Acad Sci U S A 1994;91:7787–91.
35. Good PF, Werner P, Hsu A, Olanow CW, Perl DP. Evidence for neuronal oxidative damage in Alzheimer's disease. Am J Pathol 1996;149:21–8.
36. Smith MA, Richey Harris PL, Sayre LM, Beckman JS, Perry G. Widespread peroxynitrite-mediated damage in Alzheimer's disease. J Neurosci 1997;17:2653–7.
37. Smith MA, Perry G, Richey PL, et al. Oxidative damage in Alzheimer's. Nature 1996;382:120–1.
38. Smith MA, Taneda S, Richey PL, et al. Advanced Maillard reaction end products are associated with Alzheimer disease pathology. Proc Natl Acad Sci U S A 1994;91:5710–4.
39. Pappolla MA, Omar RA, Kim KS, Robakis NK. Immunohistochemical evidence of antioxidant stress in Alzheimer's disease. Am J Pathol 1992;140:621–8.
40. Smith MA, Kutty RK, Richey PL, et al. Heme oxygenase-1 is associated with the neurofibrillary pathology of Alzheimer's disease. Am J Pathol 1994;145:42–7.
41. Schipper HM, Cissé S, Stopa EG. Expression of heme oxygenase-1 in the senescent and Alzheimer-diseased brain. Ann Neurol 1995; 37:758–68.
42. Busciglio J, Yankner BA. Apoptosis and increased generation of reactive oxygen species in Down's syndrome neurons in vitro. Nature 1995;378:776–9.
43. Nitsch RM, Blusztajn JK, Pittas AG, Slack BE, Growdon JH, Wurtman RJ. Evidence for a membrane defect in Alzheimer disease brain. Proc Natl Acad Sci U S A 1992;89:1671–5.
44. Markesbery WR, Lovell MA. Four-hydroxynonenal, a product of lipid peroxidation, is increased in the brain in Alzheimer's disease. Neurobiol Aging 1998;19:33–6.
45. Mark RJ, Lovell MA, Markesbery WR, Uchida K, Mattson MP. A role for 4-hydroxynonenal, an aldehydic product of lipid peroxidation, in disruption of ion homeostasis and neuronal death induced by amyloid β-peptide. J Neurochem 1997;68:255–64.
46. Mattson MP. Free radicals and disruption of neuronal ion homeostasis in AD: a role for amyloid β-peptide? Neurobiol Aging 1995; 16:679–82.
47. Dugan LL, Sensi SL, Canzoniero LMT, et al. Mitochondrial production of reactive oxygen species in cortical neurons following exposure to *N*-methyl-D-aspartate. J Neurosci 1995;15:6377–88.
48. Reynolds IJ, Hastings TG. Glutamate induces the production of reactive oxygen species in cultured forebrain neurons following NMDA receptor activation. J Neurosci 1995;15:3318–27.
49. Dawson VL, Kizushi VM, Huang PL, Snyder SH, Dawson TM. Resistance to neurotoxicity in cortical cultures from neuronal nitric oxide synthase-deficient mice. J Neurosci 1996;16:2463–78.
50. Montine TJ, Markesbery WR, Morrow JD, Roberts LJ. Cerebrospinal fluid F2-isoprostane levels are increased in Alzheimer's disease. Ann Neurol 1998;44:410–3.
51. Vitek MP, Bhattacharya K, Glendening JM, et al. Advanced glycation end products contribute to amyloidosis in Alzheimer disease. Proc Natl Acad Sci U S A 1994;91:4766–70.
52. Harrington CR, Colaco CALS. A glycation connection. Nature 1994;370:247–8.
53. Smith MA, Sayre LM, Monnier VM, Perry G. Radical AGEing in Alzheimer's disease. Trends Neurosci 1995;18:172–6.
54. Mattson MP, Carney JW, Butterfield DA. A tombstone in Alzheimer's? Nature 1995;373:481.
55. Yan SD, Chen X, Fu J, et al. RAGE and amyloid-β peptide neurotoxicity in Alzheimer's disease. Nature 1996;382:685–91.
56. Yan SD, Zhu H, Fu J, et al. Amyloid-β peptide-receptor for advanced glycation end product interaction elicits neuronal expression of macrophage-colony stimulating factor: a proinflammatory pathway in Alzheimer disease. Proc Natl Acad Sci U S A 1997;94:5296–301.
57. Goodman Y, Mattson MP. Ceramide protects hippocampal neurons against excitotoxic and oxidative insults, and amyloid β-peptide toxicity. J Neurochem 1996;66:869–72.
58. Mattson MP, Furukawa K, Bruce AJ, Mark RJ, Blanc EM. Amyloid cytotoxicity and Alzheimer's disease: roles of membrane oxidation and perturbed ion homeostasis. In: Brioni JD, Decker MW, eds. Pharmacological treatment of Alzheimer's disease: molecular and neurobiological foundations. New York: Wiley, 1997:239–85.
59. Barger SW, Mao X, Moerman AM, Ranganathan A. Mechanistic and metaphorical connections between NF-κB and the secreted Alzheimer's β-amyloid precursor protein. In: Patterson P, Kordon C, Christen Y, eds. Neuro-immune interactions and neurological

and psychiatric disorders. Heidelberg, Germany: Springer-Verlag, 1999:57–72.
60. Lezoualc'h F, Behl C. Transcription factor NF-κB: friend or foe of neurons? Mol Psychol 1998;3:15–20.
61. Behl C, Davis J, Cole GM, Schubert D. Vitamin E protects nerve cells from amyloid β protein toxicity. Biochem Biophys Res Commun 1992;86:944–52.
62. Behl C, Davis JB, Lesley R, Schubert D. Hydrogen peroxide mediates amyloid β protein toxicity. Cell 1994;77:817–27.
63. Sagara Y, Dargusch R, Klier FG, Schubert D, Behl C. Increased antioxidant enzyme activity in amyloid β protein-resistant cells. J Neurosci 1996;16:497–505.
64. Schubert D, Behl C, Lesley R, et al. Amyloid peptides are toxic via a common oxidative mechanism. Proc Natl Acad Sci U S A 1995; 92:1989–93.
65. Bastianetto S, Ramassamy C, Christen Y, Poirier J, Quirion R. *Ginkgo biloba* extract (EGb 761) prevents cell death induced by oxidative stress in hippocampal neuronal cell cultures. In: Packer L, Christen Y, eds. *Ginkgo biloba* extract (EGb 761) study: lesson from cell biology. Advances in EGb 761 research. Vol 7. Paris: Elsevier, 1998:85–99.
66. Pappolla MA, Sos M, Omar RA, et al. Melatonin prevents death of neuroblastoma cells exposed to the Alzheimer amyloid peptide. J Neurosci 1997;17:1683–90.
67. Bruce AJ, Malfroy B, Baudry M. β-amyloid toxicity in organotypic hippocampal cultures: protection by EUK-8, a synthetic catalytic free radical scavenger. Proc Natl Acad Sci U S A 1996; 93:2312–6.
68. Bastianetto S, Ramassamy C, Christen Y, Poirier J, Quirion E. *Ginkgo biloba* extract (EGb 761) protects in vitro rat hippocampal cells against toxicity induced by β-amyloid peptides. Soc Neurosci 1998;24:1456.
69. Dyrks T, Dyrks E, Harmann R, Masters C, Beyreuther K. Amyloidogenicity of βA4 and βA4-bearing amyloid protein precursor fragments by metal-catalyzed oxidation. J Biol Chem 1992;267:18210–7.
70. Thomas T, Thomas G, McLendon C, Sutton T, Mullan M. β-amyloid-mediated vasoactivity and vascular endothelial damage. Nature 1996;380:168–71.
71. Hensley K, Carney J, Mattson M, et al. A model for β-amyloid aggregation and neurotoxicity based on free radical generation by the peptide: relevance to Alzheimer's disease. Proc Natl Acad Sci U S A 1994;91:3270–4.
72. Butterfield DA, Hensley K, Harris M, Mattson M, Carney J. β-amyloid peptide free radical fragments initiate synaptosomal lipoperoxidation in a sequence-specific fashion: implications to Alzheimer's disease. Biochem Biophys Res Commun 1994;200:710–5.
73. McDonald DR, Brunden KR, Landreth GE. Amyloid fibrils activate tyrosine kinase-dependent signaling and superoxide production in microglia. J Neurosci 1997;17:2284–94.
74. Meda L, Cassatella MA, Szendrei GI, et al. Activation of microglial cells by β-amyloid protein and interferon gamma. Nature 1995; 374:647–50.
75. Barger SW, Horster D, Furukawa K, Goodman Y, Krieglstein J, Mattson MP. TNFα and TNFβ protect hippocampal neurons against amyloid β-peptide toxicity: evidence for involvement of a κB-binding factor and attenuation of peroxide and CaR⁺ accumulation. Proc Natl Acad Sci U S A 1995;92:9328–32.
76. Behl C, Sagara Y. Mechanism of amyloid beta protein induced neuronal cell death: current concepts and future perspectives. J Neural Transm Suppl 1997;49:125–34.
77. Lezoualc'h F, Sagara Y, Holsboer F, Behl C. High constitutive NF-κB activity mediates resistance to oxidative stress in neuronal cells. J Neurosci 1998;18:3224–32.
78. Grilli M, Pizzi M, Memo M, Spano P. Neuroprotection by aspirin and sodium salicylate through blockade of NF-κB activation. Science 1996;274:1383–5.
79. Lipton SA. Janus faces of NF-κB: neurodestruction versus neuroprotection. Nat Med 1997;3:20–2.
80. Poirier J. Apolipoprotein E in animal models of CNS injury and in Alzheimer's disease. Trends Neurosci 1994;17:525–30.
81. Miyata M, Smith JD. Apolipoprotein E allele-specific antioxidant activity and effects on cytotoxicity by oxidative insults and β-amyloid peptides. Nat Genet 1996;14:55–61.
82. Leininger-Muller B, Jolivalt C, Bertrand P, Siest G. Oxidation of human apolipoprotein E: isoforms susceptibility and protection with *Ginkgo biloba* EGb 761 extract. In: Packer L, Christen Y, eds. *Ginkgo biloba* extract (EGb 761) study: lessons from cell biology. Paris: Elsevier, 1998:57–68.
83. Beal MF. Aging, energy, and oxidative stress in neurodegenerative diseases. Ann Neurol 1995;38:357–66.
84. Mutisaya EM, Bowling AC, Beal MF. Cortical cytochrome oxidase activity is reduced in Alzheimer's disease. J Neurochem 1994; 63:2179–84.
85. Simonian NA, Hyman BT. Functional alterations in Alzheimer's disease: diminution of cytochrome oxidase in the hippocampal formation. J Neuropathol Exp Neurol 1993;52:580–5.
86. Chandrasekaran K, Giordano T, Brady DR, Stoll J, Martin JL, Rapoport SI. Impairment in mitochondrial cytochrome oxidase gene expression in Alzheimer's disease. Mol Brain Res 1994;24:336–40.
87. Parker WD, Parks JK. Cytochrome *c* oxidase in Alzheimer's disease brain: purification and characterization. Neurology 1995;45:482–6.
88. Lakis J, Glasco S, Miller SW, Thal LJ, Davis RE. Production of reactive oxygen species correlates with decreased cytochrome oxidase activity in Alzheimer's disease cybrids. Soc Neurosci 1995;21:979.
89. Beal MF. Mitochondria, free radicals, and neurodegeneration. Curr Opin Neurobiol 1996;6:661–6.
90. Swerdlow RH, Parks JK, Cassarino DS, et al. Cybrids in Alzheimer's disease: a cellular model of the disease? Neurology 1997;49:918–25.
91. Askanas V, McFerrin J, Baqué S, Alvarez RB, Sarkozi E, Engel WK. Transfer of β-amyloid precursor protein gene using adenovirus vector causes mitochondrial abnormalities in cultured normal human muscle. Proc Natl Acad Sci U S A 1996;93:1314–9.
92. Davis RE, Miller S, Herrnstadt C, et al. Mutations in mitochondrial cytochrome *c* oxidase genes segregate with late-onset Alzheimer disease. Proc Natl Acad Sci U S A 1997;94:4526–31.
93. Hirano M, Shtilbans A, Mayeux R, et al. Apparent mtDNA heteroplasmy in Alzheimer's disease patients and in normals due to PCR amplification of nucleus-embedded mtDNA pseudogenes. Proc Natl Acad Sci U S A 1997;94:14894–9.
94. Wallace DC, Stugard C, Murdock D, Schurr T, Brown MD. Ancient mtDNA sequences in the human nuclear genome: a potential source of errors in identifying pathogenic mutations. Proc Natl Acad Sci U S A 1997;94:14900–5.
95. Davis RE, Miller S, Herrnstadt C, et al. Retraction. Proc Natl Acad Sci U S A 1998;95:12069.
96. Blass JP. Cerebral metabolic impairments. In: Khachaturian ZS, Radebaugh TS, eds. Alzheimer's disease. Cause(s), diagnosis, treatment and care. Boca Raton, FL: CRC Press, 1996:187–205.
97. Shoffner JM. Oxidative phosphorylation defects and Alzheimer's disease. Neurogenetics 1997;1:13–9.
98. Sheu KF, Cooper AJL, Koike K, Koike M, Lindsay JG, Blass JP. Abnormality of the α-ketoglutarate dehydrogenase complex in fibroblasts from familial Alzheimer's disease. Ann Neurol 1994;35:312–8.
99. Rapoport SI. Positron emission tomography in Alzheimer's disease in relation to disease pathogenesis: critical review. Cerebrovasc Brain Metab Rev 1991;3:296–335.
100. Pietrini P, Dani A, Furey M, et al. Brain stimulation reverses metabolic reductions in Alzheimer disease. Biol Psychiatry 1996;39:660.
101. Rapoport SI, Hätanpää K, Brady DR, Chandrasekaran K. Brain energy metabolism, cognitive function and down-regulated oxidative phosphorylation in Alzheimer's disease. Neurodegeneration 1996;5:473–6.
102. Rapoport SI. Deux stades, réversible et irréversible, de l'insuffisance fonctionnelle dans le cerveau alzheimerien. (Two stages, reversible and irreversible of functional insufficiency in the Alzheimer brain.)

In: Besson JM, Bassant MH, Calvino B, et al, eds. De la neurophysiologie à la maladie d'Alzheimer. (The neurophysiology of Alzheimer disease.) Marseille, France: Solal, 1997:165–72 (in French).

103. Crapper McLachlan DR, Dalton AJ, Kruck TPA, et al. Intramuscular desferrioxamine in patients with Alzheimer's disease. Lancet 1991;337:1304–8.

104. Sano M, Ernesto C, Thomas RG, et al. A controlled trial of selegiline, alpha-tocopherol, or both as treatment for Alzheimer's disease. N Engl J Med 1997;336:1216–22.

105. Kanowski S, Herrmann WM, Stephan K, Wierich W, Hörr R. Proofs of efficacy of the *Ginkgo biloba* special extract EGb 761 in outpatients suffering from mild to moderate primary degenerative dementia of the Alzheimer type or multi-infarct dementia. Pharmacopsychiatry 1996;29:47–56.

106. Le Bars PL, Katz MM, Berman N, Itil T, Freedman AM, Schalzberg AF. A placebo-controlled, double-blind, randomized trial of an extract of *Ginkgo biloba* for dementia. JAMA 1997;278:1327–32.

107. Oken BS, Storzbach DM, Kaye JA. The efficacy of *Ginkgo biloba* on cognitive function in Alzheimer disease. Arch Neurol 1998;55: 1409–15.

108. DeFeudis F. *Ginkgo biloba* extract (EGb 761). From chemistry to the clinic. Wiesbaden, Germany: Ullstein-Mosby, 1998.

109. Cohen-Salmon C, Venault P, Martin B, et al. Effects of *Ginkgo biloba* extract (EGb 761) on learning and possible actions on aging. J Physiol (Paris) 1997;91:291–300.

110. Breitner JCS, Welsh KA, Helms MJ, et al. Delayed onset of Alzheimer's disease with nonsteroidal anti-inflammatory and histamine H2 blocking drugs. Neurobiol Aging 1995;16:523–30.

111. Rich JB, Rasmusson DX, Folstein MF, Carson KA, Kawas C, Brandt J. Nonsteroidal anti-inflammatory drugs in Alzheimer's disease. Neurology 1995;45:51–5.

112. Behl C, Widmann M, Trapp T, Holsboer F. 17-β estradiol protects neurons from oxidative stress-induced cell death in vitro. Biochem Biophys Res Commun 1995;216:473–82.

113. Goodman Y, Bruce AJ, Cheng B, Mattson MP. Estrogens attenuate and cortisone exacerbates excitotoxicity, oxidative injury, and amyloid β-peptide toxicity in hippocampal neurons. J Neurochem 1996; 66:869–72.

114. Chadman K, Joseph JA, Shukitt-Hale B, Prior R, Taglialatela G, Bickford PC. Diets high in antioxidant activity prevent the deleterious effects of oxidative stress on signal transduction and nerve growth factor (NGF). Soc Neurosci 1997;23:348.

115. Gale CR, Martyn CN, Cooper C. Cognitive impairment and mortality in a cohort of elderly people. BMJ 1996;312:608–11.

116. Gilman MW, Cupples LA, Gagnon D, et al. Protective effect of fruits and vegetables on development of stroke in men. JAMA 1995;273:1113–7.

117. Geula C, Wu CK, Saroff D, Lorenzo A, Yuan M, Yankner BA. Aging renders the brain vulnerable to amyloid β-protein toxicity. Nat Med 1998;4:827–31.

Alzheimer's disease: perspectives for the new millennium

Perspective

Sangram S. Sisodia, Series Editor

Department of Neurobiology, Pharmacology, and Physiology, 947 E. 58th St. Abbott 316,
The University of Chicago, Chicago, Illinois. 60637, USA.
Phone: (773) 834-2900; Fax: (773) 702-3174; E-mail: ssisodia@drugs.bsd.uchicago.edu.

This Perspectives series on Alzheimer's disease is dedicated to the late Professor Henry M. Wisniewski, whose ceaseless devotion to the field over the last four decades has paved the way for many more decades of discovery and hope in the battle against this dreadful disease.

Age-related impairments in cognition and memory have been known since ancient times, but the clinical-pathological features of the syndrome, now termed "Alzheimer's disease" (AD), were not documented in the medical literature until the first decade of this century. It is now established that AD, a complex and genetically heterogeneous disorder, is the most common type of dementia occurring in mid-to-late life, affecting 7-10% of individuals over 65 years of age and perhaps 40% of persons over 80 years of age. The prevalence of this disease, which now affects more than 4 million individuals in the US, is increasing because of very significant shifts in life expectancy and demographic parameters; it is estimated that in 2050, about 25% of the population will be over 65 years of age.

In 1907, Alois Alzheimer reported the case of a middle-aged woman who developed memory deficits and progressive loss of cognitive abilities accompanied by morbid jealousy. Autopsy disclosed the now-recognized classic pathology of AD – the presence of numerous neurofibrillary tangles (NFTs) and senile plaques (SPs) in the neocortex and hippocampus. The majority of patients with the "sporadic AD" disease exhibit clinical signs during the seventh decade, whereas individuals with inherited AD (see below) often become demented in mid-life. Affected individuals show abnormalities of memory, problem solving, language, calculation, visuospatial perceptions, judgment, and behavior; some cases show psychotic symptoms, such as hallucinations and delusions. Activities of daily living become increasingly impaired; in late stages of the disease, patients are often mute, incontinent, and bedridden and usually die of intercurrent medical illnesses.

AD selectively affects neurons in certain brain regions and neural systems, including nerve cells in the cortex, hippocampus, amygdala, anterior thalamus, basal forebrain, and several brainstem monoaminergic nuclei. Many affected nerve cells exhibit intracellular accumulations of NFTs, poorly soluble filaments comprised principally of phosphorylated tau, a microtubule-associated protein. Because phosphorylated tau binds microtubules poorly and alters their stability, modifications of tau could have effects on intracellular transport, cellular geometry, and neuronal viability. SPs, abundant in the amygdala, hippocampus, and neocortex in cases of AD, are comprised of dystrophic neurites (abnormal nerve extensions) in proximity to deposits of highly fibrillogenic 42 amino acid $A\beta$ peptides ($A\beta42$) that aggregate into β-pleated sheets. Deposition of $A\beta$ in the neural parenchyma occurs early and selectively in AD. $A\beta$ peptides are derived from β-amyloid precursor proteins (APPs) and generated by the concerted actions of as-yet unidentified proteolytic activities, termed "β-secretase" and "γ-secretase." These $A\beta$ deposits likely act to recruit additional proteins, including α_1-antichymotrypsin, components of the complement cascade, and apolipoproteins, and these complexes may attract astrocytes and microglial cells that are often conspicuous around plaques.

Over the past 5 years, we have witnessed an explosion of information pertaining to the accuracy of clinical-pathological diagnoses, genetic risk factors, mechanisms of disease, and potential therapeutic strategies. The contributors to this Perspectives series on AD highlight the extraordinary advances in these arenas and offer perspectives for future investigations in the field.

Early diagnosis and classification

AD is usually classified on the basis of criteria formulated by the NINCDS-ADRDA Joint Task Force, and these methodologies provide highly reliable clinical diagnoses that are confirmed at autopsy. From the point of view of therapeutic intervention, however, it would be highly desirable to recognize individuals at the earliest stages of AD so that maximal benefit might be obtained from agents shown to exhibit therapeutic efficacy. In this issue of the *JCI*, John C. Morris at Washington University School of Medicine offers his Perspective on the distinctions between normal aging and early AD. Morris and colleagues have used a battery of clinical and psychometric measures to establish longitudinal assessments of a large cohort of individuals who showed no detectable cognitive impairment. His clinical assessments, confirmed by quantitative postmortem studies, support the idea that healthy brain aging is possible even up to the ninth decade. Interestingly, while cognitively normal individuals exhibit minimal neocortical pathology, the extent of the lesions, primarily the burden of SPs, is significantly elevated in brains of individuals at the earliest symptomatic stages of AD. These findings offer the provocative conclusion that the beginning of AD occurs with the deposition of β-amyloid in the form of large numbers of SPs. He proposes that the presence of SPs exacerbates neurofibrillary degeneration and culminates in neuronal dysfunction and death, which in turn leads to the appearance of clinical symptoms.

Strategies to assess genetic risk for AD

Considering the demographics of the population, we anticipate a profound increase in the prevalence of age-associated disorders, none of which is more dreaded than AD. For the population over the age of 65, epidemiological and genetic linkage studies have uncovered important risk factors for AD. Examples of these phenotypic modifier alleles include apolipoprotein E4 (Apo E-ε4) and a polymorphic α2-macroglobulin (*A2M-2*) variant; additional genetic risk factors undoubtedly exist. For several years, Rudolph Tanzi and colleagues at Harvard Medical School have employed powerful family-based association studies and statistical approaches to identify genetic risk factors (*A2M-2* among them) in the aging population. In his Perspective in this issue of the *JCI*, Tanzi offers a novel genetic dichotomy model based on the concept that while rare autosomal dominant inheritance of gene mutations causes highly penetrant early-onset form of the disease, the more common forms of AD that occur with advanced age are associated with the genetic risk factors that are distributed throughout the genome in the form of common population polymorphisms (CPPs). The ultimate goal of these studies is to assemble a genotypic profile for each individual that would determine the relative risk of acquiring AD, information that would be critical for making informed decisions regarding the use of potential therapeutic agents and clinical management.

The biology of presenilins in AD

Perhaps the most active area of research in the field over the past 5 years relates to the biology of APP, presenilin 1 (PS1), and presenilin 2 (PS2), membrane proteins that are mutated in pedigrees with early-onset, autosomal dominant forms of Familial Alzheimers's Disease (FAD). In this regard, in vitro and in vivo investigations have provided compelling evidence to support the view that FAD-linked mutant APP and PS act in a dominant, gain-of-function manner, leading to the overproduction of highly fibrillogenic Aβ42 peptides. In his Perspective, which will appear in the next issue of the *JCI*, Gopal Thinakaran at the University of Chicago reviews information that has emerged over the past 4 years pertaining to the normal biology of presenilins and the cellular and physiological consequences of expressing FAD-linked mutant polypeptides. For example, analysis of mice with ablated *PS1* alleles and of lower organisms (*Drosophila* and *Caenorhabditis elegans*) lacking presenilin homologues reveals important roles for these molecules in facilitating Notch signaling and developmental cell fate decisions. In this regard, biochemical and genetic evidence indicates that PS1 plays a critical role in facilitating the proteolytic processing of several membrane proteins, including APP and Notch1. Thinakaran considers several competing models to explain how presenilins facilitate the proteolysis of these diverse substrates, and he suggests experimental strategies that may help resolve this important issue.

Modulation of Aβ production as a therapeutic strategy

Despite the wealth of evidence supporting the view that Aβ plays a critical, if not essential, role in the pathogenesis of AD, the reader should be forewarned that the field still remains somewhat divided regarding the role of amyloid deposition in AD pathogenesis. Regardless, the pharmaceutical industry is heavily invested in developing programs aimed at attenuating production and/or aggregation of Aβ peptides. In the next issue of the *JCI*, Steve Wagner at SIBIA Neurosciences Inc. provides an overview of therapeutic strategies focused on inhibiting Aβ peptide formation by modulating the still ill-defined β- and γ-secretase activities. Wagner outlines strategies to design molecules that are effective in inhibiting Aβ production in cell-based assays, as well as methods to validate the in vivo efficacy of these compounds in transgenic animals that overexpress wild-type or mutant human APP. Those compounds that prove efficacious in reducing Aβ production and amyloid burdens in transgenic mice will no doubt be introduced promptly into people with AD.

Pergamon

European Journal of Cancer Vol. 33, No. 5, pp. 703-709, 1997

Printed in Great Britain
0959-8049/97 $17.00 + 0.00

PII: S0959-8049(96)00058-0

The Biology of Replicative Senescence

J. Campisi

Department of Cancer Biology, Berkeley National Laboratory, Mailstop 70A-1118, 1 Cyclotron Road, Berkeley, California 94720, U.S.A.

Most cells cannot divide indefinitely due to a process termed cellular or replicative senescence. Replicative senescence appears to be a fundamental feature of somatic cells, with the exception of most tumour cells and possibly certain stem cells. How do cells sense the number of divisions they have completed? Although it has not yet been critically tested, the telomere shortening hypothesis is currently perhaps the best explanation for a cell division 'counting' mechanism. Why do cells irreversibly cease proliferation after completing a finite number of divisions? It is now known that replicative senescence alters the expression of a few crucial growth-regulatory genes. It is not known how these changes in growth-regulatory gene expression are related to telomere shortening in higher eukaryotes. However, lower eukaryotes have provided several plausible mechanisms. Finally, what are the physiological consequences of replicative senescence? Several lines of evidence suggest that, at least in human cells, replicative senescence is a powerful tumour suppressive mechanism. There is also indirect evidence that replicative senescence contributes to ageing. Taken together, current findings suggest that, at least in mammals, replicative senescence may have evolved to curtail tumorigenesis, but may also have the unselected effect of contributing to age-related pathologies, including cancer.

Key words: replicative senescence, cancer, ageing, somatic cells, tumour suppression

Eur J Cancer, Vol. 33, No. 5, pp. 703–709, 1997

INTRODUCTION

MOST HIGHER eukaryotic cells that can divide *in vivo* cannot do so indefinitely. Thus, normal cells are said to have a finite replicative life span. The process that limits the proliferative potential of cells has been termed cellular or replicative senescence. Replicative senescence was first described formally more than 30 years ago, for human fibroblasts in culture [1]. Since then, many cell types from a variety of animal species have been shown to have a finite replicative life span (Figure 1). Most studies of replicative senescence have used cells grown in culture. However, a limited number of studies using cells passaged in intact animals, or cells identified in intact tissues, strongly suggest that many cells undergo replicative senescence *in vivo* (reviewed in [2–4]).

Do all dividing cells senesce? Studies of simple, single-cell organisms, such as *Saccharomyces cerevisiae*, suggest that replicative senescence may be a very primitive phenotype [5]. Nonetheless, here I focus on mammalian, and principally human, cells. Replicative senescence is especially stringent in human cells which, unlike many rodent cells, rarely spontaneously immortalise (that is, spontaneously fail to senesce) [6–8]. Most proliferatively competent (that is, nonterminally differentiated) mammalian cells senesce. There are, however, three notable exceptions. First, the germ-line is capable of continuous replication. Thus, somatic cells acquire a finite replicative life span sometime during embryonic development. Very little is known about when, where or how this occurs during development. Second, some stem cells may, even in adult organisms, replicate continuously. This possibility has not yet been critically tested, and thus remains an interesting speculation, particularly for human cells. Third, many malignant tumours appear to have cells with an infinite, or immortal, replicative life span. This finding, and results discussed below, suggest that replicative senescence is a tumour suppressive mechanism, and that tumour cells often acquire mutations that allow them to overcome the proliferative constraints it imposes [6].

THE SENESCENT PHENOTYPE

The number of divisions that normal cells complete before they senesce depends on the species, age and genetic background of the donor, as well as the particular cell type (reviewed in [2, 3]). This number can be fairly large, for example, 60–80 doublings for fetal or neonatal human cells. The mechanism by which cells sense the number of divisions they have completed appears to depend, at least in part, on the length of their telomeres. However, virtually nothing is known in mammalian cells about why telomeres

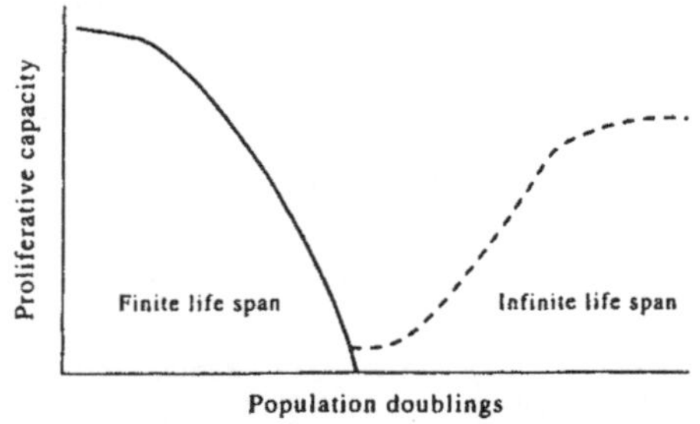

Figure 1. Replicative senescence. The proliferative capacity of most normal somatic mammalian cell populations declines more or less exponentially with each doubling. Such cells are said to have a finite replicative life span, and the process that limits the capacity for cell division has been termed cellular or replicative senescence. Replicative senescence is particularly stringent in human cells. However, oncogenic viruses or chemical or physical carcinogens may permit some human cells to escape replicative senescence and acquire an immortal or indefinite replicative life span. Cells from many rodent species may spontaneously escape senescence. In both cases, escape from senescence is a rare event, occurring with frequencies that range from 1 in 10^{-5} to 1 in 10^{-7}.

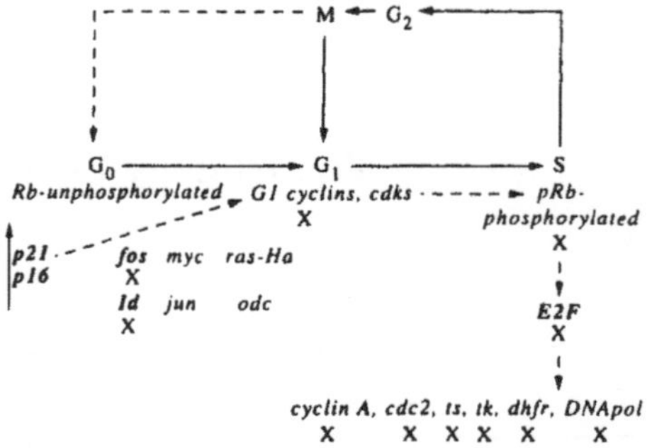

Figure 2. Blocks to cell cycle progression in senescent cells. Mitogens induce a number of growth-related genes in senescent cells. However, at least three transcriptional regulators (c-fos, Id and E2F), whose activities are important for cell proliferation, are repressed in senescent cells (indicated by an X). The repression of E2F is very likely due to the overexpression of p21, which inhibits the activity of cyclin-dependent protein kinases (CDKs). Because CDKs are responsible for phosphorylating the pRb protein, pRb remains in its growth suppressive or underphosphorylated form in senescent cells. Underphosphorylated pRb, in turn, represses E2F activity, which prevents the expression of a host of genes that are necessary or permissive for the initiation of DNA synthesis.

should influence cell proliferation or other cell phenotypes (discussed below) so strongly. Thus, before discussing the role of telomeres in replicative senescence, its genetics and the phenotypic changes that it causes will be briefly discussed.

Genetics

The finite replicative life span of cells is a dominant phenotype. This conclusion is based primarily on a large number of somatic cell fusion studies in which proliferating normal cells were fused to immortal tumour-derived cells. In the majority of cases, the hybrid cells proliferated for some time, but eventually they senesced (see [2, 3] for review). Moreover, fusions between different immortal human tumour cells often, but not always, as discussed below, produced hybrid cells that senesced [9]. Thus, replicative senescence appears to be genetically dominant, and replicative immortality appears to be recessive.

The ability of some, but not all, fusion pairs of immortal human cells to produce hybrid cells that are *not* immortal has led to the identification of four complementation groups for replicative immortality [10]. To date, three of these complementation groups have been assigned to specific human chromosomes. Thus, normal human chromosomes 4, 1 and 7 have been shown to reverse immortality (i.e. induce senescence) in multiple cell lines assigned to complementation groups B, C and D, respectively [11–13]. The genes of interest that reside on these chromosomes have not yet been identified. Taken together, these data suggest that at least four dominant-acting genetic loci act to limit the replicative potential of normal human cells.

Growth arrest

Upon completing a finite number of divisions, cell growth (used here interchangeably with proliferation) is arrested with a G1 DNA content (Figure 2). Once thus arrested, they cannot be stimulated to enter the S phase of the cell cycle by any known combination of physiological mitogens. This growth arrest, like the finite replicative life span, is a dominant trait in somatic cell hybrids (reviewed in [2, 3]). These findings suggest that once cells undergo replicative senescence they express one or more inhibitors of cell cycle progression that can act in a transdominant manner.

The growth arrest associated with cellular senescence has been studied most extensively in cultured human fibroblasts [14]. It is clear from these studies that many genes, including at least three proto-oncogenes, remain mitogen-inducible in senescent cells [15, 16] (Figure 2). Thus, senescent cells do not fail to proliferate due to a *general* breakdown of growth factor signal transduction. Rather, it is now known that replicative senescence causes the selective repression of a few positive-acting, growth regulatory genes whose expression has been shown to be important for G1 progression and DNA synthesis. In fibroblasts, these genes include the *C-FOS* proto-oncogene [16], the helix-loop-helix *ID-1* and *ID-2* genes [17], and the E2F-1 [18] and E2F-5 [19] components of the E2F transcription factor. In addition, the retinoblastoma tumour suppressor protein (pRb) remains in its growth suppressive form (that is, remains constitutively underphosphorylated) [20]. Thus, the immediate cause for the growth arrest of senescent cells appears to be a deficiency in a few, key positive-acting growth regulators, including c-fos, Id and E2F activity, and phosphorylated pRb (Figure 2).

In addition to these deficiencies in positive growth-regulators, senescent human fibroblasts overexpress two negative growth regulators: the p21 and p16 inhibitors of cyclin-dependent protein kinases (*CDKs*) [21, 22]. These high levels of p21 and p16 are very likely responsible for the accumulation of inactive CDK complexes [23], and thus the constitutive underphosphorylation of pRb [20] in senescent cells. Furthermore, because p21 can inhibit E2F by both pRb-dependent and pRb-independent mechanisms [24], the high level of p21 may well be also responsible for the lack of E2F activity in senescent cells [18, 19] (Figure 2).

p16 and p21 are obvious candidates for the dominant inhibitors of cell proliferation that are expressed by senescent cells. The mechanisms responsible for the senescence-associated increase in p16 and p21 mRNA and protein are not known. In the case of p21, the large increase in mRNA does not appear to be due to an increase in transcription (M. Nakanishi, G. Dimri, J. Campisi and J. Smith, Baylor College of Medicine and Berkeley National Laboratory, U.S.A.). It is unlikely, however, that p16 and p21 are the only growth inhibitors expressed by senescent cells. For example, indirect evidence suggests that senescent cells express at least one additional growth inhibitor, which may belong to a family of basic helix-loop-helix transcription factors [25]. Moreover, the upstream regulators of p16 and p21 expression in senescent cells, which have not yet been identified, may induce additional growth-inhibitory genes in the cells.

Other phenotypic changes

The arrest of cell proliferation is clearly the most important feature of replicative senescence with regard to its role in tumour suppression. However, in addition to this essentially irreversible arrest of cell proliferation, senescent cells display two other striking phenotypic changes.

First, senescent cells acquire resistance to apoptotic stimuli [26]. The mechanism by which senescent cells resist apoptotic death is not well understood. Nonetheless, the fact that senescent cells are in fact resistant to apoptosis emphasises the distinction, often blurred in the literature, between cell death and cell senescence [27].

Second, senescent cells show sometimes striking changes in differentiated functions. The functions that are altered by senescence depend on the cell type. For example, senescent human fibroblasts and endothelial cells overexpress the inflammatory cytokine interleukin-1α [28], senescent endothelial cells overexpress the cell-specific adhesion molecule I-CAM [29], and senescent mammary epithelial cells overexpress the β isoform of the retinoic acid receptor [30]. Very little is known about the mechanisms that alter differentiated gene expression in senescent cells. However, as is the case with growth and differentiation in many cells, the senescence-associated changes in differentiation appear to be tightly linked to growth arrest.

The altered differentiation associated with replicative senescence can have rather profound consequences for cell and, at least in principle, tissue function. For example, presenescent dermal fibroblasts express low levels of collagenase and stromelysin, metalloproteases that degrade extracellular matrix proteins. They also express high levels of the metalloprotease inhibitors TIMP-1 and TIMP-3 (tissue inhibitor of metalloproteinases 1 and 3). Upon senescence, metalloproteinase expression rises and TIMP expression falls [31–33]. Thus, in dermal fibroblasts, replicative senescence entails a clear switch from a matrix-producing phenotype to a matrix-degrading phenotype. There are now a number of examples in which senescent cells elaborate cytokines, extracellular matrix-modifying enzymes, or other molecules that can have far-reaching effects on the micro-environment of neighbouring cells.

PHYSIOLOGICAL SIGNIFICANCE: CANCER AND AGEING

A priori, a process that limits cell proliferation is a candidate tumour suppressive mechanism. Thus, the idea that replicative senescence, which entails an irreversible growth arrest, curtails tumorigenesis is intuitively appealing. In addition, there are substantial cell and molecular biological data (discussed below) to support this idea. By contrast, the senescence-associated changes in differentiation may have little to do with tumour suppression *per se*. Changes in differentiated gene expression are often coupled to changes in growth state or potential. Thus, altered differentiation may be a by-product of the senescence-associated growth arrest. However, it may not be without physiological consequence.

Altered differentiation, more than arrested growth, may be important for the role of replicative senescence in ageing [3, 4]. The idea that cell senescence has a role in both tumour suppression and ageing may seem contradictory. However, evolutionary theories suggest that some traits that were selected to optimise health during the period of reproductive fitness can have unselected, deleterious postreproductive effects (see [34] for review). Thus, replicative senescence, by curtailing cell proliferation, may contribute to the relative freedom from cancer that is seen during the first half of life. Later in life, it may be deleterious because senescent cells accumulate in aged tissue [35], where their altered differentiation may compromise tissue function and integrity [3, 4, 36]. For example, the age-dependent loss of dermal collagen may be due to senescent fibroblasts, which accumulate in aged skin [35] and express an altered balance of collagenase and TIMPs [31–33]. An additional, but highly speculative, possibility is that senescent cells may actually contribute to carcinogenesis late in life by altering the micro-environment, which is paramount for maintaining normal cell growth and differentiation [36].

TUMOUR SUPPRESSION

What is the evidence that replicative senescence is a tumour suppressive mechanism? Firstly, cells that have a finite replicative life span are orders of magnitude less likely to form tumours than replicatively immortal cells. Moreover, many tumours contain cells that have an increased division potential or are immortal (see [6–8, 37] for review). There is no evidence that immortality is required for tumorigenesis (although it may be for metastasis). Rather, there appears to be a selection for cells with an increased replicative potential in the development of cancer. Immortality, of course, permits the extensive cell division that is needed to acquire the many mutations that are the hallmark of malignant tumours.

Secondly, a number of well-recognised oncogenes appear to act, at least in part, by allowing cells to escape replicative senescence (see [3, 6, 8] for review). These include mutated or deregulated cellular genes such as *TP53* or *C-MYC*, as well as the oncogenes of certain viruses that are implicated in some human cancers (for example, the human papilloma virus genes *E6* or *E7*, or the SV40 virus large T antigen gene). Thus, mutations that lead to tumorigenesis, or the strategies of oncogenic viruses, engage mechanisms that overcome replicative senescence.

Thirdly, among the genes that are essential for establishing and maintaining replicative senescence, are two well-recognised tumour suppressors: the *TP53* and *RB* genes.

p53 and/or pRb inactivation is the primary means by which viral oncoproteins extend the replicative life span of cells (see [3, 6, 8] for review). Moreover, inactivation of *TP53* or *RB* by antisense oligonucleotides also extends the replicative life span of human fibroblasts [38]. In the absence of either *TP53* or *RB*, human fibroblasts continue to proliferate beyond the population-doubling level at which they would normally senesce, but eventually they do senesce. In the absence of both *TP53* and *RB*, human fibroblasts become genomically unstable and either die or mutate to an immortal phenotype [3, 6–8, 36, 37]. There is a striking and extremely interesting distinction between human fibroblasts and certain epithelial cells in this regard. Inactivation of only *TP53* or *RB*, for example, by the papilloma virus oncoproteins E6 or E7, is not sufficient to immortalise human fibroblasts. However, E6 or E7 alone will immortalise certain epithelial cells (for example, mammary epithelial cells) albeit at a very low frequency [39, 40]. Thus, *TP53* and *RB* are essential for the normal replicative senescence of cells. Needless to say, *TP53* and *RB* are together the most commonly lost functions in human cancers.

THE ROLE OF TELOMERES AND TELOMERASE

Telomere shortening as the cell division counting mechanism

How do cells sense the number of divisions through which they have gone? As noted above, this number can easily exceed 50 cell cycles. Moreover, for a given culture, the number of doublings at which the entire cell population senesces is fairly reproducible. At present, telomere shortening is perhaps the most viable explanation for a cell division 'counting' mechanism [8, 37, 41].

As discussed elsewhere in this Special Issue, telomeres are the ends of linear chromosomes, consisting of the repetitive sequence TTAGGG in humans and other vertebrates, and specialised proteins. The telomeric sequence and its binding proteins form a distinctive structure [42] which prevents chromosome fusions, translocation and non-dysjunctions. Thus, telomeres are essential for maintaining the stability of eukaryotic genomes [43]. Because DNA polymerases are unidirectional and require a labile primer, each round of replication leaves some 3' bases at the telomere unreplicated [8, 37, 41, 43]. Telomerase, a multimeric enzyme that adds telomeric repeats to chromosome ends *de novo* [44], is not expressed by most normal somatic cells. Thus, for most cells, the telomeres shorten with each cell cycle.

Telomere length is generally estimated from the terminal restriction fragment (TRF), which contains the telomeric TTAGGG tract plus telomere-like and non-telomeric sequences [45]. As expected, senescent human cells have shorter TRFs than their presenescent counterparts [46]. How much shorter? Germ cells, which express telomerase, have mean TRFs of 10–15 kb [45]. However, presenescent cultures from fetal or neonatal tissue generally have mean TRFs of 8–10 kb; TRFs in cells from older donors may be even shorter [46–48]. The mean TRF decreases by approximately 50 bp per doubling for cultured human fibroblasts [41, 49]. However, regardless of the starting length, the mean TRF declines to approximately 6 kb at senescence. Thus, initial mean TRF length is a good predictor of the replicative life span of cells [49]. Because TRF lengths vary substantially within a cell population, it has been proposed that senescence occurs when one or more TRF reaches a critical length [50]. This length is likely to be somewhat less than 6 kb, with perhaps most or all of the telomeric TTAGGG tract being lost from the critical chromosome.

Inactivation of *TP53* or *RB* allows cells to continue to proliferate, despite critically short TRFs [8, 38–40]. It is not known how these tumour suppressor proteins act in this capacity, but presumably they prevent cells from sensing, or allow cells to bypass, the signal(s) generated by a short telomere. As noted earlier, the majority of cells lacking functional p53 or pRb proteins do eventually senesce. There may be a minimal telomere length (TRF of approximately 2 kb) below which cell viability may be compromised. Thus, in the absence of telomerase or other mechanisms for maintaining telomere length (discussed below), cells that fail to senesce may be destined to die.

The idea that telomere shortening is the 'counting' mechanism that drives replicative senescence is attractive because it provides a plausible scheme, based on the known biochemistry of DNA replication, by which cells sense the number of completed divisions. It does not predict how a critically short telomere induces the senescent phenotype, nor does it necessarily predict that telomerase-expressing cells should not senesce. To date, however, it has not been possible to test critically this hypothesis by, for example, preventing telomere shortening in proliferating normal cells. However, certain oligonucleotides have been found to elongate the telomeres of immortal cells with relatively short telomeres; when fused to normal cells, these telomere-elongated immortal cells give rise to hybrids with a greater division potential than hybrids between normal and untreated immortal cells [51]. Because these oligonucleotides have no effect on the telomeres of normal cells, it remains to be shown that manipulation of telomere length alters the onset of normal cell senescence. Thus, despite its teleological appeal and an increasing body of indirect and correlative evidence, the telomere-shortening hypothesis of replicative senescence awaits critical testing.

The role of telomerase

As noted earlier, three cell types do not appear to undergo replicative senescence: the germ-line, certain stem cells, and many malignant tumour cells. These cells differ from most other cell types in that they express telomerase [52–58]. Thus, telomerase is undoubtedly important for unrestricted cell proliferation. However, telomerase is neither sufficient nor necessary for an unlimited or immortal replicative life span.

Firstly, normal human T-cells transiently express telomerase activity upon activation. Nonetheless, the telomeres of these cells shorten and the cells do senesce [59–62]. T-cells may not be unusual among normal somatic cells in that they express telomerase when stimulated to proliferate. Telomerase is also expressed, albeit at a low level, by cells in the proliferating basal layer of human skin [63]. However, the results in T-cells demonstrate clearly that telomerase activity *per se* can be insufficient to prevent telomere shortening and replicate senescence. The reason for this insufficiency is not known, although there are several possibilities. For example, the level of telomerase expressed by T-cells may be quantitatively inadequate to keep pace with the telomere shortening that occurs with each cell division. Alternatively, the enzyme or the telomeres may be modified such that access to the telomere or telomerase activity may be retarded or inhibited.

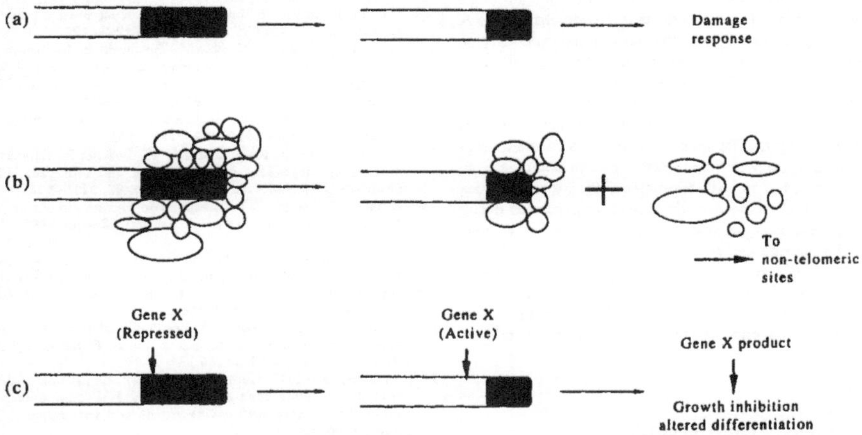

Figure 3. Three models by which a short telomere might induce the senescent phenotype. Chromosome ends are depicted, with the telomeric and immediate subtelomeric portions shown in black. (a) A shortened telomere resembles a double-strand DNA break, and thus triggers a DNA-damage response. (b) Telomeres sequester proteins that can bind to intragenomic sites, where they may activate or repress gene expression. As the telomere shortens, these proteins are freed. They may then induce or repress genes, such that growth is inhibited and differentiation is altered. (c) The field of heterochromatin surrounding the telomere shrinks as the telomere shortens. Genes within the heterochromatic region are repressed, and this repression is relieved upon telomere shortening. One or more of such genes may encode proteins that can arrest growth and/or alter differentiation.

Secondly, telomerase is not essential for maintaining telomere length or an unlimited division potential. Some human tumour cells do not express detectable telomerase activity, despite an immortal replicative life span [56, 64]. Such cells have frequently long or very heterogeneous telomeres. Moreover, some telomerase-expressing cells maintain long telomeres and an immortal phenotype even when telomerase is inhibited by reverse transcriptase inhibitors [65]. Thus, although many, if not most, tumour cells express telomerase [56], it is clear that both telomere length and an indefinite cell division can be maintained by a telomerase-independent mechanism.

Telomere shortening and the senescent phenotype

How might critically short telomeres induce the senescent phenotype? Currently, there is only speculation in mammalian cells. However, several hypotheses, based largely on experiments in lower eukaryotes, have been proposed. Of these, three are discussed here (Figure 3).

(a) A short telomere may signal a DNA damage response [14, 37, 66]. This hypothesis suggests that a critically short telomere triggers a DNA damage checkpoint from which cells cannot recover. In yeast, elimination of a telomere (from a dispensable chromosome) causes cell-cycle arrest that depends on a DNA damage–response pathway [67]. One problem with this hypothesis is that this growth arrest in yeast is transient, whereas the growth arrest of senescent mammalian cells is essentially irreversible. However, some forms of damage to mammalian cells (e.g. radiation, hydrogen peroxide) can induce permanent cell cycle arrest, although in these cases it is not known whether DNA is the critically damaged target. Moreover, the p53 protein level, which increases in the mammalian DNA damage response, is not elevated in senescent cells [68, 69]. In addition, if senescence is a damage response, immortalisation must overcome the damage-induced arrest and allow repair. However, cells reversibly immortalised by a conditional SV40 T antigen return to a growth-arrested state resembling senescence when the T antigen is inactive [66, 70, 71]. Finally, it is difficult to explain the altered differentiation of senescent cells as a simple DNA-damage response.

(b) The telomeres may bind or sequester transcription factors which can activate or repress a variety of genes. As the telomeres shorten, these factors would be available to bind intragenomic target sites, where they could act to repress or activate genes important for cell cycle progression and maintenance of differentiated function. There is a precedent for this idea in yeast, where the telomere-associated Rap1 protein has been shown to sequester silencing factors that can act at non-telomeric sites when they are not sequestered [72]. Although there is no structural homologue to Rap1 in mammalian cells, there may well be functional homologues. Although this hypothesis can in principal explain the complexity of the senescent phenotype, there is currently no molecular data to support it in mammalian cells.

(c) The heterochromatic structure of DNA near the telomere may repress regulatory genes (for example, transcription factors) that, in turn, act to arrest the growth and alter the differentiation of cells; as telomeres shorten, this field of heterochromatin should diminish, thereby derepressing these putative regulators [37, 66, 73, 74]. Again, there is strong precedent for the silencing of loci near telomeres in yeast [75, 76], although far less evidence for such effects in mammalian cells. A related hypothesis suggests that loss of gene silencing or changes in chromatin structure independent of telomere length may induce the senescent phenotype [77]. In the yeast *Saccharomyces cerevisiae*, individual cells have a finite replicative life span, but the telomeres do not shorten during the life span [5]. Despite a constant telomere length, there is nonetheless a loss of silencing near at least one, but not all, telomeres [78]. Moreover, loss of silencing at non-telomeric loci also appears to occur during replicative ageing in yeast [79]. Whether and how the senescence of yeast, in which the cell is the organism, reflects the replicative senescence of individual mammalian cells is not yet clear. The idea that chromatin structure links telomere shortening with replicative senescence in mammalian cells awaits critical testing.

CONCLUSION

Replicative senescence prevents cells from dividing indefinitely, and appears to be a powerful, albeit imperfect, tumour suppressive mechanism. This process very likely also has the unselected evolutionary consequence of contributing to ageing. As such, the growth arrest associated with senescence may be the selected phenotype that suppresses tumorigenesis, whereas the altered differentiation of senescent cells may be the linked but unselected phenotype that contributes to ageing. Very little is known at a molecular level about how the growth arrest is linked to the altered differentiation. Nonetheless, there is strong circumstantial evidence that the length of the telomeres plays an important role in timing and establishing the senescent state, and that the cell's response to a critically short telomere is mediated by the *TP53* and *RB* tumour suppressor genes.

1. Hayflick L. The limited *in vitro* lifetime of human diploid cell strains. *Exp Cell Res* 1965, 37, 614–636.
2. Stanulis-Praeger B. Cellular senescence revisited: a review. *Mech Ageing Dev* 1987, **38**, 1–48.
3. Campisi J, Dimri GP, Hara E. Control of replicative senescence. In Schneider E, Rowe J, eds. *Handbook of the Biology of Aging*. New York, Academic Press, 1996, 121–149.
4. Campisi J. Replicative senescence: an old lives tale? *Cell* 1996, **84**, 497–500.
5. Jazwinski SM. The genetics of aging in the yeast *Saccharomyces cerevisiae*. *Genetica* 1993, **91**, 35–51.
6. Sager R. Senescence as a mode of tumor suppression. *Environ Health Persp* 1991, **93**, 59–62.
7. McCormick JJ, Maher VM. Towards an understanding of the malignant transformation of diploid human fibroblasts. *Mutat Res* 1988, **199**, 273–291.
8. Shay JW, Wright WE. Defining the molecular mechanisms of human cell immortalization. *Biochim Biophys Acta* 1991, **1071**, 1–7.
9. Pereira-Smith OM, Smith JR. Evidence for the recessive nature of cellular immortality. *Science* 1983, **221**, 964–967.
10. Pereira-Smith OM, Smith JR. Genetic analysis of indefinite division in human cells: identification of four complementation groups. *Proc Natl Acad Sci USA* 1988, **85**, 6042–6046.
11. Ning Y, Weber JL, Killary AM, Ledbetter DH, Smith JR, Pereira-Smith OM. Genetic analysis of indefinite division in human cells: evidence for a senescence-related gene(s) on human chromosome 4. *Proc Natl Acad Sci USA* 1991, **88**, 5635–5639.
12. Hensler P, Annab LA, Barrett JC, Pereira-Smith OM. A gene involved in control of human cellular senescence on human chromosome 1q. *Mol Cell Biol* 1994, **14**, 2292–2297.
13. Ogata T, Ayusawa D, Namba M, Takahashi E, Oshimura M, Oishi M. Chromosome 7 suppresses indefinite division of nontumorigenic immortalized human fibroblast cell lines KMST-6 and SUSM-1. *Mol Cell Biol* 1993, **13**, 6036–6043.
14. Goldstein S. Replicative senescence: the human fibroblast comes of age. *Science* 1990, **249**, 1129–1133.
15. Rittling SR, Brooks KM, Cristofalo VJ, Baserga R. Expression of cell cycle dependent genes in young and senescent WI38 fibroblasts. *Proc Natl Acad Sci USA* 1986, **83**, 3316–3320.
16. Seshadri T, Campisi J. c-fos repression and an altered genetic program in senescent human fibroblasts. *Science* 1990, **247**, 205–209.
17. Hara E, Yamaguchi T, Nojima H, *et al*. Id related genes encoding helix loop helix proteins are required for G1 progression and are repressed in senescent human fibroblasts. *J Biol Chem* 1994, **269**, 2139–2145.
18. Dimri GP, Hara E, Campisi J. Regulation of two E2F related genes in presenescent and senescent human fibroblasts. *J Biol Chem* 1994, **269**, 16180–16186.
19. Good LF, Dimri GP, Campisi J, Chen KY. Regulation of dihydrofolate reductase gene expression and E2F components in human diploid fibroblasts during growth and senescence. *J Cell Physiol* 1996, **168**, 580–588.
20. Stein GH, Beeson M, Gordon L. Failure to phosphorylate the retinoblastoma gene product in senescent human fibroblasts. *Science* 1990, **249**, 666–669.
21. Noda A, Ning Y, Venable SF, Pereira-Smith OM, Smith JR. Cloning of senescent cell derived inhibitors of DNA synthesis using an expression screen. *Exp Cell Res* 1994, **211**, 90–98.
22. Hara E, Smith R, Parry D, Tahara H, Peters G. Regulation of p16 (CdkN2) expression and its implications for cell immortalization and senescence. *Mol Cell Biol* 1996, **16**, 859–867.
23. Dulic V, Drullinger LF, Lees E, Reed SI, Stein GH. Altered regulation of G1 cyclins in senescent human diploid fibroblasts: accumulation of inactive cyclin E-cdk and cyclin D-cdk complexes. *Proc Natl Acad Sci USA* 1993, **90**, 11034–11038.
24. Dimri GP, Nakanishi M, Desprez PY, Smith JR, Campisi J. Inhibition of E2F activity by the p21 inhibitor of cyclin-dependent protein kinases in cells expressing or lacking a functional retinoblastoma protein. *Mol Cell Biol* 1996, **16**, 2987–2997.
25. Hara E, Uzman JA, Dimri GP, Nehlin JO, Testori A, Campisi J. The helix-loop-helix protein Id-1 and a retinoblastoma protein binding mutant of SV40 T antigen synergize to reactivate DNA synthesis in senescent human fibroblasts. *Dev Genetics* 1996, **18**, 161–172.
26. Wang E. Senescent human fibroblasts resist programmed cell death and failure to suppress bcl2 is involved. *Cancer Res* 1995, **55**, 2284–2292.
27. Linskens M, Harley CB, West MD, Campisi J, Hayflick L. Replicative senescence and cell death. *Science* 1995, **267**, 17.
28. Maier JAM, Voulalas P, Roeder D, Maciag T. Extension of the life-span of human endothelial cells by an interleukin-1α antisense oligomer. *Science* 1990, **249**, 1570–1574.
29. Maier JAM, Statuto M, Ragnoti G. Senescence stimulates U037-endothelial cell interactions. *Exp Cell Res* 1993, **208**, 270–274.
30. Swisshelm K, Ryan K, Lee X, Tsou H, Peacocke M, Sager R. Down-regulation of retinoic acid receptor β in mammary carcinoma cell lines and its up-regulation in senescing normal mammary epithelial cells. *Cell Growth Diff* 1994, **5**, 133–141.
31. West MD, Pereira-Smith OM, Smith JR. Replicative senescence of human skin fibroblasts correlates with a loss of regulation and overexpression of collagenase activity. *Exp Cell Res* 1989, **184**, 138–137.
32. Millis AJ, Hoyle M, McCue HM, Martini H. Differential expression of metalloproteinase and tissue inhibitor of metalloproteinase genes in aged human fibroblasts. *Exp Cell Res* 1992, **201**, 373–379.
33. Wick M, Burger C, Brusselbach S, Lucibello FC, Muller R. A novel member of human tissue inhibitor of metalloproteinases (TIMP) gene family is regulated during G1 progression, mitogenic stimulation, differentiation and senescence. *J Biol Chem* 1994, **269**, 18953–18960.

34. Finch CE. *Longevity, Senescence and the Genome*. Chicago, Illinois, University of Chicago Press, 1990.
35. Dimri GP, Lee X, Basile G, *et al*. A novel biomarker identifies senescent human cells in culture and aging skin *in vivo*. *Proc Natl Acad Sci USA* 1995, **92**, 9363–9367.
36. Campisi J. Aging and cancer: the double-edged sword of replicative senescence. *J Am Geriatrics Soc* 1997, **45**, 1–6.
37. Harley CB, Villeponteau B. Telomeres and telomerase in aging and cancer. *Curr Opin Genet Dev* 1995, **5**, 249–255.
38. Hara E, Tsuri H, Shinozaki S, Oda K. Cooperative effect of antisense-Rb and antisense-p53 oligomers on the extension of lifespan in human diploid fibroblasts, TIG-1. *Biochem Biophys Res Comm* 1991, **179**, 528–534.
39. Shay JW, Wright WE, Brasiskyte D, Vanderhaegen BA. E6 of human papillomavirus type-16 can overcome the M1 stage immortalization in human mammary epithelial cells but not in human fibroblasts. *Oncogene* 1993, **8**, 1407–1413.
40. Wazer DE, Liu XL, Chu Q, Gao Q, Band V. Immortalization of distinct human mammary epithelial cell types by human papilloma virus 16 E6 or E7. *Proc Natl Acad Sci USA* 1995, **92**, 3687–3691.
41. Levy MZ, Allsopp RC, Futcher AB, Greider CW, Harley CB. Telomere end-replication problem and cell aging. *J Mol Biol* 1992, **225**, 951–960.
42. Tommerup H, Dousmanis A, De Lange T. Unusual chromatin in human telomeres. *Mol Cell Biol* 1994, **14**, 5777–5785.
43. Blackburn EH. Structure and function of telomeres. *Nature* 1991, **350**, 569–573.
44. Blackburn EH. Telomerases. *Ann Rev Biochem* 1992, **61**, 113–129.
45. Allshire RC, Dempster M, Hastie ND. Human telomeres contain at least three types of G-rich repeat distributed non-randomly. *Nucleic Acids Res* 1989, **17**, 4611–4627.
46. Harley CB, Futcher AB, Greider CW. Telomeres shorten during aging of human fibroblasts. *Nature* 1990, **345**, 458–460.
47. Hastie ND, Dempster M, Dunlop MG, Thompson AM, Green DK, Allshire RC. Telomere reduction in human colorectal carcinoma and with aging. *Nature* 1990, **346**, 866–868.
48. Lindsey J, McGill NI, Lindsey LA, Green DK, Cooke HJ. *In vivo* loss of telomeric repeats with age in humans. *Mutat Res* 1991, **256**, 45–48.
49. Allsopp RC, Vaziri H, Patterson C, *et al*. Telomere length predicts replicative capacity of human fibroblasts. *Proc Natl Acad Sci USA* 1992, **89**, 10114–10118.
50. Allsopp RC, Harley CB. Evidence for a critical telomere length in senescent human fibroblasts. *Exp Cell Res* 1995, **219**, 130–136.
51. Wright WE, Brasiskyte D, Piatyszek M, Shay JW. Experimental elongation of telomeres extends the lifespan of immortal × normal cell hybrids. *EMBO J* 1996, **15**, 1734–1741.
52. Kim NW, Piatyszek MA, Prowse KR, *et al*. Specific association of human telomerase activity with immortal cells and cancer. *Science* 1994, **226**, 2011–2015.
53. Counter CM, Gupta J, Harley CB, Leber B, Bacchetti S. Telomerase activity in normal leukocytes and in hematologic malignancies. *Blood* 1995, **85**, 2315–2320.
54. Hiyama K, Hirai Y, Kyoisumi S, *et al*. Telomerase activity in human peripheral blood and bone marrow cells. *J Immunol* 1995, **155**, 3711–3715.
55. Landsdorp PM. Telomere length and proliferation potential of hematopoietic stem cells. *J Cell Sci* 1995, **108**, 1–6.
56. Prowse KR, Greider CW. Development and tissue specific regulation of mouse telomerase and telomere length. *Proc Natl Acad Sci USA* 1995, **92**, 4818–4822.
57. Broccoli D, Young JW, DeLange T. Telomerase activity in normal and malignant hematopoietic cells. *Proc Natl Acad Sci USA* 1995, **93**, 9083–9086.
58. Wright WE, Piatyszek MA, Rainey WE, Byrd W, Shay JW. Telomerase activity in human germline and embryonic tissues and cells. *Dev Genetics* 1996, **18**, 173–179.
59. Vaziri H, Schachter F, Uchida I, *et al*. Loss of telomeric DNA during aging of normal and trisomy 21 human lymphocytes. *Am J Hum Genet* 1993, **52**, 661–667.
60. Buchkovich KJ, Greider CW. Telomerase regulation during entry into the cell cycle in normal human T cells. *Mol Cell Biol* 1996, **7**, 1443–1454.
61. Effros RB. Insights on immunological aging derived from the T lymphocyte cellular senescence model. *Exp Gerontol* 1996, **31**, 21–27.
62. Weng NP, Levine BL, June CH, Hodes RJ. Regulated expression of telomerase activity in human T lymphocyte development and activation. *J Exp Med* 1996, **183**, 2471–2479.
63. Harlebachor C, Boukamp P. Telomerase activity in the regenerative basal layer of the epidermis in human skin and immortal and carcinoma-derived skin keratinocytes. *Proc Natl Acad Sci* 1996, **93**, 6476–6481.
64. Bryan TM, Englezou A, Gupta J, Bacchetti S, Reddel RR. Telomere elongation in immortal human cells without detectable telomerase activity. *EMBO J* 1995, **14**, 4240–4248.
65. Strahl C, Blackburn EH. Effects of reverse transcriptase inhibitors on telomere length and telomerase activity in two immortalized human cell lines. *Mol Cell Biol* 1996, **16**, 53–65.
66. Wright WE, Shay JW. Time, telomeres and tumours: is cellular senescence more than an anticancer mechanism? *Trends Cell Biol* 1995, **5**, 293–297.
67. Sandell LL, Zakian VA. Loss of a yeast telomere: arrest, recovery and chromosome loss. *Cell* 1993, **75**, 729–739.
68. Afshari CA, Vojta PJ, Annab LA, Futreal PA, Willard TB, Barrett JC. Investigation of the role of G1/S cell cycle mediators in cellular senescence. *Exp Cell Res* 1993, **209**, 231–237.
69. Oshima J, Campisi J, Tannock CA, Martin GM. Regulation of c-fos in senescing Werner syndrome fibroblasts differs from that observed in senescing fibroblasts from normal donors. *J Cell Physiol* 1995, **162**, 277–283.
70. Wright WE, Pereira-Smith OM, Shay JW. Reversible cellular senescence: implications for immortalization of normal human diploid fibroblasts. *Mol Cell Biol* 1989, **9**, 3088–3092.
71. Radna RL, Caton Y, Jha KK, *et al*. Growth of immortal simian virus 40 ts-A transformed human fibroblasts is temperature dependent. *Mol Cell Biol* 1989, **9**, 3093–3096.
72. Marcand S, Buck SW, Moretti P, Gilson E, Shore D. Silencing of genes at nontelomeric sites in yeast is controlled by sequestration of silencing factors at telomeres by Rap1 protein. *Genes Dev* 1996, **10**, 1297–1309.
73. Shay JW, Werbin H, Wright WE. You haven't heard the end of it: telomere loss may link human aging with cancer. *Can J Aging* 1995, **14**, 511–524.
74. Wright WE, Shay JW. Telomere positional effects and the regulation of cellular senescence. *Trends Genet* 1992, **8**, 193–197.
75. Laurenson P, Rine J. Silencers, silencing and heritable transcriptional states. *Microbiol Rev* 1992, **56**, 543–560.
76. Shore D. Telomere position effect and transcriptional silencing in the yeast *Saccharomyces cerevisiae*. In Blackburn EH, Greider CW, eds. *Telomeres*. Cold Spring Harbor Press, 1995, 247–263.
77. Ogryzko VV, Harai TH, Russanova VR, Barbie DA, Howard BH. Human fibroblast commitment to a senescence-like state in responses to histone deacetylase inhibitors is cell cycle dependent. *Mol Cell Biol* 1996, **16**, 5210–5218.
78. Kim S, Villeponteau B, Jazwinski SM. Effect of replicative age on transcriptional silencing near telomeres in *Saccharomyces cerevisiae*. *Biochem Biophys Res Comm* 1996, **219**, 370–376.
79. Smeal T, Claus J, Kennedy B, Cole F, Guarente L. Loss of transcriptional silencing causes sterility in old mother cells of *S. cerevisiae*. *Cell* 1996, **84**, 633–642.

Telomerase activation, cellular immortalization and cancer

William C Hahn[1,2,4] *and Matthew Meyerson*[1,3]

The maintenance of specialized nucleoprotein structures termed telomeres is essential for chromosome stability. Without new synthesis of telomeres at chromosome ends the chromosomes shorten with progressive cell division, eventually triggering either replicative senescence or apoptosis when telomere length becomes critically short. The regulation of telomerase activity in human cells plays a significant role in the development of cancer. Telomerase is tightly repressed in the vast majority of normal human somatic cells but becomes activated during cellular immortalization and in cancers. While the mechanisms for telomerase activation in cancers have not been fully defined, they include telomerase catalytic subunit gene (*hTERT*) amplification and trans-activation of the *hTERT* promoter by the myc oncogene product. Ectopic expression of *hTERT* is sufficient to restore telomerase activity in cells that lack the enzyme and can immortalize many cell types. Understanding telomerase biology will eventually lead to several clinically relevant telomerase-based therapies. These applications include inhibiting or targeting telomerase as a novel antineoplastic strategy and using cells immortalized by telomerase for therapeutic applications.

Keywords: cancer; cell immortalization; telomerase; telomeres.

Ann Med 2001; 33: 123–129.

Telomeres protect chromosome ends

The linear structure of eukaryotic chromosomes allows for more rapid recombination and shuffling than would circular chromosomes, providing benefits for the generation of biological diversity. However, the free DNA ends of these chromosomes are susceptible to degradation and end-to-end fusions (1, 2). Moreover, the DNA-dependent DNA polymerases, responsible for almost the entirety of eukaryotic DNA replication, are unable to copy the last stretch of linear DNA on each chromosome (3–5). These DNA polymerases use small RNA primers to synthesize the lagging strand of DNA during replication. Even if an RNA primer hybridizes to the precise end of the chromosome, there would still be a gap in the DNA after the RNA is degraded. Thus, the 5' end of the lagging strand becomes shorter than the 3' end of the leading DNA strand. As terminal DNA is lost with each cycle of replication, the resulting chromosomal shortening eventually leads to chromosomal instability and cell death. This phenomenon is known as 'the end-replication problem' and is illustrated in Figure 1.

Telomeres are DNA–protein complexes at the ends of linear chromosomes. These structures protect the free ends of DNA from fusion and degradation. Furthermore, unique biochemical mechanisms have evolved to copy the telomeric DNA, solving the DNA end-replication problem. In most eukaryotes, the telomerase enzyme directs the synthesis of new telomeres; this process is described in more detail below.

The DNA component of telomeres is identical in all vertebrates: a simple repeat of the guanine-rich sequence 5'-TTAGGG-3' (7, 8). The only variation is the number of telomere repeat units at the end of each chromosome. In humans, there are roughly 1000–2000 telomeric repeats per chromosome end (7, 9). At the extreme 3' end of the telomere the G-rich strand contains several hundred base pairs of a single-stranded extension (10–13). Recent evidence suggests that telomeres form duplex loop structures called t-loops (14).

Telomeres also contain telomere-associated protein components which help to maintain telomere stability and to regulate telomere length. The TRF1 and TRF2 proteins are bound to the TTAGGG repeat sequence of vertebrate telomeres (15–17). TRF1 appears to

From the [1]Department of Adult Oncology, Dana-Farber Cancer Institute and Brigham and Women's Hospital, Boston, MA, the Departments of [2]Medicine and [3]Pathology, Harvard Medical School, Boston, MA, and the [4]Whitehead Institute for Biomedical Research, Cambridge, MA, USA.

Correspondence: William C Hahn, MD, PhD, Matthew Meyerson, MD, PhD, Department of Adult Oncology, Dana-Farber Cancer Institute, 44 Binney St, Boston, MA 02115, USA. E-mail: william_hahn@dfci.harvard.edu; matthew_meyerson@dfci.harvard.edu; Fax: +1 617 6322165.

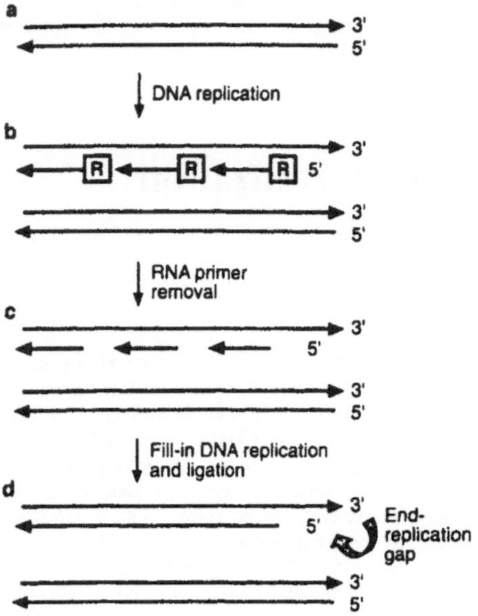

Figure 1. The end-replication problem. Conventional DNA replication with DNA-dependent DNA polymerases fails to replicate to the end of chromosomal DNA leaving a gap. Parental DNA strands are shown in grey and newly synthesized DNA strands in black. RNA primers (shown as R in the boxes, panel **b**) prime synthesis of the lagging strand. Removal of RNA primer leaves gaps (panel **c**). These gaps are copied by fill-in DNA replication, except at the extreme end of the chromosome ('End-replication gap' in panel **d**). (Adapted from (6) with permission.)

regulate telomere length by inhibiting telomere elongation once telomeres reach a critical size (18). TRF2, in contrast, suppresses end-to-end fusions between chromosomes and serves to stabilize chromosome ends (19). It is likely that these proteins form the scaffold for a larger, multiprotein complex that includes other members, such as tankyrase (20, 21) and Rap1 (22), and that act in concert to regulate telomere homeostasis.

The telomerase enzyme

The telomeric DNA of most eukaryotes, including all vertebrates, is synthesized by an unusual reverse transcriptase. This enzyme, known as telomere terminal transferase or telomerase, is composed of two core subunits, an RNA template subunit and a protein catalytic subunit. These subunits act in concert to elongate telomeres by reading from the RNA template sequence carried by the RNA subunit and synthesizing a complementary DNA strand. Mutation of this

Key messages:

- **Telomerase plays an important role in maintaining chromosome stability and determining cellular lifespan.**
- **Telomerase plays a critical role in malignant transformation.**
- **Therapeutics that target telomerase show promise as antineoplastic agents.**

RNA template leads to generation of new telomere sequences (23–26), confirming that this sequence serves as the template for telomerase. The protein subunit shares significant sequence homology with viral reverse transcriptase enzymes and is believed to catalyze the telomere synthesis reaction (27, 28). The human telomerase core subunits are *hTR*, or 'human telomerase RNA', and *hTERT*, or 'human telomerase reverse transcriptase'.

Telomerase catalytic subunits of a wide variety of species, from yeast to protozoans to humans, are highly conserved (27–31). All of these proteins share regions that are conserved in common with viral reverse transcriptase enzymes as well as other telomerase-specific conserved motifs. Mutations in conserved reverse transcriptase catalytic residues found in telomerase eliminate the enzymatic activity of telomerase (27, 28, 33–34). Other telomerase-specific motifs are also required for catalytic activity (35, 36). The similarity of telomerase to reverse transcriptases suggests that enzyme inhibitors for telomerase may be readily developed.

Cellular immortalization and telomerase

Normal human cells exhibit a limited proliferative capacity in culture and enter into an irreversible growth arrest called replicative senescence after a set number of cell divisions (37). The majority of cancer cells, in contrast, can be cultured indefinitely, suggesting that these cells have escaped the normal mechanisms that regulate lifespan. The regulation of the cell cycle is one critical molecular regulator of replicative senescence. Experimentally, replicative senescence can be bypassed by expression of viral oncoproteins, such as SV40 large T antigen (LT) or the human papilloma virus (HPV) E6 and E7 proteins, which inactivate the p53 and retinoblastoma (pRB) tumour suppressor pathways (38, 39). Such postsenescent cells then proliferate until they reach a second proliferative block, termed crisis, characterized from senescence by widespread cell death (40). Both replicative senescence and crisis limit the growth of normal human cells.

Recent evidence suggests that telomere shortening can act as a 'clock' that measures cell divisions. Telomeres progressively shorten with passage of normal human cells through multiple cycles of cell division (41). In culture, normal cells eventually reach senescence and stop dividing at a point when their telomeres have shortened considerably. As normal human cells progressively lose telomeric DNA with passage in culture, telomeric erosion has been postulated to limit cellular lifespan and to induce both senescence and crisis (42).

Neoplastic cells, however, maintain stable telomere length (43). In humans, telomerase activity is absent in most normal cells but present in the majority of cells derived from tumours (44–46). In addition, most experimentally established cell lines maintain stable telomere lengths and express telomerase. The few normal tissues that exhibit detectable telomerase activity (eg, testes, intestinal mucosa, and the lymphoid lineage) contain cell populations that require continual cell division. These observations suggest that acquisition of telomerase activity is a crucial step in attaining replicative immortality.

Although the telomerase RNA subunit is essential for telomerase activity, it is ubiquitously expressed in both normal and malignant tissues (25, 47), and thus its behaviour does not easily explain the increase in activity seen in cancer cells. In contrast, the telomerase catalytic subunit hTERT is expressed only in tissues and cells, such as tumours, that have measurable telomerase activity (29, 30, 48). Activation of hTERT may therefore represent a critical step in tumour progression and correspondingly may predispose cells to become malignant.

To investigate the role of telomerase and telomeres in the senescence, many groups have introduced *hTERT* into several types of normal human cells. Ectopic expression of hTERT in human fibroblasts (49–51), endothelial cells (52), mesothelial cells (53), and retinal pigment epithelial cells (49) conferred readily detectable telomerase activity. Remarkably, the telomerase activation was sufficient to confer a longer lifespan on these cells. Indeed, with continued passage, this forced expression of telomerase led to telomere lengthening and stabilization and immortalization without evidence of genomic instability or alterations in the cell cycle (54, 55).

However, ectopic hTERT expression and the resulting telomere elongation led to a different functional outcome in many epithelial cells. Expression of hTERT was sufficient to confer telomerase activity and stabilize telomere lengths in keratinocytes (53, 56), mammary epithelial cells (56), and lung epithelial cells (AS Lundberg, S Randell, WC Hahn, RA Weinberg, unpublished observations). However, the telomerase-expressing cells entered a period of slow growth at the passage number when control cells lacking telomerase entered senescence. This period of slow growth continued for a variable length of time in each culture, yet each culture eventually yielded a clone of rapidly dividing cells that appeared to be immortal. Molecular analysis of these emergent cells showed that in addition to expressing telomerase, they had lost the ability to produce the cyclin-dependent kinase inhibitor $p16^{INK4A}$, suggesting that immortalization required the inactivation of the pRB pathway in addition to telomere maintenance (53, 56). Corroborating these observations, co-expression of hTERT with the HPV E7 protein, which binds to and inactivates pRB, leads to immortalization in keratinocytes and mammary epithelial cells (53, 56). In human lymphocytes, it also remains unclear whether telomerase expression alone is sufficient to allow such cells to bypass senescence (57, 58). Thus, telomere maintenance plays an important role in senescence, but in many primary epithelial cells telomere maintenance through the action of telomerase requires the concomitant loss of pRB pathway regulation to achieve cellular immortalization.

Although telomeres shorten as cells approach senescence, the onset of senescence corresponds to telomere lengths that range from 4 to 8 kilobases (kb). In contrast, postsenescent cells expressing viral oncoproteins, such as LT, exhibit dramatically shorter telomere lengths (0.5–2.5 kb) as they approach crisis (43), the second proliferative barrier in human cells. Indeed, in rare cells that overcome crisis and become immortal, the *hTERT* gene and telomerase are usually spontaneously up-regulated, and telomere shortening is arrested (30, 43, 59–61). To investigate the role of telomerase in crisis, several groups have introduced hTERT into post-senescent mesenchymal and epithelial cell lines that express LT (62–64). Expression of hTERT again conferred telomerase activity, stabilized telomere lengths, and immortalized these cells regardless of their histological origin.

Taken together, these observations suggest that telomere lengths play a critical role in determining the onset of crisis. As the inactivation of the pRB and p53 tumour suppressor proteins by viral oncoproteins, such as LT, may mimic the functional loss of these pathways observed in most cancers, activation of telomerase in incipient cancer cells may allow such malignant cells to bypass crisis and become immortal. Indeed, co-expression of hTERT with LT and an oncogenic allele of Ras results in transformation of human fibroblasts, kidney epithelial cells, and mammary eptihelial cells (65, 66). These findings lend further support to the notion that the up-regulation of hTERT and telomerase activity observed in human tumours (29, 30, 32, 44, 45, 59) may be the direct cause of the ability of cancer cells to divide beyond the replicative capacity of normal somatic lineages.

Genetics of telomerase activation in cancers

Although most embryonic cells express telomerase (67), the majority of human committed somatic cells do not express detectable telomerase activity (44, 46). The observation that 85–90% of human cancer cell lines and tumour samples express telomerase thus suggests that incipient cancer cells re-acquire the ability to express hTERT during immortalization. As the regulation of hTERT appears to be largely transcriptional, most studies have focused on the regulation of the hTERT promoter, although post-translational mechanisms may also play a role in some situations (68–70). It should be noted that some cells, in particular T and B lymphocytes (71–75), appear to retain the ability to activate telomerase after specific stimulation, suggesting that physiological regulation of hTERT transcription may occur in some cell types that require the ability to have an extended lifespan.

In humans, telomerase is located on the telomeric end of chromosome 5p (5p15.33) (30, 76). Although mutations of this region are uncommon in human cancers, a recent survey of a limited number of human cancer cell lines found that 10 of 15 harboured amplifications of the *hTERT* locus (76). A second study showed *hTERT* amplification in 8 of 26 tumour cell lines and 17 of 58 primary tumours (77). These results suggest that the resulting increased *hTERT* copy number might provide one mechanism by which telomerase activation occurs in human cancers. Despite these findings, no other primary mutations of the hTERT locus have yet been documented.

Alteration of the normal transcriptional regulation of *hTERT* appears to be another mechanism for telomerase activation. One candidate activator of hTERT is c-myc. The upstream regulatory sequence of *hTERT* contains two myc consensus binding sites, and several groups have recently demonstrated that over-expression of myc in telomerase-expressing cells will drive the transcription of a reporter plasmid controlled by *hTERT* upstream regulatory sequences (78–81). As c-myc is amplified in many human cancers and has been associated with immortalization, these observations suggest that the activation of oncogenes, such as myc, may permit immortalization through transcriptional regulation of *hTERT*. Conversely, the myc antagonist, Mad, can repress transcription from the *hTERT* promoter (82, 83). However, as many human cancers do not harbour *myc* amplifications, other transcriptional regulators for *hTERT* must also exist. Notably, several other putative *hTERT* regulators have recently been reported, including activation of *hTERT* transcription by oestrogen, although this may involve a *myc* pathway (84), activation by the HPV oncoprotein E6 (85, 86), and *hTERT* repression by the Wilm's tumour protein WT-1 (87). Further work is required to determine if these activators play a physiological role for hTERT activation. In addition, Bacchetti and co-workers have reported a region of the hTERT promoter that appears to mediate transcriptional repression, implying that loss of active transcriptional repression may serve as an additional mechanism to activate telomerase during tumourigenesis (88).

Potential medical applications of telomerase and telomere biology

The observations that implicate telomerase as essential for the unlimited proliferative capacity of human cancer cells suggest that strategies that inactivate telomerase would be predicted to have potential clinical utility. Recently, several groups have tested this hypothesis by using diverse strategies. Genetic experiments in which a dominantly acting, catalytically inactive mutant of hTERT was introduced into human cancer cells resulted in complete inhibition of telomerase activity, loss of telomeres with proliferation, and eventual death of such cells through apoptosis (89, 90). Other approaches, including targeting the RNA subunit of telomerase with antisense or modified oligonucleotides complementary to the telomeric template, have also led to inhibition of telomerase activity, telomere shortening, and cell death (87–93). Taken together, these observations provide good confirmation that targeting telomerase in human cancers may be an effective antineoplastic strategy. This strategy is particularly attractive as most normal human cells lack telomerase and thus might be spared untoward effects of telomerase inhibition. Moreover, as potent inhibitors exist for viral reverse transcriptases, the prospects for developing small molecular inhibitors of the telomerase reverse transcriptase are excellent.

However, the biology of telomeres and telomerase suggests that even if such specific inhibitors are developed, they will differ significantly from traditional cytotoxic drugs. Telomere lengths differ greatly among human tumour samples, ranging from only 1 to 2 kb to more than 10 kb (9). As telomerase null cells lose approximately 50 to 100 bases for each cell division (41), significant cellular proliferation may accompany complete inhibition of telomerase activity before telomere lengths become critically short. Thus, telomerase inhibitors may need to be combined with traditional cytotoxic therapy or used after optimal tumour debulking or when patients have entered a minimal residual disease state.

An alternative approach to telomerase inhibition is to use telomerase as a target for immunotherapy. As telomerase is expressed in the majority of human tumours and appears to be essential for tumour cell proliferation, telomerase is an excellent antigen against which to raise an immunotherapeutic response. Unlike cellular receptors, such as HER2/neu, telomerase is an intracellular molecule and thus unsuitable as a target for antibodies. However, the protein component of telomerase can be processed by the intracellular proteosomes and presented in the context of major histocompatibility molecules as antigens recognized by cytotoxic T lymphocytes. Recently, several groups have shown that cytotoxic lymphocytes can be isolated from normal human donors that recognize hTERT-derived peptides (94, 95). These cytotoxic lymphocytes are then able to lyse both human cancer cell lines and primary tumour samples. In a murine immunotherapy model, mice producing such hTERT-restricted cytotoxic lymphocytes were able to reject the growth of a subcutaneous tumour without evidence of untoward autoimmunity (96). While many questions remain regarding this approach, these studies suggest that telomerase immunotherapy is an important avenue for future research.

Although an immunotherapeutic approach will avoid the problem of continued tumour proliferation after telomerase inhibition, there remain several important considerations for the development of telomerase-based antineoplastic strategies. In addition to the issue of latency, some normal cells retain the ability to activate telomerase physiologically. Because many of these types of normal cells are associated with stem cell-like behaviour, there may be significant side-effects of long-term treatment with an effective telomerase inhibitor or immunotherapy. In addition, the primary means of resistance from telomerase inhibition will likely be the activation of alternative non-telomerase-based mechanisms to maintain telomere length (97, 98). Despite these considerations, the potential utility of anti-telomerase-based cancer therapeutics warrants further development and testing of potential drugs and strategies that target telomerase. Finally, the ability of telomerase to immortalize a variety of human cell types, as described above, offers another avenue of therapy for cancer and other human diseases. Cells immortalized by telomerase may prove valuable for cell and organ transplantation and tissue regeneration. Indeed, bovine adrenal cells immortalized with telomerase and LT were recently reported to form functional adrenal glands when transplanted into adrenalectomized mice (99). Although the potential for neoplastic transformation of such immortalized cell lines remains a concern, one can easily imagine situations where the ready availability of such human cells and tissues may outweigh the long-term risks. Further investigation will undoubtably expand the potential medical applications of telomerase and telomere biology, and these possibilities are under active investigation.

References

1. McClintock B. The stability of broken ends of chromosomes in Zea mays. *Genetics* 1941; 26: 234–82.
2. Haber JE, Thorburn PC. Healing of broken linear dicentric chromosomes in yeast. *Genetics* 1984; 106: 207–26.
3. Olovnikov AM. Principle of marginotomy in template synthesis of polynucleotides (in Russian). *Dokl Akad Nauk SSSR* 1971; 201: 1496–9.
4. Olovnikov AM. A theory of marginotomy. The incomplete copying of template margin in enzymic synthesis of polynucleotides and biological significance of the phenomenon. *J Theor Biol* 1973; 41: 181–90.
5. Watson JD. Origin of concatemeric T7 DNA. *Nature New Biol* 1972; 239: 197–201.
6. Meyerson M. Role of telomerase in normal and cancer cells. *J Clin Oncol* 2000; 18: 2626–34.
7. Moyzis RK, Buckingham JM, Cram LS, Danl M, Deaven LL, Jones MD, et al. A highly conserved repetitive DNA sequence, (TTAGGG)n, present at the telomeres of human chromosomes. *Proc Natl Acad Sci U S A* 1988; 85: 6622–6.
8. Meyne J, Ratliff RL, Moyzis RK. Conservation of the human telomere sequence (TTAGGG)n among vertebrates. *Proc Natl Acad Sci U S A* 1989; 86: 7049–53.
9. de Lange T, Shiue L, Myers RM, Cox DR, Naylor SR, Killery AM, et al. Structure and variability of human chromosome ends. *Mol Cell Biol* 1990; 10: 518–27.
10. Henderson ER, Blackburn EH. An overhanging 3' terminus is a conserved feature of telomeres. *Mol Cell Biol* 1989; 9: 345–8.
11. Makarov VL, Hirose Y, Langmore JP. Long G tails at both ends of human chromosomes suggest a C strand degradation mechanism for telomere shortening. *Cell* 1997; 88: 657–66.
12. McElligott R, Wellinger RJ. The terminal DNA structure of mammalian chromosomes. *Embo J* 1997; 16: 3705–14.
13. Wright WE, Tesmer VM, Huffman KE, Levene SD, Shay JW. Normal human chromosomes have long G-rich telomeric overhangs at one end. *Genes Dev* 1997; 11: 2801–9.
14. Griffith JD, Comeau L, Rosenfield S, Stansel RM, Bianchi A, Moss H, et al. Mammalian telomeres end in a large duplex loop. *Cell* 1999; 97: 503–14.
15. Chong L, van Steensel B, Broccoli D, Erdjument-Bromage H, Hanish J, Tempst P, et al. A human telomeric protein. *Science* 1995; 270: 1663–7.
16. Broccoli D, Smogorzewska A, Chong L, de Lange T. Human telomeres contain two distinct Myb-related proteins, TRF1 and TRF2. *Nat Genet* 1997; 17: 231–5.
17. Bilaud T, Brun C, Ancelin K, Koering CE, Laroche T, Gilson E. Telomeric localization of TRF2, a novel human telobox protein. *Nat Genet* 1997; 17: 236–9.
18. van Steensel B, de Lange T. Control of telomere length by the human telomeric protein TRF1. *Nature* 1997; 385: 740–3.
19. van Steensel B, Smogorzewska A, de Lange T. TRF2 protects human telomeres from end-to-end fusions. *Cell* 1998; 92: 401–13.
20. Smith S, Giriat I, Schmitt A, de Lange T. Tankyrase, a poly(ADP-ribose) polymerase at human telomeres. *Science* 1998; 282: 1484–7.
21. Smith S, de Lange T. Tankyrase promotes telomere elongation in human cells. *Curr Biol* 2000; 10: 1299–302.

22. Li B, Oestreich S, de Lange T. Identification of human Rap1: implications for telomere evolution. *Cell* 2000; 101: 471–83.
23. Greider CW, Blackburn EH. A telomeric sequence in the RNA of Tetrahymena telomerase required for telomere repeat synthesis. *Nature* 1989; 337: 331–7.
24. Yu GL, Bradley JD, Attardi LD, Blackburn EH. In vivo alteration of telomere sequences and senescence caused by mutated Tetrahymena telomerase RNAs. *Nature* 1990; 344: 126–32.
25. Feng J, Funk WD, Wang SS, Weinrich SL, Avilion AA, Chiu CP, et al. The RNA component of human telomerase. *Science* 1995; 269: 1236–41.
26. Singer MS, Gottschling DE. TLC1: template RNA component of Saccharomyces cerevisiae telomerase. *Science* 1994; 266: 404–9.
27. Lingner J, Hughes TR, Shevchenko A, Mann M, Lundblad V, Cech TR. Reverse transcriptase motifs in the catalytic subunit of telomerase. *Science* 1997; 276: 561–7.
28. Counter CM, Meyerson M, Eaton EN, Weinberg RA. The catalytic subunit of yeast telomerase. *Proc Natl Acad Sci U S A* 1997; 94: 9202–7.
29. Nakamura TM, Morin GB, Chapman KB, Weinrich SL, Andrews WH, Lingner J, et al. Telomerase catalytic subunit homologs from fission yeast and human. *Science* 1997; 277: 955–9.
30. Meyerson M, Counter CM, Eaton EN, Ellisen LW, Steiner P, Caddle SD, et al. *hEST2*, the putative human telomerase catalytic subunit gene, is up-regulated in tumor cells and during immortalization. *Cell* 1997; 90: 785–95.
31. Harrington L, Zhou W, McPhail T, Oulton R, Yeung DS, Mar V, et al. Human telomerase contains evolutionarily conserved catalytic and structural subunits. *Genes Dev* 1997; 11: 3109–15.
32. Nakayama J, Tahara H, Tahara E, Saito M, Ito K, Nakamura H, et al. Telomerase activation by hTRT in human normal fibroblasts and hepatocellular carcinomas. *Nat Genet* 1998; 18: 65–8.
33. Weinrich SL, Pruzan R, Ma L, Ouellette M, Tesmer VM, Holt SE, et al. Reconstitution of human telomerase with the template RNA component hTR and the catalytic protein subunit hTRT. *Nat Genet* 1997; 17: 498–502.
34. Beattie TL, Zhou W, Robinson MO, Harrington L. Reconstitution of human telomerase activity in vitro. *Curr Biol* 1998; 8: 177–80.
35. Beattie TL, Zhou W, Robinson MO, Harrington L. Polymerization defects within human telomerase are distinct from telomerase RNA and TEP1 binding. *Mol Biol Cell* 2000; 11: 3329–40.
36. Bryan TM, Goodrich KJ, Cech TR. Telomerase RNA bound by protein motifs specific to telomerase reverse transcriptase. *Mol Cell* 2000; 6: 493–9.
37. Hayflick L, Moorhead PS. The serial cultivation of human diploid cell strains. *Exp Cell Res* 1961; 25: 585–621.
38. Choo CK, Ling MT, Chan KW, Tsao SW, Zheng Z, Zhang D, et al. Immortalization of human prostate epithelial cells by HPV 16 E6/E7 open reading frames. *Prostate* 1999; 40: 150–8.
39. Shay JW, Wright WE, Werbin H. Defining the molecular mechanisms of human cell immortalization. *Biochim Biophys Acta* 1991; 1072: 1–7.
40. Sedivy JM. Can ends justify the means?: telomeres and the mechanisms of replicative senescence and immortalization in mammalian cells. *Proc Natl Acad Sci U S A* 1998; 95: 9078–81.
41. Harley CB, Futcher AB, Greider CW. Telomeres shorten during ageing of human fibroblasts. *Nature* 1990; 345: 458–60.
42. Harley CB, Kim NW, Prowse KR, Weinrich SL, Hirsch KS, West MD, et al. Telomerase, cell immortality, and cancer. *Cold Spring Harb Symp Quant Biol* 1994; 59: 307–15.
43. Counter CM, Avilion AA, LeFeuvre CE, Stewart NG, Greider CW, Harley CB, et al. Telomere shortening associated with chromosome instability is arrested in immortal cells which express telomerase activity. *EMBO J* 1992; 11: 1921–9.
44. Kim NW, Piatyszek MA, Prowse KR, Harley CB, West MD, Ho PL, et al. Specific association of human telomerase activity with immortal cells and cancer. *Science* 1994; 266: 2011–5.
45. Counter CM, Hirte HW, Bacchetti S, Harley CB. Telomerase activity in human ovarian carcinoma. *Proc Natl Acad Sci U S A* 1994; 91: 2900–4.
46. Shay JW, Bacchetti S. A survey of telomerase activity in human cancer. *Eur J Cancer* 1997; 33: 787–91.
47. Avilion AA, Piatyszek MA, Gupta J, Shay JW, Bacchetti S, Greider CW. Human telomerase RNA and telomerase activity in immortal cell lines and tumor tissues. *Cancer Res* 1996; 56: 645–50.
48. Ramakrishnan S, Eppenberger U, Mueller H, Shinkai Y, Narayanan R. Expression profile of the putative catalytic subunit of the telomerase gene. *Cancer Res* 1998; 58: 622–5.
49. Bodnar AG, Ouellette M, Frolkis M, Holt SE, Chiu CP, Morin GB, et al. Extension of life-span by introduction of telomerase into normal human cells. *Science* 1998; 279: 349–52.
50. Vaziri H, Benchimol S. Reconstitution of telomerase activity in normal human cells leads to elongation of telomeres and extended replicative life span. *Curr Biol* 1998; 8: 279–82.
51. Counter CM, Meyerson M, Eaton EN, Ellisen LW, Caddle SD, Haber DA, et al. Telomerase activity is restored in human cells by ectopic expression of hTERT (hEST2), the catalytic subunit of telomerase. *Oncogene* 1998; 16: 1217–22.
52. Yang J, Chang E, Cherry AM, Bangs CD, Oei Y, Bodnar A, et al. Human endothelial cell life extension by telomerase expression. *J Biol Chem* 1999; 274: 26141–8.
53. Dickson MA, Hahn WC, Ino Y, Ronfard V, Wu JY, Weinberg RA, et al. Human keratinocytes that express hTERT and also bypass a p16(INK4a)-enforced mechanism that limits life span become immortal yet retain normal growth and differentiation characteristics. *Mol Cell Biol* 2000; 20: 1436–47.
54. Jiang XR, Jimenez G, Chang E, Frolkis M, Kusler B, Sage M, et al. Telomerase expression in human somatic cells does not induce changes associated with a transformed phenotype. *Nat Genet* 1999; 21: 111–4.
55. Morales CP, Holt SE, Ouellette M, Kaur KJ, Yan Y, Wilson KS, et al. Absence of cancer-associated changes in human fibroblasts immortalized with telomerase. *Nat Genet* 1999; 21: 115–8.
56. Kiyono T, Foster SA, Koop JI, McDougall JK, Galloway DA, Klingelhutz AJ. Both Rb/p16INK4a inactivation and telomerase activity are required to immortalize human epithelial cells. *Nature* 1998; 396: 84–8.
57. Hooijberg E, Ruizendaal JJ, Snijders PJ, Kueter EW, Walboomers JM, Spits H. Immortalization of human CD8(+) T cell clones by ectopic expression of telomerase reverse transcriptase. *J Immunol* 2000; 165: 4239–45.
58. Migliaccio M, Amacker M, Just T, Reichenbach P, Valmori D, Cerottini JC, et al. Ectopic human telomerase catalytic subunit expression maintains telomere length but is not sufficient for CD8(+) T lymphocyte immortalization. *J Immunol* 2000; 165: 4978–84.
59. Kilian A, Bowtell DDL, Abud HE, Hime GR, Venter OJ, Keese PK, et al. Isolation of a candidate human telomerase catalytic subunit gene, which reveals complex splicing patterns in different cell types. *Hum Mol Genet* 1997; 6: 2011–9.
60. Counter CM, Botelho FM, Wang P, Harley CB, Bacchetti S. Stabilization of short telomeres and telomerase activity accompany immortalization of Epstein-Barr virus-transformed human B lymphocytes. *J Virol* 1994; 68: 3410–4.

61. Klingelhutz AJ, Barber SA, Smith PP, Dyer K, McDougall JK. Restoration of telomeres in human papillomavirus-immortalized human anogenital epithelial cells. *Mol Cell Biol* 1994; 14: 961–9.
62. Counter CM, Hahn WC, Wei W, Caddle SD, Beijersbergen RL, Lansdorp PM, et al. Dissociation among in vitro telomerase activity, telomere maintenance, and cellular immortalization. *Proc Natl Acad Sci U S A* 1998; 95: 14723–8.
63. Halvorsen TL, Leibowitz G, Levine F. Telomerase activity is sufficient to allow transformed cells to escape from crisis. *Mol Cell Biol* 1999; 19: 1864–70.
64. Zhu J, Wang H, Bishop JM, Blackburn EH. Telomerase extends the lifespan of virus-transformed human cells without net telomere lengthening. *Proc Natl Acad Sci U S A* 1999; 96: 3723–8.
65. Hahn WC, Counter CM, Lundberg AS, Beijersbergen RL, Brooks MW, Weinberg RA. Creation of human tumour cells with defined genetic elements. *Nature* 1999; 400: 464–8.
66. Elenbaas B, Spirio L, Koerner F, Fleming MD, Zimonjic DB, Donaher JL, et al. Human breast cancer cells generated by oncogenic transformation of primary mammary epithelial cells. *Genes Dev* 2001; 15: 50–65.
67. Ulaner GA, Hu JF, Vu TH, Giudice LC, Hoffman AR. Telomerase activity in human development is regulated by human telomerase reverse transcriptase (hTERT) transcription and by alternate splicing of hTERT transcripts. *Cancer Res* 1998; 58: 4168–72.
68. Kharbanda S, Kumar V, Dhar S, Pandey P, Cheh C, Majumder P, et al. Regulation of the hTERT telomerase catalytic subunit by the c-Abl tyrosine kinase. *Curr Biol* 2000; 10: 568–75.
69. Kang SS, Kwon T, Kwon DY, Do SI. Akt protein kinase enhances human telomerase activity through phosphorylation of telomerase reverse transcriptase subunit. *J Biol Chem* 1999; 274: 13085–90.
70. Li H, Zhao L, Yang Z, Funder JW, Liu JP. Telomerase is controlled by protein kinase Calpha in human breast cancer cells. *J Biol Chem* 1998; 273: 33436–42.
71. Buchkovich KJ, Greider CW. Telomerase regulation during entry into the cell cycle in normal human T cells. *Mol Biol Cell* 1996; 7: 1443–54.
72. Hu BT, Lee SC, Marin E, Ryan DH, Insel RA. Telomerase is up-regulated in human germinal center B cells in vivo and can be re-expressed in memory B cells activated in vitro. *J Immunol* 1997; 159: 1068–71.
73. Lansdorp PM, Poon S, Chavez E, Dragowska V, Zijlmans M, Bryan T, et al. Telomeres in the haemopoietic system. *Ciba Found Symp* 1997; 211: 209–18.
74. Weng NP, Granger L, Hodes RJ. Telomere lengthening and telomerase activation during human B cell differentiation. *Proc Natl Acad Sci U S A* 1997; 94: 10827–32.
75. Weng N, Levine BL, June CH, Hodes RJ. Regulation of telomerase RNA template expression in human T lymphocyte development and activation. *J Immunol* 1997; 158: 3215–20.
76. Bryce LA, Morrison N, Hoare SF, Muir S, Keith WN. Mapping of the gene for the human telomerase reverse transcriptase, hTERT, to chromosome 5p15.33 by fluorescence in situ hybridization. *Neoplasia* 2000; 2: 197–201.
77. Zhang A, Zheng C, Lindvall C, Hou M, Ekedahl J, Lewensohn R, et al. Frequent amplification of the telomerase reverse transcriptase gene in human tumors. *Cancer Res* 2000; 60: 6230–5.
78. Greenberg RA, O'Hagan RC, Deng H, Xiao Q, Hann SR, Adams RR, et al. Telomerase reverse transcriptase gene is a direct target of c-Myc but is not functionally equivalent in cellular transformation. *Oncogene* 1999; 18: 1219–26.
79. Kyo S, Takakura M, Taira T, Kanaya T, Itoh H, Yutsudo M, et al. Sp1 cooperates with c-Myc to activate transcription of the human telomerase reverse transcriptase gene (hTERT). *Nucleic Acids Res* 2000; 28: 669–77.
80. Wang J, Xie LY, Allan S, Beach D, Hannon GJ. Myc activates telomerase. *Genes Dev* 1998; 12: 1769–74.
81. Wu KJ, Grandori C, Amacker M, Simon-Vermot N, Polack A, Lingner J, et al. Direct activation of TERT transcription by c-MYC. *Nat Genet* 1999; 21: 220–4.
82. Gunes C, Lichtsteiner S, Vasserot AP, Englert C. Expression of the hTERT gene is regulated at the level of transcriptional initiation and repressed by Mad1. *Cancer Res* 2000; 60: 2116–21.
83. Oh S, Song YH, Yim J, Kim TK. Identification of Mad as a repressor of the human telomerase (hTERT) gene. *Oncogene* 2000; 19: 1485–90.
84. Kyo S, Takakura M, Kanaya T, Zhuo W, Fujimoto K, Nishio Y, et al. Estrogen activates telomerase. *Cancer Res* 1999; 59: 5917–21.
85. Klingelhutz AJ, Foster SA, McDougall JK. Telomerase activation by the E6 gene product of human papillomavirus type 16. *Nature* 1996; 380: 79–82.
86. Stoppler H, Hartmann DP, Sherman L, Schlegel R. The human papillomavirus type 16 E6 and E7 oncoproteins dissociate cellular telomerase activity from the maintenance of telomere length. *J Biol Chem* 1997; 272: 13332–7.
87. Oh S, Song Y, Yim J, Kim TK. The Wilms' tumor 1 tumor suppressor gene represses transcription of the human telomerase reverse transcriptase gene. *J Biol Chem* 1999; 274: 37473–8.
88. Cong YS, Bacchetti S. Histone deacetylation is involved in the transcriptional repression of hTERT in normal human cells. *J Biol Chem* 2000; 275: 35665–8.
89. Hahn WC, Stewart SA, Brooks MW, York SG, Eaton E, Kurachi A, et al. Inhibition of telomerase limits the growth of human cancer cells. *Nat Med* 1999; 5: 1164–70.
90. Zhang X, Mar V, Zhou W, Harrington L, Robinson MO. Telomere shortening and apoptosis in telomerase-inhibited human tumor cells. *Genes Dev* 1999; 13: 2388–99.
91. Bisoffi M, Chakerian AE, Fore ML, Bryant JE, Hernandez JP, Moyzis RK, et al. Inhibition of human telomerase by a retrovirus expressing telomeric antisense RNA. *Eur J Cancer* 1998; 34: 1242–9.
92. Herbert B, Pitts AE, Baker SI, Hamilton SE, Wright WE, Shay JW, et al. Inhibition of human telomerase in immortal human cells leads to progressive telomere shortening and cell death. *Proc Natl Acad Sci U S A* 1999; 96: 14276–81.
93. Shammas MA, Simmons CG, Corey DR, Reis RJ. Telomerase inhibition by peptide nucleic acids reverses 'immortality' of transformed human cells. *Oncogene* 1999; 18: 6191–200.
94. Vonderheide RH, Hahn WC, Schultze JL, Nadler LM. The telomerase catalytic subunit is a widely expressed tumor-associated antigen recognized by cytotoxic T lymphocytes. *Immunity* 1999; 10: 673–9.
95. Minev B, Hipp J, Firat H, Schmidt JD, Langlade-Demoyen P, Zanetti M. Cytotoxic T cell immunity against telomerase reverse transcriptase in humans. *Proc Natl Acad Sci U S A* 2000; 97: 4796–801.
96. Nair SK, Heiser A, Boczkowski D, Majumdar A, Naoe M, Lebkowski JS, et al. Induction of cytotoxic T cell responses and tumor immunity against unrelated tumors using telomerase reverse transcriptase RNA transfected dendritic cells. *Nat Med* 2000; 6: 1011–7.
97. Bryan TM, Englezou A, Dalla-Pozza L, Dunham MA, Reddel RR. Evidence for an alternative mechanism for maintaining telomere length in human tumors and tumor-derived cell lines. *Nat Med* 1997; 3: 1271–4.
98. Lundblad V, Blackburn EH. An alternative pathway for yeast telomere maintenance rescues est1- senescence. *Cell* 1993; 73: 347–60.
99. Thomas M, Yang L, Hornsby PJ. Formation of functional tissue from transplanted adrenocortical cells expressing telomerase reverse transcriptase. *Nat Biotechnol* 2000; 18: 39–42.

Telomeres, Cancer, and Aging

Altering the Human Life Span

Dwayne A. Banks, PhD; Michael Fossel, MD, PhD

Population projections of the aging global society and its fiscal and social impact have depended on assumptions regarding the human life span. Until now, the assumption that the maximum human life span is fixed has been justified. Recent advances in cell biology, genetics, and our understanding of the cellular processes that underlie aging, however, have shown that this assumption is invalid in a number of animal models and suggest that this assumption may become invalid for humans as well. In vitro alteration of telomeres affects cellular senescence, and in vivo manipulation of genes and diet can increase maximum life span in animal models if these discoveries are extended to humans. We may soon be able to extend the maximum human life span and postpone or prevent the onset of diseases associated with aging. Such a possibility requires that we recognize a growing uncertainty in any attempt to project international health care costs into the next few decades. The costs may be significantly lower than projections, if life span increases and age-related disabilities are postponed or less severe, or perhaps higher, if life span increases without altering the onset and severity of disability. An appropriate uncertainty regarding the human life span undermines any attempt to accurately predict health costs in the next century.

JAMA. 1997;278:1345-1348

The same forces which operate in the birth and temporal existence of the human being operate also in his destruction and death.

Maimonides
The Guide for the Perplexed

HISTORICALLY, there has been an—as yet well-founded—assumption that although we may alter the course of age-related diseases, such as emphysema by smoking cessation and atherosclerosis by lowering cholesterol and blood pressure, the underlying process of aging cannot itself be altered.[1] The actual life span of any single individual is determined by a multitude of factors including lifestyle, socioeconomic status, diet, environmental conditions, and genetic endowments.[2,3] Some of these factors can be altered with consequent alteration of the individual (or the mean) life span, but not with any known increase in the maximum human life span. Currently, the longest anyone has ever lived is 122 years.[4] The biological origin of this maximum life span has been controversial. Is there a "clock" regulating the aging process? Or is there a biological limit that cannot be surpassed—irrespective of our genes? These are questions that gerontologists, geneticists, and molecular and cell biologists have been pondering for decades.

Recently, there has been a conceptual shift in our understanding of aging. The possibility of extending the maximum human life span has gone from legend to laboratory.[5] This change has been prompted by a growing academic literature that suggests that the aging process itself, as well as the consequent and fundamental cellular changes that occur in age-related diseases, is modifiable. This revision in views, if borne out, has profound clinical implications for the incidence of age-related diseases. Furthermore, the fiscal impact on government expenditures in the areas of health and social security will be perplexing even if only a modicum of these views hold true.

Aging is often seen from 2 vantage points: damage theories and programmed theories. Damage theories view aging as the result of accumulated errors, from free radicals for example. Programmed theories view aging as the result of genetic regulation. These paradigms are not contradictory, but complementary. Changes in gene expression that occur with cellular aging permit damage that does not occur (or accumulate) in immortal cell lines such as germ cells. The issue is not whether free radical damage underlies much of aging—it clearly does—but the more complex question of whether we may learn to control the onset and timing of such damage, which ultimately determines our health and our maximum life span.

CLONES, CANCER, AND AGING CELLS

The mean human life span has been extended remarkably over the past 2 centuries; the maximum life span has not.[6] This has resulted in a "rectangularization" of life span, in which mortality is increasingly becoming compressed against an apparently fixed end point: the maximum human life span of approximately 120 years.[7] Over the past decade, however, several laboratories have successfully extended the maximum life spans of at least 2 multicellular species genetically (*Drosophila*[8] and *Caenorhabditis elegans*[9-10]) and several more by dietary restriction.[11-13] That this increase in maximum life span can occur at all invalidates the dogma that maximum life span is fixed, and it thus invalidates the inevitability of the rectangularization model described above. That it occurs with such minimal genetic manipulation and that the effect is so substantial (2 gene mutations increase the maximum life span of *C elegans* 6-fold) raise the provocative question of what might lie in store for clinical medicine. To a first approximation, the increased life span occurs predominantly through the genetic control of free radical metabolism. This is not unexpected: cellular aging (and modifications of maximum life span)

From the Richard & Rhoda Goldman School of Public Policy & Center for the Economics and Demography of Aging, University of California, Berkeley (Dr Banks), and Michigan State University College of Human Medicine, East Lansing (Dr Fossel).
Reprints: Michael Fossel, MD, PhD, Saint Mary's Hospital, 200 Jefferson SE, Grand Rapids, MI 49503 (e-mail: Mfossel@aol.com).

operates predominantly through the damaging effects of free radicals.[14,15] Not only does free radical production increase substantially as cells age, but the free radicals produced are less well contained and less efficiently mopped up by free radical scavengers, and their consequent damage is less efficiently repaired. This simplification leaves unanswered the questions of how cancer cells and the germ cell lines avoid aging damage, though they have comparable genomes, mitochondria, and free radicals. Telomere shortening (and its consequent effects on gene expression and cell cycle mechanisms) is implicated in allowing cancer and other cell lines to avoid cell senescence.[16,17] Free radical damage appears to be the major cause of the damage that occurs in aging cells,[18,19] but the more complex issue lies in the control of the timing and release of such damage.[20]

What is the age of a cell? Unless we measure age from every new division (as we do in *Saccharomyces*, for example, in which the asexual form shows aging in that "maternal" cells have a limited number of divisions, but the number is "reset" in each new "daughter" cell), newly divided cells each inherit the same age as the single cell from which they derived. All cells—as unbroken cell lines—have, reductio ad absurdum, the same age. Life on this planet is 3.5 billion years old, and all life and every cell are equally old in a peculiar sense. But then how is the onset of free radical damage timed, if not simply by years of a cell's life? Aging does not occur for 3.5 billion years in the germ cell line, yet it clearly does so beginning sometime after fertilization. Free radical aging is not timed by the profoundly archaic age of the entire cell lineage, but rather by a clock that begins running only after fertilization (in multicellular organisms; in the asexual phase of unicellular organisms such as yeasts, aging is considered to commence at the asymmetric division of a daughter cell as she buds off the larger mother cell). Humans and other multicellular organisms derive from cells that do not show the loss of replicative ability and the morphologic and gene expression changes that are characteristic of senescent cells[1,21,22] until after the organism achieves multicellularity. Age per se does not determine aging; it is altered genetic expression that somehow permits the onset of aging to occur.

The onset and progression of aging is strongly affected by chromosomal structure[23] and gene expression.[24] Evidence from work on early aging syndromes (progerias) supports this observation. Patients with Werner syndrome have an average life span of 47 years and a known defect in helicase metabolism that results in abnormal chromosomal "unbalanced" replication.[25] Patients with Hutchinson-Gilford syndrome have an average life span of 12.7 years, and their skin fibroblasts have telomere lengths characteristic of cells from far older patients.[26,27]

The cloning of mature ovine mammary cells likewise raises the question of whether age can be reset. Wilmut et al[28] have shown that gene expression can be reset: in this case at least, the 6-year-old adult mammary cells from the donor have not yet been irreversibly aged by time per se or by the aging of the donor organism. To the contrary, successful cloning demonstrates that these cells still have sufficient intact genetic information available to allow reconstitution of an apparently healthy multicellular organism. But was Dolly—the cloned organism—6 years old at birth, or had her cells been reset to age 0? The most tempting and certainly viable hypothesis is that genes can be reset to reflect the cellular age consistent with a developing organism,[29] a hypothesis that is now (in Dolly's case) finally being tested. A considerable body of evidence suggests that cell aging does not occur because of accumulated DNA damage or poor transcription fidelity[30]; rather, the DNA is intact and transcription is intact, but transcription occurs at lower rates and in significantly altered patterns as cells age.[21,23] If Dolly's cells show normal ovine aging for her "new" age rather than being 6 years old at birth, then this will provide further support for the hypothesis that age can be reset even after it has been in progress. The nature of this putative resetting mechanism will be a prime target for further research.

Cancer research has also prompted revision of our views of cell aging.[22] The problem in oncology is not that cells age and die, but that they do not. Normal somatic cells have well-defined limits to replication. Young skin fibroblasts typically have 50 divisions before they show cellular senescence. Within the past 7 years, there has been increasing support for the model that cellular senescence in normal (noncancerous) human cells is the result of an altered (and, for the cell, dysfunctional) pattern of gene expression, and that the onset and progression of this pattern of senescent gene expression is regulated (through insufficiently understood mechanisms likely to involve heterochromatin changes) regulated by the telomeres.[31-33] Most cancer cells, however, are biologically immortal and will continue dividing indefinitely under appropriate conditions. This potential is linked to their expression of telomerase.[34-37] Additionally, it appears that most, perhaps all, cancer cells are fundamentally different with regard to their aberrant cell cycle braking systems. Typically, 60% of combined human cancers (70% of colon cancers, 40% of breast cancers) have an aberrant p53. If we include p21, D cyclins, cyclin-dependent kinases, and other parts of the cell cycle "machinery," the percentage of cell cycle abnormalities might approach 100% of cancers.[38,39] More recently, evidence is accumulating that at least 90% of cancer cells from human malignancies are also capable of resetting their telomere lengths and evading cell senescence.[40] This model has proven a powerful predictor not only of cell aging, but also of cancer cell survival.[41]

This realization—that the aging process is mutable at the cellular level and the implications of this for age-related diseases—has itself spawned more than a dozen targeted biotechnology companies over the past 4 years.[42] Not only does this realization have implications for cellular aging, but it has also prompted rethinking of the historically warranted pessimism about curing cancer. As a *Science* editorial put it at the end of 1996, predicting whether we might cure cancer, "the standard answer—'No'—may be up for revision."[43] Such optimism is not novel, but a comprehensive conceptual base, good supporting data, and promising clinical trials are new and very welcome.

The recent cloning of the protein component of human telomerase[36] will only accelerate this already rapid progress in understanding cancer and cell senescence. More importantly, to the extent that we can affect cell senescence therapeutically, we may stand to alter age-related diseases and thereby their economic, social, and human outcomes.

AN EVOLVING MODEL

Previously, aging was seen as an immutable, passive accumulation of entropic damage in the cell. The evolving paradigm, in which cell aging results from and is coordinate on altered gene expression,[29,32,33] is growing in acceptance. Although still nascent, the Senescent Gene Expression (SGE) model is a model that encompasses both the programmed and damage schools of thought on aging. It suggests that aging is a complex cascade of processes that include repeated cell division, telomere effects at the chromosome ends (due to both their lengths and their associated heterochromatin), a significantly altered pattern of senescent gene expression in "old" cells, and consequent alterations in cell metabolism in general and free radical metabolism (production, sequestration, trapping, and damage repair) in particular. In addition, even cells that do not directly demonstrate aging changes (eg, myocardial cells) depend on those that do (eg, vascular endothelial cells). Together this cascade of processes play a coordinated but complex role in aging and—far more importantly—identification of each step could provide a basis for developing therapeutic interventions to affect a single step or the whole. This model—that the process of aging in the cell can be altered or reset (as may be true in the cloning of adult sheep cells and as is demonstrably true in senescent fibroblast hybridomas[44])—undermines the historical certainty that aging and age-related diseases are immutable at the cellular level. Coordinate and similar shifts are occurring in our understanding of cancer. Differences at the heart of cancer cells, such as cell cycle errors and telomerase expression, may provide leverage to draw far clearer therapeutic lines than we now can between cancers and normal cells. While none of our currently altering views of cancer or aging promise a cure, enormous clinical potentials may be opening for us where none existed previously.[43,45,46]

These changes in our theoretical understanding of cell aging and the availability of data to support the mutability of cell aging together call into question our ability to provide accurate extrapolations of the social and economic consequences of projected aging in developed countries. In 1950, 4 years before the Salk vaccine, if we were to have predicted the medical costs (eg, iron lungs) that would accrue from polio over the next decade, we would have risked similar inaccuracies. Predictions of the costs of aging will be hollow unless they factor in the ongoing, nascent, and fundamental changes in our understanding of aging and age-related diseases.

POLICY IMPLICATIONS OF EXTENDING LIFE

We are currently incapable of altering the aging process in any meaningful way, and any assertion to the contrary is misleading and not supported by fact. It is, however, equally misleading and disingenuous as well to ignore the ongoing shift in our knowledge of cell biology as we attempt to predict the future of health care and social policy in the United States and other developed countries. The assumptions on which we base our predictions need to be clearly understood, and there exists a reasonable degree of uncertainty in assuming that aging and age-related diseases will remain immutable in the coming decades.

The fiscal and social implications of this possibility cannot be underestimated. Any increase in the life span will be accompanied by concomitant shifts in social spending on the aged, but we cannot reasonably estimate the timing, magnitude, or fiscal outcome of altering the maximum human life span and age-related diseases. Even if one denies the possibility of such alteration, the upward trend in median age of the world's population is substantial but of uncertain magnitude. Even assuming that the lifespan is fixed, the consequent rectangularization of the survival curve, described by Fries,[7] will exert fiscal and social pressures on our ability to care for the elderly.[47,48] The magnitude of these pressures, their global distribution, and their effect on well-being will depend on several factors.

First, while there exists some variation internationally in the health status of elderly cohorts,[49,50] the social costs of increasing the maximum human life span will largely depend on the level of functional impairments and chronic disabilities among the aged.[51,52] Recent research suggests that, at least in the United States, increases in the mean life span have been accompanied by a concomitant decline in the prevalence of morbidity and disability among the aged.[53,54] The magnitude of any such decline and its overall impact on health care costs, relative to prior elderly cohorts, will vary internationally as a function of a nation's health care system, preventive health measures, rate of technological diffusion, advances in the treatment of acute and chronic diseases, social and political structures, and economic systems—to mention a few. The possibility that we might delay or prevent aging and associated disabilities only increases the already considerable variance in estimating such costs. However, the impact of such innovations would be determined by the political, social, regulatory, and economic structures of each country.

Second, innovations in medical technology (therapeutic, diagnostic, or organizational) for age-related conditions need not coincide with the health care needs or necessarily promote the well-being of those in developing countries. The immediate and future medical and public health concerns of sub-Saharan Africa are not in the research, development, and diffusion of age-related technologies, but in the development of effective mechanisms for controlling the spread of infectious diseases (such as the acquired immunodeficiency syndrome,[55] malaria, and tuberculosis), and the fatal consequences of drought, famine, and civil wars, which lead to mean life spans in these countries of approximately 48 years.[56] Therefore, when considering the international fiscal effects of increasing the maximum healthy human life span and the impact of these innovations on functional impairment among the aged, it is important to acknowledge that these advances are likely to benefit developed nations preferentially.

Third, the continued decline in the level of functional impairment among the elderly populations in developed countries, along with increases in active life expectancy,[57,58] has profound implications for work and retirement years. Even though the effect of longevity on work and retirement choice remains unclear,[59] developed nations can expect years of productive life among elderly age-specific cohorts to increase. How altering the maximum human life span—by whatever magnitude—would alter this effect cannot be reliably projected, but contrary to what one might expect, increased life expectancy in the United States has been associated with an accelerated decline in the labor force participation rates of older persons for the past 50 years.[60] To sustain current levels of well-being, however, longer life spans will have to be accompanied by either correspondingly longer work years, or higher premiums to enhance private contributions to retirement accounts, or higher taxes to enhance contributions into social insurance accounts. The magnitude of the latter will be determined by the willingness of current generations to subsidize future generations, as well as the ratio of workers to retirees. Further, the structure of public and private pension funds can in themselves affect the labor force participation rates of individuals. The retirement income provided by the funds allows older workers to leave the labor force at younger ages and still support themselves in retirement years.[61] Hence, it is the interplay among mortality, health status, work, and retirement and the structure of retirement accounts that will determine the overall impact of the burgeoning aging population and any increase in the maximum human life span on the fiscal health of nations.

Finally, with increasing relative numbers of the aged, along with the growth in the proportion of the oldest old by the year 2040,[61] the international demand for long-term care could impose unexpected pressures on governments and individuals. This will be exacerbated if the proportion of offspring willing to live with their elderly parents continues to decline internationally,[62] therefore placing greater pressures on the public and private sectors for the establishment of innovative alternatives to long-term care—such as home or community-based care. The magnitude and direction of these pressures depend critically on the extent to which the maximum life span may be altered and the degree to which the prevalence of disabilities will be modified. Neither of these effects can be reliably estimated, yet current research suggests there is a reasonable possibility that both might alter within the time frame of current attempts to project future trends in international health and social welfare expenditures.

CONCLUSION

We should exercise extreme caution in projecting the future social and economic impact of an aging global society. This is particularly so because our understanding of the biology of aging is changing. The unquestioned conviction that we cannot alter aging and the cellular underpinnings of the diseases that accompany the aging process is no longer strictly tenable. The possibility that aging and its consequent clinical outcomes may be alterable at the cellular and chromosomal levels remains merely speculative and needs to be considered cautiously. However, such speculation becomes increasingly appropriate if we are to make any attempt to predict the future costs of an aging global society. It is important that we qualify and carefully define our assumptions to include the possibility that the morbidity and the mortality rates of age-related diseases, along with the maximum human life span itself, may be altered in the near future.

This research was partially supported by the National Institute on Aging-funded Center on the Economics and Demography of Aging (P20-AG-12839).

The authors gratefully acknowledge the advice of Donald Ingram, David Smith, Robert Arking, David Kirp, Ron Lee, and Jane Mauldon.

References

1. Hayflick L. *How and Why We Age*. New York, NY: Ballantine Books; 1994:313.
2. Finch CE. *Longevity, Senescence and the Genome*. Chicago, Ill: University of Chicago Press; 1990.
3. Jazwinski SM. Longevity, genes, and aging. *Science*. 1996;273:54-59.
4. Jeanne Calment, born 21 February 1875 in Arles, died 4 August 1997 in Paris, France: world's oldest woman dies at 122. *Baltimore Sun*. August 5, 1997:B5.
5. National Institute on Aging, National Institutes of Health. *In Search of the Secrets of Aging*. Bethesda, Md: National Institute on Aging, National Institutes of Health; 1997.
6. Smith DWE. *Human Longevity*. New York, NY: Oxford University Press; 1993.
7. Fries JF, Crapo LF. *Vitality and Aging: Implications of the Rectangular Curve*. San Francisco, Calif: WH Freeman; 1981
8. Lakowski B, Hekimi S. Determination of life-span in *Caenorhabditis elegans* by four clock genes. *Science*. 1996;272:1010-1013.
9. Ewbank JJ, Barnes TM, Lakowski B, Lussier M, Bussey H, Hekimi S. Structural and functional conservation of the *Caenorhabditis elegans* timing gene clk-1. *Science*. 1997;275:980-983.
10. Larsen PL, Albert PS, Riddle DL. Genes that regulate both development and longevity in *Caenorhabditis elegans*. *Genetics*. 1995;139:1567-1583.
11. Weindruch R, Walford RL. *The Retardation of Aging and Disease by Dietary Restriction*. Springfield, Ill: Charles C Thomas Publisher Inc; 1988.
12. Sohal RS, Weindruch R. Oxidative stress, caloric restriction, and aging. *Science*. 1996;273:59-63.
13. Lane MA, Ingram DK, Roth GS. Beyond the rodent model: calorie restriction in rhesus monkeys. *Age*. 1997;20:45-56.
14. Harman D. The biologic clock: the mitochondria? *J Am Geriatr Soc*. 1972;20:145-147.
15. Orr WC, Sojal RJ. Extension of life-span by overexpression of super oxide dismutase and catalase in *Drosophila melanogaster*. *Science*. 1994;263:1128-1138.
16. Lundblad V, Wright WE. Telomeres and telomerase: a simple picture becomes complex. *Cell*. 1996;87:369-375.
17. Holt SE, Wright WE, Shay JW. Regulation of telomerase activity in immortal cell lines. *Mol Cell Biol*. 1996;16:2932-2939.
18. Harman D. Aging: a theory based on free radical and radiation biology. *J Gerontol*. 1956;11:298-300.
19. Sohal RS, Weindruch R. Oxidative stress, caloric restriction, and aging. *Science*. 1996;273:59-63.
20. Fossel M. *Reversing Human Aging*. New York, NY: William Morrow & Co; 1996.
21. Dimri GP, Lee X, Basile G, et al. A biomarker that identifies senescent human cells in culture and in aging skin in vivo. *Proc Natl Acad Sci U S A*. 1995;92:9363-9367.
22. Smith JR, Pereira-Smith OM. Replicative senescence: implications for in vivo aging and tumor suppression. *Science*. 1996;273:63-67.
23. Guarente L. Do changes in chromosomes cause aging? *Cell*. 1996;86:9-12.
24. Campisi J. Replicative senescence: an old live's tale. *Cell*. 1996;84:497-500.
25. Yu CE, Oshima J, Fu YH, et al. Positional cloning of the Werner's syndrome gene. *Science*. 1996;272:258-262.
26. Allsopp RC, Vaziri H, Patterson C, et al. Telomere length predicts replicative capacity of human fibroblasts. *Proc Natl Acad Sci U S A*. 1992;89:10114-10118.
27. Oshima J, Brown WT, Martin GM. No detectable mutations at Werner helicase locus in progeria. *Lancet*. 1996;348:1106.
28. Wilmut I, Schnieke AE, McWhir J, Kind AJ, Campbell KHS. Viable offspring derived from fetal and adult mammalian cells. *Nature*. 1997;385:810-813.
29. Stewart C. An udder way of making lambs. *Nature*. 1997;385:769-771.
30. Harley CB, Pollard JW, Chamberlain JW, Stanner CP, Goldstein S. Protein synthetic errors do not increase during aging of cultured human fibroblasts. *Proc Natl Acad Sci U S A*. 1980;77:1885-1889.
31. Harley CB, Futcher AB, Greider CW. Telomeres shorten during aging of human fibroblasts. *Nature*. 1990;345:458-460.
32. Guarante L. Do changes in chromosomes cause aging? *Cell*. 1996;86:9-12.
33. Marcand S, Gasser SM, Gilson E. Chromatin: a sticky silence. *Curr Biol*. 1996;6:1222-1225.
34. Kim NW, Piatyszek MA, Prowse KR, Harley CB, West MD, Ho PLC. Specific association of human telomerase activity with immortal cells and cancer. *Science*. 1994;266:2011-2015.
35. Shay JW, Bacchetti S. A survey of telomerase activity in human cancer. *Eur J Cancer*. 1997;33:787-791.
36. Nakamura TM, Morin GB, Chapman KB, et al. Telomerase catalytic subunit homologs from fission yeast and human. *Science*. 1997;277:955-959.
37. Meyerson M, Counter C, Eaton EN, et al. hEST2, the putative human telomerase catalytic subunit gene, is up-regulated in tumor cells and during immortalization. *Cell*. 1997;90:785-795.
38. Rao RN. Targets for cancer therapy in the cell cycle pathway. *Curr Opin Oncol*. 1996;8:516-524.
39. Sherr CJ. Cancer cell cycles. *Science*. 1996;274:1672-1677.
40. Shay JW, Gazdar AF. Telomerase in the early detection of cancer. *J Clin Pathol*. 1997;50:106-109.
41. Feng J, Funk WD, Wang SS, et al. The RNA component of human telomerase. *Science*. 1995;269:1236-1241.
42. Jazwinski SM, Martin GM, Schellenberg GD, et al. Molecular and genetic strategies for treatment of age-related diseases. Presented at the National Managed Health Care Congress: Bio/technology Conferences; July 14-15, 1997; Seattle, Wash.
43. Scanning the research horizon. *Science*. 1996;274:1989. Editorial box.
44. Wright WE, Brasiskyte D, Piatyszek MA, Shay JE. Experimental elongation of telomeres extends the life span of immortal x normal cell hybrids. *EMBO J*. 1996;15:1734-1741.
45. Hartwell LH, Kastan MB. Cell cycle control and cancer. *Science*. 1994;266:1821-1828.
46. Jacks T, Weinberg RA. Cell-cycle control and its watchman. *Nature*. 1996;381:643-644.
47. Winker MA, Glass RM. Aging global population: a call for papers. *JAMA*. 1996;276:1758.
48. Marwick C. Longevity requires policy revolution. *JAMA*. 1995;273:1319-1321.
49. Manton KG, James VW. Survival after the age of 80 in the United States, Sweden, France, England, and Japan. *N Engl J Med*. 1995;333:1232-1235.
50. Sallar AM, Hogg RS, Schechter MT. Survival after age 80. *N Engl J Med*. 1996;334:537-538.
51. Lubitz J, Beebe J, Baker C. Longevity and Medicare expenditures. *N Engl J Med*. 1995;332:999-1003.
52. Schneider EL, Guralnik JM. The aging of America: impact on health care costs. *JAMA*. 1990;263:2335-2340.
53. Manton KG, Corder L, Stallard E. Chronic disability trends in the elderly United States populations: 1982-1994. *Proc Natl Acad Sci U S A*. 1997;94:2593-2598.
54. Alliance for Aging Research. *Seven Deadly Myths: Uncovering the Facts About the High Cost of the Last Year of Life*. Washington, DC: Alliance for Aging Research; May 1997.
55. Van De Walle E. The social impact of AIDS in sub-Saharan Africa. *Milbank Q*. 1990;68(1):10-32.
56. US Bureau of the Census. *Statistical Abstract of the United States, 115th Edition*. Washington, DC: US Bureau of the Census; 1995.
57. Manton KG, Stallard E, Liu K. Forecasts of active life expectancy: policy and fiscal implications. *J Gerontol*. 1993;48:11-26.
58. Sullivan DF. A single index of mortality and morbidity. *HSMHA Health Rep*. 1971;86:347-354.
59. Lee R, Tuljapurkar S. Death and taxes: longer life, consumption, and social security. *Demography*. 1997;34(1):67-81.
60. Wise DA. Retirement against the demographic trend: more older people living longer, working less, and saving less. *Demography*. 1997;34(1):83-95.
61. US Bureau of the Census. *Statistical Abstract of the United States, 1995*. Washington, DC: US Bureau of the Census; 1996.
62. Okamoto Y. Health care for the elderly in Japan: medicine and welfare in an aging society facing a crisis in long term care. *BMJ*. 1992;305:403-405.

Population Biology of Human Aging: Methods of Assessment and Sex Variation

E. KOBYLIANSKY,[1] G. LIVSHITS,[1] AND O. PAVLOVSKY[2]

Abstract Biological age, as assessed by osteographic measurements of hand bones from radiograms, was estimated in about 7000 individuals belonging to 32 ethnic groups and samples and to 2 major human groups (Europeans and Asians). Individual ages in the total group varied between 16 and 99 years old. Our biological age measure is strongly correlated with chronological age of the individual (Pearson's $r = 0.78–0.80$, $p < 0.001$), which renders the correlation amenable to fitting with a two-stage stochastic model. The following parameters of the model were estimated accordingly: t_0 is the minimal age at which initial bone changes occurred, q is the probability that an individual will first develop involutive bone change at age $t_i > t_0$, and B is the rate of aging (bone change) per unit time. The parameters varied considerably within different populations, but only a few samples showed statistically significant sex differences for all three parameters. However, B was consistently higher among women than among men. Apart from that, the distributions of the observed biological age estimates adjusted on the expected biological age values reveal no differences between men and women in any of the studied samples or in the total population.

Aging designates a developmental sequence of changes that begins in the uterus and continues throughout life (Adelman and Roth 1982). In Israel, for example, at present more than 445,000 people are aged 65 and older, which is about 12% of the total population. By the year 2000 this group is estimated to reach at least 500,000, which will be 15% of the population (*Statistical Abstracts of Israel* 1992). This plethora of the aged in modern Israel stems from a number of primary causes: enhanced life expectancy, high survival in infancy, high birth rates in the early decades of the twentieth century, and the large number of elderly immigrants entering the country in the last five years. If current trends continue, the fastest growing segment of the population will be this elderly group, which is the most vulnerable to chronic diseases, especially

[1]Research Unit—Human Population Biology, Department of Anatomy and Anthropology, Sackler Faculty of Medicine, Tel Aviv University, Tel Aviv, Israel 69978.
[2]Research Institute and Museum of Anthropology, Moscow State University, Moscow, Russia.

Human Biology, February 1995, v. 67, no. 1, pp. 87–109.

the group aged 85 and older. This latter group, which has the greatest need for health care, represented only 0.8% of the overall population in the early 1980s but nearly 22% of the population in nursing homes. Their number and proportion in nursing homes have no doubt increased since then and are likely to continue to increase in the future (Schneider and Brody 1983). This numerical imbalance between the working population responsible for creating the commodities of society and the senior population, most of which is unemployed and merely consumes these commodities, is increasing yearly and could conceivably trigger unfavorable economic pressures, even in highly developed Western communities. The problem that needs to be addressed is how to prevent elderly people from becoming a burden on the community (Shall 1975).

To resolve this problem, we first need to understand the biological processes underlying natural or biological aging in humans. According to Nakamura (1991), biological aging is the outcome of processes that usher in inevitable decline and that are rooted in heredity and circumscribed or conditioned by environment. Current dogma dictates that aging is a combination of two major determinants, namely, genetic predisposition in the family and an amalgam of exogenous social, psychological, ecological, and medical factors. Several theories propose that the aging process figures as a function in the finite life span, in which a genetic mechanism is suggested. Indeed, such a genetic basis has been elucidated on both the cellular and the familial level [e.g., Fukuchi et al. (1989), Linnane et al. (1989), Bocquet-Appel and Jacobi (1990), and Bernstein and Bernstein (1991)].

Several unresolved questions arise in this connection: (1) To what extent do genetic factors contribute to the rate of natural aging? (2) Can differences in the rate of biological aging between males and females in human populations be detected? and (3) Does the rate of aging depend on the ethnic background and the environment of an individual? In the present study we attempt to answer these questions on the basis of a large population sample comprising about 7000 individuals from 32 ethnic groups and independent samples.

In this paper, which is the first of a series, we consider a statistical model for fitting estimates of biological age and compare the rate of aging in males and females. Some results of ethnic variation are also given brief consideration, albeit ethnic differences in the rate of biological aging and its geographic trends will be dealt with mainly in the second paper. The third paper in the series will focus on a quantitative genetic analysis of biological age. First, however, we need to address the problems germane to biological age evaluation and briefly review the existing methodology.

To obtain reliable estimates of biological aging, we of course need an appropriate method for assessing the stage of aging reached by an individual. Most existing methods are based on the time-dependent changes in various physiological systems of the individual, whether cardiovascular, skeletal, endocrine, or the like.

To evaluate biological aging, Comfort (1969) recommended a comprehensive battery of tests with 60 parameters. Borkan and Norris (1980) likewise published a procedure to assess aging; their method relies on a large array of tests but mostly on biochemical and functional parameters. Subsequently, Ries and Pothig (1984) suggested an array of tests incorporating characteristics from three domains: physical, psychological, and social. More recently, Skalicky et al. (1987) proposed a battery of functional tests for measuring biological age on the basis of the cardiopulmonary system, the sense organs, testosterone level, and body surface/body mass index. The most recent system is that of Nakamura (1991), which relies on 27 variables to measure biological aging. In connection with these procedures, Borkan (1986) laid down the ground rules for selecting tests for biological age assessment: (1) The tested parameters should show a marked change with age so that alteration over a reasonably short span of years will be significant; (2) the tests should be performed with minimal disturbance of the subject; and (3) the tests should be suitable for screening a large number of people.

According to Borkan's three criteria, the aforementioned methods are not fully appropriate because most of their chosen parameters bear no direct correlation to age; they provide only a generalized picture of the functional and physiological status of an individual at any given stage of life. In addition, existing methods pose substantial financial and logistic problems, based, as they are, on performing a large number of tests on large population samples. At least part of these problems could be obviated by reliance on roentgenographic methodology.

The use of roentgenography in biological age evaluation is based on the premise that the skeletal system shows clear age-related changes, as attested by (1) the relative stability of the skeletal system's parameters on a short-term scale (weeks, months) versus the considerable changes occurring over long time intervals (years) [see Burov and Reznikov (1975)]; (2) the simplicity of evaluating various basic parameters of the bone system and the large number of criteria available for such evaluation (Kiebzak 1991); and (3) the strong correlation between skeletal age-related changes and the aging of many vital functions associated with the cardiopulmonary, visual, motor, and other systems of the organism (Ostlere and Gold 1991).

In practice, however, even within the framework of roentgenography, only select parts of the skeletal system can be evaluated for biological age. One such area is the hand, the bones of which are a convenient and useful subject for studies of aging. The use of roentgenography of the hand bones is advantageous because it offers an array of variables of age estimation. Thus one x-ray picture of the hand encompasses 27 entire bones, the distal parts of the ulna and the radius, and the sesamoid bones (Kobyliansky et al. 1985). Furthermore, age assessment from the hand bones is strongly correlated with age changes in other systems of the body, for instance, the dentition (Steel 1965) or ossification of the humerus, femur, and tibia (Walker and Lovejoy 1985; Pollitzer and Anderson 1989).

Roentgenography provides two independent types of measurements of bone biological age: osteometric measurements and osteographic measurements. Osteometric measurements are measures of length and breadth of bones, their cortical thickness, and their mineralization and density (Plato and Norris 1980; Kaur and Jit 1990). Quantitative radiology enables evaluation by two main methods: radiogrammetry and photodensitometry. Radiogrammetry entails the caliper measurement of cortical bone thickness from a simple radiograph. Although it reflects changes in endosteal bone resorption, it fails to measure intracortical and trabecular bone, which are better indexes of bone turnover (Dequeker 1976). Photodensitometry is a technique that employs a radiographic film and a standardized wedge to estimate mineral bone density proportional to the mineral mass (Ostlere and Gold 1991).

Osteographic measurements entail descriptive criteria of bone biological age, for example, the appearance of ossification centers. Pavlovsky (1987) suggests obtaining informative parameters of aging changes from a roentgenogram. Such parameters are the osteophytes or nodes of Eberdaine, exostoses, osteoporotic foci, resorption lacunae, sclerotic nuclei, sclerosis, and joint deformations. Osteographic measurements yield biological age estimates that correlate strongly with chronological age (Figure 1). Such measuring of biological age is relatively easy to perform and relatively inexpensive for large samples. Moreover, this procedure might be convenient to use for quantitative genetic analysis because it provides the normal distribution of biological age estimates for each given age. It has therefore been adopted in the present study and will also be used in a comparative ethnic study of the rate of aging or the genetic study of biological aging.

Materials and Methods

Population Samples. Data on about 7000 individuals belonging to 32 ethnic groups and samples are listed in Tables 1 and 2. The data were gathered between 1961 and 1992 in the republics of the former USSR and in India, Egypt, and Israel. All the collected data were transferred to the Research Unit on Human Population Biology of the Department of Anatomy and Anthropology at Tel Aviv University. The geographic

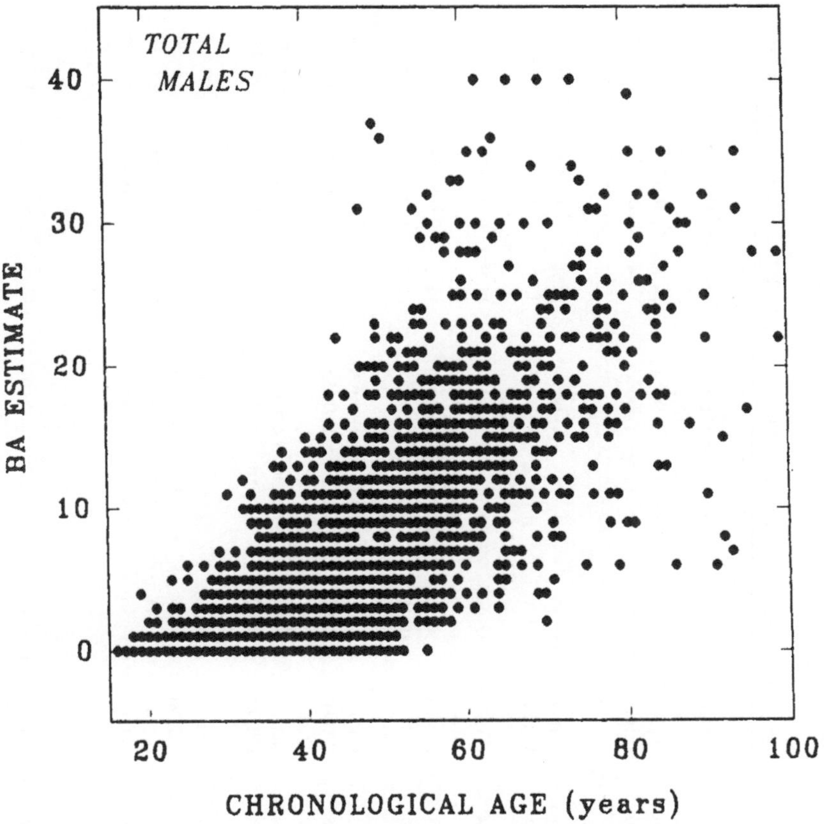

Figure 1. Relationship between estimates of biological age and chronological age in total male sample (Pearson's $r = 0.80$, $p < 0.001$). The samples are listed in Table 1.

location of our samples, their ethnic affiliation, and the specific sample sizes according to sex and chronological age are all provided in Tables 1 and 2. In 29 sampled localities data are available for both males and females. More details considering ethnic variation will be given in a subsequent paper (Livshits et al. 1995).

Variables for Biological Age Evaluation. To evaluate individual biological age scores, we performed the following analysis of hand roentgenograms in dorsopalmar projection (Pavlovsky 1987). Each phalanx was observed from all its perimeters, with attention paid to the diaphyses and proximate areas, especially at sites of tendon attachment at the middle of the phalanx, where osteophytes can be observed (Figure 2A, point

Table 1. Modeling Patterns in Bone Change with Age in Male Populations

Ethnic Group (Villages)	*Sample Size*	$t_o \pm SD$	$q \pm SD$	$B \pm SD$	*TM*	R^2	*Age Range*
Abkhazians (Chlou, Jgerda, Atara)	335	25.0 ± 0.67	0.0483 ± 0.0037	0.4683 ± 0.0033	45.7	0.5806	18–99
Abkhazians (Kaldakhvar, Aatsy)	233	28.0 ± 0.86	0.0632 ± 0.0054	0.5257 ± 0.0044	43.8	0.6701	18–95
Azerbaijanians	131	26.0 ± 1.19	0.0884 ± 0.0158	0.5129 ± 0.0045	37.4	0.7293	19–88
Bedouins (Gebeliya, Awlad/Said, Muzeina)	166	17.0 ± 0.51	0.0873 ± 0.0114	0.3793 ± 0.0074	28.4	0.6236	16–70
Buryats	125	27.0 ± 0.55	0.1176 ± 0.0140	0.6672 ± 0.0093	35.5	0.5643	20–56
Byelorussians (Veremeiki)	75	29.0 ± 1.53	0.0776 ± 0.0126	0.5859 ± 0.0074	41.9	0.7825	22–80
Chukchi	72	26.0 ± 1.09	0.0886 ± 0.0113	0.5659 ± 0.0133	37.3	0.6691	19–60
Eskimos	48	28.0 ± 1.13	0.1610 ± 0.0347	1.0000 ± 0.0000	34.2	0.8753	17–61
Georgians	97	34.9 ± 1.29	0.1519 ± 0.0240	0.4609 ± 0.0057	41.5	0.5143	16–99
Indians (North India)	191	19.2 ± 0.32	0.0937 ± 0.0155	0.4744 ± 0.0098	29.9	0.5366	19–62
Jews (Israel)	208	20.0 ± 0.56	0.0910 ± 0.0092	0.4127 ± 0.0025	31.0	0.6846	18–87
Karakalpaks (Karuzyak)	91	37.7 ± 0.99	0.1878 ± 0.0241	0.4206 ± 0.0150	43.3	0.5319	19–70
Karyaks	52	33.0 ± 1.13	0.1609 ± 0.0347	0.4656 ± 0.0180	39.2	0.8258	18–62
Kazakhs (Akkol)	142	24.0 ± 0.53	0.0599 ± 0.0079	0.4516 ± 0.0099	40.7	0.6491	19–60
Lithuanians	203	19.0 ± 0.89	0.1286 ± 0.0159	0.4360 ± 0.0043	26.8	0.7147	19–68
Mongols	99	26.0 ± 0.60	0.1066 ± 0.0131	0.3208 ± 0.0082	35.4	0.7739	23–78
Nentsi	74	25.0 ± 0.81	0.1072 ± 0.0230	0.6835 ± 0.0050	34.3	0.6812	17–78
Russians (Argoda, Burguzin, Chitkan, Kurumkan, Uro)	268	26.0 ± 0.47	0.1512 ± 0.0138	0.5229 ± 0.0058	32.6	0.3359	19–60

Russians (Azerbaijan)	208	22.0 ± 0.60	0.0994 ± 0.0114	0.5671 ± 0.0023	32.1	0.7472	17–85
Russians (Klaipeda, Palanga, Vilnius)	85	23.9 ± 0.91	0.2229 ± 0.0479	0.3962 ± 0.0058	28.4	0.6309	22–83
Russians (Kursk)	60	22.4 ± 1.22	0.1091 ± 0.0200	0.8469 ± 0.0026	31.6	0.7412	22–60
Russians (Peski)	118	23.0 ± 0.66	0.3877 ± 0.0153	0.4531 ± 0.0025	25.6	0.7729	21–90
Russians (Puttyatino)	34	22.0 ± 0.65	0.0665 ± 0.0134	0.4765 ± 0.0113	37.0	0.8552	22–84
Russians (Voshod)	79	27.0 ± 0.00	1.0000 ± 0.0000	0.4839 ± 0.0030	28.0	0.6008	20–79
Russians (Ukhta)	78	23.0 ± 0.00	1.0000 ± 0.0000	0.4366 ± 0.0048	24.0	0.6561	18–63
Tajiks (Chorku)	108	27.0 ± 0.45	0.1660 ± 0.0224	0.3672 ± 0.0113	33.0	0.4661	19–57
Tajiks (Unji)	55	28.5 ± 1.03	0.0985 ± 0.0185	0.6963 ± 0.0065	38.7	0.5009	20–78
Teke Turkmenians	94	22.0 ± 0.77	0.0490 ± 0.0082	0.5985 ± 0.0086	42.4	0.5543	19–60
Yomut Turkmenians (Kulmach, Uzun-Su)	82	26.0 ± 0.90	0.0871 ± 0.0114	0.9986 ± 0.0000	37.5	0.4356	20–50
Tuvinci (Chadan, Erzin, Toora-Jem)	173	21.0 ± 0.48	0.1287 ± 0.0159	0.5937 ± 0.0040	28.8	0.7809	19–71
Saamy	24	29.0 ± 1.10	0.0721 ± 0.0121	0.3502 ± 0.0261	42.9	0.7942	19–58
Russians (total)	930	22.0 ± 0.35	0.1110 ± 0.0064	0.5290 ± 0.0016	31.0	0.7061	17–90
Abkhazians (total)	568	25.0 ± 0.55	0.0511 ± 0.0030	0.4451 ± 0.0024	44.6	0.5975	18–95
Turkmenians (total)	176	22.0 ± 0.56	0.0639 ± 0.0062	0.6159 ± 0.0077	37.7	0.4856	19–60
Tajiks (total)	163	27.0 ± 0.47	0.1320 ± 0.0167	0.5980 ± 0.0067	34.6	0.5395	18–63

Table 2. Modeling Patterns of Bone Change with Age in Female Populations

Ethnic Group (Villages)	*Sample Size*	$t_0 \pm SD$	$q \pm SD$	$B \pm SD$	*TM*	R^2	*Age Range*
Abkhazians (Chlou, Jgerda, Atara)	302	35.0 ± 0.60	0.0820 ± 0.0084	0.5062 ± 0.0049	47.2	0.4891	17–99
Abkhazians (Kaldakhvar, Aatsy)	183	29.0 ± 1.04	0.1100 ± 0.0080	0.5666 ± 0.0043	38.1	0.6388	18–81
Azerbaijanians	164	22.0 ± 1.20	0.0693 ± 0.0110	0.4858 ± 0.0035	36.4	0.7165	19–90
Buryats	178	28.0 ± 0.54	0.1126 ± 0.0134	0.6657 ± 0.0078	36.9	0.4868	19–54
Byelorussians (Veremeiki)	96	35.0 ± 0.75	0.1855 ± 0.0162	0.6113 ± 0.0061	40.4	0.5891	20–77
Chukchi	99	27.0 ± 0.77	0.0813 ± 0.0118	0.6117 ± 0.0103	39.3	0.4845	18–57
Eskimos	59	26.0 ± 1.15	0.0566 ± 0.0111	0.6057 ± 0.0104	43.7	0.7216	19–63
Georgians	195	31.0 ± 1.04	0.0732 ± 0.0096	0.5754 ± 0.0033	44.7	0.6661	16–99
Jews (Israel)	194	24.5 ± 1.25	0.3349 ± 0.0101	0.4023 ± 0.0018	27.5	0.5463	16–86
Karakalpaks (Karuzyak)	109	38.5 ± 0.88	0.1371 ± 0.0309	0.5330 ± 0.0191	45.8	0.6457	18–78
Karyaks	42	19.0 ± 1.51	0.0444 ± 0.0071	0.5674 ± 0.0117	41.5	0.5306	18–65
Kazakhs (Akkol)	118	30.0 ± 0.86	0.0833 ± 0.0115	0.5011 ± 0.0102	42.0	0.6586	18–71
Lithuanians	163	24.0 ± 0.78	0.2088 ± 0.0294	0.4474 ± 0.0051	28.8	0.6778	20–63
Mixed group (Moscow)	87	21.0 ± 1.16	0.2079 ± 0.0629	0.4280 ± 0.0052	25.8	0.7544	18–68
Mongols	100	27.0 ± 0.79	0.0909 ± 0.0146	0.3691 ± 0.0079	38.0	0.5677	18–72
Nentsi	37	25.0 ± 0.58	0.2268 ± 0.0486	0.8078 ± 0.0042	29.4	0.6021	15–80
Russians (Argoda, Burguzin, Chitkan, Kurumkan, Uro)	274	25.0 ± 0.55	0.1154 ± 0.0096	0.5056 ± 0.0059	33.7	0.4657	18–55

Russians (Azerbaijan)	226	24.7 ± 0.76	0.1744 ± 0.0312	0.5311 ± 0.0019	30.4	0.6910	19–85
Russians (Klaipeda, Palanga, Vilnius)	115	26.9 ± 0.76	0.3539 ± 0.0278	0.4649 ± 0.0055	29.7	0.6652	22–74
Russians (Kursk)	46	23.5 ± 1.38	0.0877 ± 0.0175	0.9154 ± 0.0016	34.9	0.6974	21–73
Russians (Peski)	241	22.0 ± 0.91	0.1372 ± 0.0236	0.5301 ± 0.0019	29.3	0.7277	19–89
Russians (Puttyatino)	52	31.0 ± 2.62	0.0662 ± 0.0121	0.6613 ± 0.0049	46.1	0.6062	21–81
Russians (Voshod)	159	23.0 ± 0.91	0.3500 ± 0.0123	0.4664 ± 0.0025	25.9	0.7292	22–83
Russians (Ukhta)	201	23.0 ± 0.70	0.4358 ± 0.0377	0.4410 ± 0.0040	25.3	0.6049	19–60
Tajiks (Chorku)	90	26.0 ± 0.85	0.0752 ± 0.0107	0.6076 ± 0.0147	39.3	0.4929	18–56
Tajiks (Ünji)	88	24.0 ± 1.94	0.0263 ± 0.0163	0.7378 ± 0.0052	31.9	0.4346	19–63
Teke Turkmenians	102	32.0 ± 0.86	0.1027 ± 0.0140	0.8313 ± 0.0051	41.7	0.6141	19–60
Yomut Turkmenians (Kulmach, Uzun-Su)	78	32.0 ± 0.76	0.1118 ± 0.0166	0.8730 ± 0.0057	41.0	0.5329	19–52
Tuvincis (Chadan, Erzin, Toora-Jem)	130	25.0 ± 0.59	0.1918 ± 0.0299	0.6286 ± 0.0069	30.2	0.5729	17–60
Saamy	20	25.0 ± 0.98	0.0703 ± 0.0127	0.8816 ± 0.0044	39.2	0.6841	19–62
Russians (total)	1315	23.0 ± 0.35	0.1413 ± 0.0074	0.5234 ± 0.0013	30.1	0.7314	18–89
Abkhazians (total)	485	34.0 ± 0.49	0.1042 ± 0.0073	0.5298 ± 0.0034	43.6	0.5339	17–99
Turkmenians (total)	180	31.4 ± 0.62	0.1052 ± 0.0111	0.7964 ± 0.0053	40.9	0.5947	19–60
Tajiks (total)	178	24.0 ± 0.70	0.0805 ± 0.0097	0.7349 ± 0.0049	36.4	0.4455	18–63

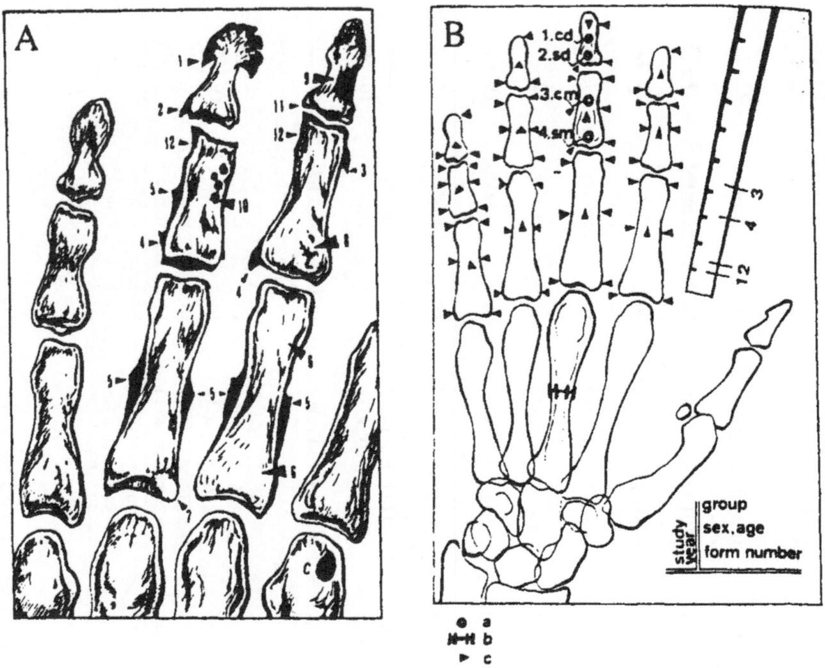

Figure 2. (A) Composite diagram of age transformation of hand bones assessed by osseographic scores. Osteophytes: (1) apostosis (not to be confused with normal variants of tuberositas unguicularis); (2) Heberden's nodes; (3) Rokhlin's nodes; (4) Bouchard's nodes; (5) exostosis. Porosis: (6) osteoporosis; (7) atrophy; (8) resorption islets. Sclerosis: (9) enostosis; (10) sclerotic nuclei; (11) narrowing of articular slit. Nontraumatic joint deformities: (12) erosion. For comparison, a normal phalanx is shown. (B) Hand radiograph indicating regions of AOS assessment. (a) Determination of mineral content: compacta (cd) and spongiosa (sd) of distal phalanx; compacta (cm) and spongiosa (sm) of medial phalanx. The scale in the upper right of the figure shows the reference aluminum cline used for the evaluation of bone density. On the third finger phalanges, 1.cd, 2.sd, 3.cm, and 4.sm designate relative density (1–4 of the reference cline) of the distal and medial phalanges compacta and spongiosa. (b) Compactness index components. (c) Elements of age-associated changes in bone structures.

1). The typical elements for bone age changes are the enlarged tuberosities at the distal phalanges and the various types of nodes, such as the nodes of Eberdaine and Bouchard in the epiphyseal areas. A few such changes are illustrated in Figure 2A (points 2–5). However, Figure 2A also shows some other involutive age changes of bone in other areas of the hand, especially those in the internal structure of the bone, where porosis or sclerosis is frequently recorded (Figure 2A, points 6–10).

Osteoporosis is associated with thinness of bone surface and/or faulty appearance of the inner structure of bone (Figure 2A, point 6); frequently there is also atrophy of the subcortical areas (Figure 2A, point 7) and deterioration of the trabecular elements (Figure 2A, point 8). Osteoporotic changes were assessed also by one of the roentgenometric methods for determining bone mineralization density (Alexeeva and Kobyliansky 1966). This was feasible because osteosclerotic changes in the bone's inner structure, for instance, are loci of high bone mineralization density and of high density of the structural elements, which appear either as cortex-like isles (Figure 2A, point 9) or as sclerotic nuclei (Figure 2A, point 10).

On surfaces of joint bones one can discern nontraumatic transformations, such as those caused by arthritis, or a narrowing of the joint cavities resulting from erosion of the periostal cartilages with age (Figure 2A, points 11 and 12). Traumatic changes, if any, have been excluded from our age-dependent assessments.

In sum, to assess the biological age rank of an individual, we recorded occurrence of (1) osteophytes in the periarticular regions of the bone and at the sites of muscle and tendon attachment, (2) manifestation of osteoporosis, (3) osteosclerosis, and (4) nontraumatic articular deformities. Only the bones of the second to fifth rays were subjected to analysis, because those of the first ray do not project directly in roentgenography.

For each biological age estimate merely the presence of a given element but not its level of development was recorded. Thus two main methodological problems were obviated. The first problem resolved is the ability to compare the data obtained by different investigators, where individual differences are expressed as a value divisible by one (equidistance principle). The second problem obviated is the discrepancy between normal and pathological aging. We took into account that morphological pathological conditions in the bones are determined not by the presence of a particular element of age rearrangement but by the extent to which substantial degenerative changes have taken place (Kellgren 1963). Only the presence of osteopathies in any joint involving two bones was marked by rank 3. The elements of rearrangement of the bone with age were recorded in toto, without differentiation into individual types, except for osteophytes.

Figure 2B shows that a maximum age-related summary score could potentially reach a rank of 80. However, in actuality, our data show that biological age estimates usually rank between 0 (no signs of the involutive changes in bone) and 40, the highest observed rank with the most advanced stage of bone change.

Stochastic Model and Statistical Analysis. The correlation between the biological age estimate and chronological age, as shown in Figure 1 for the total male sample, is high but not strictly linear. Naturally, involutive processes and detectable age-dependent changes in bone begin at a specific or particular age of an individual. The period of life between birth and the apparent onset of bone changes (t_0) was defined as the latent stage. Because after t_0 involutive changes in bone structure occur continuously throughout the remainder of life, this phase of life was designated the stage of active or visible aging.

To fit the age changes observed in biological age estimates, we developed a stochastic model that takes into account both of the stages. Let t_0 designate the age during which no visible changes occur, and let q be the probability of an individual to first develop involutive bone change at age $t_i \geq t_0$, that is, the probability for an individual to move into the visible stage of aging. It follows then that the mean age for transition from stage 1 to stage 2 is

$$TM = t_0 + \sum_{i=1}^{\infty} q(1 - q)^i i = t_0 + \frac{1}{q}. \qquad (1)$$

Let us now designate P as the probability of increasing the biological age measurement per age (time) unit, C as the amount of change, and $B = PC$ as the rate of aging per unit time. We need also X_t, a mean biological age value at time t but actually the number of changes during $t - t_0$, with the probability of each change equaling P. Now, if t is the time during which successful changes might take place, then

$$\bar{X}_{t+1} = X_t + (L - X_t) B. \qquad (2)$$

Hence

$$\bar{X}_t = L[1 - (1 - B)^t], \qquad (3)$$

$$\Delta\bar{X} = \bar{X}_{t+l} - \bar{X}_t = LB(1 - B)^t, \qquad (4)$$

where L is the maximal possible biological age value (age change) for time t, X_t is a randomly picked value of the biological age estimate, and $\Delta\bar{X}$ is the mean change of biological age during the time interval between t and $t + 1$.

Because a preliminary analysis of the relation between the biological age estimate and chronological age reveals no saturation (i.e., the biological age value does not reach a plateau) (Figure 1), we now assume that $L \rightarrow \infty$, $C \rightarrow 0$, $LC \rightarrow B \equiv$ constant. In this case $X_t = tB$ and $\Delta X \rightarrow B$, that is, linear growth with no saturation. The probability of a zero biological age estimate at any randomly picked age is $P_t(0) = 0$ for $t < t_0$ but for $t \geq t_0$ it is

$$P_t(0) = \frac{P}{P - q}(1 - q)^{t-t_0} - \frac{q}{P - q}(1 - P)^{t-t_0}. \qquad (5)$$

The parameters t_0, P, and q are not known.

To estimate parameters, we start with

$$\hat{R}_i = K_i/(n_i + 2), \tag{6}$$

where $\hat{R}_i$ is the population frequency of individuals with biological age equal to zero at ith age and K_i and n_i are the number of individuals and the total sample size at this age, respectively. The unknown parameters can now be found by the least-squares fitting procedure:

$$\Phi(t_0, P, q) = \Sigma\, [P_i(0) - \hat{R}_i]^2. \tag{7}$$

By minimizing, we get

$$\min_{\{t_0, P, q\}} \Phi(t_0, P, q) \rightarrow \hat{t}_0, \hat{P}, \hat{q}. \tag{8}$$

The mathematical expectation of the biological age estimate at any randomly picked age t is

$$M_t(X) = 0 \qquad \text{for } t < t_0, \tag{9}$$

$$M_t(X) = B(t - t_0) - \frac{B}{q}[1 - (1 - q)^{t - t_0}] \qquad \text{for } t \geq t_0. \tag{10}$$

For any given age, the mean empirical biological age estimate is

$$X_i = \frac{1}{n}\sum_{j=1}^{n} X_{ij}, \tag{11}$$

where X_{ij} is the biological age estimate in individual j at ith age. The unknown parameters B, t_0, and q can be found by the least-squares fitting procedure:

$$\min_{\{B, t_0, q\}} \sum_i [M_i(X) - \hat{X}_i]^2 \rightarrow B, t_0, q. \tag{12}$$

Estimates of the variance of the parameter means were obtained according to the method of Lloyd (1984, ch. 6).

Sex Comparisons. To compare samples, we used two statistical approaches. First, given the parameters of the model for males and females in each ethnic group and their standard deviations, we prepared a pairwise comparison using a t test with a Bonferroni correction (Sokal and Rohlf 1987). The second approach was to compare the distribution of biological age estimates in males and females with the observed biological age estimates initially adjusted for the expected relevant mean. The means for each age were obtained from the stochastic model, fitting the observed data for the respective sex and sample. Our working hypothesis was that biological age estimates between males and females belonging to the same ethnic group are not significantly different. As for the re-

Table 3. Mean and Standard Deviations of Biological Age Estimates in Total Sample

	Age (Years)						
Sample	*20–30*	*31–40*	*41–50*	*51–60*	*61–70*	*71–80*	*81–99*
Males							
$\bar{X}$	0.47	2.75	5.97	11.07	15.00	19.04	22.41
SD	1.173	2.895	4.697	6.529	7.439	6.932	7.796
N	987	878	784	464	209	122	60
Females							
$\bar{X}$	0.39	2.47	5.73	11.12	16.01	21.10	22.79
SD	1.127	2.799	4.274	6.092	7.179	8.013	7.399
N	768	1014	873	582	314	178	68

sidual obtained by this adjustment, this shall henceforth be designated as the same-sex residual (SSR).

On the assumption that there are no sex differences, we proceeded to adjust the empirical biological age estimates on females to the ones expected from the male model, and vice versa. The residual obtained here was designated the opposite-sex residual (OSR). Finally, the SSRs for males and females were combined, and their distribution was compared with the combined distribution of the OSRs. The significance of the differences between sexes was assessed by both an F test for variance ratio and the Kruskal-Wallis nonparametric analysis of variance.

Results

The mean biological age estimates and standard deviations arranged by decade for males and females are provided in Table 3. Figure 3 shows the mean values of the biological age estimates for each age in the range of the total sample; both the observed values and the expected values from the (model) fitting curve are given. The congruence between the two sets of means is very high (r = 0.920, $p < 0.001$); that is, the expected curve satisfactorily fits the observed mean estimates. However, Figure 1 shows that the variance of individual biological age values at any given age is high, and therefore the coefficient of determination R^2 is not really all embracing but accounts for only 52% of the total variation. Indeed, as will be shown mostly in the next paper in this series (Livshits et al. 1995), ethnic heterogeneity is responsible for a significant portion of this variation.

In Tables 1 and 2 we provide estimates of the major parameters of our stochastic model; these are given separately for women and men and for each ethnic group or sample. The two tables also provide the sample

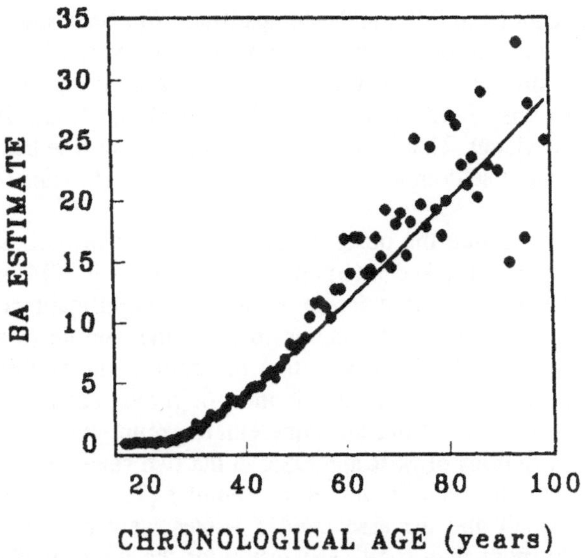

Figure 3. Observed (circles) and expected (curve) mean biological age estimates in total sample of males.

sizes, age ranges, coefficients of determination, and the expected mean chronological age (TM) for onset of visible bone aging. All samples showed high and significant correlation between the observed and expected mean biological age estimates. The R^2, however, varied among women from 43% (in Tajik of the Unji group) to 84% (in Georgians from the Georgian Republic) and among men from 34% (in Russians from the northern part of present Russia) to 87% (in Eskimos of the Chukotka Peninsula). The mean population age at the point of entry into the second stage of aging (TM) also showed wide variation between the studied populations. It reached highest values (45.6 years for males and 47.1 years for females) in Abkhazians from Chlou, Jgerda, and Atara villages and lowest values (25.3 years for females and 24 years for males) in Russians from Ukhta village.

Inspection of Tables 1 and 2 shows that model parameters in some populations are significantly different statistically for males and females, whereas in others they are not (Table 4). Parameter *B*, which estimates the rate of bone change, was significantly different between males and females in 22 of 29 samples. This value is much higher than the one expected from type 1 error (Sokal and Rohlf 1981). The results, however, do not indicate any consistent trend in sex differences. Wilcoxon's signed-rank test for two groups (Sokal and Rohlf 1981) showed no consistency in differences in the comparison of t_0 and q estimates ($p > 0.25$)

but was significant ($p < 0.05$) in the comparison of B coefficients. Women showed higher rates of bone aging more often. However, when we compared only the differences that were significant after taking into account the experiment-wise Bonferroni probability (Sokal and Rohlf 1987), no such trend was evident. That is, we cannot definitely conclude that women, more often or more seldom than men, show higher B estimates (or higher t_0 or q).

Moreover, despite the observed sex differences in parameter estimates and despite a marked interpopulation diversity in TM values, as mentioned, the TM values point to a high and significant correlation between males and females belonging to the same sample ($r = 0.787$, $p < 0.001$). This finding indicates that the combining of three model parameters results in a similar distribution of biological age estimates among men and women from the same ethnic group. Indeed, when we compared distributions of SSR and OSR in the two sexes, no significant differences were observed for any of our samples (Table 4): the F values were always small and the associated p values were higher than 0.20 and frequently more than 0.50. The results of the Kruskal-Wallis nonparametric analysis of variance and of the Kolmogorov-Smirnov two-sample tests were in good agreement with parametric analysis of variance results and therefore are not shown.

We now present some examples to illustrate the findings (Table 4). Figure 4 compares expected biological age estimates in relation to chronological age among females and males in four ethnic groups. As can be seen, the expected curves are close to one another, although some differences are evident, particularly in the combined Abkhazian population. Indeed, estimates of all three parameters are significantly different, even after Bonferroni's correction (Sokal and Rohlf 1987). However, Figure 5 demonstrates that the SSR and OSR distributions in the four groups, including Abkhazians, almost totally overlap. Thus, despite the sex differences in parameter estimates in some populations, the age-adjusted biological age distributions do not differ between males and females in any of the 29 separate samples studied or in the total sample.

Discussion

Human aging is often defined as the process of change that occurs in an individual during the course of time following maturity. This process finds its reflection in age-related changes that take place in various systems of a person, including the skeleton. Parfitt (1988) noted that bone loss in human aging is a universal, inevitable phenomenon regardless of sex, race, geographic location, socioeconomic status, or physical activity. Kiebzak (1991), in his review paper, pointed out that, despite

Table 4. Sex Comparison of Model Parameter Estimates and Age-Adjusted Distributions in Various Ethnic Groups

Ethnic Group (Villages)	*p Value of Difference*			*ANOVA*	
	t_0	q	B	*F Value*	*p Value*
Abkhazians (Chlou, Jgerda, Atara)	<0.01[a]	<0.01	<0.01[a]	0.052	0.820
Abkhazians (Kaldakhvar, Aatsy)	>0.05	<0.01[a]	<0.01[a]	0.031	0.860
Azerbaijanians	>0.05	>0.05	<0.01[a]	0.038	0.846
Buryats (Ulyuian, Agdora)	>0.05	>0.05	>0.05	1.591	0.208
Byelorussians (Veremeiki)	<0.01	<0.01	>0.05	0.130	0.718
Chukchi	>0.05	>0.05	>0.05	0.01	0.920
Eskimos	>0.05	<0.05	<0.01[a]	1.223	0.270
Georgians (Alisubani)	>0.05	<0.01	<0.01[a]	0.155	0.694
Jews (Kfar-Saba)	<0.05	<0.01[a]	<0.05	3.013	0.083
Karakalpaks (Karuzyak)	>0.05	>0.05	<0.01[a]	0.007	0.933
Karyki	<0.01[a]	<0.01	<0.01[a]	0.113	0.737
Kazakhs (Akkol)	<0.01[a]	>0.05	<0.01	0.154	0.695
Lithuanians	<0.01	>0.05	>0.05	1.237	0.267
Mongols	>0.05	>0.05	<0.01	0.782	0.377
Nentsi	>0.05	>0.05	<0.01[a]	1.603	0.207
Russians (Argoda, Burguzin, Chitkan, Kurumkan, Uro)	>0.05	>0.05	>0.05	0.131	0.718
Russians (Azerbaijan)	>0.05	>0.05	<0.01[a]	0.007	0.932
Russians (Klaipeda, Palanga, (Vilnus)	>0.05	>0.05	<0.01[a]	0.832	0.362
Russians (Kursk)	>0.05	>0.05	<0.01[a]	0.002	0.967
Russians (Peski)	>0.05	<0.01[a]	<0.01[a]	0.510	0.475
Russians (Puttyatino)	<0.01	>0.05	<0.01[a]	0.01	0.920
Russians (Voshod)	<0.01[a]	<0.01[a]	<0.01[a]	1.472	0.226
Russians (Ukhta)	>0.05	<0.01[a]	>0.05	1.225	0.269
Tajiks (Chorku)	>0.05	<0.01	<0.01[a]	1.017	0.314
Tajiks (Unji)	<0.05	>0.05	<0.01[a]	1.974	0.162
Teke Turkmenians	<0.01[a]	>0.05	<0.01[a]	0.146	0.703

Table 4. Continued

Ethnic Group (Villages)	p Value of Difference			ANOVA	
	t_0	q	B	F Value	p Value
Yomut Turkmenians (Kulmach, Uzun-Su)	<0.01[a]	>0.05	<0.01[a]	0.000	0.997
Tuvinci (Chadan, Erzin, Toora-Jem)	<0.01[a]	>0.05	<0.01[a]	0.174	0.677
Saamy	>0.05	>0.05	<0.01[a]	0.217	0.643
Russians (total)	>0.05	<0.05	>0.05	0.636	0.425
Abkhazians (total)	<0.01[a]	<0.01[a]	<0.01[a]	0.013	0.910
Turkmenians (total)	<0.01[a]	<0.05	<0.01[a]	0.201	0.654
Tadjiks (total)	<0.01	>0.05	<0.01[a]	0.528	0.468

a. $p < 0.05$ after Bonferroni correction for multiple comparison (Sokal and Rohlf 1987).

tremendous individual variation in the age at which significant bone loss becomes evident, the curve depicting loss of bone mass with age typically shows that both men and women lose bone, the process beginning at about the middle of the third decade of life and increasing gradually thereafter.

This conclusion derives mostly from postmortem investigations of bone but generally receives support from data such as our x-ray study of living individuals. Thus our biological age estimates for the total sample show high and extremely significant correlation with the chronological age of an individual, as is evident from Figure 1, where Pearson's $r = 0.800$ ($N = 3699$) for men and $r = 0.782$ ($N = 3428$) for women, with $p \ll 0.001$.

When we consider specific ethnic groups, they all show practically no visible bone changes until the third decade of life, but thereafter the changes become essentially linear with age (Figure 4). Although variation in parameter estimates between our studied groups at their different localities is considerable, the estimates nonetheless exhibit the same general pattern, with both men and women showing similar trends within each sample. As seen in Figure 5, the resulting distributions of biological age estimates adjusted for age show no sex differences. However, parameter estimates, in particular, rate of aging, consistently show significantly higher values in samples of women compared to samples of men.

Recent reviews of the literature claim that at any given age after puberty total bone mass relative to body size is always greater in women than in men (Melton et al. 1988; Parfitt 1988; Elliott et al. 1990). Yet the dynamics of bone change might not be different between the sexes, as shown by Adams et al. (1970) and also found in the present study. Elsewhere, İşcan (1991) reported sex differences in the age-related changes

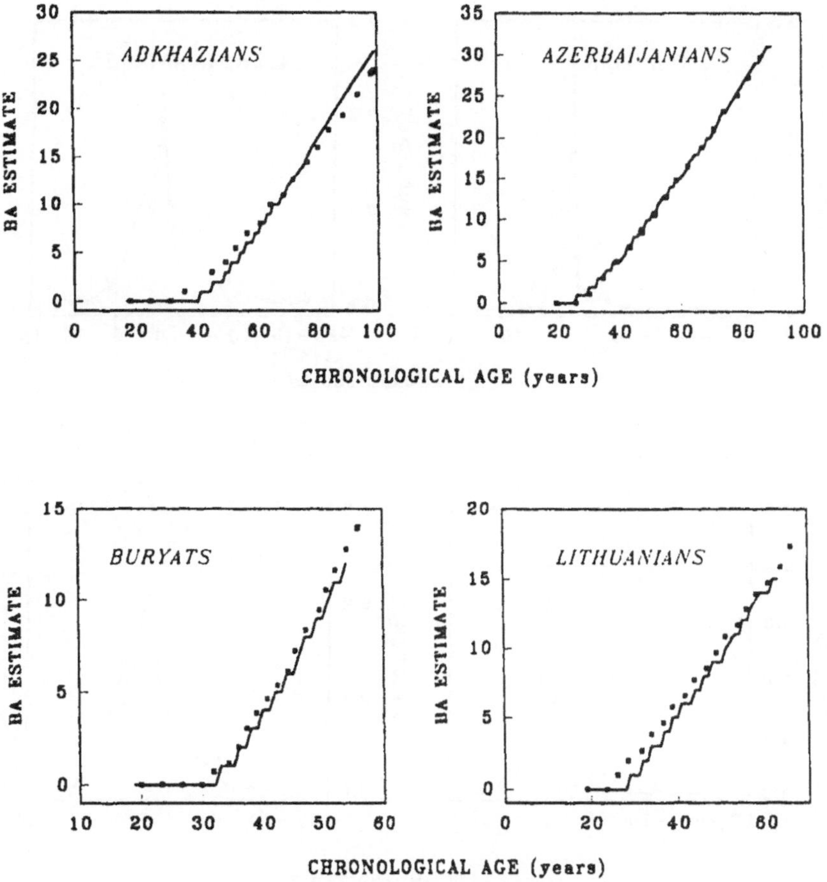

Figure 4. Expected curves of biological age estimates in male (curve) and female (squares) samples from four ethnic groups.

occurring in the costochondral junction of the ribs among white and black Americans. However, when we compared the 95% confidence levels of İşcan's (1991) data, as given in his Tables 2 and 3, we found that the data overlap greatly so that no sex differences can actually be confirmed statistically.

Linear regression of metacarpal cortical thickness, one criterion used in Milne's (1985) study, was likewise found to be the same for men and women at ages younger than 80 years, whereas the bone loss on age regression for the oldest men (>80 years) did not have a significant slope. Interestingly enough, another finding of Milne's study was that the regression lines for the index of wedging of vertebral bodies with age were parallel in men and women. Garn et al. (1992), in a 21-year follow-

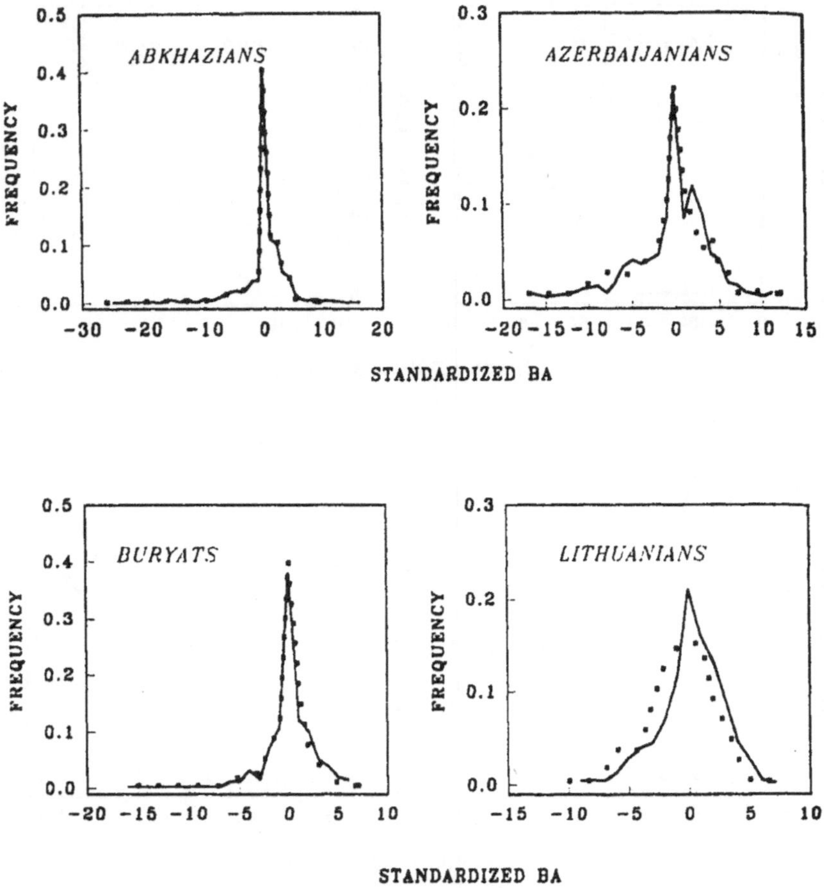

Figure 5. Same-sex residual distribution (curve) compared with opposite-sex residual distribution (squares) in four ethnic groups. For further explanations see Material and Methods section.

up study, found that although metacarpal bone loss is nearly the same in males and females, it constitutes a considerably larger percentage of the initially smaller bone mass in the female. Similarly, Kimura (1990) and Oyster (1992), who studied diverse parameters of bone change by different methods on different bones and samples, found the mean values of their parameters to be nearly always greater in men than in women, yet the regression slopes of many of these parameters on age were equal in both sexes.

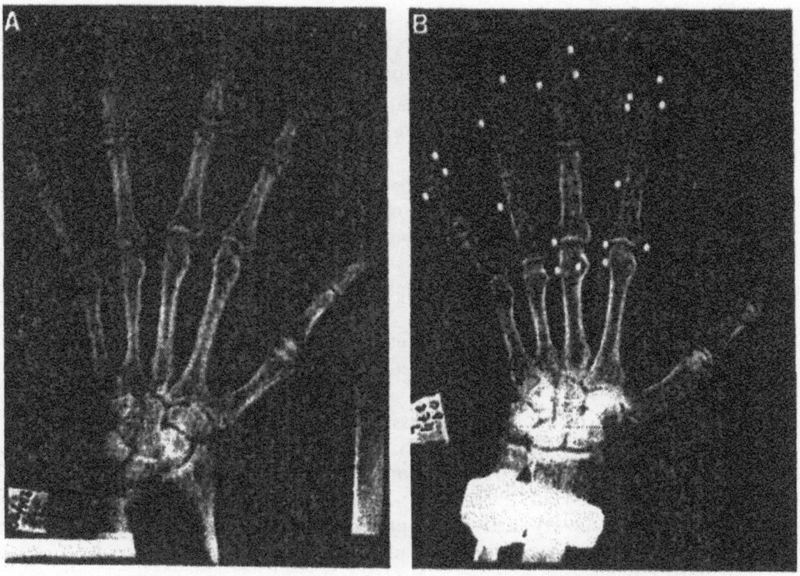

Figure 6. Radiogram of left hand bones of (A) a 20-year-old Bedouin male and (B) a 50-year-old Bedouin male.

Appendix

We present radiograms of the left hand bones of a 20- and a 50-year-old Bedouin male (Figure 6). The first specimen (Figure 6A) shows no bone change. His osseographic score is 0. The white spots on the second specimen (Figure 6B) indicate osseographic changes (see Material and Methods), with a total score of 20.

Acknowledgments We are grateful to M. Korostyshevsky for his helpful discussions and advice in the development of the stochastic model. We very much appreciate M. Crawford, R.R. Sokal, and several anonymous reviewers taking the time to critically read and comment on the earlier version of this manuscript. We thank M. Vaindler, G. Hay, and M. Goldschmidt-Nathan for computations and technical assistance during the preparation of the manuscript.

Received 27 January 1994; revision received 25 May 1994.

Literature Cited

Adams, P., G.T. Davies, and P. Sweetman. 1970. Osteoporosis and the effects of aging on bone mass in elderly men and women. *Q. J. Med.*, 601–615.

Adelman, R.C., and G.S. Roth. 1982. *Testing the Theories of Aging*. Boca Raton, FL: CRC Press.

Alexeeva, I., and E. Kobyliansky. 1966. Mineral saturation of human hand skeleton in the connection with its structure and function. *Probl. Anthropol.* 42:3–11.

Bernstein, C., and H. Bernstein. 1991. *Aging, Sex, and DNA Repair*. Orlando, FL: Academic Press.

Bocquet-Appel, J.P., and L. Jacobi. 1990. Familial transmission of longevity. *Ann. Hum. Biol.* 17:81–95.

Borkan, G.A. 1986. Biological age assessment in adulthood. In *The Biology of Human Aging*, A.H. Bittles and K.J. Collins, eds. Cambridge, England: Cambridge University Press, 81–94.

Borkan, G.A., and A.H. Norris. 1980. Assessment of biological age using a profile of physical parameters. *J. Gerontol.* 35(2):177–184.

Burov, S.A., and B.D. Reznikov. 1975. *Roentgenology in Forensic Medicine*. Saratov, Russia: Saratov University Press.

Comfort, A. 1969. Test battery to measure aging rate in man. *Lancet* 2:1411–1415.

Dequeker, J. 1976. Quantitative radiology: Radiogrammetry of cortical bone. *Br. J. Radiol.* 49:912–920.

Elliott, J.R., N.L. Gilchrist, and J.E. Wells. 1990. Effects of age and sex on bone density at the hip and spine in a normal Caucasian New Zealand population. *N.Z. Med. J.* 103:33–36.

Fukuchi, K, G.M. Martin, and R.J. Monnat. 1989. Mutator phenotype of Werner syndrome is characterized by extensive deletions. *Proc. Natl. Acad. Sci. USA* 86(15):5893–5897.

Garn, S.M., T.V. Sullivan, S.A. Decker et al. 1992. Continuing bone expansion and increasing bone loss over a two-decade period in men and women from a total community sample. *Am. J. Hum. Biol.* 4:57–68.

İşcan, M.Y. 1991. The aging process in the rib: An analysis of sex- and race-related morphological variation. *Am. J. Hum. Biol.* 3:617–623.

Kaur, H., and I. Jit. 1990. Age estimation from cortical index of the human clavicule in Northwest Indians. *Am. J. Phys. Anthropol.* 83:297–305.

Kellgren, J.H. 1963. The treatment of osteoarthrosis. *Rep. Rheum. Dis.* 15:1–12.

Kiebzak, G.M. 1991. Age-related bone changes. *Exp. Gerontol.* 26:171–187.

Kimura, K. 1990. Age and sex differences in bone density of the second metacarpal in its midshaft cross-section. *Ann. Hum. Biol.* 5:399–406.

Kobyliansky, E., I. Hershkovitz, and B. Arensburg. 1985. Use of radiograms of hand bones in predicting age and structure of Bedouin child populations, past and present. *Homo* 36:27–39.

Linnane, A.W., T. Ozawa, S. Marzuki et al. 1989. Mitochondrial DNA mutations as an important contributor to aging and degenerative diseases. *Lancet* 1:642–645.

Livshits, G., E. Kobyliansky, and O. Pavlovsky. 1995. Population biology of human aging: Ethnic and environmental factors of variation. Unpublished.

Lloyd, E., ed. 1984. *Handbook of Applicable Mathematics*. New York: Wiley.

Melton, L.J., E.Y.S. Chao, and J. Lane. 1988. Biomechanical aspects of fractures. In *Osteoporosis: Etiology, Diagnosis, and Management*. B.L. Riggs and L.J. Melton, eds. New York: Raven Press, 111–131.

Milne, J.S. 1985. *Clinical Effects of Aging: A Longitudinal Study*. London, England: Croom Helen.

Nakamura, E. 1991. A study on the basic nature of human biological aging processes based upon a hierarchical factor solution of the age-related physiological variables. *Mech. Aging Dev.* 60:153–170.

Ostlere, S.J., and R.H. Gold. 1991. Osteoporosis and bone density measurement methods. *Orthopedics* 271:149–170.

Oyster, N. 1992. Sex differences in cancellous and cortical bone strength, bone mineral content and bone density. *Age Aging* 21:353–356.

Parfitt, A.M. 1988. Bone remodeling: Relationship to the amount and structure of bone, and the pathogenesis and prevention of fracture. In *Osteoporosis: Etiology, Diagnosis, and Management*, B.L. Riggs and L.J. Melton, eds. New York: Raven Press, 45–94.

Pavlovsky, O.M. 1987. *Biological Age of a Man*. Moscow, Russia: Moscow University Press (in Russian).

Plato, C.C., and A.H. Norris. 1980. Measurements of the second metacarpal and lateral hand dominance. *Hum. Biol. Rec. Adv.* 1:159–173.

Pollitzer, W.S., and J.B. Anderson. 1989. Ethnic and genetic differences in bone mass: A review with a hereditary vs. environmental perspective. *Am. J. Clin. Nutr.* 50:1244–1259.

Ries, W., and D. Pothig. 1984. Chronological and biological age. *Exp. Gerontol.* 19:211–216.

Schneider, E.L., and J.A. Brody. 1983. Aging, natural death, and the compression of morbidity: Another view. *New Engl. J. Med.* 309:854–855.

Shall, B.A. 1975. Population structure and local differentiation. *Am. Natur.* 109:511–519.

Skalicky, M., H. Niedermuller, and G. Hofecker. 1987. Multivariate methods in the assessment of the biological age. In *Biological Age and Aging Risk Factors*. A. Ruiz-Torres, ed. Madrid, Spain: Technipublicationes, 73–82.

Smith, B.H., S.M. Garn, and P.E. Cole. 1982. Problem of sampling and inference in the study of fluctuating dental asymmetry. *Am. J. Phys. Anthropol.* 58:218–219.

Sokal, R.R., and F.J. Rohlf. 1981. *Biometry*. San Francisco, CA: W.H. Freeman.

Sokal, R.R., and F.J. Rohlf. 1987. *Introduction to Biostatistics*. New York: W.H. Freeman.

Statistical Abstracts of Israel. 1992. Jerusalem, Israel: Central Bureau of Statistics.

Steel, G.H. 1965. The relation between dental maturation and physiological maturity. *Dent. Pract.* 16(1):23–31.

Walker, R.A., and C.O. Lovejoy. 1985. Radiographic changes in the clavicule and proximal femur and their use in determination of skeletal age at death. *Am. J. Phys. Anthropol.* 68:67–78.

07019901

Some New Directions for Research on the Biology of Aging

GEORGE M. MARTIN[a]

Departments of Pathology and Genetics, University of Washington, Seattle, Washington 98195, USA

ABSTRACT: **A highly selective, eclectic, and personal view of new directions and new opportunities for research on the biology of aging is briefly outlined. Some concern is raised regarding the present emphasis on the use of centenarians for the definition of genetic loci responsible for unusually robust retention of structure and function. More progress is likely to be made were we to focus on the genetic basis for "elite" aging in middle-aged subjects examined for very specific phenotypes, as these are likely to be far less polygenic. Descriptive gerontology is entering a renaissance, given such new clinical tools as functional MRI and basic science tools such as functional genomics and proteomics. Advances in genomics should expedite answers to such questions as why some avian species have exceptionally long lifespans despite unusual loads of oxidative stress. One hopes to see renewed mechanistic studies, using such tools, at the systems levels. New methodologies are permitting the evaluation of stochastic alterations in gene structure and function in postreplicative cells. The exciting work on molecular misreading should prompt us to reexplore the Orgel hypothesis as it applies to such cell types. Epigenetic shifts in gene expression that occur in association with sexual maturation and the cessation of growth may have deleterious consequences late in the life course. It will therefore be important for gerontologists to investigate the molecular biology of pubescence. Finally, our community should investigate the impact of environmental "gerontogens," agents that accelerate specific processes of aging and specific senescent phenotypes.**

INTRODUCTION

Much of the excitement of doing science comes from the fact that we cannot easily predict the most rewarding new directions in almost any field of investigation. That dictum probably applies more to the study of aging than to most biological problems. Biogerontology interacts with all domains of biology at all levels of analysis, from molecules to populations. Add to that menu the mix of psychosocial avenues of research (which I shall make no pretense of addressing in this brief essay), and we have an almost impossible task. I can only offer a *potpourri* of my personal views of what phenotypes, which experimental approaches, and which mechanisms are especially neglected, promising, or important aspects of our science and which therefore deserve more attention as candidates for new directions for aging research.

[a]gmmartin@u.washington.edu

PHENOTYPES

Biogerontologists who work with model systems such as *Caenorhabditis elegans* and *Drosophila melanogaster* have reasonably concentrated almost exclusively upon life table parameters, particularly maximum lifespan. (Clearly, one new direction is the systematic definition of anatomic and pathophysiologic aspects of aging in these poorly understood models of aging.) We have only sampled a small subset of such model systems amenable to genetic analysis. Given the rapid advances in genomics, we should soon be able to choose from a much wider array of organisms.[1] Comparative studies across phyla can be very powerful in identifying genetic modulators of aging (and, hence, basic mechanisms) that might obtain in essentially all species ("pan-specific public modulators") or in a group of related animals such as mammals ("group-specific public modulators").[2] Medical types like me, however, have been preoccupied with the biochemical mechanisms underlying a range of late life diseases and disabilities. "Longevity genes" for people who live to be 100 but who have been demented for 15 years are not doing such individuals much good. These different approaches, of course, are not mutually exclusive. We need both.

Great progress is being made in the genetics of dementias of the Alzheimer type, but many other clinically important senescent phenotypes are crying out for more research. I shall highlight a few of these "orphan" phenotypes.[3] In terms of new directions, however, I shall emphasize a third approach, investigations of the biochemical genetic basis for unusually robust retention of structure and function. Such approaches have been neglected, with the exception of the recent flurry of interest in centenarians. Although I am happy to see this new direction of research, I am not very optimistic about our ability to interpret what I guess will be a mountain of polymorphic associations and linkage assignments, perhaps often different for different populations. Many more of us should be focusing upon rates of change of highly specific physiologic domains in middle-aged subjects. I give some arguments to support this point of view below. A recent request for applications from the U.S. National Institute on Aging highlights the importance of such research (http://grants.nih.gov/grants/guide/rfa-files/RFA-AG-99-007.html).

SOME ORPHAN PHENOTYPES OF CLINICAL INTEREST

We can pat ourselves on the back for the good progress we are making in discovering why some individuals are especially sensitive to dementias of the Alzheimer type. Not surprisingly, most of our solid information comes from the rare, "private" autosomal dominant mutations leading to early onset of disease. The hope is that this will lead us to an understanding of the vastly more common late onset "sporadic" forms of the disease. The three genes so far identified tend to support the amyloid cascade hypothesis of pathogenesis.[4] This might prove to be the case for what could be regarded as the clinically more significant "public" polymorphisms, of which only the *APOE* polymorphism has been well established.[5] It is virtually certain that other polymorphic susceptibility alleles will eventually be documented in a number of independent populations. Good candidates include the alpha 2 macroglobulin polymorphism[6] and the FE65 polymorphism.[7] Meanwhile, there are many patients waiting for help with a great variety of other geriatric ailments that impair the quality of life. I have referred to some of these as "orphan" phenotypes because of their comparative neglect by funding agencies and investigators.[3] Examples include idiopathic normal pressure hydrocephalus, trigeminal neuralgia, loss of olfactory sense, peripheral neuropathy, aging of the germ line, and degenerative intervertebral disk disease. Whereas the latter commonly presents in middle age, I consider any disability that is prevalent by about age 45 as a geriatric disorder, as any phenotype that does not reach a phenotypic threshold of expression until then will have escaped the force of natural selection.[2] I will make only a few comments about some of these research opportunities.

I was surprised to learn recently that there is not a single NIA Program-Project grant dealing with the theme of germ-line aging. I pointed out the importance of understanding ovarian aging in a review published some 22 years ago.[8] I am sure many others before and after have made the same point. If anything, it is now of greater clinical significance than ever before, because so many couples are choosing to postpone reproduction. The result has been a booming business in prenatal diagnosis to detect aneuploid pregnancies in older mothers. There is less concern for older fathers, but they too are at increased risk for the production of mutant sperm.[9] Given the recent excitement about the Bloom, Werner, and Cockayne syndromes,[10–12] all of which are likely to impact upon DNA transactions in meiosis as well as in mitosis, one would have expected a renaissance of interest in the fidelity of meiosis as it relates to aging, but I have seen no evidence of such a movement.

The apparent de-repression of integrated viral agents in late life is one of the more fascinating of the relatively neglected gerontological pathologies. Reactivation of childhood chicken pox is the classical example, but I wonder how many other such entities are waiting to be discovered. The trigeminal neuralgia form of *Herpes zoster* is among the most debilitating disorders of older people. Postherpetic neuralgia is in fact the leading cause of intractable pain in the elderly and is the single major cause of suicide in patients over the age of 70 who suffer from chronic pain.[13] Recent evidence indicates that T-cell recognition of an immediate early gene protein (IE63) may be one important component in controlling herpes reactivation from latency.[14] One would hope for more engagement of this problem by immunologists.

Normal pressure hydrocephalus is characterized by the clinical triad of gait disturbance, urinary incontinence, and dementia. A recent review came to the conclusion that "The causes and pathophysiology of idiopathic normal-pressure hydrocephalus (NPH) still remain poorly understood."[15] There are reasons to believe that the underlying pathogenesis is vascular, as arterial hypertension is a significant risk factor[16,17] and periventricular arteriosclerosis has been documented.[18] An interesting clue is that NPH can be associated with systemic lupus erythematosus.[19]

APPROACHES

Discovery Science and Hypothesis-Testing Science

Ever since I can remember, gerontologists have had trouble getting their grants approved by peer-review study groups, because they have proposed descriptive studies that were not sufficiently hypothesis-driven. But the tools for descriptive science (my colleague Lee Hood prefers the term "discovery" science) have improved dramatically. The spectacular advances in genomics will permit the rapid accumulation of our understanding of unusual gene action in new model organisms. Among these, I would give a high priority to avian genomics. Unfortunately for gerontology, the only avian species currently receiving significant attention by molecular geneticists is the domestic chicken. Evolutionary biologists have argued that species that have evolved or retained more efficient flight would be of greater interest for the study of the biology of aging.[20] These species have evolved remarkably long lifespans despite high body temperatures, high blood sugar levels, and high lifetime utilization of calories as compared to comparably sized mammals. Given these considerations, my colleague Steven N. Austad and I have embarked upon a program of avian genomics, with a focus on budgerigars (parakeets) (*Melopsittacus undulatus*) as the model organism.

We will soon have complete genomic sequences for all of the more traditional model organisms (*H. sapiens*, *M. domesticus*, *D. melanogaster*, and *C. elegans*). We will therefore see much more rapid progress on such projects as the linkage analysis of monogenic disorders of relevance to the pathobiology of human aging and the quantitative trait analysis of longevities of worms and flies. The new developments in functional genomics are also very exciting. The most recent initiative is in "proteomics," the goal being to comprehensively define networks of interacting proteins. Meanwhile, great progress has been made in the development and application of massive gene expression array technologies. The decision of *Science* magazine to publish such a study — involving a screening of changes in gene expression with aging and caloric restriction — is an example of the new kind of descriptive science that will help our field develop.[21]

The Importance of Single Cell Analytic Techniques

Studies such as the one just mentioned need to be accompanied, eventually, by single cell analytic techniques in order to address the age-old gerontological confound of shifts in population heterogeneity. For example, the aforementioned study[21] utilized skeletal muscle. In my experience, aging muscles may have more interstitial connective tissue, more adipocytes, more histiocytes, more lymphocytes, more fat cells, and relatively less skeletal muscle. Old muscle may also exhibit multiple foci of satellite cell-mediated muscle regeneration and neuritic sprouting at neuromuscular junctions. The microarray study just referred to[21] in fact provided molecular evidence of what histopathologists have observed. Techniques such as in situ polymerase chain reaction (PCR),[22] laser-assisted microdissection,[23] single cell cytoplasmic aspiration,[24] immunocytochemistry (especially accompanied by confocal microscopy),[25] and quantitative stereology[26] may help sort out these difficulties. Further studies with such tissues as skeletal muscle will also have to disentangle the relative contributions of disuse atrophy and intrinsic biological aging. Both are important to understand at the molecular levels. One way to get around this problem when studying skeletal muscle would be to include muscles like those of the tongue and diaphragm, as they are in constant use. This is also an argument for using myocardium as a model for aging in nondividing populations of cells.

A Revival of Somatic Cell Genetics

I first became involved in the study of cell aging as a frustrated postdoctoral fellow in the laboratory of Guido Pontecorvo at Glasgow University. "Ponte" had pioneered research on the parasexual cycle of fungi, using such genetic recombinational mechanisms as haploidization, mitotic nondisjunction, and mitotic crossing-over to map genes to chromosomes and to define the linear order of chromosomal genes.[27] The lab was in the process of switching from *Aspergillus nidulans* to cultivated human somatic cells. While I was eventually successful in providing some modest cytogenetic evidence for the existence of a parasexual cycle in human somatic cells,[28] my efforts at carrying out serial subcloning and replicative plating were dismal failures. This was 1961. Hayflick and Moorhead[29] had just published their famous paper on the limited replicative lifespans of such cells. But we can now establish cultures of various types of human somatic cells from individual human donors via the transgenic expression of the catalytic subunit of human telomerase.[30] We therefore now have the opportunity to revisit somatic cell genetic approaches to interesting biological questions, including those relevant to gerontology. In addition to exploitation of a parasexual cycle for genetic analysis, one can induce a variety of point mutations, conditionally express a variety of transfected genetic constructs, and fuse cells with a variety of cell types. An example of a specific application would be the use of such cultures to establish genetic haplotypes. The *APOE* chromosomal domain would be a good place to start. Another example would be the development of isogenic strains bearing various mutations for the study of the mitotic cell cycle, apoptosis, terminal differentiation, dedifferentiation, etc. One of my dreams is to turn skin fibroblasts into neuroblasts and to then terminally differentiate them into various types of glial and neural cells. This would give us immediate access to gene action in human CNS cells from individual subjects without the necessity of brain biopsy.

Some Difficulties with the Centenarian Approach

The first conceptual difficulty one faces when one initiates a genetic association study or a genetic segregation study to find genetic alleles that provide major contributions to the probability of becoming a centenarian is the fact that stochastic events clearly make a major impact upon lifespan. First of all, this is predicted by the evolutionary biological theory of aging. According to that theory, senescence is nonadaptive. It is not the result of determinative, sequential gene action. It can be viewed as a byproduct of gene action selected for entirely different purposes, the enhancement of reproductive fitness; some senescent phenotypes may also be the result of idiosyncratic constitutional mutations that escape the force of natural selection.[31,32]

Secondly, human lifespan is likely to be modulated by a large number of disease-associated genetic loci.[33] We may be misled by the reports of enhanced lifespans resulting from single gene mutations or the transfection of a single genetic construct in model organisms. These experiments[34,35] have utilized highly inbred laboratory stocks, each of which can presumed to have its special set of vulnerabilities.

A third concern is the obvious difficulty in getting DNA from members of three generations of a pedigree in order to sort out phase relations of markers.

Fourth, one would be hard put to carry out longitudinal studies in centenarians in order to establish an intrinsically slow rate of change of a putatively relevant biological marker. Moreover, any such physiologic studies would be obscured by the fact that centenarians are likely to suffer from a number of comorbid processes.

Finally, attempts to make generalizations to multiple populations should be difficult and expensive. A priori, given the likelihood of important roles for many gene-gene interactions and environmental-gene interactions, one would expect different subsets of polymorphisms to emerge in different populations.

The Genetic Basis of "Elite Aging"

I will arbitrarily define, as examples of elite aging, rates of decline of specific physiologic parameters that are slower than those observed in 99% of the population of interest. Here is one research strategy for a genetic epidemiology study that might be productive[36]: (1) Decide upon a specific physiologic function, the more specific the better, as a narrow phenotype is more likely to lead to the identification of a few loci with major effects. The assay should be relatively cheap, relatively noninvasive, highly reproducible, and, of great importance, extremely sensitive. (2) Use the assay(s) to screen a very large number of middle-aged individuals in at least two contrasting populations, preferably those that are relatively ethnically homogeneous (e.g., Japanese and Finnish). Why do I choose middle-aged subjects? First of all, as briefly noted above, by the age of 45, quantitative estimates of the rate of decline of the Malthusian parameter in age-structured human populations[2] indicate that the force of natural selection has essentially disappeared. Therefore, significant rates of functional and structural decline in various body systems should be underway by middle age. Second, people are still generally healthy in middle age, so that we have fewer concerns about comorbidity. Third, there are vast numbers of such individuals in most human populations. Fourth, one will often be able to obtain DNA from three generations (the parents and children of the middle-aged index cases). Fifth, sibs of about the same age (say within two years of the index case) should be easy to find, permitting genetic studies on sib-pairs. (3) Choose individuals who score in the upper one percentile of the population. These are the elite individuals who have remarkably robust retention of function in a given assay or assays. Carry out assays in sibs of these index cases. Define a subset of sib pairs with extreme concordance or extreme discordance for the phenotype of interest. Such pairs would provide enhanced power for genetic analysis.[37] (4) Carry out a genome wide screen for markers closely associated with the phenotype of interest. (5) Identify major genetic loci via positional cloning. Such research will be expensive and time consuming but should greatly increase our understanding of the molecular biology of aging in human beings. The challenge will be the definition of the phenotypes to be investigated and the choice of suitably sensitive assays. Perhaps this challenge will help to revive more traditional physiologic research in departments of physiology and medicine. A growing proportion of faculty members in those departments have become biochemists and molecular biologists.

Neuroimaging

Gerontologists have only begun to tap the vast potential of neuroimaging methodologies for the study of brain aging. These include magnetic resonance imaging

(MRI), functional MRI,[38] MRI spectroscopy,[39] and positron emission tomography (PET).[40]

MECHANISMS

Neuroendocrine Cascades

The concept of neuroendocrine cascades in aging was initially championed by Caleb Finch.[41] There are several reasons why I believe gerontologists will be driven to devote comparatively more effort to research on the neuroendocrine system. First, there is the exciting work in *C. elegans* on the regulation of the dauer pathway. Mutations in this pathway can lead to marked increases in lifespan. The trigger for the switch in gene expression is a pheromone related to conditions of larval crowding. This signal is mediated by a neurosecretory function.[42] An insulin-like receptor is an early player in the cascade and a downstream forkhead transcription factor is the key to the regulation of a suite of other yet-to-be-defined genes.[43] The parallels of the effects on lifespan of metabolic controls of fuel utilization in nematodes and in calorically restricted rodents[44] are striking. Second, there is increasing recognition of type 2 diabetes mellitus as a segmental progeroid syndrome. Third, there are the long-lived dwarf mice with lesions in the pituitary and associated hypothyroidism.[45] Fourth, there is the well-documented role of the thyroid hormone on metabolic rate and mitochondrial function. Fifth, there is the clinically important question of sleep disorders of the elderly, the association of sleep with lowering of core body temperatures, and the mysterious "restorative" role of sleep. The neuroendocrine basis for the setting of core body temperature deserves a great deal more research. It is still conceivable that virtually all of the reports of enhancements of animal lifespans are based on a lowering of core body temperatures or metabolic rates. This has long been known for the case of poikilotherms like fish, drosophila, and nematodes. More recent evidence has come from the study of lifespan mutations in *C. elegans,*[46] dwarf mice,[45] and calorically restricted rodents.[47]Sixth, the beautiful research from Cynthia Kenyon's lab on the coupling of reproductive biology with lifespan[48] also argues for more attention to such questions in mammals. Finally, we will probably be faced with an avalanche of clinical trials of such agents as selective estrogen and testosterone receptor agonists, growth hormone, and growth hormone release hormone. We shall need to develop the basic science underpinnings for clinicians.

Late Effects of Early Epigenetic Switches in Gene Expression

The laboratory mouse attains its full adult size at around 6 months of age, just about the time Karen Swisshelm found evidence for the beginning of a selective inactivation of a specific chromosomal site of ribosomal RNA sythesis.[49] One can assume that this mechanism evolved to partially downregulate protein synthesis after the major epoch of somatic growth. But once such a mechanism becomes engaged, it appears to have a "life of its own." The result could be a slowly progressive and insidious course of decline in protein synthesis and turnover into senescent phases of the life course.

IGF2 undergoes parental genomic imprinting. A switch from monoallelic to biallelic methylation of its promoter is said to occur "during aging" of human beings.[50] Although the existing data are scarce, they are consistent with a spreading of methylation to the promoter of the other parental allele beginning in the second decade of life. I therefore conclude that such a switch evolved as a result of strong selective pressure. Once initiated, the inactivation continues unabated, with almost complete methylation occurring at advanced ages. Here is yet another example of an adaptive switch without any built-in protection of a "brake" or "reverse gear." Given the importance of insulin-like growth factors in the pathobiology of aging, once can imagine that the late-life consequences of this strategy are substantial.

A third example is the downregulation of the expression of the NR2B subunit of the NMDA receptor at around the time of sexual maturity.[51] This epigenetic switch appears to be one of the causes of the declines in the efficiencies of memory and learning in aging mammals; the constitutive expression of additional copies of that gene leads to superior performances in several assays of memory and learning in transgenic mice.[51] But how would the downmodulation of such an apparently advantageous trait lead to enhanced reproductive fitness? Some fascinating trade-off effects remain to be discovered. We could in fact learn a great deal about the biology of aging by studying the biology of pubescence.

Alterations in Proliferative Homeostasis

Here are additional examples of phenotypes of great interest to biogerontologists that I believe also have their origins via an epigenetic switch in gene expression during early development. Unlike nematodes and fruit flies, many aged tissues from mammalian species (perhaps all mammalian species), including humans,[52] exhibit a mix of atrophy, hyperplasia, benign neoplasia, and, with considerable frequency, malignant neoplasia. Much progress has been made in the dissection of somatic mutational alterations leading to the malignant phenotype.[53] Here we have a concrete example of a stochastic mechanism leading to an important senescent phenotype. Is the hyperplasia that is found in aging tissues a precursor of neoplastic proliferation? This is a plausible scenario, as it could serve as a tumor promoter mechanism. We have almost no knowledge of how such altered homeostasis occurs, although we can guess it has to do with a loss of the fidelity of cell-cell communications and, perhaps, cell-matrix communications. One hypothesis worth testing would be that these aberrations result from asynchronous clonal attenuations[54] of cell types that participate in maintaining proliferative homeostasis.

The vast majority of research on replicative senescence has involved skin fibroblast-like cells. By contrast, there has been a paucity of research on the senescence of cells of the vascular wall (medial smooth muscle and endothelial cells).[55,56] This is unfortunate, as it may prove to be highly relevant to the pathogenesis of atherosclerosis, the major cause of morbidity and mortality in aging populations of western societies.

The question arises as to how clonal attenuation evolved. Most authors have emphasized the notion that it evolved as a mechanism to protect cells from undergoing neoplastic transformation.[57] I have argued elsewhere that perhaps a more plausible explanation is that the switch from stem cell modes of replication to tangential modes of replication was selected to permit finer control of organogenesis.[58] There is very little information as to when and how, during embryogenesis, stem cell modes of replication are abandoned and tangential types of replication are initiated.

Gerontogens

Many biologists are gainfully employed pursuing research on mutagens, carcinogens, and teratogens, but the concept of gerontogens has not yet caught on. I have defined gerontogens as exogenous agents that can accelerate aspects of biological aging.[59] One might classify these into two groups in the same way one might classify deleterious genetic alleles that impact upon aging.[33,60] Thus, there may be "segmental gerontogens" and "unimodal gerontogens." In the real world, exposure to ultraviolet light is inescapable and, therefore, various nature-nurture interactions produce different degrees of ultraviolet-induced aging of the skin. Ultraviolet-induced skin aging can therefore be considered to be an example of a unimodal gerontogen. Ultraviolet light accelerates the rates of development of such phenotypes as epidermal atrophy, basophylic degeneration of collagen, fragmentation of elastin, senile keratosis, Bowen's disease, basal cell carcinoma, squamous cell carcinoma, and malignant melanoma. These phenotypes can be presumed to result from both protein and DNA damage. Oxidative stress is widely believed to be among the more general mechanisms of aging. It will be important, therefore, to have more detailed research on the molecular biology of UV-mediated skin pathologies. An example of such relevant research is the discovery that tandem CC to TT tandem mutations are commonly found in the p53 gene of neoplasms that develop in sun-exposed skin.[61] These tandem mutations are candidate "signature mutations" for oxidative damage.[62]

A good candidate for a segmental gerontogen is tobacco smoke. It will be important to ascertain which of its numerous genotoxic agents produces DNA damage that most closely resembles that which can be observed spontaneously in various aging cell types. A study of parallel effects on proteins will also be of interest. The proneoplastic effects of tobacco smoke are exceedingly well established. Perhaps of even greater relevance to the pathobiology of aging, however, are its effects upon the vascular system, particularly the microvasculature.[63]

The Aging Brain and the Tyranny of Plaques and Tangles

Virtually every laboratory in the world with major interests in the pathogenesis, prevention, or treatment of dementia of the Alzheimer type works on aspects of the metabolism of the beta amyloid precursor protein. Small peptide derivatives of that protein are thought to be responsible for the genesis of one of the two principle histopathological lesions, the neuritic plaque. Judging from our publications (for example), our lab would be counted among the *baptists* of the scientific world. The exceptions are mostly those who are devoted to unraveling neurofibrillary tangles, the second major histopathological lesion. These colleagues are enamored of the microtubular associated protein tau, and hence are referred to as *tauists*. Hardly anyone is examining the molecular biology of the aging brain parenchyma *in between* the famous plaques and tangles. The work on molecular misreading referred to below is an exception. We need more such exceptions.

The Microvasculature

The microvasculature (capillaries, venules, and arterioles) are arguably the most important structures between the plaques and tangles discussed above. After years of relative neglect (except for the observations of congophilic angiopathy), the microvasculature of the brain has finally been getting some attention.[64] Almost nothing is known, however, about changes, in the aging organism, in the kinetics of accumulation of altered structure and function of mitochondria relative to what has been observed in unfractionated brain tissues.[65] We also need more research on the susceptibility of endothelial cells to apoptotic and necrotic injuries, the decline in the proliferative potentials of endothelial cells, and the changes in structure and function of basement membranes. Although there is a significant body of research on the blood brain barrier (BBB) in aging mammals, I am unaware of any systematic studies of individual variations in the degree to which the integrity of the BBB is maintained during aging among populations of human subjects.

Gene Instability in Postreplicative Cells

Powerful new methods for the study of mutation *in vivo* will provide us with a great deal more information about the frequencies and spectra of mutations in various cell types, including postreplicative cells such as neurons.[66] In addition to mutation, we will need to understand the basis for such processes as DNA amplifications, including increases in ploidy. The latter occurs in the aging myocardium, particularly in association with hypertension.

Molecular Misreading in Postreplicative Cells: Should We Revisit the Orgel Hypothesis?

Fred van Leewen and his colleagues[67] have provided us with exciting new evidence for the occurrence of transcriptional errors in aging neurons. There may be a general vulnerability of the genome to develop frameshifts when the complex machinery of DNA-dependent RNA polymerase engages sequences such as GAGAG. We should not ignore, however, molecular misreading at the translational level. It will be particularly important to search for age-related increases in transcriptional and translational fidelity for the case of proteins that are themselves involved in protein synthesis, as this provides the possibility of a positive feedback loop, as first envisaged by Leslie Orgel.[68] Orgel's hypothesis has been "pronounced dead" by gerontologists a number of times, but continues to be discussed.[69,70] But this conclusion was mainly based on experiments with the *in vitro* model of replicative senescence.[71] Perhaps the only test of that idea carried out in postreplicative cells *in vivo* was an experiment performed by Peter Rabinovitch and myself some 17 years ago, in which we challenged young and old mice with the encephalomyocarditis virus, an agent that replicates in neurons and myocardial cells.[72] There were two major predictions: (1) that the yield of virus should decrease in old tissues because of the theorized Orgelian exponential rise in the infidelity of protein synthesis and (2) that the frequencies of mutant viral progeny would rise for the same reasons. Prediction one was falsified, but we were unable to test prediction two because of technical reasons, including the intrinsically high mutation rates of such viral agents. Given the new experiments of van Leeuwen and colleagues, I suggest that the Orgel hypothesis was buried prematurely and that the body should be exhumed for another autopsy.

REFERENCES

1. MARTIN, G.M. & M.S. TURKER. 1988. Model systems for the genetic analysis of mechanisms of aging. J.Gerontol. **43:** B33–B39.
2. MARTIN, G.M., S.N. AUSTAD & T.E. JOHNSON. 1996. Genetic analysis of ageing: role of oxidative damage and environmental stresses. Nat.Genet. **13:** 25–34.
3. MARTIN, G.M. 1993. FRAR course on laboratory approaches to aging. "Orphan" phenotypes in gerontological research. Aging (Milano) **5:** 325–332.
4. HARDY, J.A. & G.A. HIGGINS. 1992. Alzheimer's disease: the amyloid cascade hypothesis. Science **256:** 184–185.
5. SWARTZ, R.H., S.E. BLACK & P. GEORGE-HYSLOP. 1999. Apolipoprotein E and Alzheimer's disease: a genetic, molecular and neuroimaging review. Can. J. Neurol. Sci. **26:** 77–88.
6. BLACKER, D., M.A. WILCOX, N.M. LAIRD *et al.* 1998. Alpha-2 macroglobulin is genetically associated with Alzheimer disease [see comments]. Nat.Genet. **19:** 357–360.
7. HU, Q., W.A. KUKULL, S.L. BRESSLER *et al.* 1998. The human FE65 gene: genomic structure and an intronic biallelic polymorphism associated with sporadic dementia of the Alzheimer type. Hum.Genet. **103:** 295–303.
8. MARTIN, G.M. 1977. Cellular aging-postreplicative cells. A review (Part II). Am. J. Pathol. **89:** 513–530.
9. VOGEL, F. & R. RATHENBERG. 1975. Spontaneous mutation in man. Adv. Hum. Genet. **5:** 223–318.
10. BOHR, V.A., G. DIANOV, A. BALAJEE *et al.* 1998. DNA repair and transcription in human premature aging disorders. J. Invest. Dermatol. Symp. Proc. **3:** 11–13.
11. FRY, M. & L.A. LOEB. 1998. The three faces of the WS helicase [news; comment]. Nat. Genet. **19:** 308–309.
12. YAMAGATA, K., J. KATO, A. SHIMAMOTO *et al.* 1998. Bloom's and Werner's syndrome genes suppress hyperrecombination in yeast sgs1 mutant: implication for genomic instability in human diseases. Proc. Natl. Acad. Sci. USA **95:** 8733–8738.
13. HESS, T.M., L.J. LUTZ, L.A. NAUSS & T.J. LAMER. 1990. Treatment of acute herpetic neuralgia. A case report and review of the literature [see comments]. Minn. Med. **73:** 37–40.
14. SADZOT-DELVAUX, C., P.R. KINCHINGTON, S. DEBRUS *et al.* 1997. Recognition of the latency-associated immediate early protein IE63 of varicella-zoster virus by human memory T lymphocytes. J. Immunol. **159:** 2802–2806.
15. MORI, K. & T. MIMA. 1998. To what extent has the pathophysiology of normal-pressure hydrocephalus been clarified? Crit. Rev. Neurosurg. **8:** 232–243.
16. KRAUSS, J.K., J.P. REGEL, W. VACH *et al.* 1996. Vascular risk factors and arteriosclerotic disease in idiopathic normal- pressure hydrocephalus of the elderly. Stroke **27:** 24–29.
17. PENAR, P.L., W.D. LAKIN & J. YU. 1995. Normal pressure hydrocephalus: an analysis of aetiology and response to shunting based on mathematical modeling. Neurol. Res. **17:** 83–88.
18. AKAI, K., S. UCHIGASAKI, U. TANAKA & A. KOMATSU. 1987. Normal pressure hydrocephalus. Neuropathological study. Acta Pathol. Jpn. **37:** 97–110.
19. UHL, M.D., B.E. WERNER, T.J. ROMANO & B.L. ZIDAR. 1990. Normal pressure hydrocephalus in a patient with systemic lupus erythematosus. J. Rheumatol. **17:** 1689–1691.
20. HOLMES, D.J. & S.N. AUSTAD. 1995. Birds as animal models for the comparative biology of aging: a prospectus. J. Gerontol. A Biol. Sci. Med. Sci. **50:** B59–B66.
21. LEE, C.K., R.G. KLOPP, R. WEINDRUCH & T.A. PROLLA. 1999. Gene expression profile of aging and its retardation by caloric restriction. Science **285:** 1390–1393.
22. LONG, A.A. 1998. In-situ polymerase chain reaction: foundation of the technology and today's options. Eur. J. Histochem. **42:** 101–109.

23. BANKS, R.E., M.J. DUNN, M.A. FORBES *et al.* 1999. The potential use of laser capture microdissection to selectively obtain distinct populations of cells for proteomic analysis: preliminary findings. Electrophoresis **20:** 689–700.
24. KACHARMINA, J.E., P.B. CRINO & J. EBERWINE. 1999. Preparation of cDNA from single cells and subcellular regions. Methods Enzymol. **303:** 3–18.
25. BROWN, N.L. 1999. Imaging gene expression using antibody probes. Methods Mol. Biol. **122:** 75–91.
26. WEST, M.J., L. SLOMIANKA & H.J. GUNDERSEN. 1991. Unbiased stereological estimation of the total number of neurons in the subdivisions of the rat hippocampus using the optical fractionator. Anat. Rec. **231:** 482–497.
27. PONTECORVO, G. 1958. Trends in Genetic Analysis. Columbia University Press. New York.
28. MARTIN, G.M. & C.A. SPRAGUE. 1969. Parasexual cycle in cultivated human somatic cells. Science **166:** 761–763.
29. HAYFLICK, L. & P.S. MOORHEAD. 1961. The serial cultivation of human diploid cell strains. Exp. Cell Res. **25:** 585–621.
30. BODNAR, A.G., M. OUELLETTE, M. FROLKIS *et al.* 1998. Extension of life-span by introduction of telomerase into normal human cells [see comments]. Science **279:** 349–352.
31. KIRKWOOD, T.B. & M.R. ROSE. 1991. Evolution of senescence: late survival sacrificed for reproduction. Philos. Trans. R. Soc. Lond. B Biol. Sci. **332:** 15–24.
32. ROSE, M.R. 1991. Evolutionary Biology of Aging. Ref Type: Serial (Book, Monograph). Oxford University Press.
33. Martin, G.M. 1978. Genetic syndromes in man with potential relevance to the pathobiology of aging. Birth Defects Orig. Art. Ser. **14:** 5–39.
34. LIN, Y.J., L. SEROUDE & S. BENZER. 1998. Extended life-span and stress resistance in the Drosophila mutant methuselah [see comments]. Science **282:** 943–946.
35. PARKES, T.L., A.J. ELIA, D. DICKINSON *et al.* 1998. Extension of Drosophila lifespan by overexpression of human SOD1 in motorneurons [comment] [see comments]. Nat. Genet. **19:** 171–174.
36. MARTIN, G.M. 1998. Toward a genetic analysis of unusually successful neural aging. *In* Handbook of the Aging Brain. E. Wang & D.S. Snyder, Eds.: 125–134. Ref Type: Serial (Book, Monograph). Academic Press. New York.
37. ZHANG, H. & N. RISCH. 1996. Mapping quantitative-trait loci in humans by use of extreme concordant sib pairs: selected sampling by parental phenotypes [published erratum appears in Am. J. Hum. Genet. 1997. **60:** 748–749]. Am. J. Hum. Genet. **59:** 951–957.
38. CORKIN, S. 1998. Functional MRI for studying episodic memory in aging and Alzheimer's disease. Geriatrics **53** (Suppl. 1): S13–S15.
39. ROTHMAN, D.L., N.R. SIBSON, F. HYDER *et al.* 1999. In vivo nuclear magnetic resonance spectroscopy studies of the relationship between the glutamate-glutamine neurotransmitter cycle and functional neuroenergetics [In Process Citation]. Philos. Trans. R. Soc. Lond. B Biol. Sci. **354:** 1165–1177.
40. CASELLI, R.J., N.R. GRAFF-RADFORD, E.M. REIMAN *et al.* 1999. Preclinical memory decline in cognitively normal apolipoprotein E-epsilon4 homozygotes. Neurology **53:** 201–207.
41. FINCH, C.E. 1976. The regulation of physiological changes during mammalian aging. Q. Rev. Biol. **51:** 49–83.
42. AILION, M., T. INOUE, C.I. WEAVER *et al.* 1999. Neurosecretory control of aging in *Caenorhabditis elegans*. Proc. Natl. Acad. Sci. USA **96:** 7394–7397.
43. LIN, K., J.B. DORMAN, A. RODAN & C. KENYON. 1997. daf-16: an HNF-3/forkhead family member that can function to double the life-span of *Caenorhabditis elegans* [see comments]. Science **278:** 1319–1322.
44. SOHAL, R.S. & R. WEINDRUCH. 1996. Oxidative stress, caloric restriction, and aging. Science **273:** 59–63.
45. BARTKE, A., H.M. BROWN-BORG, A.M. BODE *et al.* 1998. Does growth hormone prevent or accelerate aging? Exp. Gerontol. **33:** 675–687.
46. VAN VOORHIES, W. 1999. Genetic and environmental conditions that increase longevity in *Caenorhabditis elegans* decrease metabolic rate. Proc. Natl. Acad. Sci. USA **96:** 11399–11403.

47. TURTURRO, A. & R.W. HART. 1991. Longevity-assurance mechanisms and caloric restriction. Ann. N.Y. Acad. Sci. **621:** 363–372.
48. HSIN, H. & C. KENYON. 1999. Signals from the reproductive system regulate the lifespan of *C. elegans* [see comments]. Nature **399:** 362–366.
49. SWISSHELM, K., C.M. DISTECHE, J. THORVALDSEN *et al.* 1990. Age-related increase in methylation of ribosomal genes and inactivation of chromosome-specific rRNA gene clusters in mouse. Mutat. Res. **237:** 131–146.
50. ISSA, J.P., P.M. VERTINO, C.D. BOEHM *et al.* 1996. Switch from monoallelic to biallelic human IGF2 promoter methylation during aging and carcinogenesis. Proc. Natl. Acad. Sci. USA **93:** 11757–11762.
51. TANG, Y.-P., E. SHIMIZU, G.R. DUBE *et al.* 1999. Genetic enhancement of learning and memory in mice. Nature **401:** 63–69.
52. MARTIN, G.M. 1979. Proliferative homeostasis and its age-related aberrations. Mech.Ageing Dev. **9:** 385–391.
53. FEARON, E.R. 1995. Molecular genetics of colorectal cancer. Ann. N.Y. Acad. Sci. **768:** 101–110.
54. MARTIN, G.M., C.A. SPRAGUE, T.H. NORWOOD & W.R. PENDERGRASS. 1974. Clonal selection, attenuation and differentiation in an *in vitro* model of hyperplasia. Am. J. Pathol. **74:** 137–154.
55. MARTIN, G.M. & C.A. SPRAGUE. 1972. Clonal senescence and atherosclerosis. Lancet **2:** 1370–1371.
56. CHANG, E. & C.B. HARLEY. 1995. Telomere length and replicative aging in human vascular tissues. Proc. Natl. Acad. Sci. USA **92:** 11190–11194.
57. CAMPISI, J. 1997. The biology of replicative senescence. Eur. J. Cancer **33:** 703–709.
58. MARTIN, G. M. 1993. Clonal attenuation: causes and consequences [editorial]. J. Gerontol. **48:** B171–B172.
59. MARTIN, G.M. 1987. Interactions of aging and environmental agents: the gerontological perspective. Prog. Clin. Biol. Res. **228:** 25–80.
60. MARTIN, G.M. 1982. Syndromes of accelerated aging. Natl. Cancer Inst. Monogr. **60:** 241–247.
61. QUEILLE, S., S. SEITE, S. TISON *et al.* 1998. p53 mutations in cutaneous lesions induced in the hairless mouse by a solar ultraviolet light simulator. Mol. Carcinog. **22:** 167–174.
62. REID, T.M. & L.A. LOEB. 1993. Tandem double CC-->TT mutations are produced by reactive oxygen species. Proc. Natl. Acad. Sci. USA **90:** 3904–3907.
63. MONFRECOLA, G., G. RICCIO, C. SAVARESE *et al.* 1998. The acute effect of smoking on cutaneous microcirculation blood flow in habitual smokers and nonsmokers. Dermatology **197:** 115–118.
64. DE LA TORRE, J. C. 1997. Hemodynamic consequences of deformed microvessels in the brain in Alzheimer's disease. Ann. N.Y. Acad. Sci. **826:** 75–91.
65. CORTOPASSI, G.A. & A. WONG. 1999. Mitochondria in organismal aging and degeneration. Biochim. Biophys. Acta **1410:** 183–193.
66. DOLLE, M.E., H.J. MARTUS, M. NOVAK *et al.* 1999. Characterization of color mutants in lacZ plasmid-based transgenic mice, as detected by positive selection [In Process Citation]. Mutagenesis **14:** 287–293.
67. VAN LEEUWEN, F.W., J.P. BURBACH & E.M. HOL. 1998. Mutations in RNA: a first example of molecular misreading in Alzheimer's disease. Trends Neurosci. **21:** 331–335.
68. ORGEL, L.E. 1970. The maintenance of the accuracy of protein synthesis and its relevance to ageing: a correction. Proc. Natl. Acad. Sci. USA **67:** 1476.
69. HOLLIDAY, R. 1996. The current status of the protein error theory of aging. Exp. Gerontol. **31:** 449–452.
70. GALLANT, J., C. KURLAND, J. PARKER *et al.* 1997. The error catastrophe theory of aging. Point counterpoint [letter]. Exp. Gerontol. **32:** 333–346.
71. HARLEY, C.B., J.W. POLLARD, J.W. CHAMBERLAIN *et al.* 1980. Protein synthetic errors do not increase during aging of cultured human fibroblasts. Proc. Natl. Acad. Sci.USA **77:** 1885–1889.
72. RABINOVITCH, P.S. & G.M. MARTIN. 1982. Encephalomyocarditis virus as a probe of errors in macromolecular synthesis in aging mice. Mech. Ageing Dev. **20:** 155–163.

Bulging disks, fragile bones, fractured hips, torn ligaments, varicose veins, cataracts, hearing loss, hernias and hemorrhoids: the list of bodily malfunctions that plague us as we age is long and all too familiar. Why do we fall apart just as we reach what should be the prime of life?

The living machines we call our bodies deteriorate because they were not designed for extended operation and because we now push them to function long past their warranty period. The human body is artistically beautiful and worthy of all the wonder and amazement it invokes. But from an engineer's perspective, it is a complex network of bones, muscles, tendons, valves and joints that are directly analogous to the fallible pulleys, pumps, levers and hinges in machines. As we plunge further into our postreproductive years, our joints and other anatomical features that serve us well or cause no problems at younger ages reveal their imperfections. They wear out or otherwise contribute to the health problems that become common in the later years.

In evolutionary terms, we harbor flaws because natural selection, the force that molds our genetically controlled traits, does not aim for perfection or endless good health. If a body plan allows individuals to survive long enough to reproduce (and, in humans and various other organisms, to raise their young), then that plan will be selected. That is, individuals robust enough to reproduce will pass their genes—and therefore their body design—to the next generation. Designs that seriously hamper survival in youth will be weeded out (selected against) because most affected individuals will die before having a chance to produce offspring. More important, anatomical and physiological quirks that become disabling only after someone has reproduced will spread. For example, if a body plan leads to total collapse at age 50 but does not interfere with earlier reproduction, the arrangement will get passed along despite the harmful consequences late in life.

Had we been crafted for extended operation, we would have fewer flaws capable of making us miserable in our later days. Evolution does not work that way, however. Instead it cobbles together new features by tinkering with existing ones in a way that would have made Rube Goldberg proud.

The upright posture of humans is a case in point. It was adapted from a body plan that had mammals walking on all fours. This tinkering undoubtedly aided our early hominid ancestors: standing on our own two feet is thought to have promoted tool use and enhanced intelligence. Our backbone has since adapted somewhat to the awkward change: the lower vertebrae have grown bigger to cope with the increased vertical pressure, and our spine has curved a bit to keep us from toppling over. Yet these fixes do not ward off an array of problems that arise from our bipedal stance.

What If?

Recently the three of us began pondering what the human body would look like had it been constructed specifically for a healthy long life. The anatomical revisions depicted on the pages that follow are fanciful and incomplete. Nevertheless, we present them to draw attention to a serious point. Aging is frequently described as a disease that can be reversed or eliminated. Indeed, many purveyors of youth-in-a-bottle would have us believe that medical problems associated with aging are our own fault, arising primarily from our decadent lifestyles. Certainly any fool can shorten his or her life. But it is grossly unfair to blame people for the health consequences of inheriting a body that lacks perfect maintenance and repair systems and was not built for extended use or perpetual health. We would still wear out over time even if some mythical, ideal lifestyle could be identified and adopted.

We would look a lot different—inside and out—if evolution had designed the human body to function smoothly not only in youth but for a century or more

This reality means that aging and many of its accompanying disorders are neither unnatural nor avoidable. No simple interventions can make up for the countless imperfections that permeate our anatomy and are revealed by the passage of time. We are confident, however, that biomedical science will be able to ease certain of the maladies that result. Investigators are rapidly identifying (and discerning the function of) our myriad genes, developing pharmaceuticals to control them, and learning how to harness and enhance the extraordinary repair capabilities that already exist inside our bodies. These profound advances will eventually help compensate for many of the design flaws contained within us all.

ALL ILLUSTRATIONS BY PATRICIA J. WYNNE

WALK THIS WAY

A number of the debilitating and even some of the fatal disorders of aging stem in part from bipedal locomotion and an upright posture—ironically, the same features that have enabled the human species to flourish. Every step we take places extraordinary pressure on our feet, ankles, knees and back—structures that support the weight of the whole body above them. Over the course of just a single day, disks in the lower back are subjected to pressures equivalent to several tons per square inch. Over a lifetime, all this pressure takes its toll, as does repetitive use

Flaws

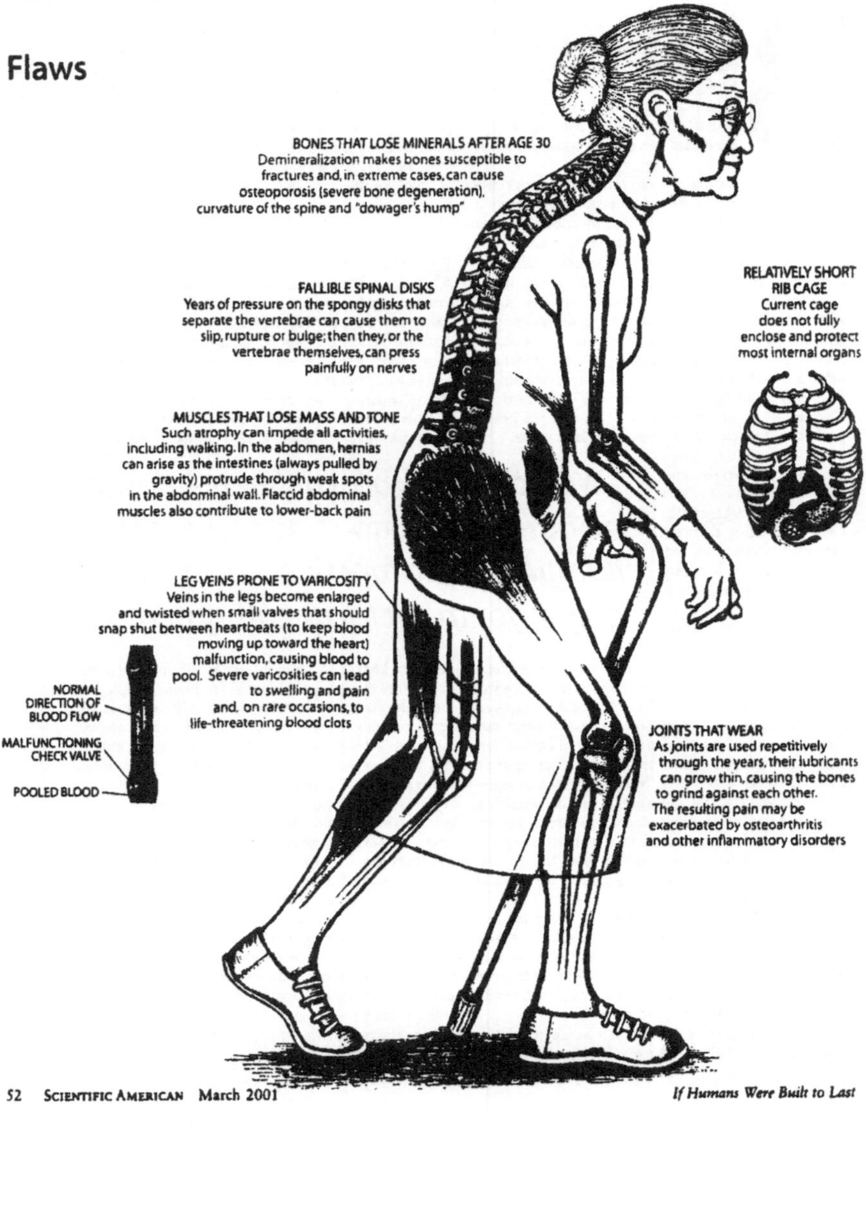

of our joints and the constant tugging of gravity on our tissues.

Although gravity tends to bring us down in the end, we do possess some features that combat its ever present pull. For instance, an intricate network of tendons helps to tether our organs to the spine, keeping them from slumping down and crushing one another.

But these anatomical fixes—like the body in general—were never meant to work forever. Had longevity and persistent good health been the overarching aim of evolution, arrangements such as those depicted below might have become commonplace.

Fixes

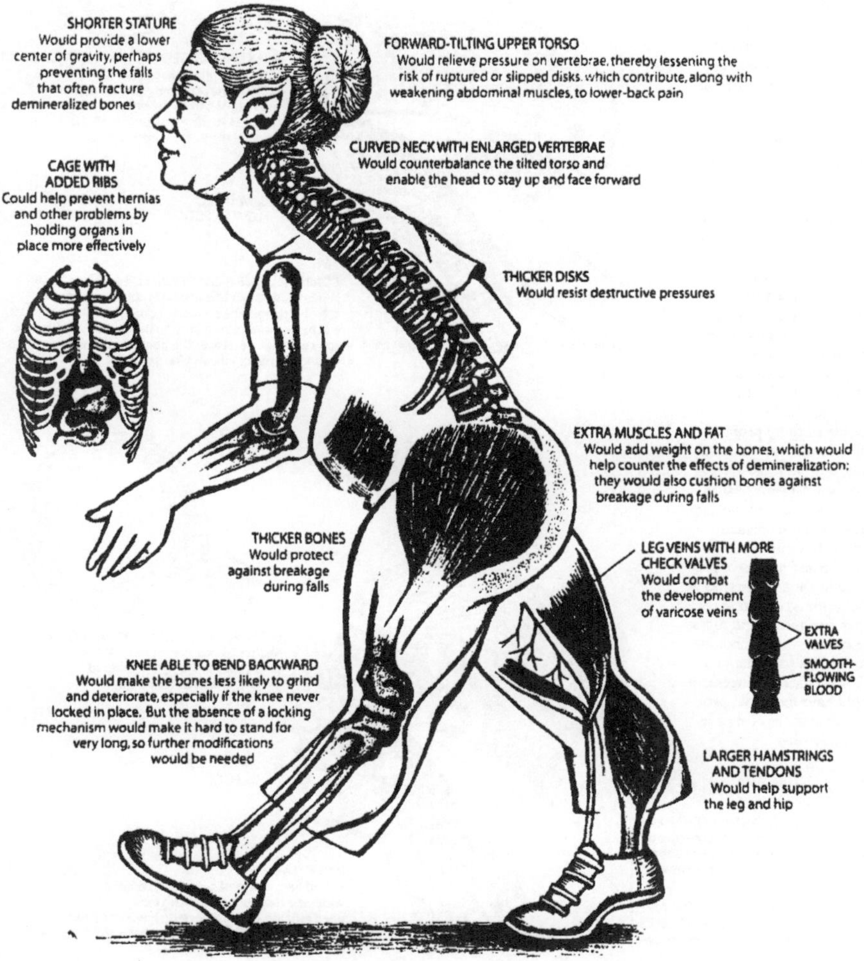

PLAN A HEAD

Various parts of the head and neck become problematic with disturbing regularity as people age. Consider the eye. The human version is an evolutionary marvel, but its complexity provides many opportunities for things to go wrong over a long lifetime.

Our vision diminishes as the protective fluid of the cornea becomes less transparent over time. The muscles that control the opening of the iris and the focusing of the lens atrophy and lose responsiveness, and the lens thickens and yellows, impairing visual acuity and color perception. Further, the retina—responsible for transmitting images to the brain—can detach fairly easily from the back of the eye, leading to blindness.

Many of those problems would be difficult to design away, but the squid eye suggests an arrangement that could have reduced the likelihood of retinal detachment. A few anatomical tweaks could also have preserved hearing in the elderly.

Suboptimal design of the upper respiratory and digestive systems makes choking another risk for older people. A simple rearrangement would have fixed that problem, albeit at the cost of severe trade-offs.

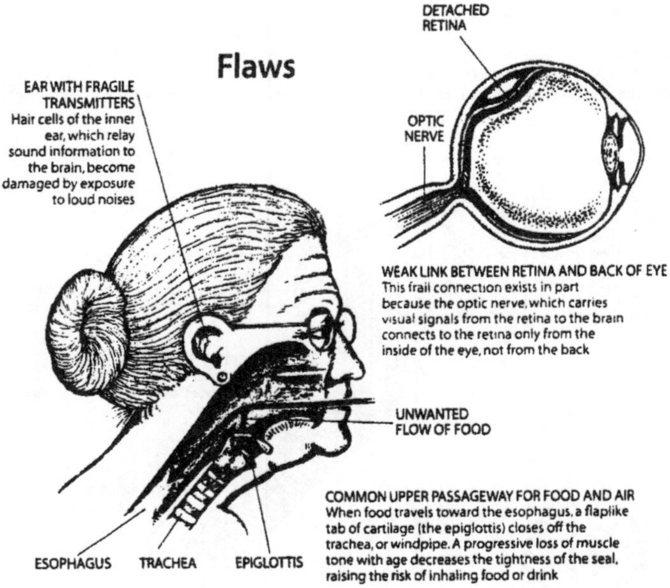

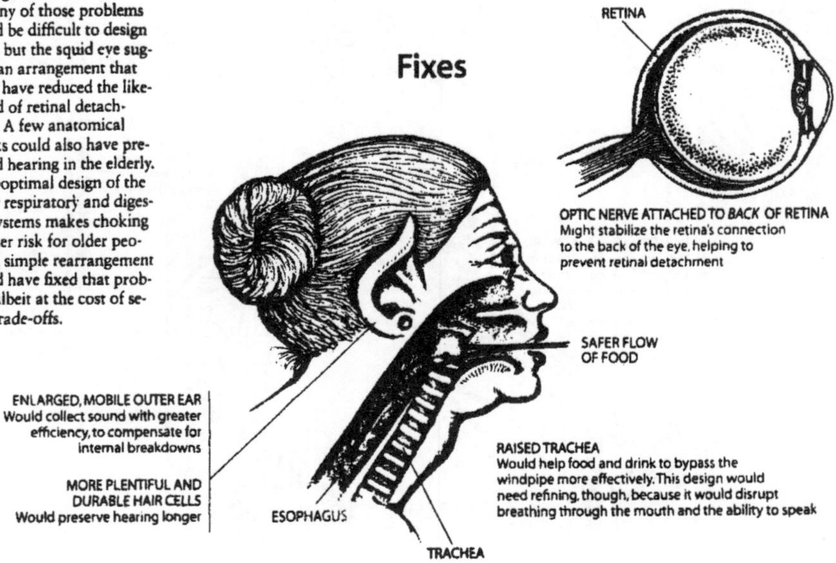

CALL A PLUMBER

An experienced plumber looking at the anatomy of a man's prostate might suspect the work of a young apprentice, because the urethra, the tube leading from the bladder, passes straight through the inside of the gland. This configuration may have as yet unknown benefits, but it eventually causes urinary problems in many men, including weak flow and a frequent need to void.

Women also cope with plumbing problems as they age, particularly incontinence. Both sexes could have been spared much discomfort if evolution had made some simple modifications in anatomical design.

MALE PROSTATE
side view

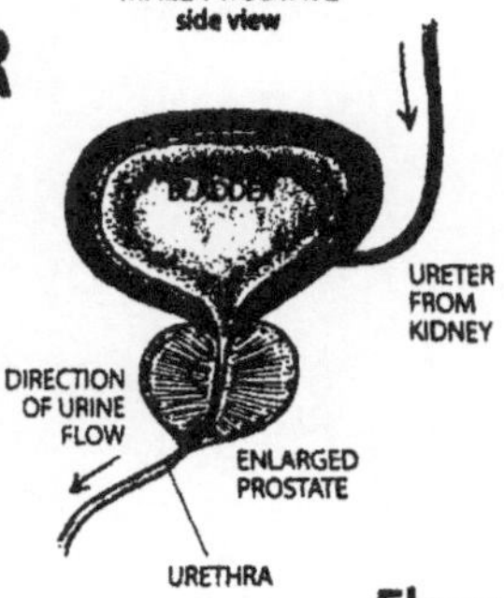

Flaw

URETHRA PRONE TO CONSTRICTION

The prostate becomes enlarged in one of every two males at some point in life. As it grows, it squeezes the urethra, potentially obstructing the flow of urine. Total obstruction can be fatal

Fix

URETHRA HUGGING OUTSIDE OF PROSTATE

Would not be squeezed if the prostate became enlarged

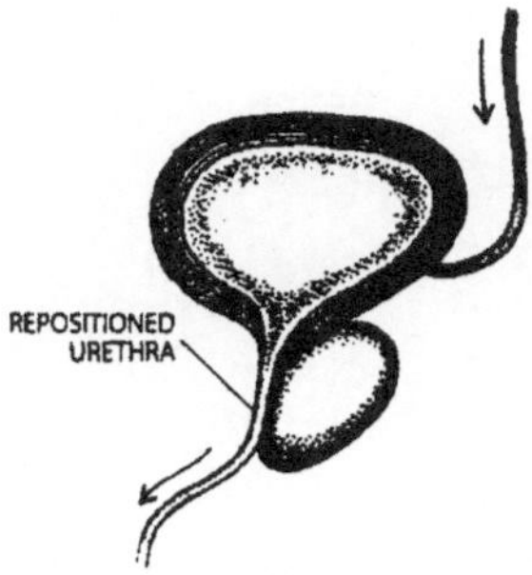

FEMALE BLADDER
front view

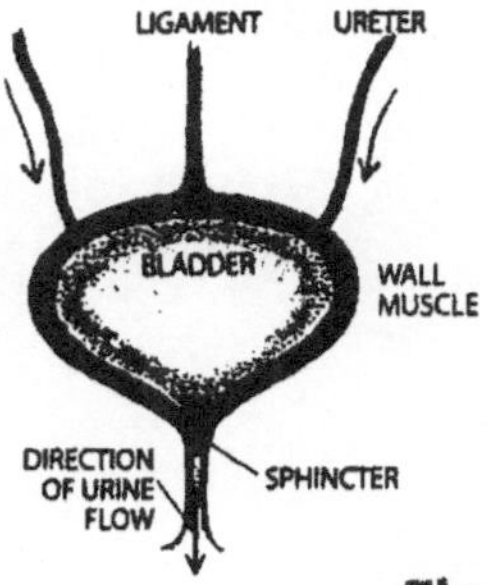

Flaw

MUSCLES AND LIGAMENTS THAT WEAKEN WITH TIME

Particularly after multiple pregnancies, the muscles of the pelvic floor and the bladder, and the ligaments that support the bladder, can sag, leading to incontinence

Fix

STRONGER SPHINCTER MUSCLES IN BLADDER AND MORE DURABLE LIGAMENTS

Would increase control over bladder function

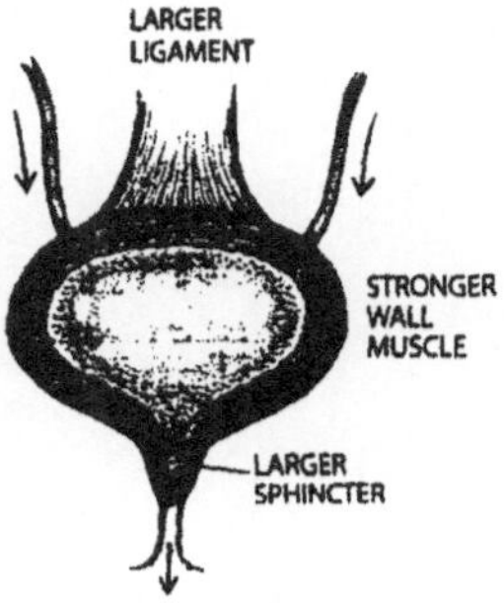

The Authors

S. JAY OLSHANSKY, BRUCE A. CARNES and ROBERT N. BUTLER all have an enduring interest in the processes that underlie human aging. Olshansky is professor in the School of Public Health at the University of Illinois at Chicago. He and Carnes, both senior research scientists at the National Opinion Research Center/Center on Aging at the University of Chicago, collaborate on studies—funded by the National Institute on Aging (NIA) and the National Aeronautics and Space Administration—of the biodemography of aging (examining the biological reasons for age-related patterns of disease and death in populations). They are co-authors of *The Quest for Immortality: Science at the Frontiers of Aging* (W. W. Norton, 2001). Butler is president of the International Longevity Center in New York City and was founding director of the NIA.

Further Information

On Growth and Form. D'Arcy Wentworth Thompson. Expanded edition, 1942. (Reprinted by Dover Publications, 1992.)

The Panda's Thumb: More Reflections in Natural History. Stephen Jay Gould. W. W. Norton, 1980.

The Blind Watchmaker: Why the Evidence of Evolution Reveals a Universe without Design. Richard Dawkins. W. W. Norton, 1986.

The Scars of Evolution: What Our Bodies Tell Us about Human Origins. Elaine Morgan. Souvenir Press, 1990. (Reprinted by Oxford University Press, 1994.)

Why We Get Sick: The New Science of Darwinian Medicine. Randolph M. Nesse and George C. Williams. Random House, 1994.

The Olshansky and Carnes Web site is www.thequestforimmortality.com

The International Longevity Center Web site is www.ilcusa.org

mechanisms of ageing and development

Mechanisms of Ageing and Development
155 (2000) 199–208

www.elsevier.com/locate/mechagedev

Meeting report

Program for testing biological interventions to promote healthy aging

Huber R. Warner [a,*], Donald Ingram [b], Richard A. Miller [c], Nancy L. Nadon [d], Arlan G. Richardson [e]

[a] *National Institute on Aging, Biology of Aging Program, 7201 Wisconsin Avenue, Gateway Building, Suite 2C231, Bethesda, MD 20892, USA*

[b] *National Institute on Aging, Gerontology Research Center, 5600 Nathan Shock Drive, Mail Box 11, Baltimore, MD 21224, USA*

[c] *University of Michigan, 5316 CCGCB, Box 0940, 1500 E. Medical Center Drive, Ann Arbor, MI 48109-0940, USA*

[d] *National Institute on Aging, Biology of Aging Program, 7201 Wisconsin Avenue, Gateway Building, Suite 2C231, Bethesda, MD 20892, USA*

[e] *The University of Texas, Health Science Center at San Antonio, 7703 Floyd Curl Drive, San Antonio, TX 78284, USA*

Received 4 April 2000; accepted 4 April 2000

Abstract

The National Institute on Aging (NIA) sponsored a workshop on September, 1999 to discuss the feasibility of establishing a program to evaluate potential intervention strategies to decelerate the rate of aging in mammals. The ultimate goal is to identify promising interventions in animals that might lead to clinical trials in humans. The participants discussed various animal models, biological endpoints and possible structure of such a program. The ability to implement such a program will require a decision by NIA staff about whether the anticipated benefits to be derived from identification of effective interventions under well controlled conditions in an animal model, in this case the mouse, would justify the anticipated cost of the testing program. © 2000 Published by Elsevier Science Ireland Ltd.

* Corresponding author. Tel.: + 1-301-4964996; fax: + 1-301-4020010.
E-mail address: warnerh@exmur.nia.nih.gov (H.R. Warner)

0047-6374/00/$ - see front matter
PII: S0047-6374(00)00118-4

Keywords: Ageing; Caloric restriction; Human clinical trials

1. Introduction

The National Institute on Aging (NIA) sponsored a workshop on September 14–16, 1999 to discuss the possibility of establishing a program to evaluate treatment strategies thought likely to decelerate the pace of aging in mammals. Similarly delaying adverse age-dependent changes in multiple cells and tissues in humans could be expected to diminish the burden of disease in old age, and subsequently increase the length of healthy life span. The specific aims of the workshop were to discuss: (1) whether there is a real need for such a program, and if so, (2) what animal model(s) should be used, (3) what endpoints should be followed in evaluating the efficacy of the intervention being tested, and (4) what strategies should be employed to optimize the program. The workshop was organized by NIA staff and was attended by staff from the NIA and FDA, as well as non-federal experts with experience in aging research. Currently, caloric restriction (CR) is the only intervention which is generally accepted to extend the maximum life span of mammalian animal models. Robust results have been consistently obtained using rats and mice, but CR also extends the life span of a wide variety of invertebrate animal models (Weindruch and Walford, 1988). More importantly, CR delays the onset of age-related pathology in rats and mice, and it is even effective when started in mature animals. It is not known whether it will work in non-human primates, but several studies in rhesus monkeys are already underway, and results so far appear to be somewhat promising (Kemnitz et al., 1994; Lane et al., 1995; Cefalu et al., 1997). However, it is unlikely that a significant portion of the human population would routinely submit to a comparable regimen, even if it were known for sure that doing so would provide 20–30 years of additional healthy life. Nevertheless, the CR phenomenon provides a paradigm for testing other interventions in animals which might be useful in human clinical trials. Considerable progress has been made identifying genes in invertebrate species which affect the rate of aging of members of these species, but less information is available about mammals. Mutations in the *dw* and *df* genes have been shown to attenuate the rate of aging in mice (Brown-Borg et al., 1996; Miller, 1999), and mice lacking the $p66^{shc}$ protein also may have an extended life span (Migliaccio et al., 1999). The testing program described below would be useful for understanding genetic alterations which extend life span in such mammalian models, although genetic intervention in humans is not the main goal of the testing program discussed here.

1.1. *Is there a need for such a program?*

Papers identifying life span and health span-extending interventions appear sporadically in the gerontological literature. A few recent examples of these include interventions to: not only reduce oxidative stress (Carney et al., 1991), but also extend life span with *N*-*t*-butyl-α-phenylnitrone (Edamatsu et al., 1995); restore biochemical and physical function with acetyl-L-carnitine (Hagen et al., 1998); prevent cardiovascular and renal pathology with aminoguanidine (Li et al., 1996); and extend life span with melatonin (Pierpaoli and Regelson, 1994). A different approach has been to transgenically extend life span by overexpression of Cu/Zn superoxide dismutase (Parkes et al., 1998; Sun and Tower, 1998).

However, many of these reports are compromised by one or more design flaws, e.g. too few animals per cohort, failure to control for possible CR, use of wrong animal model, poor housing conditions, etc. Furthermore, many research laboratories do not have either the resources or the expertise to design and conduct a scientifically robust study. For example, few of these studies have been accompanied by pathology or other biomarker assessment, which would help assess the impact of the intervention on health span. Because most of the informative endpoints in such a study may require invasive procedures, adequate testing cannot usually be done directly in humans. However, the identification of effective substances or treatments in mice could serve as a powerful spur and guide for the design of intervention studies in humans or non-human primates.

A successful program would not only achieve the above goal, but it would also provide new theories and hypotheses about basic mechanisms of aging for further research. Finally, negative results could prevent continued investment of research funds into unpromising areas of research.

A program of this kind would most likely be successful if it can bring together a group of experienced investigators and NIA staff members to design an optimized protocol for intervention evaluation that reflects substantial consensus within the research community, and can at the same time build in standardization and replicability.

2. Animal models

The criteria to be considered in selecting a test animal model were discussed at the workshop. These include:

1. External validity — the findings in the test model must be relevant to human aging processes.
2. Inexpensive — low cost is necessary to allow a large enough sample group for each test to have robust statistical power.
3. Relative ease of husbandry — this is necessary to fulfill (2).
4. Reproducible variability — the test model must have genetic diversity built in to avoid the pitfalls of genetically homogeneous populations, i.e. strain-specific pathologies, but the genetic background must be reproducible for confirmation/extension of the results. Included in this variability is the presumption that both genders must be tested.
5. Responds to caloric restriction — the argument being that because CR is the

only proven method to extend life span, any useful test model should be affected by CR. Because worms, flies, and rodents have all been shown to respond to CR with extended life span, this does not unduly limit model selection. This refers only to model selection and does not suggest responsiveness to CR as an actual test criterion.

6. To model human aging as closely as possible, the species should exhibit multiple causes of death.

Individual animal models discussed by workshop participants included rodents (Richard Miller, Richard Weindruch, Arlan Richardson); invertebrates including nematodes and fruit flies (Thomas Johnson); birds (Steven Austad); dogs (Michael Hayek); and non-human primates (Mark Lane). These discussions can be briefly summarized as follows:

2.1. Mice/rats

Although both have been used extensively in CR research, the mouse has several advantages, e.g. lower husbandry costs, a better characterized genetic system and availability of transgenic and knockout models. The larger size of rats would provide a better opportunity for repeated blood sampling, but that advantage would not outweigh the disadvantages.

Inbred strains of mice are not ideal because of their genetic homogeneity. They are also plagued by strain-specific pathologies, and have shorter life spans than genetically heterogeneous mice. F1 hybrid mice have longer life spans and fewer pathologies, but they still represent a genetically homogeneous population, quite unlike the human population. Reproducible variability could be achieved by use of hybrid animals such as 4-way cross offspring. Four-way cross mice, the product of mating two F1 hybrids so that the F2 generation mice have four different grandparents, provide another mechanism to generate genetic diversity in a reproducible manner. Four-way cross mice share a random 50% of their genes with their sibs, and they show a wide range of pathologies at death, similar to the human population.

2.2. Dogs

Although the large size of dogs would provide an excellent opportunity for multiple sampling, this advantage is largely offset by the substantial cost of using dogs.

2.3. Birds

Birds are not a good general testing model as they satisfy none of the criteria listed above. However, they do have one special advantage, i.e. they could be used to test neurological toxicity by monitoring singing behavior.

2.4. Non-human primates

Although the rhesus monkey is currently being used to evaluate whether CR retards aging in non-human primates, routine use of monkeys is prohibitive because of limited availability, high cost, and long life. However, primate assays might be suitable for advanced studies on therapeutic interventions identified by the testing program, rather than as part of the initial screening process.

Although there was strong support for the 4-way cross mouse as the primary test animal, the final decision on the best model systems will depend on definition of the exact goals of the program. One decision that needs to be made is that of large scale screening of many candidates versus confirmation of already published promising results on a few candidates. For large-scale screening, an inexpensive model such as fruit flies or nematodes might be used (see below under Large scale screening program). However, if the goal is to test the efficacy of interventions that already have given preliminary, promising findings in a published study, then a mammalian model is important so that the findings are more likely to be valid for human aging. Interventions that appear promising on theoretical grounds would also be valid candidates for consideration. The second goal that needs to be defined relates to the type of information desired, mechanistic data or therapeutic efficacy data. If it were mechanistic data, the design of the program would encompass many animal models, including transgenic and knockout mice. Although the program as originally envisioned was to focus on testing the efficacy of non-genetic therapeutic interventions to promote healthy aging, it is clear that the program could be useful for evaluating genetic interventions as well.

In summary, to develop a program to test the efficacy of putative anti-aging interventions that have been identified by investigators, a mammalian model is needed. This model must be relatively inexpensive, must provide reproducible genetic heterogeneity, and undergo aging and mortality in a manner comparable to humans. To this end, the use of 4-way cross mice seems to be the most promising, with the exception of testing genetically altered mice. Any intervention that emerges from this program with positive results could then go on to additional studies in nonhuman primates, or possibly go directly into human clinical trials.

3. Endpoints

In addition to survival data, other endpoints must be developed to provide sufficient information to conclude that an intervention may be capable of promoting healthier aging. There was general agreement that longevity is a necessary, but insufficient measurement. The requirement that health and function be preserved by the intervention indicates the need for both functional and pathology assessment. The pros and cons of various endpoints were discussed by James Nelson (survival curves), Robert Reddick (pathology assessment), Earl Stadtman and Robert Floyd (protein oxidation), Kenneth Beckman (DNA oxidation), Jan Vijg (gene expression microarrays), Don Ingram and James Joseph (behavior), Richard Miller and Simin

Meydani (immunology) and Andrew Monjan (cognition). From these discussions the outline of a minimally satisfactory panel including the following endpoints emerged.

1. Survival analysis carried out in a barrier facility and including 50–60 animals/cohort (phase 1 testing).
2. Non-invasive endpoints (for use in phase 2 testing) such as:
   locomotion (rotorod or spontaneous activity);
   cognitive function (maze learning);
   physiological function (T-cell subset analysis or immune response);
   cataracts (slit lamp exam);
3. Pathology assessment including malignancies, renal disease, and cardiovascular disease, done upon death of the animals.
4. Invasive endpoints (for possible selective use in phase 2 testing, particularly for research purposes).
5. Resistance to stress, e.g. heat.
6. Biomarkers of oxidative stress; possibilities include one or more of the following: brain glutamine synthetase (brain); mitochondrial *cis*-aconitase (muscle); other protein modifications, e.g. collagen pentosidine. Although oxidized bases in DNA have frequently been used as a biomarker, technical problems and efficient DNA repair systems make this a problematic endpoint. The measurement of urinary isoprostanes has recently been suggested as a biomarker of oxidative stress in vivo (Roberts and Morrow, 2000).
7. Gene expression — microarray technology may provide very informative endpoints once it is clear which gene probes to use on the microarray. The first major study of age-related changes using this technology was published recently (Lee et al., 1999).

Further discussion by a panel of gerontologists will be required to develop consensus about the most informative but affordable endpoints to use, and when to do the measurements.

4. Implementation strategy

4.1. Selection of interventions to test

A rigorous evaluation process should be employed before any testing is done. An Access Committee composed of investigators, both extramural and intramural, only some of whom would be testing site P.I.s, would evaluate the scientific justification for testing an intervention, based both on published results and any available unpublished data. Candidates for testing could be suggested by either a committee member or individual investigators not formally associated with the Program, such as commercial or other governmental organizations. The Access Committee membership should include the principal investigator(s) of the testing site(s), but would not include NIA staff.

The recommendations of the Access Committee would be passed on to a Scientific Advisory Committee which would determine the relative priority for testing the recommended interventions. This Committee would include NIA staff as well as at least two members from the Access Committee, including principal investigator(s) of the testing site(s).

4.2. Testing strategy

Consensus was reached at the Workshop to use mice for preliminary screening in a 2-phase program.

4.2.1. Phase 1

Proponents suggested that promising interventions could first be screened for longevity by doing a survival analysis, accompanied by longitudinal body weight measurements. Phase 1 studies would be completed within 3 years after initiating the study, because 50% survivorship would be known by then. Gross pathology assessment at time of death would also be desirable, but not essential in phase 1. Non-promising interventions would then be discarded.

4.2.2. Phase 2

Comprehensive evaluation of the remaining interventions would be done, including invasive endpoints.

The major argument for a 2-phase program is the ability to test a larger number of interventions at low cost, before commitment to an exhaustive evaluation including pathology assessment of any given intervention, although there is some risk of prematurely discarding useful interventions which extend health span but not life span.

Sentiment was also expressed to test interventions simultaneously at multiple sites, perhaps up to three. The major arguments for testing interventions at multiple sites are (1) the confidence gained from obtaining similar results at 2–3 sites; (2) the availability of site-specific expertise at different sites; (3) safety, in case one site experiences an infection or mechanical breakdown.

Each of these options would increase the overall cost of the program somewhat, thus increasing the average cost of identifying successful interventions. This would be partially offset by savings incurred when 'losers' are discarded after the relatively inexpensive phase 1 trial, and because pooling results across sites would permit the use of fewer mice per tested compound per site. This is a significant issue requiring debate by NIA staff before development of this program can proceed.

Although the emphasis in this report is on testing non-genetic interventions, it is clear that such an infrastructure would also be useful in examining mechanistic questions involving genetic differences and manipulations.

5. Cost estimates

Although detailed cost estimates depend upon implementation, preliminary calculations obtained from workshop participants indicate that a program involving three test sites, in which three proposed interventions are subjected to phase 1 trials each year, with perhaps one to two interventions proceeding to phase 2 testing each year after year 3, would cost between 1.5 and $2.0 million per year, once steady state is reached in year 3.

6. Issues to be discussed/resolved in future discussion

- The appropriate funding mechanism for such an activity would be either a contract or a cooperative agreement.
- It should not be assumed that all testing be done at primary sites. A primary site could sub-contract with other institutions to take advantage of the best expertise available. The primary concern is standardization of protocols to ensure comparability of results obtained over an extended period of time.
- Gene expression microarray analysis is an endpoint with remarkable potential to evaluate age-related changes in response to intervention. Its use would be appropriate for limited cross-sectional use, pending development of a filter or 'chip' designed specifically for this purpose.
- Cost is an obvious factor in deciding what endpoints to include in the testing plan. Any plan will require a compromise between cost and robustness.
- There must be a vehicle for publishing the results of tests conducted in the program regardless of the success of the intervention.
- Composition of Access and Scientific Advisory Committees.

7. Large scale screening program

Ultimately a comprehensive program would include not only confirmation of the efficacy of promising interventions developed from published work, but would also develop a large scale screening program to identify new potential interventions. An effective screening program must be cheap enough and rapid enough to test hundreds, or even thousands of compounds per year. It was proposed at the workshop that *Caenorhabditis elegans*, a nematode, could be used for such a purpose. Although nematodes are very distant evolutionarily from humans, there are several reasons for being optimistic about this possibility. These include:

- *C. elegans* is short-lived.
- Several *C. elegans* mutations have already been identified which extend the nematode life span by more than 2-fold; these genes have human homologs, and their gene products appear to be involved in energy metabolism in humans (Morris et al., 1996; Kimura et al., 1997).

- *C. elegans* is already being used for this purpose (B. Villeponteau, personal communication).
- Survival curves are inexpensive to do with *C. elegans*; the estimated cost is about $200 per survival curve, or about $20 000 for testing 100 interventions.

However, it could be argued that such an effort would be better if funded as a commercial activity.

8. Summary

An outline of a plan with several options for testing effective interventions to promote healthy aging in animals has been provided. If established, the program would not only identify potential human interventions, but could also provide a valuable research resource for understanding genetic factors in aging. The ability of the NIA to implement a plan will depend on the cost involved and the anticipated benefits to be derived from the discovery of effective interventions in animal models. The program as described does not include testing in humans, but is designed to identify interventions that have a reasonable chance for success in subsequent clinical trials.

A parallel large-scale screening program would be a desirable, but not a necessary part of the overall program. The cost effectiveness of such a screening program would depend on the choice of an appropriate, but inexpensive animal model for testing the effect of possibly thousands of compounds on longevity.

At a follow-up meeting convened during the 2000 Gordon Research Conference on the Biology of Aging held in Ventura, CA, a subset of the participants at the original meeting reached consensus that 4-way cross mice should be used for tests of dietary interventions. However, testing genetically altered mice, e.g. transgenic and knockout mice, would likely be done on inbred backgrounds out of necessity.

References

Brown-Borg, H.M., Borg, K.E., Meliska, C.J., Bartke, A., 1996. Dwarf mice and the aging process. Nature 384, 33.

Carney, J.M., Starke-Reed, P., Oliver, C.N., et al., 1991. Reversal of age-related increase in brain protein oxidation, decrease in enzyme activity, and loss in temporal and spatial memory by chronic administration of the spin-trapping compound *N*-tert-butyl-alpha-phenylnitrone. Proc. Natl. Acad. Sci. USA. 88, 3633–3636.

Cefalu, W.T., Wagner, J.D., Wang, Z.Q., et al., 1997. A study of caloric restriction and cardiovascular aging in cynomolgus monkeys (*Macaca fascicularis*) — a potential model for aging research. J. Gerontol. Biol. Sci. 52, B10–B19.

Edamatsu, R., Mori, A., Packer, L., 1995. The spin-trap-*N*-tert-alpha-phenyl-butylnitrone prolongs the life span of the senescence accelerated mouse. Biochem. Biophys. Res. Comm. 211, 847–849.

Hagen, T.M., Ingersoll, R.T., Wehr, C.M., et al., 1998. Acetyl-L-carnitine fed to old rats partially restores mitochondrial function and ambulatory activity. Proc. Natl. Acad. Sci. USA 95, 9562–9566.

Kemnitz, J.W., Roecker, E.B., Weindruch, R., et al., 1994. Dietary restriction increases insulin sensitivity and lowers blood glucose in rhesus monkeys. Am. J. Physiol. 266, E540–E547.

Kimura, K.D., Tissenbaum, H.A., Liu, Y., Ruvkun, G., 1997. *daf*-2, an insulin receptor-like gene that regulates longevity and diapause in *Caenorhabditis elegans*. Science 277, 942–946.

Lane, M.A., Ball, S.S., Ingram, D.K., et al., 1995. Diet restriction in rhesus monkeys lowers fasting and glucose-stimulated glucoregulatory end points. Am. J. Physiol. 268, E941–E948.

Lee, C.-K., Klopp, R.G., Weindruch, R., Prollla, T.A., 1999. Gene expression profile of aging and its retardation by caloric restriction. Science 285, 1390–1393.

Li, Y.M., Steffes, M., Donnelly, T., et al., 1996. Prevention of cardiovascular and renal pathology of aging by the advanced glycation inhibitor aminoguanidine. Proc. Natl. Acad. Sci. USA 93, 3902–3907.

Migliaccio, E., Giorgio, M., Mele, S., et al., 1999. The p66shc adaptor protein controls oxidative stress response and life span in mammals. Nature 402, 309–313.

Miller, R.A., 1999. Kleemeier award lecture: are there genes for aging? J. Gerontol. 54A, B297–B307.

Morris, J.Z., Tissenbaum, H.A., Ruvkun, G., 1996. A phosphatidylinositol-3-OH kinase family member regulating longevity and diapause in *Caenorhaditis elegans*. Nature 382, 536–539.

Parkes, T.L., Elia, A.J., Dickinson, D., et al., 1998. Extension of *Drosophila* lifespan by overexpression of human SOD1 in motorneurons. Nat. Genet. 19, 171–174.

Pierpaoli, W., Regelson, W., 1994. The pineal control of aging: the effect of melatonin and pineal grafting on aging mice. Proc. Natl. Acad. Sci. USA 91, 787–791.

Roberts, L.J., Morrow, J.D., 2000. Measurement of F2-isoprostanes as an index of oxidative stress in vivo. Free. Rad. Biol. Med. 28, 505–513.

Sun, J., Tower, J., 1998. FLP recombinase-mediated induction of Cu/Zn-superoxide dismutase transgene expression can extend the life span of adult *Drosophila melanogaster* flies. Mol. Cell. Biol. 19, 216–228.

Weindruch, R., Walford, R.L., 1988. The Retardation of Aging and Disease by Dietary Restriction. C.C. Thomas, Springfield, IL.

Acknowledgments

Hayflick, L. (1996). "Defining Aging." Chapter 1 in *How and Why We Age*. NY: Ballantine Books, 11–18. Reprinted with the permission of Ballantine Books.

Beckman, K. B. and B.N. Ames (1998). "The free radical theory of aging matures." *Physiological Reviews* 78(2): 547–81. Reprinted with the permission of the American Physiological Society.

Fossel, M. (2000). Cell senescence in human aging: a review of the theory. *In vivo* 14(1): 29–34. Reprinted with the permission of International Institute of Anticancer Research.

Barzilai, N. and I. Gabriely. (2001). Calorie Restriction: Effects on body composition, insulin signaling and aging—the role of fat depletion in the biological benefits of caloric restriction. *The Journal of Nutrition*. 131(3) S 903–906. March 2001. Reprinted with the permission of American Institute of Nutrition.

Charlesworth, B. (2000). "Fisher, Medawar, Hamilton and the evolution of aging." *Genetics*. 156(3): 927–931. Reprinted with the permission of Genetics Society of America.

Miller, R.A . and N. L. Nadon. (2000). "Principles of animal use for gerontological research." *Journals of Gerontology Series A-Biological Sciences and Medical Sciences*. 55(3): B117–23. Reprinted with the permission of Gerontological Society of America.

Vijg, J. (2000). "Somatic mutations and aging: a re-evaluation." *Mutation Research* 4470(1): 117–35. Reprinted with the permission of Elsevier Science BV.

Klinman, N. R. and G. H. Kline (1997.) "The B-cell biology of aging." *Immunological Reviews* 160:103–14. Reprinted with the permission of Blackwell Munksgaard.

Ottinger, M. A. and N. Thompson, et al. (1997). "Neurendocrine regulation of GnRH and behavior during aging in birds." *Brain Research Bulletin* 44(4): 471–7. Reprinted with the permission of Elsevier Science.

Strauss, E. (2001). Growing old together. *Science*. 292(5514), 41–43. Reprinted with permission from *Science*. Copyright © (2001) American Association for the Advancement of Science.

Macieira-Coelho, A. (2000). "Ups and downs of aging studies in vitro: the crooked path of science." *Gerontology.* 46(2): 55–63. Reprinted with the permission of Israel Gerontological Society.

Finkel, T. and N. J. Holbrook. (2000). "Oxidants, oxidative stress and the biology of ageing. *Nature.* 408(6809): 239–247. Reprinted by permission from *Nature* Vol. 408 p. 239–247. Copyright © 2000 Macmillan Magazines Ltd.

Narayanan, S. (1996). "Laboratory markers as an index of aging." *Annals of Clinical and Laboratory Science* 26(1): 50–9. Reprinted with the permission of Institute for Clinical Science.

Johnson, F. B. and D. A. Sinclair, et al. (1999). "Molecular biology of aging." *Cell* 96(2): 291–302. Reprinted with permission from Elsevier Science.

Mann, D. M. (1997) "Molecular biology's impact on our understanding of aging." *British Medical Journal* 315(7115): 1078–81. Reprinted with the permission of the British Medical Association.

Kirkwood, T. B. (1998). "Ovarian ageing and the general biology of senescence." *Maturitas* 30(2): 105–11. Reprinted with the permission of Elsevier Science Ireland Ltd.

Hensler, P. J . and O. M. Pereira-Smith (1995). "Human replicative senescence: A molecular study." *American Journal of Pathology* 147(1): 1–8. Reprinted with the permission of American Society for Investigative Pathology.

Sorkin, J. D., D. C. Muller, and R. Andres (1999). "Longitudinal change in height of men and women: implications for interpretation of the body mass index: Baltimore Longitudinal Study of Aging." *American Journal of Epidemiology.* 150(9): 969–77. Reprinted with the permission of Oxford University Press.

Harman, S. M., E. J. Metter, et al. (2001). "Longitudinal effects of aging on serum total and free testosterone levels in healthy men." Baltimore Longitudinal Study of Aging." *Journal of Clinical Endocrinology and Metabolism.* 86(2): 724–31. Reprinted with the permission of Endocrine Society.

Yaar, M. (1996) "Skin aging: observations at the cellular and molecular levels." *Israel Journal of Medical Sciences* 31(11): 1053–8. Reprinted with the permission of Israel Medical Association.

Saurat, J. H. (1995). "Retinoids and ageing." *Hormone Research.* 43(1–3): 89–92. Reprinted with the permission of S. Karger AG.

Murray, T. M. (1996). "Mechanisms of bone loss." *Journal of Rheumatology* – Supplement 45: 6–10. Reprinted with the permission of *Journal of Rheumatology*.

Frontera, W. R., Suh, D., L. S. Krivickas, V. A. Hughes, et al. (2000). "Skeletal muscle fiber quality in older men and women." *American*

Journal of Physiology. 48(3): C611–C618. Reprinted with the permission of American Physiological Society.

Seeman, T. E., B. S. McEwen, J. W. Rowe, and B. H. Singer. (2001). "Allostatic load as a marker of cumulative biological risk: MacArthur studies of successful aging." *Proceedings of the National Academcy of Sciences of the United States.* 98(8): 4770–4775. Copyright © (2001) National Academy of Sciences, U.S.A.

Swynghedauw, B., S. Besse, et al (1995). "Molecular and cellular biology of the senescent hypertrophied and failing heart." *American Journal of Cardiology* 76(13): 2D-7D. Reprinted with the permission of ESI Ltd.

Flanigan, K. M., G. Lauria, et al. (1998). "Age-related biology and diseases of muscle and nerve." *Neurologic Clinics.* 16(3): 659–69. Reprinted with the permission of W.B. Saunders Co.

Graeber, M. B., E. Grasbon-Frodl, et al. (1998) "Neurodegeneration and aging: role of the second genome." *Journal of Neuroscience Research* 52(1): 1–6. Copyright © (1998) by *Journal of Neuroscience Research*. This material is used by permission of Wiley-Liss, Inc. a subsidiary of John Wiley & Sons, Inc.

Drachman, D. A. (1997). "Aging and the brain: a new frontier." *Annals of Neurology.* 42(6): 816–828. Copyright © 1997 by *Annals of Neurology*. This material is used by permission of Wiley-Liss, Inc. a subsidiary of John Wiley & Sons, Inc.

Christen, Y. (2000). "Oxidative stress and Alzheimer disease. *American Journal of Clinical Nutrition.* 71(2) 621S–629S. Reprinted with the permission of American Society for Clinical Nutrition, Inc.

Sisodia, S. S. (1999). "Alzheimer's disease: perspectives for the new millennium." *Journal of Clinical Investigation.* 104(9): 1169–79. Reprinted with the permission of American Society for Clinical Investigation.

Campisi, J. (1997). "The Biology of Replicative Senescence." *European Journal of Cancer.* 33(5): 703–709. Reprinted with the permission of Pergamon.

Hahn, W. D. and M. Meyerson (2001). "Telomerase activation, cellular immortalization and cancer." *Annals of Medicine* 33(2): 123–9. Reprinted with the permission of Royal Society of Medicine Press Ltd.

Banks, D. A., M. Fossel. (1997). "Telomeres, cancer, and aging. Altering the human life span." *Journal of the American Medical Association.* 16: 1345–8. Copyright 1997 The American Medical Association.

Kobyliansky, E., G. Livshits, and O. Pavlovsky. (1995). "Population biology of human aging: methods of assessment and sex variation." *Human Biology* 67(1): 87–109. Reprinted with the permission of the Wayne State University Press.

G. M. Martin. (2000). "Some new directions for Research in the Biology of Aging." *Annals of the New York Academy of Sciences*. 908: 1–13. Reprinted with the permission of New York Academy of Sciences.

Olshanksy, S. J., B. A. Carnes, and R. N. Butler. (2001). "If humans were built to last." *Scientific American*. 284(3): 50–55. Reprinted with the permission of Scientific America, Inc.

Warner, H. R., D. Ingram, R. A. Miller, N. L. Nadon, and A. G. Richardson. (2000). "Program for testing biological interventions to promote healthy aging." *Mechanisms of Ageing and Development*. 115(3):199–207. Reprinted with the permission of Elsevier Science Ireland Ltd.